ADVANCES IN AGRICULTURAL MICROBIOLOGY

STUDIES in the AGRICULTURAL and FOOD SCIENCES

A series of high-level monographs which review recent research in various areas of agriculture and food science

---

Already published:

Recent Advances in Animal Nutrition 1978
Edited by W. Haresign and D. Lewis

Recent Advances in Animal Nutrition 1979
Edited by W. Haresign and D. Lewis

Recent Advances in Animal Nutrition 1980
Edited by W. Haresign

Recent Advances in Animal Nutrition 1981
Edited by W. Haresign

Recent Advances in Animal Nutrition 1982
Edited by W. Haresign

Plant Breeding for Pest and Disease Resistance
G.E. Russell

The Calf—Fourth edition
J.H.B. Roy

Energy Metabolism
Edited by Lawrence E. Mount

Growth in Animals
Edited by T.L.J. Lawrence

Mineral Nutrition of Fruit Trees
Edited by D. Atkinson, R.O. Sharples, J.E. Jackson and W.M. Waller

Sheep Breeding—Second edition
Edited by G.J. Tomes, D.E. Robertson and R.J. Lightfoot
Revised by W. Haresign

Mineral Nutrition of Animals
Edited by V.I. Georgievskii, B.N. Annenkov and V.T. Samokhin

Protein Contribution of Feedstuffs for Ruminants
Edited by E.L. Miller, I.H. Pike and A.J.H. Van Es

STUDIES IN THE AGRICULTURAL AND FOOD SCIENCES

Advances in Agricultural Microbiology

Editor
N.S. Subba Rao
Division of Microbiology, Indian Agricultural Research Institute, New Delhi

BUTTERWORTH SCIENTIFIC
London Boston Durban Singapore Sydney Toronto Wellington

First published 1982

Published jointly by
Mohan Primlani, Oxford and IBH Publishing Co., 66 Janpeth, New Delhi 110 001
and
Butterworth & Co (Publishers) Ltd 1982

British Library Cataloguing in Publication Data

Advances in agricultural microbiology.
1. Agricultural microbiology
I. Subba Rao, N.S.
630'.2'76 QR51

ISBN 0-408-10848-7

Bound in Great Britain by Mansell Bookbinders, Witham, Essex

Foreword

Though agricultural microbiology is a relatively new branch of soil science, it has emerged as a potentially very useful and distinct discipline of science, especially in the context of the current constraints of energy on the farm front. Microorganisms have relevance to agriculture in several ways—in biological nitrogen fixation, in human food and animal feed as single cell protein, as agents of insect pest control, as a source of fuel and energy, as a means to treat sewage, in converting cellulose or sugarcane juice into power alcohol, in producing new antibiotics which can control plant diseases, in generating methane or biogas, in mobilizing phosphorus to plants through endo- and ecto-mycorrhizae and so on. In fact the entire concept of bio-conversion is based on the ability of microorganisms to break down lignocelluloses.

Starting with the successful use of a bacterial insecticide preparation from *Bacillus thuringiensis* capable of killing a number of lepidopterous insect pests of crops, there has been an array of fungal, bacterial, protozoal and viral products which are being commercially made in Europe and the USA to successfully combat plant pests. Even nematode control is being envisaged by means of nematode trapping fungi. The use of microorganisms to combat plant pests is non-polluting and in fact currently some soil-borne diseases are controlled biologically by the use of antagonistic microorganisms. Similarly, spraying commercially prepared antibiotics is being routinely resorted to in Japan to control aerial diseases of plants.

On the energy front, Brazil has reduced its use of gasolene in motor cars by 10% by partially substituting it with power alcohol generated by fermenting sugarcane juice to ethyl alcohol. Biomass utilization by enzymatic conversion of nature's vast reservoir of renewable lignocelluloses into ethyl alcohol is being considered as a viable alternative to non-renewable fossil fuels.

Constant improvement of microbial processes using improved strains is possible by genetic engineering. The quick generation time of microorganisms and the ease with which their nuclear material can be handled make them ideal for microbial geneticists to 'tailor' them to yield desired products to serve mankind. In fact, it is this kind of microbial 'cell power' which is being currently harnessed by biotechnologists in developed countries to produce insulin and interferons. In the years ahead, this 'cell power' can be harnessed to improve the ability of selected microorganisms to produce more biofertilizers, to cleave cellulose more efficiently and combat plant pests with greater efficiency than ever before. In future one can foresee a tremendous upsurge of microbial technology through a greater understanding of the microbiology of agriculturally useful microorganisms.

Viewing from the above stated perspective, this book on *Advances in Agricultural Microbiology* edited by Dr. Subba Rao is very timely. The book is comprehensive in its coverage with 25 chapters arranged in three sections. It covers all areas of microbially mediated processes such as bioconversion for food, feed and fuel, mobilization of nutrients to plant growth and plant disease and pest control. I find that some of the leading world authorities in these fields have contributed chapters. This would help to get a realistic picture of the scientific situation in this field. The book is illustrated in an apt manner. The bibliography is extensive, thus making it a valuable companion to students, researchers, administrators and policy makers who foresee the need for non-polluting, less expensive and renewable sources to generate food, feed, fuel and fertilizers for feeding mankind in the years to come.

Planning Commission
Government of India
New Delhi

M.S. SWAMINATHAN*

*Currently, Director-General, International Rice Research Institute, Los Banos, Laguna, Philippines.

Preface

Agricultural microbiology is closely involved with research on organic manuring, recycling of wastes in crop cultivation and biological nitrogen fixation which are aimed at saving energy and supplementing fertilizer inputs in crop cultivation. Research on these aspects of bioconversion is crucial to the economy of developing nations which are facing the impacts of energy crisis. In advanced countries, technologies are being perfected for the bioconversion of wastes and surpluses of agricultural commodities into food, feed and fuel by microbial fermentation. Hazardous and environment-polluting pesticides are either used with caution or are being gradually replaced by biopesticides of microbial origin. The strategy, therefore, has been to rely increasingly on renewable resources on the farm through microbiological processes rather than non-renewable conventional sources not only to supplement chemical fertilizer requirements of crops but also to combat pests and plant diseases.

In this wide and varied branch of science, published literature in periodicals remains scattered and is not easily available to the majority of students and research workers in remote parts of the world who would like to gain an insight into microbial technologies in agriculture. Obviously, one individual cannot possibly summarize all the facets of agricultural microbiology and hence the need for an edited book with contributions from experts all over the world.

The contents of the book have been divided into three sections—the role of microorganisms in the mobilization of nitrogen and phosphorus to plant growth containing 13 chapters; management of pathogens, pests and weeds through microorganisms containing 5 chapters; and new strategies in bioconversion containing 7 chapters.

In the overview outline by Sylvan Wittwer, an eminent agricultural scientist of the USA, the entire gamut of biological sources of nutrients and recycling of organic wastes for food, feed and energy has been briefly covered together with statements on the future biotechnology approaches intended to augment nutrient supply to crop plants. Biotechnology in agricultural microbiology envisages gene splicing techniques with a view to transferring genes from one microorganism to another by genetic engineering to produce new microorganisms capable of speedy bioconversion or better nitrogen fixation.

Transfer of nitrogen-fixing traits (Nif genes) from a nitrogen-fixing microorganism to a non-nitrogen fixing species has been accomplished through genetic engineering. In fact, this approach has given useful results and plasmids carrying 'Nif' genes have been isolated from rhizobia and *Azotobacter*. Experiments have been carried out in which plasmids have been used as vehicles to transfer 'Nif' genes. Non-nitrogen fixing mutants of *Klebsiella* and *Azotobacter* have been made to take up 'Nif' genes to restore nitrogen fixation capacity. The first chapter is intended to highlight microbial genetics in relation to nitrogen fixation by Beringer, a specialist in this field who heads the Soil Microbiology Department, Rothamsted Experimental Station, England.

Nitrogenase is the enzyme which is responsible for the biological fixation of nitrogen and in this process nitrogen is reduced to ammonia. In the second and third chapters, the biochemistry of nitrogenase reactions and the transport mechanism of fixed nitrogen in the form of ureides in nodulated legumes have been ably discussed by Atkins and his colleague Rainbird from the botany department of the University of Western Australia, Nedlands, Australia.

It is not the prerogative of legumes to fix nitrogen since several genera of plants from non-legume families also bear root nodules which are inhabited by members of actinomycetes of the genus *Frankia*. Examples of such plants are temperate *Alnus* spp. and tropical *Casuarina* spp. These plants are very important in forest ecosystems. More and more genera of plants are being added to the list of already known nodule-bearing non-leguminous plants. The fourth chapter deals with the recent advances in this area, written by Becking from Wageningen, The Netherlands—a pioneer in this field.

Plant surfaces, especially those of roots and leaves, are congenial areas for a viable interaction between plants and microorganisms. Technically known as 'Rhizosphere' and 'Phyllosphere', these ecological microhabitats have been known to be inhabited by several nitrogen-fixing, auxin and antibiotic-producing microorganisms. Nitrogen-fixing bacteria such as *Azotobacter, Klebsiella, Beijerinckia* etc. colonize the catchment areas of plants, branches and bracts of forest trees. In fact a certain degree of intimate relationship exists between microorganisms and plants on the plant surface. This aspect has been highlighted in the fifth chapter jointly by Mukerji of the botany department, Delhi University and Subba Rao of the Indian Agricultural Research Institute, New Delhi, India.

Ever since Johanna Dobereiner of Brazil highlighted *Azospirillum* bacteria as associate symbionts in grasses, considerable work has been done on the classification, ecology, physiology, biochemistry and genetics of *Azospirillum* and other microorganisms of similar nature. The sixth chapter covers this new area of research by Patriquin of Dalhousie University, Canada who was closely associated with Dobereiner.

The blue-green alga *Anabaena azollae* lives as a microsymbiont in the aquatic water fern called *Azolla* and fixes a significant amount of nitrogen. The utility of these water-borne tiny plants in rice cultivation was known for several years in China and Korea. Considerable work has been done recently on the morphology, physiology and biochemistry of this unique symbiosis. In this connection, the work done at the Charles F. Kettering Research Laboratory, Ohio, USA is very significant. One of the research workers of this laboratory, Peters who has first-hand knowledge of this aspect and who has recently visited China tells a good deal about this fern in the seventh chapter.

During recent years in developing countries, judicious utilization of fertilizer nitrogen with organic manures is being advocated. Methods for mass scale production of *Rhizobium*, *Azotobacter, Azospirillum*, blue-green algae, phosphate solubilizing bacteria and compost accelerating microorganisms have been standardized and low cost technologies for their use have been perfected. The mass cultured microorganisms are impregnated on a carrier material such as peat, lignite or charcoal for seed or soil application. In blue-green algae and *Azolla*, methods to cultivate them in open air conditions have been devised. Such microbial preparations may be termed as 'biofertilizers' or 'microbial inoculants'. Through large-scale field experiments on the use of these biofertilizers it has been concluded that a saving of 20–30 kg N/ha is possible in cereal cultivation. This is indeed significant for developing countries where many crops do not receive even adequate fertilizer nitrogen. Quite recently, *Azospirillum* has been mass produced and used with profitable results. The eighth chapter written by Subba Rao highlights recent developments in this area.

Denitrification is a process opposite in magnitude to biological nitrogen fixation and is also mediated by soil bacteria. This process is significant in agriculture in the sense that it helps to maintain the much-needed equilibrium of nutrients in the soil. Examples of denitrifying bacteria are *Pseudomonas* and *Achromobacter*. The ninth chapter reviews research on denitrification and is written by Knowles of McGill University, Quebec, Canada who is a specialist in this area of research.

Ammonia salts are oxidized into nitrites and then to nitrates mediated by bacteria such as *Nitrosomonas* and *Nitrobacter*. These nitrates are denitrified into nitrogen gas by a variety of bacteria. Denitrification can be stemmed by nitrification inhibitors which are expensive, although inexpensive plant materials such as neem cake are also useful in controlling nitrogen losses. The tenth chapter written by Belser of New Zealand, an active worker in this area of research, is an authoritative account of the microbiology of nitrification and its inhibition processes.

Bacteria of the genus *Bacillus* and *Pseudomonas* and fungi such as *Penicillium* and *Aspergillus* solubilize bound phosphates such as tricalcium phosphate and rock phosphate and render them available for plant absorption. Application of these bacteria to seeds improves the amount of phosphate absorption by plants. The eleventh chapter written by Subba Rao briefly deals with these aspects of soil microbiology, especially in relation to low grade rock phosphate utilization.

Forest trees such as pine (*Pinus* spp.) have ectomycorrhizal associations on their roots where mycelia of fungi such as *Boletus* form an external mantle and absorb plant nutrients, especially phosphorus. Ectomycorrhizal associations are widespread in forest trees. The twelfth chapter covers the applied aspects of this branch of forest microbiology by Trappe and Molina, two U.S.D.A. experts in this field.

Examples of fungal endosymbionts forming endomycorrhizae on roots of grasses and many other plants are *Endogone* and *Glomus*. They are known as vesicular-arbuscular (V-A) mycorrhizae which help in phosphate absorption by plants. Extensive work has been done recently in this field. The thirteenth chapter written by Hayman of the Rothamsted Experimental Station, England who has exclusively worked in this area for several years, highlights the practical aspects of endomycorrhizae and makes suggestions for future developments.

The use of pesticides in agriculture has noticeable impact on microbial activities in soil. Some bacteria break down the initial structural configuration of the pesticide molecule and bring about its degradation. The fourteenth chapter covers this area of research and has been written by Walker of the Rothamsted Experimental Station, a renowned authority on the subject.

One of the splendid examples of microbial insect pathogens is *Bacillus thuringiensis* which attacks lepidopterous insects and serves as an inexpensive non-polluting bio-insecticide. Commercial preparations of this bacterium and other viral pathogens (polyhedroses and granuloses) are available for use in advanced countries. Similarly, plant pathogens have been suggested as candidates for microbial herbicides. These subjects have been covered in two chapters written by Aizawa of Japan and Freeman of the USA who have specialized in these areas.

It is possible to exploit the phenomenon of microbial antagonism in soil to achieve biological control of plant diseases. This approach is preferable to the use of chemical pesticides because microbial preparations do not pollute the environment. Nematode pathogens can be controlled by fungal predators and it is likely that even nematode problems can be solved by the application of fungi into soil. Future research in this potential area will be interesting and rewarding. The seventeenth chapter by Alan Kerr of the Waite Agricultural Research Institute, Australia, who has several years of experience in biological control of plant diseases, focusses attention on this area of research.

Commercially available antibiotics such as streptomycin, cycloheximide, aureofungin, aureomycin and blasticidin are being increasingly used in Japan and the USA to control plant diseases. The only limitation is that they are expensive. The eighteenth chapter written by Misato and Yoneyama, two experts from Japan, aims at consolidating our knowledge on the subject.

Composting is an age-old process. Cellulolytic and lignolytic bacteria and fungi are involved in degrading cellulose and lignin. *Cellulomonas* and *Trichoderma* are excellent examples of microorganisms involved in this process. Anaerobic clostridia also take part in this process. The nineteenth chapter by Subba Rao highlights the microbiological aspects of farm composting.

It is hardly necessary to emphasize the importance of microbiological aspects of methane bacteria in biogas production, and the twentieth chapter deals with these aspects and is written by Hobson of Scotland, who has devoted all his research career to the study of anaerobic digestion by microorganisms.

Direct conversion of cellulose to sugar is still in the experimental stage while the utilization of cellulosic wastes by fungi to produce a biomass which can be used as feed has no doubt potentialities for immediate exploitation and fungi are excellent candidates for growth on cellulosic materials. Similarly, surplus cereals, straw or newspaper can be fermented with yeasts to produce ethanol. Chapters 21 and 22 highlight these new strategies: Chahal and Overend of the National Research Council of Canada dwell in detail on these twin areas of obtaining feed and energy from agricultural products by microbial fermentation.

Photosynthetic bacteria grow in specialized micro-habitats and *Rhodopseudomonas* is an example of this kind which has also the ability to fix molecular nitrogen. Japanese workers have devised industrial methods for utilizing such phototrophic bacteria in agriculture. The twenty-third chapter, by Kobayashi of Japan who has spent all his research career on phototrophic bacteria, covers the developments in this area.

The use of single cell protein (SCP) from algae, bacteria and yeasts grown on easily available and inexpensive substrates—animal feed—is gaining ground because of its nutrient value. SCP has immense future possibilities in animal husbandry and Roth of West Germany, a specialist in this field, describes the role of microorganisms as animal feed in the twenty-fourth chapter.

Mushroom cultivation on logs of wood and agricultural wastes is not only a science but also an elegant art intended to supplement the increasing food needs of our growing population. From the handpicking stage of our forefathers, a stage has now come when mushroom cultivation has become a viable and stable industrial proposition and Chang and Li of Hong Kong who are authorities in this field, deal with this subject matter in the last chapter.

Most of the chapters are well illustrated and the sources of published materials which are reproduced here have been duly acknowledged at the appropriate places by the authors of the individual chapters. The authors and the editor express their gratitude to individuals and publishing companies who readily agreed to requests for reproducing such published materials. Some of the photographs presented in this book have been kindly provided by specialists in the field and the sources have been duly acknowledged at the appropriate places. To avoid delay in the release of the book, the proofs have been read by the editor. As a matter of fact, it has been possible to bring out the book in good time, thanks to the cooperation received from the publishers and the authors. I appreciate the promptness with which the individual authors wrote the manuscript and helped in many clarifications from time to time. I also wish to record my gratitude for the encouragement received in this endeavour from Dr. M.S. Swaminathan, Member, Planning Commission, Government of India, Dr. O.P. Gautam, Director-General, Indian Council of Agricultural Research, Dr. N.S. Randhawa, Deputy Director-General, Indian Council of Agricultural Research and Dr. H.K. Jain, Director, Indian Agricultural Research Institute, New Delhi. Finally, I will be failing in my duty if I do not say a word of thanks to my wife Gowri Subba Rao and my daughters Shambhavi and Shalini Subba Rao who helped in many ways for the success of this venture.

Division of Microbiology — N.S. SUBBA RAO
Indian Agricultural Research Institute
New Delhi

Contents

Section B. Management of Pathogens, Pests and Weeds through Microorganisms

Section C. New Strategies in Bioconversion

Contributors

K. Aizawa, Institute of Biological Control, Faculty of Agriculture, Kyushu University, Fukuoka 812, Japan.

C.A. Atkins, Botany Department, University of Western Australia, Nedlands, W.A. 6009, Australia.

J.H. Becking, Institute for Atomic Sciences in Agriculture, Wageningen, The Netherlands.

L.W. Belser, Cawthron Institute, P.O. Box 175, Nelson, New Zealand.

J.E. Beringer, Soil Microbiology Department, Rothamsted Experimental Station, Harpenden, Hertfordshire, AL5 2JQ, England.

H.E. Calvert, Charles F. Kettering Research Laboratory, 150 E. South College Street, Yellow Springs, Ohio 45387, USA.

D.S. Chahal, formerly of Iotech Corporation Ltd., 400-220, Laurier Avenue West, Ottawa, Ontario, Canada K1P 5Z9; 312–1800, Baseline Road, Ottawa, Ontario, Canada K2C 3N1.

S.T. Chang, Department of Biology, The Chinese University of Hong Kong, Shatin, N.T., Hong Kong.

T.E. Freeman, Plant Pathology Department, Institute of Food and Agricultural Sciences, University of Florida, Gainesville, Florida 32611, USA.

D.S. Hayman, Soil Microbiology Department, Rothamsted Experimental Station, Harpenden, Hertfordshire, AL5 2JQ, England.

P.N. Hobson, Microbial Biochemistry Department, Rowett Research Institute, Greenburn Road, Bucksburn, Aberdeen AB2958, Scotland.

A. Kerr, Department of Plant Pathology, Waite Agricultural Research Institute, University of Adelaide, Glen Osmond, Adelaide, South Australia 5064.

R. Knowles, Department of Microbiology, Macdonald Campus of McGill University, Ste Anne de Bellevue, Quebec H9X 1CO, Canada.

M. Kobayashi, Department of Agricultural Chemistry, Kyoto University, Kyoto, Japan.

S.F. Li, Department of Biology, The Chinese University of Hong Kong, Shatin, N.T., Hong Kong.

T. Misato, The Institute of Physical and Chemical Research, Wako-shi, Saitama 351, Japan.

R. Molina, United States Department of Agriculture, Forest Service, Pacific Northwest Forest and Range Experiment Station, 3200 Jefferson Way, Corvallis, Oregon 97331, USA.

K.G. Mukerji, Botany Department, University of Delhi, Delhi 110007, India.

R.P. Overend, Energy Project Biomass Convenor, National Research Council of Canada, Ottawa, Ontario, Canada K1A OR6.

D.G. Patriquin, Biology Department, Dalhousie University, Halifax, Nova Scotia, Canada B3H 4J1.

G.A. Peters, Charles F. Kettering Research Laboratory, 150 E. South College Street, Yellow Springs, Ohio 45387, USA.

R.M. Rainbird, Botany Department, University of Western Australia, Nedlands, W.A. 6009, Australia.

F.X. Roth, Institute for the Physiology of Nutrition, Technical University of Munich, D-8050 Freising-Weihenstephan, West Germany.

N.S. Subba Rao, Microbiology Division, Indian Agricultural Research Institute, New Delhi 110012, India.

J.M. Trappe, United States Department of Agriculture, Forest Service, Pacific Northwest Forest and Range Experiment Station, 3200 Jefferson Way, Corvallis, Oregon 97331, USA.

N. Walker, Rothamsted Experimental Station, Harpenden, Hertfordshire, AL5 2JQ, England.

S.H. Wittwer, College of Agriculture and Natural Resources, Michigan State University, East Lansing, Michigan 48824, USA.

K. Yoneyama, The Institute of Physical and Chemical Research, Wako-shi, Saitama 351, Japan.

S.H. WITTWER

An Overview of Agricultural Microbiology

The potential horizons for agricultural microbiology approach infinity. Only recently have we been able to envision some of the further frontiers involved in the nitrogen cycle in soils, plants, animals, and the atmosphere; the mobilization and solubilization of nutrients by mycorrhizae; the microbial possibilities for systems of integrated pest management; the role of microorganisms in the digestion of otherwise indigestible forages and the utilization of non-protein nitrogen sources consumed by ruminants for the eventual production of draft energy and of meat and milk; the bioconversion of agricultural wastes, residues and by-products for the synthesis of microbial protein useful for people and livestock, and for the generation of methane or biogas for home heating, cooking, and the generation of electricity (4). Finally, one must foresee opportunities for the transmission, through gene splicing and otherwise, of desirable characteristics from microorganisms to food, feed, and fiber crops, and to food animals.

The new focus in agricultural research must continue to shift away from fossil fuel-intensive technologies toward greater photosynthetic efficiency, more efficient nutrient and water uptake, improved biological nitrogen fixation, and greater resistance to competing biological systems and environmental stresses (23). Benefits derived from research focussed on any one or all of the above biological processes, as research areas, would be scale neutral. Agriculturally developing nations, and the developed or industrialized countries would be able to share in the same comparative benefits or advantages. These frontiers of agricultural microbiology are open to all. Advances made by one nation and shared by others can multiply the benefits of research many times (19).

We are now experiencing a renaissance in molecular biology and the primary focus is on genetic engineering of microorganisms in the greatest biological revolution of all time (22). Most of the excitement thus far has occurred in pharmacology and human medicine with the bacterial synthesis of mammalian and human proteins (insulin, interferon, and the growth hormones). These accomplishments in human physiology are providing a catalyst for work with food crops, food animals, and in food technology and processing. Only recently has the agricultural community begun to recognize the potential of microbiological techniques in plant cell research. Industrial institutions are now alerted to the possibilities of applying microbiological methods to agriculture. Dozens of new laboratories are now involved in designing microorganisms important to agriculture on the application of microbiological techniques to plant manipulation. One of the most important indicators of this global revolution in molecular biology is that most of these new laboratories have been established since 1979 (4).

In addition, many large chemical corporations and the newly formed biotechnology institutes are recruiting from academic institutions the most outstanding scientists with specialities in molecular biology, tissue culture, plant genetics, plant breeding, and microbiology. The rapid rise of new biotechnology companies has been referred to as "The Second Green Revolution." The future success of the new ventures in genetic engineering will depend on collaboration of scientists from many disciplines. There has been a literal transformation of American seed companies and those in other countries since 1978. Seed companies, with one or two exceptions, no longer exist as such, but have become integrated with or have become subsidiaries of large pharmaceutical and chemical companies. There is also the emerging issue of the future balance of sponsored agriculturally oriented microbiological research between the public and private sectors. It is now swinging rapidly toward the private sector fueled by tax write-offs and hopes of profits in the sale of patented seeds, varieties, and microorganisms on the one hand, and the lack of publicly supported microbiological research on the other (4).

The potential for advances in soil microbiology for enhancement of food production will be prominent in the future. The future resides not only with improved biological nitrogen fixation from selections of super *Rhizobium*-legume, *Azolla-Anabaena* and *Azospirillum*-grass systems, but the transfer of nitrogen fixation from prokaryotic species, bacteria, and blue-green algae to non-nitrogen fixers (3, 15). Super strains of both *Rhizobium* and *Azolla* have already been identified (4, 24). There are many systems, organisms, and crops yet to be inventoried in many parts of the world. Alfalfa may be the forage legume with the greatest potential for supplying nitrogen for succeeding crops in temperate zone agriculture. What of the rest of the world?

Billions of dollars are currently being expended globally on nitrogen fertilizers for grain and other crops (1, 4). A technology assessment is now in progress for biological nitrogen fixation in rice and maize since there are important possibilities for reducing mankind's dependence on nitrogen fertilizers manufactured from fossil fuels. There are also anticipated socioeconomic factors which may affect the widespread adoption of new biological nitrogen fixation technologies that work (18). Microorganisms are still key to all known phenomena of biological nitrogen fixation.

Another set of microbiological associations important for future agricultural productivity is the soil fungi called mycorrhizae (10). They colonize plant roots and create the equivalent of root extensions. Two significant payoffs are immediately apparent. Phosphorus and some micronutrients are made more available to plants in phosphate-poor soils by conversion into more soluble forms and by transporting them to the roots of plants; and secondly, mycorrhizae can also transport water to plants which is collected beyond the reach of the plants' root system. Thus mycorrhizal infected plants are more able to resist drought (14). The field of research on mycorrhizae constitutes one of the great opportunities in the application of microbiology for improved agricultural production on a global scale, especially for crops grown on mineral deficient soils and where soil moisture may be limited.

Losses of fertilizer and biologically fixed nitrogen applied to soils for crop production range between 50 percent in temperate zone grain production and as high as 75 percent for tropical rice culture. Two microbiologically powered processes are involved in these enormous losses—nitrification and denitrification. Denitrification is the reduction of nitrate or nitrite nitrogen to gaseous nitrogen (N_2O and N_2). It is usually catalyzed by bacteria (9, 20). Nitrification is the microbiological conversion of ammonia to form nitrate. There are nitrogen stabilizers or chemical nitrification inhibitors (2, 21) and inhibitors for denitrification (6, 7, 8, 9, 12, 16). Reduction or prevention of nitrification also reduces losses from denitrification. Fertilizer production technology (sulfur coated urea, super granules of urea and deep placement) may also reduce both nitrification and denitrification. The target of opportunity for reducing the losses of nitrogen from both nitrification and denitrification may reside in the chemical inhibition of the controlling microbiological processes (2, 20).

Improved microbial degradation of cellulose in agricultural wastes, residues, forages, and by-products must be sought after. Worldwide production of cellulose is estimated at over 100 billion tons (dry weight) per year. This is equivalent to approximately 150 pounds of cellulose produced daily for each of the earth's more than 4.4 billion inhabitants. The most important feed constituent, for ruminants in forages is cellulose. Its conversion into food on an economic basis is accomplished only by ruminants.

While cellulose is the world's most abundant organic compound lignin is next to cellulose for its prominence in nature. The biochemistry of its microbiological degradation and conversion to food and feed is much more challenging than for cellulose. Biodegradation of the ligno-cellulose fractions in plant biomass offers one of the most exciting research areas for the future (5). Not only would energy output be greatly increased, but the feed value for livestock vastly improved. The recent developments in livestock feeding involving the rumen bypass for increased protein utilization and appetite stimulation, and the use of "Rumensin" and Monensin (11) added to the rations of dairy and beef cattle to increase nutrient utilization from forages bear watching. Technologies are also available for producing food directly from microbes, once the demand becomes sufficient to warrant the cost of processing. It may now, however, require more energy to produce the synthetic and microbial foods than to utilize the primary products (13, 17).

One of the most exciting developments in agricultural microbiology is the generation of biogas (methane) from livestock wastes in China. This nation has over 300 million pigs fed in large part by aquatic plants, and in which the manure is used for the generation of methane in over seven million farm or household sized generators. Many variables are being studied in several methane gas research institutes in China. These include structural types and designs for generators, modifications in substrate, development of genetically superior organisms with special reference to low temperature tolerance and the use of solar energy as a source of heat for improved generation during winter months. The effectiveness of the entire system is the microbiological conversion of cellulose and other materials to methane under anaerobic conditions and with the production of a residue that is still effective for crop fertilization (24).

The chapters which follow cover research frontiers in agricultural microbiology which are currently grossly underfunded in view of the rising importance of renewable resources; and in view of technologies which will add to the resources of the earth; and which are nonpolluting, environmentally benign, are scale neutral, and sparing of capital, management and nonrenewable resources. Stable production at high levels must be sought after if food needs for people are to be met for an ever-increasing population growth, rising affluency of nations, and a search for improved diets.

References

1. Aldrich, S.R. *Nitrogen in Relation to Food, Environment and Energy*, Special Publication 61, Agricultural Experiment Station, College of Agriculture, University of Illinois-Urbana, Champaign, Illinois (1980).
2. American Society of Agronomy. *Nitrification Inhibitors—Potential and Limitations*, American Society of Agronomy Special Publication Number 38 (1980).
3. Bauer, W.D. Infection of legumes by rhizobia, *Annual Reviews of Plant Physiology*, *32*, 407–449 (1981).
4. Brill, W.J. Agricultural microbiology, *Scientific American*, *245* (3), 199–215 (1981).
5. Crawford, D.L. and Crawford, R.L. Microbial degradation of lignin, *Enzyme Microbiological Technology*, *29*, 11–22 (1980).

6. Delwiche, C.C. *Nitrification, Denitrification and Atmospheric Nitrous Oxide*, Academic Press, Incorporated, New York (1981).
7. Delwiche, C.C. and Bryan, B.A. Denitrification, *Annual Reviews of Microbiology*, *30*, 241–262 (1981).
8. Firestone, M.K. Denitrification, pp. 289–326 in *Nitrogen in Agricultural Soils*, Agronomy Monograph No. 22 (Edited by F.I. Stevenson, J.M. Bremmer, R.D. Hauck and D.R. Keeney), American Society of Agronomy, Madison, Wisconsin (1981).
9. Garcia, J.L. and Tiedje, J.M. Denitrification in soils, in *Microbiology of Tropical Soils* (Edited by Y. Dommergues and H.G. Diem), Martinus Nijhoff, The Hague (1981).
10. Hayman, D.S. Mycorrhiza and crop production, *Nature*, *287*, 487–488 (1980).
11. Isichei, C.D. and Bergen, W.G. The effect of monensin on the composition of abormasal nitrogen flow in steers fed grain and silage rations, Paper presented at the 32nd Annual Meeting of the Animal Science Society of America, Cornell University, New York, July (1980).
12. Knowles, R. Denitrification, pp. 323–369 in *Soil Biochemistry*, Volume 5 (Edited by E.A. Paul and J.M. Ladd), Marcel Dekker, Incorporated, New York (1981).
13. Litchfied, J.H. Microbial protein production, *Bioscience*, *30* (6), 387–396 (1980).
14. Nelsen, C.E. and Safir, G.R. Increased drought resistance in onion plants by mycorrhizal infection, *Planta* (in press) (1981).
15. Office of Technology Assessment. *Impacts of Applied Genetics: Microorganisms, Plants, and Animals*, United States Congress, Washington, D.C. (1981).
16. Payne, J.J. *Denitrification*, Wiley Interscience, New Jersey (in press) (1981).
17. Pramer, D. Microbiology and the feeding of mankind, *Bioscience*, *30* (6), 375 (1980).
18. Randolph, R.H. *A Technology Assessment of Biological Nitrogen Fixation—A Prospectus*, Resource Systems Institute, East-West Center, Honolulu, Hawaii (1981).
19. Revelle, R. Biological research and third world countries, *Bioscience*, *30*, 727 [editorial] (1980).
20. Tiedje, J.M. Denitrification, in *Methods of Soil Analysis*, second edition, American Society of Agronomy, Madison, Wisconsin (in press) (1981).
21. Varsa, E.C., Liu, S.L. and Kapusta, G. The effect of nitrification inhibitors on wheat yields and soil nitrogen retention, *Down to Earth*, *37* (3), 1–5 (1981).
22. Waldrop, M.M. *Chemical and Engineering News*, June 1, pp. 23–28 (1981).

23. Wittwer, S.H. The next generation of agricultural research, *Science*, *199*, 4327 [editorial] (1978).
24. Wittwer, S.H. Narrative account of visit to the People's Republic of China, September 8–October 6, 1980, Michigan State University, East Lansing (1980).

SECTION A

Microorganisms and Mobilization of Nutrients for Plant Growth

J.E. BERINGER

1. Microbial Genetics and Biological Nitrogen Fixation

Introduction

Genetic studies of nitrogen-fixing microorganisms have had two main aims. The first has been to make a formal genetic analysis of the genes involved in nitrogen fixation, to determine their function, their interaction with other genes and to characterize the gene products. The second has been to facilitate the manipulation of nitrogen-fixing bacteria to produce improved strains for use as inoculants on crop plants. Formal genetic studies can be split into those directly concerned with the nitrogen fixation genes (*nif* genes) of *Klebsiella pneumoniae* and studies of other nitrogen-fixing prokaryotes, such as *Rhizobium*, *Azotobacter* and the cyanobacteria. For convenience this arbitrary distinction will be used in this chapter as a basis for discussing the role of genetics in developing our understanding of nitrogen fixation and manipulating agriculturally-important bacteria.

Genetic methodology

The selection of mutants

In order to identify genes it is necessary to have strains of an organism which differ from the wild-type. These strains can be deficient in the functioning of a gene (for example, having a requirement for a metabolite; auxotrophic mutants) or have altered levels of gene expression (regulatory mutants). Either type of mutant can be obtained by screening large populations for rare spontaneous mutants or by inducing mutation.

SPONTANEOUS MUTATION. Spontaneous mutants are useful for genetic studies because they have not been produced by mutagenic procedures which are liable to have caused other defects that may or may not be detected (20). Unfortunately it is usually only possible to obtain spontaneous mutants for functions which are fairly easily selected in the laboratory. This is because spontaneous mutation rates are usually low (10^{-6} or less) and sufficiently large populations are not easy to screen unless some direct selection can be used. Therefore spontaneous mutants are often available only for genes involved in resistance to drugs or other agents and for revertants of auxotrophic mutants. This latter class of mutant is interesting because in a proportion of revertants the 'wild-type' phenotype is obtained by mutation at another site. Such 'revertants' are strains in which the mutant phenotype is suppressed; often they are regulatory mutants which over- or under-produce a product.

INDUCED MUTATION. Mutation can be induced by any treatment which alters the base composition of DNA, either directly or by interfering with the fidelity of DNA replication. Most traditionally used mutagens (e.g. nitrous acid, radiation, hydroxylamine) cause damage to DNA which results in a change of base sequence when the damaged bases are repaired or replicated. These agents have been used for most species of nitrogen-fixing prokaryotes. Variation in the efficiency of different mutagens with different strains of bacteria can occur and some mutagens can be quite ineffectual with particular species (6, 21).

Recently there has been much interest in the use of 'mutagens' which insert defined DNA sequences within the host DNA and cause major changes to the mutated gene. The first 'mutagen' of this type to be used widely was the bacteriophage Mu (33). This is a lysogenic phage which infects some members of the enterobacteriaceae and lysogenizes them by inserting into host DNA. Insertion appears to be random, and when it occurs within a gene of known function the gene product is no longer produced. Insertion mutants of this type are valuable because they inactivate the gene into which they insert and, if this gene is an operon, any genes distal to it can no longer be expressed. The terminator effect is due to the prophage terminating the transcription of the DNA (see Fig. 1).

Mu has been useful for genetic studies of *nif* genes in *Klebsiella pneumoniae* because of its ability to integrate at random and potentially produce Nif^- strains by insertion mutation or by deletion (43, 51, 52) and because of its use to define operon structure (51). Indeed insertion 'mutagens' of this type are most efficient 'tools' for defining operon structure and for finding internal promoters with operons (see Fig. 1).

Because Mu is relatively limited in the range of prokaryote species that it can infect, there has been much interest in the use of transposons as insertion mutagens (45). Transposons are sequences of DNA which are able to integrate into other DNA sequences and must do so to be replicated. Most that are in common use carry selectable drug resistances. Insertion causes the same genetic effects as Mu and they have been used in *K. pneumoniae* for the same purposes as Mu (43, 57). A number of different procedures are now available for introducing transposons into different bacterial species (7, 45, 57).

A major advantage of transposon mutation is that only one transposon is

Fig. 1. The effect of insertion mutation on the transcription of messenger RNA from an operon.

For this example the operon consists of three genes and has a weak internal promoter just before gene C. The main promoter for messenger RNA transcription is just before gene A. The same effects would be caused by transposons or phage Mu.

A. Production of messenger RNA from the 'wild-type' operon.

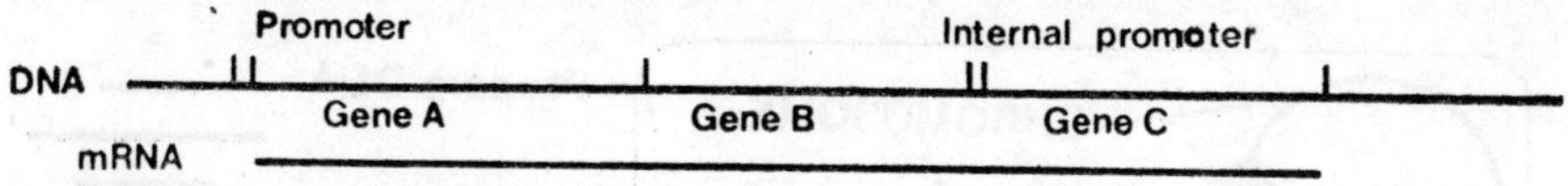

Normal promoter activity gives messenger RNA for all three genes. Therefore weak internal promoter activity within the operon is not detectable.

B. Production of messenger RNA after mutation caused by insertion of a transposon or phage Mu within gene A.

Note that in this example the mutation is due to insertion in the middle of gene A and no DNA corresponding to gene A is lost.

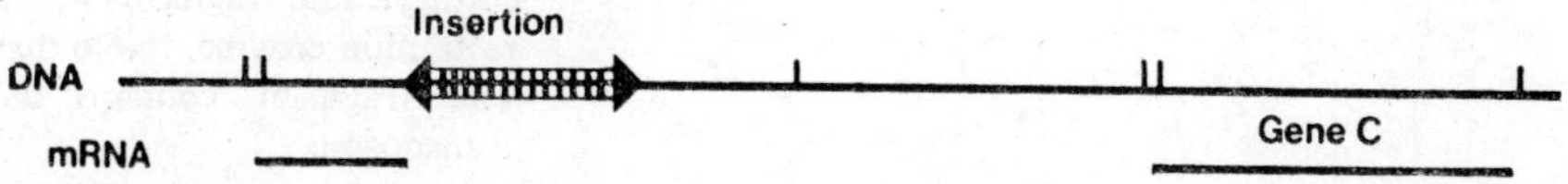

This mutation would be characterized as one in which no gene A or B product was made and only a small amount of gene B product would be detectable. If messenger RNA production were studied normal amounts of messenger for the DNA between the main promoter and the insertion of DNA would be found. No messenger RNA for the rest of gene A, none for gene B and low levels for gene C would be detected. This is because the insertion will have terminated transcription which cannot start again until another promoter occurs. Internal promoters are often weak and hence only small amounts of messenger RNA for gene C can be produced.

present and therefore the organism has undergone a single mutagenic event. Furthermore, the site of the mutation is defined by the position of the drug resistance gene carried on the transposon (4, 45, 57). Thus the mutation can be mapped by mapping the new drug resistance gene. The DNA from that region can be cloned by selecting the fragment of DNA from a total DNA preparation that hybridizes with pure transposon DNA (see Fig. 2). The fragment of DNA containing host DNA and the transposon can then be hybridized with DNA from a 'wild-type' strain and the homologous fragment identified by its ability to hybridize with the host DNA associated with the transposon. If desired the 'wild-type' gene can be re-introduced into the original mutant strain to test for its ability to suppress the mutation caused by the transposon insertion. Transposons have been used by B.G. Rolfe and J. Shine (personal communication) to produce symbiotically-defective mutants of *Rhizobium*, to isolate and clone the 'wild-type' genes and to look for gene expression in diploids. A further advantage of this type of analysis is that it enables one to look at the DNA from different strains and species to

Fig. 2. Gene isolation.

The basis for this technique is that when a gene function is lost due to transposon mutation that gene is at the site of insertion of the transposon in the host genome. It has either been inactivated due to insertion within it, or is part of an operon which has been inactivated. Pure transposon DNA is easy to prepare and can be used to 'find' other sequences of DNA which are attached to transposon DNA.

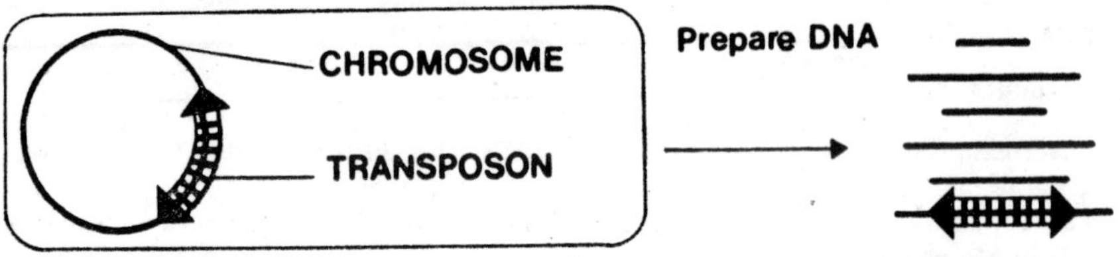

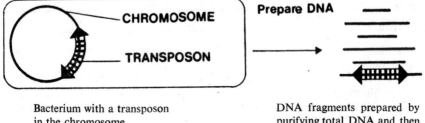

Bacterium with a transposon in the chromosome

DNA fragments prepared by purifying total DNA and then cutting it into fragments with a restriction enzyme. Note that one fragment contains the transposon.

Because the DNA fragments are of different sizes they can be separated on the basis of their migration through an agarose gel (A). One band will correspond to the fragment of DNA which contains the transposon.

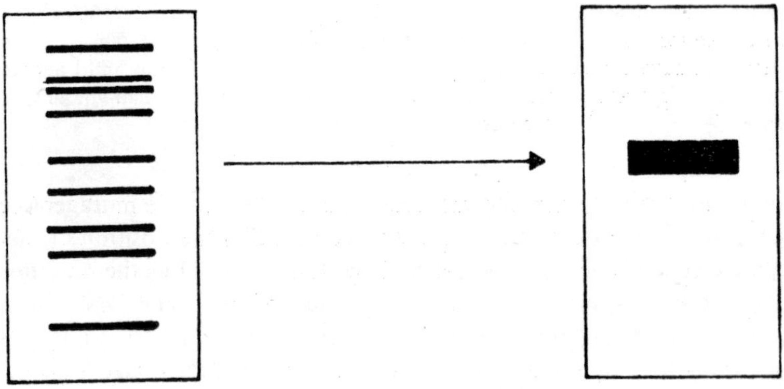

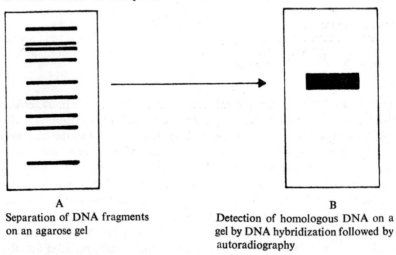

A
Separation of DNA fragments on an agarose gel

B
Detection of homologous DNA on a gel by DNA hybridization followed by autoradiography

The band which carries transposon DNA can be found by testing for the ability of purified transposon DNA to hybridize with DNA on the gel. This is done by denaturing the DNA in the gel and also a sample of purified radioactive transposon DNA. When the two are added together and allowed to renature radioactive DNA will appear in the appropriate band. It can be found by autoradiography (B). The corresponding region from another gel can be cut out and DNA isolated from it. The purified DNA can then be cloned.

The wild-type gene can be cloned by purifying DNA from the parent non-mutant strain, cutting it into fragments with restriction enzymes and separating these on an agarose gel. To find the wild-type DNA the cloned mutated gene is used as a radioactive hybridization probe in the same way as the transposon was used to isolate it. In this case the fragments of host DNA attached to the transposon provide the regions of homology to identify the gene.

Genes can also be isolated by using the same type of procedure but instead of separating DNA fragments on agarose gels they are cloned into plasmid or phage DNA. A sufficient number of cloned strains are produced so that there is a finite possibility that all the fragments have been cloned. The bacterium or phage population that carried the gene can be detected by hybridizing labelled DNA with lysed colonies of the bacterial strains or suitably treated phage plaques. Once identified the parent stock can be multiplied and used as required.

---

determine whether particular genes are conserved and to study gene function in different genetic backgrounds.

Recently a number of workers (55, 71) have shown that at least two *K. pneumoniae nif* genes are very highly conserved in nature on the basis that *K. pneumoniae nif* DNA hybridized to DNA from all the nitrogen-fixing prokaryotes that they tested. Ruvkun and Ausubel (72) have taken advantage of this DNA homology to use *K. pneumoniae nif* DNA to find and clone *Rhizobium nif* genes in preparations of *R. meliloti* DNA. Having cloned the wild-type *Rhizobium* DNA in *Escherichia coli* they were able to mutate the *nif* genes in this species using transposon mutagenesis, and then transfer the mutant genes into wild-type bacteria. Bacteria were selected in which the cloned, mutated DNA had recombined with the chromosome and replaced the chromosomal DNA. By this means they were able to produce Nif$^-$ mutants of *R. meliloti*. Because there is no theoretical limit to the number of times that DNA can be isolated and cloned, it is possible to repeatedly re-isolate flanking DNA sequences, mutate them, replace the wild-type genes with the mutant ones and thus produce a sequence of strains carrying linked mutated genes. Theoretically this technique avoids the site specificity of other mutagens. Therefore regions of DNA which have not been mutated by other means can be studied to determine whether they carry genes whose function is essential, or whether they are regions of genes with no known phenotype or of non-coding DNA. This type of directed mutagenesis should be very useful for demonstrating a linkage between closely linked genes in prokaryotes for which efficient procedures of fine structure mapping are not available.

Techniques for mapping genes

TRANSFORMATION. Transformation is a procedure in which pure DNA obtained from one strain of bacterium is added to another which can take it up and occasionally undergo recombination replacing the similar region of its own DNA with the transforming DNA. When the DNA used to transform a strain carries a gene with a phenotype different from the corresponding gene in the recipient, the recombinant will show the new phenotype. For example, transformation using DNA from a drug-resistant strain can be used to produce resistant derivatives of the recipient.

If DNA used for transformation is able to form a circle and be replicated in this form (for example plasmid DNA) recombination is not required for the establishment of donor genes. The ability to transform bacteria with plasmid DNA is very important to facilitate cloning experiments. However, for many of the genetically well-known nitrogen-fixing bacteria, transformation of the strains used for genetic studies has yet to be demonstrated convincingly. Fortunately *Klebsiella*, *Rhizobium* and *Azotobacter* are Gram-negative and can exchange DNA with *Escherichia coli* by conjugation (3, 17, 22, 26). Because *E. coli* is the organism of choice for genetic manipulation studies most of the procedures can be carried out in this species and the final cloned DNA be returned to the original host by conjugation.

Transformation has been of limited value for genetic studies of nitrogen-fixing prokaryotes to date. This is partly because the size of DNA fragment involved is small and it is, therefore, difficult to show that two genes are linked unless very large numbers of mutations and crosses are examined. There have been many reports of transformation in *Rhizobium* (2, 69, 83) and *Azotobacter* (63, 64). In the latter studies it was reported that when *Rhizobium* DNA was used to transform Nif^- *Azotobacter* strains, Nif^+ transformants were produced. Some of these Nif^+ transformants had also inherited the ability to produce polysaccharides which bound to lectins (9). Whether or not these findings can be exploited to facilitate genetic studies of *Azotobacter* species has yet to be determined. They imply that the *R. trifolii* donor had nitrogen fixation genes on a plasmid which also carried genes for the lectin-binding polysaccharide. If lectin binding really is involved in host range determination by *Rhizobium* species this is another example of the linkage of symbiotically-important genes on extrachromosomal DNA.

TRANSDUCTION. Transduction is the method of gene transfer mediated by bacteriophages. When phage particles are being assembled there is a small chance that host DNA is packaged by mistake and phages are produced which carry it (see Fig. 3). How frequently this occurs is determined by the mode of DNA replication and by the nature of the packaging process during the production of mature phages. Many of the better-known transducing phages (e.g. λ, Pl) are lysogenic phages. Such phages are advantageous for transduction because usually they do not kill the host they infect. However, virulent phages can be used, for example the transducing phage T4 used for *E. coli* (82), RL38JI used for *R. leguminosarum* and *R. trifolii* (15, 16) and DF2 used for *R. meliloti* (18, 19). With virulent phages a procedure is required to prevent the recipient bacteria being killed by the virulent nontransducing phage particles. This can be done by the use of antiserum to remove free phages (18), by the use of conditional mutations in the phage so that under non-permissive conditions (e.g. high temperature) they are unable to carry out a lytic infection (19); and by inactivating phage DNA (e.g. by irradiation) so that the proportion of virulent phages is reduced (15, 16).

Transduction can be generalized or restricted. Generalized transduction is that in which any region of the chromosome is transferred, while in restricted transduction only DNA from a specific region is transferred (Fig. 3). The specificity observed with restricted transduction is usually because the phages involved are lysogenic and occasionally transfer host DNA from the point of integration into the host genome. Bacteriophage λ is the best known example of this type of phage.

For mapping purposes and strain construction generalized transducing phages are most useful. The most widely used phage for the analysis of *nif* genes has been Pl, a lysogenic phage which does not integrate into the chromosome. Pl can infect *Klebsiella pneumoniae* and was used for early studies of the *nif* genes to demonstrate linkage of these genes to each other and to the *his* operon (80). Pl has been used recently for linkage studies of *K. pneumoniae nif* genes (42, 43) but fine structure analysis is now much more easily done by deletion analysis, cloning and the use of physical DNA studies (1, 51, 67, 68). Generalized transduction has been reported for *Rhizobium* species (15, 16, 18, 19, 48, 49, 75) though as yet it has been little used for studies of genes involved in nitrogen fixation.

Transduction, like transformation, is not a good procedure for initial mapping studies because small pieces of DNA are transferred and hence it is difficult to establish linkage maps unless very many mutants are available. However, once a rough map is available transduction is very useful for more detailed studies of the linkage of known genes, especially when they are close together and not separable by techniques such as conjugation. The transfer of small segments of DNA is important for strain construction work where it is undesirable to transfer too many genes because of the probability of introducing unknown and unwanted genes.

A further potential value for transduction is in the mapping of plasmid-borne genes. Recent studies have shown that a number of genes involved in symbiosis are plasmid borne in *Rhizobium* (4, 10, 11, 16, 39) and that these plasmids are very large (11, 59). Such plasmids cannot be mapped by conjugation using normal techniques. However, Buchanan-Wollaston *et al.* (16) showed that plasmid-borne genes could be transduced between strains and they used this procedure to demonstrate the linkage of bacteriocin and nodulation genes.

Conjugation. Conjugation is a method of gene transfer in which two bacteria form mating pairs which are capable of transferring large proportions of the donor genome. Because large fragments are transferred conjugation is the method of choice for the construction of linkage maps. In nearly all the well studied conjugation systems pair formation and DNA transfer

Fig. 3. The production of transducing bacteriophages.

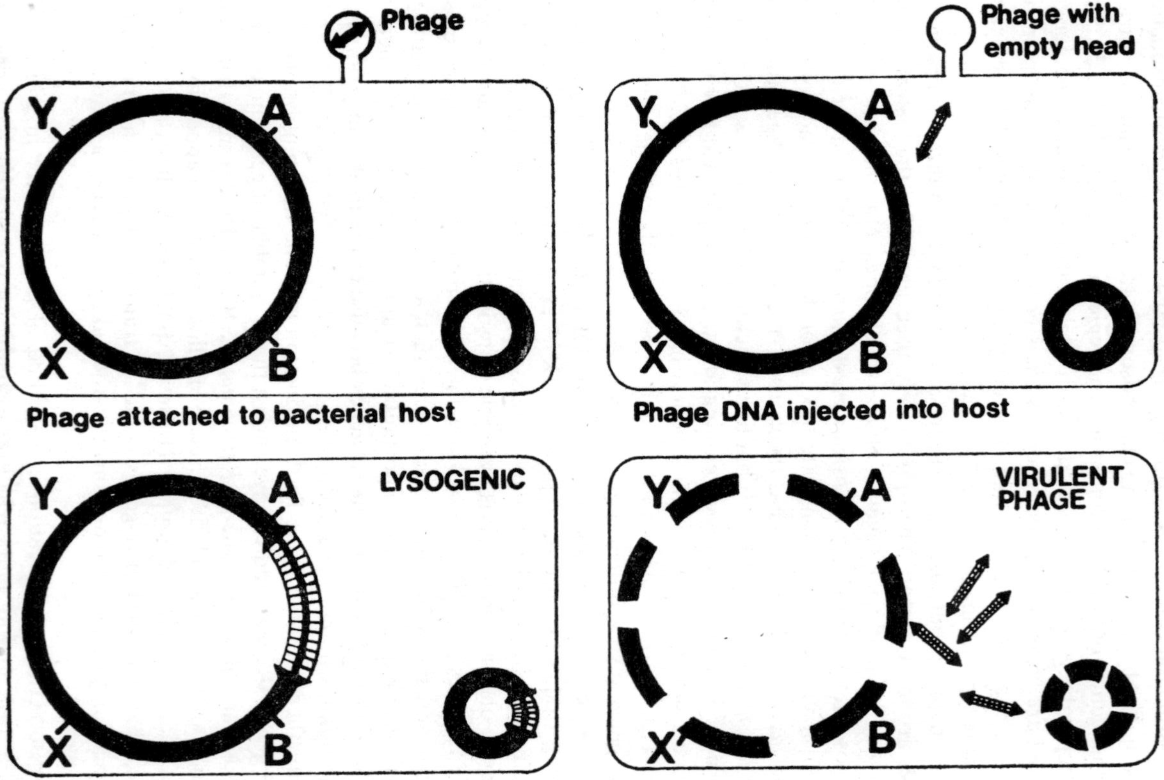

Phage DNA becomes integrated into host DNA close to gene A. Phage is now a 'prophage'. Phages are produced after induction (e.g. by treatment with ultraviolet light).

Phage DNA is replicated and viable phage particles are produced. Host DNA can be brokendown immediately.

Types of DNA packaged

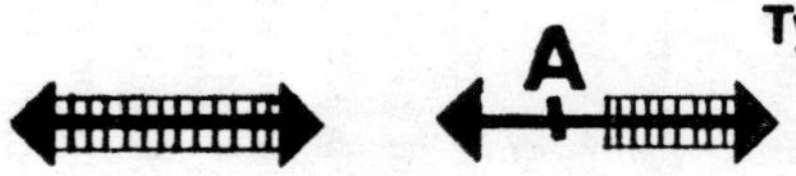

Phage DNA and consisting of phage + host DNA

Most prophages excise accurately and only phage DNA is replicated to make phage particles. Occasionally excision is imprecise and neighbouring DNA is replicated as part of the phage DNA. Therefore when phage particles are produced a small proportion carry both phage and host DNA (in this example the host gene A). Transduction by phages of this type is called restricted transduction.

Phage DNA and host DNA

A few phages carry host DNA instead of phage DNA. Their existence can only be determined by the ability of a population of phages to transduce bacteria or occasionally when the host DNA is of a different density from that of the phage DNA. In the latter case phages carrying host DNA can be separated from the rest of the population of virulent phages by centrifugation. Because any fragment of host DNA can be packaged (in this example gene Y) transduction by these phages is called generalized transduction.

are promoted by plasmids, which transfer themselves in the process. Because plasmid transfer in conjugation is usually much more frequent than the transfer of other genes it is the most common method for exchanging plasmids between strains.

The main role of conjugation for the current studies of the *K. pneumoniae nif* genes is in the exchange of derivatives of the plasmid pRD1 which carry known mutations in *nif* genes between *E. coli* and *K. pneumoniae*. Because the *nif* genes are carried on this plasmid, mutations can be induced and studied in *E. coli* and the plasmids can then be introduced into appropriate *K. pneumoniae* strains for studies of dominance and complementation (25, 26, 27, 43).

For other nitrogen-fixing prokaryotes conjugation still has an important role in the establishment of genetic maps and in the movement of indigenous plasmids. For *Klebsiella* and fastgrowing *Rhizobium* species chromosome mapping has been fairly extensive (5, 6, 18, 46, 47, 52, 53, 56), but as yet no linkage maps are available for other nitrogen-fixing species. Hopefully over the next few years maps will be produced for all the Gram-negative species on the basis of conjugation mediated by plasmids. The reason for this optimism is that there are a number of plasmids presently available which are readily transmissible between all Gram-negative bacteria (62, 72). The plasmids (e.g. RP4, R68.45) can mobilize chromosomal DNA and have been the basis of mapping studies in a number of bacterial species (5, 47, 76, 79, 81).

Not only can plasmids mobilize fragments of chromosomal DNA between strains during conjugation, but they can also mobilize other plasmids or pick-up DNA from these plasmids or the chromosome. When two plasmids fuse the product is known as a chimeric plasmid. When fragments of DNA are picked up and become integrated into plasmid DNA the new plasmids are known as plasmid primes. Chimeric plasmids can have the host range of the most promiscuous plasmid and can be used to transfer plasmids across wide taxonomic boundaries. For example, the *Agrobacterium* tumour plasmid has been transferred to *E. coli* by this means (73). Complex chimeric plasmids are often unstable. They can be used to produce plasmid primes by selecting for stable derivations. Plasmid primes can also be produced by conjugating and selecting for inheritance of donor (non-plasmid) DNA in a strain which is unable to recombine the donor DNA into its genome—e.g. *rec* A strains or other species (27, 38) or by introducing DNA into the plasmids by *in vitro* DNA manipulation techniques (34, 41, 44, 72).

All these procedures for producing plasmid primes have been used with many nitrogen-fixing bacteria and are likely to be used even more in the future. An important advantage of plasmid primes is that they facilitate studies of genes and their products in different backgrounds.

PROTOPLAST FUSION. Protoplast fusion is a technique for promoting recombination by fusing two strains which have had their cell walls removed. As yet it has not been used on a large scale for studies of nitrogen-fixing microorganisms. However, it does have the advantage that it works well with Gram-positive bacteria, which often do not have good conjugation systems. One of the best documented genetic studies using protoplast fusion has been with *Streptomyces coelicolor* (32), an actinomycete closely related to *Frankia* species. If mutant strains of *Frankia* become available it is likely that protoplast fusion will be the most useful procedure available for genetic analysis and strain construction work. Whether or not it can be used for genetic studies of Gram-negative nitrogen-fixing species has yet to be determined. The major limitation appears to be the ability of protoplasted Gram-negative bacteria to regenerate cell walls.

The great advantage of protoplast fusion for strain construction purposes is that the fused protoplasts initially contain two complete genomes. Thus with appropriate selection techniques any number of chromosomal or plasmid-borne genes may be selected. This type of gene transfer is particularly suited for the production of new strains for commercial purposes when an understanding of the basic genetics is not needed and there may be a requirement for the re-assortment of quite a large number of genes.

OTHER FORMS OF GENETIC ANALYSIS. Over the last few years a wide range of techniques have been evolved for manipulating DNA; some of these have been discussed already. Most of the techniques which have been developed not only facilitate physical studies of DNA but are also able to detect small differences in base sequence. Because details of the procedures are not necessary for an understanding of their role in genetic studies a list of techniques with a brief account of their use will be given.

a) *DNA sequencing*. Recent developments in procedures for sequencing DNA (54) have reached a point where this is a viable procedure for analysing differences between quite large molecules (e.g. plasmids). The widespread use of cloning has meant that interesting genes, such as nitrogenase structural genes, can be isolated and sequenced with comparative ease. In future, comparative studies will benefit from information about the DNA sequences of relevant genes.

b) *DNA homology*. Differences between plasmids or genes derived from different organisms can be determined by preparing single-stranded DNA from the different strains and mixing the DNA together. When conditions are altered, the DNA molecules can come together so that pairing occurs. The amount of pairing is determined by the extent that the base sequence of the DNA in the different organisms differs. Sequence homology can be estimated by measuring hybridization using spectrophotometric and other techniques or by looking at molecules under an electron microscope (heteroduplex analysis). The latter method is particularly suited to examining DNA from different closely related sources which is only expected to differ due to deletions or additions of DNA within the sequence being studied. Fairly crude estimates of hybridization are useful for taxonomic studies (29) while studies of heteroduplex formation in the electron microscope have been particularly useful for analysing plasmids and plasmid primes (23).

c) *DNA melting*. When DNA is heated the strands start to separate at temperatures which are determined by the ratio of G+C to A+T. GC : AT ratios provide a useful criterion of sequence homology for taxonomic purposes (24, 29). Much more detail can be obtained by using an electron microscope to look at DNA heated to different temperatures. Regions of DNA rich in AT base pairs melt first and thus it is possible to produce maps showing the distribution of AT-rich regions (14). Perhaps the most important use of techniques for examining regions of different % GC composition for studies of nitrogen-fixing organisms has been in the analysis of the plasmid prime pRD1, which carries the *Klebsiella nif* genes together with other chromosomal genes from the original *Klebsiella* host. Pühler *et al.* (68) showed that there were three different types of DNA present, as judged by the AT : GC ratio. These were found to correspond to plasmid DNA, *Klebsiella* chromosomal DNA and the *nif* gene region. Thus the *nif* genes had a different %GC composition to the other *Klebsiella* chromosomal DNA. It is interesting to speculate whether or not this implies that the *nif* genes came from another host in the past, particularly as recent hybridization studies have shown that the nitrogenase structural genes have very similar DNA sequences in a wide range of quite unrelated nitrogen-fixing microorganisms (55, 71).

d) *Electrophoresis of DNA*. Different sized fragments of DNA can be separated from each other in agarose gels when a current of electricity is passed through the gel (Fig. 3). Separation is dependent upon molecular weight and the structure of the DNA (e.g. linear and circular molecules migrate at different rates). This technique has been particularly useful for the analysis of plasmids and for separating fragments of DNA produced as a result of restriction enzyme digestion. At present plasmids up to a molecular weight of about 400×10^6 can be isolated from bacteria and distinguished on agarose gels. This technique has been used extensively to demonstrate the presence of plasmids in nitrogen-fixing microorganisms; for example *Rhizobium*, *Azotobacter*, *Azospirillum* and the cyanobacteria (28, 30, 31, 50, 59, 60, 66, 70). The procedure is very useful for analyzing the progeny produced as a result of crossing different strains of bacteria, to obtain information about which plasmids have been transferred, whether plasmids have been lost and whether or not recombination has occurred between plasmids (11).

Gel electrophoresis of DNA has been used extensively for separating fragments of DNA produced by restriction enzyme digestion. Because restriction enzymes cut DNA at specific points, a plasmid or chromosome will give a fixed number of bands of DNA on an agarose gel. Differences in banding patterns can be used for strain identification and for demonstrating differences between strains (58). Such patterns are particularly useful for characterizing differences between plasmids and for mapping them (37, 68).

The use of genetic markers for ecological studies

Bacteria are too small to be characterized in detail by morphological criteria. Therefore their taxonomy has been based primarily on differences in metabolic properties. While such criteria are usually (but not always) suitable for identification at the species level, they are of little value in strain identification. Serology and typing on the basis of phage and bacteriocin resistance are the methods most commonly used for strain identification, particularly with bacteria of medical importance.

One of the nitrogen-fixing microorganisms whose ecology has been studied in most detail is *Rhizobium*. This is because of the importance of these bacteria in forming nitrogen-fixing nodules on legume plants and an obvious desire to understand their distribution in the soil. Discussion of strain identification will, therefore, be based on work with *Rhizobium*, though the methods and ideas will be common to all nitrogen-fixing microorganisms.

The basis for methods of strain identification is that all organisms are the product of the functioning of many different genes. The more related the organisms are to each other, the greater the number of genes they have in common and hence the fewer the differences between them. Therefore the problem with strain identification is one of looking for a limited number of possible differences between closely related organisms. Typing on the basis of patterns of resistance and sensitivity to phages and bacteriocins, and serology are remarkably useful, though cross reactions with different strains do occur (8, 6, 5, 77, 78). The degree of confidence that can be expected depends upon the use of many different phages, bacteriocins and antisera to build up a pattern of response. Recently, differences in the intrinsic resistance to antibiotics has also been used for strain identification (8, 40).

The problem with the procedures described above is that each strain must be screened for a number of properties, which is tedious and time-consuming. However, the procedures have the advantage for ecological studies that the properties of the strains have not been altered. Strains which have had their properties altered by mutation (either spontaneous or induced) also have a role in field studies. The most attractive strains for this purpose are those which have been selected on the basis of their resistance to high levels of antibiotics. Such bacteria are easy to recognize and can be selected from large populations of sensitive bacteria by exposing the mixture to the antibiotics. Mutants of this type have been used for field (8, 12, 13) and *in vitro* studies of competition between *Rhizobium* strains (12, 13, 35, 36, 65, 74).

Traditionally the numbers of rhizobia in soil could only be estimated by adding different dilutions of soil to appropriate axenically-cultured host plants. This technique is very slow and expensive and limits the number of estimations which can be made. In many cases mutants which are resistant to high levels of antibiotics can be isolated directly from soil or the rooting medium of pot grown plants. Total numbers can then be obtained directly from counts of colonies growing on Petri dishes. The accuracy of this technique depends largely upon the absence of other bacteria which can grow on the selective medium, particularly other strains of the same species. With fast-growing rhizobia streptomycin- and spectinomycin-resistant mutants are particularly valuable (12, 36, 74). An obvious control which is required is to test a random sample of colonies for other characteristics of the resistant strain used so to confirm that it alone is being isolated. A number of other points should be borne in mind when using resistant mutants.

1) Even when a spontaneous mutant is used the strain is by definition different from the parent.

2) Some drug-resistant mutants are symbiotically-defective; mutants should be tested before use.

3) Some mutants may be either more or less competitive than the parent strain; this must be tested before comparisons are made.

4) There is always a chance that a similar mutant is present in the environment. Therefore a sample of colonies should be tested for other characteristics of the strain being examined.

5) The technique is only available to study strains which have been handled in the laboratory. To study indigenous strains, comparisons of intrinsic differences are required.

Future prospects

The future for genetic studies of all nitrogen-fixing species is very promising. Work with *Klebsiella pneumoniae* has shown that there are 17 *nif* genes. Proteins have been identified and in some cases functions for these genes are known (43, 68). Undoubtedly our knowledge of *nif* genes, and particularly information about their regulation, will advance very rapidly. Because the nitrogenase structural genes have been so highly conserved in nature (55, 71) it will be possible to find and manipulate these genes in other species. However, it remains to be demonstrated that all 17 *nif* genes are contiguous in other species. Furthermore, we do not know whether their regulation will be comparable to that of the *K. pneumoniae nif* genes or not (43, 61).

For other nitrogen-fixing species the aims are different and related to the particular properties of the organism. *Rhizobium* genetic studies will concentrate on how the symbiosis with legumes occurs and how it may be manipulated to improve nitrogen fixation. *Azotobacter* is interesting because of the role of oxygen protection mechanisms; perhaps an understanding of them could be useful for transferring the ability to fix nitrogen to other organisms. Nitrogen-fixing species of cyanobacteria are interesting because of their ability to produce the energy needed for nitrogen fixation by photosynthesis. Whether or not cyanobacteria can be exploited as producers of ammonia on an industrial scale may depend upon an adequate genetic knowledge of the relevant species.

The nitrogen-fixing symbioses are presently the most efficient sources of biologically-produced ammonia for agricultural crops. Therefore future developments in the use of biological nitrogen fixation for food production will be governed by our ability to manipulate the symbioses. Hopefully an increase in our knowledge of the genetics of *Rhizobium*, *Azospirillum*, the cyanobacteria and *Frankia* species will be useful to suggest ways in which the range and efficiency of nitrogen-fixing symbioses can be increased so that in future we need not be dependent upon the existing symbioses, but can produce them at will for desirable food plants.

References

1. Ausubel, F., Riedel, G., Cannon, F., Peskin, A. and Margolskee, R. Cloning nitrogen-fixing genes from *Klebsiella pneumoniae in vitro* and the isolation of *nif* promoter mutants affecting glutamine synthetase regulation, in *Genetic Engineering for Nitrogen Fixation* (Edited by A. Hollaender), Basic Life Sciences, Vol. 9, Plenum, New York and London (1977).
2. Balassa, G. Genetic transformation of *Rhizobium*: A review of the work of R. Balassa, *Bacteriological Reviews*, *27*, 228–241 (1963).
3. Beringer, J.E. R factor transfer in *Rhizobium leguminosarum*, *Journal of General Microbiology*, *84*, 188–198 (1974).
4. Beringer, J.E. Plasmid transfer in *Rhizobium*, pp. 358–366 in *Proceedings of the First International Symposium on Nitrogen Fixation, Vol. 2* (Edited by W.E. Newton and C.J. Nyman), Pullman, Washington State University Press (1976).
5. Beringer, J.E., Hoggan, S.A. and Johnston, A.W.B. Linkage mapping in *Rhizobium leguminosarum* by means of R plasmid-mediated recombination, *Journal of General Microbiology*, *104*, 201–207 (1978).
6. Beringer, J.E., Brewin, N.J. and Johnston, A.W.B. The genetic analysis of *Rhizobium* in relation to symbiotic nitrogen fixation, *Heredity*, *45*, 161–186 (1980).
7. Beringer, J.E., Beynon, J.L., Buchanan-Wollaston, A.V. and Johnston, A.W.B. Transfer of the drug resistance transposon Tn 5 to *Rhizobium*, *Nature, London*, *276*, 633–634 (1978).
8. Beynon, J.L. and Josey, D.P. Demonstration of heterogeneity in a natural population of *Rhizobium phaseoli* using variation in intrinsic antibiotic resistance, *Journal of General Microbiology*, *118*, 437–442 (1980).

9. Bishop, P.E., Dazzo, F.B., Appelbaum, E.R., Maier, R.M. and Brill, W.J. Intergeneric transfer of genes involved in the *Rhizobium-legume* symbiosis, *Science*, *198*, 938–940 (1977).
10. Brewin, N.J., Beringer, J.E., Buchanan-Wollaston, A.V., Johnston, A.W.B. and Hirsch, P.R. Transfer of symbiotic genes with bacteriocinogenic plasmids in *Rhizobium leguminosarum*, *Journal of General Microbiology*, *116*, 261–270 (1980).
11. Brewin, N.J., DeJong, T.M., Phillips, D.A. and Johnston, A.W.B. Co-transfer of determinants for hydrogenase activity and nodulation ability in *Rhizobium leguminosarum*, *Nature, London*, *288*, 77–79 (1980).
12. Bromfield, E.S.P. and Gareth Jones, D. The competitive ability and symbiotic effectiveness of doubly labelled antibiotic resistant mutants of *Rhizobium trifolii*, *Annals of Applied Biology*, *91*, 211–219 (1979).
13. Bromfield, E.S.P. and Gareth Jones, D. Studies of double strain occupancy of nodules and the competitive ability of *Rhizobium trifolii* on red and white clover grown in soil and agar, *Annals of Applied Biology*, *94*, 51–59 (1980).
14. Burkardt, H.J., Mattes, R., Pühler, A. and Heumann, W. Electron microscopy and computerized evaluation of some partially denatured group P resistance plasmids, *Journal of General Microbiology*, *105*, 51–62 (1978).
15. Buchanan-Wollaston, A.V. Generalized transduction in *Rhizobium leguminosarum*, *Journal of General Microbiology*, *112*, 135–142 (1979).
16. Buchanan-Wollaston, A.V., Beringer, J.E., Brewin, N.J., Hirsch, P.R. and Johnston, A.W.B. Isolation of symbiotically defective mutants in *Rhizobium leguminosarum* by insertion of the transposon Tn 5 into a transmissible plasmid, *Molecular and General Genetics*, *178*, 185–190 (1980).
17. Cannon, F.C. and Postgate, J.R. Expression of *Klebsiella* nitrogen fixation genes (*nif*) in *Azotobacter*, *Nature, London*, *260*, 271–272 (1976).
18. Casadesus, J. and Olivares, J. Rough and fine linkage mapping of the *Rhizobium meliloti* chromosome, *Molecular and General Genetics*, *174*, 203–209 (1979).
19. Casadesus, J. and Olivares, J. General transduction in *Rhizobium meliloti* by a thermosensitive mutant of bacteriophage DF2, *Journal of Bacteriology*, *139*, 316–317 (1979).
20. Cerda-Olmedo, E., Hanawalt, P.C. and Guerola, N. Mutagenesis of the replication point by nitrosoguanidine: Map and pattern of the *Escherichia coli* chromosome, *Journal of Molecular Biology*, *33*, 705–719 (1968).
21. Cunningham, D.A. Genetic studies on the nitrogen-fixing bacterium *Rhizobium trifolii*, Ph.D. thesis, University of Edinburgh, Edinburgh (1980).
22. Datta, N., Hedges, R.W., Shaw, E.J., Sykes, R.B. and Richmond, M.H.

Properties of an R Factor from *Pseudomonas aeruginosa*, *Journal of Bacteriology*, *108*, 1244–1249 (1971).

23. Davis, R.W., Simon, M. and Davidson, N. Electron microscope heteroduplex methods for mapping regions of base sequence homology in nucleic acids, pp. 413–428 in *Methods in Enzymology* (Edited by L. Grossman and K. Moldave), Vol. 21, Part D, Academic Press, New York (1971).
24. De Ley, J. DNA base composition, flagellation and taxonomy of the genus *Rhizobium*, *Journal of General Microbiology*, *41*, 85–91 (1965).
25. Dixon, R.A. and Postgate, J.R. Transfer of nitrogen-fixation genes by conjugation in *Klebsiella pneumoniae*, *Nature, London*, *234*, 47–48 (1971).
26. Dixon, R.A. and Postgate, J.R. Genetic transfer of nitrogen fixation from *Klebsiella pneumoniae* to *Escherichia coli*, *Nature, London*, *237*, 102–103 (1972).
27. Dixon, R., Kennedy, C., Kondorosi, A., Krishnapillai, V. and Merrick, M. Complementation analysis of *Klebsiella pneumoniae* mutants defective in nitrogen fixation, *Molecular and General Genetics*, *157*, 189–198 (1977).
28. Franche, C., Rosenberg, C., Quiviger, B. and Elmerich, C. Plasmids and bacteriophages of *Azospirillum*, *Society for General Microbiology Quarterly*, *8*, 136 (1981).
29. Gibbins, A.M. and Gregory, K.F. Relatedness among *Rhizobium* and *Agrobacterium* species determined by three methods of nucleic acid hybridization, *Journal of Bacteriology*, *111*, 129–141 (1972).
30. Van Den Hondel, C.A.M.J.J., Keegstra, W., Borrias, W.E. and Van Arkel, G.A. Homology of plasmids in strains of unicellular cyanobacteria, *Plasmid*, *2*, 323–333 (1979).
31. Van Den Hondel, C.A.M.J.J., Verbeeks S., Van Der Ende, A., Weisbeek, P.J., Borrias, W.E. and Van Arkel, G.A. Introduction of transposon Tn 901 into a plasmid of *Anacystis nidulans:* Preparation for cloning in cyanobacteria, *Proceedings of the National Academy of Sciences of the USA*, *77*, 1570–1574 (1980).
32. Hopwood, D.A. and Wright, H.M. Bacterial protoplast fusion: Recombination in fused protoplasts of *Streptomyces coelicolor*, *Molecular and General Genetics*, *162*, 307–317 (1978).
33. Howe, M.M. and Bade, E.G. Molecular biology of bacteriophage Mu, *Science*, *190*, 624–632 (1975).
34. Jacob, A.E., Cresswell, J.M., Hedges, R.W., Coetzee, J.N. and Beringer, J.E. Properties of plasmids constructed by the *in vitro* insertion of DNA from *Rhizobium leguminosarum* or *Proteus mirabilis* into RP4, *Molecular and General Genetics*, *147*, 315–323.
35. Johnston, A.W.B. and Beringer, J.E. Identification of the *Rhizobium* strains in pea root nodules using genetic markers, *Journal of General Microbiology*, *87*, 343–350 (1975).

36. Johnston, A.W.B. and Beringer, J.E. Pea root nodules containing more than one *Rhizobium* species, *Nature, London, 263*, 502–504 (1976).
37. Johnston, A.W.B., Bibb, M.J. and Beringer, J.E. Tryptophan genes in *Rhizobium*—their organization and their transfer to other bacterial genera, *Molecular and General Genetics, 165*, 323–330 (1978).
38. Johnston, A.W.B., Setchell, S.M. and Beringer, J.E. Interspecific crosses between *Rhizobium leguminosarum* and *R. meliloti:* Formation of haploid recombinants and of R-primes, *Journal of General Microbiology, 104*, 209–218 (1978).
39. Johnston, A.W.B., Beynon, J.L., Buchanan-Wollaston, A.V., Setchell, S.M., Hirsch, P.R. and Beringer, J.E. High frequency transfer of nodulating ability between strains and species of *Rhizobium, Nature, London, 276*, 634–636 (1978).
40. Josey, D.P., Beynon, J.L., Johnston, A.W.B. and Beringer, J.E. Strain identification in *Rhizobium* using intrinsic antibiotic resistance, *Journal of Applied Bacteriology, 46*, 343–350 (1979).
41. Julliot, J.S. and Boistard, P. Use of RP4-prime plasmids constructed *in vitro* to promote a polarized transfer of the chromosome in *Escherichia coli* and *Rhizobium meliloti, Molecular and General Genetics, 173*, 289–298 (1979).
42. Kennedy, C. Linkage map of the nitrogen fixation (*nif*) genes in *Klebsiella pneumoniae, Molecular and General Genetics, 157*, 199–204 (1977).
43. Kennedy, C., Cannon, F., Cannon, M., Dixon, R., Hill, S., Jensen, J., Kumar, S., Mclean, P., Merrick, M., Robson, R. and Postgate, J. Recent advances in the genetics and regulation of nitrogen fixation, in *Proceedings of the Fourth International Symposium on Nitrogen Fixation, Canberra* (in press) (1981).
44. Kiss, G.B., Dobo, K., Dusha, I., Breznovits, A., Orosz, L., Vincze, E. and Kondorosi, A. Isolation and characterization of an R-prime plasmid from *Rhizobium meliloti, Journal of Bacteriology, 141*, 121–128 (1980).
45. Kleckner, N., Roth, J. and Botstein, D. Genetic engineering *in vivo* using translocatable drug resistance elements: New methods in bacterial genetics, *Journal of Molecular Biology, 116*, 125–159 (1977).
46. Kondorosi, A., Kiss, G.B., Forrai, R., Vincze, E. and Banfalvi, Z. Circular linkage map of *Rhizobium meliloti* chromosome, *Nature, London, 268*, 525–527 (1977).
47. Kondorosi, A., Vincze, A.E., Johnston, A.W.B. and Beringer, J.E. A comparison of three *Rhizobium* linkage maps. *Molecular and General Genetics, 178*, 403–408 (1980).
48. Kowalski, M. Transduction in *Rhizobium meliloti, Plant and Soil, Special Volume*, pp. 63–66 (1971).
49. Kowalski, M. and Dénarié, J. Transduction d'un gène contrôlant l'expression de la fixation de l'azote chez *Rhizobium meliloti, Comptes Rendues de*

l'Academie des Sciences, Paris, 275, 141–144 (1972).

50. Lau, R.H., Sapienza, C. and Doolittle, W.F. Cyanobacterial plasmids: Their widespread occurrence, and the existence of regions of homology between plasmids in the same and different species, *Molecular and General Genetics, 178*, 203–211 (1980).
51. MacNeil, T., MacNeil, D., Roberts, G.P., Supiano, M.A. and Brill, W.J. Fine-structure mapping and complementation analysis of *nif* (nitrogen fixation) genes in *Klebsiella pneumoniae, Journal of Bacteriology, 136*, 253–266 (1978).
52. MacNeil, D., Supiano, M.A. and Brill, W.J. Order of genes near *nif* in *Klebsiella pneumoniae, Journal of Bacteriology, 138*, 1041–1045 (1979).
53. Matsumoto, H. and Tazaki, T. Genetic mapping of *aro*, *pyr* and *pur* markers in *Klebsiella pneumoniae, Japanese Journal of Microbiology, 15*, 11–20 (1971).
54. Maxam, A.M. and Gilbert, W. A new method for sequencing DNA, *Proceedings of the National Academy of Sciences of the USA, 74*, 560–564 (1977).
55. Mazur, B.J., Rice, D. and Haselkorn, R. Identification of blue-green algal nitrogen-fixation genes by using heterologous DNA hybridization probes, *Proceedings of the National Academy of Sciences of the USA, 77*, 186–190 (1980).
56. Meade, H.M. and Signer, E.R. Genetic mapping of *Rhizobium meliloti*, *Proceedings of the National Academy of Sciences of the USA, 74*, 2076–2078 (1977).
57. Merrick, M., Filser, M., Kennedy, C. and Dixon, R. Polarity mutations induced by insertion of transposons Tn 5, Tn 7 and Tn 10 into the *nif* gene cluster of *Klebsiella pneumoniae, Molecular and General Genetics, 165*, 103–111 (1978).
58. Mielenz, J.R., Jackson, L.E., O'Gara, F. and Shanmugam, K.T. Fingerprinting bacterial chromosomal DNA with restriction endonuclease *Eco* R1: Comparison of *Rhizobium* spp. and identification of mutants, *Canadian Journal of Microbiology, 25*, 803–807 (1979).
59. Nuti, M.P., Ledeboer, A.M., Lepidi, A.A. and Schilperoort, R.A. Large plasmids in different *Rhizobium* species, *Journal of General Microbiology, 100*, 421–428 (1977).
60. Nuti, M.P., Ledpidi, A.A., Prakash, R.K., Schilperoort, R.A. and Cannon, F.C. Evidence for nitrogen fixation (*nif*) genes on indigenous *Rhizobium* plasmids, *Nature, London, 282*, 533–535 (1979).
61. O'Gara, F. and Shanmugam, K.T. Regulation of nitrogen fixation by rhizobia: Export of fixed N_2 as NH_4^+, *Biochimica biophysica Acta, 437*, 313–321 (1976).
62. Olsen, R.H. and Shipley, P. Host range and properties of the *Pseudomo-*

nas aeruginosa R factor R1822, *Journal of Bacteriology*, *113*, 772–780 (1971).

63. Page, W.J. Transformation of *Azotobacter vinelandii* strains unable to fix nitrogen with *Rhizobium* spp. DNA, *Canadian Journal of Microbiology*, *24*, 209–214 (1978).
64. Page, W.J. and Von Tigerstrom, M. Optimal conditions for transformation of *Azotobacter vinelandii*, *Journal of Bacteriology*, *139*, 1058–1061 (1979).
65. Pinto, C.M., Yao, P.Y. and Vincent, J.M. Nodulating competitiveness among strains of *Rhizobium meliloti* and *R. trifolii*, *Australian Journal of Agricultural Research*, *25*, 317–329 (1974).
66. Polsinelli, M., Baldanzi, E., Bazzicalupo, M. and Gallori, E. Transfer of plasmid pRDl from *Escherichia coli* to *Azospirillum brasilense*, *Molecular and General Genetics*, *178*, 709–711 (1980).
67. Pühler, A., Burkardt, H.J. and Klipp, W. Cloning in *Escherichia coli* the genomic region of *Klebsiella pneumoniae* which encodes genes responsible for nitrogen fixation, pp. 317–325 in *Plasmids of Medical, Environmental and Commercial Importance* (Edited by K.N. Timmis and A. Pühler), Amsterdam, Elsevier/North Holland Biochemical Press (1979).
68. Pühler, A., Burkardt, H.J. and Klipp, W. Cloning of the entire region for nitrogen fixation from *Klebsiella pneumoniae* on a multicopy plasmid vehicle in *Escherichia coli*, *Molecular and General Genetics*, *176*, 17–24 (1979).
69. Raina, J.L. and Modi, V.V. Genetic transformation in *Rhizobium*, *Journal of General Microbiology*, *57*, 125–130 (1969).
70. Robson, R.L. Detection and function of indigenous plasmids of *Azotobacters*, *Society for General Microbiology Quarterly*, *8*, 136–137 (1981).
71. Ruvkun, G.B. and Ausubel, F.M. Interspecies homology of nitrogenase genes, *Proceedings of the National Academy of Sciences of the USA*, *77*, 191–195 (1980).
72. Ruvkun, G.B. and Ausubel, F.M. A general method for site directed mutagenesis in prokaryotes: Construction of mutations in symbiotic nitrogen fixation genes of *Rhizobium meliloti*, *Nature, London*, *289*, 85–88 (1980).
73. Schell, J. and Van Montagu, M. The TI-plasmid of *Agrobacterium tumefaciens*, a natural vector for the introduction of *nif* genes in plants?, pp. 159–180 in *Genetic Engineering for Nitrogen Fixation* (Edited by A. Hollaender), Plenum Press, New York (1977).
74. Schwinghamer, E.A. and Dudman, W.F. Evaluation of spectinomycin resistance as a marker for ecological studies with *Rhizobium* species, *Journal of Applied Bacteriology*, *36*, 263–272 (1973).
75. Sik, T., Horvath, J. and Chatterjee, S. Generalized transduction in *Rhizobium meliloti*, *Molecular and General Genetics*, *178*, 511–516 (1980).

76. Sistrom, W.R. Transfer of chromosomal genes mediated by plasmid R68.45 in *Rhodopseudomonas sphaeroides*, *Journal of Bacteriology*, *131*, 526–532 (1977).
77. Staniewski, R. Relationship among different *Rhizobium* strains determined by phage lysis, *Acta Microbiologica Polonica*, *19*, 3–12 (1970).
78. Staniewski, R. Typing of *Rhizobium* by phages, *Canadian Journal of Microbiology*, *16*, 1003–1009 (1970).
79. Stanisich, V.A. and Holloway, B.W. Chromosome transfer in *Pseudomonas aeruginosa* mediated by R factors, *Genetical Research*, *17*, 1–4 (1971).
80. Streicher, S., Gurney, E. and Valentine, R.C. Transduction of nitrogen fixation genes in *Klebsiella pneumoniae*, *Proceedings of the National Academy of Sciences of the USA*, *68*, 1174–1177 (1971).
81. Towner, K.J. and Vivian, A. Plasmids capable of transfer and chromosome mobilization in *Acinetobacter calcoaceticus*, *Journal of General Microbiology*, *101*, 167–171 (1977).
82. Wilson, G.G., Young, K.K.Y., Edlin, G.J. and Konigsberg, W. High frequency generalized transduction by bacteriophage T4, *Nature, London*, *280*, 80–82 (1979).
83. Zelazna-Kowalska, I. and Lorkiewicz, A. Conditions for genetical transformation in *Rhizobium meliloti*, *Acta Microbiologica Polonica Ser. A. 3*, 21–28 (1971).

C.A. ATKINS and R.M. RAINBIRD

2. Physiology and Biochemistry of Biological Nitrogen Fixation in Legumes

Introduction

Nitrogen fixation in legumes is wholly dependent on the activity of the enzyme nitrogenase, which is located within the bacteroids of the *Rhizobium* microsymbiont with energy for the reduction being totally derived from carbon substrates which pass into the bacteroids from the host cells of the nodule. The provision of energy by the host in return for reduced nitrogen from the bacteria links photosynthesis and nitrogenase activity, and in broad terms describes the symbiosis. However, the biochemistry, physiology and structure of nodules; their highly complex requirements for oxygen, water, mineral nutrients and oxidizable substrates; their close links with the vascular network of the plant, both in phloem import of sugars and xylem export of nitrogenous solutes; and the interaction of nodular metabolism with the nitrogen and carbon demands of other plant organs suggests that, in legumes, nitrogen fixation is truly an integrated "plant process". This chapter will, therefore, deal with the physiology and biochemistry of fixation at the cellular level, at the level of the nodule and at the level of the whole plant. It is not possible, within the space available, to exhaustively review all aspects of symbiotic nitrogen fixation and the reader is referred to several monographs and reviews which have recently been written on the subject (38, 53, 54, 56, 82, 83, 85, 91, 92, 103 and 120).

Fixation at the cellular level

Biochemistry of nitrogenase

The nitrogenase complex consists of two dissociating proteins—one termed nitrogenase reductase or Fe protein, which contains four iron and four acid labile sulphur atoms, while the other, the MoFe protein or nitrogenase, which

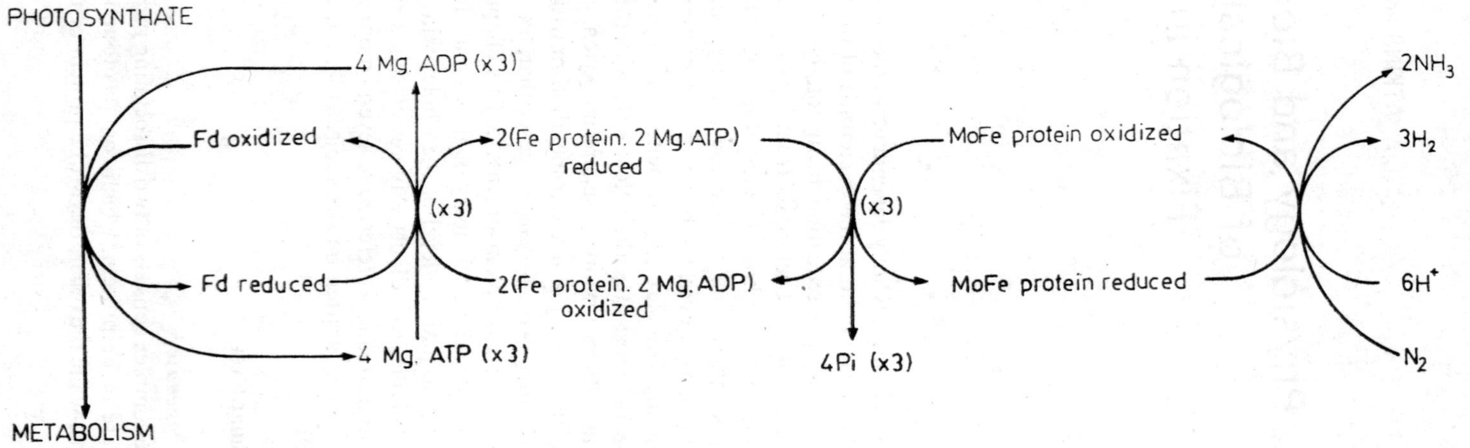
PHOTOSYNTHATE
METABOLISM
4 Mg. ADP (x3)
Fd oxidized
Fd reduced
4 Mg. ATP (x3)
(x3)
2(Fe protein. 2 Mg. ATP) reduced
2(Fe protein. 2 Mg. ADP) oxidized
(x3)
4Pi (x3)
MoFe protein oxidized
MoFe protein reduced
2NH3
3H2
6H+
N2

Fig. 1.

contains two molybdenum, 28–32 iron, and approximately 28 acid labile sulphur atoms. Together these two proteins catalyze the reduction of N_2 to NH_3, acetylene to ethylene, H^+ to H_2, and a number of other reactions. The source of reductant for these electron transfers is believed to be reduced ferredoxin or flavodoxin (145). While these low potential proteins may indeed serve as the ultimate source of electrons in all nitrogen-fixing organisms (33) the terminal steps of electron transport leading to their reduction, particularly in *Rhizobium* bacteroids, have not been defined.

The passage of electrons through the nitrogenase complex is particularly relevant to current understanding of the energetics of N_2 fixation. The sequence of transfers through the complex to reducible substrates, as suggested by Mortenson (81), is outlined in Fig. 1.

Electrons flow from ferredoxin to the Fe protein which, when complexed with ATP, serves to reduce the MoFe protein with the release of inorganic phosphate. It is the MoFe protein that eventually passes electrons to the reducible substrate. Based on inhibitor studies Hwang *et al.* (60) proposed five sites (or modified sites) for the nitrogenase complex: (A) N_2 and H_2 site, (B) acetylene site, (C) azide, cyanide, and methylisocyanide site, (D) CO site, (E) H^+ site. Studies of the partitioning of electrons between N_2, H^+ and acetylene by Hageman *et al.* (51) have shown that the allocation of electrons to these substrates depends upon their relative concentrations and upon the electron flux through the complex. At low electron flux, N_2 competes poorly with acetylene and H^+ as a substrate, while reduction of H^+ to H_2 was shown to be dominant. At a high electron flux N_2 becomes a very effective substrate and the dominant electron acceptor. Furthermore, the electron flux can be altered by changing the concentration of Mg ATP or the ratio of the Fe protein to MoFe protein (51). In the intact nitrogen-fixing system the supply of both ATP and reductant depends on the availability of oxidizable carbon substrates for respiration and it might be predicted that under limiting conditions, hydrogen evolution will predominate.

As shown in Fig. 1, ATP hydrolysis is associated with the transfer of electrons from nitrogenase reductase to nitrogenase. However, the amount of ATP hydrolyzed for each electron pair transferred is uncertain, with estimates varying from two to four or five (48). The variation is associated with different preparations of the enzyme, changes in temperature, different ratios of the two nitrogenase proteins and the ADP/ATP ratio. Orme-Johnson *et al.* (88) proposed a futile cycle in which only a single electron is passed on to the MoFe protein. Under these conditions it is possible that the electron will simply return to the Fe protein, causing ATP hydrolysis but no reduction. Under optimal conditions with the purified complex a value of two molecules of ATP hydrolysed per electron passing to reducible substrates has been widely accepted (88).

Reduction of molecular nitrogen may thus be represented as:

$$N_2 + 6e^- + 12ATP + 8H^+ \xrightarrow{Mg^{2+}} 2NH_4^+ + 12ADP + 12Pi \quad (1)$$

In vitro as well as in the intact nitrogen-fixing organism nitrogenase reduces protons to evolve H_2 gas concomitant with ammonia formation. The exact relationship between N_2 reduction and H_2 evolution is not clear (117), the proportioning of electrons between the two substrates varying as described above. Measurements using the isolated enzyme complex indicate, however, that for every N_2 molecule reduced a minimum of one molecule of H_2 is evolved.

$$2H^+ + 2e^- + 4ATP \xrightarrow{Mg^{2+}} H_2 + 4ADP + 4Pi \quad (2)$$

Thus the overall model equation for nitrogenase functioning in legume nodules may be represented as:

$$N_2 + 16ATP + 8e^- + 10H^+ \xrightarrow{Mg^{2+}} 2NH_4^+ + H_2 + 16ADP + 16Pi \quad (3)$$

Sources of energy for nitrogenase activity

Carbohydrate, principally sucrose, formed in leaf photosynthesis is delivered in the downward-moving phloem stream to the root system and nodules of the legume (93). Oxidation of these energy-yielding substrates by the respiratory metabolism of the nodule ultimately produces ATP and reduced ferredoxin or flavodoxin. The exact nature of the carbon substrates passing from the host plant cytosol of the nodule to the bacteroids has, however, not been defined. While some enzymes of hexose utilization have been detected in bacteroids from a number of legume species (63), the absence of the key glycolytic enzyme, fructose 1-6-diphosphate aldolase (113) indicates that a functional glycolytic sequence is not present. Furthermore, Ronson and Primrose (113) have shown that *Rhizobium trifolii* mutants defective in sucrose, glucose and fructose catabolism may still form fully-effective nodules. Using isolated bacteroids from which endogenous substrates were leached, Bergersen and Turner (21) found that high rates of nitrogen fixation were supported by succinate and to a lesser extent by fumarate and pyruvate (see also 106). Further evidence in support of organic acids as the source of reduced carbon comes from experiments by Wilcockson and Werner (139), who showed that *Rhizobium japonicum* grew and fixed nitrogen more effectively on media containing an organic acid and a pentose sugar than on media containing only one of these carbon sources. Also Ronson *et al.* (112) have recently, with the aid of mutants defective in C_4-dicarboxylate transport, shown that the supply of C_4-dicarboxylic acids is essential for nitrogen fixation in clover nodules.

If organic acids are indeed the form of energy-yielding substrate transferred from the host to the bacteroid then partial oxidation of sugars in the host cell cytosol would be necessary. Wooi and Broughton (144) examined the metabolism of carbon substrates by bacteroid-containing host cell protoplasts isolated from nitrogen-fixing *Vigna unguiculata* nodules and concluded that the cytosol of these cells contained functional glycolytic and Kreb's cycle pathways. In addition, the plant cell cytosol of nodules from a number of species shows high levels of phospho-enol pyruvate (PEP) carboxylase activity (36, 68) and while ^{14}C-labelling studies indicate that some of the C_4 acids produced in this way are exported from the nodule as asparagine, aspartate or malate, their transfer to, and oxidation by bacteroids may be significant. Studies of the metabolism of ^{14}C-organic acids by isolated soybean bacteroids (125) indicate the presence of a functional TCA cycle.

Bacteroids contain a number of unique cytochromes (3) and show at least two terminal cytochrome C oxidase systems (6). One of these has optimal activity at very low stabilized oxygen tensions (0.1 μM) and is associated with the production of the high bacteroid ATP levels required for nitrogen fixation (22). The other, "low efficiency", system may be important in "protective" respiration. Recent observations by Laane *et al.* (64, 65) have led to the suggestion that the generation of reductant for nitrogenase in bacteroids is regulated by the electrical component of a proton-motive force across the bacterial plasma membranes. While this is consistent with current concepts of membrane-bound oxidative phosphorylation processes (80), details in the flow of electrons to nitrogenase are not clear.

Regulation of nitrogenase

The discovery that free-living cultures of a number of *Rhizobium* strains exhibit functional nitrogenase when grown under defined conditions of nitrogen limitation and reduced oxygen level (18, 47) has eliminated previous theories for the strict regulation of nitrogenase biosynthesis by the host cell (39). While diffusible factors produced from plant cells may modulate *Rhizobium* nitrogenase activity (109) the transformation of vegetative cells to nitrogen-fixing bacteroids is apparently a response to the low oxygen tensions of the developing nodule. Ammonia represses nitrogenase synthesis in free-living nitrogen fixers (48, 136) as well as in free-living *Rhizobium* cultures (24, 27). In both cases adenylylation of the ammonia-assimilating enzyme glutamine synthetase is apparently a positive control element (23, 27, 70, 135) in the regulation. High levels of glutamine formed as a result of the increased ammonia activate the adenylylation mechanism and inactivate glutamine synthetase, which in turn represses genes coding for the nitrogenase enzyme complex (128). In nodules, where ammonia is excreted from the bacteroids (17, 65) and is assimilated by glutamine synthetase in the host plant cytosol (72), control of nitrogenase synthesis by adenylylation of bacteroid glutamine synthetase (32) has not been observed (27).

In addition to factors affecting biosynthesis of nitrogenase, enzyme activity may also be regulated by positive or negative effectors.

Changes in ATP level and in "energy-charge" (35, 81) have been recognized in this respect. ADP is a potent inhibitor of all known reductions by the enzyme complex (134) as is Mg^{2+} in ratios greater or lower than 1:1 in respect of ATP (62, 130). While ammonia may affect nitrogenase synthesis neither ammonia (48) nor amino compounds have a direct effect on nitrogenase activity. In contrast, carbamyl phosphate, which has not been recognized as a major product of fixation in legume nodules, inhibits nitrogen reduction but not H_2 evolution by isolated nitrogenase (81). Oxygen is a potent inhibitor of the enzyme complex *in vitro* but is unlikely to play a significant role in the intact nodule where a complex protection mechanism involving leghaemoglobin maintains a low free oxygen level in the tissues.

Indirect controls of nitrogenase activity may also arise from changes in the partitioning of electrons between ammonia and hydrogen production, impaired diffusion of oxygen to respiratory sites (122) or the availability of photosynthate from the host plant (55).

Hydrogen and nitrogenase activity

While hydrogen production is apparently a constant feature of nitrogenase activity, both as an isolated enzyme and *in vivo* in the intact nodule, in a number of symbioses, the presence of a bacteroid-located uptake hydrogenase utilizes part or all of the H_2 produced (40, 41, 117), and is generally considered to yield ATP from the oxidation (40, 41). Recent studies by Emerich *et al.* (44) have provided evidence which suggests that this hydrogenase may function to scavenge oxygen and thus protect nitrogenase from inactivation; to prevent inhibition of nitrogenase by hydrogen gas (60); as well as to recover at least part of the energy which is lost to hydrogen evolution.

Some nodules maintain a relatively constant ratio of H_2 evolved to N_2 reduced, while in some this ratio varies with change in environmental factors and with nodule and plant development (26). As discussed earlier, variation in the allocation of electrons between these two substrates may reflect the association of the nitrogenase proteins and the electron flux through the complex. However, judging from published values for gas exchange of H_2 and N_2 and hydrogenase assays using labelled or unlabelled hydrogen or the stimulation of H_2 evolution by nodules held in N_2-free atmospheres (103, 114, 117, 146), an electron partitioning to proton reduction from 25 to around 60 per cent would appear to be a likely range of values for nitrogenase functioning. Based on theoretical considerations of the biochemical costing of the reactions involved, Pate *et al.* (94) estimated that if uptake hydrogenase is absent or inoperative, nitrogenase will utilize from 28–53 mol ATP/mol N_2 fixed; with complete recycling of the evolved H_2 by hydrogenase the nitrogenase will then utilisc 25–39 mol ATP/mol N_2. That is a saving of from 11 to 26 per cent of the energy required for nitrogenase functioning. While nitrogenase activity is only one of the energy-dependent processes in nodules, it is the major cost of respired carbon (68, 94; see also Table 1) and a conserving reaction such as that of uptake hydrogenase could, on theoretical grounds, cause a marked increase in the efficiency of photosynthate use in the nodules of those symbioses formed with naturally-occurring or induced mutants of *Rhizobium* which are H_2-uptake-positive (Hup^+). That this potential for increasing legume yield may be realized is shown in the field trials of Albrecht *et al.* (1) in which soybean, nodulated with a number of Hup^+ and Hup^- *R. japonicum* strains showed significant increase in plant dry weight and in total plant nitrogen in those with the uptake hydrogenase.

Fixation at the nodule level

Assimilation of ammonia

There is general agreement that ammonia is the first stable product of nitrogen fixation in legume nodules (21, 61) and that this ammonia is released from the bacteroids to the host cell where organic forms of nitrogen are synthesized (17, 65). Early reports using $^{15}N_2$ (7, 16, 61) concluded that the initial assimilatory reaction was the formation of glutamate by glutamate dehydrogenase (49). However, more recent studies of the short-term labelling of nodules with $^{13}N_2$ (76) and $^{15}N_2$ (87) have indicated that glutamine is the initial product. This is consistent with the observation that glutamine synthetase activity is prominent in the plant cytosol fraction of nodules from a wide range of symbioses (30, 57, 72). The plant cytosol fraction of nodules also contains an NADH-dependent glutamate synthase (30, 57, 104) which, together with glutamine synthetase, constitutes the "high-affinity" pathway for ammonia assimilation proposed by Miflin and Lea (77, 78) to be widespread in plant tissues effectively utilising inorganic nitrogen. Nodules also contain significant levels of glutamate dehydrogenase in the plant cell cytosol (11, 13, 49) and Boland (28) has proposed a mechanism in which glutamate dehydrogenase, glutamate synthase and glutamine synthetase interact in the assimilation of ammonia. As discussed earlier, bacteroids also contain active glutamine synthetase and Brown and Dilworth (32) demonstrated that glutamate synthetase activity was also present. While bacteroid glutamine synthetase could function in the regulation of nitrogenase synthesis, a more general function for the two enzymes in the synthesis of bacteroid protein during nodule growth might be indicated, particularly in view of the observation that a substantial fraction of the nitrogen incorporated into a growing nodule tissue is derived directly from nitrogen fixed in the nodule (68, 86).

While glutamine, formed from fixed nitrogen, may be exported from the nodule it is rarely the major nitrogenous solute of the xylem leaving the root system. Secondary reactions in the nodule in which the amide and amino groups of glutamine are transferred lead to the formation of the exported solutes. Current information indicates that effectively-nodulated legumes produce either asparagine as the major end product of fixation or the ureides, allantoin and allantoic acid (see also Chapter 3). While this rather arbitrary division of legumes into "amide" or "ureide" producers may require modification in the light of a more extensive examination of species, at present "amide" producers are temperate species while those producing ureides are of tropical origin (92, 123).

Both an aspartate-amino transferase and a glutamine-dependent asparagine synthetase have been detected in the plant fraction of legume nodules (10, 110, 119). The amino transferase is also found in the bacteroids and the non-bacteroid-containing cortical cells (10, 40), whereas the asparagine synthetase is restricted to the cytosol of the bacteroid-containing cells (119) and is thus in close proximity to the site of ammonia assimilation. While short-term studies with $^{13}N_2$ (76) showed little accumulation of labelled asparagine in detached soybean nodules, possibly because of a low nitrogen partial pressure used in the experiments (see 78), labelling in asparagine was extensive from ^{15}N (amide) glutamine or $^{15}NH_3$ supplied to nodules in the studies of Fujihara and Yamaguchi (46). Evidence for the formation of ureides from fixed nitrogen and the metabolic pathways proposed for the synthesis of these molecules in nodules are reviewed extensively in the next chapter. While the detailed reactions of ureide biosynthesis have not been defined, they are apparently formed from the oxidation of purine base and utilise both amide and amino nitrogen formed as initial products of ammonia assimilation in the plant cell cytosol (8, 9, 10, 57).

A limited number of studies have investigated the control of cytosol ammonia assimilation in nodules (108). Both glutamine synthetase (72) and asparagine synthetase (69) appear to be sensitive to inhibition by AMP and ADP and may thus, like nitrogenase, be modulated by changes in "energy charge" in nodule tissues. Glutamate synthase also showed complex product inhibition with both glutamate and NAD (29), while one of the key enzymes of ureide synthesis, uricase, is extremely sensitive to inhibition by ammonia, glutamine and xanthine (105). Even though studies with isolated enzymes must be interpreted with caution in extrapolating to the *in vivo* situation, complex regulation of the pathways of ammonia assimilation is probably important in maintaining the observed constant composition of exported solutes and to ensure rapid and efficient assimilation of ammonia.

Whereas assimilation of ammonia by the glutamine synthetase/glutamate synthase/glutamate dehydrogenase system probably represents the major pathway in many symbioses, other enzymes might contribute in part. The xylem stream from some nodules contains a significant amount of arginine, e.g. *Vigna radiata* (94), and indeed this amino acid is the major fixed product in some legume (31) while the product is citrulline in non-legume nodules (31, 141); in each case the formation of carbamyl phosphate might represent a direct route for the incorporation of newly-fixed ammonia. Likewise the two amide transferase enzymes which participate in the formation of the purine ring precursors of the ureides might utilize ammonia directly, rather than amide nitrogen as is the case in other tissues (also see Chapter 3). Nodules from a number of species have been shown to contain significant alanine dehydrogenase activity (42, 129), which is restricted to the bacteroids (129) and like glutamine synthetase could function in the nitrogen nutrition of the microsymbiont. In support of this, Werner and Stripf (138) found that in free-living *Rhizobium japonicum* derepressed for nitrogenase, alanine dehydrogenase activity increased more than four-fold compared to cultures grown with an exogenous nitrogen supply and in which nitrogenase remained repressed.

Gaseous diffusion in nodules and the functioning of leghaemoglobin

Nodules are dense organs possessing significant endodermal investments (see 37) which protect the tissues against desiccation in the relatively dry environment of the soil. These features are, however, likely to impede gaseous diffusion and the extensive network of intercellular spaces (20) and surface lenticels (37, 90) found in some nodules are probably adaptations aiding gas exchange (123). While there is no direct evidence indicating nitrogen fixation being limited by the availability of dissolved nitrogen gas to the bacteroids, the supply of oxygen to the central tissues of the nodule is viewed as critical to nodule functioning (131). Indeed, a number of experiments, particularly with soybean nodules (see 18), indicate that at ambient atmospheric concentrations (0.2 atm), oxygen is a limiting factor for nitrogen fixation. Treatments, such as waterlogging, which reduce oxygen diffusion result in prompt inhibition of nitrogenase activity in nodules (122). However, oxygen levels above 0.5 atm inhibit fixation (34), presumably due to the inactivation of nitrogenase by the oxygen (142).

The maintenance of a low partial pressure of free dissolved oxygen in the nodule tissues while allowing at the same time a high flux of oxygen to the bacteroidal respiratory sites is believed to be achieved by the unique nodule pigment, leghaemoglobin (4, 5, 18). Experiments in which purified oxyleghaemoglobin was added to dense, oxygen-limited, bacteroid suspensions resulted in increased respiration, ATP formation and nitrogenase activity (6, 22, 140) and led to the leghaemoglobin-carrier-facilitated diffusion concept (5). Significantly oxidative phosphorylation in bacteroids is most efficient at dissolved oxygen concentration of 10–100 nM (6) and, as discussed above. involves a unique "high-oxygen-affinity" terminal oxidase. Based on measurements of the dissociation constant for oxyleghaemoglobin (3) and the content of the pigment in soybean nodules, Bergersen and Goodchild (20) estimated the free-dissolved oxygen at the bacteroid surface to be around 10 nM. This estimate was based on the localization of leghaemoglobin within the peribacteroid spaces of the infected cells and thus in close contact with the bacteroid membrane (20). Recent studies (111, 137) suggested, however, that the peribacteroid membrane may separate the bacteroid from the oxygen buffering system although a precise cellular location for leghaemoglobin was not established.

While an important function for leghaemoglobin seems obvious for legume nodules the pigment is not necessary for effective nitrogen fixation in all symbioses. Nodules formed between actinomycetes of the genus *Frankia* and non-legumes lack haemoglobin (15), as do those of the subtropical species *Parasponia andersonii* which forms effective nodules with some 'slow-growing' *Rhizobium* isolates (132). The presence of other oxygen buffering systems in these nodules has not been reported and it is perhaps significant that in *Parasponia*, unlike many legumes, the optimum pO_2 for fixation is very narrow with activity falling rapidly above and below 0.2 atm O_2 (133).

Transport of fixed nitrogen from the site of reduction in nodules

The movement of ammonia from the site of synthesis in the bacteroid necessitates transfer across three membranes to the host cell cytosol. Two of these are bacterial and the third, the peribacteroid membrane, surrounding one or a number of bacteroids, is probably of host plasmalemma origin (37). Although it is not established whether neutral ammonia or cationic ammonium is the mobile species transferred to the host cell, recent experiments by Laane *et al.* (64, 65) suggest that ammonium is actively pumped out of *R. leguminosarum* bacteroids. Soybean bacteroids exhibit a respiration-driven proton translocation (106) indicating that the periplasmic space is likely to be acidic and that any ammonia moving through this compartment is in the cationic form. Subsequent assimilation of the ammonium ion to form amino groups by the host cytoplasm would then result in the release of protons (107). The consequences of proton imbalance and the formation of pH gradients on the transport of ammonia and the exchange of other ionic species, such as dicarboxylic organic acids, between bacteroids and host cytoplasm remain to be evaluated.

In other plant tissues the principal enzymes involved in ammonia assimilation (glutamine synthetase, glutamate synthase) are compartmented into plastids (78). Preliminary results indicate that this might also be the case for legume nodules especially for glutamate synthase (12). Some of the enzymes of ureide formation in soybean nodules are also located in organelles (52), so that a complex set of membrane transfers may be required before fixed nitrogen is finally assimilated into the exported forms.

The structure and functional significance of the tissue arrangement and vasculature of nodules in relation to solute export has been recently reviewed (37, 89, 92). Nitrogenous solutes probably traverse a symplastic route to the centrifugally-located vascular elements where they may be actively secreted to the apoplast. This secretion of solutes is then considered to generate an osmotic flux of water into the vascular apoplast from that of the non-vascular compartments of the nodule, thus flushing fixation products out in the xylem stream (50). Xylem exudate collected from detached nodules of a number of species show solute concentrations ten times or more greater than those in bleeding sap from the whole nodulated root system; with asparagine levels of 124–145 mM (50, 93, 95, 143) and 38–90 mM ureides (127; Rainbird, Atkins and Pate, unpublished). Calculations based on the daily export of nitrogen to the host plant indicate that xylem export of solutes at the concentrations above would require an amount of water only two times the total volume of the exporting nodules (77, 90) and that at least half of this could be derived from the mass flow of phloem fluid into the nodule. Sprent (122) has shown that lateral exchange of tritiated water from the root to the nodule occurs in soybean and this might constitute an additional source of water for xylem export. In any case, the necessity for water absorption from the soil through the relatively impervious and small area of nodule surface might not be great.

In general, nodules from symbioses in which ureides constitute the major exported solutes of nitrogen, are of a determinate growth habit and have a vascular system in which the strands are fused apically to form a closed loop (37, 92, 123). By comparison, those from "amide-exporting" symbioses are indeterminate and have an open vascular system in which the ends of the traces do not meet. Measurements of water flow through nodules of each type have not been made but the difference in structure suggests that there would be a much lower resistance to flow in determinate nodules. Thus the transpiration stream from the roots of determinate plants might more easily flow through the nodule, allowing higher rates of water flux and more dilute solutions in xylem than might be possible in nodules of indeterminate habit (92, 123). These differences may be related to the solubility and stability of the two classes of exported compound, as well as to the formation of specialized exporting mechanisms (e.g. transfer cells (95, 123)).

Energy costing of nodule functioning

THEORETICAL. Considerable experimental evidence supports the concept that the capacity of effective legume-*Rhizobium* symbioses to fix nitrogen is limited by the energy supply from the host plant (2, 25, 55, 66, 126). A large research effort has therefore been directed towards studies of the energetics of symbiotic fixation to define the potential for improvement by more efficient use of plant energy resources. Thus a clear understanding of the relationship between theoretical energy costs and experimentally-measured energy use has been sought.

Thermodynamic considerations of the chemistry of gaseous nitrogen reduction suggest that reduction of 1 mol N_2 requires energy equivalent to that provided by 0.22 mol glucose (14, 75). Knowledge of the biochemistry of nitrogenase functioning (see equation 1, page 28) however indicates that, *in vitro* at least, 2 mol ATP are hydrolyzed per electron passing through the complex, so that overall reduction of 1 mol N_2 would require the equivalent of 21 mol ATP or 0.55 mol glucose (equating each $2e^-$ required for reduction with 3 ATP, i.e. oxidative phosphorylation with $P/2e^- = 3$; and assuming a yield of 38 mol ATP/mol glucose oxidized); representing a maximum efficiency for the biological process of 40 per cent. Production of 1 mol H_2 concomitant with N_2 reduction (equations 2 and 3, page 28) would require the equivalent of a further 7 mol ATP or 0.18 mol glucose giving a total of 28 mol ATP or 0.73 mol glucose. Bacterial oxidative phosphorylation could occur with a $P/2e^-$ ratio less than 3 (124), so that *in vivo* these costings in terms of glucose consumption would be increased. As discussed earlier, uptake hydrogenase located in the bacteroids is likely to recycle some or all of the H_2 evolved from nitrogenase and yield ATP from the oxidation (43). Uncertainty about the partitioning of electrons between proton and nitrogen reduction and the extent to which uptake hydrogenase recycles electrons does not allow an estimate of the theoretical costs for nitrogenase *in vivo* except within defined limits (94). Costs would range from 22 mol ATP/mol N_2 fixed (10 per cent electron allocation to H^+ reduction and complete recovery of H_2 with a yield of 3 ATP/mol H_2 oxidized) to 70 mol ATP/mol N_2 (70 per cent electron allocation and zero recovery of H_2). As an example the measured evolution of H_2 in a *Lupinus albus* symbiosis during early vegetative growth was slightly greater than 3 mol H_2/mol N_2 fixed (68). This could be accommodated by a 60 per cent electron allocation to H^+ reduction with oxidation of 30 per cent of the evolved H_2; or by a 50 per cent electron allocation to H^+ but with no hydrogenase activity. The cost in the first case would be 48 mol ATP/mol N_2 and in the second 42 mol ATP/mol N_2. The range of theoretical costs for nitrogenase/hydrogenase activity cited above has suggested that in terms of carbohydrate consumption the biological process would require 0.66–1.38 mol glucose/mol N_2; representing an efficiency of 16–33 per cent compared to the estimate based on thermodynamic considerations.

A second element in the energy cost of fixation is that associated with the assimilation of ammonia. Since nodules of different symbioses vary in the types and amounts of nitrogen-containing compounds which they form and export (92), estimates of cost based on observed xylem sap composition are useful. Using known and postulated metabolic pathways an early study (9) found that whether ureides or amides were produced as the predominant export, the costs for ammonia assimilation in nodules of *Vigna*, *Lupinus* and *Pisum* were not significantly different and averaged 6.6 mol ATP or 0.17 mol glucose/mol N_2 fixed. However, further estimates (94) based on complete analysis of xylem sap collected from detached nodules coupled with the recent knowledge on the pathway of ureide synthesis (see Chapter 3) and the operation of PEP carboxylase in providing C_4 carbon skeletons (36, 68) have shown small differences between symbioses. For *Lupinus albus* in which 75 per cent of exported molecules were amides (glutamine and asparagine) the net cost was 5.3 mol ATP equivalents/mol N_2, for *Vigna unguiculata* in which 37 per cent of exported solutes were ureides and 30 per cent amides the cost was 5.5 mol ATP/mol N_2 and for *Vigna radiata* in which 50 per cent exported solutes were ureides and 17 per cent amides the cost was greatest at 6.1 mol ATP/mol N_2 (94). In comparison, the energy required for the functioning of nitrogenase is greater than that needed for ammonia assimilation although the differences among the different symbioses are slight.

A third component in the cost of symbioses is the growth and maintenance of the nodule itself. These costs are difficult to estimate and measurements of each have not been possible. Using calculated coefficients of growth and maintenance respiration for plant tissues in general (101, 102) or from a partitioning of respiration components by the procedure of McCree (71) as applied to measurements of nodulated root gas exchange by Mahon (73), estimates as low as 7 mol ATP or 0.2 mol glucose/mol N_2 fixed (94) or as high as 25 mol ATP or 0.7 mol glucose/mol N_2 fixed (75) have been made.

While energy costs associated with transport might normally be classified as part of "maintenance" the considerable activity of nodules in this function warrants its consideration as a separate item in the 'costs' of fixation. As discussed earlier, the number of membrane transfers of ammonia and nitrogenous solutes within nodules is difficult to estimate. Furthermore, the association of ATP-dependent reactions with any or all of these transfers has not been established. However, assuming that at least two transfers of ammonia and one of exported solute occur and that each requires hydrolysis of 1 ATP, then for a symbiosis exporting ureides, 5 mol ATP or 0.13 mol glucose/mol N_2 would be involved which is essentially the same for a symbiosis exporting amides (94).

A summation of the above theoretical estimates of the costs for nodule functioning is shown in Table 1. The range of total values results largely from the extremes of nitrogenase function and emphasizes the need for a clearer understanding of the energy requirements of this enzyme *in vivo*.

Table 1. Theoretically-based cost estimates for the energy required by component processes in actively fixing legume nodules[1]

Item of functioning	mol glucose/mol N_2 fixed
Nitrogenase/Hydrogenase	0.66–1.38
Ammonia assimilation and related carbon metabolism	0.14–0.16
Transport of fixed nitrogen	0.13
Growth and maintenance of nodule	0.2–0.7
Total	1.13–2.37

[1]The assumptions used in making these estimates are described in the text.

EXPERIMENTAL. A number of different experimental approaches have been used to assess the cost to the plant of nitrogen fixation by nodules (9, 94, 103, 118). These range from a comparison of the dry weight gain or of the CO_2 efflux by the root systems of nodulated and non-nodulated plants; measurement of CO_2 efflux from nodulated roots treated with combined nitrogen to eliminate nitrogenase activity; the use of a multi-component model of respiration to separate the CO_2 efflux due to different functions in the nodulated root system; and measurements of CO_2 efflux of attached nodule clusters, root systems before and after nodule removal or of detached nodules. The shortcomings in all these approaches have been recognized and discussed (9, 94). The values obtained from a wide range of symbioses vary from as low as 0.4 to as high as 7.8 mol glucose/mol N_2 fixed (103). Most recent measurements involving CO_2 efflux determinations indicate values from around 1 mol glucose/ mol N_2 fixed for cowpea (68) and 1.4 for *Lupinus albus* (94), to around 2.5 for a wide range of symbioses with different species (73, 74, 115, 116).

A mean value of 1.7 mol glucose/mol N_2 fixed has been suggested by Mahon (75). While this value clearly falls within the limits suggested from theoretical considerations of the component energy costs (Table 1) real variations between symbioses have been observed using the same technique (9, 68 73, 115, 116) thereby showing that differences in the efficiency with which N_2 is fixed to exist in nodulated plants.

One factor contributing to the variation in measured values may be the recycling of respired CO_2. Nodules from a number of legumes contain active carboxylase enzymes which may effectively trap some of the respired CO_2 (36, 67, 68) and so change the assumed proportionality between CO_2 efflux and energy utilization. Measurement of CO_2 fixation against a net efflux from nodules, even with considerably elevated atmospheric levels of CO_2, is however difficult and probably seriously underestimates fixation. Assessment based on the maximum anapleurotic inputs of CO_2 via PEP carboxylase to form the C_4 compounds exported from nodules of particular symbioses (aspartate, asparagine and malate) suggests that an amount of carbon equivalent to 0.8 mol glucose/mol N_2 fixed in cowpea and 1.5 in lupin could be contributed in this way (94). A clearer understanding of such carbon conservation awaits unambiguous measurements of the components of net CO_2 efflux from nodules.

In view of the many uncertainties in both the theoretical estimates and the measured gas exchanges of nodules, it is probably unwise to place great emphasis on the apparent 'fit' between the theoretical and measured energy costs. The schemes outlined are approximate and undoubtedly reflect somewhat simplistic views of nodule functioning.

THE CARBON AND NITROGEN ECONOMY OF NODULES. In addition to the oxidation of incoming carbohydrate to provide reductant and ATP for the component reactions of nodules, assimilates are also used to provide carbon for nodule growth and the carbon skeletons of exported solutes of nitrogen. Detailed carbon and nitrogen economies of a number of symbioses have been constructed over the whole life cycle of the plant (9) or for short periods during vegetative growth (68, 94) in which each of these components has been measured. These balance studies show that nodules vary considerably in their total import of carbon from the host plant (see Chapter 3)—for example, *Lupinus albus: Rhizobium* WU425 utilized 6.85 g C/g N as N_2 fixed, whereas *Vigna unguiculata: Rhizobium* 176-A-27 utilized 5.42 g C/g N as N_2 fixed (94). The greater economy shown by cowpea was due to slightly less C lost as CO_2 in respiration, and significantly less required for nodule growth and for the export of ureides (C:N = 1) when compared to the export of asparagine (C:N = 2) in the lupin. Similarly this proportioning markedly changes during nodule development (9) and could be expected to change with environmental conditions.

Fixation at the whole plant level

There is a strong interdependence between photosynthesis and nitrogen assimilation in plants. Treatments which increase the rate of photosynthesis or reduce the proportion of carbon flux to photorespiration result in increased nitrogen fixation and yield in some field-grown grain legumes (55). However, selection of cultivars for increased photosynthetic rates, for reduced photorespiration or for 'C4' photosynthetic characteristics from naturally-occurring populations or from cross-breeding programmes have not, to date, identified desirable variants. Similarly, changes in ambient CO_2 or O_2 concentration in the field are, at present, not practicable.

In addition to the overall rates of CO_2 exchange, the partitioning of assimilated carbon and the efficiency with which it is used are important determinants of productivity (45), particularly in relation to N assimilation. A quantitative description of translocation and utilization of C, N and H_2O has been attempted for a number of grain legumes (58, 79, 99). This has allowed the development of empirically-based models which describe assimilate partitioning in these plants and which can be used for comparisons between modes of N assimilation (97), different stress conditions or differently yielding cultivars (98). For a review of these studies the reader is referred to Pate (91), Pate and Atkins (92) and Atkins, Herridge and Pate (9).

Carbon and nitrogen balance studies which have been carried out for the whole growth cycle of nodulated legumes and which have included respiratory carbon lossess as CO_2 in their measurement of the amount of net photosynthate available during growth have provided estimates of the proportion of the plant's resources utilized in nodule functioning (see Table 2). In each case the efficiency of carbon use by nodules varied with plant development as did the proportion of photosynthate partitioned to nodule functioning. Consis-

Table 2. Estimates of the plant's net photosynthetic resources allocated to nodule functioning during three periods of growth for three symbioses

Stage of growth	Vegetative	Flowering and fruit set	Pod-filling
Lupinus albus: Rhizobium WU425[1]			
Days after sowing	0–49	50–94	95–135
Total C imported/N fixed (mg/mg) in nodules	6.5	6.1	7.5
Total C imported by nodules as % of net photosynthesis	31.8	16.9	14.9
Total C imported minus C exported from nodules/N fixed (mg/mg)	4.5	3.9	5.0
Total C utilized in nodule functions as a % of net photosynthesis	21.8	10.9	9.9
Vigna unguiculata: Rhizobium CB756[2]			
Days	0–61	62–78	79–120
mg/mg	3.5	2.9	3.5
%	14.8	9.1	4.7
mg/mg	2.3	1.3	2.0
%	9.9	4.0	2.7
Vigna radiata: Rhizobium CB756[3]			
Days	0–36	37–58	59–98
mg/mg	8.0	4.2	2.8
%	23.1	10.0	6.3
mg/mg	5.4	2.8	1.4
%	15.6	6.7	3.2

[1]Values derived from data in Pate and Herridge (96); Atkins *et al.* (9); Layzell *et al.* (68).

[2]Values derived from data in Herridge and Pate (58); Atkins *et al.* (9); Layzell *et al.* (68).

[3]From unpublished data of R.M. Rainbird, C.A. Atkins and J.S. Pate.

tent with a pattern of increasing allocation of resources to fruit development as plants age, nodules received a decreasing proportion of available carbon. For the three associations the overall use of photosynthate was around 10 per cent of what was available. While the operation of nitrogenase is an expensive reaction in terms of energy demand, the allocation of photosynthate to nitrogen reduction is only likely to be 60 per cent of the total used in nodules (Table 1). Manipulation of the efficiency with which nitrogenase/hydrogenase utilizes ATP and reductant is therefore only likely to release 0.5–2.0 per cent of total net photosynthate from nodule function for other uses in the plant.

The high demand by developing fruits of grain legumes is satisfied by N mobilized from vegetative tissues as well as by currently assimilated N from N_2 fixation of nodules or from soil N (9). Mobilization of leaf protein in turn results in reduced photosynthesis and a consequent limitation of photosynthate supply for N assimilation (121). The literature indicates that grain legume species can differ widely in the pattern of nitrogen flow to their fruits (9, 100). In *Lupinus albus* and some *Vigna unguiculata* cultivars, more than 25 per cent of the nitrogen fixed during growth occurs in late pod-filling (9), while in other cowpea cultivars and *Pisum* species fixation declines abruptly after flowering with only 10–16 per cent of the plant's nitrogen contributed during seed maturation (100). The interactions between different organs for assimilate demand, both in terms of C and N, coupled with hormonal control of leaf senescence (84) and its association with flowering or seed growth are poorly understood. While undoubtedly there is a link between potosynthetic activity in the shoot and the quantity of fixed nitrogen exported from nodules, successful manipulation of this relationship may lie in areas other than those of increasing photosynthetic rate, especially late in plant development when a programmed series of changes in the assimilating functions of the legume appear to be directed by the reproductive structures.

References

1. Albrecht, S.L., Maier, R.J., Hanus, F.J., Russell, S.A., Emerich, D.W. and Evans, H.J. Hydrogenase in *Rhizobium japonicum* increases nitrogen fixation by nodulated soybeans, *Science, 203*, 1255–1257 (1979).
2. Allison, F.E. Legume nodule development in relation to available energy supply, *Journal of the American Society of Agronomy, 31*, 149–158 (1939).
3. Appleby, C.A. Properties of leghaemoglobin *in vivo*, and its isolation as ferrous oxyleghaemoglobin, *Biochimica et Biophysica Acta, 188*, 222–229 (1969).
4. Appleby, C.A. Leghaemoglobin, pp. 521–554 in *The Biology of Nitrogen Fixation* (Edited by A. Quispel), North Holland, Amsterdam (1974).
5. Appleby, C.A., Bergersen, F.J., Macnicol, P.K., Turner, G.L., Wittenberg, B.A. and Wittenberg, J.B. The role of leghaemoglobin in symbiotic N_2 fixation, pp. 274–292 in *Proceedings of First International Symposium on Nitrogen Fixation, Vol. 1* (Edited by W.F. Newton and C.J. Nyman), Washington State University Press, U.S.A. (1976).

6. Appleby, C.A., Turner, G.L. and Macnicol, P.K. Involvement of oxyleghaemoglobin and cytochrome P-450 in an efficient oxidative phosphorylation pathway which supports nitrogen fixation in *Rhizobium*, *Biochimica et Biophysica Acta*, *387*, 461–474 (1975).
7. Aprison, M.H., Magee, W.E. and Burris, R.H. Nitrogen fixation by excised soybean root nodules, *Journal of Biological Chemistry*, *208*, 29–39 (1954).
8. Atkins, C.A. Metabolism of purine nucleotides to form ureides in nitrogen fixing nodules of cowpea (*Vigna unguiculata* L. Walp.), *Federation of European Biochemical Societies Letters* (in press) (1981).
9. Atkins, C.A., Herridge, D.F. and Pate, J.S. The economy of carbon and nitrogen in nitrogen-fixing annual legumes—experimental observations and theoretical considerations, pp. 211–242 in *Isotopes in Biological Dinitrogen Fixation* (Edited by C.N. Welsh), FAO/IAEA Advisory Group Conference, Vienna (1978).
10. Atkins, C.A., Rainbird, R.M. and Pate, J.S. Evidence for a purine pathway of ureide synthesis in N_2-fixing nodules of cowpea (*Vigna unguiculata* (L.) Walp.), *Zeitschrift für Pflanzenphysiologie*, *97*, 249–260 (1980).
11. Awonaike, K.O., Lea, P.J. and Miflin, B.J. The effect of added nitrate on the enzymes of nitrogen assimilation in *Phaseolus* root nodules, pp. 91–95 in *Nitrogen Assimilation in Plants* (Edited by E.J. Hewitt and C.V. Cutting), Academic Press, New York (1979).
12. Awonaike, K.O., Lea, P.J. and Miflin, B.J. The localisation of ammonia-assimilating enzymes in the organelles of *P. vulgaris* nodules, *Proceedings of the Fourth International Symposium on Nitrogen Fixation*, Canberra (1981).
13. Awonaike, K.O., Lea, P.J., Miflin, B.J. and Day, J.M. The development of the enzymes involved in nitrogen assimilation in the root nodules of *Phaseolus vulgaris*, *Plant Physiology*, *59*, 52 (1977).
14. Bayliss, N.S. The thermochemistry of biological nitrogen fixation, *Australian Journal of Biological Science*, *9*, 364–370 (1956).
15. Becking, J.H. Dinitrogen-fixing associations in higher plants other than legumes, pp. 185–275 in *A Treatise on Dinitrogen Fixation, Section III, Biology* (Edited by R.W.F. Hardy and W.S. Silver), Wiley, New York (1977).
16. Bergersen, F.J. Ammonia—an early stable product of nitrogen fixation by soybean root nodules, *Australian Journal of Biological Science*, *18*, 1–9 (1965).
17. Bergersen, F.J. Biochemistry of symbiotic nitrogen fixation in legumes, *Annual Review of Plant Physiology*, *22*, 121–140 (1971).
18. Bergersen, F.J. Physiological chemistry of dinitrogen fixation by legumes, pp. 519–555 in *A Treatise on Dinitrogen Fixation, Section III*,

Biology (Edited by R.W.F. Hardy and W.S. Silver), Wiley, New York (1977).

19. Bergersen, F.J. and Goodchild, D.J. Aeration pathways in soybean root nodules, *Australian Journal of Biological Sciences*, *26*, 729–740 (1973).
20. Bergersen, F.J. and Goodchild, D.J. Cellular location and concentration of leghaemoglobin in soybean root nodules, *Australian Journal of Biological Science*, *26*, 741–756 (1973).
21. Bergersen, F.J. and Turner, G.L. Nitrogen fixation by the bacteroid fraction of breis of soybean root nodules, *Biochimica et Biophysica Acta*, *141*, 507–515 (1967).
22. Bergersen, F.J. and Turner, G.L. Leghaemoglobin and the supply of O_2 to nitrogen-fixing root nodule bacteroids: Presence of two oxidase systems and ATP production at low free O_2 concentrations, *Journal of General Microbiology*, *91*, 345–354 (1975).
23. Bergersen, F.J. and Turner, G.L. Activity of nitrogenase and glutamine synthetase in relation to availability of oxygen in continuous cultures of a strain of cowpea *Rhizobium* sp. supplied with excess ammonium, *Biochimica et Biophysica Acta*, *538*, 406–416 (1978).
24. Bergersen, F.J., Turner, G.L., Gibson, A.H. and Dudman, W.F. Nitrogenase activity and respiration of cultures of *Rhizobium* spp. with a special reference to concentration of dissolved oxygen, *Biochimica et Biophysica Acta*, *444*, 164–174 (1976).
25. Bethlenfalvay, G.J. and Phillips, D.A. Interactions between symbiotic nitrogen fixation, combined-N application and photosynthesis in *Pisum sativum*, *Physiologie Plantarum*, *42*, 119–123 (1978).
26. Bethlenfalvay, G.J. and Phillips, D.A. Variation in nitrogenase and hydrogenase activity of Alaska pea root nodules, *Plant Physiology*, *63*, 816–820 (1979).
27. Bishop, P.E., Guevara, J.G., Engelke, J.A. and Evans, H.J. Relation between glutamine synthetase and nitrogenase activities in the symbiotic association between *Rhizobium japonicum* and *Glycine max*, *Plant Physiology*, *57*, 542–546 (1976).
28. Boland, M.J. Ammonia assimilation in nodules: Computer model of a multienzyme system, in *Proceedings of the Fourth International Symposium on Nitrogen Fixation, Canberra* (1981).
29. Boland, M.J. and Benny, A.G. Enzymes of nitrogen metabolism in legume nodules: Purification and properties of NADH—dependent glutamate synthase from lupin nodules, *European Journal of Biochemistry*, *79*, 355–362.
30. Boland, M.J., Fordyce, A.M. and Greenwood, R.M. Enzymes of nitrogen metabolism in legume nodules: A comparative study, *Australian Journal of Plant Physiology*, *5*, 553–559 (1978).

31. Bollard, E.G. Translocation of organic nitrogen in the xylem, *Australian Journal of Biological Science*, *10*, 292–301 (1957).
32. Brown, C.M. and Dilworth, M.J. Ammonium assimilation by *Rhizobium* cultures and bacteroids, *Journal of General Microbiology*, *86*, 38–48.
33. Burns, R.C. and Hardy, R.W.F. Nitrogen fixation in bacteria and higher plants, pp. 1–189 in *Molecular Biology, Biochemistry and Biophysics 21* (Edited by A. Kleinzer, G.F. Springer, H.G. Wittman), Springer-Verlag, Berlin (1975).
34. Burris, R.H., Magee, W.E. and Bach, M.K. The pN_2 and the pO_2 function for nitrogen fixation by excised soybean nodules, *Annales Academiae Scientiarum Fennicae Series A2 Chemica*, *60*, 190–199 (1955).
35. Ching, T.M. Regulation of nitrogenase activity in soybean nodules by ATP and energy change, *Life Sciences*, *18*, 1071–1076 (1976).
36. Christeller, J.T., Laing, W.A. and Sutton, W.D. Carbon dioxide fixation by lupin root nodules. I. Characterization, association with phosphoenolpyruvate carboxylase, and correlation with nitrogen fixation during nodule development, *Plant Physiology*, *60*, 47–50 (1977).
37. Dart, P.J. Infection and development of leguminous nodules, pp. 367–472 in *A Treatise on Dinitrogen Fixation*, *Section III*, *Biology* (Edited by R.W.F. Hardy and W.S. Silver), Wiley, New York (1977).
38. Dilworth, M.J. Dinitrogen fixation, *Annual Review of Plant Physiology*, *25*, 81–114 (1974).
39. Dilworth, M.J. and Parker, C.A. Development of the nitrogen-fixing system in legumes, *Journal of Theoretical Biology*, *25*, 208–218 (1969).
40. Dixon, R.O.D. Hydrogen uptake and exchange by pea root nodules, *Annals of Botany*, *31*, 179–193 (1967).
41. Dixon, R.O.D. Nitrogenase-hydrogenase inter-relationships in rhizobia, *Biochemie*, *60*, 233–236 (1978).
42. Dunn, S.D. and Klucas, R.V. Studies on possible routes of ammonium assimilation in soybean root nodule bacteroids, *Canadian Journal of Microbiology*, *19*, 1493–1499 (1973).
43. Emerich, D.W., Albrecht, S.L., Russell, S.A., Ching, T.M. and Evans, H.J. Oxyleghaemoglobin-mediated hydrogen oxidation by *Rhizobium japonicum*, USDA 122 DES Bacteroids, *Plant Physiology*, *65*, 605–609 (1980).
44. Emerich, D.W., Ruiz-Argüeso, Ching, T.M. and Evans, H.J. Hydrogen-dependent nitrogenase and ATP formation in *Rhizobium japonicum* bacteroids, *Journal of Bacteriology*, *137*, 153–160 (1979).
45. Evans, L.T. Beyond photosynthesis—the role of respiration, translocation and growth potential in determining productivity, pp. 501–507 in *Photosynthesis and Productivity in Different Environments* (Edited by J.P. Cooper), Cambridge University Press, London.
46. Fujihara, S. and Yamaguchi, M. Asparagine formation in soybean

nodules, *Plant Physiology*, *66*, 139–141 (1980).
47. Gibson, A.H., Scowcroft, W.R. and Pagan, J.D. Nitrogen fixation in plants: An expanding horizon? pp. 387–417 in *Proceedings of the Second International Symposium on Nitrogen Fixation, Salamaca* (Edited by W.E. Newton, J.R. Postgate and C. Rodrigues Barrueco), Academic Press, New York (1977).
48. Gordon, J.K. and Brill, W.J. Depression of nitrogenase synthesis in the presence of excess NH_4^+, *Biochemical and Biophysical Research Communications*, *59*, 867–971 (1974).
49. Grimes, H. and Fottrell, P.F. Enzymes involved in glutamate metabolism in legume root nodules, *Nature*, *212*, 295–296 (1966).
50. Gunning, B.E.S., Pate, J.S., Minchin, F.R. and Marks, I. Quantitative aspects of transfer cell structure in relation to vein loading in leaves and solute transport in legume nodules. Transport at the cellular level, *Symposium of the Society of Experimental Biology*, *28*, 87–126 (1974).
51. Hageman, R.V., Orme-Johnson, W.H. and Burris, R.H. Role of magnesium adenosine 5'-triphosphate in the hydrogen evolution reaction catalyzed by nitrogenase from *Azotobacter vinelandii*, *Biochemistry*, *19*, 2333–2342 (1980).
52. Hanks, J.F., Tolbert, N.E. and Schubert, K.R. Localisation of enzymes of ureide biosynthesis in peroxisomes and microsomes of nodules, *Plant Physiology* (in press) (1981).
53. Hardy, R.W.F., Bottomley, F. and Burns, R. *A Treatise on Dinitrogen Fixation, Sections I and II, Inorganic and Physical Chemistry and Biochemistry*, John Wiley & Sons Inc., New York (1979).
54. Hardy, R.W.F. and Gibson, A.H. *A Treatise on Dinitrogen Fixation, Section IV, Agronomy and Ecology*, John Wiley & Sons Inc., New York (1977).
55. Hardy, R.W.F. and Havelka, U.D. Nitrogen fixation research. A key to world food? *Science*, *188*, 633–643 (1975).
56. Hardy, R.W.F. and Silver, W.S. *A Treatise on Dinitrogen Fixation, Section III, Biology*, John Wiley & Sons Inc., New York (1977).
57. Herridge, D.F., Atkins, C.A., Pate, J.S. and Rainbird, R.M. Allantoin and allantoic acid in the nitrogen economy of the cowpea (*Vigna unguiculata* (L.) Walp.), *Plant Physiology*, *62*, 495–498 (1978).
58. Herridge, D.F. and Pate, J.S. Utilization of net photosynthate for nitrogen fixation and protein production in an annual legume, *Plant Physiology*, *60*, 759–764 (1977).
59. Hollander, A. (Editor). Genetic engineering for nitrogen fixation, *Basic Life Sciences*, *Vol. 9*, Plenum Press, New York (1977).
60. Hwang, I.C., Chen, C.H. and Burris, R.H. Inhibition of nitrogenase, *Biochimica et Biophysica Acta*, *292*, 256–270 (1973).
61. Kennedy, I.R. Primary products of symbiotic nitrogen fixation. I.

Short-term exposures of serradella nodules to $^{15}N_2$, *Biochimica et Biophysica Acta*, *130*, 285–294 (1966).

62. Kennedy, I.R., Morris, J.A. and Mortenson, L.E. N_2 fixation by purified components of the N_2-fixing system of *Clostridium pasteurianum*, *Biochimica et Biophysica Acta*, *153*, 777–786 (1968).
63. Kidby, D.K. Carbon metabolism in legume root nodules, Ph.D. Thesis, University of Western Australia (1967).
64. Laane, C., Krone, W., Konings, W., Haaker, H. and Veeger, C. The involvement of the membrane potential in nitrogen fixation by bacteroids of *Rhizobium leguminosarum*, *Federation of European Biochemical Societies Letters*, *103*, 328–332 (1979).
65. Laane, C., Krone, W., Konings, W., Haaker, H. and Veeger, C. Short-term effect of ammonium chloride on nitrogen fixation by *Azotobacter vinelandii* and by bacteroids of *Rhizobium leguminosarum*, *European Journal of Biochemistry*, *103*, 39–46 (1980).
66. Lawn, R.J. and Brun, W.A. Symbiotic nitrogen fixation in soybeans. I. Effect of photosynthetic source-sink manipulations, *Crop Science*, *14*, 11–16 (1974).
67. Lawrie, A.C. and Wheeler, C.T. Nitrogen fixation in the root nodules of *vicia faba* L. in relation to the assimilation of carbon. II. The dark fixation of carbon dioxide, *New Phytologist*, *74*, 429–436 (1975).
68. Layzell, D.B., Rainbird, R., Atkins, C.A. and Pate, J.S. Economy of photosynthate use in N-fixing legume nodules: Observations on two contrasting symbioses, *Plant Physiology*, *64*, 888–891 (1979).
69. Lea, P.J. and Fowden, L. The purification and properties of glutamine-dependent asparagine synthetase isolated from *Lupinus albus*, *Proceedings of the Royal Society* (*London*) *B.*, *192*, 13–26 (1975).
70. Ludwig, R.A. and Signer, E.R. Glutamine synthetase and control of nitrogen fixation in *Rhizobium*, *Nature*, *267*, 245–248 (1977).
71. McCree, K.J. An equation for the rate of respiration of white clover plants grown under controlled conditions, pp. 221–229 in *Prediction and Measurement of Photosynthetic Productivity* (Edited by I. Setlik), Pudoc, Wageningen (1970).
72. McPharland, R.H., Guevara, J.C., Becker, R.R. and Evans, H.J. The purification and properties of the glutamine synthetase from the cytosol of soybean root nodules, *Biochemical Journal*, *153*, 597–606 (1976).
73. Mahon, J.D. Respiration and the energy requirements for nitrogen fixation in nodulated pea roots, *Plant Physiology*, *60*, 817–821 (1977).
74. Mahon, J.D. Environmental and genotypic effects on the respiration associated with symbiotic nitrogen fixation in peas, *Plant Physiology*, *63*, 892–897 (1979).
75. Mahon, J.D. Energy relationships, in *Nitrogen Fixation, Vol. 3* (Edited by W.J. Broughton), Oxford University Press, Oxford (in press) (1981).

76. Meeks, J.C., Wolk, C.P., Schilling, N., Shaffer, P.W., Avissar, Y. and Chien, W.S. Initial organic products of fixation of [^{13}N] dinitrogen by root nodules of soybean (*Glycine max*), *Plant Physiology, 61*, 980–983 (1978).
77. Miflin, B.J. and Lea, P.J. Amino acid metabolism, *Annual Review of Plant Physiology, 28*, 299–329 (1977).
78. Miflin, B.J. and Lea, P.J. Ammonia assimilation, pp. 169–202 in *The Biochemistry of Plants, Vol. 5*, Academic Press, New York (1980).
79. Minchin, F.R. and Pate, J.S. The carbon balance of a legume and the functional economy of its root nodules, *Journal of Experimental Botany, 24*, 259–271 (1973).
80. Mitchell, P. Chemiosmotic coupling in oxidative and photosynthetic phosphorylation, *Biological Reviews, 41*, 445–502 (1966).
81. Mortenson, L.E. Regulation of nitrogen fixation, pp. 179–227 in *Current Topics in Cellular Regulation, Vol. 13* (Edited by B.L. Horecker and E.R. Stadtman) (1978).
82. Mortenson, L.E. and Thorneley, R.N.F. Structure and function of nitrogenase, *Annual Review of Biochemistry, 48*, 387–418 (1979).
83. Newton, W., Postgate, J.R. and Rodriguez-Barrueco, C. (Editors). *Recent Developments in Nitrogen Fixation*, Academic Press, London (1977).
84. Nooden, L.D. Regulation of senescence, pp. 139–151 in *World Soybean Research Conference II. Proceedings* (Edited by F.T. Corbin), Westview Press Co., U.S.A. (1980).
85. Nutman, P.S. (Editor). Symbiotic nitrogen fixation in plants. International Biological Programme 7, Cambridge University Press, London, New York, Melbourne (1976).
86. Oghoghorie, C.G.O. and Pate, J.S. Exploration of the nitrogen transport system of a nodulated legume using ^{15}N, *Planta, 104*, 35–49 (1972).
87. Ohyama, T. and Kumazawa, K. Nitrogen assimilation in soybean nodules, II. $^{15}N_2$ assimilation in bacteroid and cytosol fractions of soybean nodules, *Soil Science and Plant Nutrition, 26*, 205–213 (1980).
88. Orme-Johnson, W.H., Davis, L.C., Henzl, M.T., Averill, B.A., Orme-Johnson, N.R., Munk, E. and Zimmerman, R. Pathways of biological N_2 reduction, pp. 131–178 in *Recent Developments in Nitrogen Fixation* (Edited by W. Newton, J.R. Postgate and C. Rodriguez-Barrueco), Academic Press, London (1977).
89. Pate, J.S. Transport in symbiotic systems fixing nitrogen, pp. 278–803 in *Transport in Plants, II. Part B. Tissues and Organs, Encyclopedia of Plant Physiology, New Series, Vol. 2* (Edited by U. Lüttge and M.G. Pitman), Springer, New York (1976).
90. Pate, J.S. Functional biology of dinitrogen fixation by legumes, pp. 473–517, Ch. 9, in *A Treatise on Dinitrogen Fixation, Section III, Biology*

(Edited by R.W.F. Hardy and W.S. Silver) (1977).

91. Pate, J.S. Transport and partitioning of nitrogenous solutes, *Annual Review of Plant Physiology*, *31*, 313–340 (1980).
92. Pate, J.S. and Atkins, C.A. Nitrogen uptake, transport and utilisation, in *Nitrogen Fixation, Vol. 3* (Edited by W.J. Broughton), Oxford University Press, U.K. (in press) (1981).
93. Pate, J.S., Atkins, C.A., Hamel, K., McNeil, D.L. and Layzell, D.B. Transport of organic solutes in phloem and xylem of a nodulated legume, *Plant Physiology*, *63*, 1082–1088 (1979).
94. Pate, J.S., Atkins, C.A. and Rainbird, R.M. Theoretical and experimental costings of nitrogen fixation and related processes in nodules of legumes, in *Proceedings of the Fourth International Symposium on Nitrogen Fixation, Canberra 1980* (1981).
95. Pate, J.S., Gunning, B.E.S. and Briarty, L.G. Ultrastructure and functioning of the transport system of the leguminous root nodule, *Planta*, *85*, 11–34 (1969).
96. Pate, J.S. and Herridge, D.F. Partitioning and utilization of net photosynthate in a nodulated annual legume, *Journal of Experimental Botany*, *29*, 401–412 (1978).
97. Pate, J.S., Layzell, D.B. and Atkins, C.A. Economy of C and N in a nodulated and non-nodulated (NO_3-grown) legume, *Plant Physiology*, *64*, 1083–1088 (1979).
98. Pate, J.S., Layzell, D.B. and Atkins, C.A. Transport exchanges of carbon, nitrogen and water in the context of whole plant growth and functioning—case history of a nodulated annual legume, *Berichte Deutsche Botanisches Gesellschaft*, *93*, 243–255 (1980).
99. Pate, J.S., Layzell, D.B. and McNeil, D.L. Modelling the transport and utilization of C and N in a nodulated legume, *Plant Physiology*, *63*, 730–738 (1979).
100. Pate, J.S. and Minchin, F.R. Comparative studies of carbon and nitrogen nutrition of selected grain legumes, pp. 105–114 in *Proceedings of the International Legume Conference, Royal Botanic Gardens* (Edited by R.J. Summerfield and A.H. Bunting), Kew, England (1980).
101. Penning de Vries, F.W.T. The cost of maintenance processes in plant cells, *Annals of Botany*, *39*, 77–92 (1975).
102. Penning de Vries, F.W.T., Brunsting, A.H.M. and van Laar, H.H. Products, requirements and efficiency of biosynthesis. A quantitative approach, *Journal of Theoretical Biology*, *45*, 339–377 (1974).
103. Phillips, D.A. Efficiency of symbiotic nitrogen fixation in legumes, *Annual Review of Plant Physiology*, *31*, 29–49 (1980).
104. Planqué, K., Kennedy, I.R., de Vries, G.E., Quispel, A. and van Brussel, A.A.N. Location of nitrogenase and ammonia-assimilatory enzymes in bacteroids of *Rhizobium leguminosarum* and *Rhizobium lupinii*, *Journal*

of General Microbiology, *102*, 95–104 (1977).

105. Rainbird, R.M. and Atkins, C.A. Purification and some properties of urate oxidase from nitrogen-fixing nodules of cowpea, *Biochimica et Biophysica Acta* (in press) (1981).
106. Ratcliffe, H.D., Drozd, J.W., Bull, A.T. and Daniel, R.M. Energy coupling in soybean bacteroids, *Federation of European Microbiology Societies Microbiology Letters*, *8*, 111–115 (1980).
107. Raven, J.A. and Smith, F.A. Nitrogen assimilation and transport in vascular land plants in relation to intracellular pH regulation, *New Phytologist*, *76*, 415–431 (1976).
108. Rawsthorne, S., Minchin, F.R., Summerfield, R.J., Cookson, C. and Coombs, J. Carbon and nitrogen metabolism in legume root nodules, *Phytochemistry*, *19*, 341–355 (1980).
109. Reporter, M. Synergetic cultures of *Glycine max* root cells and rhizobia separated by membrane filters, *Plant Physiology*, *57*, 651–655 (1976).
110. Reynolds, P.H.S. and Farnden, K.J.F. The involvement of aspartate amino transferases in ammonium assimilation in lupin nodules, *Phytochemistry*, *18*, 1625–1630 (1979).
111. Robertson, J.G., Warburton, M.P., Lyttleton, P., Fordyce, A.M. and Bullivant, S.B. Membranes in lupin root nodules. II. Preparation and properties of peribacteroid envelope inner membranes from developing lupin nodules, *Journal of Cell Science*, *30*, 151–174 (1978).
112. Ronson, C.W., Lyttleton, P. and Robertson, J.G. Ineffective mutants of *Rhizobium trifolii* defective in C_4-dicarboxylate transport, in *Proceedings of the Fourth International Symposium of Nitrogen Fixation, Canberra* (1981).
113. Ronson, C.W. and Primrose, S.B. Carbohydrate metabolism in *Rhizobium trifolii*: Identification and symbiotic properties of mutants, *Journal of General Microbiology*, *112*, 77–88 (1979).
114. Ruiz-Argueso, T., Hanus, J. and Evans, H.J. Hydrogen production and uptake by pea nodules as affected by strains of *Rhizobium leguminosarum*, *Archives of Microbiology*, *116*, 113–118 (1978).
115. Ryle, G.J.A., Powell, C.E. and Gordon, A.J. The respiratory costs of nitrogen fixation in soybean, cowpea and white clover. I. Nitrogen fixation and the respiration of the nodulated root, *Journal of Experimental Botany*, *30*, 135–144.
116. Ryle, G.J.A., Powell, C.E. and Gordon, A.J. The respiratory costs of nitrogen fixation in soybean, cowpea and white clover. II. Comparisons of the cost of nitrogen fixation and the utilization of combined nitrogen, *Journal of Experimental Botany*, *30*, 145–153 (1979).
117. Schubert, K.R. and Evans, H.J. Hydrogen evolution, a major factor affecting the efficiency of nitrogen fixation in nodulated symbionts, *Proceedings of the National Academy of Science, U.S.A.*, *73*, 1207–1211 1976).

118. Schubert, K.R. and Ryle, G.J.A. The energy requirements for nitrogen fixation in nodulated legumes, pp. 85–96 in *Advances in Legume Science* (Edited by R.J. Summerfield and A.H. Bunting), Royal Botanic Gardens, Kew, U.K. (1980).
119. Scott, D.B., Robertson, J. and Farnden, K.J.F. Ammonia assimilation in lupin nodules, *Nature*, *262*, 703–705 (1976).
120. Shanmugam, K.T., O'Gara, F., Andersen, K. and Valentine, R.C. Biological nitrogen fixation, *Annual Review of Plant Physiology*, *29*, 263–276 (1978).
121. Sinclair, T.R. and de Wit, C.T. Photosynthate and nitrogen requirements for seed production by various crops, *Science*, *189*, 565–567 (1975).
122. Sprent, J.I. The effects of water stress on nitrogen-fixing root nodules. IV. Effects on whole plants of *Vicia faba* and *Glycine max*, *New Phytologist*, *71*, 603–611 (1972).
123. Sprent, J.I. Root nodule anatomy, type of export product and evolutionary origin in some *Leguminoseae*, *Plant Cell and Environment*, *3*, 35–43 (1980).
124. Stouthamer, A.H. and Bettenhausen, C. Utilization of energy for growth and maintenance in continuous and batch cultures of microorganisms, *Biochimica et Biophysica Acta*, *301*, 53–70 (1973).
125. Stovall, I. and Cole, M. Organic acid metabolism by isolated *Rhizobium japonicum* bacteroids, *Plant Physiology*, *61*, 787–790 (1978).
126. Streeter, J.G. Growth of two soybean shoots on a single root. Effect of nitrogen and dry matter accumulation by shoots and on the rate of nitrogen fixation by nodulated roots, *Journal of Experimental Botany*, *25*, 189–198 (1974).
127. Streeter, J.G. Allantoin and allantoic acid in tissues and stem exudate from field-grown soybean plants, *Plant Physiology*, *63*, 478–480 (1979).
128. Streicher, S. and Valentine, R.C. The genetic basis of dinitrogen fixation in *Klebsiella pneumoniae*, pp. 623–656 in *A Treatise on Dinitrogen Fixation, Section III, Biology* (Edited by R.W.F. Hardy and W.S. Silver), Wiley, New York (1977).
129. Stripf, R. and Werner, D. Differentiation of *Rhizobium japonicum*. II. Enzymatic activities in bacteroids and plant cytoplasm during the development of nodules of *Glycine max*, *Zeitschrift für Naturforschung 33c*, 373–381 (1978).
130. Thornley, R.N.F. and Willinson, K.R. Nitrogenase of *Klebsiella pneumoniae:* Inhibition of acetylene reduction by magnesium ion explained by the formation of an inactive dimagnesium-adenosine triphosphate complex, *Biochemical Journal*, *139*, 211–214 (1974).
131. Tjepkema, J.D. and Yocum, C.S. Respiration and oxygen transport in soybean nodules, *Planta*, *115*, 59–72 (1973).

132. Trinick, M.J. Symbiosis between *Rhizobium* and the non-legume *Trema aspera*, *Nature*, *244*, 459–460 (1973).
133. Trinick, M.J. The effective *Rhizobium* symbiosis with the non-legume *Parasponia andersonii*, in *Proceedings of the Fourth International Symposium on Nitrogen Fixation, Canberra* (1981).
134. Tso, M.Y.W. and Burris, R.H. The binding of ATP and ADP by nitrogenase components from *Clostridium pasteurianum*, *Biochimica et Biophysica Acta*, *309*, 263–270 (1973).
135. Tubb, R.S. Regulation of nitrogen fixation in *Rhizobium* species, *Applied and Environmental Microbiology*, *32*, 483–488 (1976).
136. Tubb, R.S. and Postgate, J.R. Control of nitrogenase synthesis, *Journal of General Microbiology*, *79*, 103–117 (1973).
137. Verma, D.P.S. and Bal, A.K. Intracellular site of synthesis and localization of leghaemoglobin in root nodules, *Proceedings of the National Academy of Science* (*U.S.A.*), *73*, 3843–3847 (1976).
138. Werner, D. and Stripf, R. Differentiation of *Rhizobium japonicum*. I. Enzymatic comparison of nitrogenase repressed and derepressed free-living cells and of bacteroids, *Zeitschrift für Naturforschung*, *33c*, 245–252 (1978).
139. Wilcockson, J. and Werner, D. Organic acids and prolonged nitrogenase activity by non-growing, free-living *Rhizobium japonicum*, *Archives of Microbiology*, *122*, 153–159 (1979).
140. Wittenberg, J.B., Bergersen, F.J., Appleby, C.A. and Turner, G.L. Facilitated oxygen diffusion. The role of leghaemoglobin in nitrogen fixation by bacteroids isolated from soybean root nodules, *Journal of Biological Chemistry*, *249*, 4057–4066 (1974).
141. Wolfgang, H. and Hothes, K. Papierchromatographische Untersuchungen an pflanzlichen Blutungssäften, *Naturwissenschaften*, *40*, 606 (1953).
142. Wong, P.P. and Burris, R.H. Nature of oxygen inhibition of nitrogenase from *Azotobacter vinelandii*, *Proceedings of the National Academy of Science* (*U.S.A.*), *69*, 672–675 (1972).
143. Wong, P.P. and Evans, H.J. Poly-β-hydroxybutyrate utilization by soybean (*Glycine max* Merr.) nodules and assessment of its role in maintenance of nitrogenase activity, *Plant Physiology*, *47*, 750–755 (1971).
144. Wooi, K.C. and Broughton, W.J. Isolation and metabolism of *Vigna unguiculata* root nodule protoplasts, *Planta*, *145*, 487–495 (1979).
145. Yoch, D.C. and Valentine, R.C. Ferredoxins and flavodoxins of bacteria, *Annual Review of Microbiology*, *26*, 139–162 (1972).
146. Zabtolowicz, R.M., Russell, S.A. and Evans, H.J. Effect of the hydrogenase system in *Rhizobium japonicum* on the nitrogen fixation and growth of soybeans at different stages of development, *Agronomy Journal*, *72*, 555–559 (1980).

C.A. ATKINS

3. Ureide Metabolism and the Significance of Ureides in Legumes

Introduction

The ureides are nitrogenous organic compounds which contain one or more of the ureido grouping (NH_2-CO-NH-). The most important in biological systems are the ureides of glyoxylic acid, allantoin (5-ureido-hydantoin) and allantoic acid (diureido acetic acid) (structures, Fig. 1). While allantoin is an important excretory product in many primates and in some reptiles and allantoic acid serves a similar function in some teleost fishes, both are common plant constituents and have been found in species from twenty-one higher plant families (169). Allantoin and allantoic acid occur especially in relatively large amounts in members of the Aceraceae, Borraginaceae, Hippocastanaceae and Plantanaceae (95). As stored nitrogenous solutes of the stems and underground organs in *Symphytum* spp. or *Acer* spp. they may be mobilized in the spring and transferred in xylem to growing tips where they constitute a major source of nitrogen for new growth (94).

Allantoin and allantoic acid have also been detected in many species of the leguminoseae, particularly in seeds or in young seedlings (169) and also in xylem sap (Table 1) where ureides may be the predominant nitrogenous solutes, especially of effectively nodulated plants (89, 118). Current interest in these compounds in legumes stems from the realization that they are produced in nodules and exported to the host plant as the major products of nitrogen fixation (59) and also because many of the species which show a 'ureide-based' nitrogen economy are agriculturally important tropical or subtropical crop plants (see Table 1). To a lesser extent allantoin and allantoic acid may serve as forms of translocated nitrogen in young legume seedlings following hydrolysis of nucleic acids stored in the cotyledons (48).

The other important ureide in plants is citrulline (structure, Fig. 1) which is an intermediate in the pathway of arginine biosynthesis and also appears to serve a similar function to allantoin and allantoic acid in being a major solute for storage and translocation of nitrogen in the Betulaceae and Juglandaceae (25, 132). However, citrulline is also a significant constituent of the xylem stream in some legumes (25), of the nodules and xylem of

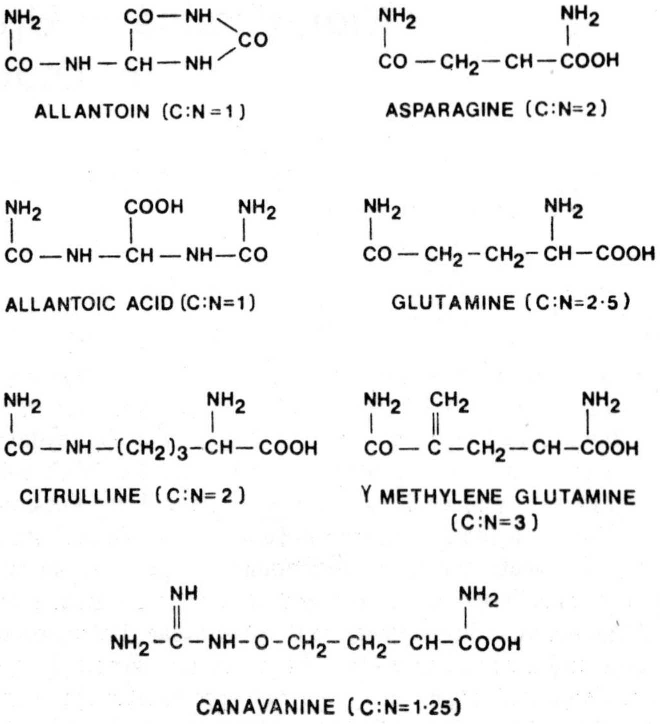

Fig. 1. Molecular structure of some nitrogenous solutes found in xylem sap.

nitrogen-fixing non-legumes such as *Alnus* spp. and *Myrica* spp. (25, 92, 178) and of the translocating strands of corraloid roots of *Macrozamia riedlii* infected with nitrogen-fixing blue-green algae (54).

The subject matter of this chapter will stress the formation and significance of ureides, especially of allantoin and allantoic acid, in legumes. However, details of the metabolic relationships between nitrogen assimilation and ureide synthesis, the transport of ureides from their site(s) of formation to those of utilization and the metabolic pathways of ureide breakdown and transfer of nitrogen for protein synthesis in legumes are, at this stage, largely speculative. However, information from other plants, microorganisms and animal systems in which these compounds are of importance and which have been more extensively studied is utilized to provide a cohesive picture and to unfold the importance and consequences of ureide synthesis in nitrogen fixation.

Table 1. Occurrence of ureides (allantoin and allantoic acid) in xylem sap of nodulated legumes

Species in which ureides are major[1] solutes of N in xylem sap	Species in which ureides are minor solutes of N in xylem sap	Species in which analysis has not detected ureides in xylem sap
Albizia lophanth[2, 3] (25)	*Cicer artietinum*[4]	*Lathyrus cicera*[4]
Arachis hypogea[4]	*Lens esculenta*[4]	*L. sativus*[4]
Cajanus cajan[4]	*Pisum arvense* (123, 124)	*Lupinus albus* (116)
Cyamopsis tetragonoloba (118)	*Vicia ervilia*[4]	*L. angustifolius*[4]
Glycine max[5] (85, 89, 118, 152)	*V. sativa*[4]	*L. cosentinii*[4]
Macrotyloma uniflorum (118)		*L. mutabilis*[4]
Phaseolus vulgaris (36, 113)		*Pisum sativum* (52)
Psophocarpus tetragonolobus[5] (118)		*Trifolium repens* (37)
Vigna angularis (118)		*Vicia calcarata*[4]
V. mungo (118)		*V. faba* (120)
V. radiata (118)		
V. triloba (118)		
V. unguiculata[5] (59, 118)		
V. umbellata (118)		

[1]Greater than 30% of xylem-borne N.
[2]Stem tracheal sap collected for *A. lophanth* and root bleeding sap for all other species.
[3]Also contained citrulline.
[4]R.M. Rainbird, unpublished data.
[5]Ureides also found as the major solutes of N in xylem sap of detached nodules (152; R.M. Rainbird, J.S. Pate and C.A. Atkins, unpublished; 59).
Note: References to published information are given in parentheses.

Occurrence and significance of ureides in legumes

The products of nitrogen fixation are transported from the nodule to the organs of the host plant in the transpiration stream and there is good evidence to suppose that analyses of xylem sap bleeding from cut, nodulated roots of detached nodules will identify the transported forms of fixed nitrogen (114). While the concentration of nitrogenous solutes in bleeding xylem sap (0.1–1 mg N ml^{-1}) of root, forced out of the decapitated plant under root pressure is likely to be considerably higher than that which would normally be moving in the intact transpiration stream (113, 151), the distribution of nitrogen among different solutes is probably little changed (113).

Present information suggests that effectively nodulated legumes fall within the classes of being either amide-exporting or ureide-exporting species. In the former group asparagine, glutamine or substituted amides, such as γ-methylene glutamine (structure, Fig. 1), constitute the bulk of fixed nitrogen exported from the root system. In *Lupinus albus,* for example, asparagine accounts for 60–80 per cent of xylem nitrogen throughout plant development (116) and a similar situation applies to a large number of other temperate species (115). In the latter group ureides are frequently 80 per cent or more of xylem nitrogen throughout development (59, 89). In both cases analysis of xylem sap bleeding from detached nodules confirms that the major solutes in root bleeding sap are contributed by the nodules. These exudates contain 1–6 mg N ml^{-1}, that is more than ten times the concentration found in root bleeding sap, and show 124–140 mM asparagine in *Pisum sativum* (52), *Vicia faba* (120) and *Lupinus albus* (116), 38–90 mM for ureides in *Glycine max* (152) and *Psophocarpus tetragonolobus* (R.M. Rainbird, C.A. Atkins and J.S. Pate, unpublished) and 3–5 times more allantoin and allantoic acid in nodule sap compared to root bleeding sap in *Vigna unguiculata* (59).

Analysis of xylem sap collected from above and below the nodulated zone of the root system of young, actively-fixing cowpea plants shows that more than 98% of the xylem-borne nitrogen entering the shoot is due to the contribution made by nodules and that most of the nodule's output is in the form of ureides and, to a lesser extent, glutamine (Table 2). Provision of [^{15}N] nitrogen gas to the root system of nodulated plants results in rapid labelling of ureides in nodules (88, 103) and other organs (104) of soybean and in nodules and xylem sap of cowpea (59). Similarly, allantoin and allantoic acid

Table 2. Composition of nitrogenous solutes in root bleeding xylem sap collected below and above the nodulated zone of the root system of young (40 days old) cowpea (*Vigna unguiculata* L. Walp) plants

Nitrogenous solute	Below nodules	Above nodules
	(μg N ml^{-1})	
Allantoin + allantoic acid	2.4	439.5
Aspartic acid	2.7	2.9
Threonine	tr[1]	1.0
Serine	0.8	1.4
Asparagine	ND[2]	3.0
Glutamic acid	1.3	1.8
Glutamine	ND	42.6
Alanine	0.8	1.0
Valine	ND	2.1
Isoleucine + leucine	1.1	1.4
Arginine	ND	6.0
Total	9.1	502.7

[1]Less than 0.1 μg N/ml.
[2]Not detectable.

are readily labelled from [^{15}N] nitrogen gas in detached nodules of ureide-forming species (106).

High concentrations of ureide nitrogen in the xylem stream is usually associated with effective nodulation and high rates of fixation (118) although non-nodulated or nodulated plants utilising combined nitrogen, as nitrate supplied to the roots, also contain ureides in xylem. The levels are frequently low compared to symbiotically-nourished plants and the predominant nitrogenous solutes in xylem are amides and free nitrate (8, 36, 64, 85, 89, 118, 152). In field grown plants where both soil nitrogen and N_2 are used xylem ureide levels vary during the season (152) reflecting the relative activity of fixation and nitrate uptake and assimilation at different stages of plant development.

In *Phaseolus vulgaris* however, significant levels of xylem-borne ureides are found in nodulated plants supplied with ammonium or ammonium nitrate and allantoic acid is the predominant organic nitrogenous compound or solute (163). Similarly, the concentrations of allantoin or allantoic acid in the vegetative and reproductive organs of nodulated cowpea or soybean are generally higher than in the organs of comparable non-nodulated plants or nodulated plants supplied with combined nitrogen (47, 87, 88, 118). Matsumoto *et al.* (88) showed that the concentration of allantoin in soybean plants was in fact negatively correlated with nodule mass. However, the pattern of ureide storage in non-nodulated plants appears to be different from that in symbiotic plants (118). In soybean, for example, the ureide levels in stems, roots and pods of a non-nodulating isoline supplied with ammonium nitrogen were lower than those of a comparable nodulated line but in the leaves of the non-nodulating line ureide levels were in fact higher (47). In cowpea, while stems and petioles were the major sites of storage for ureides in both symbiotically-dependent plants and nodulated plants supplied with high levels of nitrate, in non-nodulated plants, ureide was, however, stored in greater amounts in leaflets and roots than in stems or petioles (118).

Ureides are found in seeds or in young seedlings of many species (169) and allantoin levels may increase markedly following germination (31, 48). In this latter case massive hydrolysis of 'storage RNA' (15, 109) in cotyledons releases purine bases which are in turn oxidized to ureides. These serve as a transportable nitrogen source which is apparently readily metabolized to release nitrogen for protein synthesis in early seedling growth (48).

Biosynthesis of allantoin and allantoic acid

General

The formation of allantoin in plants is probably through the oxidation of purines (16, 33, 66, 83, 99, 131, 153) by a series of reactions similar to those f animal and microbial metabolism (7, 175, 177). Suggestions that allantoic

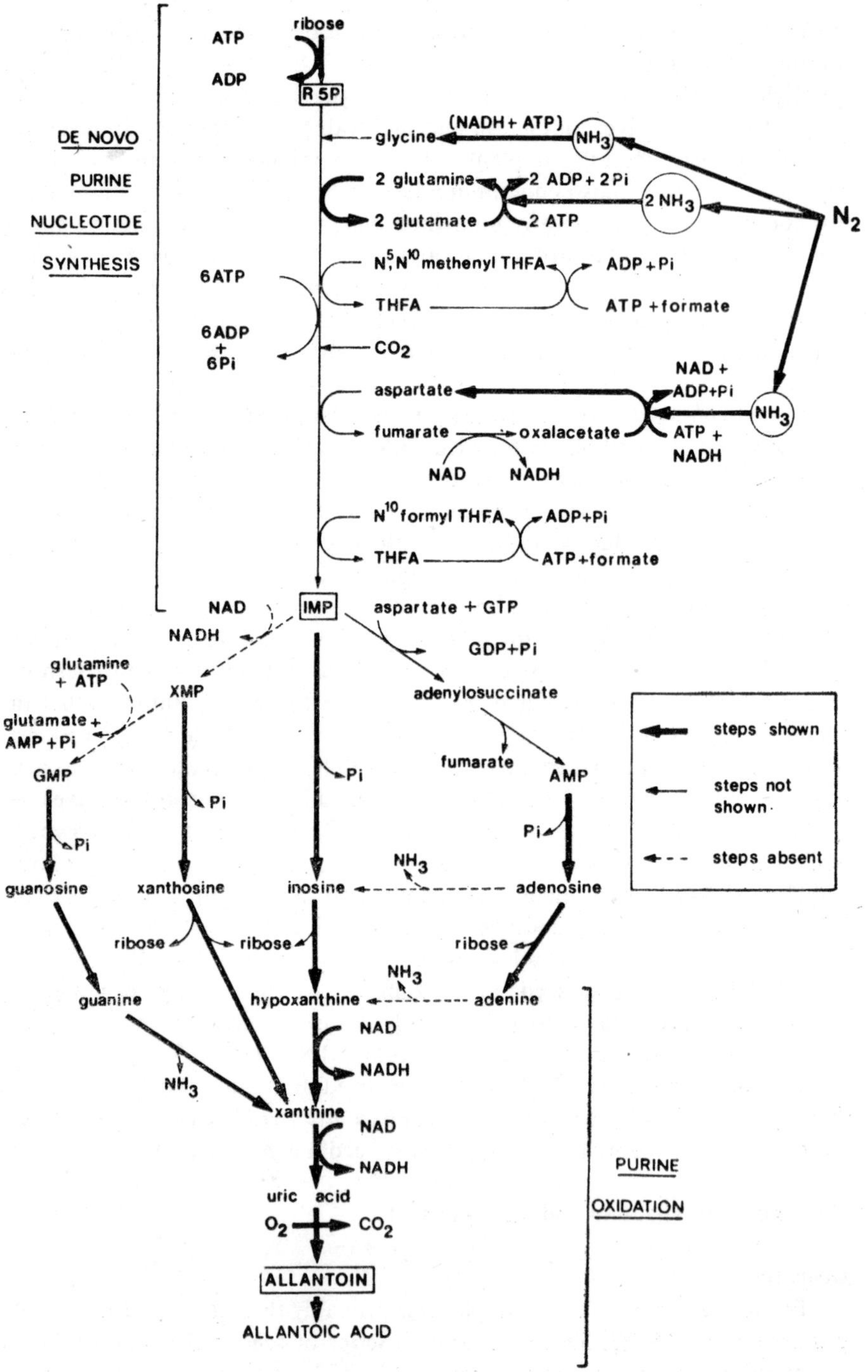
DE NOVO PURINE NUCLEOTIDE SYNTHESIS
ATP
ADP
ribose
R 5P
glycine
(NADH + ATP)
NH_3
2 glutamine
2 glutamate
2 ADP + 2 Pi
2 ATP
2 NH_3
N_2
N^5,N^{10} methenyl THFA
THFA
ADP + Pi
ATP + formate
6ATP
6ADP + 6Pi
CO_2
aspartate
fumarate
oxalacetate
NAD
NADH
NAD + ADP+Pi
ATP + NADH
NH_3
N^{10} formyl THFA
THFA
ADP+Pi
ATP+formate
IMP
NAD
NADH
aspartate + GTP
GDP+Pi
glutamine + ATP
glutamate + AMP + Pi
XMP
GMP
adenylosuccinate
fumarate
AMP
Pi
Pi
Pi
Pi
steps shown
steps not shown
steps absent
NH_3
guanosine
xanthosine
inosine
adenosine
ribose
ribose
ribose
NH_3
guanine
hypoxanthine
adenine
NAD
NADH
NH_3
xanthine
NAD
NADH
uric acid
O_2
CO_2
ALLANTOIN
ALLANTOIC ACID
PURINE OXIDATION

acid might be formed from the condensation of urea and a two carbon organic acid such as glyoxylate (26, 149) have generally received little recent experimental support (131). In rapidly growing embryonic axes of soybean (1) in germinating pollen of *Pinus mugo* (102) and isolated mesophyll protoplasts of tobacco leaves (14) and in all instances of higher plant tissues in which high rates of nucleic acid synthesis are occurring, enzymological studies and [^{14}C] labelling tracer experiments have shown that the pathways of *de novo* purine synthesis are broadly similar to those established in other organisms (32). The purine ring structure is believed to be derived from glycine, two amide group of glutamine, the amino group of aspartate and two activated C_1 units (see Fig. 2). The metabolic sequence produces inosine monophosphate (IMP) which serves as the precursor for the adenine and guanine nucleotides (see Fig. 2) in normal nucleic acid synthesis. Plants also show many of the reactions of purine salvage (14, 53, 144, 153) which result in extensive interconversion of purine bases, nucleosides and nucleotides similar to those occurring in animal tissues (97). Xanthine and hypoxanthine in plants serve not only as precursors for ureides formed as a result of nucleic acid hydrolysis (48) or from deamination of exogenously supplied adenylates (99) but also serve as substrates for the synthesis of secondary purine derivatives (154).

Nodulated legumes

The detailed metabolic pathways linking nitrogen fixation, purine synthesis and ureide production in legume root nodules have not been established despite a number of recent studies within this area and the clear demonstrations that [^{15}N] labelled ureides result from fixation of [^{15}N] nitrogen gas by intact cowpea (59) and soybean (103, 104) plants. There is general agreement that ammonia is the first stable product of nitrogen fixation in nodules (22, 68, 69, 91), that it is released from the bacteroids (21, 75) and that assimilation to form organic solutes of nitrogen occurs in the cytoplasm and organelles of the surrounding host cell. Early reports tracing the labelling of amino compounds in nodules supplied [^{15}N] with nitrogen gas concluded that reductive amination to form glutamate was the initial reaction involved in ammonia assimilation (4, 20, 68, 69). However, studies of the short-term

Fig. 2. Possible metabolic pathways of linked *de novo* purine synthesis and purine oxidation indicating how reduced nitrogen might be incorporated into the final products, allantoin and allantoic acid. Heavy arrows indicate pathways shown to occur in legume nodules, lighter arrows are postulated steps not yet shown in legume nodules and broken arrows are steps in which assays for the enzymes in legume nodules have been negative. Abbreviations used: R5-P, Ribose 5 phosphate; ATP, adenosine triphosphate; ADP, adenosine diphosphate; AMP, adenosine monophosphate; IMP, inosine monophosphate; XMP, xanthosine monophosphate; GMP, guanosine monophosphate; NAD, nicotinamide adenine dinucleotide; THFA, tetrahydrofolate; Pi, orthophosphate.

labelling of nodules with [^{13}N] (91) or with [^{15}N] nitrogen gas (105, 106) have recently indicated that glutamine, not glutamate, is the initial organic product of nitrogen fixation. Enzymes of the 'high affinity' pathway for ammonia assimilation (93), namely glutamine synthetase (EC 6.3.1.2) and NADH: glutamate synthase (EC 2.6.1.53), are active in cytosol extracts of many legumes including those producing ureides (13, 24, 30, 127, 143) and although active NADH: glutamate dehydrogenase (EC 1.4.1.3) is also found in nodule extracts (51, 143) a metabolic sequence shown in Fig. 3 seems likely. Both aspartate aminotransferase (EC 2.6.1.1) and glutamine-dependent asparagine synthetase (EC 6.3.5.4) have been demonstrated for the cytosol of nodules (12, 128, 134, 142) which together with labelling data showing [^{15}N] asparagine synthesis from [^{15}N] ammonia or [amide—^{15}N] glutamine in soybean nodules (50) supports the scheme shown in Fig. 3.

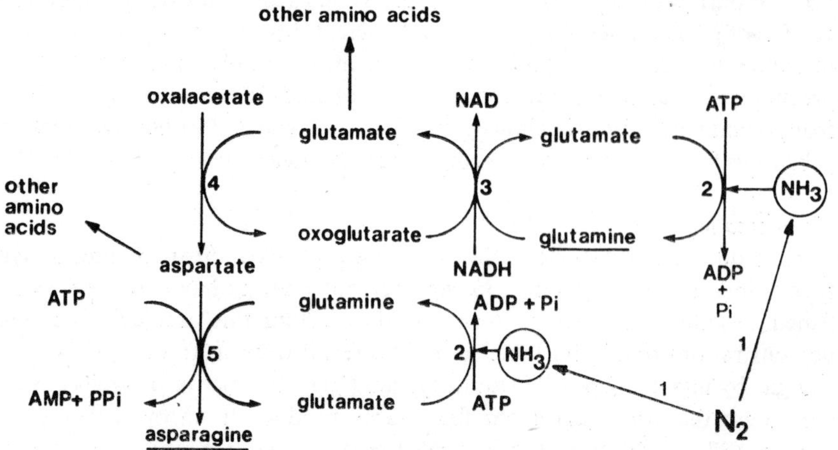

Fig. 3. Metabolic routes for the primary assimilation of ammonia derived from nitrogen fixation in legume root nodules. Abbreviations as in Fig. 2 and PPi, pyrophosphate.

Thus the sources of nitrogen required for *de novo* purine synthesis (see Fig. 2) are readily formed in the nodules of ureide-producing species. However, the only enzymes of the pathway leading to IMP which have been detected in nodule extracts are phosphoribosyl pyrophosphate (PRPP) synthetase (EC 2,7.6.1) (140) and one of the glutamine-dependent amidotransferases (glutamine: PRPP amidotransferase; EC 2.4.2.14) (130; C.A. Atkins, unpublished). The nature of the folate derivatives utilized in the reaction sequence and the enzymology of their synthesis have not been studied; those indicated in Fig. 3 have been described for animal and microbial systems (32).

Within one hour of supplying [^{14}C] carbon dioxide to illuminated leaves of cowpea plants, labelled allantoin and allantoic acid are detected in xylem sap leaving the root system (59) indicating utilization of sucrose currently delivered to the nodule for ureide synthesis. Slices of cowpea nodule tissue showed conversion of [U—^{14}C] glucose and [1, 2—^{14}C] glycine to [^{14}C] allantoin and allantoic acid (13). Glycine was the better precursor, consistent with its more probable direct utilization in purine synthesis. However exogenously supplied substrates are not extensively metabolized by whole nodules or nodule slices and more direct evidence for the incorporation of precursors, such as glycine, by the *de novo* purine nucleotide pathway awaits the reconstruction of the metabolism in a cell-free preparation. In other tissues these enzymes have been recovered as a functional multi-enzyme complex (135) and if this proves to be the case in legume nodules the activities can be expected to be tightly regulated and quite labile to isolation.

The presence of significant activities of NAD : xanthine oxidoreductase (EC 1.2.1.37) (13, 171), O_2 : xanthine oxidoreductase (EC 1.2.3.2) (48, 155), urate oxidase (EC 1.7.3.3) (59, 155, 183), allantoinase (EC 3.5.25) (59, 155) and guanine aminohydrolase (EC 3.5.4.3) (181) in nodules of ureide-producing legumes indicates active oxidation of the products of *de novo* purine synthesis. The activity of these enzymes in nodules of an asparagine-exporting species, *Lupinus albus* (116), is very low (R.M. Rainbird and C.A. Atkins, unpublished). Furthermore, in cowpea they (enzymes) are more active in the cytosol of the bacteroid-containing tissue than in either the bacteroids or the cytosol of the cortex of the nodule (12, 59) and are thus largely restricted to the tissue active in ammonia assimilation.

More direct evidence of ureide formation from purines comes from studies using cell-free extracts of nodules of soybean (171) and cowpea (13, 181, 182). Both IMP and XMP are readily metabolized to form ureides at rates commensurate with those of N_2 fixation (3–6 μmol N_2 fixed$\cdot h^{-1}\cdot g^{-1}$ fresh weight nodule); however GMP is only slowly utilized and AMP not at all (13, 180). The absence of NAD (P) : IMP oxidoreductase (EC 1.2.1.14), glutamine: GMP synthetase (EC 6.3.4.1) and enzymes catalyzing the deamination of adenylates results in restriction of IMP metabolism to a direct route leading to formation of allantoin (13). Direct metabolism of relatively low levels of IMP (apparent Km = 30 μM IMP; 13) in this way leads to effective transfer of nitrogen assimilated in *de novo* purine nucleotide synthesis and avoids release and reassimilation of ammonia which would result if the branched pathways of purine nucleotide formation (Fig. 2) also contributed to ureide synthesis.

Analysis of the products of the metabolism of labelled and unlabelled purines in cell-free extracts (13, 180–182) supports the direct sequence shown in Fig. 2. NAD reduction due to xanthine oxidoreductase activity accompanies ureide synthesis (13, 171) although in extracts of soybean nodules oxygen-dependent xanthine oxidation has been demonstrated (48, 155). While the overall synthesis of ureides by extracts of cowpea nodules requires molecular oxygen (182) this requirement has been completely accounted for by the activity and kinetic properties of urate oxidase (129, 182).

Considerable support for this pathway operating *in vivo* comes from experiments involving application of the hypoxanthine analogue, allopurinol (4-hydroxy pyrazolo [3, 4-d] pyrimidine), to nodulated plants or to extracts of nodules (12, 13, 48, 171). Addition of 0.5 mM allopurinol to the nutrient solution bathing nodulated roots of intact cowpea plants causes a rapid fall in xylem-borne ureides but little effect on levels of amino compounds in xylem, while a concomitant decrease is observed in ureide levels of the nodules and a parallel increase in the pool size of xanthine (12). Concentrations of the inhibitor as low as 10μM also effectively arrest ureide synthesis and NAD reduction in cell free extracts utilizing purine nucleotides or bases as substrates (13, 171, 181).

Oxidation of purine nucleotides in nodules apparently occurs outside the bacteroids (13, 59, 171). Extracts from soybean bacteroids prepared using a French pressure cell (171) showed limited urate oxidase and allantoinase activities but xanthine oxidation, in contrast to earlier reports (48, 155, 157), was absent. In other plant tissues urate oxidase has been located, together with catalase, in microbodies (19, 63, 111, 112, 136, 159, 160) and this has led to the suggestion that such an arrangement allows effective dissimilation of the hydrogen peroxide generated by urate oxidation. Allantoinase has also been shown in microbodies from castor bean endosperm (160) and together with allantoinase activity in the appendices of *Arum* spp. (112) indicating that a significant portion of the purine oxidation pathway in plants might be located in this cellular compartment. Separation of organelles from nodule tissue by the usual sucrose density gradient techniques is complicated by a very extensive zone of bacteroids which interferes with the resolution of other organelle zones and which overlaps with the expected position of microbodies. However, a recent study (55) using young soybean nodules and non-linear sucrose density gradients, in which the resolution of organelles in the higher density region of the gradient was increased, has shown that urate oxidase is associated with catalase at a density consistent with the location of both enzymes in microbodies. Xanthine oxidoreductase was, however, completely soluble and allantoinase was recovered in a microsomal fraction following disruption of the tissue. Purified urate oxidase is sensitive to inhibition by the initial products of nitrogen fixation (ammonia, glutamine) and by xanthine (129), in each case at concentrations likely to be present in nodules during normal functioning (6–10 μM), so that compartmentation in microbodies may serve to isolate the enzyme from these solutes.

A most significant property of both crude preparations and highly purified urate oxidase from cowpea nodules, and one which is likely to regulate the activity of the enzyme *in vivo*, is its relatively low affinity for oxygen (Km = 28 μM dissolved oxygen; 129, 182). Measurements based on the dissociation constant for oxyleghaemoglobin and the content of leghaemoglobin in soybean nodules suggest that the concentration of dissolved free oxygen is likely to be 10–200 nM (2). The 'high oxygen affinity' oxidase of bacteroids functions effectively at 10–100 nM dissolved oxygen (3) and recent measurements of hydrogen oxidation and respiration by soybean bacteroids showed an apparent Km around 10 nM oxygen (41). Thus, unless the kinetic properties of urate oxidase are different *in vivo* or the enzyme uses a source of oxygen other than the general tissue supply, the microaerophilic environment maintained in nodules (168) is likely to severely limit ureide synthesis through uric acid oxidation. That this limitation might be a normal feature of nodule functioning in ureide-producing plants is supported by differential effects of reduced partial pressure of oxygen (pO_2) in the rhizosphere on the nature of solutes formed from [^{15}N] nitrogen gas by intact soybean plants (107). When pO_2 was changed from 0.2 to 0.1 atm. the labelling of allantoin was more severely inhibited when compared to labelling of ammonia or amino compounds.

Non-nodulated legumes

Even though in non-nodulated legumes ureide levels in xylem and in tissues are generally lower than in effectively nodulated plants, the fact that ureides are formed at all indicates that pathways must exist for the channelling of newly assimilated combined nitrogen into allantoin and allantoic acid. Labelling experiments supplying [^{15}N] ammonium or [^{15}N] nitrate salts to non-nodulated soybean plants (49, 104) shows some enrichment of allantoin nitrogen in roots and stem tissues. However, compared to the enrichment of amino compounds the labelling in allantoin is slight and indicates that the rate of synthesis of ureides in these plants is very low.

In non-nodulated *Phaseolus vulgaris*, where significant levels of allantoic acid in xylem have been detected (163, 164, 165), ureide synthesis results from prolonged supply of ammonium to the plants. In similar plants provided with nitrate in the rooting medium considerably lower levels of ureide occur in xylem while appreciable free nitrate is transported to the tops (163, 164). Mothes (94) has investigated the site of synthesis of ureides in non-nodulated plants of the closely similar species, *Phaseolus coccineus*. In rooted leaf cuttings treated with auxin and supplied with nitrogen as ammonium nitrate in liquid culture, a very large root system is supported by a single photosynthesizing leaf. Under these conditions of carbohydrate limitation allantoin is the principal nitrogenous solute of the leaf blade (96), reaching up to 100 times the concentration in rooted leaf cuttings grown without added auxin and in which a more balanced ratio of root to shoot is maintained. By supplying ammonia either to leaf blades or to roots, experiments indicated that allantoin is synthesized in the roots and not in the leaves (94). However, under the more severe conditions of 'carbohydrate starvation' which occurs in detached leaves held in darkness and at elevated temperature (32°C; 94) allantoin is formed as a product of protein and possibly also of nucleic acid degradation whereas at lower temperatures (18 or 4°C) senescence yields mainly amides with only small quantities of ureide.

In *Symphytum uplandicum* [8–^{14}C] hypoxanthine is readily converted to allantoin in both excised roots and leaf discs whereas only in roots is [2–^{14}C] glycine converted to allantoin (33). Similarly, in leaves of other species exogenously supplied purines are readily metabolized to form ureides (16, 73, 74, 99, 154) indicating that the enzymes of purine oxidation are functional. However, those of *de novo* purine synthesis may be low in activity or inhibited (94) in leaves but functional in roots. Synthesis of allantoin through the oxidation of purines in roots is further supported by the close correlation between external nitrogen supply and urate oxidase activity in *Phaseolus coccineus* (161). Pathways similar to those thought to occur in nodules (Fig. 2) might be expected also in roots although some differences in the enzymology between the two sites of ureide formation have been noted. Urate oxidase from radicles of young non-nodulating soybean differs from the enzyme of nodules in its affinity for urate, pH optimum and in requiring a low molecular weight cofactor for activity (155, 156). Like the enzyme from nodules the urate oxidase of *Phaseolus* root is also located in microbodies (161) together with catalase.

As suggested by Mothes (94) the formation of ureides in non-nodulated plants appears to be part of a physiological response to a limiting supply of carbohydrates and that inorganic nitrogen, taken up by the plant as a consequence of transpiration, is assimilated and stored with the greatest economy of carbon use by forming molecules with a narrow C : N ratio (Fig. 1). This applies especially to ammonia which is not readily translocated as such but is assimilated in the root system, where, even under normal conditions of growth, carbohydrate supply is likely to be limited. The nitrate may be stored in tissues at relatively high concentration as such (137) or may be reduced following xylem transport to shoot and assimilated by a pathway which probably directly utilizes the products of photosynthesis (6, 18, 34). In the latter case nitrate assimilation in leaves might be viewed as a process which does not require respiratory oxidation of substrates to provide reductant, ATP or carbon skeletons and so is not limited by carbohydrate supply.

In some non-nodulated legumes the principal sites of nitrate assimilation are in the root system [e.g. *Lupinus albus*, (9,10); *Pisum* spp. and *Vicia faba*, (113)], while in the three ureide producers studied [*Glycine max*, (162); *Vigna unguiculata*, (8); *Phaseolus vulgaris* (163, 164)] the principal sites are in the above-ground organs. It remains to be seen whether it is a general rule that legumes of tropical origin, which produce ureides as nodulated plants, exhibit a shoot-dominated pattern of nitrate reduction whereas temperate species, many of which have an amide-based nitrogen economy, show the opposite of this. As nodule functions are likely to be limited by phloemz delivered photosynthate (56) the formation of ureides might be regarded as a response to limiting carbohydrate supply, which, together with leaf-based nitrate reductase, also a response to limiting carbohydrate supply, are coincident in legumes which have evolved in tropical environments.

Biosynthesis of citrulline

This ureide may constitute a significant proportion of xylem-borne nitrogen in some legumes (Table 1) and probably a more extensive examination of species will result in more examples with a 'citrulline based' nitrogen economy being recognized. There have been no studies of citrulline synthesis in relation to nitrogen fixation in legumes and most information in this regard relies on limited results with the nodulated non-legumes *Alnus* spp. and *Myrica* spp. (80, 81, 92, 133, 178).

Labelling studies using [^{15}N] nitrogen gas with *Alnus* have shown a pattern in nodules consistent with formation of ammonia as an intermediate in fixation with assimilation to form glutamate, citrulline and aspartate. While it seems reasonable to suppose that in *Alnus* and *Myrica* nodules, ammonia is assimilated by the combined activity of glutamine synthetase and glutamate synthase (see above) the earlier labelling studies (81) as well as results from a more recent study of [^{13}N] ammonium assimilation (138) suggest that glutamate dehydrogenase and glutamine synthetase are both involved. Consistent with a role for citrulline in the transport and storage of nitrogen, [^{15}N]-labelled arginine was not readily formed from [^{15}N] nitrogen gas (80). Furthermore, the carbamino-nitrogen of citrulline was more heavily labelled than both the ornithine nitrogen of citrulline and the glutamate of the nodule indicating direct incorporation of fixed nitrogen by carbamoyl phosphate synthetase (EC 2.7.2.5) and condensation of ornithine and carbamoyl phosphate as in Fig. 4. While the detailed enzymology of this sequence has not been described for nitrogen-fixing nodules, these reactions, which form part of the pathway for arginine biosynthesis, have been demonstrated in a number of diverse plant tissues (17, 26, 28, 131) and might be expected to occur also in nodules.

Translocation of ureides

Export from the nodule

Although nodules of different legumes vary in their size, shape and pattern of growth, all show a rather sparse network of peripherally-located vascular tissue, through which nitrogenous solutes are exported to the host plant in the xylem. Vascular tissue generally occupies less than 5 per cent of the total volume of a nodule, and, solutes exchanged between the conducting elements and the central cells of the bacteroid-containing tissue must traverse

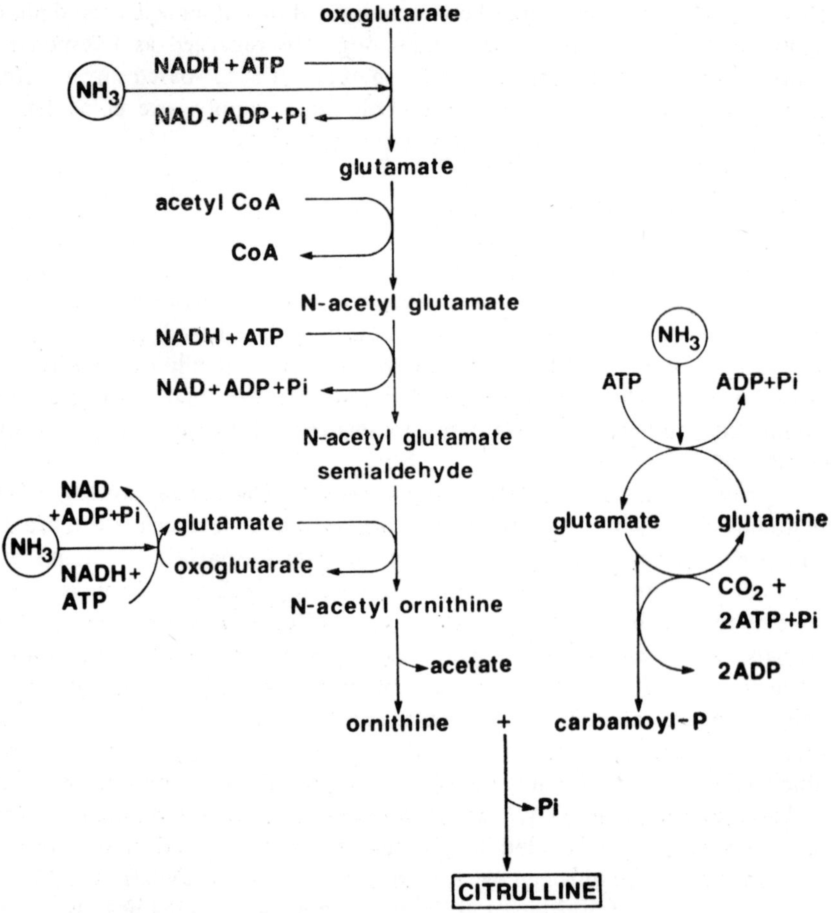

Fig. 4. Possible metabolic route for the synthesis of citrulline in relation to ammonia assimilation following nitrogen fixation. Abbreviations as in Fig. 2.

0.5–2 mm (120). The existense of Plasmodesmatal connections, observed at all cellular boundaries from the phloem to the bacteroid-containing cells, support the view (119) that phloem-delivered sugars travel a symplastic route to their sites of utilization with solutes of fixed nitrogen travelling the same route in the opposite direction. These solutes (amides or ureides) are then considered to be actively secreted to the apoplast of the vascular tissue (52) and by generating an osmotic flux of water into the xylem from the surrounding apo plast of the non-vascular tissue are voided out of the nodule.

Many of the ureide-forming legumes belong to the predominantly tropical tribes Phaseoleae and Glycineae and form nodules which are of essentially determinate growth (148) and which have a vascular system in which the strands fuse apically to comprise a series of closed loops. The association between ureide production and a determinate pattern of nodule growth may be more than a coincidental one and relate to the limited solubility of ureides as solutes during the transport of fixed nitrogen (148). With low resistance to water flow in the 'closed' vasculature of such nodules, the voiding of fixation products in the xylem might more easily aid in the usage of the transpiration stream to sweep solutes from the nodule, thus achieving higher rates of water flux and resulting in more dilute solutions in xylem than would be possible in indeterminate nodules which show open-ended strands of vascular tissue. The amides asparagine and glutamine which serve as the principal exported fixation products from nodules of plants which produce little or no ureide (Table 1) (115) are several times more soluble in water than either allantoin or allantoic acid, and nodules using amides could be expected to transport fixed nitrogen with economy in water use (148).

The concentration of allantoin plus allantoic acid found in nodule bleeding sap (38–90 mM in soybean, 152; 66 mM in winged bean, R.M. Rainbird, C.A. Atkins and J.S. Pate, unpublished) shows that indeed nodules might operate close to the aqueous solubility limit of these compounds (about 100 mM for allantoin plus allantoic acid at pH 5; 148) and the rate at which fixation products are exported to the host plant could be limited by such factors. The pH of xylem might also be critical as the solubility and stability of ureides, particularly stability of allantoic acid, are markedly affected by pH (148).

Distribution to plant parts

Transport in xylem from the nodulated root to the shoot suggests that in fully expanded leaves, the major sites of transpiration are likely to be the main regions to which ureides would be initially distributed. The significant level of ureides in stems of both cowpea (59) and soybean (87) indicates however that these tissues have a greater capacity to abstract solutes from the xylem than might be predicted from their transpirational loss of water (78).

Most detailed studies of the fate of xylem-borne nitrogenous solutes in legumes have been made using *Lupinus albus*, a species which has an 'asparagine based' nitrogen economy (116) but which bleeds from phloem (116, 121), allowing the metabolism and transfer of either [^{14}C] or [^{15}N]-labelled substrates to phloem to be readily examined. Analysis of phloem sap at fruit tips or at petioles (10, 11, 90) provides information on the extent to which particular solutes are metabolized or loaded directly onto phloem for redirection to the weakly or non-transpiring organs of the plant (apices, fruits, roots). In lupin the major xylem solute, asparagine, is efficiently transferred to phloem largely in unchanged form (11) whereas lesser constituents of xylem (glutamine, aspartate, glutamate) are metabolized extensively by leaves (122) providing nitrogen for protein synthesis and for the export in phloem of amino compounds formed in photosynthesis.

While experiments in which [^{14}C], [^{15}N] double-labelled ureides are supplied to cut fruiting shoots of cowpea confirm that ureides are indeed utilized for protein synthesis in the shoot (C.A. Atkins, unpublished), the sites at which ureide nitrogen becomes amino nitrogen have not been established. Unfortunately a legume with a ureide-based nitrogen economy which bleeds spontaneously from cut phloem has not been found so that the extent to which allantoin and allantoic acid participate in direct xylem to phloem transfer in leaves and petioles could be determined. Further, it is tempting to suppose that like asparagine in lupin, the neutral molecule allantoin is transferred to phloem without metabolism whereas analogous to the dicarboxylic acids in lupin (10, 117), allantoic acid preferentially enters the metabolism of the leaf. Phloem-fed organs in both soybean and cowpea appear to have enzymic mechanisms for ureide utilization and although it might be inferred from this that ureides are translocated in phloem there has been no direct demonstration that these compounds are indeed mobile in phloem of legumes. Perhaps techniques of indirect phloem analysis such as the use of chelating agents (70) which have been successfully applied to sucrose translocation in soybean (44), or the 'abraded spot' method used to examine the loading of nitrogenous solutes onto phloem in soybean leaves (61, 62) can be used to examine this question.

In a number of non-legume species allantoin and allantoic acid have been found in natural phloem sap (39, 158) and in sampled sieve tube sap (184, 185) where the concentrations range from 5 to 30 mM and these compounds constitute a significant proportion of phloem-borne nitrogen. While the presence of ureides in phloem does not necessarily mean their loading in leaves or stems (43), it does indicate that the compounds are mobile in phloem and that they could serve as sources of nitrogen for non-transpiring organs. Citrulline has also been detected in the phloem sap of a number of non-legume plants (42, 67, 185).

Utilization of ureides for protein synthesis

During periods of intense nitrogen fixation ureides commonly carry 70–90 per cent of the nitrogen exported from the root to the shoot of cowpea (7, 59), soybean (88) and mung bean (R.M. Rainbird, C.A. Atkins and J.S. Pate, unpublished). Presumably this is also the case for the other legume species in which high concentrations of ureides in xylem have been detected (Table 1). Although a significant amount of allantoin and allantoic acid accumulates in all organs during vegetative growth (59, 87, 151) and in pods during reproductive growth (59, 87) balance sheets for the utilization of the compounds during different stages of the growth cycle show that, on a whole shoot basis, most (75–90 per cent) of the incoming ureide is promptly metabolized (59).

Ureides which do accumulate in roots, stems and especially in developing pods of soybean are metabolized late in growth (87) providing an additional source of nitrogen for seed filling. The extent to which ureides remain in non-reproductive plant parts at harvest probably reflects the relative demand of protein synthesis for nitrogen. In a lightly fruiting cultivar of cowpea (cv. caloona) considerable leaf nitrogen, including ureides, was mobilized during seed filling but in stems the high concentration of ureide maintained throughout growth was retained in these tissues at maturity (59). Concentrations of stored allantoin and allantoic acid, especially in stems and petioles, may exceed their expected solubility in the tissue water present in these organs (148) suggesting that ureides might occur as insoluble deposits of their salts.

Allantoinase, the enzyme degrading allantoin to allantoic acid, has been detected in stems, leaves and nodules of soybean (155, 174, 176) and in all organs of cowpea during vegetative and reproductive growth (59). In both cowpea and soybean fruits allantoinase in the pod walls declines with development but increases markedly in seeds as the rate of accumulation of storage protein increases towards maturity (59, 166). While this pattern of enzyme development is consistent with allantoin serving as a major source of nitrogen for seed development there is, at present, no evidence that the ureides are delivered in quantity to fruits. Very active allantoinases have also been detected in dry seeds of soybean (82, 176) and in young seedlings of *Lablab* spp. (84), *Lathyrus sativus* (101) and *vigna radiata* (98, 176). In these cases the enzyme functions in hydrolysis of allantoin derived from purines which are released by the breakdown of nucleic acids in cotyledons during germination (48, 109, 110).

The metabolic sequence involved in the further utilization of allantoic acid with release of nitrogen in a form readily assimilated into amino acids and proteins has not been defined. Singh (145) has described allantoicase (EC 3.5.3.4) in germinating cotyledons of peanut and demonstrated the formation of both glyoxylate and urea as products of the reaction. In an earlier study (31) enzyme-catalyzed breakdown of allantoic acid was described for extracts of mature soybean seed but the products formed were not characterized. The existence of this enzyme in plants has been inferred (164, 169) but not unambiguously demonstrated in extracts of the actively growing tissues of ureide-producing legumes. Allantoic acid in solution is labile, especially at neutral or slightly acidic pH, and in the presence of plant extracts breaks down readily in a non-enzymic reaction to yield glyoxylate and urea (169). Allantoicase apparently accounts for allantoic acid utilization in a number of bacteria in which ureidoglycolate (see Fig. 5) is formed as an intermediate in the metabolism (175). While cleavage of ureidoglycolate to form glyoxylate and urea might be catalyzed by allantoicase, in number of bacteria, optically-specific ureidoglycolases (EC 4.3.2.3) have been demonstrated (175). Urea formed in this sequence (Fig. 5) would be readily hydrolyzed by urease (EC

3.5.1.5) yielding ammonia for utilization by the well-established pathways of ammonia assimilation which have been described in many plant tissues (93). However, even though urease has been demonstrated in a number of legumes, especially in mature seeds (26, 169), its presence has yet to be confirmed for the actively growing tissues of the many ureide-producing legumes. In cowpea, urease activity is present in nodules, root tips, leaves, stems plus petioles and in developing fruits (C.A. Atkins, unpublished) indicating that in this species at least the capacity for urea hydrolysis is present in all vegetative tissues and is not confined to the mature seeds.

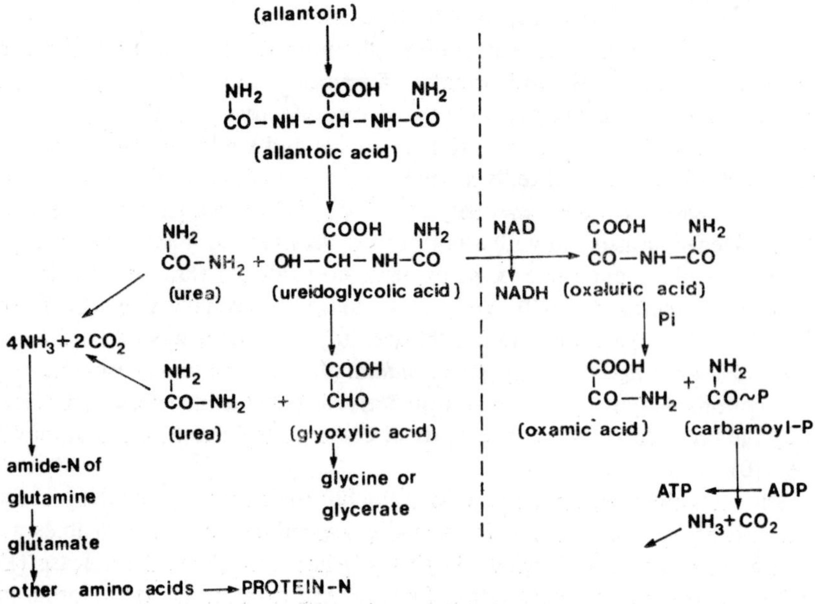

Fig. 5. Possible pathways for the enzymatic degradation of allantoin to yield ammonia for amino acid synthesis. Abbreviations as in Fig. 2. See text for details.

Apparently a number of plant tissues which utilize urea lack urease (26) and a reversal of reactions of the urea cycle has been suggested as a mechanism for urea catabolism in these plants (131). However more direct mechanisms of urea breakdown might also operate in plants lacking urease. An NADP-dependent urea oxidoreductase has been reported for some higher plant tissues (108) and the enzymes of the allophanate pathway of urea breakdown, found in some microorganisms (175) have been demonstrated in green algae (167). An alternative mechanism of ureidoglycolate breakdown (Fig. 5) which has been described in *Streptococcus allantoicus* and a number of Enterobacteriaceae (175) involves reduction of NAD and formation of oxalurate which is metabolized to form oxamate and carbamoyl phosphate. Operation of this pathway would yield NADH and ATP in addition to NH_3, oxamate and urea as end products (Fig. 5). While such a sequence seems unlikely in higher plants significant quantities of oxamate together with allantoin have been reported from analyses of sugar beet (71, 72).

Urease activity may decline rapidly following tissue disruption unless precautions to prevent denaturation are observed (131) so that negative urease assays should be interpreted with caution. Furthermore, urease appears to be a substrate-induced enzyme in higher plants (27, 86, 146) suggesting that in ureide-producing legumes activity might remain relatively low unless the plants are effectively nodulated and indeed producing ureides.

Just as the assimilation of ammonia released from amide breakdown in leaves is likely to occur in chloroplasts and be linked directly to photosynthetic generation of ATP and reductant (93) so too will be the case with ammonia released from ureide catabolism. However earlier steps of ureide breakdown in photosynthetic tissue might also be linked to chloroplast reactions. In the moss protonema of *Funaria hydrometrica* both the uptake and metabolism of allantoin are controlled by light (57, 58). Although allantoin is hydrolyzed to allantoic acid in both light and dark further metabolism to form urea can only occur in the light (57).

Significance of ureides in the carbon economy of nodules

The efficiency with which photosynthate, delivered in phloem to nodules, is utilized for nodule functions varies between different legume host : *Rhizobium* associations (7). These differences are exemplified by comparing an amide-exporting symbiosis and one exporting ureides (Table 3, data taken from 79). Items of carbon use are expressed in relation to nitrogen fixed. The more than two-fold greater requirement for carbon in the lupin symbiosis is due to greater carbon use in all three items of their budget. The larger requirement of carbon used for export of nitrogen reflects the greater C : N ratio of asparagine in lupin (Fig. 1) compared to that of ureides in cowpea (Fig. 1) and contributes significantly to the overall difference in carbon requirement. Quantitatively the most important difference between nodules of the two legumes is in the amount of carbon lost as carbon dioxide due to respiration (Table 3) and a number of other factors might also contribute to this disparity.

One factor probably contributing significantly to the lower carbon dioxide evolution of cowpea was the negligible evolution of hydrogen gas from this particular symbiosis (0.02 ± 0.01 mol H_2/mol N_2 fixed, 79) compared to the substantial hydrogen production by that of lupin (3.1 ± 0.7 mol H_2/mol N_2 fixed, 79). The factors responsible for such differences in hydrogen exchange have not been clearly defined (40, 139, 141) and so estimating the likely impact

Table 3. Economy of carbon use in nodules of two legume : *Rhizobium* symbioses

Item in carbon budget	Cowpea : *Rhizobium* CB 756	Lupin : *Rhizobium* WU 425
	(g atom C g atom^{-1} N)	
C lost as CO_2 in respiration	3.6 ± 0.6	8.5 ± 0.7
C exported with fixed N in xylem	1.1 ± 0.1	2.3 ± 0.2
C incorporated into nodule mass	0.8 ± 0.3	1.6 ± 0.4
Total C imported in phloem	5.5 ± 1.0	12.4 ± 1.3

Note: Measurements were made over a two-week period in early vegetative growth and plants were grown in minus N water culture with cuvettes attached to the nodulated zone of their primary roots for collection of CO_2. Nitrogen fixation was measured as incremental nitrogen increase by Kjeldahl and the carbon attached to exported nitrogen by analysis of nitrogenous solutes in root bleeding xylem sap. Data are taken from a previous study (79).

of net hydrogen evolution on the carbon economy of the nodule is not possible. A second factor likely to alter the carbon dioxide loss by nodules is the extent to which carbon dioxide fixation contributes carbon to nodule growth and export of nitrogen. The 'cytosol' fraction from a number of nodules of different species contain significant phosphoenolpyruvate (PEP) carboxylase (EC 4.1.1.31) activity (12, 35, 79, 125) and a functional role for this enzyme in nodules in diverting PEP to form the four carbon skeletons of asparagine for export has been proposed (7, 35, 76, 77). While rates of such an 'internal carbon dioxide retrieval' mechanism are difficult to measure against net efflux from the tissue, even at greatly elevated external carbon dioxide levels, the potential for such a mechanism is sufficient (35) to suppose that all exported asparagine in amide-producing species might be derived in this manner.

In ureide-exporting plants four carbon compounds constitute only a minor fraction of xylem exports (Table 2). However, PEP carboxylase activities in such species can be considerably greater than those found in extracts of nodules of amide-exporting species (79) and [^{14}C] carbon dioxide fixation by nodulated roots results in prompt labelling of ureides in xylem sap (36: K. Hillman and C.A. Atkins, unpublished) indicating that formation of oxalacetate might be linked in some way to allantoin synthesis. An obvious linkage would be through the glyoxylate bypass of the Kreb's Cycle with glyoxylate being either aminated to form glycine or producing formate (38) for activation to the tetrahydrofolate derivatives required for *de novo* synthesis of purines (Fig. 2). However, while a partial glyoxylate bypass has been inferred for isolated soybean bacteroids (150), isocitrate lyase (EC 4.1.3.1) is absent from the bacteroids and cytosol of soybean and cowpea nodules active in nitrogen fixation (65: C.A. Atkins, unpublished) and the enzyme appears only in soybean nodules during senescence or when poly-β-hydroxybutyrate utilization is initiated (179).

A consideration of the requirements of asparagine and ureide synthesis for ATP and reductant (see Figs. 2 and 3) shows that asparagine requires the equivalent of 7 ATP and 4 carbon atoms for its formation whereas allantoin requires 10 ATP and 4 carbon atoms. Simply equating carbon atoms with their potential as sucrose to yield energy in oxidative phosphorylation and expressing the cost on a nitrogen basis, asparagine might be expected to require 15.5 ATP equivalents per nitrogen and ureides 8.5. This difference in energy cost would also contribute to a lower respiratory burden in ureide-exporting plants and might in part account for the lower carbon dioxide evolution in cowpea versus lupin (Table 3). For comparison citrulline formed according to the pathway in Fig. 4 and costed on the same basis as above requires the equivalent of 17 ATP per nitrogen.

Transport of ureide nitrogen might also require less energy and therefore involves less supporting respiration compared to that required for transport of amide nitrogen if membrane-bound ATP-ases (60) are involved in the secretion of solutes from the symplast of the nodule tissues to xylem. The number of membrane barriers traversed in this way have not been defined but conservation of respiratory products by transporting ureides versus other nitrogenous solutes is likely to have only a minor effect on the economy of carbon use by nodules.

As a proportion of the total net photosynthate generated in nodulated legumes throughout their growth and development, that used in nodule functioning accounts for 9–12 per cent (7). While this may seem to be a relatively minor proportion of the plant's photosynthate resource, small variations in the allocation of photosynthate between organs or variations in the rate of photosynthesis may have quite significant effects on nitrogen fixation (23, 56, 126). Thus the differences in economy of carbon use afforded by synthesis of ureides may be significant in determining the potential of nitrogen fixation in relation to increasing yield in future grain or pasture legume improvement programmes.

Possible use of ureides as indicators of nitrogen fixation in legumes

Formation of ureides and their presence in xylem sap as dominant nitrogenous solutes is closely related to the presence of effective nitrogen-fixing nodules (8, 36, 64, 85, 87, 89, 118) suggesting that the relative content of allantoin and/or allantoic acid in xylem sap or in particular tissues might reflect the current rates of nitrogen fixation and be used as a convenient and simple assay for the nodulatino status of field grown plants.

Even though a leaf punch or stem segment assay would be a useful non-destructive field technique, the complexity of the response of ureides in tissues of nodulated plants to combined nitrogen (36, 87, 118), the relatively small pools of ureide compared to total tissue nitrogen, the possible diurnal variation in these pools and the fact they may simply reflect stored solutes indicates that tissue analysis is unlikely to be a useful assay of current fixation. Since the nitrogenous solutes found in xylem sap are probably representative of current products of nitrogen assimilation in the root system and since ureide as a proportion of total sap nitrogen is closely correlated with the proportion of plant nitrogen derived from fixation (Fig. 6) sap analysis might provide a more predictable assay method. While this correlation, made using a [^{15}N] nitrate feeding technique to estimate fixation, is restricted to pot-grown, nodulated cowpea (118) a similar response in xylem composition to supplied nitrate has been shown for eight other ureide-producing legume

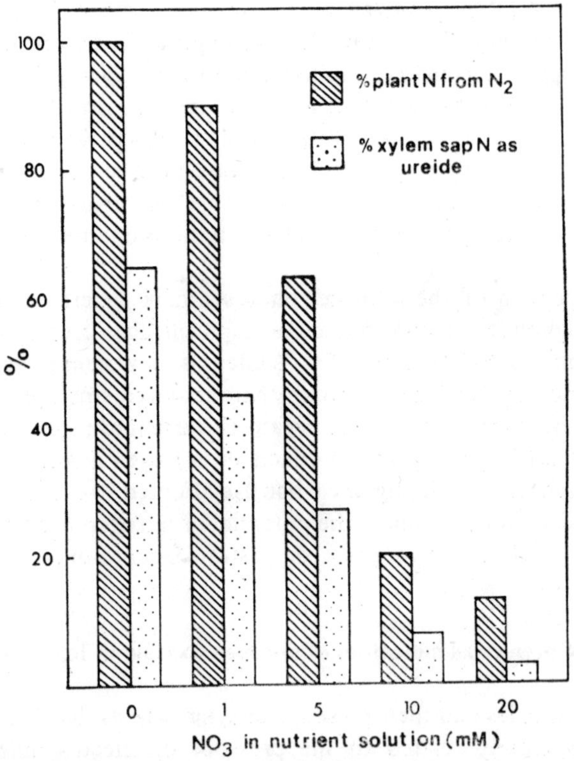

Fig. 6. Relationships between percentage of xylem-borne nitrogen and percentage of plant nitrogen from molecular nitrogen in nodulated *Vigna unguiculata* plants grown with a range of concentrations of nitrate. Nitrogen fixation was estimated from a [^{15}N] nitrate isotope dilution technique.

species (118) and a positive correlation between percentage total sap nitrogen as ureide and nitrogen fixation estimated by acetylene reduction has also been found for soybean (64).

The concentration of ureides present in xylem sap of most ureide-forming legumes studied to date (Table 1) is so sufficient that small samples (5–50μl) may be quantitatively assayed directly using a number of reliable and simple colorimetric assays (29, 100, 170) or detected qualitatively in a spot test on paper using Ehrlich's reagent (147). Allantoin and allantoic acid may be separated and quantitated using cellulose thin layer (59) or ion exchange column chromatography (5).

The ratio of the two amides, asparagine and glutamine, in xylem sap of nodulated cowpea plants supplied with increasing levels of nitrate, like ureide content, is also closely correlated with the extent to which fixed nitrogen contributes to total plant nitrogen (118). Expressed as the ratio of asparagine-N to glutamine-N this value is negatively correlated with nitrogen fixation in a range of other ureide-forming legumes (118). Such ratios as well as the relative abundance of other xylem-borne solutes might also be useful as indicators of nitrogen fixation. Examples of such solutes are citrulline (Table 1), γ-methylene glutamine (structure, Fig. 1; 45, 46) and canavanine (structure, Fig. 1; 172, 173) and probably a greater survey of translocated solutes of nitrogen in legume species will reveal others.

Concluding remarks

In their review of urea, ureides and guananidines in plants in 1962, Reinbothe and Mothes (131) raised many of the questions concerning the synthesis, breakdown and physiological role of these compounds which remain unanswered at present. Since 1962 however the realization that ureides function, not only in the transport and storage of nitrogen in a wide range of agriculturally important legume species, but that these compounds are quantitatively the most important products of nitrogen fixation and constitute the major source of protein nitrogen in these plants, has added considerable impetus to research in this area. While the cost-benefit analysis of plants utilizing allantoin or allantoic acid in their primary metabolism of nitrogen compared with those using amides is far from complete, considerations of solute transport and the solubility of translocated substances, nodule carbon economy and the provision of photosynthate to nitrogen fixation indicate that there may be significant physiological consequences of having a 'ureide-based' nitrogen economy. Increasing legume yields in the tropical and subtropical regions of the world will place increasing demands on the capacity and efficiency of the host plant: *Rhizobium* symbiosis to supply nitrogen, most of which will be assimilated to form ureides which will subsequently yield seed protein.

References

1. Anderson, J.D. Purine nucleotide metabolism of germinating soybean embryonic axes, *Plant Physiology*, *63*, 100–104 (1979).
2. Appleby, C.A. Leghaemoglobin, pp. 521–544 in *The Biology of Nitrogen Fixation* (Edited by A. Quispel), North Holland, Amsterdam (1974).
3. Appleby, C.A., Turner, G.L. and Macnicol, P.K. Involvement of oxyleghaemoglobin and cytochrome P-450 in an efficient oxidative phosphorylation pathway which supports nitrogen fixation in *Rhizobium*, *Biochimica et Biophysica Acta*, *387*, 461–474 (1975).
4. Aprison, M.H., Magee, W.E. and Burris, R.H. Nitrogen fixation by excised soybean root nodules, *Journal of Biological Chemistry*, *208*, 29–39 (1954).
5. Atkins, C.A. and Canvin, D.T. Analysis of ^{14}C-labelled acidic photosynthetic products by ion-exchange chromatography, *Photosynthetica*, *5*, 341–351 (1971).
6. Atkins, C.A. and Canvin, D.T. Nitrate, nitrite and ammonia assimilation by leaves; effect of inhibitors, *Planta*, *123*, 41–51 (1975).
7. Atkins, C.A., Herridge, D.F. and Pate, J.S. The economy of carbon and nitrogen in nitrogen-fixing annual legumes—experimental observations and theoretical considerations, pp. 211–242 in *Isotopes in Biological Dinitrogen Fixation* (Edited by C.N. Welsh), FAO/IAEA Advisory Group Conference, Vienna (1977).
8. Atkins, C.A., Pate, J.S., Griffiths, G.J. and White, S.T. Economy of carbon and nitrogen in nodulated and non-nodulated (NO_3-grown) cowpea (*Vigna unguiculata* (L.) Walp.), *Plant Physiology*, *66*, 978–983 (1980).
9. Atkins, C.A., Pate, J.S. and Layzell, D.B. Assimilation and transport of nitrogen in non-nodulated (NO_3-grown) *Lupinus albus* L., *Plant Physiology*, *64*, 1078–1082 (1979).
10. Atkins, C.A., Pate, J.S. and McNeil, D.L. Phloem loading and metabolism of xylem-borne amino compounds in fruiting shoots of a legume, *Journal of Experimental Botany*, *31*, 1509–1520 (1980).
11. Atkins, C.A., Pate, J.S. and Sharkey, P.J. Asparagine metabolism—key to the nitrogen nutrition of developing legume seeds, *Plant Physiology*, *56*, 807–812 (1975).
12. Atkins, C.A., Rainbird, R.M. and Pate, J.S. Evidence for a purine pathway of ureide synthesis in N_2-fixing nodules of cowpea (*Vigna unguiculata* (L.) Walp.), *Zeitschrift für Pflanzenphysiologie*, *97*, 249–260 (1980).
13. Atkins, C.A. Metabolism of purine nucleotides to form ureides in nitrogen-fixing nodules of cowpea (*Vigna unguiculata* (L.) Walp.), *Federation of European Biochemical Societies Letters*, *125*, 89–93 (1981).
14. Barankiewicz, J., and Paszkowski, J. Purine metabolism in mesophyll protoplasts of tobacco (*Nicotiana tabacum*) leaves, *Biochemical Journal*, *186*, 343–350 (1980).

15. Barker, G.R. and Douglas, T. Function of ribonuclease in germinating peas, *Nature*, *188*, 943–944 (1960).
16. Barnes, R.L. Formation of allantoin and allantoic acid from adenine in leaves of *Acer saccharinum* L., *Nature*, *184*, 1944 (1959).
17. Beevers, L. *Nitrogen Metabolism in Plants*, pp. 38–41, Edward Arnold, London (1976).
18. Beevers, L. and Hageman, R.H. The role of light in nitrate metabolism in higher plants, *Photophysiology*, *7*, 85–113 (1972).
19. Berger, C. and Gerhardt, B. Charakterisierung der Microbodies aus Spadix-Appenices von *Arum maculatum* L. und *Sauromatum guttatum* Schoff, *Planta*, *96*, 326–338 (1971).
20. Bergersen, F.J. Ammonia—an early stable product of nitrogen fixation by soybean root nodules, *Australian Journal of Biological Sciences*, *18*, 1–9 (1965).
21. Bergersen, F.J. Biochemistry of symbiotic nitrogen fixation in legumes, *Annual Review of Plant Physiology*, *22*, 121–140 (1971).
22. Bergersen, F.J. and Turner, G.L. Nitrogen fixation by the bacterial fraction of breis of soybean root nodules, *Biochimica et Biophysica Acta*, *141*, 507–515 (1967).
23. Bethlenfalvay, G.J. and Phillips, D.A. Interactions between symbiotic nitrogen fixation, combined-N application and photosynthesis in *Pisum sativum*, *Physiologia Plantarum*, *42*, 119–123 (1978).
24. Boland, M.S., Fordyce, A.M. and Greenwood, R.M. Enzymes of nitrogen metabolism in legume nodules. A comparative study, *Australian Journal of Plant Physiology*, *5*, 553–559 (1978).
25. Bollard, E.G. Translocation of organic nitrogen in the xylem, *Australian Journal of Biological Sciences*, *10*, 292–301 (1957).
26. Bollard, E.G. Urease, urea, and ureides in plants, pp. 304–328 in *Utilization of Nitrogen and its Compounds by Plants*, Symposia of the Society of Experimental Biology, Vol. 13, Cambridge University Press, Cambridge.
27. Bollard, E.G., Cook, A.R. and Turner, H.A. Urea as sole source of nitrogen for plant growth. I. The development of urease activity in *Spirodela oligorrhiza*, *Planta*, *83*, 1–12 (1968).
28. Bone, D.H. Metabolism of citrulline and ornithine in mung bean mitochondria, *Plant Physiology*, *34*, 171–175 (1959).
29. Borchers, R. Allantoin determination, *Analytical Biochemistry*, *79*, 612–613 (1979).
30. Brown, C.M. and Dilworth, M.S. Ammonium assimilation by *Rhizobium* cultures and bacteroids, *Journal of General Microbiology*, *86*, 38–48 (1975).
31. Brunel, A. and Échevin, R. La Présence, L'origine et le role Physiolo-

gique des Ureides Glyoxyliques dans les germinations de *Soja hispida* munch, *Revue Géneral de Botanique, 50*, 73–93 (1938).

32. Buchanan, J.M. and Hartman, S.C. Enzymic reactions in the synthesis of the purines, *Advances in Enzymology, 21*, 199–261 (1959).
33. Butler, G.W., Ferguson, J.D. and Allison, R.M. The biosynthesis of allantoin in *symphytum*, *Physiologia Plantarum, 14*, 310–321 (1961).
34. Canvin, D.T. and Atkins, C.A. Nitrate, nitrite and ammonia assimilation by leaves: Effect of light, carbon dioxide and oxygen, *Planta, 116*, 207–224 (1974).
35. Christeller, J.T., Laing, W.A. and Sutton, W.D. Carbon dioxide fixation by lupin root nodules. I. Characterization, association with phosphoenolpyruvate carboxylase, and correlation with nitrogen fixation during nodule development, *Plant Physiology, 60*, 47–50 (1977).
36. Cookson, C., Hughes, H. and Coombs, J. Effects of combined nitrogen on anapleurotic carbon assimilation and bleeding sap composition in *Phaseolus vulgaris* L., *Planta, 148*, 338–345 (1980).
37. Copeland, R., and Pate, J.S. Nitrogen metabolism of nodulated white clover in the presence and absence of nitrate nitrogen, in *White Clover Research*, British Grassland Society, *6*, 71–77 (1970).
38. Corbett, J.R. and Davies, D.D. The decarboxylation of glyoxylic acid, *Plant Physiology, 40* S, 68 (1965).
39. Die Van, J. and Tammes, P.M.L. Phloem exudation from monocotyledonous axes, pp. 196–222 in *Encyclopedia of Plant Physiology, New Series, Vol. I. Transport in Plant Phloem* (Edited by M.H. Zimmerman and J.A. Milburn), Springer-Verlag, Berlin (1975).
40. Dixon, R.O.D. Nitrogenase-hydrogenase inter-relationships in rhizobia, *Biochemie, 60*, 233–236 (1978).
41. Emerich, D.W., Albrecht, S.L., Russell, S.A., Ching, T.M. and Evans, H.J. Oxyleghaemoglobin-mediated hydrogen oxidation by *Rhizobium japonicum* USDA 122 DES bacteroids, *Plant Physiology, 65*, 605–609 (1980).
42. Eschrich, W. Der Phloemsaft von *Cucubita ficifolia*, *Planta, 60*, 216–224 (1963).
43. Eschrich, W. and Heyser, W. Biochemistry of phloem constituents, pp. 101–136 in *Encyclopedia of Plant Physiology, New Series, Vol. I. Transport in Plant Phloem* (Edited by M.H. Zimmerman and J.A. Milburn), Springer-Verlag, New York (1975).
44. Fellows, R.J., Egli, D.B. and Leggett, J.E. A pod leakage technique for phloem translocation studies in soybean (*Glycine max* [L] Merr.), *Plant Physiology, 62*, 812–814 (1978).
45. Fowden, L. The nitrogen metabolism of groundnut plants: The role of γ-methylene glutamine and γ-methylene glutamic acid, *Annals of Botany, 18*, 417–440 (1954).

46. Fowden, L. The non-protein amino acids of plants, *Endeavour*, *21*, 35–42 (1962).
47. Fujihara, S., Yamamoto, K. and Yamaguchi, M. A possible role of allantoin and the influence of nodulation on its production in soybean plants, *Plant and Soil*, *48*, 233–242 (1977).
48. Fujihara, S. and Yamaguchi, M. Effects of allopurinol [4-hydroxypyrazolo (3, 4-d) pyrimidine] on the metabolism of allantoin in soybean plants, *Plant Physiology*, *62*, 134–138 (1978).
49. Fujihara, S. and Yamaguchi, M. Probable site of allantoin formation in nodulating soybean plants, *Phytochemistry*, *17*, 1239–1243 (1978).
50. Fujihara, S. and Yamaguchi, M. Asparagine formation in soybean nodules, *Plant Physiology*, *66*, 139–141 (1980).
51. Grimes, H. and Fottrell, P.F. Enzymes involved in glutamate metabolism in legume root nodules, *Nature*, *212*, 295–296 (1966).
52. Gunning, B.E.S., Pate, J.S., Minchin, F.R. and Marks, I. Quantitative aspects of transfer cell structure in relation to vein loading in leaves and solute transport in legume nodules. Transport at the cellular level, *Symposia of the Society of Experimental Biology*, *28*, 87–126 (1974).
53. Guranowski, A. and Barankiewicz, J. Purine salvage in cotyledons of germinating lupin seeds, *Federation of European Biochemical Societies Letters*, *104*, 95–98 (1979).
54. Halliday, J. An interpretation of seasonal and short term fluctuations in nitrogen fixation, Ph.D. Thesis, University of Western Australia, pp. 71–73 (1975).
55. Hanks, J.F., Tolbert, N.E. and Schubert, K.R. Localisation of enzymes of ureide biosynthesis in peroxisomes and microsomes of nodules, *Plant Physiology* (1981) (in press).
56. Hardy, R.W.F. and Havelka, U.D. Nitrogen fixation research. A key to world food? *Science*, *188*, 633–643 (1975).
57. Hartman, V.E. and Arnold, G. Ureide metabolism in moss protonema of *Funaria hygrometrica* L. (Sibth). III. Influence of light and darkness on the uptake and the metabolism of allantoin-7-^{14}C, *Biochemie und Physiologie den Pflanzen*, *166*, 57–72 (1974).
58. Hartman, V.E. and Geissler, G. Ureide metabolism in the moss protonema of *Funaria hygrometrica* L. (Sibth). I. The influence of varicoloured irradiations, *Biochemie und Physiologie den Pflanzen*, *164*, 614–622 (1973).
59. Herridge, D.F., Atkins, C.A., Pate, J.S. and Rainbird, R. Allantoin and allantoic acid in the nitrogen economy of the cowpea (*Vigna unguiculata* (L.) Walp), *Plant Physiology*, *62*, 495–498.
60. Hodges, T.K. ATPases associated with membranes of plant cells, in *Encyclopedia of Plant Physiology, New Series, Transport in Plants. II. Part A, Cells* (Edited by U. Luttge and M.G. Pitman), Springer-Verlag,

Berlin, 260–283 (1976).

61. Housley, T.L., Peterson, D.M. and Schrader, L.E. Long distance translocation of sucrose, serine, leucine, lysine and CO_2 assimilates. I. Soybean, *Plant Physiology*, *59*, 217–220 (1977).
62. Housley, T.L., Schrader, L.E., Miller, M. and Setter, T.L. Partitioning of (^{14}C) photosynthate and long distance translocation of amino acids in pre-flowering and flowering nodulated and non-nodulated soybeans, *Plant Physiology*, *64*, 94–98 (1979).
63. Huang, A.H.C. and Beevers, H. Localization of enzymes within microbodies, *Journal of Cell Biology*, *58*, 379–389 (1973).
64. Israel, D.W. and McClure, P.R. Nitrogen Translocation in the xylem of soybeans, pp. 111–128 in *World Soybean Research Conference, II. Proceedings* (Edited by F.T. Corbin), Westview Press, U.S.A. (1980).
65. Johnson, G.V., Evans, H.J. and Ching, T.M. Enzymes of the glyoxylate cycle in rhizobia and nodules of legumes, *Plant Physiology*, *41*, 1330–1336 (1966).
66. Kapoor, M. and Waygood, E.R. Initial steps of purine biosynthesis in wheat germ, *Biochemical and Biophysical Research Communications*, *9*, 7–10 (1962).
67. Kating, H. and Eschrich, W. Aufnahme Einbau und Transport von ^{14}C in *Curcubita ficifolia* II. Applikation von Bicarbonat—^{14}C uber die Wurzel, *Planta*, *60*, 598–611 (1964).
68. Kennedy, I.R. Primary products of symbiotic nitrogen fixation. I. Short-term exposures of *Serradella* nodules to $^{15}N_2$, *Biochimica et Biophysica Acta*, *130*, 285–294 (1966).
69. Kennedy, I.R. Primary products of symbiotic nitrogen fixation. II. Pulse-labelling of *Serradella* nodules with $^{15}N_2$, *Biochimica et Biophysica Acta*, *130*, 295–303 (1966).
70. King, R.W. and Zeevaart, J.A.D. Enhancement of phloem exudation from cut petioles by chelating agents, *Plant Physiology*, *53*, 96–103, (1974).
71. Kḿinek, M. Studies of the oxalogenic substances in sugar beets. I. Allantoin studies and determinations, *Listy cukrovarnické*, *54*, 461–465 (1936).
72. Kḿinek, M. Studies of the oxalogenic substances in sugar beets. II. A new non-sugar, oxamic acid isolated from sugar beets, *Listy cukrovarnické*, *54*, 469–479 (1936).
73. Krupka, R.M. and Towers, G.H.N. Studies of the keto acids of wheat. II. Glyoxylic acid and its relation to allantoin, *Canadian Journal of Botany*, *36*, 179–186 (1958).
74. Krupka, R.M. and Towers, G.H.N. Studies of the metabolic relations of allantoin in wheat, *Canadian Journal of Botany*, *37*, 539–545 (1959).
75. Laane, C., Krone, W., Konings, W., Haaker, H. and Veeger, C. Short-

term effect of ammonium chloride on nitrogen fixation by *Azotobacter vinelandii* and by bacteroids of *Rhizobium leguminosarum*, *European Journal of Biochemistry*, *103*, 39–46 (1980).

76. Laing, W.T., Christeller, J.T. and Sutton, W.D. Carbon dioxide fixation by lupin root nodules. II. Studies with ^{14}C-labelled glucose, the pathway of glucose catabolism and the effects of some treatments that inhibit nitrogen fixation, *Plant Physiology*, *63*, 450–454 (1979).
77. Lawrie, A.C. and Wheeler, C.T. Nitrogen fixation in the root nodules of *Vicia faba* L. in relation to the assimilation of carbon. II. The dark fixation of carbon dioxide, *New Phytologist*, *74*, 429–436 (1975).
78. Layzell, D.B., Pate, J.S., Atkins, C.A. and Canvin, D.T. Partitioning of carbon and nitrogen and the nutrition of root and shoot apex in a nodulated legume, *Plant Physiology*, *67*, 30–36 (1981).
79. Layzell, D.B., Rainbird, R., Atkins, C.A. and Pate, J.S. Economy of photosynthate use in N-fixing legume nodules: observations on two contrasting symbioses, *Plant Physiology*, *64*, 888–891 (1979).
80. Leaf, G., Gardner, I.C. and Bond, G. Observations on the composition of metabolism of the nitrogen-fixing root nodules of *Alnus*, *Journal of Experimental Botany*, *9*, 320–331 (1958).
81. Leaf, G., Gardner, I.C. and Bond, G. Observations on the composition and metabolism of the nitrogen-fixing root nodules of *Myrica*, *Biochemical Journal*, *72*, 662–667 (1959).
82. Lee, K.W. and Roush, A.H. Allantoinase assays and their application to yeast and soybean allantoinases, *Archives Biochemistry and Biophysics*, *108*, 460–467 (1964).
83. Martin, P. Verteilung von Stickstaff aus Spross und Wurzel bei jungen Bohnen pflanzen nach der Aufnahme von NO_3^- und NH_4^+, *Zeitschrift für Pflanzenernährung und Bodenkunde*, *2*, 181–193 (1976).
84. Mary, A. and Sastri, K.A. An unusual allantoinase from *Dolichos biflorus*, *Phytochemistry*, *17*, 397–399 (1978).
85. Matsumoto, T., Yamamoto, Y. and Yatazawa M. Role of root nodules in the nitrogen nutrition of soybeans. II. Fluctuation in allantoin concentration of the bleeding sap, *Journal of the Science of Soil and Manure Japan*, *47*, 463–469 (1976).
86. Matsumoto, H., Yasuda, T., Kobayashi, M. and Takahashi, E. The inducible formation of urease in rice plants, *Soil Science and Plant Nutrition*, *12*, 239–244 (1966).
87. Matsumoto, T., Yatazawa, M. and Yamamoto, Y. Distribution and change in the contents of allantoin and allantoic acid in developing nodulating and non-nodulatin soybean plants, *Plant and Cell Physiology*, *18*, 353–359 (1977a).
88. Matsumoto, T., Yatazawa, M. and Yamamoto, Y. Incorporation of ^{15}N into allantoin in nodulated soybean plants supplied with $^{15}N_2$, *Plant*

and Cell Physiology, 18, 459–462 (1977b).

89. McClure, P.R. and Israel, D.W. Transport of nitrogen in the xylem of soybean plants, *Plant Physiology, 64*, 411–416 (1979).
90. McNeil, D.L., Atkins, C.A. and Pate, J.S. Uptake and utilization of xylem-borne amino compounds by shoot organs of a legume, *Plant Physiology, 63*, 1076–1081 (1979).
91. Meeks, J.C., Wolk, C.P., Schilling, N., Shaffer, P.W., Avissar, Y. and Chien, W.S. Initial organic products of fixation of [^{13}N] dinitrogen by root nodules of soybean (*Glycine* max), *Plant Physiology, 61*, 980–983 (1978).
92. Miettinen, J.K. and Virtanen, A.I. The free amino acids in the leaves, roots and root nodules of the alder (*Alnus*), *Physiologia Plantarum, 5*, 540–547 (1952).
93. Miflin, B.S. and Lea, P.J. Amino acid metabolism, *Annual Review of Plant Physiology, 28*, 299–329 (1977).
94. Mothes, K. The metabolism of urea and ureides, *Canadian Journal of Botany, 39*, 1785–1807 (1961).
95. Mothes, K. and Engelbrecht, L. Über Allantoinsäure und Allantoin. I. Ihre Rolle als Wanderformdes Stickstoffs und ihre Beziehungen zum Eiweisstoffwechsel des Ahorns, *Flora, 139*, 586–616 (1952).
96. Mothes, K. and Engelbrecht, L. Über den Stickstoffumsatz in Blattsteeklingen, *Flora, 143*, 428–472 (1956).
97. Murray, A.W. The biological significance of purine salvage, *Annual Review of Biochemistry, 40*, 811–826 (1971).
98. Nagai, Y. and Funahashi, S. Allantoinase from mung bean seedlings, *Agricultural and Biological Chemistry, 25*, 265–268 (1961).
99. Nguyen, J. Effect of light on deamination and oxidation of adenylic compounds in cotyledons of *Pharbitus nil, Physiologia Plantarum, 46*, 255–259 (1979).
100. Nirmala, J. and Sastry, K.S. Modified method for analysis of glyoxylate derivatives in biological materials, *Analytical Biochemistry, 47*, 218–227 (1972).
101. Nirmala, J. and Sastry, K.S. The allantoinase of *Lathyrus sativus, Phytochemistry, 14*, 1971–1973 (1975).
102. Nygaard, P. Nucleotide metabolism during pine pollen germination, *Physiologia Plantarum, 28*, 361–371 (1973).
103. Ohyama, T. and Kumazawa, K. Incorporation of ^{15}N into various nitrogenous compounds in intact soybean nodules after exposure to $^{15}N_2$ gas, *Soil Science and Plant Nutrition, 24*, 525–533 (1978).
104. Ohyama, T. and Kumazawa, K. Assimilation and transport of nitrogenous compounds originated from $^{15}N_2$-fixation and $^{15}NO_3$-absorption, *Soil Science and Plant Nutrition, 25*, 9–19 (1979).
105. Ohyama, T. and Kumazawa, K. Nitrogen assimilation in soybean no-

dules. I. The role of GS/GOGAT system in the assimilation of ammonia produced by N_2-fixation, *Soil Science and Plant Nutrition, 26*, 109–115 (1980a).

106. Ohyama, T. and Kumazawa, K. Nitrogen assimilation in soybean nodules. II. $^{15}N_2$ assimilation in bacteroid and cytosol fractions of soybean nodules, *Soil Science and Plant Nutrition, 26*, 205–213 (1980b).
107. Ohyama, T. and Kumazawa, K. Nitrogen assimilation in soybean nodules. III. Effects of rhizosphere pO_2 on the assimilation of $^{15}N_2$ in nodules attached to intact plants, *Soil Science and Plant Nutrition, 26*, 321–324 (1980c).
108. Omura, H., Osajima, Y. and Tsukamoto, T. Properties of urea dehydrogenase in tissues of higher plants, *Enzymologia, 31*, 129–154 (1966).
109. Oota, Y., Fujii, R. and Osawa, S. Changes in chemical constituents during the germination stage of a bean, *Vigna sesquipedalis, Journal of Biological Chemistry, 40*, 649–661 (1953).
110. Oota, Y. and Takata, K. Changes in microsomal ribonucleoprotein in the time course of the germination stage as revealed by electrophoresis, *Physiologia Plantarum, 12*, 518–525 (1959).
111. Parish, R.W. Urate oxidase in peroxisomes from maize root tips, *Planta, 104*, 247–251 (1972).
112. Parish, R.W. Peroxisomes from the *Arum italicum* appendix, *Zeitschrift für Pflanzenphysiologie, 67*, 430–442 (1972).
113. Pate, J.S. Uptake, assimilation and transport of nitrogen compounds by plants, *Soil Biology and Biochemistry, 5*, 109–119 (1973).
114. Pate, J.S. Transport and partitioning of nitrogenous solutes, *Annual Review of Plant Physiology, 31*, 313–340 (1980).
115. Pate, J.S. and Atkins, C.A. Nitrogen uptake, transport and utilisation, in *Ecology of Nitrogen Fixation, Vol. 3. Legumes* (Edited by W.J. Broughton), Oxford University Press, United Kingdom (in press) (1981).
116. Pate, J.S., Atkins, C.A., Hamel, K., McNeil, D.L. and Layzell, D.B. Transport of organic solutes in phloem and xylem of a nodulated legume, *Plant Physiology, 63*, 1082–1088 (1979).
117. Pate, J.S., Atkins, C.A., Herridge, D.F. and Layzell, D.B. Synthesis, storage and utilization of amino compounds in white lupin (*Lupinus albus* L.), *Plant Physiology, 67*, 37–42 (1981).
118. Pate, J.S., Atkins, C.A., White, S.T., Rainbird, R.M. and Woo, K.C. Nitrogen nutrition and xylem transport of nitrogen in ureide-producing grain legumes, *Plant Physiology, 65*, 961–965 (1980).
119. Pate, J.S. and Gunning, B.E.S. Transfer cells, *Annual Review of Plant Physiology, 23*, 173–196 (1972).
120. Pate, J.S., Gunning, B.E.S. and Briarty, L.G. Ultrastructure and functioning of the transport system of the leguminous root nodule, *Planta, 85*, 11–34 (1969).

121. Pate, J.S., Sharkey, P.J. and Lewis, O.A.M. Phloem bleeding from legume fruits. A technique for study of fruit nutrition, *Planta, 120*, 229–243 (1974).
122. Pate, J.S., Sharkey, P.J. and Lewis, O.A.M. Xylem to phloem transfer of solutes in fruiting shoots of a legume studied by a phloem bleeding technique, *Planta, 122*, 11–26 (1975).
123. Pate, J.S., Walker, J. and Wallace, W. Nitrogen-containing compounds in the shoot system of *Pisum sativum* L. II. The significance of amino acids and amides released from nodulated roots, *Annals of Botany, 29*, 475–493 (1965).
124. Pate, J.S. and Wallace, W. Movement of assimilated nitrogen from the root system of the field pea (*Pisum arvense* L.), *Annals of Botany, 28*, 80–99 (1964).
125. Peterson, J.B. and Evans, H.J. Phosphoenolpyruvate carboxylase from soybean nodule cytosol. Evidence for isoenzymes and kinetics of the most active component, *Biochimica et Biophysica Acta, 567*, 445–452 (1979).
126. Phillips, D.A. Efficiency of symbiotic nitrogen fixation in legumes, *Annual Review of Plant Physiology, 31*, 29–49 (1980).
127. Planque, K., Kennedy, I.R., de Vries, G.E., Quispel, A. and von Brussel, A.A.N. Location of nitrogenase and ammonia-assimilatory enzymes in bacteroids of *Rhizobium leguminosarum* and *Rhizobium lupinii*, *Journal of General Microbiology, 102*, 95–104 (1977).
128. Radyukina, N.A., Puskin, A.V., Evstigneeva, Z.G. and Kretovich, V.L. *Doklady Akademiya nauk SSSR, 234*, p. 1209 (1977).
129. Rainbird, R.M. and Atkins, C.A. Purification and some properties of urate oxidase from nitrogen-fixing nodules of cowpea, *Biochimica et Biophysica Acta, 659*, 132–140 (1981).
130. Rao, K.P., Blevins, D.G. and Randall, D.D. Glutamine-phosphoribosylpyrophosphate amidotransferase from soybean nodules, *Plant Physiology, 65*, p. 110 (1980).
131. Reinbothe, H. and Mothes, K. Urea, ureides and guanidines in plants, *Annual Review of Plant Physiology, 13*, 129–150 (1962).
132. Reuter, G. Über den Stickstoffhaushalt der Betulaceaen und anderer Laub- und Nadelhölzer, *Flora, 144*, 420–446 (1957).
133. Reuter, G. and Wolffgang, H. Vergleichende Untersuchungen über den charakter der Stickstoff-Verbindungen von Baumblutungssaften bei Betulaceaen und anderen Holzarten, *Flora, 142*, 146–155 (1954).
134. Reynolds, P.H.S. and Farnden, K.J.F. The involvement of aspartate aminotransferases in ammonium assimilation in lupin nodules, *Phytochemistry, 18*, 1625–1630 (1979).
135. Rowe, P.B., McCairns, E., Madsen, G., Sauer, D. and Elliott, H. *De novo* purine synthesis in avian liver. Co-purification of the enzymes and

properties of the pathway, *Journal of Biological Chemistry*, *253*, 7711–7721 (1978).
136. Ruis, H. Isolation and characterisation of peroxisomes from potato tubers, *Hoppe-Seyler's Zeitschrift für Physiologishes Chemie*, *352*, 1105–1112 (1971).
137. Schrader, L.E. Uptake, accumulation, assimilation and transport of nitrogen in higher plants, pp. 101–141 in *Nitrogen in the Environment, Vol. 2* (Edited by D.R. Nielsen and S.G. MacDonald), Academic Press, New York (1978).
138. Schubert, K.R., Firestone, R. and Coker, G.T. Ammonia assimilation in root nodules of *Alnus glutinosa:* Tracer studies using [^{13}N] NH_4^+, *Plant Physiology*, *65*, S, 113.
139. Schubert, K.R., Jennings, N.T. and Evans, H.J. Hydrogen reactions of nodulated leguminous plants. II. Effects on dry matter accumulation and nitrogen fixation, *Plant Physiology*, *61*, 398–401 (1978).
140. Schubert, K.R. and de Shone, G.M. Enzymes of purine biosynthesis and catabolism in soybean root nodules: Role in ureide biosynthesis, *Plant Physiology*, *65*, S, 111, (1980).
141. Schubert, K.R. and Ryle, G.J.A. The energy requirements for nitrogen fixation in nodulated legumes, pp. 85–96 in *Advances in Legume Science* (Edited by R.J. Summerfield and A.H. Bunting), Royal Botanic Gardens, Kew, United Kingdom (1980).
142. Scott, D.B., Robertson, J. and Farnden, K.J.F. Ammonia assimilation in lupin nodules, *Nature*, *262*, 703–705 (1976).
143. Sen, D. and Schulman, H.M. Enzymes of ammonia assimilation in the cytosol of developing soybean root nodules, *New Phytologist*, *85*, 243–250 (1980).
144. Silver, A.V. and Gilmore, V. The metabolism of purines and their derivatives in seedlings of *Pisum sativum*, *Phytochemistry*, *8*, 2295–2299 (1969).
145. Singh, R. Evidence for the presence of allantoicase in germinating peanuts, *Phytochemistry*, *7*, 1503–1508 (1968).
146. Skokut, T.S. and Filner, P. Slow adaptive changes in urease levels of tobacco cells cultured on urea and other nitrogen sources, *Plant Physiology*, *65*, 995–1003 (1980).
147. Smith, I. *Chromatographic Techniques*, William Heinemann Medical Book Ltd., London (1958).
148. Sprent, J.I. Root nodule anatomy, type of export product and evolutionary origin in some *Leguminoseae*, *Plant Cell and Environment*, *3*, 35–43 (1980).
149. Steward, F.C. and Pollard, J.K. Nitrogen metabolism in plants: Ten years in retrospect, *Annual Review of Plant Physiology*, *8*, 65–114 (1957).
150. Stovall, I. and Cole, M. Organic acid metabolism by isolated *Rhizobium*

japonicum bacteroids, *Plant Physiology*, *61*, 787–790 (1978).

151. Streeter, J.G. Nitrogen nutrition of field-grown soybean plants. I. Seasonal variations in soil nitrogen and nitrogen composition of stem exudate, *Agronomy Journal*, *64*, 311–312 (1972).
152. Streeter, J.G. Allantoin and allantoic acid in tissues and stem exudate from field-grown soybean plants, *Plant Physiology*, *63*, 478–480 (1979).
153. Suzuki, T. and Takahashi, E. Metabolism of xanthine and hypoxanthine in the tea plant (*Thea Sinensis* L.), *Biochemical Journal*, *146*, 79–85 (1975a).
154. Suzuki, T. and Takahashi, E. Biosynthesis of caffeine by tea leaf extracts. Enzymic formation of theobromine from 7-methylxanthine and of caffeine from theobromine, *Biochemical Journal*, *146*, 87–96 (1975b).
155. Tajima, S. and Yamamoto, Y. Enzymes of purine catabolism in soybean plants, *Plant and Cell Physiology*, *16*, 271–282 (1975).
156. Tajima, S. and Yamamoto, Y. Regulation of uricase activity in developing roots of *Glycine max*, non-nodulating variety A62–2, *Plant and Cell Physiology*, *18*, 247–253 (1977).
157. Tajima, S., Yatazawa, M. and Yamamoto, Y. Allantoin production and its utilization in relation to nodule formation in soybeans. Enzymatic studies, *Soil Science and Plant Nutrition*, *23*, 225–235 (1977).
158. Tammes, P.M.L. and van Die, J. Studies on phloem exudation from *Yucca flaccida* Haw. I. Some observations on the phenomenon of bleeding and the composition of the exudate, *Acta Botanica Neerlandica*, *13*, 76–83 (1964).
159. Theimer, R.R. Enzymatische Untersuchungen über die Stickstoffeassimilation in Wurzelen von *Phaseolus coccineus*, *Hoppe-Seyler's Zeitschrift für Physiologiches Chemie*, 354, p. 1251 (1973).
160. Theimer, R.R. and Beevers, H. Uricase and allantoinase in glyoxysomes, *Plant Physiology*, *47*, 246–251 (1971).
161. Theimer, R.R. and Heidinger, P. Control of particulate urate oxidase activity in bean roots by external nitrogen supply, *Zeitschrift für Pflanzenphysiologie*, *73*, 360–370 (1974).
162. Thibodeau, P.S. and Jaworski, E.G. Patterns of nitrogen utilization in the soybean, *Planta*, *127*, 133–147 (1975).
163. Thomas, R.J., Feller, U. and Erisman, K.H. The effect of different inorganic nitrogen sources and plant age on the composition of bleeding sap of *Phaseolus vulgaris* (L.), *New Phytologist*, *82*, 657–669 (1979).
164. Thomas, R.J., Feller, U. and Erisman, K.H. Allantoic acid metabolism in non-nodulated bushbeans, *Plant Physiology*, *63*, S, 50 (1979).
165. Thomas, R.J., Feller, U. and Erisman, K.H. Ureide metabolism in nonnodulated *Phaseolus vulgaris* (L.), *Journal of Experimental Botany*, *31*, 409–418 (1980).

166. Thomas, R.J., Khandavilli, V. and Schrader, L.F. Allantoinase activity in soybean leaves and fruits during development, *Plant Physiology, 65,* p. 55 (1980).
167. Thompson, J.F. and Muenster, A.M.E. Separation of the *Chlorella* ATP: Urea amidolyase into two components, *Biochemical and Biophysical Research Communications, 43,* 1049–1055 (1971).
168. Tjepkema, J.D. and Yocum, C.S. Respiration and oxygen transport in soybean nodules, *Planta, 115,* 59–72 (1973).
169. Tracey, M.V. Urea and ureides, pp. 119–141 in *Modern Methods of Plant Analysis, Vol. 4* (Edited by K. Paech and M.V. Tracey), Springer-Verlag, Berlin (1955).
170. Trijbels, F. and Vogels G.D. Degradation of allantoin by *Pseudomonas acidovorans, Biochimica et Biophysica Acta, 113,* 292–301 (1966).
171. Triplett, E.W., Blevins, D.G. and Randall, D.D. Allantoic acid synthesis in soybean root nodule cytosol via xanthine dehydrogenase, *Plant Physiology, 65,* 1203–1206 (1980).
172. Tschiersch, B. Über Canavanin, *Flora, 147,* 405–416 (1959).
173. Tschiersch, B. Über das Vorkommen von Canavanin, *Flora, 150,* 87–94 (1961).
174. Vogels, G.D. and van der Drift, C. Allantoinases from bacterial, plant and animal sources. II. Effect of bivalent cations and reducing substances on the enzymic activity, *Biochimica et Biophysica Acta, 122,* 497–509 (1966).
175. Vogels, G.D. and van der Drift, C. Degradation of purines and pyrimidines by microorganisms, *Bacteriological Reviews, 40,* 403–468 (1976).
176. Vogels, G.D., Trijbels, F. and Uffink, A. Allantoinases from bacterial, plant and animal sources. I. Purification and enzymic properties, *Biochimica et Biophysica Acta, 122,* 484–496 (1966).
177. Wang, D. and Waygood, E.R. Enzymes of synthesis of purine and pyrimidine nucleotides, pp. 421–447 in *Modern Methods of Plant Analysis, Vol. 7* (Edited by H.F. Linskens, B.D. Sanwal and M.V. Tracey), Springer-Verlag, Berlin (1964).
178. Wolfgang, H. and Mothes, K. Papierchromatographische Untersuchungen an pflanzlichen Blutungssäften, *Naturwissenschaften, 40,* 606 (1953).
179. Wong, P.P. and Evans, H.J. Poly-β-hydroxybutyrate utilization by soybean (*Glycine max*) nodules and assessment of its role in maintenance of nitrogenase activity, *Plant Physiology, 47,* 750–755 (1971).
180. Woo, K.C., Atkins, C.A. and Pate, J.S. Ureide synthesis in cell-free extracts of cowpea (*Vigna unguiculata* (L.) Walp.) nodules, *Plant Physiology,* 65, S, 57 (1980).
181. Woo, K.C., Atkins, C.A. and Pate, J.S. The biosynthesis of ureides from purines in a cell-free system from nodule extracts of cowpea (*Vigna unguiculata* (L.) Walp.), *Plant Physiology, 66,* 735–739 (1980).

182. Woo, K.C., Atkins, C.A. and Pate, J.S. Ureide synthesis in a cell-free system from cowpea (*Vigna unguiculata* (L.) Walp.) nodules. Studies with O_2, pH and purine metabolites, *Plant Physiology* (in press) (1981).
183. Yamamoto, Y. and Yatazawa, M. Formation and functions of nitrogen-fixing nodules in soybean. Allantoin formation in the symbiotic condition, pp. 25–34 in *Nitrogen Fixation and Nitrogen Cycle* (Edited by H. Takahashi), JIBP Synthesis, University of Tokyo Press (1975).
184. Ziegler, H. Nature of transported substances, pp. 59–100 in *Encyclopedia of Plant Physiology, New Series, Vol. 1* (Edited by M.H. Zimmermann and J.A. Milburn), Springer-Verlag, Berlin (1975).
185. Ziegler, H. and Schnabel, M. Über Hannstoffderivate in Siebröhrensaft, *Flora, 150,* 306–317 (1961).

J.H. BECKING

4. Nitrogen Fixation in Nodulated Plants other than Legumes

Introduction

Non-leguminous plants possessing root nodules with nitrogen-fixing capacity occur in a large number of phylogenetically unrelated families and genera of dicotyledonous angiosperms. They have different morphological appearances ranging from small, prostrate herbs (e.g. *Dryas*) to small shrubs (e.g. *Ceanothus* and *Colletia*) and tree-like woody species (e.g. *Alnus* and *Casuarina*).

With the exception of *Rubus ellipticus* (Rosaceae) most of these nodulated plants are of no agricultural value, but they often play a prominent role in plant succession in natural ecosystems disturbed by soil erosion, landslides, fire or volcanic activity, and human interference which include road constructions, forest clearings, etc. Most of these species are therefore pioneer colonizers in natural plant succession or in forests denuded of timber. They are usually short-lived, especially the smaller species, and because of their nitrogen-fixing capacity, they assist the initial growth of other more aggressive plant species but soon become eliminated. Due to this reason, these non-leguminous nodulated plants are rarely found in climax vegetation but may survive in such a situation merely due to extreme environmental conditions, e.g. *Myrica javanica* as a component of the crater vegetation on Java and *Parasponia* on lahar streams produced by volcanic activity on Mt. Kelud (Java, Indonesia).

Because of their colonizing habit, most nodulated non-leguminous plants have particular significance for land reclamation in forestry. Besides their ability to contribute nitrogen by way of fixation, the more woody species like *Alnus* and *Casuarina* also provide suitable timber and firewood for human beings.

Nodulated species

The older literature dealing with non-leguminous nitrogen-fixing dicotyledons can be found in the reviews of Becking (13, 14, 16, 19, 21) and Bond

(29). The present communication covers only the more recent contributions in this field. A complete list of the non-leguminous nodulated species including the more recent discoveries is presented in Table 1. As evident from this table the nodulated plants comprise 7 orders, 8 families, 17 genera and 173 species of decotyledonous plants.

Table 1. Classification of non-leguminous dinitrogen fixing angiosperms with *Frankia* symbioses

Order	Family	Tribe	Genus	Number of nodulated species (in parentheses, total number of species)[a]
Casuarinales	Casuarinaceae	—	*Casuarina*	24 (45)
Myricales	Myricaceae	—	*Myrica*	26 (35)
			Comptonia	1 (1)
Fagales	Betulaceae	Betuleae	*Alnus*	33 (35)
Rhamnales	Elaeagnaceae	—	*Elaeagnus*	16 (45)
			Hippophae	1 (3)
			Shepherdia	3 (3)
	Rhamnaceae	Rhamneae	*Ceanothus*	31 (55)
		Colletieae	*Discaria*	6 (10)
			Colletia	3 (17)
			Trevoa	1 (6)
Coriariales	Coriariaceae	—	*Coriaria*	13 (15)
Rosales	Rosaceae	Rubieae	*Rubus*[b]	1 (250) (429)
		Dryadeae	*Dryas*	3 (4)
			Purshia	2 (2)
		Cercocarpeae	*Cercocarpus*	4 (20)
Cucurbitales	Datiscaceae	—	*Datisca*	2 (2)

[a]Taxonomic estimates mainly based on Willis (87).

[b]According to Focke (39) 429 *Rubus* species occur world-wide, but Willis (87) gives as estimate 250 *Rubus* species.

With regard to their root-nodule development, two types of symbioses can be distinguished. Most of these symbioses are actinomycetous (or actinorhizal) symbioses caused by an actinomycete of the genus *Frankia* belonging to the Frankiaceae (15, 18). However, in one genus of dicotyledons (i.e. Genus, *Parasponia*) a true bacterium (Eubacteriales or the Procaryotes) is involved in the symbioses. This bacterium belongs to the genus *Rhizobium* of the Rhizobiaceae and it is to a certain degree promiscuous as it can also nodulate and effectively fix nitrogen in a number of tropical legumes such as cow-pea (*Vigna* spp.) and other leguminous plants.

Actinomycetous symbioses

A recent discovery of this type of symbioses is root nodulation of the genus *Datisca* in the family Datiscaceae of the Order Cucurbitales. Uptil now, Cucurbitales was not in the picture among nodulated species of non-leguminous plants and had no affinity to other nodulated orders. Nodulation and nitrogen fixation (acetylene reduction) was observed in two representatives of the genus *Datisca*—*D. cannabina* (see Plate 1) having a mediterranean distribution ranging from the Indo-Arabian region to the Himalayas and Central Asia, and *D. glomorata* found in S.W. United States (California) and N.W. Mexico. In both the species, root nodulation and nitrogen fixation was recently discovered and confirmed by Chaudhary (35, 36).

Root nodulation in *Datisca cannabina* was, however, reported earlier more than 55 years ago by Severini (71), who also did some experiments on the growth of nodulated plants in nitrogen-deficient medium to demonstrate their nitrogen-fixing ability. In spite of the fact that root nodulation of *Datisca cannabina* was cited by Metcalfe and Chalk (59) without any reference to the possibility of nitrogen fixation, this observation somehow escaped notice of research workers until recently.

Root nodulation in Rosaceae was already known in three *Cercocarpus* species (i.e., *C. betuloides*, *C. montanus* and *C. paucidentatus*); recently nodulation was also observed in *Cercocarpus ledifolius* growing as pioneer species in *Pinus flexilis* stands at higher altitudes in California. Recently root nodulation and nitrogen fixation (acetylene reduction) has been confirmed in *Rubus ellipticus* (Plate 2) from Java, Indonesia (24). Although the presence of root nodules in this species was earlier reported by Mrs. Sri Soemartono as part of the I.B.P. Survey (29), this observation was received with some surprise and even suspicion. The genus *Rubus* is unrelated to other nodulating species of non-legumes and no tests for nitrogen fixation of the *Rubus* material were made during the I.B.P. survey. Moreover, in all non-leguminous nodulated plants so far investigated, nodulation is a generic character and root nodulation habit in one species among the 250 to 429 species of the genus *Rubus* (tribe Rubiae) (39, 87) is indeed very remarkable. Perhaps in the genus *Rubus*, other nodulating species will eventually be found by extensive survey. *Purshia tridentata* is an important browse plant in arid regions of western U.S.A. and root nodulation and nitrogen fixation in this little-known plant was studied by Bond (30).

Root nodulation habit in the genus *Colletia* of the Rhamnaceae of the Order Rhamnales was first observed by Bond in a specimen of *C. paradoxa* (syn. *C. cruciata*) growing in the Glasgow Botanical Gardens and reported the same in the final conference of I.B.P. Section PP-N in Edinburgh 1973 (29). Subsequently, Medan and Tortosa (58) reported nodulation in plants of *C. paradoxa* and *C. spinosissima* growing in the Botanical Gardens of Buenos Aires, Argentina and in four species of the genus *Discaria* also belonging to the Rhamnaceae, i.e., *Discaria americana*, *D. serratifolia*, *D. trinervis* and *D. nana*, occurring locally in natural vegetations. So far nodulation in *Dis-*

caria was only observed in the species *D. toumatou* endemic to New Zealand (61) and in an unclassified species growing in the botanical gardens of Edinburgh, raised from seed collected from the wilderness in Chile (29). Root nodulation habit was also observed in *Colletia*, and nitrogen fixation (acetylene reduction) measured in *C. paradoxa* (syn. *C. cruciata*) and *C. armata* (syn. *C. spinosa*) growing in the Cibodas Mountain Gardens (altitude 1450 m) and in Mt. Pangrango-Gedeh, W. Java, Indonesia (24). Added to these observations, nodulation was also recorded in *C. paradoxa* and *C. armata* plants growing in botanical gardens of Amsterdam, Cologne, Munich and Nantes (Becking, unpublished).

Very recently, in the genus *Trevoa* of the family Rhamnaceae root nodulation was observed. The species *Trevoa trinervis* closely related to *Colletia*, was found to bear root nodules and acetylene reduction tests in natural habitat indicated the ability of the plant to fix nitrogen (68) which was comparable to that observed in *Ceanothus* and *Cercocarpus* when measured under similar conditions (43, 38, 47). *Trevoa trinervis* is an important matorral shrub in Chile occurring especially on disturbed sites, more or less in the same way as *Ceanothus* and *Cercocarpus* species which occur as xerophytic chaparral shrubs in California. Unlike the above-mentioned Californian species, the leaves of *Trevoa trinervis*, however, are drought decidous. Therefore, with the onset of drought stress, the leaves of *Trevoa* are rapidly shed, but photosynthesis is maintained by chlorophyllous tissue of young stems and spines. Similarly, *Trevoa* closely resembles *Colletia* species (Plate 3) in which the leaves are also shed during the unfavourable season, but phyllocladioid stems and spines continue the photosynthetic activity and provide indispensable carbohydrates to the nodules.

Root nodulation and the possibility of nitrogen fixation was claimed by Allen *et al.* (5) in *Arctostaphylos uva-ursi* of the Ericaeae occurring in natural habitats in Alaska. As pointed out by Becking (21), the observed root nodulation in Ericaceous species such as *Arctostaphylos* and *Calluna* is very probably mycorrhizal and that there is no evidence of nitrogen fixation for such eukaryotic systems. Recently Tiffney *et al.* (78) re-examined the nodular structures of *Arctostaphylos uva-ursi* in material from Massachusetts, U.S.A. (Nantucket and Cape Cod) and some other localities in the United States. They observed nodular structures on the rhizomes which simulate root nodules but failed to show nitrogen fixation when tested by acetylene reduction tests. They found no evidence of mycorrhizal growth associated with these structures and they suggested that such structures are simply aggregates of latent buds initiated by physical damage to the plant due to human interference caused by motor tracks or footpaths or by fire.

Rhizobium symbioses

This kind of symbioses in non-leguminous plants is only found in the genus *Parasponia* of the Ulmaceae belonging to the Urticales. Evidence of root nodulation in the Ulmaceae by a *Rhizobium* species was highlighted by Trinick (82). This report, however, named the plant as *Trema aspera* occurring in the Pangia District of Papua (New Guinea). Later, this species was considered to be *Trema cannabina* var. *scabra* (85) but recently it has been classified as a species of *Parasponia* (2, 3). The confusion between *Trema* and *Parasponia* is understandable because representatives of both the genera are morphologically very similar and obviously have caused confusion resulting in many re-identifications and changes of names on labels of herbarium specimens deposited in the Herbarium Bogoriensis, Bogor, Indonesia or Rijksherbarium, University of Leiden, The Netherlands. Minor morphological characteristics can only differentiate representatives of both the genera. *Parasponia* can be distinguished from *Trema* by its imbricate perianth lobes of the male flowers and the presence of interpetiolar connate stipulates enclosing the terminal bud (7, 77).

The presence of root nodules in *Parasponia* was known at a very early date in this century. In a symposium on "Green manure in Indonesia" Ham (41) drew attention to the fact that the non-leguminous tree called "anggrung" (in the Javanese language of Central and East Java) or "kuray" (in the Sundanese language of Western Java) bore root nodules and indicated that these root nodules are probably able to "collect nitrogen" like so many leguminous plant species.

The vernacular names "anggrung" and "kuray" are used indiscriminately for *Trema* and *Parasponia* in Java. Later, Backer and van Slooten (8) could confirm root nodulation in *Trema* in one case, but not in other cases. Since at that time, no rigid discrimination was made between representatives of the genera *Trema* and *Parasponia*, it is likely that both species have been confused and that the tree bearing root nodules was in fact *Parasponia*. Finally, Clason (37) in an extensive study of the regeneration of the flora of volcanic ash soils of Mt. Kelud (East Java, Indonesia) clearly described the pioneer habit of *Parasponia* on these virgin soils and observed profuse root nodulation of this species. In addition, he stated very clearly that "nitrogenous food being possibly obtained in this way" (37). Dried herbarium specimens of root nodules of *Parasponia* collected at that time by Clason on Mt. Kelud are still present in the herbarium material of the Herbarium Bogoriensis at Bogor, Indonesia.

However, Clason was not the first to draw attention to the property of *Parasponia* as a pioneer among colonizers. Earlier Junghuhn cited by Clason (37) described *Parasponia* as a pioneer vegetation at places cleared by volcanic activity and flow of solidified lava on Mt. Merapi, Central Java, and remarked that the plant is an active colonizer of bare, virgin soils at altitudes between 1500 and 1800 m. Both Junghuhn and Clason (37) referred to the plant species as *Parasponia parviflora* Miq., but based on herbarium material obtained from these localities the species has recently been reclassified as *Parasponia rugosa* Bl (77). Trinick (82) found the same species in Papua having a distribution from Central Java over the Lesser Sunda islands and some Greater Sunda Islands (Sulawesi) and the Philippines extending up to Papua.

Root nodulation in *Parasponia* has now been recorded in three among the five species of this genus—*P. rugosa* (38, 82), *P. parviflora* (2, 23, see Plate 4 a–c) and *P. andersonii* (83). It is likely that the two other unexamined *Parasponia* species, which are confined to New Guinea, are also nodulated.

It is noteworthy that *Trema* have a more western distribution i.e. western parts of Malaysia (Western Indonesia and Malay Peninsula), Africa and the American continent, whereas *Parasponia* species are typically Asiatic and more common in the eastern parts of Malaysia (i.e. mainly Central and Eastern Indonesia) and New Guinea.

Rhizobial root nodulation has also been reported in some xerophytic desert plants belonging to the Zygophyllaceae. Sabet (69) observed root nodulation in a number of species of the genera *Zygophyllum* (*Z. coccineum*, *Z. album*, *Z. decumbens* and *Z. simplex*), *Fagonia* (*F. arabica*) and *Tribulus* (*T. altus*) growing in the poor sandy soils of the Egyptian deserts. Mostafa and Mahmoud (62) reported the isolation of *Rhizobium*-like strains from these root nodules, which can also produce effective root nodules on roots of some cultivated legumes such as *Trifolium alexandrinum* and *Arachis hypogaea*. They reported, however, that some of these Zygophyllaceous species apparently had their own nodule causing bacterium. Both Sabet (69) and Mostafa and Mahmoud (62) mentioned that nodulated plants of Zygophyllaceae showed more vigorous growth in unsterilized soil or in sterilized soil after inoculation than non-nodulated plants and the latter in due course of time suffered from symptoms of nitrogen deficiency. They concluded from these experiments that root nodules supplied the host with nitrogen. The presence of root nodules on Zygophyllaceous plants was confirmed by Athar and Mahmood (6) by recording nodulation in *Zygophyllum simplex*, *Fagonia cretica* and *Tribulus terrestris* growing in dry habitats of sandy soils, low in nutrition near Karachi in W. Pakistan. From morphological observations and staining techniques, they suggested that the endophyte within the root nodules was a *Rhizobium* species. The authors examined *Zygophyllum coccineum* plants in Egyptian desert soil on the sides of the road connecting Cairo and Suez. Nodular structures were observed on the roots (Plate 5b) but acetylene reduction field tests *in situ* with root nodules did not reveal any nitrogenase activity (Becking, unpublished).

Biology of the symbioses

Actinomycetous symbioses

About 170 plant species belonging to 17 genera, 8 families and 7 orders of worldwide distribution have been reported to bear actinomycete-induced root nodules with nitrogen-fixing capacity (Table 1). The actinomycete involved has been classified as the genus *Frankia* of the family Frankiaceae (15, 18).

Actinomycetous root nodules are morphologically and anatomically distinct from legume root nodules. Two main types can be distinguished (Plate 6 a–c). In the *Alnus*-type of root nodule, by dichotomous branching, a coralloid root nodule is formed. The nodule lobes are in the axil of lateral roots whose apical meristem grows slowly or is completely inhibited. In the *Myrica/Casuarina* type of root nodules, the apex of each nodule lobe produces, however, a normal but negative geotropic root. In this way, the root nodule mass gets covered with upward growing rootlets.

As pointed out by Becking (13, 14, 16, 19) the very initial development of these actinomycetous root nodules is different from that of leguminous plants. In actinomycetous symbioses firstly, a pre-nodule is formed, which is apparent only as a slight thickening of the main root. In longitudinal section pre-nodules appear to consist only of a few host-cell layers and a restricted number of cortical cells invaded by the endophyte which invades the cortex from the root hair. From this pre-nodule stage, the true root nodule is formed by a lateral root primordium developed by meristematic proliferation of the pericycle and midcortical cells of the main root. At a later stage of root-nodule development, the infection progresses into the lateral root initiated within the pre-nodule. In the pre-nodule stage, the endophyte is present in its vesicular form forming a cortical parenchyma cell-layer of only three to four cells thick (14, 19; Plate 7), even though these nodules do not fix nitrogen. The pre-nodule stage is accompanied by an extensive curling of the initially straight root hairs at the site of infection (14, 16, 19). This stimulus is probably due to the production of auxin or other growth substances by the endophyte. At the infection site at least one root hair becomes infected with the endophyte. Torrey (79), Callaham and Torrey (34) and Torrey and Callaham (81) stated that only one or very few root-hair infections give rise to pre-nodule formation, which situation is apparently different from that of leguminous root nodules where usually many root hair infections and abortive infections are associated with the root-nodule formation. Penetration is the result of a physiological interaction between the plant and the endophyte, possibly a hormonal one.

A study of the fine structure of the endophyte revealed the presence of hyphae, vesicle structures and spore-like particles mainly in the *Alnus* endophyte as seen by transmission and scanning electron microscopy (25, 17, 18, 19, 20). A scanning electron micrograph of the host cells of *Alnus glutinosa* containing vesicular endophyte structures is shown in Plate 8 and mature spore-like endophytic structures as seen by transmission electron microscopy are presented in Plate 9. Gardner (40) has also studied the fine structure of other non-leguminous root nodules such as those of *Hippophae rhamnoides*, *Myrica gale* and *M. cerifera*, *Ceanothus velutinus* and *Casuarina cunninghamiana*.

Strand and Laetsch (75, 76) very thoroughly studied the cell and endophytic structures of the root nodules of *Ceanothus integerrimus*. Lalonde and Knowles (50, 51), Lalonde and Devoe (49) and Lalonde *et al.* (52) did detailed cytological studies of the material ensheathing the endophyte in the *Alnus crispa* var. *mollis* root nodules and investigated the origin of the membrane envelopes by using cytological staining techniques combined with transmission electron microscopy and freeze-etching microscopy.

Studies on root initiation and growth of the *Myrica/Casuarina* type of root nodules have been extensively carried out by Torrey and his group at Harvard Forest, Petersham, U.S.A. (79, 32, 34, 63, 80, 81). In the *Myrica/Comptonia* symbiosis, as in the *Alnus* symbiosis root hair infection results in the formation of a pre-nodule. Subsequently, three different stages of root nodule development can be distinguished—first nodule-lobe formation, then a transitional or arrested stage of variable duration, and finally nodule-root-development. As in *Alnus* root nodules, the primary nodule lobe is a lateral root primordium originating endogenously within the pre-nodule and its formation involves pericycle, endodermis and cortical cell derivatives. The nodule lobe develops slowly while the cortical parenchyma cells are invaded by the actinomycetous endophyte. After a dormant period ranging from a few days to several weeks, the nodule-lobe meristem begins to grow and form an elongate nodule root which undergoes slow but continuous growth about 3 to 4 cm length. New nodule lobe primordia are initiated endogenously at the base of the existing nodule lobes, and ultimately form a cluster of nodule roots. Each nodule root has an apical meristem, a reduced root-cap and a modified root structure possessing an elaborate cortical intercellular space system and a reduced central vascular cylinder. The endophyte is restricted to cortical cells of the nodule lobe and is totally absent from tissues of the nodule root.

A probable function of the nodule roots is to facilitate gas diffusion to the nitrogen-fixing endophyte in the nodule lobe, under low oxygen tension. An anatomical study of nodule formation in *Casuarina* showed essentially the same sequence of events as the one described earlier for *Myrica/Comptonia* root nodules. Anatomical analysis of nodule formation in *Casuarina* showed that the large number of nodule lobes formed are due to repeated endogenous lateral root initiations, one placed upon another in a complexly branched and truncated root system. The endophyte-infected cortical tissues derived from successive root primordia form the swollen nodular mass (79).

Rhizobium symbioses

The internal structure of the root nodule of *Parasponia rugosa* (formerly classified as *Trema aspera* or *T. cannabina* var. *scabra*) has been described by Trinick (82, 83) and Trinick and Galbraith (85). Initially the nodule structure was described to have some resemblance to that of leguminous plants, but later more affinities with non-leguminous root nodules were found. In a

Plate 1. *Datisca cannabina* L. (Datiscaceae, Cucurbitales). (a) Young plant, $\times 0.5$; (b) Twig of mature plant, $\times 0.35$; (c) Root nodules, $\times 1.2$. (Root nodules probably abnormal because they showed no acetylene reduction activity.)

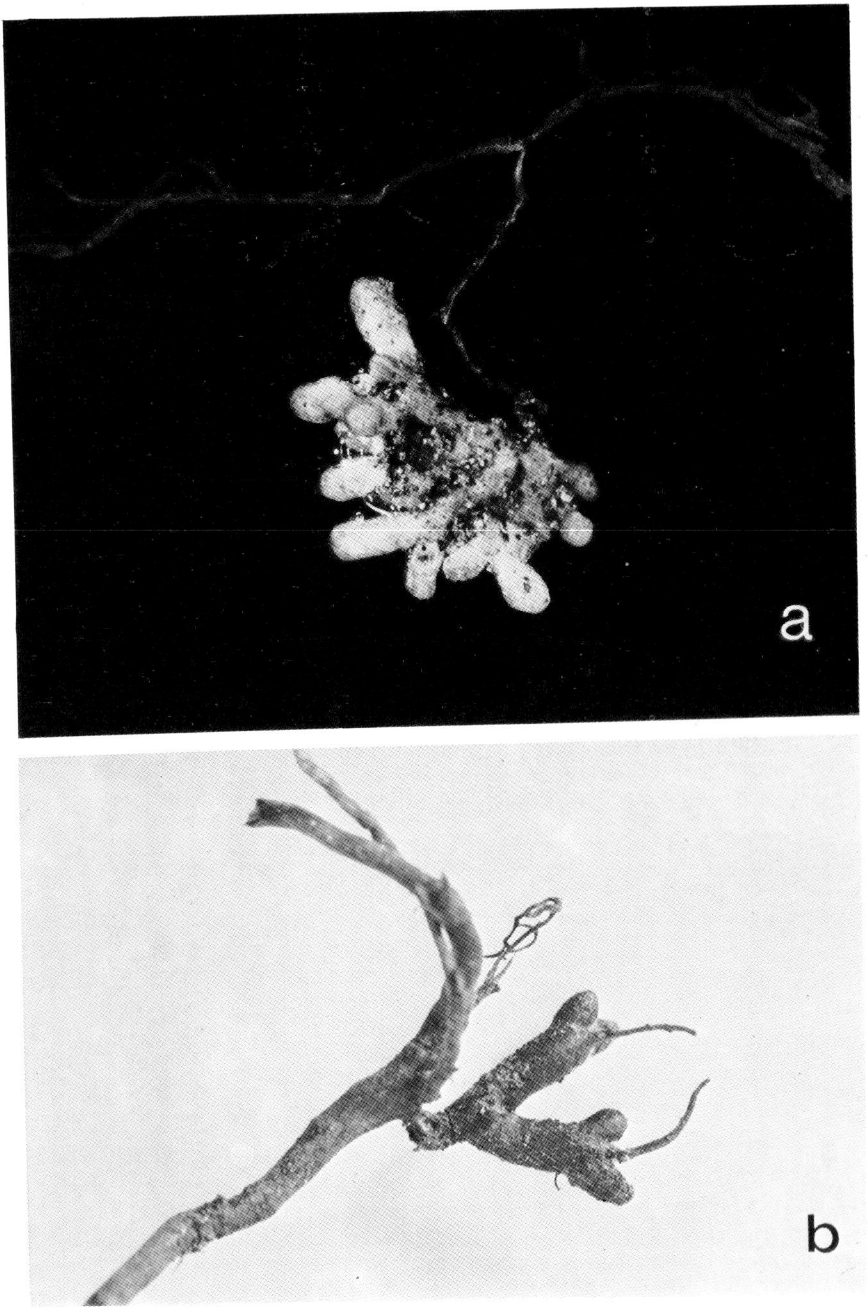

Plate 2. Actinomycetous root nodules of some Rosaceae species. (a) *Dryas drummondii* Richardson, × 4.5; (b) *Rubus ellipticus* J.E. Smith, × 3.0.

Plate 3. *Colletia paradoxa* (Spreng.) Escalante (syn. *C. cruciata* Gillies ex Hook.). (a) Twig of a mature plant showing phyllocladioid stems with spines and flowers, × 1.2; (b) Root nodules, × 1.5.

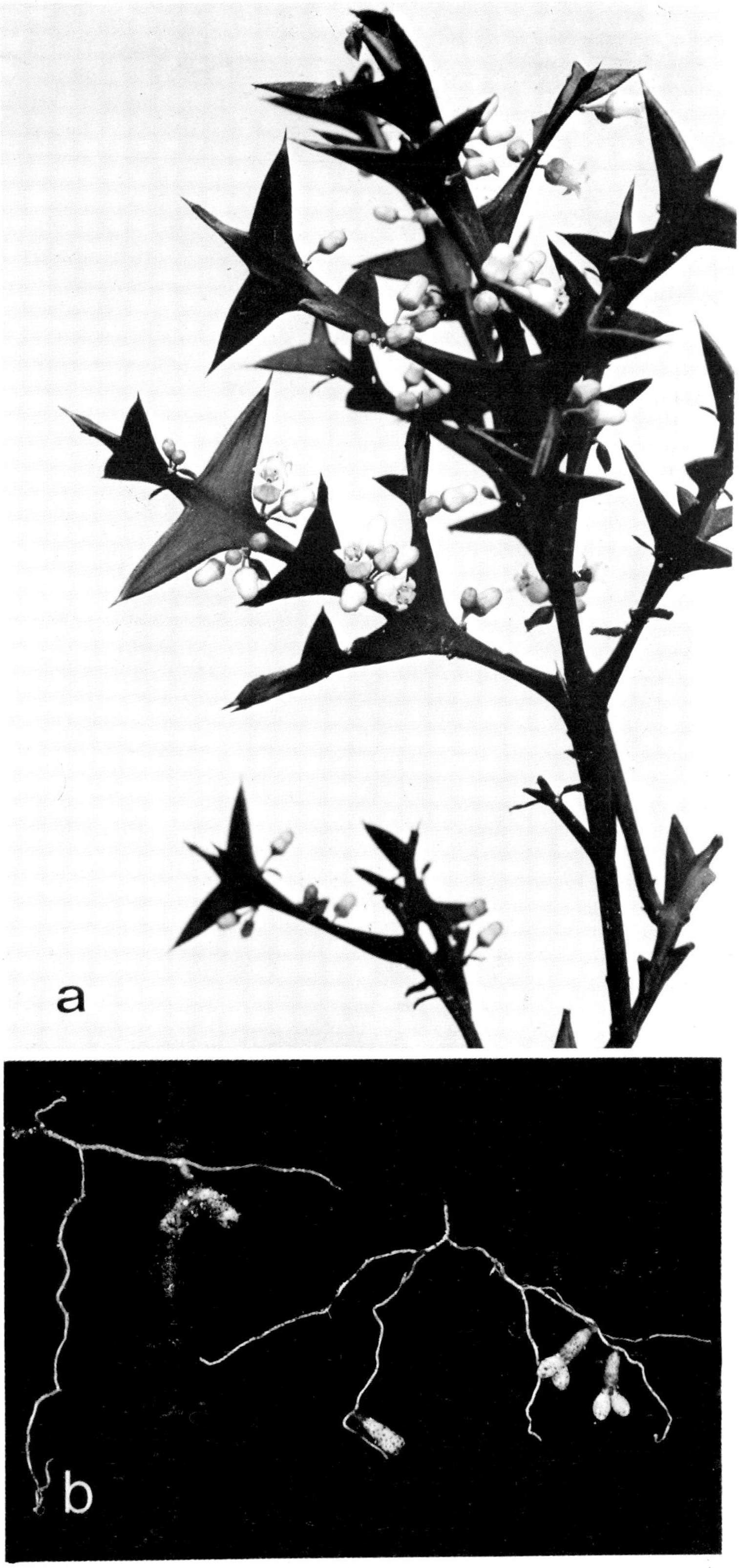
a
b

Plate 4. Rhizobial root nodules of *Parasponia* (Ulmaceae, Urticales). (a) Young nodulated plant of *Parasponia parviflora* Miq., × 0.63; (b) and (c) Root nodules; × 3.2 and × 5.0, respectively.

Plate 5. *Zygophyllum coccineum* L. (Zygophyllaceae, Malpighiales). (a) Young plant, × 1.3; (b) Root nodules, × 1.4.

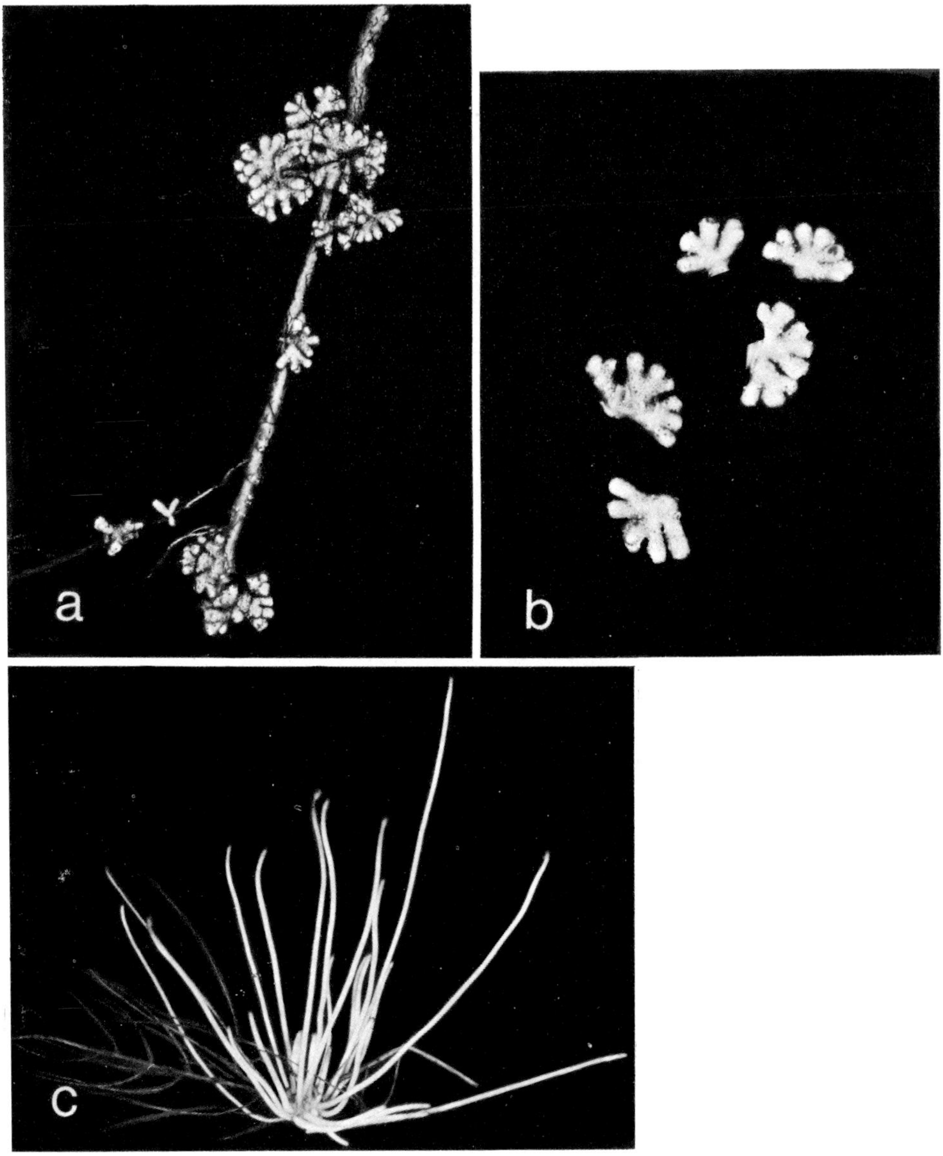

Plate 6. *Alnus* and *Myrica/Casuarina* type of root nodules. (a) Coralloid root nodule of *Alnus glutinosa* L. (Gaertn.), × 0.9; (b) Detached and divided root nodules of *Alnus glutinosa* L. (Gaertn.), showing dichotomous branching of nodule lobes, × 1.5; (c) *Casuarina equisetifolia* L. root nodules showing the feature that the apex of each nodule lobe gives rise to a negative geotrophic root, × 1.5.

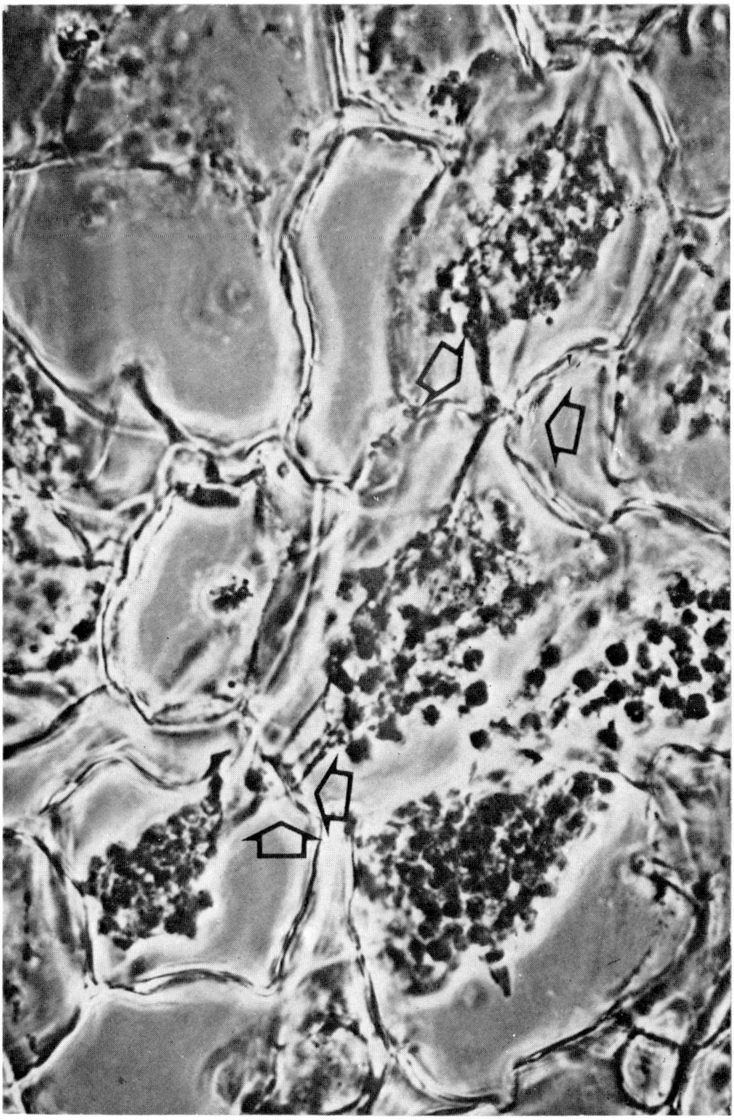

Plate 7. Paraffin-wax section through a root nodule of *Alnus glutinosa* L. (Gaertn.) showing the perforation of the host-cell wall by the hyphae penetrating the successive host cells. In the centre of the host cell vesicular endophytic structures are visible. × 970.

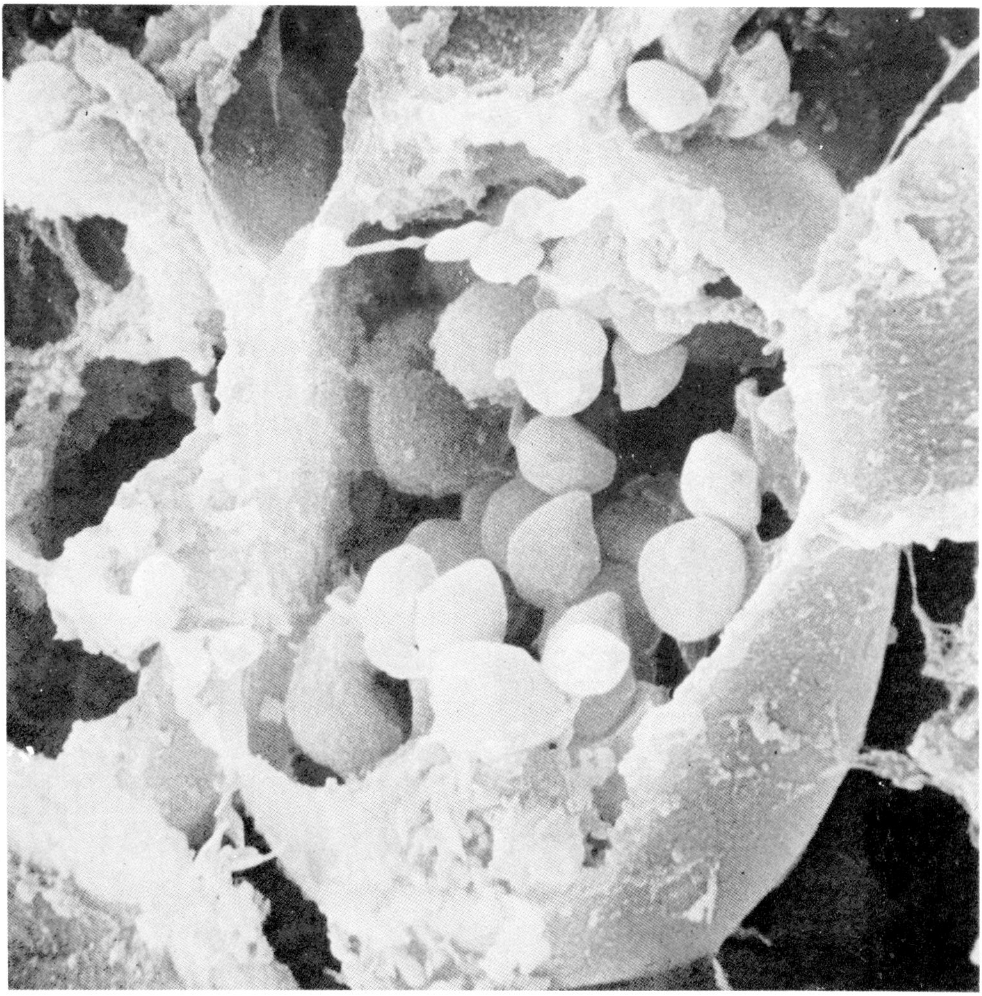

Plate 8. Scanning electron micrograph of an opened host cell of *Alnus glutinosa* L. (Gaertn.) showing the vesicular structures of the endophyte at the tip of the hyphae. × 5250.

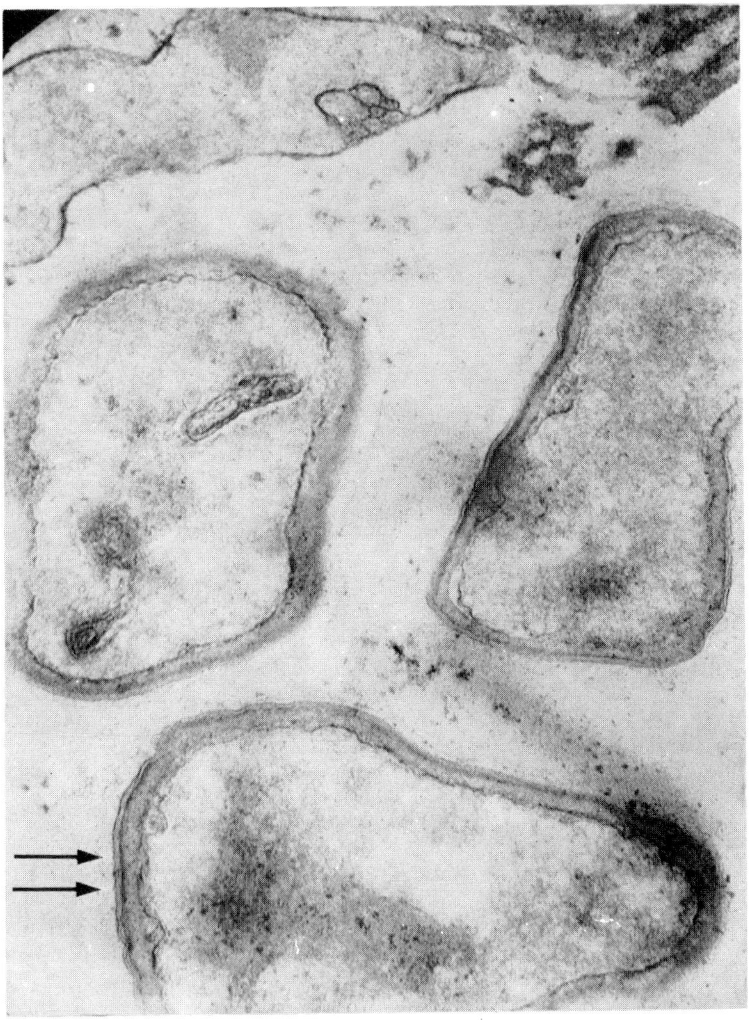

Plate 9. Transmission electron micrograph of mature spores of the endophyte within the host cell. Note the dead host cell devoid of cytoplasmic content. The very thick endophyte cell wall of the spore is surrounded by a double-layered cytoplasmic membrane (see arrows) of probably host origin. × 68,000.

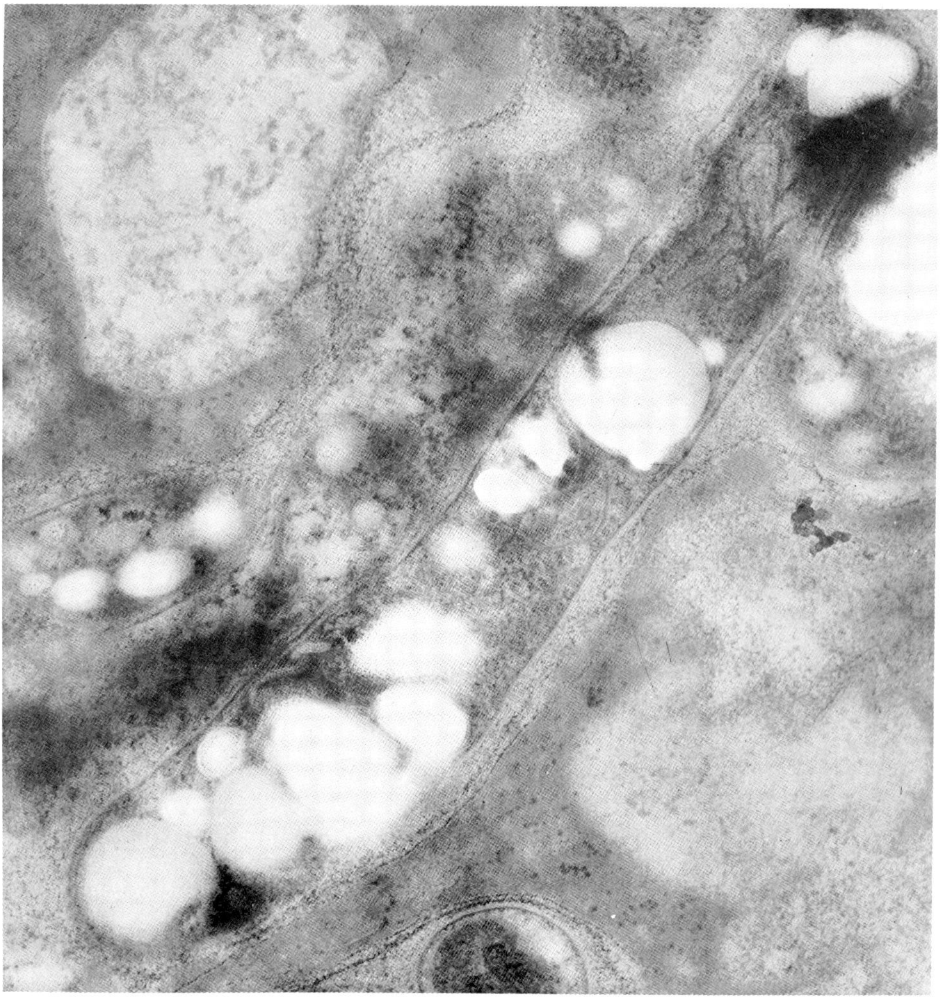

Plate 10. Transmission electron micrograph of a free *Rhizobium* bacteroid in the cytoplasm of a *Parasponia parviflora* Miq. root nodule cell. × 70,000. The reserve material inside the bacterium is most probably poly- ß -hydroxybutyrate.

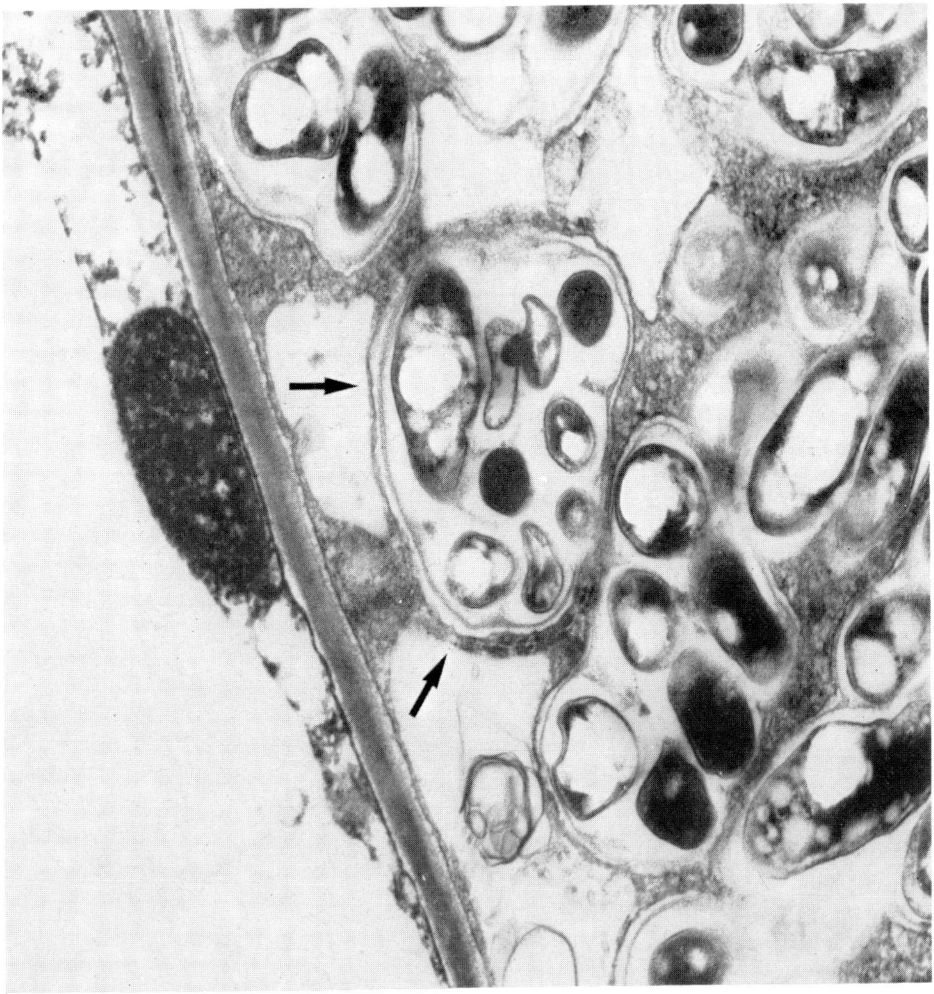

Plate 11. Transmission electron micrograph. Transverse section through the membrane envelope enclosing *Rhizobium* cells in a *Parasponia parviflora* Miq. host cell. Note the living cytoplasm of the host cell and the double-layered cytoplasmic structure of the envelopes (see arrows). × 35,000.

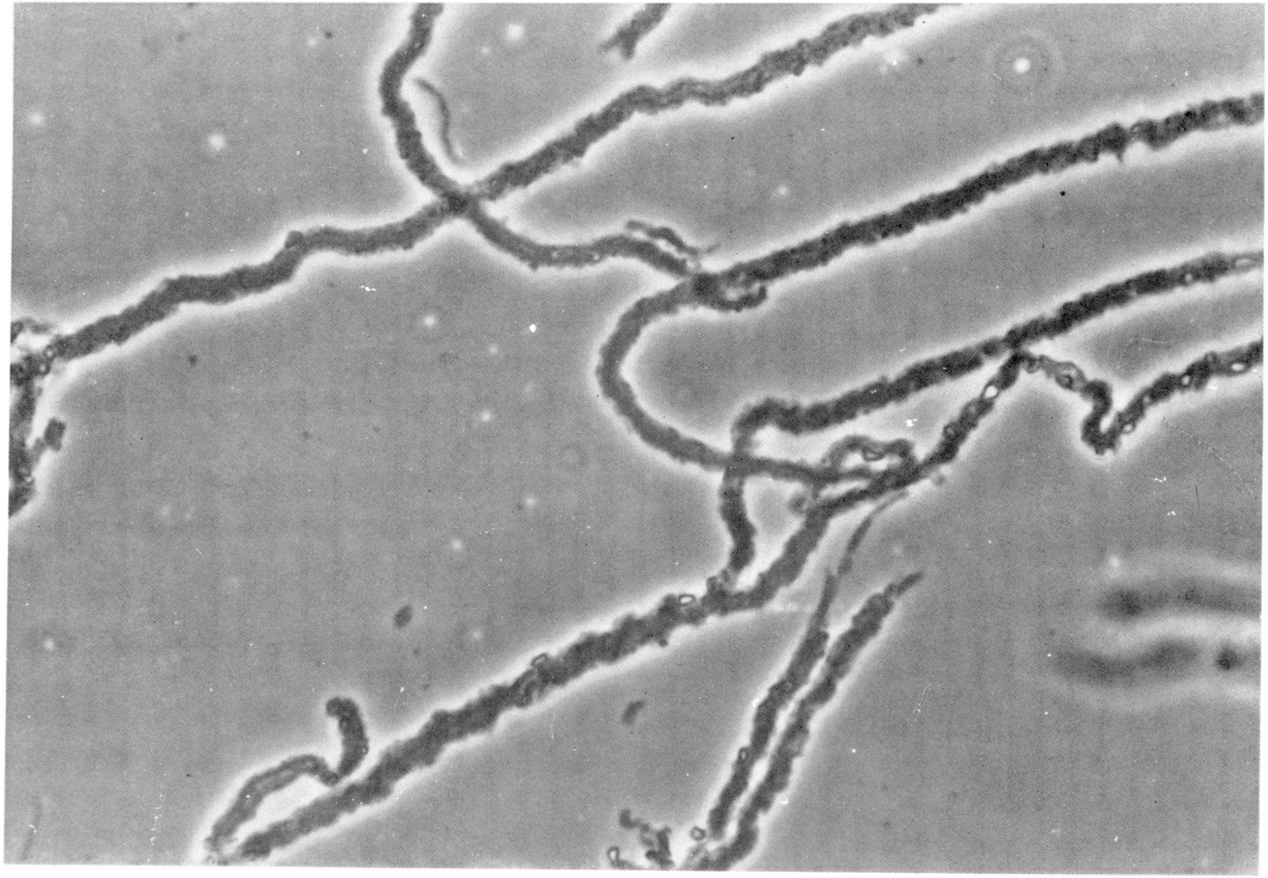

Plate 12. Thread-like structures in *Parasponia parviflora* Miq. host cells enclosing *Rhizobium* bacteria. Note the variable thickness of the threads. Phase contrast micrograph ×1650.

transection, the root nodules show a central vascular bundle surrounded by a horse-shoe-shaped bacteroid host cell tissue. The central vascular stele is a typical feature of non-leguminous root nodules. As in other non-leguminous root-nodules, *Parasponia* nodules possess an apical meristematic zone which provides scope for the continuous elongation of the nodules. The infection thread persists and penetrates the newly formed host cells immediately behind the apical meristem. The thread can pass directly through the host-cell wall to infect other cells, or it can enter young cells from the intercellular spaces. As in legumes and other non-legumes nodulated by *Frankia* spp., the host cells of *Parasponia* invaded by the *Rhizobium* endophyte also become hypertrophied. However, unlike the infection threads of leguminous root nodules, the *Rhizobium* cells in *Parasponia* are very rarely released from the infection threads. Usually the infection thread continues to grow until it has filled completely the host cell. The wall of the infection thread was observed to be continuous with the host-cell wall and therefore probably contains cellulose and pectins. The persistence of infection threads within host cells appears to be a peculiar feature of *Parasponia* root nodules and because more than two–thirds of the infected host cells of the nodule had these structures, it was presumed that nitrogen fixation occurred in the bacteria within the thread-like structures. Hence, bacteria within the infection threads might have properties similar to bacteroids enclosed in a membrane envelope of host cells of leguminous root nodules.

A study of the internal structure of *Parasponia andersonii* root nodules showed that the infection-thread structures in nodules of this plant were even more dominant than those of the preceding species, since rhizobia were never found to be released from the infection threads into the host cytoplasm (84). The persistence of the thread structures was confirmed by examining serial sections of host cells containing little and very extensive infection. The absence of "released" rhizobia in sections of root nodules of *Parasponia* when seen in the light microscope may be due to either the improved resolution obtained in the Araldite-embedding material compared to wax used in the study (85) or to an intrinsic difference between the two species—*P. andersonii* and *P. rugosa.* Transmission electron micrographs showed that the walls of the threads varied greatly in thickness and often thread structures were observed without rigid wall and enclosed only by a plasmic membrane. In view of the observed differences between the two *Parasponia* species, it was surmised that the two species represent probably different evolutionary stages of the *Rhizobium* infection (84).

In a preliminary study of *Parasponia parviflora* root nodules from Java, Indonesia, Becking (23) showed that they were of the coralloid type, but different from the *Alnus* root-nodules type of *Frankia* symbioses, because nodule branching is more irregular and not strictly dichotomous. Moreover, the larger root nodules of the *Alnus*-type are usually sessile on the thicker roots, whereas the *Parasponia* root nodules showed thin bases (Plate 4, b and c). Bacteroids are released into the host cytoplasm, but when free these structures (Plate 10) are not distorted in shape as observed usually in bacteroids in nodules of leguminous plants. Moreover, a large proportion of the rhizobial cells are found in thin-walled membrane envelopes consisting of a double layered plasmic membrane (Plate 11). These membrane structures are long and thin and the longitudinal axis of these envelopes is often orientated in the same direction as the host cell wall. In extreme cases, thread-like structures may develop (Plate 12). In transection, generally about 5–25 bacteria can be counted in these elongated membrane structures (23) although sometimes only one bacterial cell is visible. The thread-like structures are usually very uneven and variable in size and it is likely that among them the thicker ones are either composite structures of many individual threads or they become so thick by some internal growth process (Plate 12).

Isolation of the endophyte

Actinomycetous symbioses

Most of the earlier literature concerning isolation of the endophyte has been summarized by Becking (14, 16, 19, 20, 21). One of the noteworthy attempt is that of Pommer (64), in *Alnus glutinosa* where the isolate was in general morphologically very similar to the symbiotic endophyte observed *in vivo*; moreover, the isolated endophyte could induce root nodulation in aseptically grown *Alnus glutinosa* seedling within four to six weeks. The isolate exhibited slow growth under normal atmospheric conditions on agar plates containing a simple glucose-asparagin agar supplied with *Alnus* root-nodules extract. Unfortunately Pommer's isolate was subsequently lost (Pommer unpublished) and the result could not be reproduced by others as cited by Quispel (65) and this unpublished work of the author.

Becking (12, 14, 16, 19) obtained bacteria-free endophyte from nodules of *Alnus glutinosa* by culturing tissue of root-nodule using a complex nutrient medium used for the *in vitro* cultivation of plant tissues. The actinomycetous endophyte, however, did not grow on the medium without the plant tissue. The endophyte invaded the newly formed callus tissue only rather very slowly and because of this growth characteristic, the endophyte was often lost in the subsequent callus explants. The isolated endophyte nodulated *Alnus glutinosa* seedlings in axenic cultures but the nodules proved to be ineffective as revealed by poor nitrogenase activity.

Lalonde *et al.* (53) reported the isolation of the endophyte from *Alnus crispa* var. *mollis* on a rather complex tissue culture medium (42). The isolate, however, lost resemblance to the nodular endophyte *in vivo* after several subcultures and also failed to nodulate aseptically grown seedlings of *Alnus crispa* var. *mollis*. Later, extensive immuno-labelling tests were done to demonstrate that the isolate was in reality an ineffective strain of the nodular endophyte but such immuno-labelling reactions could also be characteristic of other actinomycete species as already shown by the same authors. Therefore, a definite and unambiguous proof for establishing the identity of the isolate of Lalonde *et al.* (53) endophyte cannot be taken for granted. On the other hand, it is rather unlikely that it is the true endophyte.

Using micro-dissection and enzyme degradation techniques, Callaham *et al.* (33) could isolate a slow growing actinomycete from the root nodules of *Comptonia peregrina* in a medium containing sucrose, mineral salts, amino acids, and vitamins or in a yeast mannitol/sucrose medium supplemented with growth substances (thiamine-HCl, nicotinic acid, pyridoxin-HCl, etc.) in standing liquid medium in petri dishes. The isolate grew very slowly in axenic culture in this medium under these conditions.

It was, however, observed that somewhat better growth could be obtained in yeast extract medium (Difco Yeast Extract, 0.5% v/v) in standing test tube cultures with 5–6 cm deep medium. The isolate appeared to be microaerophilic because the growth was better at the bottom of the tubes in liquid medium but could not grow under complete anaerobiosis. The isolate could induce root nodulation in *Comptonia* seedlings in sand culture in the green house. The endophyte could be re-isolated from the nodules and thus according to the authors, could fulfill the Koch's postulates. It is, however, a pity that this experiment was not conducted in monoxenic culture with axenic *Comptonia* seedlings. Moreover, the fulfilment of Koch's postulates as stated by the authors is disputable because a small number of nodules were also induced in the uninoculated controls probably due to contamination. Again, the independent evaluation tests performed by Lalonde (48) also suffer from the above-mentioned challengeable immuno-labelling reactions and therefore may be regarded as inconclusive. Although surface sterilized *Comptonia* seeds were used at the start of the experiment, no tests were described by the authors to verify the extent to which monoxenic conditions of the cultures were maintained.

The endophyte of actinomycetous root nodules have been described by Becking (15) to belong to the genus *Frankia* of the family Frankiaceae (15, 18). The taxonomic criteria for this classification included the structure and dimensions of the hyphae, spherical vesicular bodies or club-shaped structures at the tip of the hyphae and the presence or absence of spore-like endophytic cells, sometimes also called bacteria-like cells, bacteroids or "granulae". Moreover, like the speciation in the genus *Rhizobium*, cross-inoculation grouping was also attempted to define species in *Frankia*. Attempts have also been made to determine cell-wall constituents of the endophyte, because the classification of actinomycetes is based on major cell-wall types (55).

Originally, the endophyte was described as an obligate symbiont (15) because of the unsuccessful attempt to isolate the root nodule organism.

Later, it was, however, observed that there was considerable growth of the endophyte as hyphae in monoxenic cultures in the rhizosphere of *Alnus glutinosa* seedlings (19) and that there could be a free stage or viable form of the endophyte in soil during which the organism could induce root nodulation in axenic *Alnus* seedlings. For this reason, the term "obligate" with regard to the *Alnus* association has been purposefully omitted for this group in the *Bergery's Manual of Determinative Bacteriology*. In this manual it is clearly stated that there is a free stage of the endophyte in soil and the endophyte is probably micro-aerophilic (18). The micro-aerophilic nature of the endophyte was indeed proved to be true when the organism was finally isolated (33).

The vesicular or club-shaped structures at the tip of the hyphae in *Frankia* species are probably degenerated sporangia modified under the influence of the host cytoplasm and probably associated with nitrogen fixation, since nodule slices containing the endophyte predominantly in vesicular form showed more nitrogen fixation (acetylene reduction) than those containing the endophyte in the form of hyphae or spore-like structures and few vesicles (18, 22). Moreover cell-wall analyses for diaminopimilic acid (DAP) by paper chromatography and chemical methods revealed the presence of meso-DAP (but not LL-DAP), arabinose, galactose and glycine in the *Alnus* endophyte. The cell-wall preparations were obtained by a filtration technique of nodule tissue homogenates using filter cloth of variable mesh. By this technique, isolated hyphal and vesicular endophyte clusters were sedimented on a cloth of 10 mm mesh. These endophyte clusters were carefully freed from adhering host-cell components by repeated washings with phosphate buffer. The endophytic cells were subsequently ruptured by ultrasonic treatment and the sub-cellular fragments were thoroughly washed with buffer to remove cytoplasmic constituents prior to hydrolysis and cell-wall analyses (18, 19, 20). Although we could find DAP in the *Alnus* endophyte, we did not detect it in the *Casuarina* endophyte and similarly Quispel (65) reported its absence in the *Myrica gale* endophyte.

Like *Rhizobium* species, the cross inoculation grouping of the *Frankia* species has been criticized, since a number of cases of promiscuity were observed between representatives of different groups. Rodriguez-Barrueco and Bond (67) observed that inoculum (crushed nodules) of *Alnus glutinosa* induced root nodulation in *Myrica gale*, but that the reciprocal combination of *Myrica gale* inoculum and *Alnus glutinosa* plants did not result in nodulation. Recently Miguel *et al.* (60) notably succeeded in nodulating *Alnus glutinosa* by a *Myrica gale* inoculum but also observed that a *Hippophae rhammoides* inoculum induced root nodulation in *Myrica gale, Coriaria myrtifolia* and *Elaeagnus angustifolia* although it did not nodulate *Alnus glutinosa.* Most of these experiments were, however, performed in water cultures under non-axenic conditions using an inoculum from water or field-grown plants. Therefore, unambiguous proof for these interchanges cannot be taken for granted because the involvement of other actinomycetes cannot be excluded.

Furthermore, it is necessary to distinguish normal and abnormal (artificial) associations between symbionts among non-leguminous plants possessing root nodulation habit. For instance, when *Alnus glutinosa* endophyte was artificially inoculated to a *Myrica gale* host, it resulted in the formation of root nodules possessing an endophyte morphologically different from the endophyte resulting from a normal association between *M. gale* and its true symbiont and greatly resembled the endophyte of *Alnus glutinosa* formed by its true symbiont (Rodriguez-Barrueco, personal communication). Likewise, the nitrogen-fixing potentiality was greatly reduced in the abnormal symbioses in comparison with its normal one. Besides this fact, the normal endophyte of *Alnus* and *Myrica* cannot be identical because the *Alnus* endophyte contains DAP and the *Myrica* endophyte lacks it and as is well known the presence or absence of DAP in the cell walls is a diagnostic feature in the taxonomy of actinomycetes (56, 54). Undoubtedly, analyses of the cell-wall components of the free-living acti-nomycete isolates from root nodules of plants grown in axenic culture must receive increased attention (9, 33), and studies in that direction will give an insight into the interrelations among *Frankia* species.

Rhizobium symbioses

The isolation of the rhizobial endophyte from non-leguminous root nodules (*Parasponia* spp.) does not pose any problem. Homogenates of surface-sterilized root nodules which have been earlier washed with ethanol 70 per cent (v/v), and subsequently with hydrogen peroxide 6 per cent (v/v) when plated on agar medium containing mannitol and yeast extract produce good rhizobial growth. This medium has the following composition: distilled water 1000 ml; yeast extract, 1.0 g; mannitol, 10 g; K_2HPO_4, 0.5 g; $MgSO_4 \cdot 7\ H_2O$, 0.5 g; NaCl, 0.1 g; $CaCO_3$, 3.0 g; trace-element solution 1 ml; agar, 12 g (pH 6.8–7). On these plates, the *Rhizobium* strains often appear as the sole organism and in other cases (with less severe surface disinfection) it appears at least as the dominant organism.

Biochemical progress

Nitrogenase activity

Nitrogen-fixing activity in nodule breis from non-leguminous nodulating plants was first reported using $^{15}N_2$ and C_2H_2 reduction methods by Sloger and Silver (73, 74) and Sloger (72). The nitrogenase activity of the nodule breis was much lower than that of intact nodules. The presence of a reducing agent (sodium dithionite) and the absence of O_2 was essential during homogenization, but O_2 was required, presumably for the production of ATP during exposure of the homogenate.

Reasonable nitrogenase activity in nodule homogenates was reported by Van Straten *et al.* (86) and Akkermans *et al.* (4) as revealed by acetylene reduction assay, if 0.3 M sucrose and 100 mM dithionite ($Na_2S_2O_4$) were present during anaerobic homogenization. The addition of Na-dithionite was essential because the nodule material contains large amounts of phenolic compounds which after oxidation inhibit nitrogenase activity. The reaction was found to be ATP-dependent, but strange enough the addition of an ATP-generating system [creatine phosphate (Cr$\sim$P)/creatine phosphokinase or phospho (enol) pyruvate (PEP)/pyruvate kinase (PK)] decreased the nitrogenase activity during short-term experiments. Apparently, Cr$\sim$P and PEP inhibited acetylene reduction, but the nature of this inhibition could not be resolved. The observed nitrogenase activity was ascribed mainly to the vesicle clusters because 60 per cent of the homogenate activity was recovered by the resuspended residue left behind from a 10 μm filter. The filtrate passing 10 μm consisting of hyphal fragments and disrupted vesicle clusters accounted for only 29 per cent of the total activity of the homogenate. The latter fraction of the activity was all particle bound. The remaining 11 per cent of the activity was lost during the experimental procedure. These findings suggest again that the sites of nitrogen fixation are the vesicles confirming the earlier results which showed a strong tetrazolium reducing activity of the vesicles (1) and those which demonstrated that tissue slices of nodules containing host cells filled with vesicles had higher nitrogenase (C_2H_2 reduction) activity than nodule slices containing the other forms of the endophyte (18, 22).

Nitrogenase activity in cell-free preparations of actinomycetous root nodules of *Alnus glutinosa* was obtained by Benson *et al.* (26) who disrupted nodules in liquid nitrogen to release the actinomycetal endophyte and suspended it in 100 mM potassium phosphate buffer (pH 7.4) containing 10 mM dithionite. After centrifugation for four times in fresh phosphate buffer the homogenate was washed in tris-buffer (20mM tris-HCl, pH 7.2) with 20 mM ascorbate and 2 mM dithionite to free the homogenate from most of the inhibitory phenolic compounds released during homogenization. Nitrogenase (C_2H_2 reduction) activity was stable even when degassed buffer alone was used in the later washings, indicating that dithionite is not needed in late washing to retain activity. Cell-free preparations were made by mixing the sedimented material with solid polyvinylpolypyrrolidone (PVP) and sonication for 1.5–2 min. However, prolonged sonication rapidly inactivated the C_2H_2-reducing activity. The sonicated material was centrifuged down and the supernatant containing the cell-free extract had an initial activity of 0.5–1 μ mole of C_2H_2 reduced per g of nodules per hour, or about half the activity of the particulate homogenates. The cell-free enzyme is somewhat unstable after isolation as shown by the non-linear time course in 60 minutes. The cell-free nitrogenase required ATP, Mg_2^+ and Na_2SO_4 for acetylene reduction.

Hydrogenase activity

Nitrogenase-dependent hydrogen evolution has been observed in detached legume and non-legume root nodules and in reaction mixture containing cell-free nitrogenase.

An evaluation of the magnitude of this hydrogen evolution seen as energy loss in terms of the efficiency of electron transfer to dinitrogen, via nitrogenase, indicated that hydrogen production may severely reduce nitrogen fixation in leguminous plants. For most leguminous plants it was shown that 40–60 per cent of the energy of the electron flow was lost through hydrogen evolution, but in some tropical legumes (i.e. *Vigna sinensis* and *Phaseolus aureus*) and in some non-leguminous plants (i.e. *Alnus rubra, Purshia tridentata, Elaeagnus angustifolia, Ceanothus velutinus* and *Myrica californica*), the relative efficiency was much higher being 70–90 per cent, because they have evolved a mechanism for minimizing net hydrogen production by recycling hydrogen (70).

Benson *et al.* (26) studied hydrogen production in crude homogenates and cell-free extracts of *Alnus glutinosa* root nodules. The H_2 evolution was ATP dependent and occurred at a rate comparable to the rate of C_2H_2 reduction. Since intact actinomycetous root nodules evolve hydrogenase, there must be a highly active or tightly coupled uptake hydrogenase in the system. As explained above, the relation between the uptake hydrogenase and ATP dependent H_2 production by nitrogenase is of special interest, because part of the ATP expended for H_2 production can be recovered by coupled hydrogenase-catalyzed re-oxidation of H_2 thereby increasing the efficiency of the system.

Other researches

Wheeler (88) showed in translocation studies with ^{14}C-labelled photosynthate to the root nodules of first-year *Alnus glutinosa* plants grown under natural illumination and constant temperature, that a maximum influx of new photosynthate occurred at the time of the midday peak in nitrogen fixation. Analysis of fluctuations in the levels of the main free sugars present in the nodules suggested that a substantial part of the nodule carbohydrate is unavailable for nitrogen fixation and that maximal rates of fixation are only attained when new photosynthates are entering the nodules. Similar diurnal changes in the nitrogen fixation rate were observed by Bond and Mackintosh (31) in *Casuarina cunninghamiana* using the N-15 method.

Studies of hormones in non-leguminous root nodules have revealed the presence of cytokinin-like substances in *Alnus-glutinosa* and *Myrica gale* root nodules (66, 28). Analyses of the cytokinin extracts of different plant parts showed that a zeatin-9-glucoside-like substance was the prominent cytokinin in nodules and leaves of *Alnus*, but that a zeatin riboside-like substance was the major cytokinin present in the roots and root pressure sap (44). Cytokinin levels determined by bioassay were also observed in other non-leguminous root nodules such as *Purshia tridentata, Myrica gale, Hippophae rhamnoides* and *Colletia paradoxa* (45). In addition, gibberellin-like (GA-like) substances were detected in various parts of young nodulated non-leguminous plants as estimated by the lettuce hypototyl bioassay (46).

References

1. Akkermans, A.D.L. Nitrogen fixation and nodulation of *Alnus* and *Hippophae* under natural conditions, Ph.D. thesis, University of Leiden, The Netherlands (1971).
2. Akkermans, A.D.L., Abdulkadir, S. and Trinick, M.J. N_2-fixing root nodules in Elmaceae: *Parasponia* or (and) *Trema* spp.? *Plant and Soil, 49,* 711–715 (1978a).
3. Akkermans, A.D.L., Abdulkadir, S. and Trinick, M.J. Nitrogen-fixing root nodules in Elmaceae, *Nature, London, 274,* 190 (1978b).
4. Akkermans, A.D.L., Straten, J. van and Roelofsen, W. Nitrogenase activity of nodule homogenates of *Alnus glutinosa*: A comparison with the *Rhizobium*-pea system, pp. 591–603 in *Recent Developments in Nitrogen Fixation* (Edited by N. Newton, J.R. Postgate and C. Rodriguez-Barrueco), *Proceedings of the Second International Symposium on Nitrogen Fixation*, Salamanca, Academic Press, London (1977).
5. Allen, E.K., Allen, O.N. and Klebesadel, L.J. An insight into symbiotic nitrogen-fixation plant associations in Alaska, *Proceedings of the 14th Alaskan Science Conference*, Anchorage, Alaska, Publ. Alaska Division, American Association for Advancement of Science, 54–63 (1964).
6. Athar, M. and Mahmood, A. Root nodules in some members of Zygophyllaceae growing at Karachi University Campus, *Pakistan Journal of Botany, 4,* 209–210 (1972).
7. Backer, C.A. and Bakhuizen van den Brink, R.C. Flora of Java (Spermatophytes only), Vol. 2, Angiospermae, Families, pp. 111–160, Noordhoff, Groningen, The Netherlands (1965).
8. Backer, C.A., and van Slooten, D.F. *Geïllustreerd Handboek der Javaansche Theeonkruiden en hunne beteekenis voor de cultuur*, Algemeen Proefstation voor Thee. Drukkerijen Ruygrok & Co., Batavia, Indonesia (1924).
9. Baker, D. and Torrey, J.G. The isolation and cultivation of actinomycetous root nodule endophyte, pp. 38–56 in *Symbiotic Nitrogen Fixation in the Management of Temperate Forest* (Edited by J.C. Gordon, C.T. Wheeler and D.A. Perry), Forest Research Laboratory, Oregon State University, Corvallis, U.S.A. (1979).
10. Becker, B., Lechevalier, M.P., Gordon, R.E. and Lechevalier, H.A. Rapid differentiation between *Nocardia* and *Streptomyces* by paper chromatography of whole-cell hydrolysates, *Applied Microbiology, 12,* 421–423 (1964).

11. Becker, B., Lechevalier, M.P. and Lechevalier, H.A. Chemical composition of cell-wall preparations from strains of various form genera of aerobic actinomycetes, *Applied Microbiology, 13*, 236–243 (1965).
12. Becking, J.H. *In vitro* cultivation of Alder root-nodule tissue containing the endophyte, *Nature, London, 207*, 885–887 (1965).
13. Becking, J.H. Interactions nutritionelles plantes-actinomycetes. Rapport Général, *Annales de l'Institut Pasteut Suppl. 111*, 211–246 (1966).
14. Becking, J.H. Nitrogen fixation by non-leguminous plants, Symposium on Nitrogen in Soil, Groningen, Stikstof, *Dutch Nitrogenous Fertilizer Review, 12*, 47–74 (1968).
15. Becking, J.H. Frankiaceae fam. nov. (Actinomycetales) with one new combination and six new species of the genus *Frankia* Brunchorst 1886, 174, *International Journal of Systematic Bacteriology, 20*, 201–220 (1970a).
16. Becking, J.H. Plant-endophyte symbiosis in non-leguminous plants. 2nd Conference on Global Impact of Applied Microbiology, Addis Ababa, Ethiopia, *Plant and Soil, 32*, 611–654 (1970b).
17. Becking, J.H. Biological fixation of atmospheric nitrogen: Other systems. 4th Conference on Global Impact of Applied Microbiology, Sao Paulo, Brazil, pp. 421–460, *Publication of Brazilian Microbiological Society* (1973).
18. Becking, J.H. Key of Family Frankiaceae and Genus *Frankia*, pp. 702–706, 871–872 in *Bergey's Manual of Determinative Bacteriology* (8th edition) (Edited by R.E. Buchanan and N.E. Gibbons), Williams & Wilkins, Baltimore, U.S.A. (1974).
19. Becking, J.H. Root nodules in non-legumes, pp. 507–566 in *The Development and Function of Roots*, Third Cabot Symposium, Harvard University, Massachusetts, U.S.A. (Edited by J.G. Torrey and D.T. Clarkson), Academic Press, London (1975).
20. Becking, J.H. Actinomycete symbioses in non-legumes, pp. 581–591 in *Proceedings of the First International Symposium on Nitrogen Fixation*, Pullman, Washington (Edited by W.E. Newton and C.J. Nyman), Washington State University Press, U.S.A. (1976).
21. Becking, J.H. Nitrogen-fixing associations in higher plants other than legumes, pp. 185–275 in *Dinitrogen (N_2) Fixation, Vol. 2*, Chapter 6 (Edited by R.W.F. Hardy and W.S. Silver), John Wiley & Sons, Inc., New York, U.S.A. (1977a).
22. Becking, J.H. Endophyte and association establishment in non-leguminous nitrogen-fixing plant. 2nd International Symposium on N_2 fixation, Salamanca, Spain, pp. 551–567 in *Recent Developments in Nitrogen Fixation* (Edited by W. Newton, J.R. Postgate and C. Rodriguez-Barrueco), Academic Press, London (1977b).
23. Becking, J.H. Root nodule symbioses between *Rhizobium* and *Parasponia* (Ulmaceae), *Plant and Soil, 51*, 289–296 (1979a).

24. Becking, J.H. Nitrogen-fixation by *Rubus ellipticus* J.E. Smith, *Plant and Soil, 53*, 541–545 (1979b).
25. Becking, J.H., De Boer, W.F. and Houwing, A.L. Electron microscopy of the endophyte of *Alnus glutinosa, Antonie van Leeuwenhoek, Journal of Microbiology and Serology, 30*, 343–376 (1964).
26. Benson, D.R., Arp, D.P. and Burris, R.H. Cell-free nitrogenase and hydrogenase from actinorhizal root nodules, *Science, 205*, 688–689 (1979).
27. *Bergey's Manual of Determinative Bacteriology* (8th edition), (Edited by R.E. Buchanan and N.E. Gibbons), Williams & Wilkins Baltimore, U.S.A.
28. Bermudez de Castro, F., Canizo, A., Costa, A., Miguel, C. and Rodriguez-Barrueco, C. Cytokinins and nodulation of the non-legumes *Alnus glutinosa* and *Myrica gale*. 2nd International Symposium on Nitrogen Fixation, Salamanca, Spain, pp. 539–550 in *Recent Developments in Nitrogen Fixation* (Edited by W. Newton, J.R. Postgate and C. Rodriguez-Barrueco), Academic Press, London (1977).
29. Bond, G. The results of the IBP survey of root nodule formation in non-leguminous angiosperms, pp. 443–474 in *Symbiotic Nitrogen Fixation in Plants* (Edited by P.S. Nutman), Cambridge University Press, Cambridge, England (1976a).
30. Bond, G. Observations on the root nodules of *Purshia tridentata, Proceedings of Royal Society London B, 193*, 127–135 (1976b).
31. Bond, G. and Mackintosh, A.H. Diurnal changes in nitrogen fixation in the root nodules of *Casuarina, Proceedings of Royal Society London B, 192*, 1–12 (1975).
32. Bowes, B., Callaham, D. and Torrey, J.G. Time-lapse photographic observations of morphogenesis in root nodules of *Comptonia peregrina* (Myricaceae), *American Journal of Botany, 64*, 516–525 (1977).
33. Callaham, D., Del Tredici, P. and Torrey, J.G. Isolation and cultivation *in vitro* of the actinomycete causing root nodulation in *Comptonia, Science, 199*, 899–902 (1978).
34. Callaham, D. and Torrey, J.G. Prenodule formation and primary nodule development in roots of *Comptonia* (Myricaceae), *Canadian Journal of Botany, 55*, 2306–2318 (1977).
35. Chaudhary, A.H. The discovery of root nodules in new species of non-leguminous angiosperms from Pakistan and their significance, in *Limitations and Potentials for Biological Nitrogen Fixation in the Tropics* (Edited by J. Dobereiner, R.H. Burris and A. Hollaender), Plenum Press, New York (1978).
36. Chaudhary, A.H. Nitrogen-fixing root nodules in *Datisca cannabina, Plant and Soil, 51*, 163–165 (1979).
37. Clason, E.W. The vegetation of the Upper Badak region of Mount Kelud (East Java), *Bulletin Jardin Botanique, Buitenzorg* (Bogor), *13*, 509–518 (1935).

38. Delwiche, C.C., Zinke, P.J. and Johnson, C.M. Nitrogen fixation by *Ceanothus, Plant Physiology, 40,* 1045–1047 (1965).
39. Focke, W.O. *Die natürlichen Pflanzenfamilien* (Edited by A. Engler and P. Prantl), Teil 111, 3. Abteilung Rosaceae, pp. 1–61, Verlag W. Engelmann, Leipzig (1894).
40. Gardner, I.C. Ultrastructural studies of non-leguminous root nodules, pp. 485–496 in *Symbiotic Nitrogen Fixation in Plants* (Edited by P.S. Nutman), Cambridge University Press, Cambridge, England (1976).
41. Ham, S.P. Discussion in "Groene Bemesting" ("Green Manure"), Handelingen 10de Congress Ned., *Indisch Landbouw Syndicaat 11* (2) *26*, pp. 25–27 (1909).
42. Harvey, A.E. Tissue culture of *Pinus monticola* on a chemically defined medium, *Canadian Journal of Botany, 45,* 1783–1787 (1967).
43. Hellmers, H. and Kelleher, J.M. *Ceanothus leucodermis* and soil nitrogen in southern California mountains, *Forest Science, 5,* 275–278 (1959).
44. Henson, I.E. and Wheeler, C.T. Hormones in plants bearing nitrogen-fixing root nodules: Distribution and seasonal changes in levels of cytokinins in *Alnus glutinosa* L. Gaertn. *Journal of Experimental Botany, 28,* 205–214 (1977a).
45. Henson, I.E. and Wheeler, C.T. Hormones in plants bearing nitrogen-fixing root nodules: Cytokinin levels in root and root nodules of some non-leguminous plants, *Zeitschrift für Pflanzen-physiologie, 84,* 179–182 (1977b).
46. Henson, I.E. and Wheeler, C.T. Hormones in plants bearing nitrogen-fixing root nodules: Gibberellin-like substance in *Alnus glutinosa* L. Gaertn. *New Phytologist, 78,* 373–381 (1977c).
47. Kummerow, J., Alexander, J.V., Neel, J.W. and Fishbeck, K. Symbiotic nitrogen fixation in *Ceanothus* roots, *American Journal of Botany, 65,* 63–69 (1978).
48. Lalonde, M. Confirmation of the infectivity of a free-living actinomycete isolated from *Comptonia peregrina* root nodules by immunological and ultrastructural studies, *Canadian Journal of Botany, 56,* 2621–2635 (1978).
49. Lalonde, M., and Devoe, I.W. Origin of the membrane envelope enclosing the *Alnus crispa* var. *mollis* fern root nodule endophyte as revealed by freeze-etching microscopy, *Physiological Plant Pathology, 8,* 123–129 (1976).
50. Lalonde, M. and Knowles, R. Ultrastructure of the *Alnus crispa* var. *mollis* Fern. root nodule endophyte, *Canadian Journal of Microbiology, 21,* 1058–1080 (1975a).
51. Lalonde, M. and Knowles, R. Ultrastructure, composition and biogenesis of the encapsulation material surrounding the endophyte in *Alnus crispa* var. *mollis, Canadian Journal of Botany, 53,* 1951–1971 (1975b).
52. Lalonde, M., Knowles, R. and Devoe, I.W. Absence of "void area" in

freeze-etched vesicles of the *Alnus crispa* var. *mollis* Fern root nodule endophyte, *Archiv für Microbiologie, 107*, 263–267 (1976).
53. Lalonde, M., Knowles, R. and Fortin, J.A. Demonstration of the isolation of non-infective *Alnus crispa* var. *mollis* Fern nodule endophyte by morphological immunolabelling and whole cell composition studies, *Canadian Journal of Microbiology, 21*, 1901–1920 (1975).
54. Lechevalier, M.P. and Feteke, E. *Chemical Methods as Criteria for Separation of Actinomycetes into Genera*, Workshop Subcommittee on Actinomycetes of the American Society for Microbiology, Rutgers University, New Brunswick, New Jersey (1971).
55. Lechevalier, M.P. and Lechevalier, H.A. Chemical composition as criterion in the classification of aerobic actinomycetes, *International Journal of Systematic Bacteriology, 20*, 435–443 (1970).
56. Lechevalier, H.A., Lechevalier, M.P., Becker, B. Comparison of the chemical composition of cell walls of Nocardiae with that of other aerobic actinomycetes, *International Journal of Systematic Bacteriology, 16*, 151–160 (1966).
57. Lepper, M.G. and Fleschner, M. Nitrogen fixation by *Cercocarpus ledifolius* (Rosaceae) in pioneer habitats, *Oecologia* (Berlin), *27*, 333–338 (1977).
58. Medan, D. and Tortosa, R.D. Nodulos radicales en Discaria y Colletia (Rhamnaceae), *Boletin de Socieded Argentina de Botánica, 17*, 323–336 (1976).
59. Metcalfe, R.C. and Chalk, L. *Anatomy of the Dicotyledons, Vol. 1* (2nd edition), Clarendon Press, Oxford (1957).
60. Miguel, C., Cañizo, A., Costa, A. and Rodriguez-Barrueco, C. Some aspects of the *Alnus* type root nodule symbiosis, pp. 121–133 in *Limitations and Potentials for Biological Nitrogen Fixation in the Tropics* (Edited by J. Dobereiner, R.H. Burris and A. Hollaender), Plenum Press, New York (1978).
61. Morrison, T.M. and Harris, G.P. Root nodules in *Discaria toumatou* Raoul Choix, *Nature, London, 182*, 1746–1747 (1958).
62. Mostafa, M.A. and Mahmoud, M.Z. Bacterial isolates from root nodules of Zygophyllaceae, *Nature, London, 167*, 446–447 (1951).
63. Newcomb, W., Peterson, R.L., Callaham, D. and Torrey, J.G. Structure and host-actinomycete interaction in developing root nodules of *Comptonia peregrina, Canadian Journal of Botany, 56*, 502–531 (1978).
64. Pommer, E.H. Über die Isolierung des Endophyten aus den Wurzelknöllchen von *Alnus glutinosa* Gaertn. und über erfolgreiche Re-Infektionsversuche, *Berichte der Deutschen der Deutschen Botanischen Gesellschaft, 72*, 138–150 (1959).
65. Quispel, A. The endophyte of the root nodules in non-leguminous plants, pp. 599–520 in *The Biology of Nitrogen Fixation* (Edited by A. Quispel),

North-Holland Publishing Company, Amsterdam (1974).

66. Rodrigues-Barrueco, C. and Bermudez de Castro, F. Cytokinin-induced pseudonodules on *Alnus glutinosa*, *Physiologia Plantarum*, *29*, 277–280 (1973).
67. Rodriguez-Barrueco, C. and Bond, G. A discussion of the results of cross-inoculation trials between *Alnus glutinosa* and *Myrica gale*, pp. 561–565 in *Symbiotic Nitrogen Fixation in Plants* (Edited by P.S. Nutman), International Biological Programme, Vol. 7, Cambridge University Press, Cambridge, England (1976).
68. Rundel, P.W. and Neel, J.W. Nitrogen fixation by *Trevoa trinervis* (Rhamnaceae) in the Chilean Matorral, *Flora*, *167*, 123–132 (1978).
69. Sabet, Y.S. Bacterial root nodules in the Zygophyllaceae, *Nature, London*, *157*, 656–657 (1946).
70. Schubert, K.R. and Evans, H.J. Hydrogen evolution: A major factor affecting the efficiency of nitrogen fixation in nodulated symbionts, *Proceedings of the National Academy of Sciences, U.S.A.*, *73*, 1207–1211 (1976).
71. Severini, C. Sui tubercoli radicali di *Datisca cannabina*, *Annali di Botanica*, *15*, 29–51 (1922).
72. Sloger, C. Nitrogen fixation by tissues of leguminous and non-leguminous plants, Ph.D. thesis, University of Florida, U.S.A. (1968).
73. Sloger, C. and Silver, W.S. Note on nitrogen fixation by excised root nodules and nodular homogenates of *Myrica cerifera* L., pp. 299–302 in *Non-heme Iron Proteins: Role in Energy Conversion* (Edited by A. San Pietro), Antioch Press, Yellow Springs, Ohio, U.S.A. (1965).
74. Sloger, C. and Silver, W.S. Nitrogen fixation by excised root nodules and nodular homogenates of *Myrica cerifera* L. Abstracts of the 9th International Congress on Microbiology, Moscow, p. 285 (1966).
75. Strand, R. and Laetsch, W.M. Cell and endophyte structure of the nitrogen-fixing root nodules of *Ceanothus integerrimus*. I. Fine structure of the nodule and its endosymbiont, *Protoplasma*, *93*, 165–178 (1977a).
76. Strand, R., and Laetsch, W.M. Cell and endophyte structure of nitrogen fixing root nodules of *Ceanothus integerrimus*. II. Progress of the endophyte into young cells of the growing nodule, *Protoplasma*, *93*, 179–190 (1977b).
77. Soepadmo, E. Ulmaceae, pp. 31–76 in *Flora Malesiana*, Ser. I (Edited by C.G.G.J. van Steenis), Vol. 8, Noordhoff-Kolf N.V., Jakarta (1977).
78. Tiffney, W.N., Benson, D.R. and Eveleigh, D.E. Does *Arctostaphylos uva-ursi* (Bearberry) have nitrogen fixing root nodules? *American Journal of Botany*, *65*, 626–628 (1978).
79. Torrey, J.G. Initiation and development of root nodules of *Casuarina* (Casuarinaceae), *American Journal of Botany*, *63*, 335–344 (1976).
80. Torrey, J.G. and Callaham, D. Determinate development of nodule roots in actinomycete-induced root nodules of *Myrica gale* L., *Canadian*

Journal of Botany, *56*, 1357–1364 (1978).

81. Torrey, J.G. and Callaham, D. Early nodule development in *Myrica gale*, *Botanical Gazette*, *140* (Suppl.), 10–14 (1979).
82. Trinick, M.J. Symbiosis between *Rhizobium* and the non-legume *Trema aspera*, *Nature*, *London*, *244*, 459–460 (1973).
83. Trinick, M.J. *Rhizobium* symbiosis with a non-legume, pp. 507–517 in *Proceedings of the First International Symposium in Nitrogen Fixation* (Edited by W.E. Newton and G.J. Nyman), Pullman, Washington, Washington State University Press, U.S.A. (1976).
84. Trinick, M.J. Structure of nitrogen fixing nodules formed by *Rhizobium* on root of *Parasponia andersonii*, *Canadian Journal of Microbiology*, *25*, 565–578 (1979).
85. Trinick, M.J. and Galbraith, J. Structure of root nodules formed by *Rhizobium* on the non-legume *Trema cannabina* var. *Scabra*, *Archiv für Microbiologie*, *108*, 159–166 (1976).
86. Van Straten, J., Akkermans, A.D.L. and Roelofsen, W. Nitrogenase activity of endophyte suspensions derived from root nodules of *Alnus*, *Hippophae*, *Shepherdia* and *Myrica* spp., *Nature*, *London*, *266*, 257–258 (1977).
87. Willis, J.C. *A Dictionary of the Flowering Plants and Ferns* (8th edition), (Revised by H.K. Airy Shaw), Cambridge University Press, Cambridge, England (1973).
88. Wheeler, C.T. The causation of the diurnal changes in nitrogen fixation in the nodules of *Alnus glutinosa*, *New Phytologist*, *70*, 487–495 (1971).

K.G. MUKERJI and N.S. SUBBA RAO

5. Plant Surface Microflora and Plant Nutrition

Introduction

The surface of the aerial and underground parts of plants growing in soil under natural conditions harbours a large and varied population of microorganisms, potentially parasitic as well as non-parasitic ones (38, 39, 46, 47, 85, 88, 140). Changes in the environmental conditions and in the physiological status of the plants lead to changes in their surface microflora (106). Some of these microorganisms grow on the surface of healthy plants while others grow on the surface of senescent or damaged plants. Different parts of the plant represent specific ecological niches for colonization by various, and to some extent specific types of microorganisms. The subject of the microbial ecology of plant surface is so fascinating that it has attracted the attention of many workers all over the world (48, 112, 140) with emphasis on the following: the microbial population of plant surface as influenced by environmental changes; the influence of foliar application of a fungicide, or of any chemical such as growth hormones, inhibitors or nutrients on the biology of plant surface microorganisms; the interactions among microbial populations on the plant surface; the interaction between the plant and the surface microorganisms; pollution as a factor influencing the surface colonizers; the role of plant parts which can have a carryover effect on the spread of a particular plant disease; leaching of leaves as a physiological process in relation to plant surface colonizations, and lastly, the mechanism of succession of microflora on plant surface.

Such studies have provided clues to answering certain essential questions regarding: (i) the role of plant surface microorganisms in the biological control of certain epidemic diseases; (ii) the role of microorganisms as active decomposers of different dead and decaying plant parts in relation to soil fertility; (iii) the evaluation of the role of a pathogen in the absence of visible symptoms; (iv) the establishment of the cause of certain allergies common to mankind; and (v) the utility of plant surface nitrogen-fixing microorganisms as a source of nitrogen nutrition of plants. The first four aspects have been well reviewed and discussed elsewhere and the reader is advised to consult these papers for further information (38, 39, 48, 105, 112, 138–143). The purpose of this chapter is to highlight the last aspect and to elaborate the role of plant surface microorganisms in plant nutrition.

Several workers have from time to time described the microflora on plant surface under different terminologies such as rhizosphere, rhizoplane, gemmisphere, spermatosphere, spermosphere, phyllosphere, phylloplane and so on. However, in this chapter, it is proposed to consider plant surface microorganisms by restricting descriptions to the following headings: rhizosphere, rhizoplane and mycorrhiza (root region); phyllosphere, phylloplane and leaf nodulation (leaf region) and caulosphere and cauloplane (the stem region).

Rhizosphere and rhizoplane

Hiltner (69) found that the soil around the roots supported more microbial activity than the soil which was at a distance from the roots. He named this zone of intense microbial activity as the "rhizosphere". Clark (29) suggested the term "rhizoplane" as the zone of actual root influence and defined it as the external surface of plant roots with closely adhering soil particles. The influence of the root on soil microorganisms starts immediately after seed germination which increases as the plant grows and reaches a maximum when plants have reached the peak of their vegetative growth (78). Rovira (123) considered rhizosphere as a "poorly defined zone of soil with a microbiological gradient in which the maximum changes to the microflora occur in the soil adjacent to the root and decline with distance away from the roots." He also felt that other terms like outer rhizosphere and root surface (rhizoplane) were vague. Similarly, whether the term rhizosphere could be used in connection with ectomycorrhizeae has been a debatable point (39, 149).

The roots exert a selective action on microorganisms, stimulating the growth of some while suppressing that of others. Simultaneously, associative and antagonistic phenomena among microorganisms are also initiated. According to Stenton (153), the roots of higher plants provide a "virgin ecological niche to soil microorganisms" within the soil. Sorokina (151) emphasized the presence of specific microflora complexes around different plant root systems. Mishra (97) and Mishra and Srivastava (98) also reported that different plant species grown in the same type of soil can harbour different fungal flora in the rhizosphere. Elkan (56) working with nodulating and genetically related non-nodulating soybean lines reported differences in the total numbers and nutritional requirements of the bacterial isolates. Clark (30) observed a greater

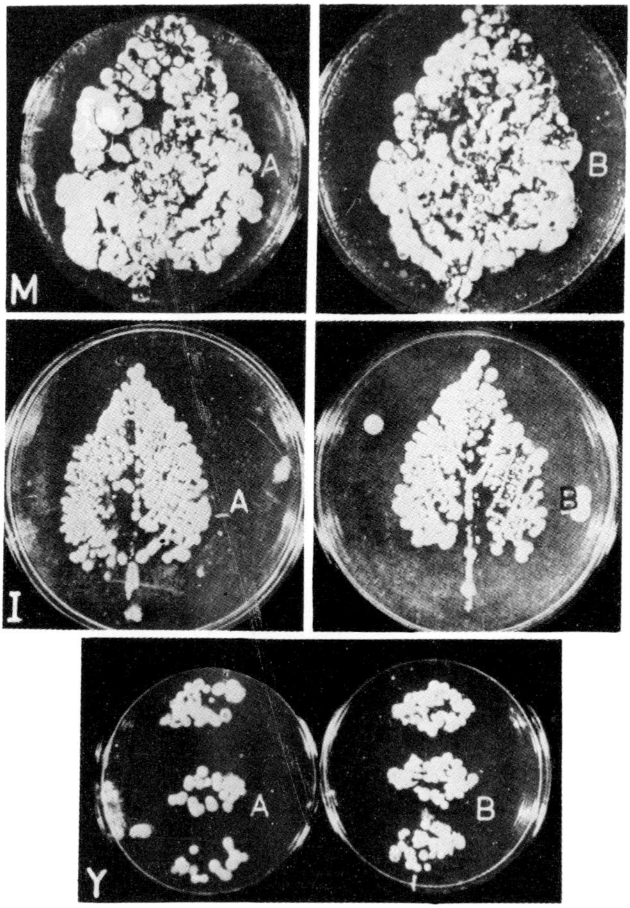

Plate 1. Showing distribution of bacterial colonies on ventral (A) and dorsal (B) surface of mature (M), immature (I), and young (Y) leaves of *Lantana Camara*.

Plate 2. Two leaves of *Psychotria mucronata* showing distinct leaf nodules (courtesy: Dr. Muriel E. Rhodes-Roberts).

number of microorganisms in the rhizosphere of nodulating soybeans, than in the near isogenic, non-nodulating lines. Neal *et al.* (107) described changes in the rhizosphere microflora of spring wheat induced by disomic substitution of a chromosome. Ranga Rao and Mukerji (116) found that the total counts of fungi differed appreciably in the rhizoplane and rhizosphere of four cultivars of wheat, and changed with the age of the plant.

Roots initially have little or no microbial colonization but as the plants grow in the soil, the root exudates composed of a mixture of nearly 18 amino acids, 10 sugars, 10 organic acids, mucilage and other substances together with sloughed-off root cap and other cells (66, 126) exert influence on microbial colonization. These nutrients allow the dormant spores to germinate, and there is a zone of enhanced activity in which the ruderal species are very evident (113). The ruderal species (67) generally have a short life expectancy, a large reproductive capacity and the ability to exploit habitats that are potentially productive and intermittently favourable for growth. Examples of this group of microorganisms are the fungi belonging to mucorales of the class phycomycetes. Normally, the root surface becomes colonized by such fungi within a period of a few days which grow and traverse up and down the root until they encounter the next colonist.

Bowen (22) has pictured a conceptual model of the rhizosphere biology. This is based on three main ideas: (i) the plant is the main driving force for the system and thus factors affecting its growth will also affect microbial growth; (ii) the distribution of roots and the microbial movement to them are essential considerations in understanding the composition of the root microflora; (iii) microbial growth at the root surface can itself affect nutrient losses from the root and may also affect the substrates available for microorganisms away from the root surface.

The generation times of different microorganisms on the root surface, rhizosphere, and soil away from the roots are different. The generation times (growth rates) for a *Pseudomonas* sp. and a *Bacillus* sp. on the roots of *Pinus radiata* in the soil was found to be 5.2 hours and 39 hours respectively in comparison to 77 hours and more than 100 hours in the nearby soil (24). According to Bowen (22), the generation time will be more useful in understanding the biology of the rhizosphere rather than the usual root/soil (R/S, rhizosphere effect) ratios, because the generation time concept (i) is a more dynamic parameter than the static measurement usually made with rhizosphere studies, (ii) brings rhizosphere biology into line with other fields of population biology, and (iii) is internally consistent and a more valid measurement than growth comparisons. Apart from the production of growth stimulating or inhibiting substances by the rhizosphere microflora, microorganisms in the rhizosphere also influence the inhibition or stimulation of soil-borne plant pathogens, root-hair production and root morphology (125). Rhizosphere microflora is also considered important as it may control the growth and spread of pathogens by antibiosis (21, 108). However, the evidence for the control of soil-borne pathogens by antibiosis in soil rests largely on the correlation between disease suppression and increase in number of antibiotic-producing organisms (71, see also Chapter 18 under Section C). The rhizosphere microorganisms are also known to influence plant growth by controlling the availability and uptake of nutrients (90, 92, 124, 156).

Microorganisms in the root region and nitrogen fixation

Several bacteria in the rhizosphere help in nitrogen fixation. Nitrogen-fixing *Azospirillum* strains have been isolated from tropical and some temperate grass root surfaces (49, 51, 53, 158). Some species of *Azotobacter* form root-associations with seasonal grasses and are host-specific e.g. *A. paspali* associates with the roots of *Paspalum notatum* and fixes about 100 kg N/ha soil (52, 53, 115). The maximum benefit from root-nitrogen-fixing bacteria association can be obtained in agricultural systems, such as grassland or the systems with crop plants having the C-4 photosynthetic pathway and where nitrogen-fertilizer use is restricted by cost or availability (43, 50). Several species of *Azotobacter*, *Beijerinckia*, *Derxia*, *Bacillus*, *Klebsiella* and *Clostridium* are now known to fix nitrogen under field conditions in association with roots of plants (36, 82).

Nitrogen fixation has been observed in the rhizome and roots of fresh-water plants, *Glyceria* and *Typha* under anaerobic conditions and it was about 60 kg N/ha/yr with *Glyceria* (25). Significant nitrogen-fixing activity has also been reported in the rhizoplane-rhizosphere of certain tropical (*Thalasia*, *Syringodicem* and *Diplanthera*) and temperate (*Zostera*) marine angiosperms whose roots were associated with species of *Azotobacter* (109).

Wallace and Romney (160), Wallace *et al.* (161) and Hunter *et al.* (72) surveyed the desert flora at the Nevada for nitrogen fixation, and found measurable activity associated with the roots of the following plants: *Artemisia, spinescens*, *A. tridentata*, *Hymenoclea salsola*, *Tetradymia canescens* of the compositae; *Coleogyne ramosissima* of the rosaceae; *Atriplex canescens* and *A. confertifolia* of the chenopodiaceae; *Krameria parvifolia* of the Krameriaceae; *Larrea tridentata* of the zygophyllaceae; *Lycium pallidum* and *L. shockleyi* of the solanaceae; *Menodora spinescens* of the oleaceae; *Thamnosa montana* of the rutaceae; *Yucca schidigera* of the liliaceae; *Bromus rubens* and *Stipa speciosa* of the poaceae. Snyder and Wullstein (150) found similar activity of *Oryzopsis hymenoides*. Since root nodules were not obvious, this type of fixation might well involve such microorganisms as *Azotobacter*, *Clostridium* (32, 91) and *Azospirillum* living in the rhizosphere. All are heterotrophic depending on outside carbon sources to fix nitrogen (74, 117, 118); thus a loose or parasymbiotic condition or the occurrence of endophytic flora could well be involved in the root or the soil-root interface (59).

Nitrogen fixation was found to be associated with sand grain root sheaths (rhizosheaths) occurring on the following xeric grasses—*Oryzopsis hymenoides*, *Agropyron daystachyum*, *Stipa comata* and *Aristida purpurea*. Nitrogenase activity was found to be associated with the rhizosheath and *Bacillus polymyxa*-like nitrogen-fixing microorganisms were isolated from the rhizosheaths of each grass (168).

Mycorrhizal associations and root nodulation

Asai (2) first suggested that mycorrhizae were a necessary precondition for effective nodulation in many legumes. Nitrogen-fixing microorganisms can increase soil nitrogen while mycorrhizal fungi effectively augment the absorbing surface of the roots (102, 103). Successful nodulation is dependent on an adequate supply of phosphorus. Many tropical soils are so phosphorus-deficient that they cannot respond to nitrogen until this deficiency is corrected. It is here that endomycorrhiza takes on added significance, as it enhances the absorption of phosphorus. Endomycorrhizae do not interact directly with nitrogen-fixing bacteria. Stimulation of nitrogen-fixing systems by endomycorrhizae can be replaced by adding extra phosphorus (28).

There are several reports on the stimulation of nodulation by *Rhizobium* in legumes by endomycorrhizae (23, 100). Crush (31) and Daft and El Giahmi (33, 34) found that the weight of nodules, amount of nodular tissue, nitrogen and phosphorus content of the plant, concentration of leghaemoglobin and rates of acetylene reduction were greater in mycorrhizal plants than in non-mycorrhizal ones. Gates (64) and Mosse *et al.* (104) found similar results with *Trifolium*, *Stylosanthes* and *Centrosema* spp. and noted that only mycorrhizal plants were able to nodulate in severely P-deficient soils. The growth and nodulation of *Pueraria phaseoloides* and *Stylosanthes guyanensis* (growth as ground cover for rubber) was severely retarded unless plants were infected with endomycorrhiza or given large doses of rock phosphate (159). Combined inoculation of *Glomus mosseae* and *Rhizobium meliloti* on lucerne in calcareous soil enhanced the activity of *Rhizobium* and increased the yield by 211 per cent (4). Bagyaraj *et al.* (6, 7) reported increased yield in soybean by the dual inoculation of endomycorrhizal fungus and *Rhizobium*. Asimi *et al.* (3) reported better nodulation and higher nitrogenase activity in soybeans with *Glomus mosseae* in a sterile unamended soil (see also Chapter 14).

Endomycorrhizal interaction with other nitrogen-fixing microorganisms has also been reported. Rose (120) reported that of the 25 species of actinomycete-nodulated angiosperms examined, 23 were infected with endomycorrhizae.

Several workers have indicated the possibility of nitrogen fixation by mycorrhizeae in desert plants (57, 79, 81, 148, 163, 165, 166). Mycorrhizae are found abundantly in association with desert shrubs and herbs of many families around the world. Eleusenova and Selivanov (55) have shown that there is considerable seasonal variation in the abundance of these fungal associates, with the greatest abundance noted during the wet seasons. Since nitrogen is often the nutrient most limiting for plant growth in deserts (164), it might be possible that mycorrhizae depending on rhizospheral carbon and vitamins are able to fix nitrogen under conditions of minimal moisture. For instance, Williams and Aldon (165) have reported improvements in growth responses of *Atriplex canescens* when it was inoculated with *Endogone mosseae*, an endomycorrhiza-forming organism.

Mycorrhizae and their effect on rhizosphere biology

Bagyaraj and Menge (5) reported that a synergistic or additive interaction seems to exist between *Glomus fasciculatus* and *Azotobacter*. Large populations of bacteria and actinomycetes were recovered from the rhizosphere of tomato plants inoculated with *Glomus fasciculatus* and *Azotobacter chroococcum* either individually or in combination. Plants inoculated with both the microorganisms had higher numbers of microorganisms in the rhizosphere than plants inoculated with either of the two. *Azotobacter* enhanced spore production and infection by *Glomus* which may have been responsible for the better yield of the crop.

Marx (94) and Davy (40) reported the effects of ectotrophic mycorrhiza on root-borne pathogens and rhizosphere organisms. Endomycorrhizeae in the root would have a direct effect on the exudations of substances from the roots. This in turn, would affect the distribution of microorganisms in the rhizosphere and might be responsible for rendering the mycorrhizal roots susceptible or resistant to the attack by pathogens.

In the rhizosphere there might be competition for getting adjusted to the root surface between two members of the endogonaceae which are the causative agents of endomycorrhizal associations. Ross and Ruttencutter (122) observed an interaction between *Glomus macrocarpus* var. *geosporus* and *Gigaspora gigantea*. His results indicated that *Glomus macrocarpus* var. *geosporus* does not compete well with *Gigaspora gigantea* as the population of the former increases rapidly in the absence of the latter. It has often been observed that some other fungi or bacteria dominate the members of the endogonaceae and limit populations of endomycorrhizal fungi in soil (11, 12, 65, 122, 133, 152).

There have been several reports of endomycorrhizal fungi interacting with other microorganisms in the rhizosphere, notably with the root pathogens (102) and thus provide a distinct possibility that the colonization of roots by endomycorrhizal fungi confers resistance to invasion by other root pathogens. For example, root infection of tomato by *Fusarium oxysporum* f. *lycospersici* gets diminished and wilt symptoms get abated if roots become mycorrhizal.

Similarly, the disease index of cotton and tobacco infected by *Thielaviopsis basicola* was lowered in mycorrhizal plants as compared to non-mycorrhizal plants (8, 9, 134). Schonbeck and Dehne (135) reported that diseases caused by *Helminthosporium sativum* (=*Drechslera sorokiniana*) and *Erysiphe graminis* on barley, *Colletotrichum lindemuthianum* and *Uromyces phaseoli* on bean, *Botrytis cinerea* on *lettuce*, *Erysiphe chicoracearum* on cucumber and TMV on tobacco increased and *Olpidium brassicae* on lettuce and tobacco decreased on inoculation with *Glomus mosseae*. Schenck *et al.* (132), Sikora and Schonbeck (146) and Sikora (145) reported that mycorrhizal infection had an adverse effect on the nematode *Meloidogyne incognita* during several phases of its life cycle probably as a result of changes in the physiology of the host. Similarly, it was observed that stunting of cotton caused by *Meloidogyne incognita* could be nullified by endomycorrhizeae (119). Inoculation of tomato roots with root knot nematodes (*Meloidogyne incognita* and *M. javanica*) and *Glomus fasciculatus* enhanced infection and sporulation of the latter and reduced significantly the number and size of the root knot galls (7). Khan and Khan (80) reported the resistance of mycorrhizal wheat roots to invasion by *Urocystis tritici*. Fox and Spasoff (62) reported that *Heterodera solanacearum* and *Gigaspora gigantea* mutually suppressed reproduction of one another. Iqbal *et al.* (73) has reported that although the damping off of *Brassica napus* seedlings due to *Rhizoctonia solani* occurred in the presence of mycorrhiza, infection however, did not take place in mycorrhizal seedlings of *Brassica campestris*. Conversely *Glomus fasciculatus* had no effect upon the infection and subsequent disease development by *Phytophthora cinnamomi* on avacado (95).

Dehne and Schonbeck (45) have reported that *Glomus mosseae* induced higher chitinase activity and arginine accumulation in the roots of tomato, *Phaseolus vulgaris*, maize, and tobacco. They have discussed a possible relation between arginine accumulation, chitinase activation and the development of resistance in mycorrhizal plants. Davis *et al.* (42) reported that infection by *Verticillium dahliae* was more severe in mycorrhizal cotton plants inoculated with *Glomus fasciculatus* than non-mycorrhizal ones at 20 μg P/g soil although at 300 μg P/g soil, the infection was equally severe in both. Therefore, it seems that adequate phosphorus nutrition in cotton results in severe wilt infection. The possible reasons for this as given by these authors are: (i) an increase in the number of avenues for penetration since the production of chlamydospores in the cortex might cause ruptures; (ii) dilution of potassium concentration in plants; (iii) larger population of the pathogen due to the improved nutrient status of the host; (iv) greater amount of tissue leads to more transpiration and hence to greater movement of the conidia. Davis and Menge (41) reported that in soils fertilized with less than 15 μg P/g soil, the roots of sweet orange seedlings infected with *Phytophthora parasitica* and the mycorrhizal fungus *Glomus fasciculatus* were healthier and weighed more than roots infected with *Phytophthora* alone. However, the beneficial effect of

Glomus was eliminated when 56 μg P/g soil was added. These results suggest that tolerance to *Phytophthora parasitica* root rot in citrus infected with *Glomus fasciculatus* is caused by the ability of the mycorrhizal roots to absorb more phosphorus and possibly other minerals as compared to non-mycorrhizal roots.

More TMV was detected by immunofluorescence in mycorrhizal tomato roots when compared to non-mycorrhizal ones and increased virus multiplication was observed in cells containing the mycorrhizal fungus in its arbuscular stage (35, 136).

Phyllosphere and phylloplane

Leaf surface carries a heterogenous population of microbes, which grow, reproduce, and multiply on leaves in dynamic equilibrium with the existing micro- and macro-environment. The microbial succession on leaves has been the subject matter for investigations since 1955 (48, 68, 84, 111, 112, 127, 138–140, 143). The leaf surface microbes are important in several ways. For instance, some of them are known to fix atmospheric nitrogen for the benefit of higher plants, have antagonistic action against fungal parasites, degrade plant surface waxes and cuticles, produce plant hormones, decompose plant material after leaf fall, activate plants to produce phytoalexins, have toxic effects on cattle, act as a source of allerginic air-borne spores, and influence the growth behaviour and root exudation of plants (144).

The age and position of a leaf on the plant is an important factor for microbial colonization of its surface (Fig. 1). The physiological and biochemical status of the leaf greatly influences the population and composition of the microflora (Fig. 2). The earliest colonizers on newly formed leaves have to face almost no competition as they are devoid of any microbes, and in fact, they receive a potential supply of surface nutrients. But as they get established, they face a relatively hostile environment (114) because of widely fluctuating temperatures and the incidence of ultraviolet radiations (113). They may immediately grow utilizing the fresh supply of substrates present on the leaf surface or lie dormant and inactive until the leaf becomes senescent.

The leaf surface medium comprises exudates, chemical compounds resulting from biological activity of various microbes including nitrogen fixers and components resulting from atmospheric pollution. The structure and chemistry of the leaf surface influence the occurrence of the plant surface colonizers (60, 70, 93) apart from physical factors like temperature, relative humidity, light and wind velocity which also interact with the leaf surface community in various ways (19).

Leaves at the seedling stage of plants usually harbour the least number of microbes which increases as the plants age, reaching the maximum population only on leaves which start yellowing at maturity (46, 143). The youngest

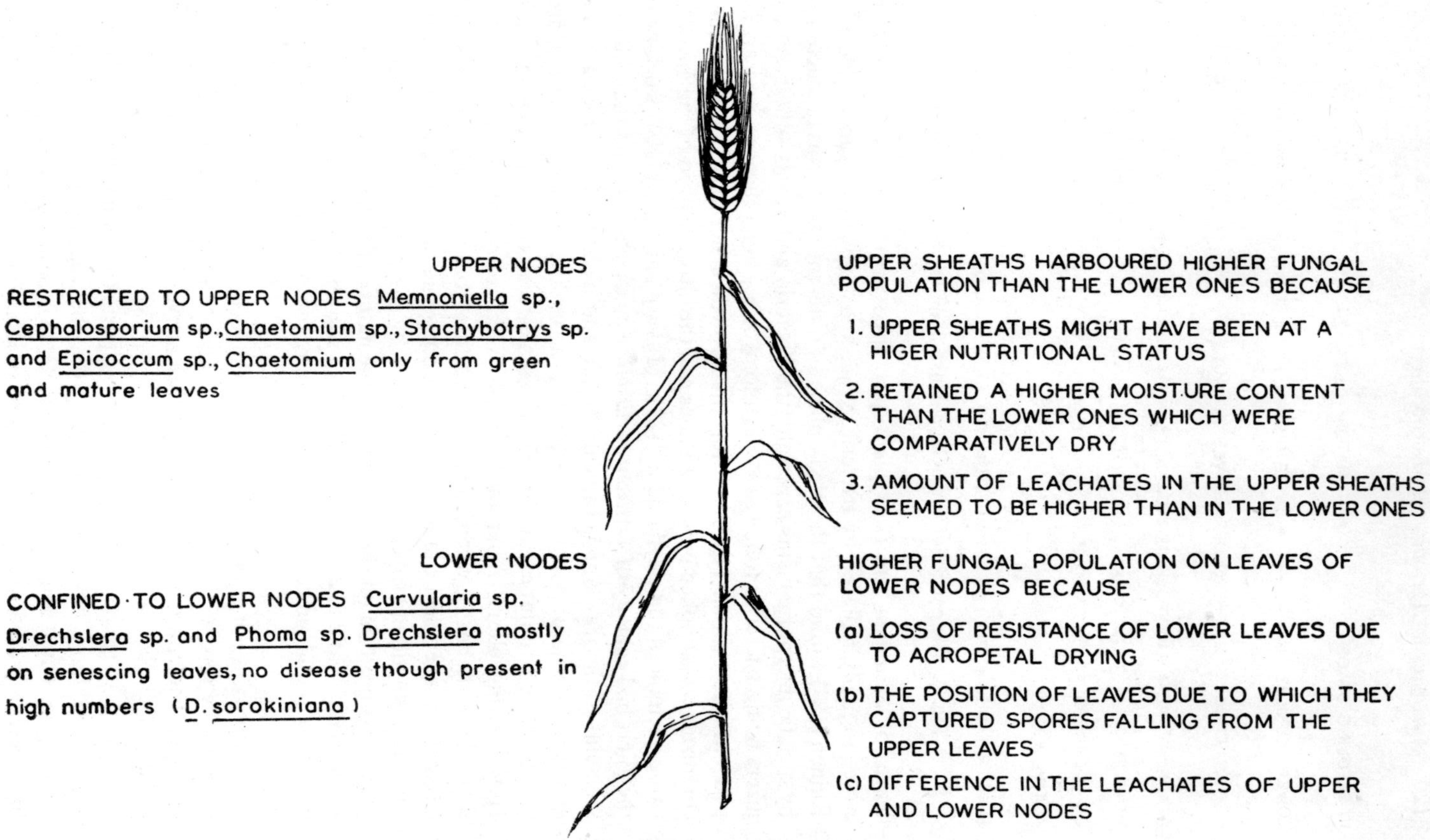

UPPER NODES
RESTRICTED TO UPPER NODES Memnoniella sp., Cephalosporium sp., Chaetomium sp., Stachybotrys sp. and Epicoccum sp., Chaetomium only from green and mature leaves
LOWER NODES
CONFINED TO LOWER NODES Curvularia sp. Drechslera sp. and Phoma sp. Drechslera mostly on senescing leaves, no disease though present in high numbers (D. sorokiniana)
UPPER SHEATHS HARBOURED HIGHER FUNGAL POPULATION THAN THE LOWER ONES BECAUSE
1. UPPER SHEATHS MIGHT HAVE BEEN AT A HIGER NUTRITIONAL STATUS
2. RETAINED A HIGHER MOISTURE CONTENT THAN THE LOWER ONES WHICH WERE COMPARATIVELY DRY
3. AMOUNT OF LEACHATES IN THE UPPER SHEATHS SEEMED TO BE HIGHER THAN IN THE LOWER ONES
HIGHER FUNGAL POPULATION ON LEAVES OF LOWER NODES BECAUSE
(a) LOSS OF RESISTANCE OF LOWER LEAVES DUE TO ACROPETAL DRYING
(b) THE POSITION OF LEAVES DUE TO WHICH THEY CAPTURED SPORES FALLING FROM THE UPPER LEAVES
(c) DIFFERENCE IN THE LEACHATES OF UPPER AND LOWER NODES

Fig. 1. Distribution of microfungi on the leaves of a wheat plant.

IN SEEDLING STAGE → LOWER POPULATION

MATURE LEAVES → INTERMEDIATE POPULATION

SENESCENT LEAVES → MAXIMUM POPULATION

LOWER POPULATION ⟶ INTERMEDIATE POPULATION ⟶ MAXIMUM POPULATION

MORE MICROBES ON THE OLDER LEAVES BECAUSE:

1) LEACHING FROM THE FOLIAGE INCREASES WITH AGEING. OLD LEAVES LEACH NUTRIENTS MORE RAPIDLY THAN THE YOUNGER ONES THESE LEACHATES INVITE GREATER NUMBER OF MICROBES AT THE SENESCENT PHASE WHICH IN TURN REPRESENTS HIGH MICROBIAL ACTIVITY

2) SUBSTRATES (e.g. POLLEN GRAINS) ON THE LEAF SURFACE MAY ALSO ALTER MICROFLORA

Fig. 2. The pattern of microbial population on leaves at different ages.

leaves of *Sesamum* and *Gossypium* harboured actinomycetes at first which were gradually replaced by bacteria, yeast and filamentous fungi (139, 141). Ruinen (127) found that the phylloplane of different plants was characterized by specific microörganisms especially the bacteria and yeasts—the first colonizers being bacteria, followed by actinomycetes, fungi, lichens, and arthropods in succession. The food for the growth and reproduction of these colonizers was supplied by leaf exudates and the nitrogen fixed by the nitrogen-fixing bacteria of the phylloplane (127). Mukerji and Rikhy (106) working on the microbial ecology of *Triticum aestivum* (wheat) leaves found that different segments of the leaf lamina (apical, middle and basal) had different amounts and types of microorganisms. In general the middle leaf segments harboured the maximum number of fungal propagules and the leaf apex the least. However, unicellular fungi and bacteria were recovered in larger numbers from the leaf sheath, when compared to those from other leaf segments. The circumstantial evidence indicates that the differences in the quality and quantity of epiphytic microbes are determined by the physiological status of the organs, its position on the plant and by the nature of substrates present on its surface (Fig. 3 and Plate 1). These findings corroborate those of Ruinen (130) who also found more unicellular microorganisms in the sheath region of grasses.

The presence of epiphytic bacteria on the surface of aquatic plants has been reported by several workers. Strzelczyk and Mielczarek (155) claimed that epiphytic bacteria are metabolically more active than planktonic or benthic isolates.

Both pathogenic and saprophytic bacteria (including nitrogen fixers), yeasts, and fungi have been isolated from the vegetative buds of tropical woody plants, and some annuals (37, 86, 87, 88). Some plant pathogenic fungi have been reported to have their source of inoculum in the buds, e.g. *Tilletia*

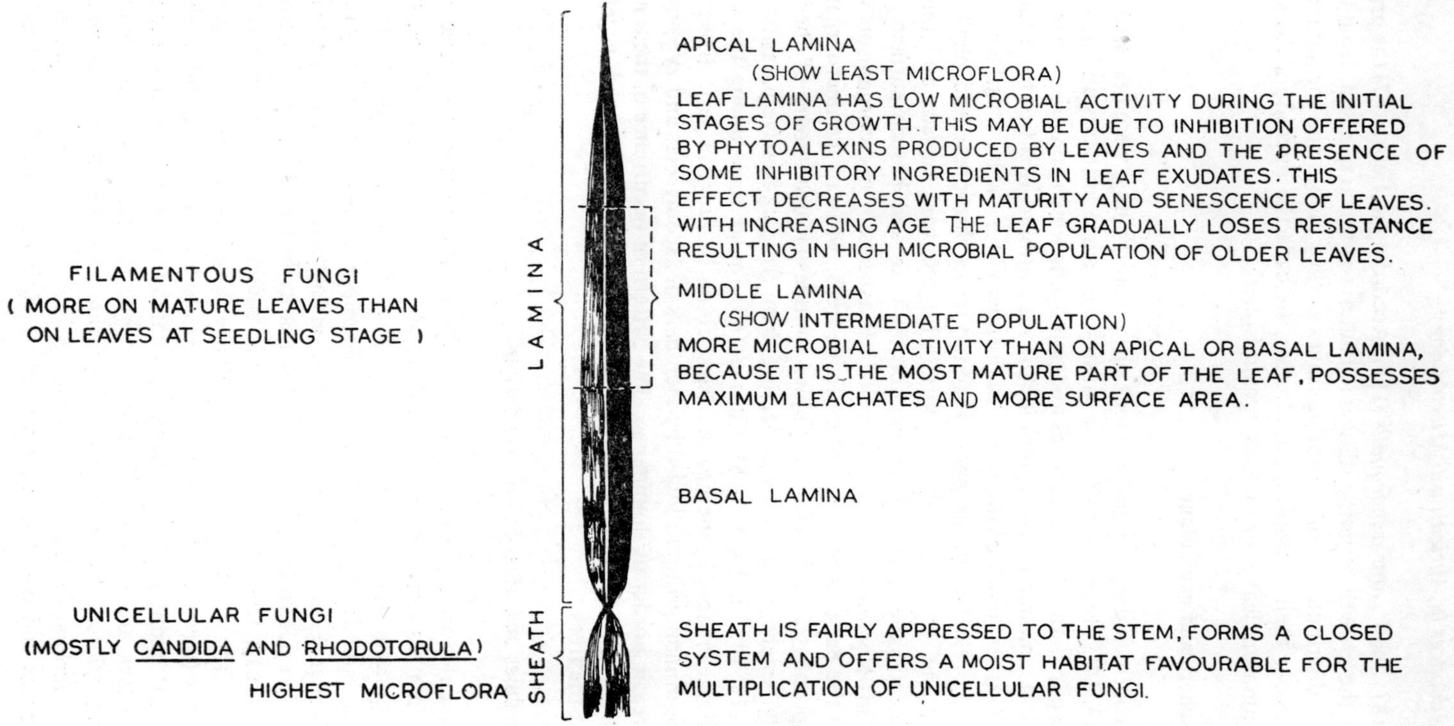

FILAMENTOUS FUNGI
(MORE ON MATURE LEAVES THAN
ON LEAVES AT SEEDLING STAGE)
UNICELLULAR FUNGI
(MOSTLY CANDIDA AND RHODOTORULA)
HIGHEST MICROFLORA
LAMINA
SHEATH
APICAL LAMINA
(SHOW LEAST MICROFLORA)
LEAF LAMINA HAS LOW MICROBIAL ACTIVITY DURING THE INITIAL STAGES OF GROWTH. THIS MAY BE DUE TO INHIBITION OFFERED BY PHYTOALEXINS PRODUCED BY LEAVES AND THE PRESENCE OF SOME INHIBITORY INGREDIENTS IN LEAF EXUDATES. THIS EFFECT DECREASES WITH MATURITY AND SENESCENCE OF LEAVES. WITH INCREASING AGE THE LEAF GRADUALLY LOSES RESISTANCE RESULTING IN HIGH MICROBIAL POPULATION OF OLDER LEAVES.
MIDDLE LAMINA
(SHOW INTERMEDIATE POPULATION)
MORE MICROBIAL ACTIVITY THAN ON APICAL OR BASAL LAMINA, BECAUSE IT IS THE MOST MATURE PART OF THE LEAF, POSSESSES MAXIMUM LEACHATES AND MORE SURFACE AREA.
BASAL LAMINA
SHEATH IS FAIRLY APPRESSED TO THE STEM, FORMS A CLOSED SYSTEM AND OFFERS A MOIST HABITAT FAVOURABLE FOR THE MULTIPLICATION OF UNICELLULAR FUNGI.

Fig. 3. Distribution of microfungi on different segments of a wheat leaf.

caries (87) and *Venturia inaequalis* (111). According to Leben (87) the primary site for the growth of non-pathogenic bacteria is the active bud, from where they are distributed to the maturing leaves as they unfold and the latter act as the sites for their multiplication. In fact, he calls the microflora associated with immature leaves or buds as "gemmisphere".

Caulosphere and cauloplane

The bark (including adjoining wood) surface microfungi were first studied by Bier (13, 14). He thought that this substrate is rich in nutrients and may carry a large population of saprophytes and pathogens which comprise a part of its biological community and determine the susceptibility of a tree to certain diseases. Garner (63) used the term "caulosphere" for the bark surface and the non-living cells within the bark. Like the terms rhizoplane and phylloplane "cauloplane" may be more appropriate for the bark surface and this will also include the microbial population on twig and branch surface. Relatively, little is known about the ecological aspects of microorganisms which inhabit the bark surface of healthy living trees. Most of the microorganisms are saprophytes and cause no injury to trees, instead some of them can actually benefit the trees by protecting them from aggressive parasitic organisms (13, 14, 15, 16, 17, 18). Garner (63) found that among the several microfungi inhabiting the healthy bark, those of fungi imperfecti were the most dominant group followed by members of ascomycetes and zygomycetes in decreasing order of dominance. The pattern of occurrence of these microorganisms is mainly influenced by environmental factors and vegetation around the trees.

Leaf nodules and other foliar associations

Leaf nodules are known in more than 400 species, belonging to three genera of two families—myrsinaceae and rubiaceae. The genera involved are, *Ardisia* of myrsinaceae and *Pavetta* and *Psychotria* of rubiaceae (Plate 2). The bacterial leaf nodule associations have been comprehensively reviewed by Lersten and Horner (89) and Fletcher (60). Various aspects of the development and symbiotic cycle in *Psychotria* have been described by Fletcher and Rhodes-Roberts (61). During the life cycle, the bacterial endophyte in *Psychotria*, grows on the surface of the plant. Bacteria present in the mucilage at the stem apex (terminal vegetative bud) moves to leaf primordia and establishes in the leaf nodules (Fig. 4). Bacteria isolated from non-nodulated and nodulated leaves are similar and include several genera, namely, *Streptomyces*, *Bacillus*, *Micrococcus* or *Sarcina*; small Grampositive rods, *Mycoplana rubra*, *Flavobacterium*, *Enterobacter*, *Erwinia*, *Klebsiella*, *Pseudomonas fluorescens*, *Alcaligenes* and many *Agrobacterium*- or *Rhizobium*- or *Chromobacterium*-

like organisms. These bacteria have been repeatedly isolated by Becking (10) and Fletcher and Rhodes-Roberts (61).

Ruinen (128, 131) found large number of microbes on tropical foliage which amounted to more than 10^7 *Beijerinkia*/cm^2 in cacao. In the leaf sheath

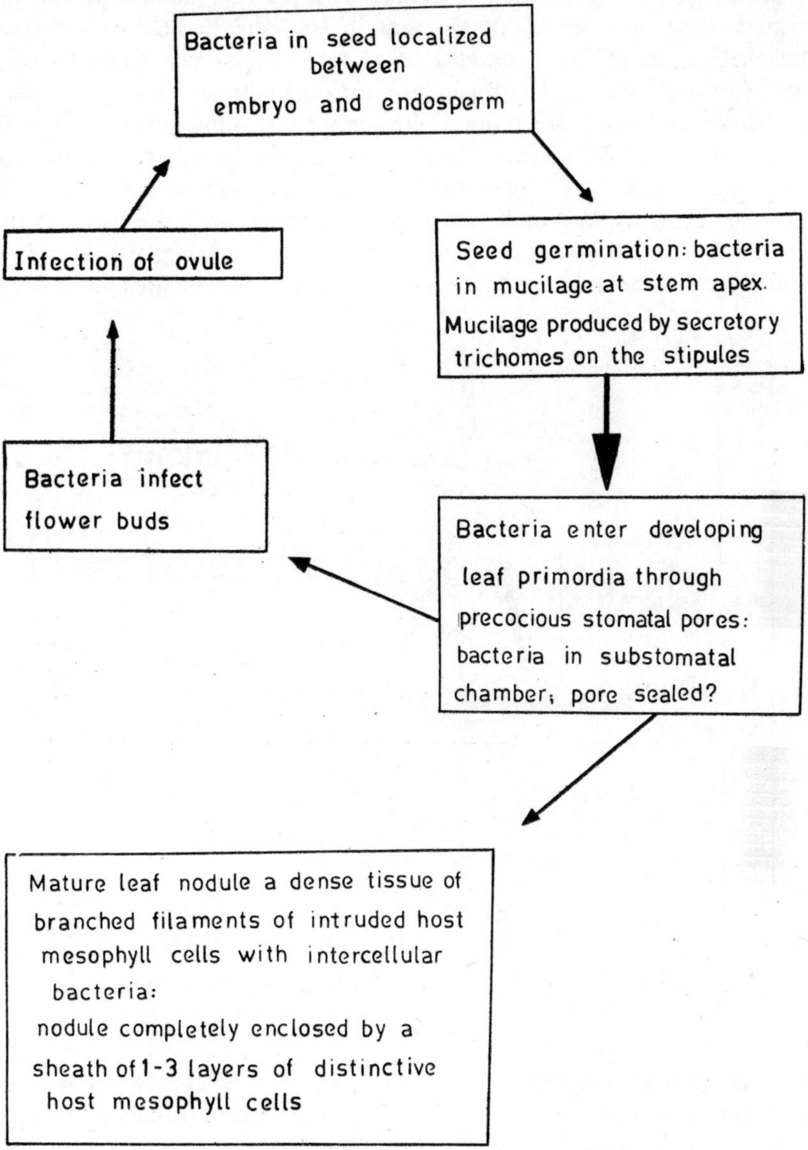

Fig. 4. Life cycle of the bacterial endophyte in *Psychotria* [from Fletcher and Rhodes-Roberts 61)].

of grasses, such as *Tripsacum*, the number went up to 10^{10} microbial cells/ml of fluid in the sheath of mature leaves. In Surinam, Indonesia, she found that dew and the run-off water from leaves during rain contained measurable amounts of carbohydrates and amino acids. She reported carbohydrate concentrations of the order of 14–200 ppm in rain and dew, 465 ppm in stem flow, and up to 10,000 ppm in leaf sheath fluid of *Tripsacum*. Detached leaves of *Coffea arabica*, *Phaseolus vulgaris* and *Gossypium* were floated on the surface of nitrogen-free, mineral salts solution in illuminated petri dishes and left in green house for some days. Subsequent analysis provided evidence of an increase in nitrogen content of the leaves which was attributed to fixation by bacteria originally present on the leaves and activated under the experimental conditions, nourished by carbohydrate leachates from the leaf. Ruinen observed an increase of nitrogen in the range of 6–48 μg N/mg total N/day. Inoculation of *Azotobacter* on leaves stimulated the nitrogen

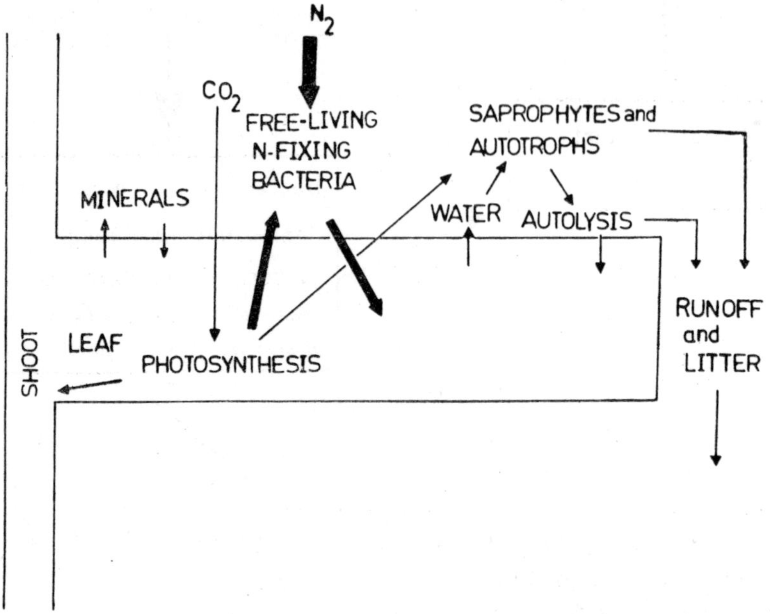

Fig. 5. Diagrammatic representation of possible interactions between phylloplane microorganisms and leaf [after Ruinen (128); Knowles (82)].

gain. On the basis of accumulated circumstantial evidence, she proposed that the plant-microorganism association in the phylloplane formed a loose symbiotic nitrogen-fixing system (Fig. 5). Photosynthesis supplied energy substrates for the heterotrophic diazotrophs which in turn provided the fixed nitrogen for plant nutrition either through the leaf surface or via run-off and root uptake (20). Jones (75) found significant nitrogen-fixation by unidentified phylloplane bacteria, of about 0.03–0.34 μg N/mg total N/day equivalent to about 19–64 g N/ha/day.

Ruinen (127) was the first to realize the significance of leaf surface bacteria in relation to natural vegetation in a nitrogen-poor soil. According to her the widespread occurrence of nitrogen-fixing bacteria on the leaf surface in the humid tropics serves as a "factory" for the production of organic nitrogen, and contribute to the nitrogen economy of the system. This nitrogen fixation ultimately results in the increase of available soil nitrogen. Several workers have reported an increase of soil nitrogen under grass cover (101, 110, 129, 130, 167). Jones (76) also reported nitrogen-fixing bacteria on leaves of all types of trees present in Grizedale forest in the English Lake district, U.K. and Jones *et al.* (77) found the annual rate of fixation for *Pseudotsuga menziensii* (Douglas fir) in Grizedale forest to be 20.82 mg N/g of leaf nitrogen. The amount of nitrogen fixed varied in different conditions and ranged from 4.43 to 14.60 kg N/ha. Denison (44) thinks that in addition to bacteria, lichens also fix nitrogen in the phylloplane of Douglas fir. The amount of nitrogen fixed varied (Tables 1 and 2) from season to season (77). Experiments with nodulated

Table 1. Monthly variation in the amount of nitrogen fixed on the leaves of Douglas fir [from Jones *et al.* (77)]

Months	Amount of nitrogen fixed (mg N/g leaf N)
January	0.00
February	0.00
March	0.68
April	3.47
May	5.21
June	1.02
July	0.00
August	0.68
September	1.68
October	1.74
November	2.34
December	2.79

leaves of *Psychotria* and *Pavetta* have been done using ^{15}N and acetylene reduction techniques but no clearcut evidences have come forth to attribute nitrogen-fixing abilities to leaf nodules. Alternate explanations for the significance of leaf nodule symbiosis based on growth factor (cytokinin-like substances) production have been put forward by recent workers in the field (10).

Very little is known about the mechanism of nitrogen fixation in the

Table 2. Percentage of shoots or leaf buds supporting nitrogen-fixing bacteria on larch—*Larix eurolepis* [from Jones *et al.* (77)]

Months	Percentage	Plant surface
January	88	buds only
February	100	buds only
March	31	leaf flush but mainly buds
April	29	leaves only
May	49	leaves only
June	35	leaves only
July	46	leaves only
August	25	leaves only
September	26	leaves only
October	7	leaves only—brown
November	63	buds only
December	63	buds only

phylloplane of arid and semi-arid plants. Nodules have been observed on the leaves of some tropical plants (99, 137, 147, 154). The extensive leaf nodulation or galls found on *Artemisia* may be a possible locus of nitrogen fixation in semi-arid environments (58). The phylloplane microflora is also considered to be quite effective and important in the nitrogen nutrition of forest trees (59).

Acknowledgements

Thanks are due to Mr. Kishen Lal and Mr. Harshavardhan for the diagrams and photographs and to Dr. (Miss) Darshan Kaur, Mr. Inderjeet Singh Bakshi and Mrs. Mita Sharma for help in literature collection.

References

1. Ames, R.N. and Linderman, R.G. The growth of easter lily (*Lilium longiflorum*) as influenced by vesicular-arbuscular mycorrhizal fungi, *Fusarium oxysporum* and fertility level, *Canadian Journal of Botany*, *56*, 2773–2780 (1978).
2. Asai, T. Über die Mycorrhizenbildung der leguminosen Pflanzen, *Japanese Journal of Botany*, *13*, 463–485 (1944).

3. Asimi, S., Gianinazzi-Pearson, V. and Gianinazzi, S. Influence of increasing soil P levels on interaction between vesicular-arbuscular mycorrhizae and *Rhizobium* in soybeans, *Canadian Journal of Botany, 58*, 2200–2205 (1980).
4. Azcon, G., De Aguilar, C., Azcon, R. and Barea, J.M. Endomycorrhizal fungi and *Rhizobium* as biological fertilisers for *Medicago sativa* in normal cultivation, *Nature, 279*, 325–327 (1979).
5. Bagyaraj, D.A. and Menge, J.A. Interaction between a VA mycorrhiza and *Azotobacter* and their effects on rhizosphere microflora and plant growth, *New Phytologist, 80*, 567–574 (1978).
6. Bagyaraj, D.A., Manjunath, A. and Patil, R.B. Interaction between a vesicular-arbuscular mycorrhiza and *Rhizobium* and their effects on soybean in the field, *New Phytologist, 82*, 141–145 (1979a).
7. Bagyaraj, D.A., Manjunath, A. and Reddy, D.D.R. Interaction of vesicular-arbuscular mycorrhiza with root knot nematodes in tomato, *Plant and Soil, 51*, 397–404 (1979b).
8. Baltruschat, H. and Schonbeck, F. The influence of endotrophic mycorrhiza on the infestation of tobacco by *Thielaviopsis basicola, Phytopathologische Zeitschrift, 84*, 172–188 (1976).
9. Baltruschat, H., Sikora, R.A. and Schonbeck, F. Effect of VA mycorrhiza (*Endogone mosseae*) on the establishment of *Thielaviopsis basicola* and *Meloidogyne incognita* in tobacco, Abstract, *Second International Congress of Plant Pathology*, 661 (1973).
10. Becking, J.H. The physiological significance of the leaf nodules of *Psychotria, Plant and Soil,* Special Volume, 361–374 (1971).
11. Bhattacharjee, M. and Mukerji, K.G. The SEM structure of *Sclerocystis coremio ides, Nova Hedwigia* (in press) (1981).
12. Bhattacharjee, M., Mukerji, K.G., Tewari, J.P. and Skoropad, W.P. The structure and hyperparasitism of a new species of *Gigaspora, Transactions of the British Mycological Society* (in press) (1981).
13. Bier, J.E. Tissue saprophytes and the possibility of biological control of some tree diseases, *Forest Chronicle, 39*, 82–84 (1963a).
14. Bier, J.E. Further effect of bark saprophytes on *Hypoxylon* canker, *Forest Science, 9*, 263–269 (1963b).
15. Bier, J.E. The relation of some bark factors to canker susceptibility, *Phytopathology, 54*, 250–253 (1964).
16. Bier, J.E. and Rowat, M.H. The relation of bark moisture to the development of canker diseases caused by native facultative parasites. VII. Some effects of the saprophytes on the bark of polar and willow on the incidence of *Hypoxylon* canker, *Canadian Journal of Botany, 40*, 61–69 (1962a).
17. Bier, J.E. and Rowat, M.H. The relation of bark moisture to the development of canker diseases caused by native facultative parasites. VIII.

Ascospore infection of *H. pruinatum* (Klotzsch) Cke., *Canadian Journal of Botany*, *40*, 879–901 (1962b).

18. Bier, J.E. and Rowat, M.H. Some inoculum and substrate factors in the cultural inhibition of *Hypoxylon pruinatum* (Klotzsch) Cke. by pyrocatechol, *Canadian Journal of Botany*, *41*, 1585–1596 (1963).
19. Blakeman, J.P. and Parbery, D.G. Stimulation of appresorium formation in *Colletotrichum acutatum* by phylloplane bacteria, *Physiological Plant Pathology*, *11*, 313–325 (1977).
20. Bond, G. Some biological aspects of nitrogen fixation, in *Recent Aspects of Nitrogen Metabolism in Plants* (Edited by E.J. Hewitt and C.V. Cutting), Academic Press, London, 15–25, (1968).
21. Bowen, G.D. Dysfunction and shortfalls in symbiotic responses, pp. 231–256 in *Plant Disease*, Vol. 3 (Edited by J.G. Horsfall and E.B. Cowling), Academic Press, New York (1978).
22. Bowen, G.D. Misconceptions, concepts and approaches in rhizosphere biology, pp. 283–304 in *Contemporary Microbial Ecology* (Edited by D.C. Ellwood, J.N. Hedger, M.J. Latham, J.M. Lynch and J.H. Slater), Academic Press, London (1980).
23. Bowen, G.D. The effects of mycorrhizas on nitrogen uptake by plants, in *Terrestrial Nitrogen Cycles: Processes, Ecosystem Strategies and Management Impacts* (Edited by F.E. Clark and T. Rosswall), Ecological Bulletin 33, Swedish National Science Research Council, Stockholm, Sweden (1981).
24. Bowen, G.D. and Rovira, A.D. Microbial colonisation of plant roots, *Annual Review of Plant Pathology*, *14*, 121–144 (1976).
25. Bristow, J.M. Nitrogen fixation in the rhizosphere of freshwater angiosperms, *Canadian Journal of Botany*, *52*, 217–221 (1964).
26. Brown, M.E. Plant growth substances produced by microorganisms of soil rhizosphere, *Journal of Applied Bacteriology*, *35*, 443–451 (1972).
27. Brown, M.E. Rhizosphere microorganisms—opportunists, bandits and benefactors, pp. 221–238 in *Soil Microbiology—A critical review* (Edited by N. Walker), Butterworths, London (1975).
28. Carling, D.E., Riehle, W.G., Brown, M.F. and Johnson, D.R. Effects of a vesicular-arbuscular mycorrhizal fungus on nitrate reductase and nitrogenase activities in nodulating and non-nodulating soybeans, *Phytopathology*, *68*, 1590–1596 (1978).
29. Clark, F.E. Soil microorganisms and plant roots, *Advances in Agronomy*, *1*, 241–288 (1949).
30. Clark, F.E. Nodulation responses of two near isogenic lines of soybeans, *Canadian Journal of Microbiology*, *3*, 113–123 (1957).
31. Crush, J.R. Plant growth responses to vesicular-arbuscular mycorrhiza. VIII. Growth and nodulation of some herbage legumes, *New Phytologist*, *73*, 743 (1974).

32. Dalton, H. The cultivation of diazotrophic microorganisms, pp. 1–64 in *Methods for Evaluating Biological Nitrogen Fixation* (Edited by F.J. Bergersen), John Wiley & Sons (1980).
33. Daft, M.J. and El Giahmi, A.A. Effect of *Endogone*-mycorrhiza on plant growth. VII. Influence of infection on growth and nodulation in French bean (*Phaseolus vulgaris*), *New Phytologist*, *73*, 1139–1147 (1974).
34. Daft, M.J. and El Giahmi, A.A. Studies on nodulated and mycorrhizal peanuts, *Annals of Applied Biology*, *83*, 273–276 (1976).
35. Daft, M.J. and Okusanya, B.O. Effect of *Endogone*-mycorrhiza on plant growth. V. Influence of infection on the multiplication of viruses in tomato, petunia and strawberry, *New Phytologist*, *72*, 975–983 (1973).
36. Dart, P.J. and Day, J.M. Non-symbiotic nitrogen fixation in the field, pp. 225–251 in *Soil Microbiology—A critical review* (Edited by N. Walker), Butterworths, London (1975).
37. Davenport, R.R. Epiphytic yeasts associated with the developing grape vine, M.Sc. thesis, University of Bristol, England (1970).
38. Davenport, R.R. The distribution of yeasts and yeast-like organisms in an English vineyard, Ph.D. dissertation, University of Bristol, England (1975).
39. Davenport, R.R. Ecological concepts in studies of microorganisms in aerial plant surfaces, pp. 199–215 in *Microbiology of Aerial Plant Surfaces* (Edited by C.H. Dickinson and T.F. Preece), Academic Press, London (1976).
40. Davey, C.B. Nonpathogenic organisms associated with mycorrhizae, in *Mycorrhizae* (Edited by E. Hacskaylo), United States Department of Agriculture, Miscellaneous Publication, *1189*, 114–121 (1971).
41. Davis, R.M. and Menge, J.A. Influence of *Glomus fasciculatus* and soil phosphorus on *Phytophthora* root rot of *Citrus*, *Phytopathology*, *70*, 447–452 (1980).
42. Davis, R.M., Menge, J.A. and Erwin, D.C. Influence of *Glomus fasciculatus* and soil phosphorus on *Verticillium* wilt of cotton, *Phytopathology*, *69*, 453–456 (1979).
43. Day, J.M. Nitrogen-fixing association between bacteria and tropical grass roots, in *Biological Nitrogen Fixation in Farming Systems of the Tropics* (Edited by A. Ayanaba and P.J. Dart), John Wiley, Toronto, *23*, 273–288 (1977).
44. Denison, W.C. Life in tall trees, *Scientific American*, *228*, 74–80 (1973).
45. Dehne, H.W. and Schonbeck, F. Untersuchungen zum Einfluss der endotrophen Mycorrhiza auf Pflanzenkrankheiten. 3. Chitinase aktivitata und Ornithinzyklus, *Zeitschrift für Pflanzenkrankheiten und Pflanzenschutz*, *85*, 666-678 (1978).
46. Dickinson, C.H. Fungal colonisation of *Pisum* leaves, *Canadian Journal of Botany*, *45*, 915–925 (1967).

47. Dickinson, C.H. Fungi on the aerial surfaces of higher plants, pp. 293–324 in *Microbiology of Aerial Plant Surfaces* (Edited by C.H. Dickinson and T.F. Preece), Academic Press, London (1976).
48. Dickinson, C.H. and Preece, T.F. (Editors). *Microbiology of Aerial Plant Surfaces*, Academic Press, London (1976).
49. Dobereiner, J. Rhizosphere associations between grasses and nitrogen-fixing bacteria, Effect of oxygen on nitrogenase activity in the rhizosphere of *Paspalum notatum*, *Soil Biology and Biochemistry*, *5*, 157–159 (1973).
50. Dobereiner, J. Present and future opportunities to improve the nitrogen nutrition of crops through biological fixation, pp. 1, 3–12 in *Biological Nitrogen Fixation in Farming Systems of the Tropics* (Edited by A. Ayanaba and P.J. Dart), John Wiley, Toronto (1977).
51. Dobereiner, J. and Day, J.M. Associative symbiosis in tropical grasses: Characterisation of microorganisms and dinitrogen fixing sites, in *Symposium on Nitrogen Fixation* (Edited by W.E. Newton and C.J. Nyman), Washington State University Press, Pullman, U.S.A. (1976).
52. Dobereiner, J., Day, J.M. and Dart, J.M. Nitrogenase activity and oxygen sensitivity of the *Paspalum notatum-Azotobacter paspali* association, *Journal of General Microbiology*, *71*, 103–116 (1972).
53. Dobereiner, J., Day, J.M. and vön Bulow, J.F.W. Association of nitrogen-fixing microorganisms with roots of forage grass and grain species, *Proceedings of International Winter Wheat Conference*, Zagreb, Yugoslavia (1975).
54. Doetsch, R.N. and Cook, T.M. *Introduction to Bacteria and their Ecobiology*, University Park Press, Baltimore, U.S.A., *4*, 133–120 (1973).
55. Eleusenova, N.G. and Selivanov, A. Seasonal changes in mycorrhizae of desert plants, *Mikology and Fitopathology*, *9*, 473–476 (1975).
56. Elkan, G.H. Comparison of rhizosphere microorganisms of genetically related nodulating and non-nodulating soybean lines, *Canadian Journal of Microbiology*, *8*, 79–87 (1962).
57. Fanelli, C. and Albonetti, S.G. Nitrogen fixation linked with mycorrhizas, *Annals of Botany* (Rome), *31*, 175–186 (1972).
58. Farnsworth, R.B. Nodulation and nitrogen fixation in shrubs, pp. 32–71 in *Proceedings of Symposium and Workshop on Wildland Shrubs* (Edited by H.C. Stutz), Brigham Young University Press, Provo, Utah, U.S.A. (1975).
59. Farnsworth, R.B., Romney, E.M. and Wallace, A. Nitrogen fixation by microfloral-higher plant associations in arid to semi-arid environments, in *Nitrogen in Desert Ecosystems* (Edited by N.E. West and J.J. Skuijins), US/IBP Synthesis Series 9, Dowden. Hutchinson and Rose Inc., U.S.A., *2*, 17–19 (1978).
60. Fletcher, L.M. Bacterial symbiosis in the leaf nodules of Myrsinaceae

and Rubiaceae, in *Microbiology of Aerial Plant Surfaces* (Edited by C.H. Dickinson and T.F. Preece), Academic Press, London, 465–485 (1976).

61. Fletcher, L.M. and Rhodes-Roberts, M.E. The bacterial leaf nodule association in *Psychotria*, pp. 99–118 in *Plant Pathogens* (Edited by D.W. Lovelock), Society of Applied Bacteriology, Technical Series No. 12, Academic Press, London (1979).
62. Fox, J.A. and Spasoff, L. Introduction of *Heterodera solanacearum* and *Endogone gigantea* on tobacco, *Journal of Nematology*, *4*, 224–225 (1972).
63. Garner, J.H.B. Some notes on the study of bark fungi, *Canadian Journal of Botany*, *45*, 540–541 (1967).
64. Gates, C.T. Nodule and plant development in *Stylosanthes humulis* HBK: Symbiotic responses to phosphorus and sulfur, *Australian Journal of Botany*, *22*, 45–55 (1974).
65. Godfrey, R.M. Studies on British species of *Endogone*. II. Fungal parasites, *Transactions of the British Mycological Society*, *40*, 136–144 (1957).
66. Griffin, G.J., Hale, M.G. and Shay, F.J. Nature and quantity of sloughed organic matter produced by roots of axenic peanut plants, *Soil Biology and Biochemistry*, *8*, 29–32 (1976).
67. Grime, J.P. *Plant Strategies and Vegetation Processes*, John Wiley, New York (1976).
68. Hallam, N.D. and Juniper, B.E. The anatomy of leaf surface, pp. 3–37 in *Ecology of Leaf Surface Microorganisms* (Edited by T.F. Preece and C.H. Dickinson), Academic Press, London (1971).
69. Hiltner, L. Über neuere Erfahrungen und Probleme auf dem Gebiet der Bodenbakteriologie und unter besonderer Berucksichtigung der Grundungung und Brache, *Arb. Deut. Landw. Ges.*, *98*, 59–78 (1904).
70. Hollowway, P.J. The chemical and physical characteristics of leaf surfaces, pp. 39–53 in *Ecology of Leaf Surface Microorganisms* (Edited by T.F. Preece and C.H. Dickinson), Academic Press, London (1971).
71. Hornby, D. Microbial antagonisms in the rhizosphere, *Annals of Applied Biology*, *89*, 97–100 (1978).
72. Hunter, R.B., Wallace, A., Romney, E.M. and Wieland, P.A.T. Nitrogen transformations in Rock Valley and adjacent areas of Mojave Desert, *US/IBP Desert Biome Research Memoirs*, 75–85, Utah State University, Logan, p. 8 (1975).
73. Iqbal, S.H., Qureshi, K.S. and Jabed, S.A. Influence of vesicular-arbuscular mycorrhiza on damping off caused by *Rhizoctonia solani* in *Brassica napus*, *Biologia* (Lahore), *23*, 197–208 (1977).
74. Ishizawa, S., Suzuki, T. and Araragi, M. Ecological study of free living nitrogen fixers in paddy soils, pp. 41–49 in *Nitrogen Fixation and Nitro-*

gen Cycle (Edited by H. Takahashi), Japanese IBP Synthesis, Vol. 12, University of Tokyo Press, Japan (1975).

75. Jones, K. Nitrogen fixation in the phyllosphere of the Douglas fir, *Pseudotsuga douglasii*, *Annals of Botany*, *34*, 239–244 (1970).
76. Jones, K. Nitrogen-fixing bacteria in the canopy of conifers in a temperate forest, pp. 451–463 in *Microbiology of Aerial Plant Surfaces* (Edited by C.H. Dickinson and T.F. Preece), Academic Press, London (1976).
77. Jones, K., King, E. and Eastlick, M. Nitrogen fixation by free living bacteria in the soil and in the canopy of Douglas fir, *Annals of Botany*, *38*, 765–772 (1974).
78. Katznelson, H. Nature and importance of the rhizosphere, pp. 187–209 in *Ecology of Soil-borne Plant Pathogens—Prelude to Biological Control* (Edited by K.F. Baker and W.C. Snyder), University of California Press, Los Angeles, Berkeley (1965).
79. Khan, A.G. The occurrence of mycorrhizas in halophytes, hydrophytes and xerophytes and of *Endogone* spores in adjacent soils, *Journal of General Microbiology*, *81*, 7–14 (1974).
80. Khan, S. and Khan, A.G. Role of vesicular-arbuscular mycorrhizal fungus in the resistance of wheat roots to infection by *Urocystis tritici*, *Proceedings of Pakistan Science Congress*, *25*, 1–11 (1974).
81. Khudiari, A.K. Mycorrhiza in desert soils, *Bio-Science*, *19*, 598–599 (1969).
82. Knowles, R. The significance of asymbiotic dinitrogen fixation by bacteria, in *A Treatise on Dinitrogen Fixation, Section IV. Agronomy and Ecology* (Edited by R.W.F. Hardy and A.H. Gibson), John Wiley and Sons, New York, *2*, 33–49 (1977).
83. Krieg, N.R. and Tarrand, J.J. Taxonomy of root associated nitrogen fixing bacterium, *Spirillum lipoferum*, in *Limitations and Potentials for Biological Nitrogen Fixation in the Tropics* (Edited by J. Dobereiner, R.H. Burris, A. Hollaender, A.A. Franco, C.A. Neyra and D.B. Scott), Plenum Press, New York, *10*, 317–333 (1977).
84. Last, F.T. Seasonal influence of *Sporobolomyces* on cereal leaves, *Transactions of the British Mycological Society*, *38*, 221–239 (1955).
85. Last, F.T. and Warren, R.C. Non-parasitic microbes colonising green leaves: Their form and functions, *Endeavour*, *31*, 143–150 (1972).
86. Leben, C. Colonisation of soybean buds by bacteria: Observations with the scanning electron microscope, *Canadian Journal of Microbiology*, *15*, 319–320 (1969).
87. Leben, C. The bud in relation to the epiphytic microflora, pp. 117–127 in *Ecology of Leaf Surface Microorganisms* (Edited by T.F. Preece and C.H. Dickinson), Academic Press, London (1971).
88. Leben, C., Schroth, M.N. and Hildebrand, D.C. Colonization and

movement of *Pseudomonas syringe* on healthy bean seedlings, *Phytopathology, 60*, 677–680 (1970).

89. Lersten, N.R. and Horner, H.T. Jr. Bacterial leaf nodule symbiosis in angiosperms with emphasis on Rubiaceae and Myrsinaceae, *The Botanical Review, 42*, 145–214 (1976).
90. Loutit, M.W. and Brooks, R.R. Rhizosphere organisms and molybdenum concentrations in plants, *Soil Biology and Biochemistry, 2*, 131–135 (1970).
91. Mahmoud, S.A.Z., Sbou El-Fadl, M. and El-mofty, M.K. Studies on the rhizosphere microflora of a desert plant, *Folia microbiologia, 9*, 1–8 (1964).
92. Martin, J.K. The influence of rhizosphere microflora on the availability of P32-myoinositol hexaphosphate phosphorus to wheat, *Soil Biology and Biochemistry, 5*, 473–483 (1973).
93. Martin, J.K. and Juniper, B.E. *The Cuticles of Plants*, Edward Arnold (Publishers) Ltd. (1970).
94. Marx, D.H. Ectomycorrhizae as biological deterrents to pathogenic root infections, pp. 81–96 in *Mycorrhizae* (Edited by E. Hacskaylo), United States Department of Agriculture, Miscellaneous Publication No. 1189 (1971).
95. Matare, R. and Hattingh, M.J. Effect on mycorrhizal status of avocado seedlings on root rot caused by *Phytophthora cinnamomi, Plant and Soil, 49*, 433–436 (1976).
96. Mathur, M. and Mukerji, K.G. Antagonistic behaviour of *Cladosporium spongiosum* against *Phyllactibia dalbergiae* on *Dalbergia sissoo, Angewandte Botanik* (in press) (1981).
97. Mishra, R.R. Nature of rhizosphere fungal flora of certain plants, *Plant and Soil, 37*, 162–166 (1967).
98. Mishra, R.R. and Srivastava, V.B. Rhizosphere fungal flora of certain legumes, *Annales de l'Institut Pasteur, 117*, 717–723 (1969).
99. Mishustin, E.N. and Shilnikova, V.K. *Biological Fixation of Atmospheric Nitrogen*, Pennsylvania State University Press, University Park, (1971).
100. Moiroud, A., Capellano, A. and Bartschi, H. Fixation d'agite ches les especes ligneuses symbiotiques. I. Ultrastructure des nodules, mycorrhizes a vesicules et a arbuscules et activite reductrice de C_2H_2 de jeunes plants de *Robinia pseudoacacia* cultives au laboratoire, *Canadian Journal of Botany, 59*, 481–490 (1981).
101. Moore, A.W. *Soils and Fertilizers, 29*, 113–128 (1966).
102. Mosse, B. Advances in the study of vesicular-arbuscular mycorrhiza. *Annual Review of Phytopathology, 11*, 171–196 (1973).
103. Mosse, B. A microbiologist's view of root anatomy, pp. 339–366 in *Soil Microbiology—A critical review* (Edited by N. Walker), Butterworths, London, (1975).

104. Mosse, B., Powell, C.L.L. and Hayman, D.S. Plant growth responses to vesicular-arbuscular mycorrhiza. IX. Interactions between VA mycorrhiza, rock phosphate and symbiotic nitrogen fixation, *New Phytologist*, *76*, 331 (1976).
105. Mukerji, K.G. The phylloplane—Pathogens versus non-pathogens, pp. 20–25 in *Current Trends in Plant Pathology* (Edited by S.P. Ray Chaudhuri and J.P. Verma), Parnassus Publishers, New Delhi (1974).
106. Mukerji, K.G. and Rikhy, M. The ecology of microfungi on leaves of *Triticum aestivum* cv Kalyan sona, *Abstract*, III International Symposium on the Microbiology of Leaf Surfaces, Aberdeen, U.K. (1980).
107. Neal, J.L., Atkinson, T.G. and Larson, R.I. Changes in the rhizosphere microflora of spring wheat induced by disomic substitution of a chromosome, *Canadian Journal of Microbiology*, *16*, 153–158 (1970).
108. Newman, E.I. Root microorganisms: Their significance in the ecosystem, *Biological Reviews*, *53*, 511–554 (1978).
109. Partquin, D. and Knowles, R. *Marine Biology* (Berlin), *16*, 49 (1972).
110. Pate, J.S. Uptake, assimilation and transport of nitrogen compounds by plants, *Soil Biology and Soil Biochemistry*, *5*, 109–119 (1973).
111. Preece, T.F. Micro-exploration and mapping of apple scab infections, *Transactions of the British Mycological Society*, *46*, 523–529 (1963).
112. Preece, T.F. and Dickinson, C.H. (Editors). *Ecology of Leaf Surface Microorganisms*, Academic Press, London (1971).
113. Pugh, G.J.F. Strategies in fungal ecology, *Transactions of the British Mycological Society*, *75*, 1–14 (1980).
114. Pugh, G.J.F. and Buckley, N.G. The leaf surface as a substrate for colonisation by fungi, pp. 431–445 in *Ecology of Leaf Surface Microorganisms* (Edited by T.F. Preece and C.H. Dickinson), Academic Press, London (1971).
115. Raju, P.N., Evans, H.S. and Seidler, R.S. An asymbiotic nitrogen-fixing bacterium from the root environment of corn, *Proceedings of the National Academy of Sciences*, USA, *69*, 3474–3478 (1972).
116. Ranga Rao, V. and Mukerji, K.G. Fungi in the root zone of four cultivars of wheat, *Annales de l'Institut Pasteur*, *121*, 533–544 (1971).
117. Richards, B.N. Nitrogen fixation in the rhizosphere of conifers, *Soil Biology and Biochemistry*, *5*, 149–152 (1973).
118. Richards, B.N. *Introduction to the Soil Ecosystems*, Longman Group Limited, U.K., 9, 214–247 (1974).
119. Roncadori, R.W. and Hussey, R.S. Interaction of the endomycorrhizal fungus *Gigaspora margarita* and root-knot nematode on cotton, *Phytopathology*, *67*, 1507–1512 (1977).
120. Rose, S.L. Mycorrhizal associations of some actinomycete nodulated nitrogen-fixing plants, *Canadian Journal of Botany*, *58*, 1449–1454 (1980).
121. Rose, S.L. and Youngberg, C.T. Tripartite associations in snow brush

(*Ceanothus velutinus*): Effect of vesicular-arbuscular mycorrhizae on growth, nodulation and nitrogen fixation, *Canadian Journal of Botany*, *59*, 34–39 (1981).

122. Rose, J.P. and Ruttencutter, R. Population dynamics of two vesicular-arbuscular endomycorrhizal fungi and the role of hyperparasitic fungi, *Phytopathology*, *67*, 490 (1977).
123. Rovira, A.D. Interactions between plant roots and soil microorganisms, *Annual Review of Microbiology*, *19*, 241–266 (1965).
124. Rovira, A.D., Bowen, G.D. and Foster, R.C. The nature of the rhizosphere and the influence of the rhizosphere microflora and mycorrhizas on plant nutrition, in *Encyclopedia of Plant Physiology*, Vol. 12 (Edited by A. Lauchi and R.L. Bieleski), Springer-Verlag, Berlin (1980).
125. Rovira, A.D. and Campbell, R. A scanning electron microscope study of interactions between microorganisms and *Geaumannomyces graminis* (Syn. *Ophiobolus graminis*) on wheat roots, *Microbial Ecology*, *2*, 177–185 (1975).
126. Rovira, A.D., Foster, R.C. and Martin, J.K. Notes on the terminology, origin, nature and nomenclature of the organic materials in the rhizosphere, pp. 1–4 in *The Soil Root Interface* (Edited by J.L. Harley and R.S. Russell), Academic Press, London (1979).
127. Ruinen, J. The phyllosphere. III. Nitrogen fixation in the phyllosphere, *Plant and Soil*, *22*, 375–394 (1956).
128. Ruinen, J. The phyllosphere. I. An ecologically neglected milieu. *Plant and Soil*, *15*, 81–109 (1961).
129. Ruinen, J. The phyllosphere. V. The grass sheath, a habitat for nitrogen-fixing microorganisms, *Plant and Soil*, *33*, 661–671 (1970).
130. Ruinen, J. The grass sheath as a site of nitrogen fixation, pp. 567–579 in *Ecology of Leaf Surface Mircoorganisms* (Edited by T.F. Preece and C.H. Dickinson), Academic Press, London (1971).
131. Ruinen, J. Nitrogen fixation in the phyllosphere, pp. 121–167 in *The Biology of Nitrogen Fixation* (Edited by A. Quispel), North Holland Publishing Co., Amesterdam (1974).
132. Schenck, N.C., Kinloch, R.A. and Dickson, S.W. Interaction of endomycorrhizal fungi and root-knot nematode on soybean, pp. 607–618 in *Endomycorrhizas* (Edited by F.E. Sander, B. Mosse and P.B. Tinker), Academic Press, London (1975).
133. Schenck, N.C. and Nicolson, T.H. A zoosporic fungus occurring on species of *Gigaspora margarita* and other vesicular-arbuscular mycorrhizal fungi, *Mycologia*, *69*, 1049–1053 (1977).
134. Schonbeck, F. Einfluss der endotrophen Mycorrhiza auf die Krankheitenresistenz hoherer Pflanzen, *Zeitschrift für Pflanzenkrankheiten und Pflanzenschutz*, *85*, 191–196 (1978).
135. Schonbeck, F. and Dehne, H.N. Untersuchungen zum Einfluss der endo-

trophen Mycorrhiza auf Pflanzenkrankheiten 4, Pilzliche Sprossparasiten *Olipidium brassicae*, TMV, *Zeitschrift für Pflanzenkrankheiten und Pflanzenschutz*, *86*, 103–112 (1979).

136. Schonbeck, F. and Spengler, G. Nachweis von TMV in Mycorrhiza-haltigen zillen der Tomate mit Hilfe der Immunofluoreszenz, *Phytopathologische Zeitschrift*, *94*, 84–86 (1979).

137. Schwartz, W. Bacterian and actinomycetan Symbiosen, *Encyclopedia of Plant Physiology*, *11*, 560–572 (1959).

138. Sharma, K.R. and Mukerji, K.G. Succession of fungi on cotton leaves, *Annales de l'Institut Pasteur*, *122*, 425–454 (1972a).

139. Sharma, K.R. and Mukerji, K.G. Prevalence of *Candida albicans* on *Gossypium* leaves, *Journal of the Indian Botanical Society*, *51*, 291–297 (1972b).

140. Sharma, K.R. and Mukerji, K.G. Microbial colonization of aerial parts of plants—A review, *Acta Phytopathologica Academiae Scientiarum Hungaricae*, *8*, 425–461 (1973).

141. Sharma, K.R. and Mukerji, K.G. *Candida albicans*, a natural inhabitant of the phyllosphere, *Japanese Journal of Ecology*, *24*, 60–63 (1974a).

142. Sharma, K.R. and Mukerji, K.G. Incidence of pathogenic fungi on leaves, *Indian Phytopathology*, *27*, 558–566 (1974b).

143. Sharma, K.R. and Mukerji, K.G. Microbial ecology of *Sesamum orientale* L. and *Gossypium hirsutum* L., pp. 375–390 in *Microbiology of Aerial Plant Surfaces* (Edited by C.H. Dickinson and T.F. Preece), Academic Press, London (1976).

144. Sharma, K.R. and Mukerji, K.G. The ecology of microfungi on leaves, in *Recent Advances in the Biology of Microorganisms* (Edited by K.S. Bilgrami and K.M. Vyas), Bishen Singh Mahendra Pal Singh, Dehradun, *1*, 69–79 (1980).

145. Sikora, R.A. Einfluss der endotrophen Mykorrhiza (*Glomus mosseae*) auf das wirt-parasit-verhaltnis von *Meloidogyne incognita* in Tomaten, *Zeitschrift für Pflanzenkrankheiten und Pflanzenschutz*, *85*, 197–202 (1978).

146. Sikora, R.A. and Schonbeck, F. Effect of vesicular-arbuscular mycorrhiza (*Endogone mosseae*) on the population dynamics of the root-knot nematodes (*Meloidogyne incognita* and *M. hapla*), *Proceedings of the Eighth International Plant Protection Congress*, 158–166 (1975).

147. Silver, W.S., Centifanto, J.M. and Nicholas, D.J.D. Nitrogen fixation by leaf nodule endophyte of *Psychotria bacteriophila*, *Nature*, London, *199*, 396–397 (1968).

148. Skujins, J. Nitrogen cycling in arid ecosystems, in *Terrestrial Nitrogen Cycles, Processes, Ecosystems, Strategies and Management Impacts* (Edited by F.E. Clark and T. Rosswall), Swedish National Science Research Council, Sweden (1981).

149. Snell, W.H. and Dick, E.A. *A Glossary of Mycology*, Harvard University Press, Cambridge, Massachusetts (1971).
150. Snyder, J.M. and Wullstein, L.H. The role of desert cryptogams in nitrogen fixation, *Ann. Midl. Nat.*, *90*, 257–265 (1973).
151. Sorokina, T.A. The specificity of the microflora complex of the plant root system, *Proceedings of Symposium on Relationship between Soil Microorganisms and Plant Roots, 1963*, 42–47 (1965).
152. Sparrow, P.K. A *Rhizidiomycopsis* on azygospores of *Gigaspora margarita*, *Mycologia*, *69*, 1053–1058 (1977).
153. Stenton, H. Colonisation of roots of *Pisum sativum* by fungi, *Transactions of the British Mycological Society*, *41*, 74–80 (1958).
154. Stevenson, G.B. Bacterial symbiosis in some New Zealand plants, *Annals of Botany*, *17*, 343 (1953).
155. Strzelczyk, E. and Mielczarek, A. Comparative studies on metabolic activity of planktonic, benthic and epiphytic bacteria, *Hydrobiologia*, *38*, 67–77 (1971).
156. Subba Rao, N.S., Bidwell, R.G.S. and Bailey, D.L. The effect of rhizosphere fungi on the uptake and metabolism of nutrients by tomato plants, *Canadian Journal of Botany*, *39*, 1759–1764 (1961).
157. Tien, T.M., Gaskins, M.H. and Hubbell, D.H. Plant growth substances produced by *Azospirillum brasiliense* and their effect on the growth of Pearl millet (*Pennisetum americanum* L.), *Applied Environment Microbiology*, *37*, 1016–1024 (1979).
158. vön Bulow, J.F.W. and Dobereiner, J. Potential for nitrogen fixation in maize genotypes in Brazil, *Proceedings of the National Academy of Sciences*, USA, *72*, 2389–2393 (1975).
159. Waidyanatha, V.P., De, S., Yagaratnam, W. and Ariyaratne, W.A. Mycorrhizal infection on growth and nitrogen fixation of *Pueraria* and *Stylosanthes* and uptake of phosphorus from two rock phosphates, *New Phytologist*, *82*, 147–152 (1979).
160. Wallace, A. and Romney, E.M. *Radioecology and Ecophysiology of Desert Plants at the Nevada Test Site*, United States Atomic Energy Commission, TID-26954, Office of Information Services, Springfield, Va. (1972).
161. Wallace, A., Romney, E.M., Cha, J.W. and Soufi, S.M. Nitrogen transformations in Rock Valley and adjacent areas of the Mojave Desert, *US/IBP Desert Biome Research Memoirs*, 74–76, Utah State University, Logan, p. 25 (1974).
162. Webley, D.M. and Duff, R.B. The incidence in soils and other habitats of microorganisms producing 2-ketogluconic acid, *Plant and Soil*, *22*, 307–313.
163. Went, F.W. and Stark, N. Mycorrhiza. *Bio-Science*, *18*, 1035–1039 (1968).

164. West, N.E. Nutrient cycling in desert ecosystems, in *Arid Ecosystems* (Edited by R.A. Perry and D.W. Goodall), Cambridge University Press, London (1980).
165. William, S.E. and Aldon, E.F. Growth of *Atriplex canescens* (Pursh) Nutt. improved by formation of vesicular-arbuscular mycorrhizae, *Endogone mosseae*, *Soil Science Society of America, Proceedings*, *38*, 962–965 (1974).
166. William, S.E. and Aldon, E.F. Endomycorrhizal (vesicular-arbuscular) associations of some arid zone shrubs, *South West Nat.*, *20*, 437–444 (1976).
167. Woldendorp, J.W. *Meded. Landbhogesch. Opzoekstns. Gent*, *63*, 1–100, (1963).
168. Wullstein, L.H., Bruening, M.L. and Bollen, W.B. Nitrogen-fixation associated with sand grain root sheaths (rhizosheaths) of certain xeric grasses, *Physiologia Plantarum*, *46*, 1–4 (1979).

D.G. PATRIQUIN

6. New Developments in Grass-Bacteria Associations

Introduction

The existence of N_2-fixing grass-bacteria associations, or "associative symbioses", has been recognized or suspected for over 50 years (17), but practically all of our knowledge of the nature of these systems has been gained in the last 15 years. The latter developments are attributable to the introduction of the acetylene reduction assay for N_2 fixation (89, 188). Moore (127) in 1966 compiled literature dating back to 1928 which suggested that annual gains of N under grasses were frequently in the range of 30 to 60, and ran as high as 375, pounds of N per hectare. Yeasts, fungi, and even higher plants were suspected N_2-fixing agents, and the N_2 fixation process was considered to be physiologically exothermic. By 1974 it was clear that nitrogenase is restricted to prokaryotes, the enzyme complex had been purified, its requirement for ATP established, and the N_2-fixing genes had been transferred to *E. coli* (158). In that year, Dobereiner and Day (62) reported their discovery of a novel type of diazotroph (N_2-fixing organism), *Azospirillum* (syn. *Spirillum lipoferum*, 191), in association with tropical grasses, and they speculated that its apparently unique physiology facilitated the formation of a primitive sort of symbiosis.

It is not surprising that the intense scrutiny that this organism and its association with grasses received over the ensuing two years (reviewed in Neyra and Dobereiner, 131), led to some questioning and modification of the methodologies employed, and to speculations related to grass-diazotroph associations; it was also accompanied by some disappointment that there were not much immediate benefits to be had from inoculating temperate crops with *Azospirillum* (16). Throughout that period Dobereiner maintained that it was premature to expect immediate applications, that there was far more to be learned about these associations and their functioning before agronomic applications could be expected (59). This viewpoint has been vindicated by the developments since the publication of the comprehensive review of grass diazotroph associations by Neyra and Dobereiner in 1977 (131); it is these developments that are reviewed herein.

I will make frequent reference to salt marsh cord grass, *Spartina alterniflora.* This plant is a C-4 type emergent halophyte which occurs in marshes of subtropical and temperate zone estuaries in eastern North America. Fixation of $^{15}N_2$ in the rhizosphere and incorporation of some of that N by marsh grasses was reported by Jones in 1974 (96); subsequent studies have shown that *Spartina alterniflora* could provide a convenient model for studying associative N_2 fixation.

Evidence for N_2 fixation associated with grasses

C_2H_2 assays

Many of the earlier assessments of nitrogenase activity associated with grasses were based on acetylene-reducing activities (ARA) in bottles or flasks containing roots that had been "preincubated" overnight at low pO_2. For *Paspalum notatum*, the first grass for which a specific agent of N_2 fixation (*Azotobacter paspali*) was identified, Dobereiner *et al.* (63) observed immediate and roughly constant ARA by plant-soil systems contained in cores. When the soil was removed and roots were excised and assayed, ARA was initiated only after a lag period of about 8 hours. They suggested that the "lag" was due to disturbance of the root-*Azotobacter* association during sampling, possibly through an effect on the oxygen status of nitrogenase, or on root exudation. This postlag ARA was lower than that observed for the whole plant system, and therefore seemed to give a conservative indication of whole plant nitrogenase activity. Surveys of plants using the "preincubated excised root technique" resulted in the discovery of the grass-*Azospirillum* associations (62).

It was subsequently found that for certain plant species or under certain conditions, post-lag excised root ARA was much greater than immediate ARA measured in cores or *in situ* (16, 116, 194, 196), and it has now been generally accepted that post-lag ARA is not a reliable indicator of *in situ* levels of nitrogenase activity. Nevertheless, if its limitations are appreciated, the preincubated excised root technique is a convenient tool in surveys of plants for associations with diazotrophs (183), and post-lag ARA may reveal ecologically (150, 183), physiologically (32), or genetically (81) sensible trends. Thus post-lag excised root ARA might best be regarded as a parameter of, rather than a direct measure of, the *in situ* nitrogenase activity.

There have been various explanations for "the lag", and the enhanced ARA, compared to *in situ* ARA, that occurs after the lag. Thirty-fold or greater proliferation of diazotrophs occurs during the preincubation of roots excised and washed from corn (16, 138, 195), but not during the preincubation of unwashed roots (195). Post-lag ARA in the washed roots is higher than in unwashed roots (91, 195), except in *Spartina* (144) for which proliferation of diazotrophs was not observed during the preincubation of washed roots (197) or of roots and soil (53). Thus proliferation may result in the enhancement of ARA, but does not appear to be the cause of the lag. The lag is of longer duration for roots taken from N-fertilized plants than from unfertilized plants (195). In *Spartina* the lag decreases in length through the season (144) and may disappear altogether as reproductive growth is approached (54). Hence it has been proposed that the absence of nitrogenase activity or very low nitrogenase activity during the lag is due to the suppression of the same by ammonia or nitrate (53). As the season progresses, either less combined N is present (116) or bacterial activity is higher, resulting in progressively shorter periods being required to consume the mineral N (53).

Dicker and Smith (53) reported that the effects of adding ammonium to root (*Spartina*) and soil samples were mimicked by chloramphenicol, while the effects of nitrate, which caused amore immediate and complete suppression of ARA than did ammonium, were mimicked by chlorate; chlorate is reduced to chlorite, possibly by the same system responsible for the reduction of nitrate. On a molar basis, three times as much chlorate as nitrate were required to suppress ARA to a similar degree, which is in reasonable agreement with a value of 4, which would be expected if the suppression were due to competition between nitrogenase and the nitrate/chlorate reductase systems for electrons. They propose that *in situ* nitrogenase activity (and accordingly, immediate ARA by excised roots, and the length of the lag) are controlled by two processes: (i) repression and derepression of nitrogenase synthesis, mediated by ammonia, and (ii) competition for reducing power and energy between the processes of nitrate reduction and nitrogen fixation.

van Berkum and Sloger (198) observed immediate ARA for excised roots from *Spartina alterniflora*, *Scirpus olneyii*, and rice growing in salt marsh soils, and for maize growing in upland soils. ARA was immediately linear when roots were kept under N_2, but accelerating rates of acetylene reduction were observed when roots of grasses or of immature soybeans were handled under air, or when nodulated roots of mature soybeans were exposed to high pO_2 prior to assay. These authors propose that acetylene reduction without an extended lag period should occur if nitrogen fixation is associated with the plant in the field.

These various observations and proposals do not completely resolve the uncertainties surrounding the lag. For example, one would expect that the addition of reduced carbon to excised roots would shorten the lag by reducing competition for electrons, and possibly by immobilizing N; however such is not the case (144). Rice plants contained in dessicators exhibit immediate, linear (constant) ARA when the gas phase is evacuated and back-filled with argon-C_2H_2, but roots excised from the same plants exhibit no ARA for 6 hours (92). The possibility of oxygen inactivation of nitrogenase during the sampling of certain plant systems cannot be ruled out. Shaking of *Spartina* roots that exhibited ARA in air induced an 8 hours lag (144) and van Berkum's and

Sloger's data (198) clearly demonstrate a transient inactivation of nitrogenase by air, in that case for only about 1 hour, but it might be longer for plants taken from reduced soils. Thus, while it is reasonable to conclude that immediate, linear ARA by excised roots does reflect nitrogenase activity *in situ*, the absence of nitrogenase activity within 6 to 8 hours after excision of roots may not, in all cases, reflect the absence of nitrogenase activity *in situ*.

Rates of N_2 fixation that have been estimated from ARA in lag-free systems are in general considerably lower than rates estimated from post-lag ARA of excised roots (Table 1). *In situ*, C_2H_2 assays involving the insertion of a cylinder around plants (7, 147), or enclosure of leaves alone in a cylinder (147) entail less disturbance of plants than with core techniques. When these are assayed for nitrogenase activity sometime after they are taken from the field (e.g., 194, 196), they may be prone to the same enhancement phenomena as are excised roots. Also, the growth of blue-green algae occurs rapidly in confined, high moisture environments as occurring in watered pots in greenhouses.

Several chemicals have been found to effectively suppress phototrophic nitrogenase activity without inhibiting heterotrophic nitrogenase activity; these include streptomycin (85), "propanil" (3′, 4′-dichloropropionanilide) (24, 86), and zinc ion with manganese ethylene bisdithiocarbamate (213). Propanil inhibits nitrogenase activity of blue-green algae but not of bacteria, and thus in combination with the exclusion of light from the soil surface, can be used to distinguish between N_2 fixation by blue-green algae, photosynthetic bacteria and heterotrophic bacteria (86).

In waterlogged soils, it may not be possible to saturate the nitrogenase with C_2H_4, and C_2H_2 can become occluded within the soil (122, 147). Growing plants or transferring field grown plants to water culture and assaying these in two phase shoot-root systems overcomes some of these problems (27, 207), but of course excludes the rhizosphere system.

Acetylene inhibits the reduction of nitrous oxide to dinitrogen by denitrifying enzymes, and the accumulation of nitrous oxide in the presence of C_2H_2 indicates denitrifying activity (216). Assays for denitrification in plant-soil systems can be carried out simultaneously with assays for nitrogenase activity (153, 216), but are subject to same sorts of limitations as are C_2H_2 assays of nitrogenase activity.

Problems related to various side effects of C_2H_2 on microbial activities are reviewed by Knowles (106). For grass systems, the most serious problem with the C_2H_2 assay is the one resulting from the inhibition of oxidation of ethylene by acetylene (210). Since the potential rate of aerobic decomposition of ethylene greatly exceeds its anaerobic production (41, 210), ethylene does not normally accumulate in aerobic-anaerobic systems, i.e. in "controls" without C_2H_2. However, when C_2H_2 is introduced, the oxidation of ethylene is inhibited and ethylene accumulates resulting in "acetylene-dependent, nitrogenase-independent ethylene production." The troublesome feature of this phenomenon is that the conditions for its occurrence are very similar to those which favor root associated nitrogenase activity: the presence of anae-

Table 1. Daily rates of N_2 fixation estimated from acetylene-reducing activities in lag-free systems, and from that of "preincubated" excised roots

Method of measurement, plant and site	Rate in lag-free system (g N per hectare per day[a])	Rate from assays of preincubated excised roots (g N per hectare per day[a])	Reference
Bags around plants in containers			
Maize, different cultivars in non-sterile soil, inoculated with *Azospirillum*, Oregon, U.S.A.	0.1–2.3	1–151	16
Rice in waterlogged soil, Trinidad	78	112	24
Bluegrass in artificial soil inoculated with *Klebsiella pneumoniae*, Nebraska, U.S.A.	11.5		213
Chambers, in situ			
Rice, different cultivars in waterlogged soil at heading stage, Philippines	17–61		206
Wild grasses, Nova Scotia, Canada	4.8–60		183
Spartina alterniflora in salt marsh, Canada, during reproductive growth	62.1	377	116
Cores			
Forage grasses, over 12 weeks, Australia	46.4–151		209
Forage grasses, Texas, U.S.A.	50[b]–165		208
Forage grasses, fertilized, Brazil, averages for different cultivars	89–975		176
Digitaria decumbens	16.1	36	196
Paspolum notatum	28.2	291	196
Sorghum vulgare in Brazil	14.7	347	
Soil+roots extruded under Argon			
Turtle grass (*Thalassia testudinum*, submerged aquatic)			
Florida and Bahamas	51–243		33
Barbados	376	265–1325	33, 151

[a]I have made use of only reports in which data are given allowing estimation of rates on a per unit area basis; a 3:1 ratio was used in converting moles C_2H_2 reduced to moles N_2 fixed. Values are means of more than one assay.

[b]Many values were greater than 50.

robic-aerobic interfaces, plant roots, high moisture and low soil nitrate all stimulate soil ethylene production (181) and nitrogenase activity.

Witty (210) modified the flame ionization detector on his gas chromatograph so as to minimize the "dead" volume and to enable the collection of effluent CO_2 produced by oxidation of individual hydrocarbons as they passed through the detector. By incubating samples with ^{14}C-labelled C_2H_2 and determining the specific activity of the ethylene produced, he was able to estimate the proportion of ethylene production that was due to nitrogenase activity. That proportion was close to 100 per cent for root nodules, cell suspensions of *Bacillus polymyxa* and *Clostridium pasteurianum*, and for a very active soil core containing the grass *Lolium perenne* (N_2 fixation estimated from C_2H_2 dependent C_2H_4 production was over 300 g N ha^{-1} day^{-1}). However, it was only 43 per cent for a core of *Lolium perenne* with low activity (estimated N_2 fixation less than 100 g N ha^{-1} day^{-1}), and Witty (210) suggests that if acetylene reduction techniques are to be used for assaying such systems, ^{14}C-acetylene controls should be included. Many of the reported acetylene reducing activities in grass systems (Table 1) are in the range in which Witty found substantial acetylene-dependent, nitrogenase-independent ethylene production.

As an alternative to $^{14}C_2H_2$ controls, van Berkum and Sloger (197) considered that the rate of consumption of C_2H_4 added to excised root systems lacking C_2H_2 would indicate a maximum possible rate of endogenous ethylene production because the rate of ethylene oxidation normally exceeds the rate of production. The validity of this sort of "control" still needs to be investigated. The production of ethylene in strictly anaerobic systems (without C_2H_2) may give some indication of whether endogenous production of ethylene is likely to be a problem. However, low rates of ethylene production in anaerobic controls might not indicate that real endogenous ethylene production is low, because that could be dependent on interactions between aerobic and anaerobic zones (181). In short, all estimates of nitrogenase activity in grasses that are based on C_2H_2 reduction assays must be regarded as suspect, presumptive, or at best, semiquantitative unless they are verified through the use of appropriate controls, or by alternative means of measuring N_2 fixation.

^{15}N tracer techniques

Measurement of $^{15}N_2$ fixation remains the only completely unequivocal means of demonstrating N_2 fixation. Incorporation of recently and heterotrophically fixed $^{15}N_2$ in plant tissues has been demonstrated for the salt marsh gasses *Puccinellia maritima* and *Spartina anglica* (96), for the tropical forage grasses *Digitaria decumbens* and *Paspalum notatum* (51), sugar cane (171) and rice (215) growing in soil, and for field grown rice maintained in a hydroponics system (94). Rates of $^{15}N_2$ fixation associated with rice (94) were similar to values of N_2 fixation estimated from C_2H_2 assays of plants in the field (206). The latter, integrated over the season, indicated N_2 fixation of 4.8 to 5.9 kg per hectare, which is equivalent to about 10 per cent of the total N accumulated in shoots. Much higher N_2 fixation was evident for rice in soil, $^{15}N_2$ fixation over 13 days extrapolating to 8.1 to 25.6 kg per hectare in one month. In one experiment 74 per cent was recovered in the soil (Table 2) and 16 per cent in the leaves and panicle (215). These results illustrate the quantitative significance of rhizosphere N_2 fixation, and show that a good proportion of that N_2 fixation may be recovered in plants within a short period. Large amounts of ^{15}N have been recovered in aboveground parts of rice and sugar cane (Table 2). Shoots of rice (207) and *Spartina* (154) reduce acetylene, and diazotorphs have been isolated from stem (99, 117, 149, 206); thus shoots (leaf sheaths, lower parts of leaves or stems) may contain sites of N_2 fixation.

Substantial non-rhizosphere N_2 fixation occurs in rice soils. Charyulu and Rao (36) reported that 0.9 to 4.4 mg $^{15}N_2$ were fixed per kg flooded rice soil over 28 days in the dark; in non-flooded soils, the rates were 0.4 to 2.2 mg N_2 fixed. Addition of cellulose resulted in increased N_2 fixation (to 19 and 14 mg N per kg flooded and non-flooded soil respectively), while addition of ammonium sulfate suppressed N_2 fixation. Fixation of 4 mg N per kg soil is equivalent to about 8 kg N per hectare to 15 cm depth.

Because it entails enclosing samples with a gas, use of $^{15}N_2$ to measure N_2 fixation is restricted to relatively small samples and to short intervals. The classical "nitrogen difference" method, and the recently introduced "isotope dilution" technique are not subject to these limitations. In the difference method, N_2 fixation is estimated as the difference between total N accumulation in N_2-fixing plants and non-N_2-fixing control plants. The isotope dilution technique [or the modified A value technique (73)], which is mathematically identical to the isotope dilution technique (161) is based on the principle that if a plant is growing in a medium that has a higher isotopic $^{15}N/^{14}N$ ratio than that of the atmosphere, then the isotopic composition of plant N will reflect the proportions of N derived from the medium and from the atmosphere, i.e. in the plant, the medium ^{15}N enrichment will be diluted in direct proportion to the amount of N_2 fixed from the atmosphere. The isotopic ratio of the medium is determined from the isotopic composition of non-N_2-fixing control plants growing in that medium. When ^{15}N enriched fertilizer (usually 5% atom excess or greater) is added, dilution of the fertilizer-N in the non-fixing control indicates the relative use of soil and fertilizer N, and the dilution of ^{15}N in the fixing plant compared to the non-fixing plant indicates the proportion of N derived from the atmosphere by the fixing plants. To measure N_2 fixation at the zero level of added fertilizer, ^{15}N may be incorporated into soil organic matter prior to the experiment (115); or where there is sufficient accuracy and replication or precision in sampling, variation in the natural isotopic ratios can be used to estimate N_2 fixation (3, 50, 160) since soil organic matter is naturally enriched in ^{15}N relative to the atmosphere (3, 50, 160, 178). The

Table 2. Recovery of recently fixed N in different tissues or in soil

Plant system (and reference)	Distribution (%) of recovered ^{15}N						
	Soil	Roots	Rhizomes	Leaves	Panicle	"shoots"	Other
Rice							
in soil[a] (215)	74.1	10.2		6.7	8.9		
in hydroponics[b] (94)		47.3		0.6	0.1		Outer leaf sheath, 43.3 Inner leaf sheath, 8.6
in hydroponics[c] (94)		27.5		8.0	2.6		Outer leaf sheath, 46.0 Inner leaf sheath, 9.4 Basal node, 6.4
Digitaria decumbens[d] (51)		33.1	49.3	6.3			
Paspalum notatum[d] (51)		47.5	43.6	8.6			
Sugar cane[e] (171)		21.4				78.6	

[a]In flowering stage, 13 days exposure to $^{15}N_2$.
[b]IR 26, 7 days exposure at preheading stage.
[c]Latisil, 7 days exposure at preheading stage.
[d]In pots for 2 weeks after transplanting from field, 72 hours exposure.
[e]3-month-old plants grown from seed in soil; 40 hours exposure.

significance of fractionation of N isotopes within the plant still needs to be examined more closely before the natural isotope variation can be used more generally for estimating N_2 fixation. It has been found that ^{15}N contents of senescent leaves and nodules are not representative of whole plants (178).

Comparison of N difference methods with isotope dilution methods applied to legumes (3, 161, 173) indicates that the former method may not give reliable estimates of N_2 fixation, apparently because of differences in total uptake of soil N between N_2-fixing and non-N_2-fixing plants. In soybean and lupins, nodulation seems to suppress the use of soil N to some extent, resulting in underestimation of N_2 fixation by the difference method (3). In a laboratory study, Rennie (163) found that isotope dilution and N difference methods gave similar estimates of N_2 fixation in maize inoculated with *Azospirillum*.

The major problem confronting workers attempting to evaluate associative N_2 fixation in the field using these techniques is the uncertainty concerning what types of plants should be used as non-N_2-fixing controls. The various types of diazotrophs implicated in associative N_2 fixation are ubiquitously distributed, and there are no obvious structures to indicate when they are fixing N_2 in association with plants. In this context, the search for good non-N_2-fixing plants [possibly such as those identified by Rennie (166)] becomes as important as the search for N_2-fixing plants.

Surveys of natural isotopic ratios may be of value in identifying N_2-fixing plants. Legumes are typically less enriched in ^{15}N than are non-legumes (50, 178). Rennie and Larson (166) consider that N_2 fixation has to contribute more than 10 to 15 percent of the total plant N in order to be detected by variation in natural isotopic ratios. They surveyed vegetation at a mine reclamation site (low soil N); the soil had a Δ_a ^{15}N of 7 (which corresponds to a 0.7 percent enrichment in ^{15}N relative to the atmosphere), and most of the grasses had a value in the region of 5; legumes had values (-0.22 to 1.90) approaching those of the atmosphere. Three grasses (*Elymus angustus, Agropyron elongatum* and *Agropyron dasystachyum*) had values in the range 2.31 to 3.04, suggesting that they might receive up to 50 percent of their total plant N from N_2 fixation. Similarly, on the basis of variations in natural isotopic ratios, Vose *et al.* (204) considered that the percentage of N in some varieties of 8-month-old sugar cane at one site could have been as high as 30 percent.

Nitrogen budgets

App *et al.* (4) measured changes in total plant and soil N over intervals equivalent to four to six crops in pots containing flooded rice soil ± rice plants, ± exposure of soil to light, ± iron and phosphorus fertilizer, and ± blue-green algal or *Azolla* inoculant. A negative balance was observed for rice soil without plants in the dark, no change for soil exposed to light but without plants, and positive balances for all treatments with plants. Gains in N were equivalent to approximately 14–18, 40, and 60 to 70 percent of total crop N for pots with plants in darkened soil, pots with plants in soil exposed to light, and for pots with plants in soil exposed to light and fertilizer added with or without inoculant. The first figures (14–18 percent) are attributed to heterotrophic N_2 fixation; in one experiment the gain was not significantly reduced by removing the stubble after each crop. Actual N_2 fixation could have been higher than this because without rice plants there was a negative balance; or the difference between pots with and without rice plants (soil darkened) could have been due to both reduced losses of N and increased N_2 fixation in the presence of plants. The rice plant apparently stimulated phototrophic N_2 fixation, and this was further enhanced by the addition of iron and phosphorus fertilizer, but not very greatly by addition of blue-green algal and *Azolla* inocula. Gains in pots with plants and soil exposed to light extrapolated to 21 kg N per crop. Long-term field balances for plots receiving no fertilizer indicated gains of about 37 kg N per crop per hectare (49).

Schank *et al.* (176) estimated the amount of N withdrawn from the soil by 30 cultivars of tropical forage grasses in Brazil from the difference between soil N to 25 cm depth before and after 13 harvests (28 day intervals). Of the 602 kg N per hectare recovered in forage, 200 could be accounted for in applied fertilizer and 346 by uptake from the soil leaving 56 kg N per hectare unaccounted for; N_2 fixation estimated from C_2H_2 reduction assays of cores was 38.6 kg per hectare.

Inoculation experiments

Increased yields and N contents resulting from inoculation of plants with diazotrophs cannot automatically be attributed to N_2 fixation by the inoculants. Microorganisms are well known to influence plant growth both negatively and positively through the production of hormones or by effecting changes in the rhizosphere populations. "Bacterization" of seeds or roots can result in highly significant but usually variable increases in yields (30). Kloepper *et al.* (104) believed that much of the variation in results from inoculation with "plant growth promoting bacteria" is related to differences in root colonization by the applied bacteria. Inoculation of potato plants with two *Pseudomonas* strains shown to be aggressive colonizers of roots resulted in growth increases up to fivefold in young plants in the greenhouse, and in final yield increases in the field of 17 percent. Brown (31) found that inoculation of *Paspalum notatum* with *Azotobacter paspali*, or addition of supernatant from *A. paspali* cultures, caused significant increases in plant weight before there was development (in older plants) of root nitrogenase activity. Culture fluids from *A. paspali* contained indole acetic acid, and gibberelin- and cytokinin-like substances. Tryptophan, a precursor of auxin is converted to auxin by *Azospirillum* (167). Inoculation of Pearl Millet and Guinea Grass with *Azospirillum brasilense* SP 7 results in more mucigel, root hairs and lateral roots than in controls (75). Such observations illustrate that it is necessary to correlate increases in yield or total N resulting from inoculation with other measurements of nitrogenase activity, such as C_2H_2 or $^{15}N_2$ reduction, if the increases are to be attributed to N_2 fixation.

For the most part, inoculation of grasses with diazotrophs has given rather inconclusive results, or significant increases in both yield and total N have not been observed except at very low levels of total plant N (2, 11, 15, 16, 26, 32, 163). Statistically significant increases in total N (18 to 30%) and yield were observed in two of five experiments with maize (40, 163, 2, 15, 32) and were correlated with other measures of N_2 fixation, but the total N uptake was low (2–21 mg N per plant at five to eight weeks) and not agronomically significant. Substantial increases in yield were reported following the inoculation of rye, barley and wheat with *Azospirillum* when plants were fertilized with 40 kg N per hectare but there were no equivalent increases at other levels of fertilizer-N (190). Hormonal effects have been implicated as responsible for at least part of the yield increases by forage grasses following inoculation with *Azospirillum* (185, 93).

The most encouraging results to date are those of Rennie and Larson (165, 166) obtained with disomic chromosome substitution lines of spring wheat inoculated with *Azospirillum brasilense* SP 7 or with a *Bacillus* sp. isolated from wheat. Single chromosome pairs ($2n=42$) were reciprocally exchanged between a root rot resistant line (Cadet) and a root rot susceptible line (Rescue). Chromosome 5B from Cadet conferred resistance to the Rescue line, and chromosome 5B from Rescue conferred susceptibility to the Cadet line. In

Table 3. Effects of inoculation with diazotrophs on N accumulation in disomic chromosome substitution lines of spring wheat [from Rennie and Larson (165)]

Line	Root rot suscepti-bility	Plant N in control[a] (mg/plant)	Plant N, inoc. plants (as fraction of control)		% of total N from N_2 fixation	
			Bacillus	*Azospirillum*	*Bacillus*	*Azospirillum*
Cadet		14.9	1.27	1.51	21.4	33.9
C-R2A		12.9	1.38	2.17	27.7	54.0
C-R2D		19.0	1.71	1.71	41.4	41.6
C-R5B	+	14.8	1.08[b]	1.29	0	22.4
C-R5D		15.1	1.90	1.33	47.5	24.8
Rescue	+	20.1	0.57	0.90[b]	0	0
R-C2A	+	21.8	0.77	0.93[b]	0	0
R-C2D	+	17.1	1.05[b]	1.49	0	33.0
R-C5B		18.1	2.48	1.47	59.6	32.3
R-C5D	+	14.4	2.44	1.45	59.0	31.0

[a]Plants were grown for five weeks in Leonard jar assemblies, inoculations made at two weeks, four plants per treatment.

[b] Not significantly different from control values.

Leonard jars, most root rot susceptible lines exhibited no gains in N, while all root rot resistant lines exhibited gains in N following inoculation (Table 3). Independent measurements of N_2 fixation were not reported, but C_2H_2 reduction has been reported for similar substitutions in another study (95 cited in 165). Some lines responded more to inoculation with *Bacillus*, and others to inoculation with *Azospirillum*. Most important, the absolute levels of the N gains were in the range that could be of agronomical significance: for R-C5B inoculated with *Bacillus*, the gains (Table 3) were equivalent to field gains of about 27 kg N per hectare, assuming 300 plants per m^2, over a three-week period. Inoculation of seeds with the *Bacillus* sp. resulted in similar, bit less pronounced, effects in the fields, with gains in N being of the order of ten percent of shoot N (166).

Bacteriology

Commonly, more than one species of diazotroph can be isolated from grasses or other non-nodulated plant exhibiting nitrogenase activity, and at least 10 species are commonly associated with grasses (Table 4). Counts of various types or groups of organisms associated with roots have generally been based on characteristics observed in enrichment cultures, rather than in pure cultures. Rennie (164) noted that common N_2-fixing bacteria can grow on several carbon sources; therefore presumptive identification of diazotrophs, and particularly of microaerophils, from the occurrence of C_2H_2 reduction in single carbon source media is prone to error. For isolating diazotrophs from nature, he used a medium including three carbon sources (sucrose, mannitol and sodium lactate), yeast extract and basal salts. Representatives from eight commonly isolated genera of diazotrophs grew well on the medium, and 75 percent of isolates from the higher dilutions in the medium exhibited C_2H_2 reduction. The use of premixed microtube systems for conducting biochemical tests such as are used with clinical materials, facilitates routine identification of diazotrophs isolated from non-specific media. Rennie (162) developed a computer assisted scheme for the identification of nine genera of diazotrophs based on interpretation of 70 biochemical tests of the API 20E and 50E systems (Analytab Products Inc.), plus additional tests for C_2H_2 reduction, nitrate and nitrite reduction, catalase, oxidase, motility and growth on MacConkey's bile salt medium. In most cases only the API 20E tests and the seven additional tests were required. His identifications of isolates from plants of temperate and tropical zones probably give a reasonable indication of the relative frequencies of occurrence of known diazotrophs with grasses (Table 4).

Beijerinckia was not included in Rennie's (162) data base and is not included in Table 3. However, except for Dobereiner's reports of its occurrence with sugar cane (58), this organism has not been reported in other studies to be commonly associated with plants, or with sugar cane at other locales (90).

Table 4. Identification of N_2-fixing (C_2H_2-reducing) isolates from various grasses by Rennie (162)

Organism[a]	No. of grass species from which organism was isolated[b]	
	temperate spp. (n=8)	tropical spp. (n=13)
Aerobes		
Azotobacter vinelandii	4	3
Derxia gummosa		3
Microaerophils		
Azospirillum brasilense	3	10
Azospirillum lipoferum		3
Azospirillum spp.		4
Facultative anaerobes		
Bacillus polymyxa	1	4
Bacillus spp.		1
Enterobacter cloacae	5	5
Enterobacter spp.		1
Erwinia herbicola	7	1
Klebsiella pneumoniae	1	3

[a]Data base for non-enteric types included in addition to those listed above, *Bacillus macerans*, *B. cereus*, *B. subtilis*, *Clostridium pasteurianum*, *Azotobacter chroococcum*, *Rhodospirillum rubrum*, *Agrobacterium tumefaciens*; identifications were based on standard tests for Enterobacteriaceae plus seven additional tests as described in text.

[b]These were not collected systematically and the numbers should be regarded as simply giving a general impression of the frequency of occurrence of the different diazotrophs.

Beijerinckia isolate BC from rice soil seemed not to be stimulated in the rhizosphere of rice (56). High numbers of *Azotobacter* are found in some rhizosphere (90) and non-rhizosphere (55, 119) soils, but *Paspalum notatum* remains the only grass for which a functional relationship and specificity in association with an aerobic diazotroph (in this case, *Azotobacter paspali*) has been demonstrated. Others undoubtedly occur. The effects of pO_2 on ARA by roots excised from the graminaceous weed *Eragrostis ferruginea* (134) are similar to those observed for *Paspalum notatum* (64), and an organism resembling *Azotobacter*, but not identical to any described species was isolated from this grass (134).

Rennie's (162) and other studies suggest that *Azospirillum* species (84, 90, 203, 206, 212, 135) and facultative anaerobies of the Enterobacteriaceae (84, 135, 155) or *Bacillus* spp. (135, 214) are more commonly or more closely (90) associated with grasses than are Azotobacteriaceae. Facultative anaerobes are found without co-occurrence of *Azospirillum* (90, 214) but where both groups have been looked for, facultative anaerobes have been consistently

Table 5. Characteristics useful in distinguishing diazotrophic microaerophils [from Baldani and Dobereiner; McClung and Patriquin and Tarrand *et al.* (10, 125, 191)]

Character	*Azospirillum lipoferum*	*Azospirillum brasilense* nir+	*Azospirillum brasilense* nir−	*Campylobacter* sp[b]
Growth in semisolid low N medium[a]	microaerophilic	microaerophilic	microaerophilic	microaerophilic
Growth on plates containing combined N, under air	good	good	good	poor
Denitrification[a]	+	+	−	−
Cell form in alkaline medium[a]	large polymorph	very motile normal form (ca. 1 μm)	very motile normal form (ca. 1 μm)	very motile, small (0.3–0.6 μm)
Growth in semisolid medium with glucose as sole carbon source	good	poor	poor	poor
Mol% G+C	69–70	69–70	69–70	32
PHB in N_2-fixing conditions	+	+	+	−
Biotin requirement	+	−	−	−
Growth on citrate	+	+	+	−
Fermentation of glucose or fructose	acidification and scant growth	−	−	−
Media with ribose, mannitol or sorbitol	usually acidified	−	−	−

[a]Tests used by Baldani and Dobereiner (10) to identify *Azospirillum* isolates.
[b]The isolate from *Spartina* has a salt requirement (minimally 2 g/1 NaCl).

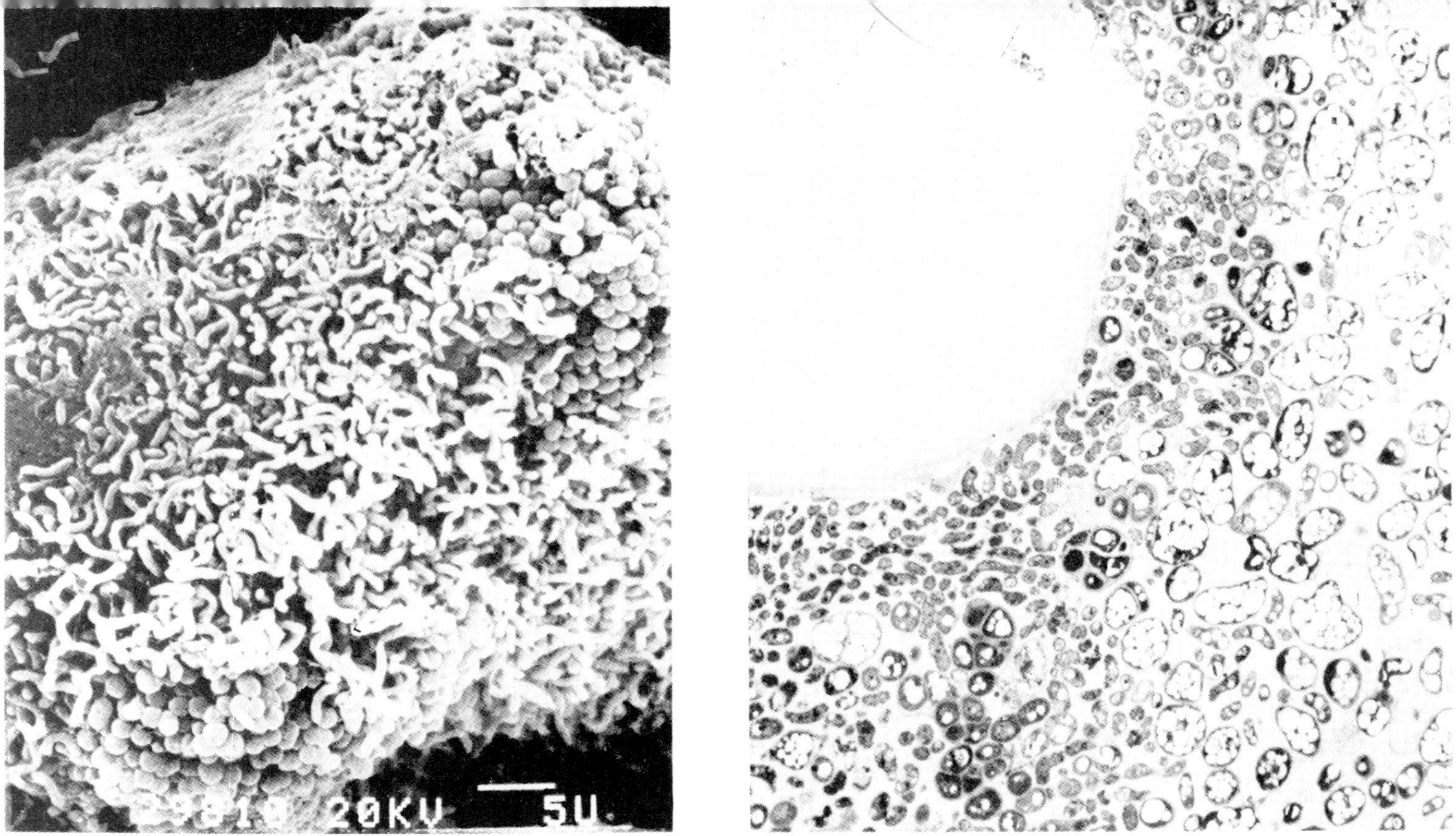

Plate 1. Wild type, vibrioid forms, and pleomorphic (spherical) forms of *Azospirillum brasilense* SP 7 growing on surface of sugar cane callus (left), and in intercellular space of callus (right). Left: scanning electron micrograph, × approx. 2000; Right: transmission electron micrograph, × approx. 3300. See Berg *et al.* (20) for details. Photographs courtesy of Dr. Indra Vasil. Note presence of PHB in, and the multicellular nature of certain of, the large spherical forms at right.

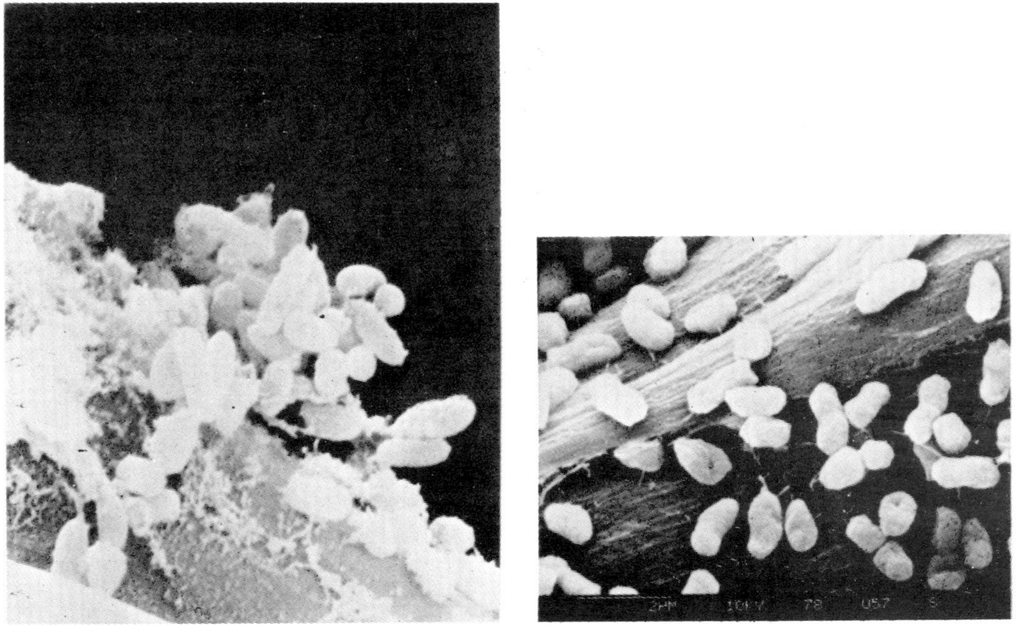

Plate 2. Scanning electron micrographs of *Azospirillum brasilense* SP 7 adsorbed to root hairs. Left: on pearl millet, × approx. 9250, see Garcia *et al.* (75) for details, photograph courtesy of Dr. Frank Dazzo. Right: on maize, × approx. 6000, from McClung (124). Note granular material on root hair at left, and at right, note fibrillar extensions from azospirilla to root hair.

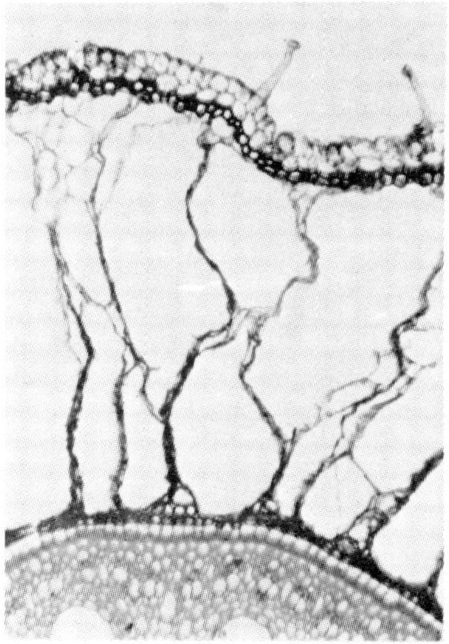

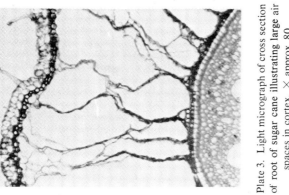

Plate 3. Light micrograph of cross section of root of sugar cane illustrating large air spaces in cortex. × approx. 80.

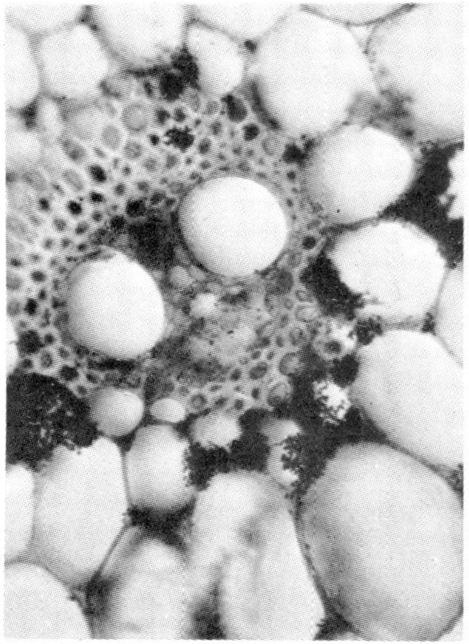

Plate 4. Light micrograph of cross section of stem of sugar cane illustrating tetrazolium-reducing bacteria in intercellular spaces of stem parenchyma and in protoxylem and sclerenchymatous cells of vascular bundle (149); × approx. 520.

found associated with plants harboring azospirilla or other microaerophils (90, 154, 206), and in approximately equal or greater numbers (154, 206). Nitrogenase negative strains of both *Azospirillum* and the facultative anaerobes are also apparently commonly associated with grasses (162). The possibility that *Rhizobium* and actinomycetes are involved in grass N_2 fixation should probably be considered as strains of both of these types of organisms fix N_2 *in vitro* (18, 101, 108, 76).

Two species of *Azospirillum* (syn. *Spirillum lipoferum*) are now recognized on the basis of DNA homologies, and several other characteristics correlate with these homologies and can be used to distinguish the two species (Table 5). Some isolates do not appear to belong to either of these two species (135). *Azospirillum lipoferum* can grow on glucose, but *A. brasilense* cannot; both species grow best on organic acids, and can utilize fructose (191). *Azospirillum lipoferum*, but not *A. brasilense*, is capable of autotrophic growth on H_2 (174). The previously designated Group I and Group III strains of *Spirillum lipoferum* (175) which are respectively nitrite reductase (denitrification) positive and negative are both included as *A. brasilense*, although Dobereiner still considers them as separate subspecies (10).

We have isolated an obligately microerophilic diazotroph, identified as a *Campylobacter* species, from roots of *Spartina* (125). On superficial examination this organism resembles *Azospirillum*, so it seems probable that more critical examination of isolates from other plants will reveal this organism elsewhere. The G+C content, the absence of PHB granules, and the obligately microaerophilic habit distinguish this genus from *Azospirillum*.

Fluorescent antibody (FA) techniques have been used to identify inoculant strains of *Azospirillum* and *Beijerinckia* BC in field samples. Primary antisera prepared against *A. brasilense* strains 13t, 51e, 84 and JM 125A2 from Brazil were tested against various soil organisms (177); all showed no reaction with non-azospirilla. There was variable cross reactivity between strains but it was possible to distinguish the Brazilian from local (Florida) strains. *Panicum maximum* that had been inoculated for three successive years was found to harbor significantly higher numbers of inoculant (13t) strains than of indigenous strains. Antiserum prepared against 13t bound to surface cross reactive antigens in *Panicum maximum* (46). The fluorescent antibody prepared against *Beijerinckia* BC was highly specific, as shown by negative reactions with six *Azotobacter* species, four *Beijerinckia* species and 44 unidentified isolates (57).

Dobereiner and associates (10, 60) attempted to use low level streptomycin resistance as a marker for *Azospirillum* strains used as inoculants but this marker was not distinctive because of the high proportion of organisms amongst the normal rhizosphere and root microflora that exhibited low level resistance to streptomycin and other antibiotics (60). Presumably, higher level resistance could be used as a marker.

Physiology of *Azospirillum*

Oxygen

Azospirillum fixes N_2 under microaerophilic conditions but will grow aerobically if a combined N source is available. Doubling times of approximately one and six hours have been reported for growth on NH_4^+ under air and on N_2 under low pO_2, respectively (140). Nelson and Knowles (130) obtained steady state continuous cultures of *A. brasilense* SP 7 at pO_2s between 0.0000 and 0.0150 atm; calculated efficiencies of N_2 fixation were maximal in the range 0.005 to 0.0075 atm. Little ARA occurred at pO_2s above 0.01 atm. There was no evidence of respiratory protection of the sort that occurs in *Azotobacter*.

"C-forms" of *Azospirillum* are observed when the organism is growing on plates (and fixing N_2) under air, and in association with sugar cane callus (20). These spherical C-forms are distinguished from the typical vibrioid "V-forms" by the presence of a capsule external to the lipopolysaccharide layer, reduced motility, formation of encapsulated clusters and more PHB than usually observed for V forms as shown in Plate 1 (19, 20). This capsulation may provide protection from oxygen under N_2-fixing conditions (19).

According to Okon *et al.* (139), *Azospirillum brasilense* is aerotactically attracted to microerophilic conditions even when not fixing N_2; they suggest that the microerophilic habitat provided by plant roots attracts *Azospirillum*.

Nitrate

Nitrate is assimilated (reduced to ammonium) by all isolates of *Azospirillum* that have been examined (191). *Azospirillum* was the first diazotroph reported to exhibit dissimilatory reduction of nitrate (132); subsequently this denitrifying activity was found in some strains of *Rhizobium* (218). All *A. lipoferum* and about half of *A. brasilense* strains will reduce nitrate to N_2 under oxygen limited conditions; those not producing gas (nitrite reductase negative strains, nir$^-$) reduce nitrate to nitrite which accumulates (132). Neyra and van Berkum (133) reported that nitrate stimulated nitrogenase activity of *A. brasilense* SP 7 under anaerobic conditions, but Nelson and Knowles (130) could find no evidence of nitrate supported anaerobic nitrogenase activity in continuous culture systems. At low pO_2, C_2H_2 reduction occurred only when nitrate concentration dropped below 10 μg per milliliter.

Magalhaes *et al.* (118) obtained nitrate reductase (NR$^-$) mutants of *A. lipoferum* and *A. brasilense* by the selection of chlorate resistant mutants in low oxygen, deep agar tubes. Both assimilatory and dissimilatory nitrate reductases seemed to be lost, and 10 mM nitrate inhibited nitrogenase activity of parent strains but not of NR$^-$ mutants. Nitrite inhibited nitrogenase activity of all parent strains and of mutants including those that were nir$^-$. The authors suggested that "Construction of proper mutants may lead to systems capable of nitrogen fixation in the presence of high nitrate levels and thus permit complementation of the two processes."

Hydrogen

In the absence of other substrates, nitrogenase catalyzes an irreversible, ATP dependent reduction of hydrogen ion to H_2. This process is nearly completely inhibited by acetylene, but not by N_2. It is suspected that H_2 evolution in the presence of N_2 is an intrinsic part of the N_2 reduction process; loss of hydrogen gas produced by nitrogenase is considered to be energetically wasteful (168).

In addition to the H_2 evolving properties of nitrogenase, all N_2-fixing organisms possess conventional hydrogenases (168). In obligate and facultative anaerobes, reversible hydrogenases are present and seem to function primarily in the reduction of protons as terminal electron acceptors. Many aerobic diazotrophs contain a unidirectional H_2-oxidizing hydrogenase which is capable of recycling H_2 evolved from nitrogenase. This recycling may serve to enhance respiratory protection of nitrogenase and/or provide ATP or reducing power. The autotrophic growth by hydrogen uptake positive (Hup^+) strains of *Rhizobium japonicum* (88) confirms that it can function to provide both ATP and reducing power. Hup^- strains of *R. japonicum* are common in nature, and appear to be less efficient N_2 fixers than are Hup^+ strains (1).

Neither *Azotobacter chroococcum* (184) nor *Azospirillum brasilense* SP 7 (35) release H_2 while growing on N_2. It has been shown through selective inhibition of the uptake hydrogenase (184) or by use of the tritium exchange assay (35) that H_2 evolved by these organisms *in vivo* under N_2-fixing conditions is efficiently scavenged by the uptake hydrogenase. Hydrogen-dependent growth and autotrophic CO_2 fixation have recently been demonstrated for four strains of *Derxia gummosa* (156), and for *A. lipoferum* but not for *A. brasilense* (174; Dr. H. Schlegel, personal communication).

Efficiencies of N_2 fixation

Values equivalent to the highest efficiencies of N_2 fixation first reported by Dobereiner and Day for *Azospirillum* (63), 115 mg N per gram lactate, have not been observed in other studies. For example Okon *et al.* (137) reported values of 20 to 24, and Nelson and Knowles (130), 4.7 to 28. Dobereiner (61) reported that efficiencies increase with increasing age of the culture, reaching values of 92 mg N per g lactate and 49 mg N per hour glucose for *A. brasilense* and *A. lipoferum* respectively in an early stationary phase; at this stage hydrogenase activity is highest and specific nitrogenase activity is carbon limited (61).

Methylotrophic growth

Azospirillum brasilense SP 7, *A. lipoferum* SP 208 and *Derxia gummosa* can be grown with methane, ethanol or formate as the sole carbon source (174).

Sites and processes of association

Rhizoplane

Non-specific bacterial colonization of grasses and other non-legumes is described as patchy (5, 148, 217), increasing away from the root tip (5, 117) with bacterial colonies initially present in regions of elongation (117), occurring along grooves between adjacent epidermal cells (5, 117, 145, 148, 217) and associated with sloughing cells (117, 141, 217), fibrillar mucigel (5, 217), ruptured epidermal cells (5), or concentrated where epidermal cells have been sloughed off (71, 141). The root tip is typically free of bacteria (5, 199), possibly because of high (root) growth rates, or of secretion of antibiotics at the root tip (199). The fungal population becomes progressively more important as root cells become moribund (143). Percentage cover of root surfaces is typically of the order of 1 to 10 percent or less (5, 148, 170).

Distribution of FA labelled *A. brasilense* in field grown grasses was seen to be generally similar to that of the much more numerous unlabelled bacteria (177). Likewise, the distribution of inoculated *Azospirillum* cells of roots of axenically grown wheat and sorghum seedlings was similar to that of tetrazolium-reducing bacteria (which include diazotrophs and non-diazotrophs) on roots of field grown plants (148). Such observations suggest that the colonization of roots by *Azospirillum* does not differ from that of non-diazotrophic bacteria normally associated with the roots.

Following the report (25) that fluorescein labelled soybean lectin preparations bound to *Rhizobium japonicum* but not to rhizobia in other cross inoculation groups, considerable attention has been focused on the role of lectins in the recognition and attachment process in legumes. Dazzo (43) suggests that there is a non-specific mechanism which allows all rhizobia to adhere to roots in low numbers, and an additional specific mechanism which results in the selective adhesion of much larger numbers. The clover lectin "trifolliin" specifically agglutinates *R. trifolii* and accumulates at the tips of clover root hairs, which are also the primary locales of specific adhesion by *R. trifolii* (44). Dazzo and Hubbell (45) proposed a "cross binding model" in which a multivalent lectin is bound by unique cross reactive antigens on the rhizobia and on the plant cell wall. Addition of 2-deoxyglucose, a component of the antigenic surface polysaccharide of *R. trifolii* and of clover roots, to clover root-*R. trifolii* systems reduces adsorption by a factor of ten; other sugars inhibit adsorption in other cross inoculation groups (187). The receptor site for the lectin on rhizobia has alternately been proposed to lie in the LPS and EPS layers (211, 45). Using gold labelled lectin, Bal and Shantharam (6) observed binding sites in both the capsule and the cell wall; they propose that capsular receptors are responsible for the attachment of bacteria to the root hair surface, and that host specificity and recognition are mediated by receptors on the cell wall LPS layer.

The large plasmid pWZ2 is required for specific adsorption of *R. trifolii* to clover (219). In several legumes, the attachment of rhizobia is polar (187). Soybean lectin-binding sites are also aggregated at poles of *R. japonicum* (187). Receptors on rhizobia for lectins are transient (22, 48) in nature. As cultures of *R. japonicum* mature from early to late log phase, the percentage of cells which are capsulated and which bind lectins declines sharply and there are also changes in the composition of capsular polysaccharides (128). Fibrillar appendages, which are observed on rhizobia after prolonged incubations with roots, may stabilize attachment after the initial specific recognition and symbiotic contact have occurred (129).

Certain features of the rhizobial attachment process seem to be operative in the associative systems. Garcia *et al.* (75) used the Fahreus slide assemblies to study the attachment of *Azospirillum* strains to roots of small seeded grasses. After short-term incubations, most of the azospirilla adsorbed from cultures were associated with granular material on root hairs (Plate 2) and with fibrillar mucigel accumulating on surfaces of old spidermal cells. Addition of nitrate resulted in degranulation of the root hairs and inhibited attachment of azospirilla. Pearl millet excreted a protease-sensitive, non-dialyzable substance which bound azospirilla and promoted their adherence to root hairs. McClung (124) likewise observed *Azospirillum* to be associated with granular material on root hairs of maize, and fibrillar appendages were also observed (Plate 2).

Shimshick and Hebert (180) examined the dynamics of binding of *Rhizobium* spp. and of *Azotobacter vinelandii* to roots of wheat and rice. The rhizobia exhibited time dependent attachment and detachment, and the steady state adsorption was described by a Langmuir adsorption isotherm. The adsorption constant (or equilibrium constant for binding) for a soybean *Rhizobium* was five times that of a cowpea *Rhizobium*. Numbers of bound *Azotobacter* increased with increasing unattached concentration, but electron microscope observations showed that adsorption did not involve monolayer binding, and the authors did not attempt to describe it quantitatively. The adsorbed *Azotobacter* but not the adsorbed rhizobia exhibited C_2H_2 reduction; rates were proportional to the numbers adsorbed, and were low unless exogenous reduced carbon was supplied. Streptomycin inhibited the binding of rhizobia (suggesting the involvement of protein), but not of *Azotobacter*. The authors suggested that practical root-surface nitrogen fixing systems in non-legumes may be feasible, and that physical models of this sort could indicate appropriate conditions of temperature and cell concentration for inoculation.

Along these lines Garcia *et al.* (75) reported the numbers of adsorbed cells per 200 μmeter long root hairs of Pearl millet incubated with standardized concentrations of various bacteria. The numbers were highest for three strains of *A. brasilense* (24 to 32), low for *Azotobacter vinelandii, Klebsiella pneumoniae* and *E. coli* (less than 0.7), and were intermediate for *Pseudomonas fluorescens* and *Rhizobium trifolii* (11 to 14) suggesting some selective advantage for association with *A. brasilense.* The authors noted that *P. fluorescens* and *R. trifolii* share certain biochemical and serological traits with *Azospirillum.*

Gotz and associates reported polar attachment and nitrogenase activity by cowpea *Rhizobium* sp. 32Hl associated with root of *Petunia* (82) and with wheat seedlings (83). The polar attachment is curious in the light of the indications that lectins are involved in polar binding and specificity of binding of rhizobia to legumes (above).

Internal

In 1974 Dobereiner and Day (62) reported that root pieces of *Digitaria decumbens* exhibiting high nitrogenase activity were characterized by strongly reducing cells within the cortex, where cells packed with *S. lipoferum*-like, tetrazolium-reducing bacteria were observed. They suggested that these bacteria-filled cells may represent a primitive root cell-diazotroph symbiosis.

Concurrent with that report, and subsequently, there have been a number of reports of non-specific bacterial invasion of the cell lumen, or of intercellular spaces. Scanning electron microscopy revealed holes of 0.5 to 2 μmeters diameter on the outer epidermal surfaces of dune grasses and transmission electron microscopy indicated that such perforations were associated with bacteria (141). Bacteria-filled epidermal or outer cortical cells, surrounded by intact, "clean" cells have been observed in rice (5), *Spartina* (145), maize (148) and in sorghum and rice grown gnotobiotically with *Azospirillum* (148). It seems unlikely that such cells remain functional. This condition probably represents, if not a pathogenic condition, an initial stage in the decomposition of the outer region of the root.

Non-disruptive associations (which leave the plant cell membrane intact) may occur where bacteria are embedded in the inner region of the cell wall but lie outside of the cell membrane (141), or where they invade the intercellular spaces. In *Panicum maximum* grown gnotobiotically with *Azospirillum*, Garcia *et al.* (74) noted progressive sloughing of epidermal cells on older tissues, and concurrently, bacterial invasion of the exposed middle lamellar regions of the underlying cortex cells; the bacteria also gained entrance at points of emergence of lateral roots.

FA-labelled *Azospirillum* cells were observed at the periphery of cortical cells in field grown *Cynodon dactylon* (177). The ability of *Azospirillum* to penetrate the middle lamella may be related to the pectinase activity by this organism (74). *Beijerinckia* became established on the lacunal (air space) walls of basal regions of roots of rice seedlings grown under gnotobiotic conditions with this organism for 30 days. High numbers of diazotrophs were recovered after surface sterilization of basal zones, but not of younger regions of roots, of field grown plants (56). Lacunal walls potentially provide a much greater surface area for bacterial colonization (Plate 3) than is provided by the external surface of the root; this could account for the recovery of greater numbers of organisms from the "histosphere" of rice than from the rhizoplane (206).

A *Bacillus* sp. proliferates in the rhizoplane, and invades the intercellular spaces of the root cortex of gnotobiotically grown wheat (112), forming a functional N_2-fixing association (165). Intercellular spaces are larger in inoculated than in axenic plants (112).

It is not clear whether these cortical invasions differ fundamentally from the progressive, radial invasion and decomposition of roots by bacteria such as has been described for wheat (71, 121). The latter are initiated while the stele is functional, may be preceded by some autolytic degeneration of the epidermal and cortical cells, progress radially from the root surface inwards, and can lead to the invasion of the endodermis and death of the plant or severance of the root. A non-specific and apparently common sort of invasion which clearly differs from the progressive, radial type of invasion, was described by Patriquin and Dobereiner (148). Tetrazolium-reducing bacteria were observed in the inner cortex and inside of the stele of maize and of other tropical grasses in the absence of significant bacterial colonization or collapse of outerlying tissues or of disruption of the endodermis, and in which the stele remained functional. They were found mostly within the intercellular spaces, and within xylem vessels. Those observations referred to bacteria producing prominent formazan crystals in the presence of 2, 3, 5 triphenyltetrazolium-chloride; such bacteria include diazotrophs, but not exclusively so. Lakshmi *et al.* (109) and Patriquin and Dobereiner (148) observed almost identical distributions of bacteria within grasses that were grown gnotobiotically with *Azospirillum*; they noted longitudinal distributions of bacteria in the cortex, bacteria filled root hairs (109, 148), and bacteria inside of xylem vessels (109).

Bacteria in the stele of field grown plants studied by Patriquin and Dobereiner (148) remained viable after a 6-hour treatment of root pieces with 1 percent chloramine-t, and included *Azospirillum*. The common occurrence and longitudinal distribution of bacteria in the inner cortex and occasionally in the stele of branches, and a radiating pattern of colonization of the stele of main roots in the region of branches suggested that the initial colonization of these inner areas occurs in the regions of branches, and then spreads longitudinally. Entry into the roots could occur through disrupted cortical tissues where the branches emerge from the main roots; this area is believed to be the principal venue of infection of some legumes by rhizobia (42), and one venue of invasion of gnotobiotically grown grasses by *Azospirillum* (74). However, it is difficult to understand how the invasion of roots via these cortical disruptions could result in invasion of the stele without disrupting the endodermis. Megalhaes *et al.* (117) observed the colonization of field grown maize by tetrazolium-reducing bacteria within 2 cm of root apices. In cross sections just behind the root apices tetrazolium-reducing bacteria were observed in the cortex, and in the adjacent region of the stele. They suggested that initial entry into the stele occurs at apical regions where there is no appreciable thickening of the endodermal walls or of the Casparian strip, or where the stele has not differentiated as such. This process may resemble that involving vascular nematodes whose larva penetrate the root in the region of cell elongation prior to cell differentiation (200). Presumably, some of the mechanisms that a plant normally uses to resist bacterial invasion of roots are not operative at the root apex. It is of interest that initial events in the colonization of soybean roots by *Rhizobium* take place at the root apex before there is visible differentiation of root hairs (23).

Evidence that colonization of root tissues takes place via root apices, and/or at points of emergence of lateral roots, is important because this seems to explain various correlations between the root nitrogenase activity and the branching of roots (62, 154). Dobereiner and Day (62) noted that excised roots exhibiting high nitrogenase activity were "all of the same gross morphology: medium thick brown roots with many laterals and an intact cortex." It is implicit in such an explanation that excised root nitrogenase activity is largely internal in nature. But do these internally located bacteria exhibit nitrogenase activity *in situ*?

Spartina alterniflora has a well developed lacunal system in roots and leaves which, like that in rice, functions in the transport of oxygen to roots situated in anoxic environments. When hydroponically maintained plants are incubated with C_2H_2 in a two-phase leaf-root system, ethylene is evolved in both the upper and lower phases following at most a 1 or 2-hour lag. Measurements of diffusion of C_2H_4 from around the roots to the upper phase, and of nitrogenase activity by leaves alone showed that diffusion of C_2H_4 from the lower phase and leaf associated nitrogenase activity could account for only a fraction of C_2H_4 in the upper phase. The latter was therefore attributed to the reduction of acetylene by diazotrophs located in close proximity to the lacunal system. The addition of mercuric chloride to seawater in the lower phase resulted in an immediate 75 percent reduction in its C_2H_4 evolution, but did not affect that of the upper phase. The upper phase C_2H_4 evolution was also much less sensitive than that of the lower phase to the addition of ammonium to the latter. The proportion of total plant ARA made up by "endorhizal" (upper phase) ARA was highest (75%) in plants with the highest total plant activities (27). van Berkum and Sloger (197) enclosed leaves of *Spartina* in a plastic bag, added C_2H_2, and measured the evolution of C_2H_4 in the lacunae by inserting the needle of a syringe directly into them. These approaches to measuring nitrogenase activity by internally located or endorhizal diazotrophs may be applicable to certain tropical grasses which do not normally grow in waterlogged soils, but which do possess well developed lacunae, e.g. *Paspalum notatum*, and sugar cane (Plate 3). We have suggested also that the preincubation of excised roots in ammonium and/or mercuric chloride could be a means of screening plants for internal sites of nitrogenase activity (27). Using such techniques and a two-phase system as above, Boyle and Patriquin (28) concluded that nitrogenase activity associated with experimentally grown lowland rice at Seropédica, Brazil, was mainly external (or "exorhizal") in nature.

Specificity of invasive bacteria

The bacterial population within defined zones of the root can be characterized by determinative studies of the population recovered after treatment with surface-sterilizing agents. The radial penetration of these agents can be assessed by observing the radial distribution of tetrazolium reduction in plant tissues and by bacteria after similar treatment (148). The use of strong sterilizing agents or long periods of exposure, the cutting of roots into small root pieces, and the shaking of roots while in these agents may be necessary to ensure effective surface sterilization, and the extent of penetration may vary with the age of the roots (117). For *Spartina*, studies of this sort reveal a greater "relative enrichment" of facultative anaerobic and microaerophilic diazotrophs than that of microaerophils inside of the roots (Table 6).

Table 6. Proportion of diazotrophs in bacterial population of soil, washed roots, and surface sterile roots from *Spartina alterniflora* [from Boyle and Patriquin and Patriquin and McClung (29, 154)]

Sample	Total number peg wet weight (plate count)	diazotrophs: glucose-using anaerobes	diazotrophs: malate-using micro-aerophils	Relative enrichment[b] of diazotrophs in root interior: anaerobes	Relative enrichment[b] of diazotrophs in root interior: miro-aerophils
		(nos. as % of total)			
Field samples					
Non-rhizosphere soil	1.09×10^6	0.012	0.022		
Rhizosphere soil	28.8×10^6	3.2	0.32		
Washed roots	34.7×10^6	2.6	0.081		
Surface sterile roots[a]	0.10×10^6	5.4	1.3	2.0	16
Plants in hydroponics[c]					
Washed roots	16.1×10^6	2.9	0.022		
Surface sterile roots	0.69×10^6	6.7	0.41	2.3	19

[a]1% chloramine-t for 2 hours.

[b]% of total in surface sterile roots divided by % of total in washed roots.

[c]Field plants transferred to flowthrough seawater-hydroponics; counts after 6 mos. 590000 *Azotobacter*-like organisms per gram washed roots; none from surface sterile roots (29).

Non-diazotrophic bacteria make 90 percent or more, numerically, of the population recovered after surface sterilization of the roots of *Spartina* (Table 6), wheat (155), and stems of sugar cane (149), and of roots of maize that had received *Azospirillum* inoculant (138). Diazotrophs may be relatively more numerous in rice (205). *Azotobacter*, or *Azotobacter*-like organisms have variously been reported to be excluded (27, 191) or included (21) in the bacterial flora received from surface sterilized roots.

Evidence for invasiveness of roots by *Azospirillum* spp. led Dobereiner and associates to look for evidence of host plant specificity in the associations. In greenhouse and field experiments in which soils were inoculated to ensure the presence of *A. lipoferum* and of nir^+ and nir^- strains of *A. brasilense*, more than 50 percent of the isolates from chloramine-t treated roots of maize were *A. lipoferum*, and more than 88 percent of isolates from roots of wheat and rice were *A. brasilense* nir^- strains (10, 60). Examination of other C-4 and C-3 type grasses suggests that *A. lipoferum* strains are selected by C-4 type plants, and *A. brasilense* nir^- strains by C-3 type plants. An exception is sugar cane (C-4 type) which harbored *A. brasilense* nir^- strains.

They also discovered during attempts to recover low level streptomycin resistant (str^r) inoculant strains from field plants, that there is a natural enrichment in roots and rhizosphere of *Azospirillum* and other bacterial types that are resistant to low levels of streptomycin. The percentage of isolates (diazotrophs and non-diazotrophs) that were str^r was less than 0.1 percent for nonrhizosphere soil, approximately 1 percent for rhizosphere soil, and greater than 50 percent for the population recovered from chloramine-t treated roots (10). Root isolates were in general more resistant to a number of antibiotics than were soil isolates, but the differences were most pronounced with respect to streptomycin (60). The proportion of str^r strains in populations recovered from roots varied markedly with the growth stage of maize (60). Dobereiner suggests that low level streptomycin resistance could be related to the presence of actinomycetes on roots, and to the relative stability of streptomycin once taken up by the roots. While this property (str^r) seems to confer some selective advantage to colonizers of rhizosphere soil and roots, it is not in itself the basis of strain specificity (60).

Rhizosphere

To some extent, the rhizosphere soil has been overlooked as a possibly important site of N_2 fixation, perhaps because of difficulties in assaying rhizosphere N_2 fixation separately from the plant, and perhaps because it is assumed that the potential for N_2 fixation and efficiency of transfer of that N to the plant would be lower than for N_2 fixation that is more intimately associated with the plant. The $^{15}N_2$ studies with rice, cited above, illustrate that the rhizosphere can be quantitatively very important, and that a substantial portion of the N fixed there may be almost immediately available to the plant.

Wullstein *et al.* (214) found nitrogenase activity associated with "rhizosheaths" of certain xeric grasses. These sheaths of sand grains appear to increase the efficiency of uptake or of the retention of water, and contain as much as 0.7 percent carbon on root-free basis, possibly derived from lysis of root cortical cells. The rhizosheaths were found to harbor a N_2-fixing *Bacillus*; other types of diazotrophs were looked for but not found.

Soil around roots of prairie grasses exhibited ten times the ARA on a per plant basis as did the roots (193). Seasonally, we found that ARA of washed sods of *Spartina* amounted to 79 percent of that in soil slices (roots + soil), but on individual days, the soil slices exhibited as much as three times the ARA of washed sods (154). Differential responses of roots versus rhizosphere soil to additions of various carbon compounds and inorganic nitrogen suggest that there are physiologically distinctive populations in the two sites—with organic acid using microaerophils being of greatest importance in N_2 fixation in the roots, and sugar-using organisms, in the rhizosphere soil (27). Dicker and Smith (55) found *Azotobacter*, *Desulfovibrio*, and *Clostridium* to be present in numbers of the order of 10^7 to 10^8, 10^3 to 10^4, and 10^2 to 10^4, respectively per gram of marsh soil. Pasteurization reduced ARA by 92 percent and the addition of $BaCl_2$ (to precipitate sulfate) reduced it by 50 percent suggesting that sulfate reducers were responsible for a major fraction of rhizosphere ARA. Because of the more limited supply of sulfate, sulfate reducers are not likely to be important N_2 fixers in rice paddies (67).

The plant may exert a determining effect on the qualitative composition of the rhizosphere (98) or even be a source of rhizosphere organisms (149). Kavimandan *et al.* (98) reported that relative to non-rhizosphere soil, numbers of *Azotobacter* are enriched and those of *Clostridium* are reduced in the rhizosphere of maize and sorghum, while the reverse is true in rhizosphere soil of wheat.

Cyclic colonization of sugar cane

The colonization of roots as described by Patriquin and Dobereiner (148) is essentially a longitudinal rather than a radial process; bacteria enter the roots at apices or where laterals emerge and spread longitudinally through intercellular spaces, or in the xylem. It is not surprising in this context that bacteria are found in intercellular spaces and in the xylem in stems of sugar cane (Plate 4), and that diazotrophs have been isolated from surface sterile stems of wheat (99), sugar cane (149) and maize (117). Their occurrence in stems of maize and wheat is probably of little significance. In cane, bacteria are concentrated in peripheral nodal regions where the "set" roots originate and following sprouting of cuttings move into the rhizosphere soil where they exhibit ARA (149); cuttings and the roots themselves exhibit little or no ARA at this stage but do so after two to three months (172). Thus there appears to be a cyclic colonization or infection of cane in a sequence soil—roots—

stems—(following sprouting)—soil. The significance of this phenomenon for N_2 fixation in cane through its life cycle is not determined. It may offer a convenient means of selecting varieties with high associated N_2 fixation, and for introducing appropriate strains of diazotrophs with new varieties, or in new locales.

Tissue culture-diazotrophs associations

The simplification of the plant component afforded by plant cell or tissue culture is appropriate for studying certain aspects of plant-diazotroph associations. Thus the observation of nitrogenase activity by cowpea rhizobia growing on the surfaces of, or in intercellular spaces, or separated but in close proximity to, callus of various non-leguminous plants (38, 186) demonstrated that neither nodulation, nor intracellular infection, or even legume hosts were essential for nitrogenase activity by these rhizobia.

Child and Kurz (38) used tissue culture systems to compare nutritional requirements of *Rhizobium* sp. 32Hl and *Azospirillum brasilense* SP 7. The addition of a 5-carbon sugar (arabinose) enhanced nitrogenase activity of both organisms grown in the proximity to calli of legumes and non-legumes, and this nitrogenase activity was further enhanced by the addition of succinate (sucrose was the principal carbon compound in the medium). In the same medium without callus, arabinose and a tricarboxylic acid were required for high levels on nitrogenase activity by *Azospirillum.* Wheat plant cell cultures appeared to "activate" certain carbohydrates (e.g. fructose), which by themselves did not support high nitrogenase activity by *Azospirillum.*

Calson and Chalef (34) "forced" an association between *Azotobacter vinelandii* and plant cells by inoculating carrot cell suspension with an adenine-requiring strain of *Azotobacter*, and plating the mixture on solidified N-free medium. One in 10^4 to 10^5 carrot cells produced callus in which *Azotobacter* grew intercellularly or in the underlying agar. No calli were formed without *Azotobacter.* The associations reduced acetylene, and were maintained for over 18 months.

Vasil *et al.* (201) observed marked inhibition of callus growth after inoculation with *Azospirillum brasilense* SP 7 for pearl millet, tobacco and centipede grass, but they were able to maintain slow-growing sugar cane callus-azospirilla associations for over 18 months (subculturing every five to six weeks). These associations reduced acetylene, but did not grow faster than calli without *Azospirillum.*

Azospirillum

Azospirillum grew over the surfaces of calli and in intercellular spaces of older parts of the calli (Plate 1); a significant portion of the azospirilla were pleomorphic, capsulated forms (Plate 1). *Azospirillum* reisolated from sugar cane callus had higher specific nitrogenase activity than did organisms from stock cultures (19). They were able to regenerate plantlets with roots and shoots from the calli and to grow these in soil; bacteria cells were absent in shoots and roots within intact cells, tissues or organs but vascular elements in the old callus tissues retained bacteria in the intercellular spaces (201).

Carbon interactions

Aeration and carbon metabolism

Diazotrophs most commonly associated with roots of grasses include aerobes, microaerophils and facultative anaerobes which fix N_2 most efficiently at pO_2s in the region of 4, 0.4 and 0.00 kPa, respectively (64, 130, 105). Thus the precise nature of the oxygen gradient surrounding and within roots can be expected to influence the distribution activity and types of associated diazotrophs. Bacterial metabolism in turn may modify O_2 gradients. The O_2 gradient, through its effects on the metabolism of potentially fermentative microorganisms or plant tissues, and on the "leakiness" of plant cells, may affect the availability and types, as well as the efficiency of use, of carbon substrates by diazotrophs.

Higher plants are obligate aerobes. "Flood tolerant" species, typically, have a well developed aerenchyma (system of air spaces or lacunae in cortex and sometimes in the stele) which functions in the internal transport of O_2 to root tissues. Aerenchyma can arise "lysigenously" (by collapse of cells) or "schizogenously" (by separation of cells) (69). Lysigenous aerenchyma develops in adventitious roots of many mesophytic species in response to the reduced availability of oxygen associated with waterlogging. In sunflower and probably maize, O_2 deficiency triggers the production of ethylene which causes increased cellulase activity and the development of aerenchyma (100). Inadequate availability of nitrogen can lead to cavity formation in maize under aerated conditions (107). In many hydrophytes and even in some species which are not generally hydrophytic in nature such as sugar cane and *paspalum notatum* (146), well developed aerenchyma are normal, constitutive features of roots (Plate 3). Aerenchymal O_2 concentrations of 2 to 8 percent have been reported for hydrophytes (142, 192).

Cytochrome oxidase activity is near maximal at a pO_2 of 0.1 kPa less (142). Due to the restricted diffusion of O_2 in aqueous phases between and in cells, the pO_2 external to the root or in aerenchyma must be in the region of 2 to 5 kPa for near maximal O_2 consumption. Part of the decline in O_2 consumption that occurs when pO_2 falls below these levels, and the greater sensitivity of roots of some species to pO_2 may be due to the activity of other oxidases which have lower affinities for O_2 than cytochrome oxidase (142). As pO_2 declines, the respiratory quotient (R.Q. = CO_2 evolved/O_2 consumed) may rise above 1 due to fermentative CO_2 production. The R.Q. of roots under air increases with the increasing diameter of roots, apparently due to fermentation in the center of the root (136). Roots commonly produce alcohol under low pO_2; the autotoxic effects of alcohol may contribute to the limited tolerance of many species to waterlogging of soils. Many flood-tolerant species reoxidize NADH by producing a predominance of other less toxic reduced end products including malate, shikimate, glycerol, amino acids and lactate (142). Largely on the basis of their observations on the occurrence and activity of alcohol dehydrogenase (ADH), McMannon and Crawford (126) proposed the following theory of flooding tolerance (Fig. 1):

> In intolerant roots, on flooding, normal respiration is blocked, and glycolysis proceeds to the production of acetaldehyde and ethanol. Acetaldehyde induces alcohol dehydrogenase activity which, together with a reduction in apparent K_M value, accelerates glycolysis. Malate present is decarboxylated by 'malic enzyme' to pyruvate and thence to acetaldehyde, contributing further to ethanol production. Oxaloacetate and hence malate may be formed by the carboxylation of phosphoenolpyruvate, but the malate will not accumulate. Ethanol and acetaldehyde do accumulate, and contribute to the poisoning of metabolism.
>
> In 'tolerant' roots on flooding, normal respiration is at least partially blocked, and glycolysis may proceed to the production of acetaldehyde and ethanol, but the former fails to induce ADH activity, the apparent K_M value remains unchanged, and no acceleration of glycolysis ensues. The malate present is not decarboxylated, because malic enzyme is absent. Oxaloacetate and hence malate are produced by the carboxylation of phosphoenolpyruvate, and malate accumulates. This is non-toxic, and may remain without harm to the plant until aerobic conditions are restored.

Such adaptations are not universal; rice for example carries out a predominantly alcoholic fermentation. It has been suggested that the thin roots of rice permit ready loss of the alcohol by diffusion into the surrounding water (142). Smith and Ap Rees (182) questioned the McMannon-Crawford theory because for several flood-tolerant species they found (i) appreciable activity of malic enzyme, (ii) increased ADH, (iii) high rates of alcohol production, and (iv) no accumulation of malate. Their observations were on the apical 2 cm of roots, while those of McMannon and Crawford were on whole roots; meristematic regions have an intrinsically fermentative metabolism because of the incomplete development of mitochondria (142). The whole situation requires detailed study.

In legumes, nitrogenase activity is related to sucrose concentration (189), but organic acids rather than sugars appear to be used by bacteroids to provide ATP and reductant for N_2 fixation (169). Interestingly, ADH, malate dehydrogenase (MDH) and phosphoenolpyruvate carboxylase (PEPC) activities, relative to those of isocitrate dehydrogenase, are reported to be 4, 30, and 9 times, respectively higher in nodules than in root tissues of *Pisum*

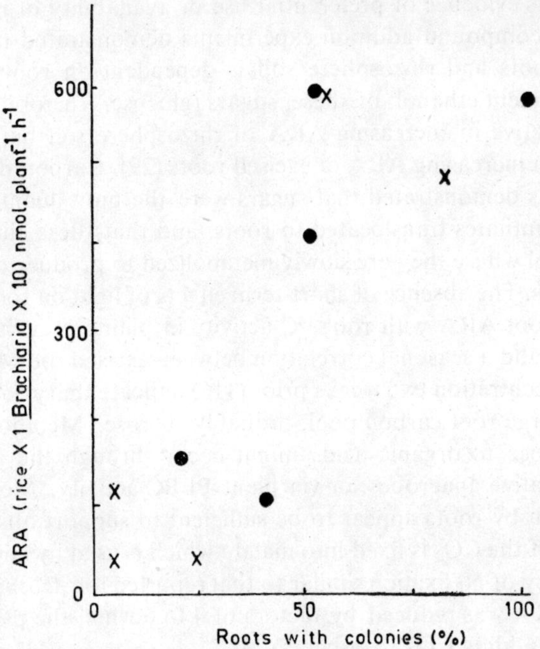

Fig. 1. Acetylene-reducing activities of seedlings in relation to the colonization of roots by diazotrophs. ●: rice; ×: *Brachiaria mutica*. Plants were grown gnotobiotically with various diazotrophs, and assayed as described by Patriquin and Dobereiner (148). Following assays, plants were incubated in tetrazolium solution and the entire surfaces of all roots (n=5 to 12) on each plant were examined for bacterial colonies. Individual roots were either heavily colonized or were not colonized at all. Each point represents one plant. *Brachiaria* plants were inoculated with various *Azospirillum* isolates. Rice plants, in order of decreasing ARA in figure, were inoculated with *Derxia*, *A. lipoferum*, *Beijerinckia*, *A. brasilense* nir$^+$, *A. brasilense* nir$^-$ (Dobereiner and Patriquin, unpublished observations).

sativum, a notably flood intolerant species; very low, if any activity of malic enzyme was detected. De Vrieset *et al.* (52) attribute these increases to O_2 deficiency in the nodules, and contend that the organic acids produced by PEPC activity serve both as precursors for amino acid synthesis (39) and as carbon and energy sources for the bacteroids. Isolated bacteroids exhibited high rates of malate and succinate accumulation at concentrations of these organic acids equivalent to those in the nodule tissue (52).

Dobereiner and Day (63) suggested that the inferred internal location of *spirillum lipoferum* in grasses was related in some way to the ability of that organism to use organic acids; they suggested that there may be some connection with the C-4 photosynthetic metabolism of tropical grasses. We investigated this 'malate connection' in *Spartina alterniflora*. While we could not find any evidence for a direct interrelationship with C-4 photosynthesis *per se*, there was evidence of preferential use or availability of malate in the roots. Carbon compound addition experiments demonstrated that ARA of both excised roots and rhizosphere soil is dependent on root compounds soluble in 80 percent ethanol; of these, sugars (glucose, fructose and sucrose) were most effective in increasing ARA of rhizosphere soil but malate was most effective in increasing ARA of excised roots (29). Carbon dioxide labelling experiments demonstrated that sugars were the only quantitatively important photosynthates translocated to roots, and that these entered a large root carbon pool where they are slowly metabolized to produce organic acids and amino acids. The absence of short-term effects of light on root ARA (29), correlation of root ARA with root ^{14}C activity in plants previously exposed to $^{14}CO_2$ (154), and a seasonal correlation between excised root ARA and the root sugar concentration two weeks prior (116) indicate that root ARA is dependent on a large root carbon pool, probably sucrose. Metabolic transformation of sucrose to organic acids might occur through the fermentative action of facultative anaerobes, or via plant PEPC activity. Measured rates of $^{14}CO_2$ fixation by roots appear to be sufficient to support observed nitrogenase activity if the CO_2 is fixed into malate which is used by microaerophils with an efficiency of N_2 fixation similar to that reported for *Azospirillum* (138); excised root ARA was reduced by factors of 4 to 6 when the pCO_2 in flasks was reduced by adding CO_2 absorbents (29).

Rhizosphere carbon

Sugars are the predominant form in which recent photosynthates are excreted into the rhizosphere of *Spartina* (29). Matsumoto *et al.* (123) documented seasonal patterns and types of carbon compounds excreted by hydroponically grown maize. On a fresh weight of plant basis, all products were maximal at the first sampling (7 days, approximately 13 μg ethanol insoluble N, 23 μg ethanol insoluble sugar (polysaccharides), 37 μg ethanol soluble N, 16 μg ethanol soluble sugars, and 4.2 μg organic acids per gram fresh weight per day) and declined by factors of five- to tenfold by 30 to 40 days; plant fresh weight increased about tenfold during the same period. Thereafter, levels remained low except for ethanol soluble sugars which exhibited a second peak of production at 40 days, just before heading. Ninety percent of the soluble N was amino nitrogen and glutamate was the most common amino acid throughout plant growth, followed by alanine. Lactate was the predominant organic acid. Stachyose (a tetrasaccharide with two molecules of galactose and one each of fructose and glucose) was the main soluble sugar, except at the preheading stage when glucose and fructose were also abundant.

Lee *et al.* (114) found no sharp seasonal fluctuations in levels of free sugars in rice soils. Rhizosphere soil nitrogenase activity associated with *Spartina* appears to be the highest (154) when root sugar levels are also highest (116) presumably because roots are leakiest at such times. Since the addition of alcohol insoluble extracts of roots did not stimulate nitrogenase activity of diazotrophs in rhizosphere soil from carbon-starved plants whereas the alcohol soluble extracts did so (29), we think it is unlikely that materials such as sloughed off root cells and mucigel are a major reduced carbon source for rhizosphere diazotrophs. Wullstein *et al.* (214) suggested that materials released from lysed cortical cells of xeric grasses supported nitrogenase activity in rhizosheaths.

Potential for N_2 fixation

Dobereiner (58) noted that *Rhizobium* bacteroids can fix maximally about 10 times as much N per gram carbohydrate consumed (she cited 310 mg N perg ram carbohydrate; more recent estimates (57) are about half of this) as *Azotobacter*, but that grasses have only half the N content of legumes and that the C-4 pathway in tropical grasses might provide more carbon than the strictly C-3 metabolism of legumes; "under the circumstances it does not seem too exagerated to admit the theoretical possibility of at least half of the plant's nitrogen requirement being met by root exudates, as long as the larger part of the root exudates is used by nitrogen-fixing bacteria."

Barber and Martin (13) found that between 5 and 10 percent, and 12 and 18 percent of photosynthetically fixed carbon was released from wheat and barley grown under sterile and non-sterile conditions respectively for three weeks. Martin (121) studied plants over the period from three to eight weeks which is probably more relevant in considerations of associative N_2 fixation (see below); the loss of ^{14}C from roots (as $^{14}CO_2$ and soil ^{14}C) was on the average 17.3 percent of the total ^{14}C recovered from pots. The total loss was not effected by the presence of soil microorganisms but 58 percent of the total loss was respired in the non-axenic systems, versus 29 percent in the gamma-radiated system. Losses of carbon from roots to soil were attributed to the autolytic degeneration of root cortical cells. Conceivably, a substantial fraction of even the respired carbon might be alternatively used by an internally located bacterial population because a large fraction of root-respired carbon is respired via the cyanide-insensitive pathway which seems to function in burning off excess root carbon (110); the proportion of respiration accounted for by the cyanide-insensitive pathway is smaller in nodulated than in non-nodulated legumes, possibly because *Rhizobium* acts as an alternative sink for this excess carbon (111).

Barber and Lynch (12) suggested that with rates of exudation cited by Barber and Martin (13), and an efficiency of N_2 fixation of 10 to 15 mg N per gram carbohydrate, at most, 15 percent of the N content of wheat might be derived from associative N_2 fixation. Because non-diazotrophs normally make up the bulk of the rhizosphere population, they suggested that the real value is likely to be lower, and that this process could cause an appreciable increase in the N supply to temperate cereals only if greater amounts of carbohydrate were exuded by roots, or efficiencies of fixation were higher. It seems not unrealistic to expect that efficiencies of N_2 fixation could be two or three times higher than those cited by Barber and Lynch (12), and that the overall efficiency of fixing N and transferring N to the plant would be higher in an endorhizal association that in a strictly rhizosphere association. The highest calculated contribution of bacterial N_2 fixation to total plant N in the inoculated wheat plants studied by Rennie and Larson (165) was 59 percent (Table 3). N_2 fixation was highest in non-leaky plants, and there was evidence that the *Bacillus* involved became established in the intercellular spaces of the root cortex (112). We have suggested that bacteria in intercellular spaces might induce a localized leakage of carbon compounds from plant cells, and that even in a mixed bacterial population, most of this carbon might be oxidized by the N_2-fixing microaerophils if they were located in a region of slightly higher pO_2 than other organisms, or if they are able to utilize oxygen more effectively at low pO_2 than other organisms (146).

In summary, both theoretical and empirical considerations suggest that the "potential" for N_2 fixation by associative symbioses is of the order of 50 percent of the total N taken up by shoots. Declining efficiencies of N_2 fixation as carbon substrate concentration is increased might limit the absolute amount of N_2 fixation, i.e. result in the maximum theoretical contribution being lower in plants or at sites of very high productivity (152).

Environmental factors

Nitrogen

High levels of ammonium or nitrate inhibit the nodulation of legumes (72). After nodulation, the inorganic nitrogen suppresses nitrogenase activity by causing a reduction in the translocation of carbon substrates to roots rather than by the regulation of nitrogenase synthesis (113). There is evidence that in the N_2-fixing state, rhizobial NH_4-assimilating enzymes are non-functional. In free-living diazotrophs, the repression of nitrogenase synthesis by ammonium is mediated by the ammonium-assimilating system (159).

ARA of roots excised from various species in a Nova Scotian saltmarsh were inversely correlated with groundwater ammonium concentrations (150); ARA was high where ammonium was less than 3 μM, and low where it was above 33 μM (0.46 ppm). The latter concentration is much lower than the "threshold" value of 40 ppm ammonium-N on a soil weight basis for the suppression of nitrogenase activity in rice by added ammonium (9). However it is a true solution value (this was a quaking mat marsh in which there was free circulation of water) and is in the range (10 to 25 μM) in which repression/derepression of nitrogenase occurs in continuous cultures of *Azotobacter vinelandii* (103). In several studies, nitrate has been observed to cause a more complete and immediate suppression of nitrogenase activity by roots and soil from *Spartina* than does ammonium (53, 87, 197); this is attributed to competition by nitrate for electrons (53). When *Spartina* is maintained hydroponically with low ammonium and transferred to 200 μM ammonium solution, "exorhizal" nitrogenase activity is suppressed by 64 percent but "endorhizal" activity is not affected, at least for several days (27). Hanson (87) reported that monthly additions of ammonium nitrate to a saltmarsh stimulated nitrogenase activity, possibly by increasing primary production. Application of fertilizer-N to rice in mid-season caused an initial depression and later enhancement in nitrogenase activity (81). In several experiments, the addition of N has been reported to enhance the positive effects of inoculation (185, 190). Garcia *et al.* (75) reported that nitrate inhibited the attachment of *Azospirillum* to root hairs. McClung (124) found that attachment of *A. brasilense* SP 7 to root hairs of maize was inhibited by nitrate or ammonium, but the attachment of two strains originally isolated from maize was not inhibited by combined N. He also found that relative numbers of diazotrophs (MPN estimates of diazotrophs/colony forming units on combined N plates) increased as nitrogenase activity of three plant species increased along natural gradients of moisture and inorganic N; this suggests that environmental factors influence the colonization of roots by diazotrophs *in situ*.

Temperature, pH, light and moisture

Initially, *Azospirillum* was isolated only from tropical soils, and was considered to be rare or non-existent in temperate soils. Soil temperatures below 25°C may limit nitrogenase activity associated with tropical grasses (62), but the occurrence of *Azospirillum* and association of nitrogenase activity with roots of temperate zone plants such as *Festuca rubra* illustrates that temperature *per se* is not a limiting factor for the formation of these types of associations in temperate regions (84).

Like *Azotobacter* (58), *Azospirillum* is sensitive to low pH, but is found associated with plants in soils in which the pH (less than 5.5) is not favorable to its growth (65).

Pronounced diurnal fluctuations in nitrogenase activity in whole plant systems (8), or light/dark effects on whole plant nitrogenase activity (63) have been attributed to short term effects of changing light on the supply of photosynthate to roots and associated bacteria. Alternatively, such effects could be due to activity of phototrophic diazotrophs, or to changes in pO_2, temperature or stomatal opening. Under controlled laboratory conditions we could not find a pronounced short-term effect of light on nitrogenase activity associated with *Spartina*; this was consistent with other evidence that recent photosynthates enter a large root carbon pool which would buffer fluctuations in the supply of photosynthates to roots (29), and is similar to observations on legumes (68).

Typically, ARA is reported to increase with increasing soil moisture (183,

209); these increases could be due in part to an increasing endogenous production of ethylene.

Seasonal variations in nitrogenase activity

Nitrogenase activity of roots excised from grasses (116, 131), and of intact plants (114, 116, 205) is typically maximal at some stage during reproductive growth. Lee *et al.* (114) found no relation between seasonal patterns of nitrogenase activity and either exchangeable ammonium or sugar concentration in the soil. In *Spartina,* the seasonal maximum in excised root nitrogenase activity, and in whole plant nitrogenase activity follows the accumulation of sugars in the roots (116). In annuals, nitrogenase activity is generally low during the early vegetative growth, except in some instances immediately following germination (66), and reaching the maximum, drops more rapidly than in *Spartina.* In addition to the seasonal accumulation of sugars in early reproductive phases (116), seasonal variations in nitrogenase activity and differences in the patterns between plants may be related to one or more of the following: (i) high concentration of excreted carbon compounds in the seedling stage (123); (ii) presence of inhibitory compounds at certain stages (97, 149); (iii) suppression of nitrogenase activity in the early part of the season by combined N (53, 116); (iv) invasion of the root interior by diazotrophs (117); and (v) sharp decline in root sugars following the maximum in annuals (70) versus continued accumulation of sugars in perennials (116).

Goals and strategies of future research

There is a variety of evidence indicating that both the plant and bacterial genomes influence associative N_2 fixation, and hence that both have to be appropriately selected in order to maximize N_2 fixation. The inoculation experiments of Rennie and Larson (Table 3) demonstrate most explicitly plant/bacteria interactions. They suggested that since root rot susceptible lines are considered to be "leaky", and the resistant lines not so, and since the latter consistently supported N_2-fixing activity while the former did not, non-leaky roots rather than leaky roots is the feature to be sought in the plant genome. We came to the same conclusion independently, based on considerations of carbon metabolism and on the evidence for internal localization of nitrogenase activity in *Spartina* (146). Correlations between whole plant ARA and root mass of different rice cultivars (92, 114), and between whole plant or excised root ARA and root sugar concentration (116), and carbon-14 activity of roots from plants previously exposed to $^{14}CO_2$ (154) suggest that there is likely to be some trade off between increasing root nitrogenase activity and increasing harvest index, at least where the latter are grains. It is not surprising in this context that there is evidence that older varieties of rice support greater N_2 fixation than do modern varieties (81).

There is evidence of various sorts suggesting that nitrogenase activity in grasses is correlated with the absolute numbers or relative proportions of diazotrophs in the root microflora (154, 181; Fig. 1). Thus increasing the proportion or numbers of diazotrophs associated with roots by the selection of highly competitive strains, heavy inoculation, or by the introduction of N_2-fixing genes to other root associated bacteria (102), should result in increased N_2 fixation.

The first apparently successful example of "engineering" a new type of N_2-fixing intracellular, eukaryote-prokaryote association is that in which cells of *Azotobacter vinelandii* were introduced into the mycorrhizal fungus *Rhizopogon* (78, 79, 80). The modified fungus was capable of acetylene reduction, and on being reassociated with its host, *Pinus radiata*, giving rise to increased levels of nitrogen and more vigorous growth in N-deprived seedlings (79). Because most major crops form mycorrhizal associations, this approach seems to hold great potential, but it is presently not applicable to grasses because of the difficulties in manipulating their associated mycorrhiza in culture (77).

Giles and Whitehead (79) were at first concerned that their new fungal strain was pathogenic. This turned out not to be the case (77), but it illustrates the need to be concerned about such a possibility. There is also concern that conventional plant breeding for increased nitrogen fixation could produce plants that are more susceptible to certain pathogens (202).

Though a combination of the above—increasing root carbon, increasing numbers of diazotrophs associated with roots, and introduction of N_2-fixing mycorrhiza—and complementation of N_2 fixation and use of combined N by appropriate timing of fertilizer-N applications (120), or by use of nitrate reductase negative mutants (117), it seems reasonable to expect that grasses could fix 50 percent of their total N at moderate to high levels of production. The research required to achieve such a goal is still very basic in nature. We have major gaps in our understanding of the functioning of the associative systems, and hence of the plant and microbial characteristics that must be selected to maximize N_2 fixation and the transfer of that N to the plant.

On a more immediate basis, I suggest that the best prospects for using associative N_2 fixation lie with the perennial forage and biomass crops, and particularly with the latter, notably with sugar cane, for the following reasons: (1) They have large root systems. (2) The perennial habit facilitates inoculation and maintenance of high levels of associated diazotrophs. (3) The product of interest (in sugar cane) is also a carbon substrate for diazotrophs; while the amount of N required per unit area may be large, there is a large amount of carbon produced per unit of N assimilated and the diversion of energy toward N_2 fixation can be expected to involve less of a trade off with yield than in cereals. (4) Submaximal yields may be acceptable, if inputs are substantially reduced by associative N_2 fixation. (5) In a perennial system, the precise site of N_2 fixation is not as critical as in an annual plant system. Provided that the overall efficiency of use of soil N is high, which it usually is in a perennial plant system (leaching and denitrification losses are lower than in an annual system), then most of the N_2 fixed, regardless of where it is fixed, will ultimately be available to the plant. In this context associative N_2 fixation needs to be examined in relation to the N dynamics of the whole system, and the methodologies employed to study strictly symbiotic N_2-fixing systems may not be the most appropriate ones. For example, the isotope dilution technique provides an estimate of the amount of N_2 fixed that is incorporated in the plant in the current growing season, but it does not estimate the total amount of N_2 fixed which could be several-fold higher. In the whole system context, it is just as important to look at the input of recently fixed N_2 from rhizosphere and non-rhizosphere sources to soil humus, and at the subsequent release of that N. The potential for reducing fertilizer-N inputs in sugar cane by maximizing associative N_2 fixation through plant breeding (172), irrespective of the immediate fate of that N, and by managing plant residues in such a way as to maximize non-rhizosphere N_2 fixation, is probably high and certainly more readily achievable than is the prospect of achieving the same reductions through development of legume-like symbioses.

Summary

The developments of the last four to five years have borne out Dobereiner's contention that there was much more to be learned about the associative systems before applied benefits could be reasonably expected; at the same time, they have verified many of her basic concepts about these associations. The major developments are as follows.

1) We have more critical methodologies for assessing N_2 fixation, identifying microorganisms, and determining the parameters of the associative systems; still, the problem of assaying N_2 fixation is a difficult one.

2) Rhizosphere soil N_2 fixation has been shown to be substantial and of immediate significance to the plant.

3) Facultatively anaerobic diazotrophs have been recognized as important agents of N_2 fixation in many associative systems.

4) The physiology of *Azospirillum* and the process of association of this organism with plants have proved to be remarkably similar to those of rhizobia with legumes; differences between the various types of associations are beginning to look to be more of a quantitative, rather than of a qualitative, nature.

5) Internal colonization of roots by diazotrophs has been verified, but this sort of colonization is not limited to diazotrophs, nor is it intracellular.

6) There is clear evidence of plant/bacteria interactions on N_2 fixation,

and some of the determinants of these interactions have been identified.

7) Positive and significant inoculation effects that are attributable to N_2 fixation have been reported; it is evident also that certain inoculation responses are not due to N_2 fixation.

8) N_2-fixing genes or whole organisms have been incorporated in non-diazotrophic root associates.

The real potential of associative N_2 fixation to supply plant N would appear to be of the order of 50 percent of total plant N. To achieve such a potential will probably require increasing the root carbon poll, and increasing the proportion and numbers of diazotrophs associated with roots. Such a goal is still a matter of basic rather than applied research. The perennial grasses seem to offer more immediate prospects for substituting biological N_2 fixation for fertilizer-N than do cereals because there is less trade off between providing carbon to roots and carbon to harvested portions, and because the precise sites of N_2 fixation and the proportion of N_2 fixed that is transferred to the plant within one season are of less consequence.

References

1. Albrecht, S.L., Maier, R.J., Hanus, F.J., Russell, S.A., Emerich, D.W. and Evans, H.J. Hydrogenase in *Rhizobium japonicum* increases nitrogen fixation by nodulated soybeans, *Science, 203,* 1255–1257 (1979).
2. Albrecht, S.L., Okon, Y. and Burris, R.H. Effect of light and temperature on the association between *Zea mays* and *Spirillum lipoferum, Plant Physiology, 60,* 528–531 (1978).
3. Amarger, N., Mariotti, A., Mariotti, F., Durr, J.C. Bourguignon, C. and Lagacherie, B. Estimate of symbiotically fixed nitrogen in field grown soybeans using variations in ^{15}N natural abundance, *Plant and Soil, 52,* 269–280 (1979).
4. App, A.A., Watanabe, I. and Alexander, M. Nonsymbiotic nitrogen fixation associated with the rice plant in flooded soils, *Soil Science, 130,* 283–289 (1980).
5. Asanuma, S., Tanaka, H., and Yatazawa, M. Rhizoplane microorganisms of rice seedlings as examined by scanning electron microscopy, *Soil Science and Plant Nutrition, 25,* 539–551 (1979).
6. Bal, A.K. and Shantharam, S. Localization of primary and secondary lectin receptor sites in *Rhizobium japonicum,* in *Proceedings of the Fourth International Symposium on N_2 Fixation,* Canberra, December 1980 (Edited by A.H. Gibson), Australian Academy of Sciences (1981).
7. Balandreau, J. and Dommergues, Y. Assaying nitrogenase (C_2H_2) activity in the field, *Ecological Bulletins* (Stockholm), *17,* 247–254 (1973).

8. Balandreau, J.P., Millier, C.R. and Dommergues, Y.R. Diurnal variations of nitrogenase activity in the field, *Applied Microbiology, 27*, 662–665 (1974).
9. Balandreau, J., Rinaudo, G., Fares-Hamed, I. and Dommergues, Y. Nitrogen fixation in the rhizosphere of rice plants, in *Nitrogen Fixation by Free-Living Microorganisms* (Edited by W.D.P. Stewart), Cambridge University Press, Cambridge (1975).
10. Baldani, V.L.D. and Dobereiner, J. Host-plant specificity in the infection of cereals with *Azospirillum* spp., *Soil Biology and Biochemistry, 12*, 433–439 (1980).
11. Baltensperger, A.A., Schank, S.C., Smith, R.L., Littell, R.C., Bouton, J.H. and Dudeck, A.E. Effect of inoculation with *Azospirillum* and *Azotobacter* on turf-type Bermuda genotypes, *Crop Science, 18*, 1043–1045 (1978).
12. Barber, D.A. and Lynch, J.M. Microbial growth in the rhizosphere, *Soil Biology and Biochemisty, 9*, 305–308 (1977).
13. Barber, D.A. and Martin, J.K. The release of organic substrates by cereal roots in soil, *New Phytologist, 76*, 69–80 (1976).
14. Barber, L.E. and Evans, H.J. Characterization of a nitrogen-fixing bacterial strain from the roots of *Digitaria sanguinalis, Canadian Journal of Microbiology, 22*, 254–260 (1976).
15. Barber, L.E., Russell, S.A. and Evans, H.J. Inoculation of millet with *Azospirillum, Plant and Soil, 52*, 49–57 (1979).
16. Barber, L.E., Tjepkema, J.D., Russell, S.A. and Evans, H.J. Acetylene reduction (nitrogen fixation) associated with corn inoculated with *Spirillum, Applied and Environmental Microbiology, 32*, 108–113 (1976).
17. Becking, J.H. Putative nitrogen fixation in other symbioses, in *The Biology of Nitrogen Fixation* (Edited by A. Quispel), North-Holland Publishing Co., Amsterdam (1974).
18. Bedmar, E.J. and Olivares, J. Nitrogen fixation (acetylene reduction) by free-living *Rhizobium meliloti, Current Microbiology, 56*, 181–193 (1979).
19. Berg, R.H., Tyler, M.E., Novick, N.J., Vasil, V. and Vasil, I.K. Biology of *Azospirillum*-sugar cane association: Enhancement of nitrogenase activity, *Applied and Environmental Microbiology, 39*, 642–649 (1980).
20. Berg, R.H., Vasil, V. and Vasil, I.K. The biology of *Azospirillum*-sugarcane association II. Ultrastructure, *Protoplasma, 101*, 143–163 (1979).
21. Bhide, V.P. and Purandare, A.G. Occurrence of *Azotobacter* within the root cells of *Cynodon dactylon, Current Science, 48*, 913–914 (1979).
22. Bhuvaneswari, T.V., Pueppke, S.G. and Bauer, W.D. Role of lectins in plant-microorganism interactions I. Binding of soybean lectin to rhizobia, *Plant Physiology, 60*, 486–491 (1977).
23. Bhuvaneswari, T.V., Turgeon, G. and Bauer, W.D. Early events in the infection of soybean (*Glycine max* L. Merr) by *Rhizobium japonicum* I.

Localization of infectible root cells, *Plant Physiology*, *66*, 1027–1031 (1980).

24. Boddey, R.M., Quilt, P. and Ahmad, L. Acetylene reduction in the rhizosphere of rice: Methods of assay, *Plant and Soil*, *50*, 567–574 (1978).
25. Bohlool, B.B. and Schmidt, E.L. Lectures: A possible basis for specificity in the *Rhizobium*-legume root nodule symbiosis, *Science*, *185*, 269–271 (1974).
26. Bouton, J.H. and Zuberer, D.A. Response of *Panicum maximum* Jacq. to inoculation with *Azospirillum brasilense*, *Plant and Soil*, *52*, 585–590 (1979).
27. Boyle, C.D. and Patriquin, D.G. Endorhizal and exorhizal acetylene-reducing activity in a grass (*Spartina alterniflora* Loisel.)-diazotroph association, *Plant Physiology*, *66*, 276–280 (1980).
28. Boyle, C.D. and Patriquin, D.G. Acetylene-reducing activity (ARA) by endorhizosphere diazotrophs, in *Proceedings of International Workshop on Associative N_2-fixation*, Piracicaba, Brazil, July 1979 (Edited by P.B. Vose and A.P. Ruschel), CRC Press, West Palm Beach, Florida (1981).
29. Boyle, C.D. and Patriquin, D.G. Carbon metabolism of *Spartina alterniflora* Loisel. in relation to that of associated N_2-fixing bacteria, *New Phytologist* (in press).
30. Brown, M.E. Seed and root bacterization, *Annual Review of Phytopathology*, *12*, 181-197 (1974).
31. Brown, M.E. Role of *Azotobacter paspali* in association with *Paspalum notatum*, *Journal of Applied Bacteriology*, *40*, 341–348 (1976).
32. Burris, R.H., Okon, Y. and Albrecht, S.L. Properties and reactions of *Spirillum lipoferum*, *Ecological Bulletins*, *26*, 353–363 (1978).
33. Capone, D.G. and Taylor, B.F. N_2 fixation in the rhizosphere of *Thalassia testudinum*, *Canadian Journal of Microbiology*, *26*, 998–1005 (1980).
34. Carlson, P.S. and Chalef, R.S. Forced association between higher plant and bacterial cells *in vitro*, *Nature*, *252*, 393–395 (1974).
35. Chan, Y.K., Nelson, L.M. and Knowles, R. Hydrogen metabolism of *Azospirillum brasilense* in nitrogen-free medium, *Canadian Journal of Microbiology*, *26*, 1126–1131 (1980).
36. Charyulu, P.B.B.N. and Rao, V.R. Nitrogen fixation in some Indian rice soils, *Soil Science*, *128*, 86–89 (1979).
37. Child, J.J. Nitrogen fixation by a *Rhizobium* sp. in association with non-leguminous plant cell cultures, *Nature*, *253*, 350–351 (1975).
38. Child, J.J. and Kurz, W.G.W. Inducing effect of plant cells on nitrogenase activity by *Spirillum* and *Rhizobium in vitro*, *Canadian Journal of Microbiology*, *24*, 143–148 (1978).
39. Christeller, J.T., Laing, W.A. and Sutton, W.D. Carbon dioxide fixation by lupin root nodules, *Plant Physiology*, *60*, 47–50 (1977).

40. Cohen, E., Okon, Y., Kigle, J., Nur, I. and Henis, Y. Increase in dry weight and total nitrogen content in *Zea mays* and *Setaria italica* associated with nitrogen-fixing *Azospirillum* spp., *Plant Physiology, 66*, 746–749 (1980).
41. Cornforth, I.S. The persistence of ethylene in aerobic soils, *Plant and Soil, 42*, 85–96 (1975).
42. Dart, P. Infection and development of leguminous modules, in *A Treatise on dinitrogen fixation, Section III, Biology* (Edited by R.W.F. Hardy and W.S. Silver), John Wiley and Sons, New York (1977).
43. Dazzo, F.B. Determinants of host specificity in the *Rhizobium*-clover symbioses, in *Nitrogen Fixation*, Vol. II (Edited by W.E. Newton and W.H. Orme-Johnson), University Park Press, Baltimore (1980).
44. Dazzo, F.B. and Brill, W.J. Bacterial polysaccharide which binds *Rhizobium trifolii* to clover root hairs, *Journal of Bacteriology, 137*, 1362–1373 (1979).
45. Dazzo, F.B. and Hubbell, D.H. Antigenic differences between infective and noninfective strains of *Rhizobium trifolii*, *Applied Microbiology, 30*, 172–177 (1975).
46. Dazzo, F.B. and Milam, J.R. Serological studies of *Spirillum lipoferum*, *Proceedings of Soil and Crop Science Society of Florida, 35*, 121–126 (1976).
47. Dazzo, F.B., Napoli, C.A. and Hubbell, D.H. Adsorption of bacteria roots as related to host specificity in the *Rhizobium*-clover symbiosis, *Applied and Environmental Microbiology, 32*, 166–171.
48. Dazzo, F.B., Urbano, M.R. and Brill, W.J. Transient appearance of lectin receptors on *Rhizobium trifolii*, *Current Microbiology, 2*, 15–20 (1979).
49. DeDatta, S.K. and Gomez, K.A. Changes in soil fertility under intensive rice cropping with improved varieties, *Soil Science, 120*, 361–366 (1975).
50. Delwiche, C.C., Zinke, P.J. Johnson, C.M. and Virginia, R.A. Nitrogen isotope distribution as a presumptive indicator of nitrogen fixation, *Botanical Gazette, 140* (Supplement), 565–569 (1979).
51. De-Polli, H., Matsui, E., Dobereiner, J. and Salati, E. Confirmation of nitrogen fixation in two tropical grasses by $^{15}N_2$ incorporation, *Soil Biology and Biochemistry, 9*, 119–123 (1977).
52. de Vries, G.E., In'T Veld, P. and Kijne, J.W. Production of organic acids in *Pisum sativum* root nodules as a result of oxygen stress, *Plant Science Letters, 20*, 115–123 (1980).
53. Dicker, H.J. and Smith, D.W. Physiological ecology of acetylene reduction (nitrogen fixation) in a Delaware salt marsh, *Microbial Ecology, 6*, 161–171 (1980).
54. Dicker, H.J. and Smith, D.W. Acetylene reduction (nitrogen fixation) in

a Delaware, USA salt marsh, *Marine Biology*, *57*, 241–250 (1980).
55. Dicker, H.J. and Smith, D.W. Enumeration and relative importance of acetylene-reducing (nitrogen-fixing) bacteria in a Delaware salt marsh, *Applied and Environmental Microbiology*, *39*, 1019–1025 (1980).
56. Diem, H.G., Rougier, M., Humad-Fares, I., Balandreau, J.P. and Dommergues, Y.R. Colonization of rice roots by bacteria, *Ecological Bulletins* (Stockholm), *26*, 305–311 (1978).
57. Diem, H.G., Schmidt, E.L. and Dobereiner, J. The use of the fluorescent-antibody technique to study the behaviour of a *Beijerinchis* isolate in the rhizosphere and spermosphere of rice, *Ecological Bulletins* (Stockholm), *26*, 312–318 (1978).
58. Dobereiner, J. Nitrogen-fixing bacteria in the rhizosphere, in *The Biology of Nitrogen Fixation* (Edited by A. Quispel), North Holland Publishing Co., Amsterdam (1974).
59. Dobereiner, J. N_2 fixation associated with non-leguminous plants, in *Genetic Engineering for Nitrogen Fixation* (Edited by A. Hollaender), Plenum Press, New York and London (1977).
60. Dobereiner, J. and Baldani, V.L.D. Selective infection of maize roots by streptomycin-resistant *Azospirillum lipoferium* and other bacteria, *Canadian Journal of Microbiology*, *25*, 1264–1269 (1979).
61. Dobereiner, J. and Boddey, R.M. Nitrogen fixation in association with Graminae, in *Proceedings of Fourth International Symposium on N_2 Fixation*, Canberra, December, 1980 (Edited by A.H. Gibson), Australian Academy of Sciences (in press).
62. Dobereiner, J. and Day, J.M. Associative symbioses in tropical grasses: Characterization of microorganisms and dinitrogen-fixing sites, in *Proceedings of the 1st International Symposium on Nitrogen Fixation* (Edited by W.E. Newton and C.J. Nyman), Washington State University Press, Pullman (1976).
63. Dobereiner, J., Day, J.M. and Dart, P.J. Nitrogenase activity and oxygen sensitivity of the *Paspalum notatum-Azotobacter paspali* association, *Journal of General Microbiology*, *71*, 103–116 (1972).
64. Dobereiner, J., Day, J.M. and Dart, P.J. Rhizosphere associations between grasses and nitrogen-fixing bacteria: Effect of O_2 on nitrogenase activity in the rhizosphere of *Paspalum notatum*, *Soil Biology and Biochemistry*, *5*, 157–159 (1973).
65. Dobereiner, J., Marriel, I.E. and Nery, M. Ecological distribution of *Spirillum lipoferum* Beijerinck, *Canadian Journal of Microbiology*, *22*, 1464–1473 (1976).
66. Dommergues, Y., Balandreau, J., Rinaudo, G. and Weindhard, P. Nonsymbiotic nitrogen fixation in the rhizospheres of rice, maize and different tropical grasses, *Soil Biology and Biochemistry*, *5*, 83–89 (1975).
67. Durbin, K.J. and Watanabe, I. Sulfate reducing bacteria and nitrogen

fixation in flooded rice soil, *Soil Biology and Biochemistry*, *12*, 11–14, (1980).

68. Eckart, J.F. and Raguse, C.A. Effects of diurnal variation in light and temperature on acetylene reduction activity (nitrogen fixation) of subterranean clover, *Agronomy Journal*, *72*, 519–523 (1980).
69. Esau, K. *Plant Anatomy*, John Wiley, New York (1965).
70. Evans, L.T. and Wardlaw, I.F. Aspects of the comparative physiology of grain yield in cereals, *Advances in Agronomy*, *28*, 301-359 (1976).
71. Foster, R.C. and Rovira, A.D. Ultrastructure of wheat rhizosphere, *New Phytologist*, *76*, 343–352 (1976).
72. Fred, E.B., Baldwin, I.L. and McCoy, E. *Root nodule bacteria and leguminous plants*, The University of Wisconsin Press, Madison (1932).
73. Fried, M. and Middleboe, V. Measurement of amount of nitrogen fixed by a legume crop, *Plant and Soil*, *47*, 713–715 (1977).
74. Garcia, M. Umali, Hubbell, D.H. and Gaskins, M.H. Process of infection of *Panicum maximum* by *Spirillum lipoferium*, in *Environmental Role of Nitrogen-fixing Blue-green Algae and Asymbiotic Bacteria* (Edited by U. Granhall), *Ecological Bulletins* (Stockholm), *26*, 373–379 (1978).
75. Garcia, M. Umali, Hubbell, D.H., Gaskins, M.H. and Dazzo, F.B. Association of *Azospirillum* with grass roots, *Applied and Environmental Microbiology*, *39*, 219–226 (1980).
76. Gauthier, D., Diem, H.G. and Dommergues, V. *In vitro* nitrogen fixation by two Actinomycete strains isolated from *Casuarina* nodules, *Applied and Environmental Microbiology*, *41*, 306–308 (1981).
77. Giles, K.L. and Vasil, I. Nitrogen fixation and plant tissue culture, in *Perspectives in Plant Cell and Tissue Culture*, Supplement *11B* (Edited by I.K. Vasil), Academic Press, New York, London (1980).
78. Giles, K.L. and Whitehead, H.C.M. The transfer of nitrogen-fixing ability to a eukaryotic cell, *Cytobios*, *14*, 49–61 (1975).
79. Giles, K.L. and Whitehead, H.C.M. Reassociation of a modified mycorrhiza with the host plant roots (*Pinus radiata*) and the transfer of acetylene reduction activity, *Plant and Soil*, *48*, 143–152 (1977).
80. Giles, K.L. and Whitehead, H. The localization of introduced *Azotobacter* cells within the mycelium of a modified mycorrhiza (*Rhizopogon*) capable of nitrogen fixation, *Plant Science Letters*, *10*, 367–372 (1977).
81. Gilmour, J.T., Gilmour, G.M. and Johnston, T.H. Nitrogenase activity of rice plant root systems. *Soil Biology and Biochemistry*, *10*, 261–264. (1978).
82. Gotz, E.M. Attachment to plant root surface and nitrogenase activity of rhizobia associated with *Petunia* plants, *Zeitschrift für Pflanzenphysiologie*, *98*, 465–470 (1980).
83. Gotz, E.M. and Hess, D. Nitrogenase activity induced by wheat plants, *Zeitschrift für Pflanzenphysiologie*, *98*, 453–458 (1980).

84. Haahtela, K., Wartioraara, T., Sundmain, V. and Skujins, J. Root-associated N_2 fixation (acetylene reduction) by Enterobactaceae and *Azospirillum* strains in cold-climate spodosols, *Applied and Environmental Microbiology*, *41*, 203–206 (1981).
85. Habte, M. and Alexander, M. Use of streptomycin for suppressing blue-green algal nitrogenase activity during the assessment of nitrogenase activity in the rice rhizosphere, *Soil Science Society of America Journal*, *44*, 756–760 (1980).
86. Habte, M. and Alexander, M. Nitrogen fixation by photosynthetic bacteria in lowland rice culture, *Applied and Environmental Microbiology*, *39*, 342–347 (1980).
87. Hanson, R.B. Comparison of nitrogen fixation activity on tall and short *Spartina alterniflora* salt marsh soils, *Applied and Environmental Microbiology*, *33*, 596–602 (1977).
88. Hanus, F.J., Maier, R.J. and Evans, H.J. Autotrophic growth of H_2-uptake-positive strains of *Rhizobium japonicum* in an atmosphere supplied with hydrogen gas, *Proceedings of National Academy of Sciences* (*U.S.A.*), *76*, 1788–1792 (1979).
89. Hardy, R.W.F., Holsten, R.D., Jackson, E.K. and Burns, R.C. The acetylene-ethylene assay for N_2 fixation: Laboratory and field evaluation, *Plant Physiology*, *43*, 1185–1207 (1968).
90. Hegazi, M.A., Eid, M., Farag, R.S. and Monib, M. Asymbiotic N_2 fixation in the rhizosphere of sugar cane planted under semi-arid conditions of Egypt, *Revue Ecologie et Biologie du sol*, *16*, 23–37 (1979).
91. Hegazi, M.A., Monib, M. and Vlassak, K. Effect of inoculation with N_2-fixing spirilla and *Azotobacter* on nitrogenase activity on roots of maize grown under subtropical conditions, *Applied and Environmental Microbiology*, *38*, 621–625 (1979).
92. Hirota, Y., Fujii, T., Sano, Y. and Iyama, S. Nitrogen fixation in the rhizosphere of rice, *Nature*, *276*, 416–417 (1978).
93. Hubbell, D.H., Tien, T.M., Gaskins, M.H. and Lee, J.K. Physiological interaction in the *Azospirillum*-grass root association, in *Proceedings of International Workshop on Associative N_2-fixation*, Piracicaba, Brazil July, 1979 (Edited by P.B. Vose and A.P. Ruschel), CRC Press, West Palm Beach, Florida (in press).
94. Ito, O., Cabrera, D. and Watanabe, I. Fixation of dinitrogen-15 associated with rice plants, *Applied and Environmental Microbiology*, *39*, 554–558 (1980).
95. Johnson, V.A. and Mattern, P.J. Genetic improvement of productivity and nutritional quality of wheat, Report Research Findings U.S. Agency of International Development, Contract number AID/taC-1093 (1977).
96. Jones, K. Nitrogen fixation in a salt marsh, *Journal of Ecology*, *62*, 583–595 (1974).

97. Kandasamy, D. and Prasad, N.N. Colonization by rhizobia of the seed and roots of legumes in relation to exudation of phenolics, *Soil Biology and Biochemistry*, *11*, 73–75 (1979).
98. Kavimandan, S.K., Kumari, M.L. and Subba Rao, N.S. Non-symbiotic nitrogen-fixing bacteria in the rhizosphere of wheat, maize and sorghum, *Proceedings of the Indian Academy of Science*, *87B*, 299–302 (1978).
99. Kavimandan, S.K., Subba Rao, N.S. and Mohrir, A.V. Isolation of *Spirillum lipoferum* from the stems of wheat and nitrogen fixation in enrichment cultures, *Current Science*, *47*, 96–98 (1978).
100. Kawase, M. Role of cellulose in aerenchyma development in sunflower, *American Journal of Botany*, *66*, 183–190 (1979).
101. Keister, D.L. Acetylene reduction by pure culture of rhizobia, *Journal of Bacteriology*, *123*, 1265–1268 (1975).
102. Kleeberger, A. and Klingmuller, W. Plasmid-mediated transfer of nitrogen-fixing capability to bacteria from the rhizosphere of grasses, *Molecular and General Genetics*, *180*, 621–627 (1980).
103. Kleiner, D. Quantitative relations for the repression of nitrogenase synthesis in *Azotobacter vinelandii*, *Archives of Microbiology*, *101*, 153–159 (1974).
104. Kloepper, J.W., Schroth, M.M. and Miller, T.D. Effects of rhizosphere colonization by plant growth-promoting rhizobacteria on potato plant development and yield, *Phytopathology*, *70*, 1078–1082 (1980).
105. Klucas, R. Nitrogen fixation by *Klebsiella* grown in the presence of oxygen, *Canadian Journal of Microbiology*, *18*, 1845–1850 (1972).
106. Knowles, R. The measurement of nitrogen fixation, in *Proceedings of Fourth International Symposium on N_2 Fixation*, Canberra, December, 1980 (Edited by A.H. Gibson), Australian Academy of Sciences (in press).
107. Konings, H. and Verschuren, G. Formation of aerenchyma in roots of *Zea mays* in aerated solutions, and its relation to nutrient supply, *Physiologia Plantarum*, *49*, 265–270 (1980).
108. Kurz, W.G.W. and LaRue, T.A. Nitrogenase activity in rhizobia in absence of plant host, *Nature*, *256*, 407–409 (1975).
109. Lakshmi, V., Rao, A.S., Vijayalakshmi, K., Lakshmi-Kumari, M. Tilak, K.V.B.R. and Subba Rao, N.S. Establishment and survival of *Spirillum lipoferum*, *Proceedings of the Indian Academy of Science*, *86B*, 397–404 (1977).
110. Lambers, H. The physiological significance of cyanide-resistant respiration in higher plants, *Plant, Cell and Environment*, *3*, 293–302 (1980).
111. Lambers, H., Layzell, D.B. and Pate, J.S. Efficiency and regulation of root respiration in a legume: Effects of the N source, *Physiologia Plantarum*, *50*, 319–325 (1980).
112. Larson, R.I. and Neal, J.L. Selective colonization of the rhizosphere of

wheat by nitrogen-fixing bacteria, in *Environmental Role of Nitrogen-fixing Blue-green Algae and Asymbiotic Bacteria* (Edited by U. Granhall), *Ecological Bulletins* (Stockholm), *26*, 331–342 (1978).

113. Latimore, M., Giddens, J. and Ashley, D.A. Effect of ammonium and nitrate nitrogen upon photosynthate supply and nitrogen fixation by soybeans, *Crop Science*, *17*, 399–404 (1977).

114. Lee, K.K., Castro, T. and Yoshida, T. Nitrogen fixation throughout growth, and varietal differences in nitrogen fixation by the rhizosphere of rice planted in pots, *Plant and Soil*, *48*, 613–619 (1977).

115. Legg, J.O. and Sloger, C.A. Tracer method for determining symbiotic nitrogen fixation in field studies, in *Proceedings of Second International Conference on Stable Isotopes*, Oak Brook, Illinois (1976).

116. Livingstone, D.C. and Patriquin, D.G. Nitrogenase activity in relation to season, carbohydrates and organic acids in a temperate zone root association, *Soil Biology and Biochemistry*, *12*, 543–546 (1980).

117. Magalhaes, L.M.S., Patriquin, D. and Dobereiner, J. Infection of field grown maize with *Azospirillum* spp., *Revista Brasileira de Biologia*, *39*, 587–596 (1979).

118. Magalhaes, L.M.S., Neyra, C.A. and Dobereiner, J. Nitrate and nitrite reductase negative mutants of N_2-fixing *Azospirillum* spp., *Archives of Microbiology*, *117*, 247–252 (1978).

119. Mahmoud, S.A.Z., El-Sawy, M., Ishac, Y.Z. and El-Safty, M.M. The effects of salinity and alkalinity on the distribution and capacity of N_2-fixation by *Azotobacter* in Egyptian soils, *Ecological Bulletins* (Stockholm), *26*, 99–109 (1978).

120. Mahon, J.D. and Child, J.J. Growth response of inoculated peas (*Pisum sativum*) to combined nitrogen, *Canadian Journal of Botany*, *57*, 1687–1693 (1979).

121. Martin, J.K. Factors influencing the loss of organic carbon from wheat roots, *Soil Biology and Biochemistry*, *9*, 1–7 (1977).

122. Matsuguchi, T., Shimonura, T., Lee, S.K. Factors regulating acetylene reduction assay for measuring heterotrophic nitrogen fixation in waterlogged soils, *Soil Science and Plant Nutrition*, *25*, 323–336 (1979).

123. Matsumoto, H., Okada, K. and Takahashi, E. Excretion products of maize roots from seedling to seed development stage, *Plant and Soil*, *53*, 17–26 (1979).

124. McClung, C.R. The effects of environmental factors on nitrogenase activity and root colonization by diazotrophic bacteria in non-nodulated angiosperms, M.Sc. thesis, Dalhousie University, Halifax, Canada (1980).

125. McClung, C.R. and Patriquin, D.G. Isolation of a nitrogen fixing *Campylobacter* species from the roots of *Spartina alterniflora* Loisel., *Canadian Journal of Microbiology*, *26*, 881–886 (1980).

126. McMannon, M. and Crawford, R.M.M. A metabolic theory of flooding tolerance: The significance of enzyme distribution and behaviour, *New Phytologist, 70*, 299–306 (1971).
127. Moore, A.W. Non-symbiotic nitrogen fixation in soil and soil-plant systems, *Soils and Fertilizers, 29*, 113–128 (1966).
128. Mort, A.J. and Bauer, W.D. Composition of the capsular and extracellular polysaccharides of *Rhizobium japonicum*, *Plant Physiology, 66*, 158–163 (1980).
129. Napoli, C.A., Dazzo, F.B. and Hubbell, D.H. Production of cellulose microfibrils by *Rhizobium*, *Applied Microbiology, 30*, 123–131 (1975).
130. Nelson, L.M. and Knowles, R. Effect of oxygen and nitrate on nitrogen fixation and denitrification by *Azospirillum brasilense* grown in continuous culture, *Canadian Journal of Microbiology, 24*, 1395–1403 (1978).
131. Neyra, C.A. and Dobereiner, J. Nitrogen fixation in grasses, *Advances in Agronomy, 29*, 1–38 (1977).
132. Neyra, C.A., Dobereiner, J., LaLande, R. and Knowles, R. Denitrification by N_2 fixing *Spirillum lipoferum*, *Canadian Journal of Microbiology. 23*, 300–305 (1977).
133. Neyra, C.A. and van Berkum, P. Nitrate reduction and nitrogenase activity in *Spirillum lipoferum*, *Canadian Journal of Microbiology, 23*, 306–310 (1977).
134. Nioh, I. Nitrogen fixation and a nitrogen-fixing bacterium from the roots of *Eragrostis ferruginea*, *Journal of General and Applied Microbiology, 25*, 261–271 (1979).
135. Nur, I., Okon, Y. and Henis, Y. Comparative studies of nitrogen-fixing bacteria associated with grasses in Israel with *Azospirillum brasilense*, *Canadian Journal of Microbiology, 26*, 714–718 (1980).
136. Obroucheva, N.V. Physiology of growing root cells, in *The Development and Function of Roots* (Edited by J.G. Torrey and D.T. Clarkson), Academic Press, New York, London (1975).
137. Okon, Y., Albrecht, S.L. and Burris, R.H. Methods for growing *Spirillum lipoferum* and for counting it in pure culture and in association with plants, *Applied and Environmental Microbiology, 33*, 85–87 (1977).
138. Okon, Y., Albrecht, S.L. and Burris, R.H. Factors affecting growth and nitrogen fixation of *Spirillum lipoferum*, *Journal of Bacteriology, 127*, 1248–1254 (1976).
139. Okon, Y., Cakmakci, L., Nur, I. and Chet, I. Aerotaxis and chemotaxis of *Azospirillum brasilense*: A note, *Microbial Ecology, 6*, 277–280 (1980).
140. Okon, Y., Houchins, J.P., Albrecht, S.L. and Burns, R.H. Growth of *Spirillum lipoferum* at constant partial pressures of oxygen, and the properties of its nitrogenase in cell-free extracts, *Journal of General Microbiology, 98*, 87–93 (1977).

141. Old, K.M. and Nicolson, T.H. Electron microscopal studies of the microflora of roots of sand dune grasses, *New Phytologist*, *74*, 51–58 (1975).
142. Opik, H. *The Respiration of Higher Plants*, Edward Arnold, London (1980).
143. Parkinson, D., Taylor, G.S. and Pearson, R. Studies on fungi in the root region. I. The development of fungi on young roots, *Plant and Soil*, *19*, 332–349 (1963).
144. Patriquin, D.G. Factors affecting nitrogenase activity (acetylene-reducing activity) associated with excised roots of the emergent halophyte *Spartina alterniflora* Loisel., *Aquatic Botany*, *4*, 193–210 (1978).
145. Patriquin, D.G. Nitrogen fixation (acetylene reduction) associated with cord grass, *Spartina alterniflora* Loisel., *Ecological Bulletins* (Stockholm), *26*, 20–27 (1978).
146. Patriquin, D.G., Boyle, C.D., Livingstone, D.C. and McClung, C.R. Physiology of the associative symbiosis in salt marsh cord grass, *Spartina alterniflora* Loisel., in *Proceedings of International Workshop on Associative N_2 -fixation*, Piracicaba, Brazil, July 1979 (Edited by P.B. Vose and A.P. Ruschel), CRC Press, West Palm Beach, Florida (1981).
147. Patriquin, D.G. and Denike, D. *In situ* acetylene reduction assays of nitrogenase activity associated with the emergent halophyte *Spartina alterniflora* Loisel.: Methodological problems, *Aquatic Botany*, *4*, 211–226 (1978).
148. Patriquin, D.G. and Dobereiner, J. 1978. Light microscopy observations of tetrazolium-reducing bacteria in the endorhizosphere of maize and other grasses in Brazil, *Canadian Journal of Microbiology*, *24*, 734–742 (1978).
149. Patriquin, D.G., Gracioli, L.A. and Ruschel, A.P. Nitrogenase activity of sugar cane propagated from stem cuttings in sterile vermiculite, *Soil Biology and Biochemistry*, *12*, 413–417 (1980).
150. Patriquin, D.G. and Keddy, C. Nitrogenase activity (acetylene reduction) in a Nova Scotian salt marsh: Its association with angiosperms and the influence of some edaphic factors, *Aquatic Botany*, *4*, 227–244 (1978).
151. Patriquin, D.G. and Knowles, R. Nitrogen fixation in the rhizosphere of marine angiosperms, *Marine Biology*, *16*, 49–58 (1972).
152. Patriquin, D.G. and Knowles, R. Effects of oxygen, mannitol and ammonium concentrations on nitrogenase (C_2H_2) activity in a marine skeletal carbonate sand, *Marine Biology*, *32*, 49–62 (1975).
153. Patriquin, D.G., MacKinnon, J.C. and Wilkie, K.I. Seasonal patterns of denitrification and leaf nitrate reductase activity in a corn field, *Canadian Journal of Soil Science*, *58*, 283–285 (1978).
154. Patriquin, D.G. and McClung, C.R. Nitrogen accretion, and the nature

and possible significance of N_2 fixation (acetylene reduction) in a Nova Scotia *Spartina alterniflora* stand, *Marine Biology*, *47*, 227–242 (1978).
155. Pedersen, W.L., Chakrabarty, K., Klucas, R.V. and Vidaver, A.K. Nitrogen fixation (acetylene reduction) associated with roots of winter wheat and sorghum in Nebraska, *Applied and Environmental Microbiology*, *35*, 129–135 (1978).
156. Pedrosa, F.O., Dobereiner, J. and Yates, M.G. Hydrogen dependent growth and autotrophic carbon dioxide fixation in *Derxia*, *Journal of General Microbiology*, *119*, 547–551 (1980).
157. Phillips, D.A. Efficiency of symbiotic nitrogen fixation in legumes, *Annual Review of Plant Physiology*, *31*, 29–49 (1980).
158. Quispel, A. (Editor). *The Biology of Nitrogen Fixation*, North-Holland Publishing Co., Amsterdam (1974).
159. Rawsthorne, S., Minchin, F.R., Summerfield, R.J., Cookson, C. and Coombs, J. Carbon and nitrogen metabolism in legume root nodules, *Phytochemistry*, *19*, 341–355 (1980).
160. Rennie, R.J., Paul, E.A. and Johns, L.E. Natural nitrogen-15 abundance of soil and plant samples, *Canadian Journal of Soil Science*, *56*, 43–50 (1976).
161. Rennie, R.J. Comparison of ^{15}N-aided methods for determining symbiotic dinitrogen fixation, *Revue Ecologie et Biologie de Sol*, *16*, 455–463 (1979).
162. Rennie, R.J. Dinitrogen-fixing bacteria: Computer assisted identification of soil isolates, *Canadian Journal of Microbiology*, *26*, 1275–1283 (1980).
163. Rennie, R.J. ^{15}N-Isotope dilution as a measure of dinitrogen fixation by *Azospirillum* associated with maize, *Canadian Journal of Botany*, *58*, 21–24 (1980).
164. Rennie, R.J. A single medium for the isolation of acetylene-reducing (dinitrogen-fixing) bacteria from soils, *Canadian Journal of Microbiology*, *27*, 8–14 (1981).
165. Rennie, R.J. and Larson, R.I. Dinitrogen fixation associated with disomic chromosome substitution lines of spring wheat, *Canadian Journal of Botany*, *57*, 2771–2775 (1979).
166. Rennie, R.J. and Larson, R.I. Dinitrogen fixation associated with disomic chromosome substitution lines of spring wheat in the phytotron and in the field, in *Associative Dinitrogen Fixation* (Edited by P.B Vose and A.P. Ruschel), CRC Press, Miami (1981).
167. Reynders, L. and Vlassak, K. Conversion of tryptophan to indoleacetic acid by *Azospirillum brasilense*, *Soil Biology and Biochemistry*, *11*, 547–548 (1979).
168. Robson, R.L. and Postgate, J.R. Oxygen and hydrogen in biological nitrogen fixation, *Annual Review of Microbiology*, *34*, 183–207 (1980).
169. Ronson, C.W. and Primrose, S.B. Carbohydrate metabolism in *Rhizo-*

bium trifolii: Identification and symbiotic properties of mutants, *Journal of General Microbiology, 112,* 77–88 (1979).

170. Rovira, A.D., Newmen, E.I., Bowen, H.J. and Campbell, R. Quantitative assessment of the rhizoplane microflora by direct microscopy, *Soil Biology and Biochemistry, 6,* 211–216 (1976).
171. Ruschel, A.P., Henis, Y. and Salati, E. Nitrogen-15 tracing of N-fixation with soil-grown sugar cane seedlings, *Soil Biology and Biochemistry, 7,* 181–182 (1975).
172. Ruschel, A.P. and Ruschel, R. Varietal differences affecting nitrogenase activity in the rhizosphere of sugar cane, *Proceedings of International Society of Sugarcane Technologists XVI Congress, 2,* 1941–1947 (1977).
173. Ruschel, A.P., Vose, P.B., Victoria, R.L. and Salati, E. Comparison of isoline techniques and non-nodulating isolines to study the effect of ammonium fertilization on dinitrogen fixation in soybean, *Glycine max., Plant and Soil, 53,* 513–525 (1979).
174. Sampaio, M.J.A.M., daSilva, E.M.R., Dobereiner, J., Yates, M.G. and Pedrosa, F.O. Autotrophy and methylotrophy in *Derxia gummosa, Azospirillum lipoferum* and *A. brasilense,* in *Proceedings of Fourth International Symposium on* N_2 *Fixation,* Canberra, December (Edited by A.H. Gibson), Australian Academy of Sciences (1981).
175. Sampaio, M.J.A.M., de Vasconcelos, L. and Dobereiner, J. Characterization of three groups within *Spirillum lipoferum* Beijerinck, *Ecological Bulletins* (Stockholm), *26,* 364–365 (1978).
176. Schank, S.C., Day, J.M. and de Lucas, E.D. Nitrogenase activity, nitrogen content, *in vitro* digestibility and yield of 30 tropical forage grasses in Brazil, *Tropical Agriculture* (Trinidad), *54,* 119–125 (1977).
177. Schank, S.C., Smith, R.L., Weiser, G.C., Zuberer, D.A., Bouton, J.H., Quesenberry, K.H., Tyler, M.E. and Littel, R.C. Fluorescent antibody technique to identify *Azospirillum brasilense* associated with roots of grasses, *Soil Biology and Biochemistry, 11,* 287–295 (1979).
178. Shearer, G. and Kohe, D.H. ^{15}N abundance in N-fixing and non-N-fixing plants, in *Recent Developments in Mass Spectrometry in Biochemistry and Medicine* (Edited by A. Frigerio), Plenum Press, N.Y. (1978).
179. Shearman, R.C., Petersen, W.L., Klucas, R.V. and Kinbacher, E.J. Nitrogen fixation associated with "Park" Kentucky bluegrass (*Poa pratensis* L.), *Canadian Journal of Microbiology, 25,* 1197–1200 (1979).
180. Shimshick, E.J. and Hebert, R.R. Binding characteristics of N_2-fixing bacteria to cereal roots, *Applied and Environmental Microbiology, 38,* 447–453 (1979).
181. Smith, A.M. Ethylene in soil biology, *Annual Review of Phytopathology, 14,* 53–73 (1976).
182. Smith, A.M. and Ap Rees, T. Pathways of carbohydrate fermentation in the roots of marsh plants, *Planta, 146,* 327–334 (1979).

183. Smith, D. and Patriquin, D.G. A survey of angiosperms in Nova Scotia for rhizosphere nitrogenase (acetylene-reduction) activity, *Canadian Journal of Botany*, *56*, 2218–2223 (1978).
184. Smith, L.A., Hill, S. and Yates, M.G. Inhibition by acetylene of conventional hydrogenase in nitrogen-fixing bacteria, *Nature*, *262*, 209–210 (1976).
185. Smith, R.L., Bouton, J.H., Schank, S.C., Queensbury, K.H., Tyler, M.E., Milam., J.R., Gaskins, M.H. and Littell, R.C. Nitrogen fixation in grasses inoculated with *Spirillum lipoferum*, *Science*, *193*, 1003–1005 (1976).
186. Snowcroft, W.R. and Gibson, A.H. Nitrogen fixation by *Rhizobium* associated with tobacco and cowpea cell cultures, *Nature*, *253*, 351–352 (1975).
187. Stacey, G., Paau, A.S. and Brill, W.A. Host recognition in the *Rhizobium*-soybean symbiosis, *Plant Physiology*, *66*, 609–614 (1980).
188. Stewart, W.D.P., Fitzgerald, G.P. and Burris, R.H. *In situ* studies on N_2 fixation using the acetylene reduction technique, *Proceedings of the National Academy of Science U.S.A.*, *58*, 2071–2078 (1967).
189. Streeter, J.G. and Bosler, M.E. Carbohydrates in soybean nodules: Identification of compounds and possible relationships to nitrogen fixation, *Plant Science Letters*, *7*, 321–329 (1976).
190. Subba Rao, N.S., Tilak, K.V.B.R., Singh, C.S. and Lakshmi-Kumari, M. Response of a few economic species of graminaceous plants to inoculation with *Azospirillum brasilense*, *Current Science*, *48*, 133–134 (1979).
191. Tarrand, J.J., Krieg, N.R. and Dobereiner, J. A taxonomic study of the *Spirillum lipoferum* group, with descriptions of a new genus, *Azospirillum* gen. nor. and two species *Azospirillum lipoferum* (Beijerinck) comb. nov. and *Azospirillum brasilense* sp. nov., *Canadian Journal of Microbiology*, *24*, 967–980 (1978).
192. Teal, J.M., Kanwisher, J.W. Gas transport in the marsh grass *Spartina alterniflora*, *Journal of Experimental Botany*, *17*, 355–361 (1966).
193. Tjepkema, J.D. and Burris, R.H. Nitrogenase activity associated with some prairie grasses, *Plant and Soil*, *45*, 81–94 (1976).
194. Tjepkema, J. and van Berkum, P. Acetylene reduction by soil cores of maize and sorghum in Brazil, *Applied and Environmental Microbiology*, *33*, 626–629 (1977).
195. van Berkum, P. Evaluation of acetylene reduction by excised roots for the determination of nitrogen fixation by grasses, *Soil Biology and Biochemistry*, *12*, 141–145 (1980).
196. van Berkum, P. and Day, J.M. Nitrogenase activity associated with soil cores of grasses in Brazil, *Soil Biology and Biochemistry*, *12*, 137–140, (1980).

197. van Berkum, P. and Sloger, C. Immediate acetylene reduction by excised grass roots not previously preincubated at low oxygen tensions, *Plant Physiology*, *64*, 739–743 (1979).
198. van Berkum, P. and Sloger, C. Comparing time course profiles of immediate acetylene reduction by grasses and legumes, *Applied and Environmental Microbiology*, *41*, 184–189 (1981).
199. van Egeraat, A.W.S.M. Root exudates of pea seedlings and their effect upon *Rhizobium leguminosarum*, in *Microbial Ecology* (Edited by M.W. Loutit and J.A.R. Miles), Springer-Verlag, Berlin, Heidelberg, New York (1978).
200. van Gundy, S.D. and Freckman, D.W. Phytoparasitic nematodes in below ground agroecosystems, in *Soil Organisms and Components of Ecosystems* (Edited by U. Lohm and T. Persson), *Ecological Bulletins* (Stockholm), *23*, 320–329 (1977).
201. Vasil, V., Vasil, I.K., Zuberer, D.A. and Hubbell, D.H. The biology of *Azospirillum*-sugarcane association I. Establishment of the association, *Zeitschrift für Pflanzenphysiologie*, *95*, 141–147 (1979).
202. Viands, D.R., Barnes, D.K. and Frosheiser, F.I. An association between resistance to bacterial wilt and nitrogen fixation in alfalfa, *Crop Science*, *20*, 699–705 (1980).
203. Vlassak, K., Reynders, L. Association of free-living nitrogen-fixing bacteria with plant roots in temperate regions, in *Microbial Ecology* (Edited by M.W. Loutit and J.A.R. Miles), Springer-Verlag, New York (1980).
204. Vose, P.B., Ruschel, A.P. and Salati, E. Determination of N_2-fixation, especially in relation to the employment of nitrogen-15 and of natural isotope variation, *Abstracts 89–90*, *Proceedings of the 2nd Latin American Botanical Congress* (1978).
205. Watanabe, I. and Barraquio, W.L. Low levels of nitrogen required for isolation of free-living N_2-fixing organisms from rice, *Nature*, *277*, 565–566 (1979).
206. Watanabe, I., Barraquio, W.L., de Guzman, R. and Cabrera, D. Nitrogen-fixing (acetylene reduction) activity and population of aerobic heterotrophic nitrogen-fixing bacteria associated with wet land rice, *Applied and Environmental Microbiology*, *37*, 813–819 (1979).
207. Watanabe, I. and Cabrera, D. Nitrogen fixation associated with the rice plant grown in water culture, *Applied and Environmental Microbiology*, *37*, 373–378 (1979).
208. Weaver, R.V., Wright, S.F., Varanka, M.W., Smith, O.E. and Holt, E.C. Dinitrogen fixation (C_2H_2) by established forage grasses in Texas, *Agronomy Journal*, *72*, 965–968 (1980).
209. Weir, K.L. Nitrogenase activity associated with three tropical grasses growing in undisturbed soil coves, *Soil Biology and Biochemistry*, *12*, 131–136 (1980).

210. Witty, J.F. Acetylene reduction assay can overestimate nitrogen-fixation in soil, *Soil Biology and Biochemistry*, *11*, 209–210 (1979).
211. Wolpert, J.S. and Albersheim, P. Host-symbiont interactions. I. The lectins of legumes interact with the O-antigen-containing lipopolysaccharides of their symbiont rhizobia, *Biochemical and Biophysical Research Communications*, *70*, 729–737 (1976).
212. Wong, P.P. and Stenberg, N.E. Characterization of *Azospirillum* isolated from nitrogen-fixing roots of harvested sorghum plants, *Applied and Environmental Microbiology*, *38*, 1189–1191 (1979).
213. Wood, L.V., Klucas, R.V. and Shearman, R.C. Nitrogen fixation (acetylene reduction) by *Klebsiella preumonial* in association with 'Park' Kentucky bluegrass (*Poa pratensis* L.), *Canadian Journal of Microbiology*, *27*, 52–56 (1980).
214. Wullstein, L.G., Bruening, M.L. and Bollen, W.B. Nitrogen fixation associated with sand grain root sheaths (rhizosheaths) of certain xeric grasses, *Physiologia Plantarum*, *46*, 1–4 (1979).
215. Yoshida, T. and Yoneyama, T. Atmospheric dinitrogen fixation in the flooded rice rhizosphere as determined by the N-15 isotope technique, *Soil Science and Plant Nutrition*, *26*, 551–559 (1980).
216. Yoshinari, T., Hynes, R. and Knowles, R. Acetylene inhibition of nitrous oxide production and measurement of denitrification and nitrogen fixation in soil, *Soil Biology and Biochemistry*, *9*, 177–183 (1977).
217. Zuberer, D.A. and Silver, W.S. N_2-fixation (acetylene reduction) and the microbial colonization of mangrove roots, *New Phytologist*, *82*, 467–471 (1979).
218. Zablotowicz, R.M., Eskew, D.L. and Focht, D.D. Denitrification in *Rhizobium*, *Canadian Journal of Microbiology*, *24*, 757–760 (1978).
219. Zurkowski, W. Specific adsorption of bacteria to clover root hairs, related to the presence of the plasmid pWZ2 in cells of *Rhizobium trifolii*, *Microbios*, *27*, 27–32 (1980).

G.A. PETERS and H.E. CALVERT

7. The Azolla-Anabaena Symbioses*

Azolla, a genus of heterosporous floating aquatic ferns established by Lamarck in 1873 (66), is usually included with *Salvinia* in the Salviniaceae. However, the two genera are markedly different and some authors have placed *Azolla* in a separate family, the *Azollaceae* (see 25, 27, 29, 58). Species delineation is based primarily upon reproductive structures. The genus comprises four new world species in the subgenus Euazolla, and two old world species the subgenus Rhizosperma (29). The Euazolla species include *A. caroliniana* Willdenow, *A. filiculoides* Lamarck (type species), *A. mexicana* Presl and *A. microphylla* Kaulfuss. *A. pinnata* R. Brown and *A. nilolotica* DeCaisne comprise the subgenus Rhizosperma. The various species are widely distributed in tropical and temperate fresh water ecosystems throughout the world. A world distribution map of the individual species based on literature citations has recently been presented (27).

All species normally contain a heterocystous cyanobacterium, designated *Anabaena azollae* Strasburger, in specialized leaf chambers. Although there is as yet no information for *A. nilotica* or *A. microphylla*, N_2 fixed by the endophyte in the other species can provide the associations with their total N requirements (47). Thus at least these members of the genus, and presumably the other two as well, are able to grow in environments deficient in combined N. Since wind and wave action as well as other forms of turbulence cause fragmentation and drastically diminished growth, *Azolla* species are not found on large lakes or swiftly moving waters. Rather, their growth is confined to the more placid surfaces found in ponds, marshes, canals, drainage ditches and significantly rice paddies.

Azolla has a long history of use as an alternative N source for rice in Vietnam and the Peoples' Republic of China (11, 26, 29). During 1975–1976 detailed field studies of *Azolla* with rice were initiated at the International Rice Research Institute in the Philippines, the Central Rice Research Institute at Cuttack, India and at the University of California, Davis. Subsequently studies

*Contribution No. 739 from the C.F. Kettering Research Laboratory.

have been initiated at ORSTOM, Dakar, Senegal and, in conjunction with projects such as the International Network on Soil Fertility and Fertilizer Efficiency in Rice (INSFFER), in numerous other countries including Thailand, Bangladesh, Nepal and other locations in India. While *Azolla* is by no means a panacea, the outcome of the studies to date is generally promising and consistent in demonstrating the potential of these associations as an alternative N source for rice.

This chapter will first address the *Azolla* growth mode and life cycle along with aspects of laboratory culture and nutritional requirements. Next the developmental morphology, whole plant and developmental physiology, and biochemistry of *A. caroliniana* are described. This is followed by a section on the history, current use, assets and shortfalls of *Azolla* as an alternative N source for rice. Emphasis is placed upon biomass, N production, and yield response of rice to *Azolla* N versus commercial fertilizer N. Aspects of management practices are considered and some future research goals addressed.

Growth mode and life cycle

Sporophytes *A. caroliniana* Willd., *A. filiculoides* Lam., *A. mexicana* Presl (two isolates) and *A. pinnata* R. Br. are shown in Plate 1, A. The sporophytes exhibit a multibranched floating stem which bears deeply bilobed leaves and adventitious roots. The roots occur at branch points (nodes) on the ventral side of the rhizome. Mature portions of the root are covered with root hairs and a sheath and root cap cover the young portion. The anatomy of the root of *A. pinnata* has been studied in detail (16). The alternately arranged leaves occur in two lateral rows and may overlap to such an extent that they hide the rhizome. Each leaf comprises a dorsal and ventral lobe. The ventral lobes are thinner than the dorsal lobes and float on the water surface. The dorsal lobes are aerial and contain the endophytic *Anabaena* in ellipsoidal cavities.

Vegetative propagation is rapid and constitutes the more common means of multiplication. It is facilitated by an abscission layer occurring at the point of root and lateral branch attachment (53). *Azolla* also reproduces via a sexual cycle, producing sporocarps which enables the association to overwinter and, to at least some extent, withstand desiccation. The life cycle and/or sporocarp development have been described in maximum detail for *A. filiculoides* (5, 8) and *A. pinnata* (25, 53). The literature on this subject has recently been summarized (27). With the exception of *A. nilotica* (12), sporocarps are borne in pairs on short stalks which originate from the first ventral leaf lobe initial of a lateral branch. The pair may be two megasporocarps, two microsporocarps or one of each. The microsporocarps are appreciably larger than the megasporocarps (Plate 1, B). At maturity the sporocarps dehise from the sporophyte. The microsporocarps soon disintegrate releasing massulae with glochidia. The glochidia anchor the massulae to the megaspores. The megaspores germinate producing gametophytes which bear one or more archegonia and the microspores germinate producing prothalli which differentiate antheridia. Antherozoids released from antheridia fertilize the egg cells in the archegonia creating the zygote which leads back to the sporophyte.

The symbiosis is retained throughout the sexual cycle. Spores or spore-like filaments of the endophyte are invariably found within the megasporocarp. The *Anabaena* spores apparently germinate during or shortly after formation of the female gametophytes within the megasporocarp, producing undifferentiated filaments. These filaments become associated with the shoot apex of the developing sporophyte, perpetuating the symbiosis.

Laboratory culture and nutrients

Azolla species collected from natural environments can usually be easily cultured in the laboratory. These photosynthetic, N_2-fixing floating aquatic plants require light energy, air, water, macronutrients other than combined N, and micronutrients. The use of a medium lacking an N source will necessarily hold the growth of most other organisms in check. It does not, however, prevent growth of free-living N_2-fixing cyanobacteria. Furthermore, unless epiphytic contaminants such as eukaryotic algae and cyanobacteria are removed, it is impossible to conduct meaningful physiological studies on medium containing a combined N source. Thus, we routinely treat field-collected *Azolla* as described previously (43). In brief, the plants are washed 20 times, with vigorous agitation in large volumes of distilled water, surface sterilized with a solution containing 0.12 percent sodium hypochorite and 0.01 percent Triton X-100 as a wetting agent for 20 to 30 minutes, washed a minimum of six times with large volumes of sterilized distilled water, and transferred to sterilized nutrient media with and without a combined N source. This treatment causes death of some of the fronds but subcultures of small, healthy plants can usually be obtained within a week. This treatment does not free the *Azolla* of the endophyte. The absence of any growth of other photosynthetic organisms during repeated subcultures onto medium containing combined N is indicative of the plants being free of epiphytic contaminants.

A variety of inorganic nutrient solutions have been used for *Azolla* culture, including N-free modifications and dilutions of Crone's, Hoagland's, or Knop's solution. Becking (4) summarizes the composition of numerous media, comparing the cation and anion concentrations as well as the initial pH. Peters *et al.* (47) determined biomass increase, C and N content, nitrogenase activity (C_2H_2 reduction), percentage dry weight and Chlorophyll a/b ratios for clones of *A. caroliniana*, *A. filiculoides*, *A. mexicana* and *A. pinnata* as a function of nutrient solution, pH, light intensity, temperature and photoperiod in a controlled environment. Using the optimum growth temperature (25°C for *A. caroliniana* and *A. filiculoides*, 30°C for *A. mexicana* and *A. pinnata*), a

16 hours light period, a photon flux density of at least 400 $\mu E/m^2/sec$ and the nutrient solution of Yoshida *et al.* (78) as modified by Watanabe *et al.* (75), buffered at either pH 6 with 10 mM MES (2 (N-Morpholino) ethane sulfonic acid) or pH 8 with TRICINE (N-Tris (hydroxymethyl) methyl glycine), all species were shown to double their biomass in two days or less and contain 5–6 percent N on a dry weight basis. No significant difference was observed in growth on media buffered at pH 5, 6, 7 or 8 but growth decreased at pH 9. The inclusion of buffer is not essential. The decrease in pH of the unbuffered nutrient solutions during growth on N-free medium can be corrected by frequent transfer and/or daily pH adjustment. Cultures are maintained in gauze-cotton capped Fernbach or Erlenmeyer flasks and while not axenic, handled with sterile technique. For maximal growth, filtered humidified air is introduced to the culture to maintain the pCO_2. Stock cultures can be conveniently maintained for several months between transfers on the nutrient solution solidified with 2 percent agar.

While there are reports that *Azolla* growth and nitrogenase activity decline at light intensities greater than one-half of full sunlight, about 1000 $\mu E \cdot m^{-2} \cdot sec^{-1}$ photosynthetically active radiation (2–4), others have not observed such effects (47, 67, 68, 74). It has recently been reported (71) that in Malaysia floodwaters, inhibition of growth by full sunlight was eliminated when the water was amended with 20 ppm phosphorus as $Ca(H_2PO_4)_2$. Except for nitrogen, which can be supplied in total by N_2 fixation, the macronutrients essential to the *Azolla-Anabaena* association are basically the same as those of other photoautotrophs.

In a study of the symptoms and effects of P, K, Ca, Mg and Fe deficiencies on strains of *A. filiculoides*, *A. mexicana* and *A. pinnata*, the appearance of deficiency symptoms were Ca > Fe > P > Mg > K, but magnesium and calcium deficiencies produced the most adverse effects on growth (13). In a recent study with *A. imbricata* (*pinnata*), the threshold levels of P, K, Mg and Ca for growth were about 0.03, 0.4, 0.4 and 0.5 $mmol \cdot liter^{-1}$, respectively, and for nitrogenase activity, 0.03, 0.6, 0.5, and 0.5 $mmol \cdot liter^{-1}$, respectively (77). Using a continuous flow system the threshold level P content in *Azolla* appeared to be 0.25 percent, since below this plant growth was proportional to P content (65). Plants grew normally with 0.06 ppm P in the medium but biomass and N contents were diminished at 0.03 ppm P in the medium.

In regard to micronutrients, the deficiency of Fe, Mn, Co, Zn, Cu, Mo and B had unfavorable effects on growth and N_2 fixation in *A. imbricata* (*pinnata*) (77). Threshold levels of Fe, Mn, Mo and B were 50, 20, 0.3 and 30 $\mu g \cdot liter^{-1}$, respectively, for growth and 20, 10, 1 and 20 $\mu g \cdot liter^{-1}$, respectively for nitrogenase activity. Earlier studies had shown that cobalt (23) and molybdenum (6) were essential micronutrients for the symbiotic association grown in the absence of combined nitrogen sources.

In a given situation, any of the macro- or micronutrients might be limiting to growth and N_2 fixation. However, Fe and P are most often limiting to *Azolla* growth in natural waters. In Danish lakes *Azolla* growth was limited when manganese and ferrous ions or iron-humus complexes were not available in suitable concentrations (33). The availability was determined by pH and calcium concentration since high Ca concentration reduces absorption of ferrous ions. There is a diminished availability of iron as iron citrate to *Azolla* at pH 7.5 (75) and the Fe/mg dry weight of *Azolla* plants grown in medium buffered at pH 6 was found to be about three times greater than that in plants grown in the same medium buffered at pH 9 (47). In California rice paddies with water at pH 8–8.5, supplemental iron is required to support *Azolla* growth (69). Under field conditions it is phosphorus which is most often limiting to *Azolla* yields (60, 61, 67–69, 71, 74, 75). While *Azolla* can store phosphate and potassium (76, 77), water soluble phosphate, as superphosphate, is rapidly absorbed by paddy soils and becomes unavailable to *Azolla* plants unless their roots can touch the soil. In fact, early Vietnamese observations that *Azolla* would not grow unless their roots came into contact with mud (57) may well have been the result of inadequate soluble phosphate in the waters. While normally free-floating, the roots of *Azolla* are capable of penetrating into water-laden soils and plants may become "rooted" in shallow water. Whether uptake of nutrients occurs only via the roots or through the floating stem and ventral leaf lobes as well is unresolved.

Morphology and development

The growth and development of the *Azolla* and the *Anabaena* are synchronous (19, 20). *Azolla* shoot tips are supported above the water surface by their stem axes and a small colony of *Anabaena* filaments is associated with every apical meristem. These undifferentiated *Anabaena* filaments occupy an opening between the upcurved meristem and the overarching young leaves (Plate 2, A). Partitioning of *Anabaena* filaments begins in the young leaves contiguous with the apical *Anabaena* colony and is completed by the time a leaf emerges as a distinct organ visible to the naked eye (Plate 2, B).

The cavity begins as a depression in the adaxial epidermis of the young leaves associated with the *Anabaena* filaments in the apical region. As each forming leaf is displaced from the apical meristem, some of the *Anabaena* filaments associated with the apex are separated from it and become associated with the forming leaf cavities (Plate 2, A). The partitioning of the *Anabaena* filaments into the forming cavity involves a specialized epidermal hair called the primary branched hair (7) (Plates 2, A, 3, B). There is a single primary branched hair (PBH) in each leaf cavity. It originates from the axil of the forming dorsal lobe and its growth is directed into the apical *Anabaena* colony. The terminal cells of this hair exhibit transfer cell ultrastructure (34), and may thus be metabolically interactive with the *Anabaena* filaments. The *Anabaena* filaments in the apical region are clearly attracted to the PBHs.

As a developing leaf is displaced from the shoot tip by continued growth, *Anabaena* filaments entangled with its PBH are separated from the apical *Anabaena* mass. Subsequently, cells around the opening of the forming cavity become meristematic. They produce epidermal cells which form the cover of the cavity, engulfing the PBH with its associated *Anabaena* filaments. As this closure of the leaf cavity commences, the *Anabaena* filaments begin to differentiate heterocysts and many additional epidermal hairs begin to form on the cavity walls. Another branched hair, termed the secondary branched hair (7), develops on the wall of each forming cavity and numerous simple hairs also begin to emerge from the cavity walls (Plate 3, B). The simple hairs constitute a population distinct from the two branched hairs with twenty or more occurring in a mature cavity (7). They are also morphologically distinct from the branched hairs, being elongate and consisting of

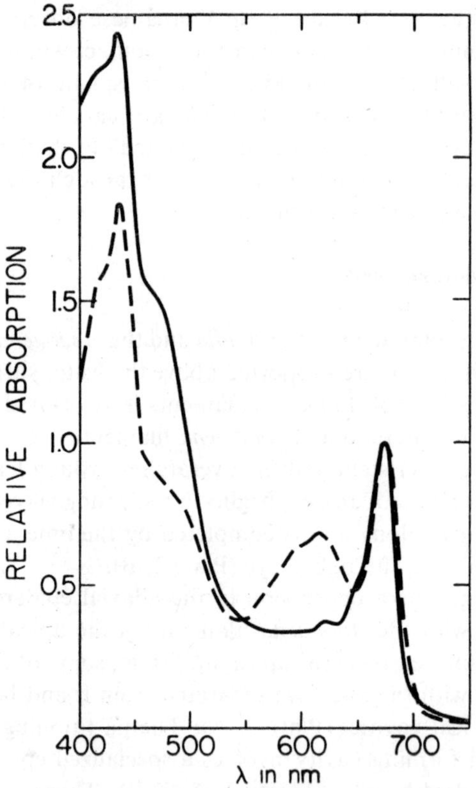

Fig. 1. Absorption spectra of cell-free extracts from the endophyte-free *Azolla* (———) and the endophytic *Anabaena* (– – –). Endophyte-free *Azolla* was obtained using antibiotics [from Peters and Mayne (43)]. The spectra have been normalized at 673 nm [after Rap *et al.* (54)].

a stalk cell and a terminal cell. The two branched hairs are located in similar positions in every leaf cavity, along the path of the dorsal lobe foliar trace, while the location of the simple hairs appears to be random.

During the maturation of the leaf cavity, the *Anabaena* filaments multiply and increase in size. In mature cavities heterocyst frequencies reach 25–30 percent and the *Anabaena* filaments are localized along the periphery of the cavity walls (Plate 3, A). The vast majority of the 20–25 simple hairs (Plate 3, B) exhibit transfer cell ultrastructure in their terminal cells. The terminal cells of the branched hairs have senesced and newly differentiated cells between the stalk cell and terminal cells exhibit transfer cell ultrastructure.

Whole plant physiology and biochemistry

Photosynthesis

The complementation of light harvesting pigments of the partners is shown in Fig. 1. *Azolla* chloroplasts contain chlorophylls *a* and *b*, the *Anabaena* filaments contain chlorophyll *a* and phycobiliproteins, and both contain carotenoids. The absorption spectra of the purified phycobiliproteins

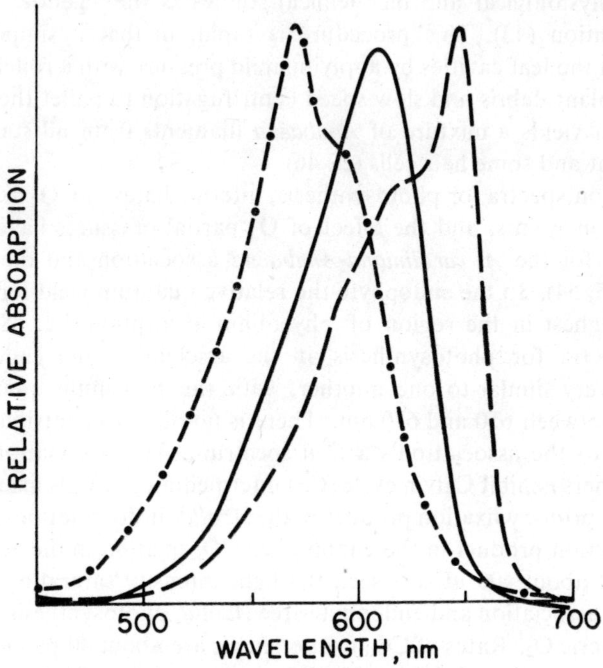

Fig. 2. Absorption spectra of the purified phycobiliproteins from the endophytic *Anabaena*. Phycoerythrocyanin (–·–), Phycocyanin (—), and Allophycocyanin (– –). Phycocyanin accounts for about 70% of the endophytes phycobiliprotein [from Tyagi *et al.* (72)].

are shown in Fig. 2. *In vivo*, phycocyanin accounts for about 70 per cent of the total phycobiliprotein while phycoerythrocyanin and allophycocyanin account for about 17 percent and 13 percent respectively (72). In *A. caroliniana*, the endophyte accounts foı 10–20 percent of the association's total chlorophyll (38, 44, 55) and about 16 per cent of its total protein, with phycobiliproteins accounting for 4–10 percent of the *Anabaena azollae* protein (55). The *Azolla* pigments obscure the endophyte's absorption in the intact associations. *Azolla* may also contain anthocyanins (21, 51) and a variety of environmental factors can trigger their production. Their presence causes *Azolla* to change from green to a light orange, red, or even maroon.

Azolla chloroplasts and *Anabaena* filaments, separated on sucrose density gradients, have been characterized with respect to partial reactions of photosynthesis, delayed light emission and P700 content (43). Furthermore, packets of the endophyte, corresponding in size and contour to the individual leaf cavities (Plate 3, C) can be isolated using cellulolytic enzymes to digest the leaf tissue (36, 49) and filaments of the endophyte can also be removed from individual leaves under the dissection microscope (38, 45). However, the method of preference for obtaining sufficient quantities of the endophyte for most physiological and biochemical studies is the "gentle" *Anabaena azollae* isolation (43). This procedure is rapid, in that it simply involves rupturing of the leaf cavities by applying mild pressure with a roller, filtration to remove plant debris and slow speed centrifugation to pellet the *Anabaena* filaments. It yields a mixture of *Anabaena* filaments from all stages of leaf development and some hair cells (cf. 46).

The action spectra for photosynthesis, intermediates of CO_2 fixation, CO_2 compensation points, and the effect of O_2 partial pressures have also been determined for the *A. caroliniana-Anabaena* association and the individual partners (45, 54). In the endophyte the relative quantum yield for photosynthesis is highest in the region of phycobilin absorption, i.e. 580–640 nm. Action spectra for photosynthesis in the association and endophyte-free *Azolla* are very similar to one another, with the maximum quantum yield occurring between 650 and 670 nm. There is no obvious contribution by the endophyte to the association's action spectrum. The association and individual partners exhibit Calvin cycle (C3) intermediates of CO_2 fixation. While sucrose is a primary fixation product in the *Azolla*, it does not occur as a ^{14}C-labeled reaction product in the endophyte. CO_2 fixation in the association is saturated at about 400 $\mu E \cdot m^2 \cdot sec.$, the light intensity saturating for growth (47). In the association and endophyte-free *Azolla*, photosynthesis is inhibited by atmospheric O_2. Rates of CO_2 fixation in air are about 40 percent less than those at 2 percent O_2 and the aerobic CO_2 compensation point is about 40 ppm CO_2. Photosynthesis by the endophytic *Anabaena* is not inhibited by atmospheric O_2 and the CO_2 compensation point is about 4 ppm CO_2 at 20 percent and 2 percent O_2. Results obtained with the isolated *Anabaena* demonstrate its capabilities, but do not necessarily extrapolate to the symbiotic state. The actual contribution of the endophyte to the association's total photosynthetic capability and probable interaction in fern-endophyte carbon metabolism have not been resolved.

Nitrogen fixation

Acetylene reduction, $^{15}N_2$ fixation and ATP dependent H_2 production have been used to study nitrogenase activity in the *Azolla caroliniana-Anabaena azollae* association and in the isolated *Anabaena*. The anaerobic dark reduction of all substrates is negligible. Dark aerobic reductions are dependent upon the endogenous supplies of reductant accumulated during prior photosynthesis and rates are usually no more than 40% of those obtained aerobically in the light. The reduction of all substrates is maximal under anaerobic or microaerobic conditions in the light.

With a purified nitrogenase and saturating levels of acetylene, all electron flow through the enzyme is used for the reduction of C_2H_2 to C_2H_4. This is not the case with the natural substrate (N_2) since some reduction of protons always accompanies the reduction of N_2 resulting in the formation of two products, H_2 and ammonia. Thus with the purified enzyme the amount of C_2H_4 produced at saturating C_2H_2 will equal the amount of H_2 production in the absence of any reducible substrate other than protons, i.e., under an argon atmosphere. However, there will be more than three times as much C_2H_2 reduced as N_2 reduced when the enzyme is provided with saturating C_2H_2 versus atmospheric (or higher) levels of N_2. For example, if 25 percent of the electron flow under N_2 goes to H_2 production, C_2H_4, will equal NH_3+H_2 and a conversion factor of 4 C_2H_2 reduced/N_2 reduced would be attained.

In vivo determinations are subject to many additional factors, one of which is the occurrence of a unidirectional hydrogenase. This enzyme oxidizes the H_2 produced by the nitrogenase, recycling electrons and ATP into the system. Thus part of the energy and reducing power expended in H_2 production is recovered. There is, however, a cost for the recycling. If one *assumes* that 25 percent of the electron flow *in vivo* is lost to H_2 production, and the unidirectional hydrogenase recycles it with an efficiency of 75 percent, one would predict a C_2H_2/N_2 conversion factor of 3.25 moles C_2H_2 reduced per mole of N_2 fixed. While values of less than 3 are theoretically impossible, such values have been obtained in *in vivo* determinations.

The endophyte's nitrogenase activity is saturated at 10–15 percent C_2H_2 and there is little or no detectable nitrogenase catalyzed H_2 production (39, 40, 48). The amount of H_2 which can be measured under argon or various N_2 partial pressures is dependent upon the amount of unidirectional hydrogenase activity present in the endophyte (37, 40, 48). H_2 production under argon is always lower than acetylene reduction in the N_2 grown plants or the *Anabaena* isolated from them. Rates of N_2 production are variable, but an average value is about 60 percent of the rate of C_2H_2 reduction (39, 48). When the association is grown on a nutrient solution containing nitrate, a significant nitrogenase activity is retained but the rates of H_2 production under argon may equal the rates of acetylene reduction and H_2 production under an air atmosphere which is up to 40 percent of that measured under argon (30, 37, 39). The occurrence of unidirectional hydrogenase activity necessarily complicates the determinations of the inhibition of H_2 production by N_2. However, H_2 production is markedly inhibited by N_2 concentrations well below those of air (48).

The C_2H_2/N_2 conversion factor for the *Azolla-Anabaena* association and isolated endophyte has been determined as a function of N_2 partial pressures (48), using $^{15}N_2$ enriched air atmospheres (24, 46), and for the association, as a function of photoperiod, with determinations of this relationship at the midpoint of the light and dark cycles (40, 41). In the latter, a value of 4.08 was attained at the midpoint of the 16 hours light and 4.4 at the midpoint of the 8 hours dark cycle and, during a 24 hours period, 81 percent of the N input was estimated to occur during the light period. While the conversion factor has ranged from about 1.7 to 5.1 in individual studies, most values fall between 3 and 4. For further details and more technical information on the reduction of C_2H_2, N_2, and protons, the relationships among them, effects of combined N sources and electron balance studies (i.e., electron allocation coefficients), the reader is referred to specific publications (36, 37, 39, 40, 44–46, 48).

Relationship of photosynthesis and N_2 fixation

Photosynthesis is necessarily the ultimate source of all reductant and ATP for nitrogenase activity in these associations. If one assumes that reductant is generated in the same manner in the light and dark, the diminished rates of N_2 fixation under aerobic dark conditions versus those obtained under aerobic light conditions with DCMU (35, 36) imply that: dark respiratory-driven nitrogenase activity may be ATP limited; and, that increased activities in the light are the result of photosynthetically generated ATP. The cumulative results of a number of studies (35, 36, 44, 46) strongly imply that reductants provided by prior photosynthesis and cyclic photophosphorylation are the primary driving forces of nitrogenase activity in the light.

The action spectra for nitrogenase catalyzed acetylene reduction have been determined for the association and the endophyte (73), further showing the interaction of photosynthesis and N_2 fixation. Surprisingly, in both the association and the isolated *Anabaena* the relative rates of C_2H_2 reduction per incident quantum are nearly equivalent in the regions of phycobiliprotein absorption and chlorophyll absorption (73 and unpublished results). The most pronounced difference in the action spectra for photosynthesis

Plate 1. (A) Sporophytes of *A. caroliniana* (a), *A. filiculoides* (b), *A. mexicana* (c and d) and *A. pinnata* (e). Details of the sources, identification, and growth characteristics are provided in Peters *et al.* (47). (B) Ventral view of sporulating *A. mexicana* from a laboratory culture showing both microsporocarps (mi) and megasporocarps (me).

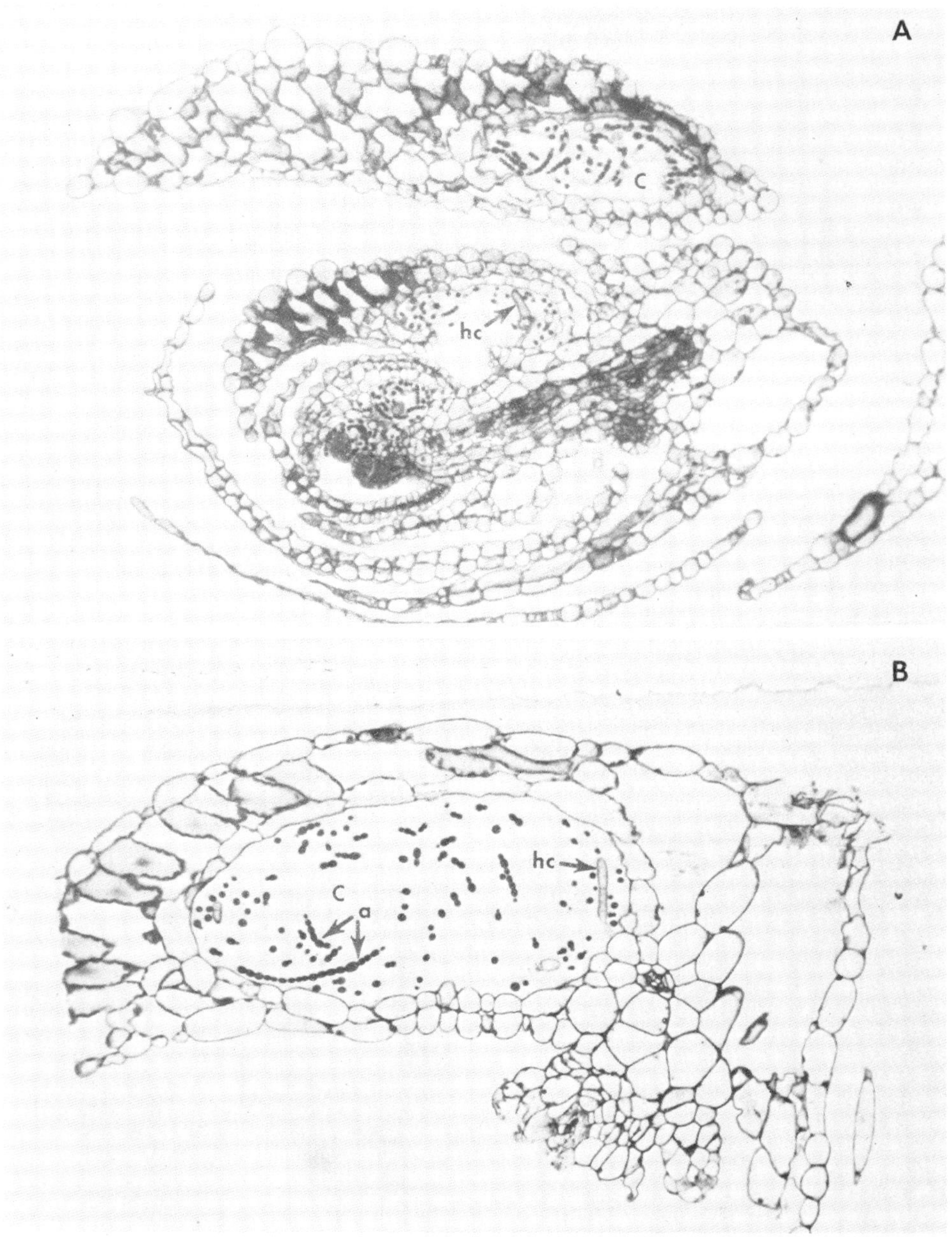

Plate 2. (A) Light micrograph of a section through the stem tip of *A. caroliniana* depicting the generative *Anabaena azollae* filaments in the apical region and several sequential stages in the development of the leaf cavities (c) in the dorsal lobe. Note the prominent primary branched hair (hc) to the right of the cavity opening in the next to the oldest leaf and the manner in which the apical region is protected by the overarching of both dorsal and ventral leaf lobes [figure reproduced from Peters *et al.* (46)]. (B) Light micrograph of a section through the second dorsal leaf lobe of *A. caroliniana.* (The oldest leaf shown in 2A being designated leaf number one in our dissections.) Filaments of *Anabaena* (a) bearing a few heterocysts are distributed throughout the cavity (c) and there is a prominent simple hair [figure reproduced from Peters *et al.* (46)].

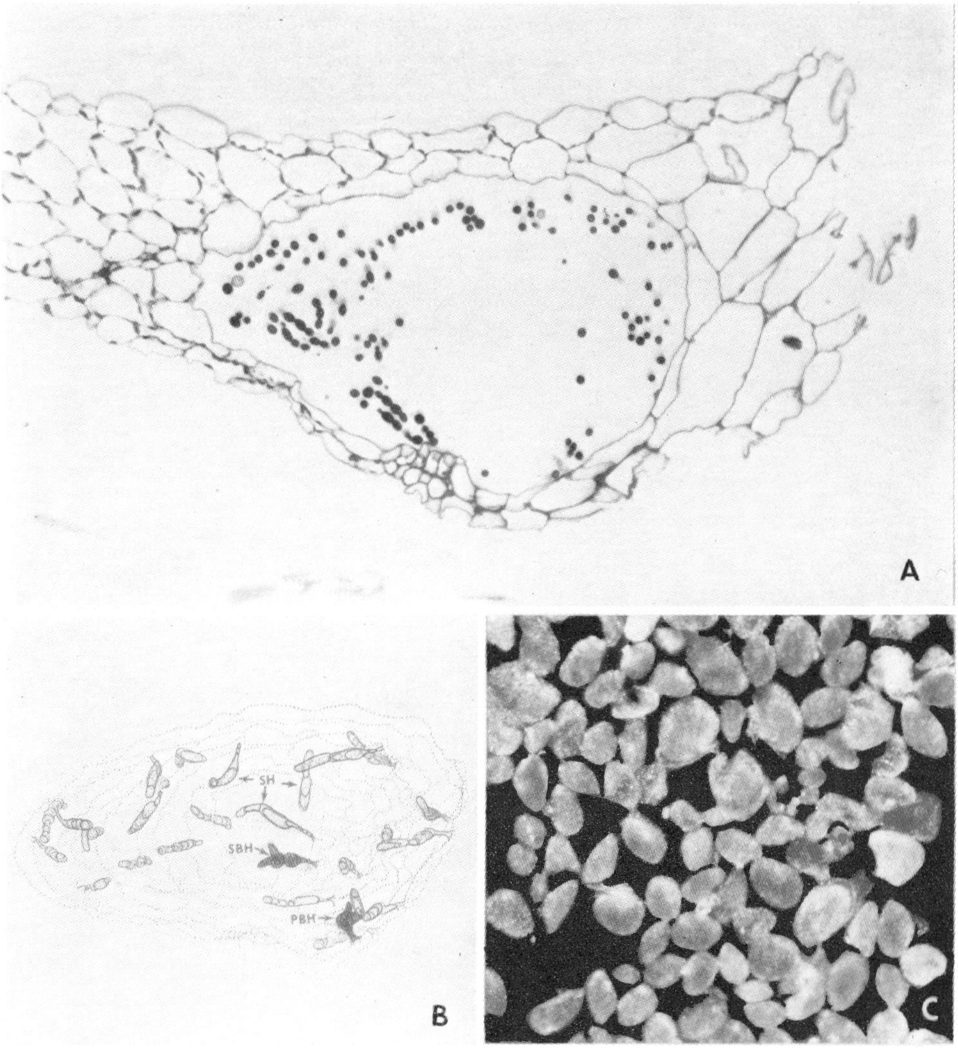

Plate 3. (A) Light micrograph of a section through an older dorsal leaf lobe (leaf 12 counting back from the oldest leaf in Plate 1A) in which the filaments of the endophyte are associated with the periphery of the leaf cavity. (B) A serial reconstruction of a mature leaf cavity showing the number and distribution of hair types. While present in the leaf sections, *Anabaena* filaments are not shown in the reconstruction since they obscure the hairs. Note the single primary branched hair (PBH), single secondary branched hair (SBH) and numerous simple hairs (SH). (C) A purified preparation of "algal packets" obtained by digesting away the fern leaves with cellulolytic enzymes. The individual packets, which correspond in size and shape to the leaf cavities, contain the *Anabaena* filaments and cavity hairs. The reconstruction of the cavity from serial sections through the leaf in 3B is indicative of the hairs in a packet from a mature cavity [figure reproduced from Peters (36)].

(PSII) and N_2 fixation (PSI) is that the action spectrum for photosynthesis exhibits the characteristic red drop in the region of chlorophyll *a* absorption while that for C_2H_2 reduction it does not (73). Although phycobiliproteins are generally considered to be accessory pigments for PSII, and to be depleted or absent in heterocysts, phycobiliprotein fluorescence in heterocysts of the endophyte is nearly equivalent to that of the vegetative cells. The retention of phycobiliproteins in heterocysts and their apparent association with PSI may be an adaptation by the endophyte enabling it to effectively harvest solar energy not absorbed by the fern pigments for N_2 fixation. However, similar results have been reported for the free-living cyanobacterium *Anabaena variabilis* (50).

Transfer of fixed nitrogen from the *Anabaena* to the *Azolla*

In the absence of combined nitrogen, N_2 fixed by the *Anabaena* provides the total N requirement of the association. When isolated from the *Azolla* and exposed to $^{15}N_2$, filaments of endophytic *Anabaena* release fixed N_2 into the incubation medium as ammonium (37, 46). These and related studies, which compared supernatants after incubating the endophyte under Ar-$^{14}CO_2$ and N_2-$^{14}CO_2$ (46, 55) indicated little or no release of organic N compounds such as amino acids. These results are in general accord with the report that the free ammonia accounted for up to 50 percent of the total pool N in *A. caroliniana*, the remainder being contributed by the amino acids, especially glutamine and that the intracellular (or perhaps intracavity) ammonia N pool was 5 to 10 times higher in N_2 grown plants than in those lacking the endophyte and/or provided with combined N (31). It should be noted, however, that the free ammonium accounts for a small fraction of the total plant N.

Release of fixed N as ammonia by the endophyte and the occurrence of free ammonia in the tissues and/or leaf cavity of the N_2 grown association led to a study of the enzymes involved in ammonia assimilation. Both partners exhibit glutamine synthetase (GS), glutamate synthase (GOGAT) and glutamate dehydrogenase (GDH) activities (55). However, it has been estimated that about 90 percent of the association's total GS activity and 80 percent of its total GDH activity are attributable to the *Azolla*. (These values are an average of the activities and distribution of these enzymes in all developmental stages of both partners. The ammonia assimilating enzymes and their portioning between the partners as a function of leaf development is considered later.) In accordance with the preceding incubation of the association under $^{15}N_2$ enriched air followed by chase periods with air (45) showed a low percentage of the total ^{15}N in the ammonia fraction with a rapid incorporation into ethanol soluble compounds (amino acids, etc.) followed by incorporation into ethanol insoluble compounds (proteins, etc.) (Fig. 3).

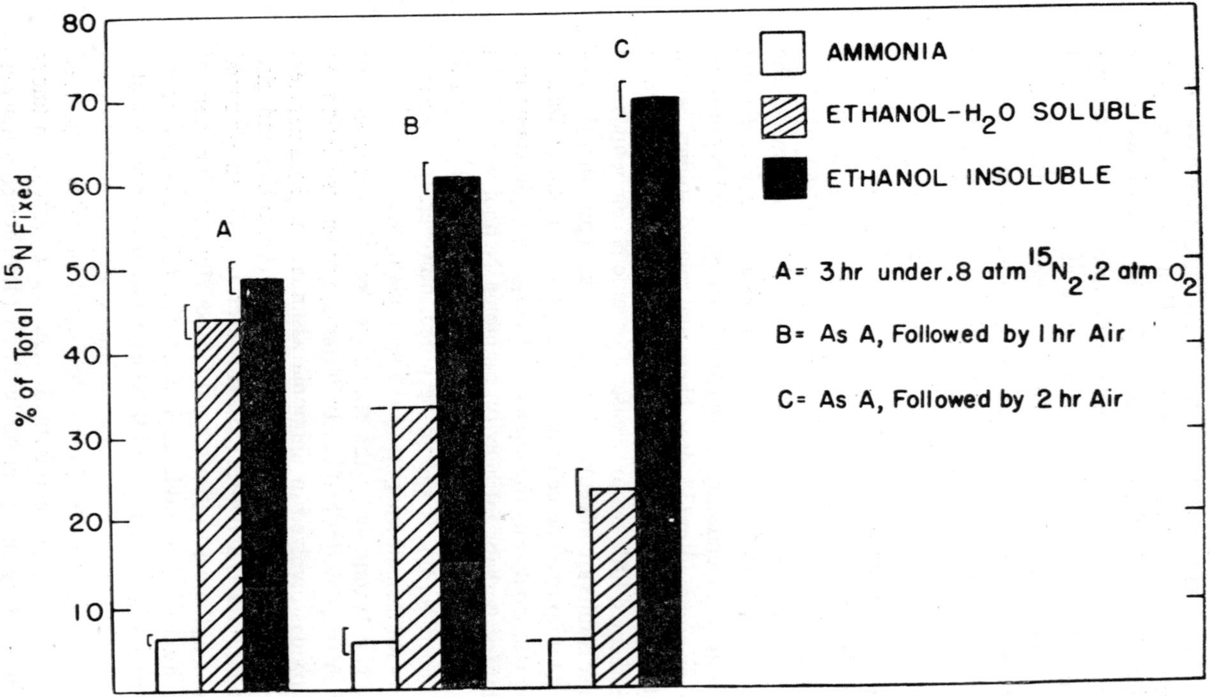

Fig. 3. The distribution of ^{15}N among three fractions obtained from the *Azolla caroliniana-Anabaena* arsociation after assimilation under steady state conditions and as a function of chasing the ^{15}N with $^{14}N_2$ in air. Three sets of flasks (A, B, and C) containing *Azolla* plants grown in the absence of combined N were filled with 0.8 atm N_2 (containing approximately 33 atom % ^{15}N) and 0.2 atm O_2 and incubated in the light 700 ft-c, 26°. At the end of 3 hours the plants in set A were extracted with hot 80% ethanol. Sets B and C were opened to room air and extracted after one (B) and two (C) additional hours of incubation. Samples were fractionated into ethanol insoluble, ethanol-H_2O soluble, and microdiffusible NH_3 portions. The ethanol insoluble (protein) and residual ethanol H_2O soluble (amino acid) fractions, after removing the ethanol and microdiffusion of ammonia, were subjected to Kjeldahl digestion and distillation. Total N of the individual samples and atom % excess ^{15}N were determined as described previously (37, 48) [after Peters *et al.* (45)].

Developmental physiology

Anabaena filaments associated with the plant apex (Plate 2, A) lack heterocysts and nitrogenase activity. As the endophyte occupies the leaf cavities there is a rapid differentiation of heterocysts and a rapid increase in nitrogenase activity. This developmental profile, originally described by Hill (19, 20) using *A. filiculoides* has been studied in considerable detail in *A. caroliniana* (7, 24, 46). This approach uses dissected main stem axes, stem segments and/or individual leaves.

The absence of nitrogenase activity in the apical *Anabaena* filaments requires a transport of the N_2 fixed by the *Anabaena* in mature leaf cavities to the apical region to meet the nitrogen requirements of both the plant tissues and the generative *Anabaena* filaments. Transport of fixed N has recently been demonstrated in main stem axes using a pulse-chase approach with $^{15}N_2$ (24). The transported compound(s) have not been resolved and there is as yet no direct evidence that N is transferred from the *Azolla* to the generative *Anabaena* filaments. However, the N content and dry matter were found to decrease with increasing leaf age while the C/N ratio increased (24). These findings are consistent with the suggestion that filaments of the endophyte in mature cavities might have diminished capability to metabolize ammonia (55) with the host exerting a control or regulation of metabolic processes in the endophyte as a function of the developmental gradient. While the host causes the endophyte to rapidly differentiate a disproportionate number of heterocysts and to serve as an ammonium production facility, the factors responsible for diminished cell division greatly increased heterocyst differentiation and diminished ability to assimilate the ammonia from N_2 fixation during the developmental profile in the *Azolla* endophyte are not yet resolved. It has been suggested that the GS activity found in isolates of the endophyte might be associated primarily with the undifferentiated apical filaments (55). While GS levels in the endophytic *Anabaena* were half of those found in Newton's free-living isolate and *A. cylindrica*, the endophyte preparations contained epidermal hairs of the *Azolla* (55). Thus it was stated that due to a possible contribution from the hairs, activities attributed to the endophyte might be slightly high. Haselkorn *et al.* (18) employed an antibody against the purified GS from *Anabaena* 7120 and found that the antigen levels of the endophytic *Anabaena* were only 5–10 percent of those observed in Newton's free-living isolate (32). Further, in the association the antigen concentration was greatest in the endophyte associated with younger leaves. Thus, there is reason to suspect a gradient in the endophyte's GS with a decrease in the endophyte's GS, paralleling the differentiation of heterocysts and epidermal hairs in the leaf cavities. This is in contrast to free-living cyanobacteria where the GS is preferentially associated with heterocysts and it has been suggested that in *Azolla* and other plant-cyanobacterial symbioses, the host might produce an effector substance which inhibits the endophyte's GS activity or synthesis (17, 64).

While GDH is not found in appreciable quantities in free-living N_2 fixing cyanobacteria or other symbiotic forms this enzyme has an appreciably lower affinity for ammonia than does glutamine synthetase. It has been postulated (55) that in the *Azolla-Anabaena* association the endophyte's GDH might be associated with those filaments occupying mature cavities and actively fixing N_2. In this case the ammonia released by the endophyte would normally be assimilated by the *Azolla* GS, with the endophyte's GDH providing a regulatory role, enabling the endophyte to effectively reassimilate released ammonia at high intra-cavity ammonia concentrations. The amount of the fixed N_2 which is utilized by the individual partners and N metabolism in the association as a function of the developmental profile are subjects of current research.

The relative contribution of the individual partners to the association's total photosynthetic capability and the extent of interaction in fern-endophyte carbon metabolism are also largely unknown. However, it is highly probable that they also vary as a function of the developmental profile. It is our opinion that to sustain functional heterocysts at frequencies of 15–20 percent or more there is probably an absolute requirement for an exogenous carbon source to maintain levels of reducing power. Thus in the *Azolla-Anabaena* association we would postulate a transition from photoautotrophic metabolism in generative filaments to a photoheterotrophic or mixotrophic mode of metabolism with increasing differentiation of heterocyst. Ray *et al.* (54) suggested that in mature leaf cavities oxidative metabolism of sucrose from the *Azolla* could conceivably provide reductant for N_2 fixation. Although $^{14}CO_2$ time-course studies showed the endophytic *Anabaena* did not synthesize sucrose (54), this sugar, along with glucose and fructose, have recently been identified as the major soluble di- and monosaccharides in the endophyte (42). This necessarily implies crossfeeding with that sucrose found in the endophyte having been synthesized by the *Azolla*.

Azolla as a weed

Azolla species are capable of very rapid vegetative growth, with *A. caroliniana*, *A. filiculoides*, *A. mexicana* and *A. pinnata* capable of doubling their biomass in two days or less under optimal laboratory conditions (47). The rapid vegetative propagation of *Azolla* enables it to completely cover moderately protected water surfaces. Thus, depending upon where it occurs, *Azolla* may be considered a noxious weed or an important green manure crop. For example, in New Jersey, *A. caroliniana* reportedly covered a canal for five miles with a mat of vegetation so dense that it was impossible to row a boat through it (10) and nutrient rich water in an agricultural drainage ditch in the Sacramento Valley, California, rice growing region supported an *A. filiculoides* cover containing 2.2 Tdr wt/hectare (approximately 38T fr wt/hectare) of biomass with an N content of 150 kg N/hectare and P content of 12 kg/hectare (69). A similar situation was described in South Africa (2) but with more serious implications. *A. filiculoides* was established in several slow-flowing feeder streams and ponds behind farm dams, and upon completion of the Headrik Verwood Dam Catchment it was feared that like *Salvinia*, *Azolla* might colonize the large expanse of open waters created by the new dam. Although large mats of *Azolla* were indeed washed from the feeder streams into the main body of water by annual summer floods, it was found that *Azolla* was not a threat to the open waters of the lake since wind and wave action resulted in fragmentation and the plants were unable to survive. Interestingly, only slightly more than a decade ago, *Azolla* was considered as a weed in most rice paddies and, excepting a report that one farmer used *Azolla* to suppress other weed growth (14), in taro patches as well (29). We shall now consider the traditional use of *Azolla* and its recent and continuing metamorphosis from water weed status to a potentially important alternative N source for rice culture in many other countries.

Traditional and current use of *Azolla* with rice

Those traits which caused *Azolla* to be considered a weed are essentially the same as those which have been exploited in its traditional use as a fodder and/or green manure. In Vietnam, there are several legends about its domestication. According to Dao and Tran (11) the use of *Azolla* can be traced back many centuries and the popular legend is that the cultivation of *Azolla* was promoted by a Buddhist monk in the eleventh century. In another legend (15, 27, 29) *Azolla* was domesticated by a peasant woman, Ba Heng, in the village of Le Van, Thai Binh Province. In 1955 the use of *Azolla* in Vietnam was restricted to about 40,000 ha in the Red River delta. A decade later, during which techniques for both its multiplication and its dual cropping with winter rice were popularized, its use had been extended to 320,000 hectares, (11). After the introduction of higher yielding rice varieties in 1965 *Azolla* was grown primarily as a separate crop in rotation with rice (11) and in 1976 about 40 percent of the rice field area in northern Vietnam was used for *Azolla* culture after the summer rice harvest (74). The use of *Azolla* in Vietnam has leveled off and/or declined slightly at present (Dao The Tuan, personal communication IRRI workshop, 1980).

A book on agricultural techniques written in 540 A.D. entitled, "The Art of Feeding the People" (Chih Min Tao Shu) reportedly describes the cultivation and use of *Azolla* in rice fields and, at the end of the Ming Dynasty (beginning of 17th century) there were many "local" records of *Azolla's* use as a green manure (26). However, the use of *Azolla* in the Peoples' Republic of

China did not expand markedly until 1962. An example of the interest and type of studies conducted with *Azolla* from 1962 to 1974 is contained in a handbook on the "Cultivation, Propagation and Utilization of *Azolla*" (1). Having established various methods to maintain vegetative *Azolla* during unfavorable climatic conditions, the use of *Azolla* in the Peoples Republic of China has now been extended to approximately 1.34 million hectares in total (26).

Rice is a major source of food calories and can be considered a pivotal crop in mankind's efforts to stave off famine. Nitrogen is the nutrient most often limiting crop yield, especially with new high yield varieties. Since the production of commercial fertilizer N is intimately linked to non-renewable energy sources, the use of *Azolla* as an alternative N source for rice is now receiving considerable interest in many countries.

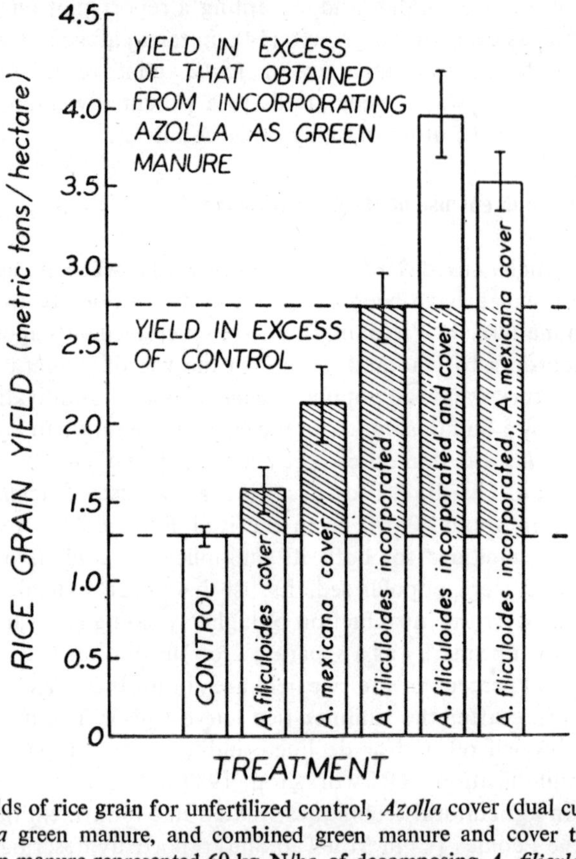

Fig. 4. Yields of rice grain for unfertilized control, *Azolla* cover (dual culture with rice), *Azolla* green manure, and combined green manure and cover treatments. *Azolla* green manure represented 60 kg N/ha of decomposing *A. filiculoides* cover grown in dual culture with rice representing 31 kg N/ha. Cover for *A. mexicana* represented 38 kg N/ha. Experiments conducted between May 3 and Oct. 30, 1976 [after Talley *et al.* (69)].

Field studies were initiated five to six years ago at the International Rice Research Institute (IRRI) in the Philippines (75), at the Central Rice Research Institute (CRRI), Cuttack, India (59) and at the University of California, Davis (69) to further determine the problems and the applicability of *Azolla* in rice production. These initial studies clearly demonstrated the potential of *Azolla* as an alternative or supplemental nitrogen source for rice in tropical and temperate regions. *Azolla* is most effective when grown as a green manure during the fallow season of rice and incorporated into the soil but nitrogen is also provided when *Azolla* is grown in dual culture, as a cover crop with rice (60, 61, 63, 69). *A. mexicana* was more effective in increasing rice yields when grown in dual culture than *A. filiculoides* (52, 67, 69), and a combination of *A. mexicana* or *A. filiculoides* in dual culture following the

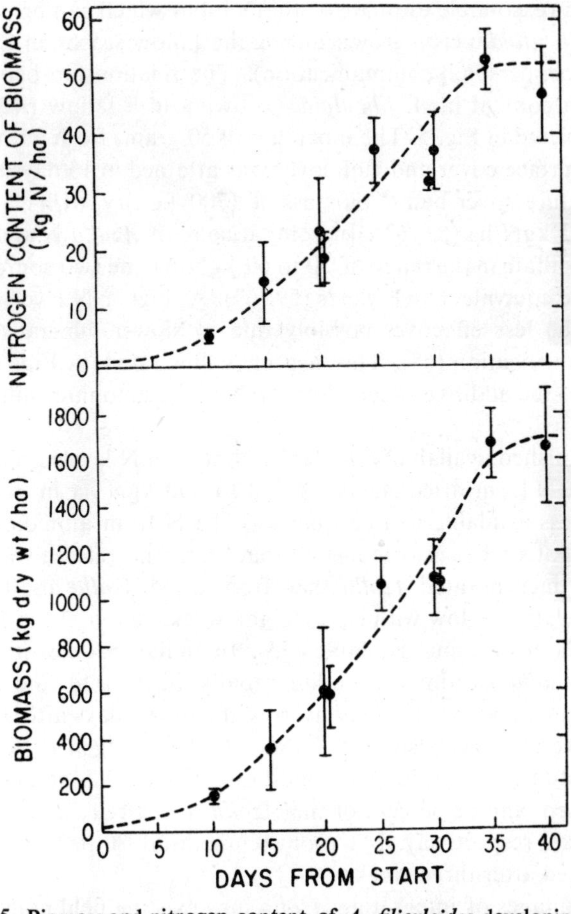

Fig. 5. Biomass and nitrogen content of *A. filiculoides* developing in a fallow rice field (Spring 1977) [after Talley and Rains (67)].

incorporation of *A. filiculoides* grown during the fallow season increased yields relative to that obtained with the incorporated *Azolla* alone (Fig. 4). In addition, as reported for taro (14), under appropriate management practices the establishment of an *Azolla* cover in dual culture has the added benefit of suppressing weed growth (52, 69). As much as 100 kgN/ha has been provided to a single rice crop at IRRI in the Philippines using the Chinese dual row method and a more labor-intensive approach in which the *Azolla* is incorporated into the soil at intervals through the growing season after briefly draining the paddy. Other studies at IRRI have shown an accumulation of 330 kgN/ha over 220 days (75) and 460 kgN/ha from 22 consecutive crops of *A. pinnata* over 330 days with an average daily accumulation of 1.4 kgN/hr (74). Inoculation with *A.* filiculoides equivalent to 1.2 kgN/ha, onto fallow flooded paddies has given yields between 33 and 93 kgN/ha (68) and 60 kgN/ha appears to be a reasonable estimate of the N input which can be expected from a single *A. filiculoides* crop grown during the fallow season in California (67, 68; and Rains, personal communication). The relationship between biomass and nitrogen content in *A. filiculoides* growing in a fallow rice field in California is depicted in Fig. 5. The inoculum of 50 grams fresh wt/m^2 gave about 10 percent surface cover and full cover was attained in 15 days. Twenty days later the mature cover had a biomass of 1700 kg dry wt/ha and a nitrogen content of 52 kgN/ha (52, 67). In comparisons of *Azolla* N and commercial ammonium sulfate in the range of 30 to 60 kgN/ha the two sources have been found to give equivalent rice yields (59, 69). At higher N levels the *Azolla* N appears to be less effective, possibly due to slower mineralization or loss through denitrification (68). This aspect is illustrated in Fig. 6 which also demonstrates the additive effect of *Azolla* N and ammonium sulfate on paddy yield.

The diminished availability of *Azolla* N at high N levels is consistent with a report that N from dried *Azolla* at 250 to 500 kgN/ha in pot studies was 40 percent less available to rice than was the N from ammoniacal fertilizer (75). These pot studies also indicated that ammonia release was more rapid from freshly incorporated *Azolla* than from dried *Azolla* and that mineralization was relatively slow with six to eight weeks required for 75 percent of the *Azolla* N to become available (75). In Indian paddies it was reported that fresh *Azolla* decomposes quite rapidly in the soil and rice seedlings should be planted either immediately or within a few days after incorporation (60). However, it was also stated that *Azolla* nitrogen is released rather slowly and studies under flooded conditions showed that at 24 ± 2°C, 34 percent, 56 percent and 82 percent of the *Azolla* N was released after one, three and six weeks, respectively. [At room temperature 80 percent of the *Azolla* N was released after three weeks (60).]

The advantages of integrating *Azolla* into existing field practices for lowland rice are well illustrated by current studies in India (59–63). The annual production of the Central Rice Research Institute farm, based on an average for three years, is reported to be about 347 tons fresh wt/ha which contains 868 kgN (62, 63). This is based on 12 to 15 crops harvested at 20 to 30 day intervals. The research, being carried out under a collaborative program initiated by the Indian Council of Agricultural Research, has shown that a single crop of *A. pinnata* incorporated into rice fields can provide the equivalent of at least 20 to 30 kgN/ha of chemical fertilizer N. When several *Azolla* crops or dual cropping are used the equivalent of 50 kgN/ha is added and rice yields increased up to 40 percent, 0.5 to 2 tons/ha versus unfertilized controls (60–63). Moreover, *Azolla* retains nitrogenase activity in the presence of combined N fertilizer (37, 40, 44), its N contribution is additive to the chemical fertilizer N (52, 60, 61, 63), and it adds large quantities of organic matter to the soil thus both enriching it and improving its texture.

In the last few years the International Rice Research Institute has provided impetus to *Azolla* field research. It has brought together individuals from around the world to discuss *Azolla*, beginning with a Nitrogen and Rice Symposium in 1978, and sponsored the compilation of an International Bibliography of *Azolla* with 671 citations from the world literature between 1873 and 1979 (9). Workers at IRRI have accepted and maintained clones of *Azolla* from numerous locations around the world for distribution elsewhere and the Institute continues to sponsor research and training programs on *Azolla* and its utilization with rice. International collaborative projects were finalized in 1979 involving field trials on the effect of phosphorus on growth, N, and P contents of several *Azolla* isolates in areas including Thailand, Nepal, Bangladesh and India and the first year's results compiled (22, 63). At present field trials include developing management practices and, whereas traditional use and most earlier trials were restricted to the endemic species, current trials are comparing various species as well as populations, varieties or strains of a given species. The isolates from widely separated geographical areas are being characterized in regard to temperature tolerances, pest resistance, and N input, selecting those best suited for a specific environment.

It is of interest to note that prior to 1977 *Azolla* was simply considered as a curiosity in West Africa. However, in field studies conducted at ORSTOM (Office de La Recherche Scientifique et Technique Outre-Mer) DaKar, Senegal in 1979 and 1980, the latter carried out according to the recommendations of the International Network on Soil Fertility and Fertilizer Efficiency (INSFFER), use of *Azolla* resulted in higher average rice yields than were observed in INSFFER trials in Asia during 1979 (22, 56). The West African Rice Development Association (WARDA) is currently studying the adaptation of a collection of *Azolla* strains from various parts of the world. It is probable that with the support of the Food and Agriculture Organization (FAO) and collaboration of ORSTOM, WARDA and IITA (International Institute of Tropical Agriculture), *Azolla's* use as a green manure in West Africa will be established in the next decade (56).

It is obvious that *Azolla* certainly has potential as an alternative N source for rice in many areas. However, much remains to be determined and *Azolla* is by no means a panacea. Considerable attention has been focused on the phosphate requirement of *Azolla*. While it is important to remember that this may well vary with the locale, overall management practices and other factors, the following are noted. Split applications of 1 kg P_2O_5/ha at two to four-day intervals was sufficient to maintain high productivity of *Azolla* in field studies at IRRI and it was calculated that *Azolla* produced 1.9 kg additional N with an application of 1 kg P_2O_5 (76). In California, P was supplied at 7.2 kg/ha as KH_2PO_4 at weekly intervals over a four-week period and it was estimated that each kg of P resulted in 5 kg of additional N in the *Azolla* biomass 35 days after inoculation (69). Vietnamese workers recommended the addition of 1–2 kg P_2O_5 every five days and concluded that 1 kg P_2O_5 increased *Azolla* yields equivalent to 2.2 kg of N (11) and 4–6 kg P_2O_5/ha week was recommended in Indian paddies (59). One of the problems with adding phosphate to the *Azolla*, and a reason for split applications, is that the phosphate rapidly becomes unavailable to the floating *Azolla*. That the actual phosphate requirement of *Azolla* is considerably lower than that which it is necessary to add to the paddy water is suggested by the finding that *Azolla* grows normally with P supplied at 0.06 ppm in a flow through system (65). Thus, the phosphate requirement of *Azolla* might best be met by the advent of a slow release source which either floats or has a specific attraction to the *Azolla* sporophyte. There are, however, two other points worth noting. First, *Azolla* strains which have even lower phosphate requirements may be found through screening procedures. It has recently been reported that an *A. pinnata* isolate from Vietnam grows well in paddies at CRRI without the application of phosphorus (62). Second, it is important to remember that while added phosphorus rapidly becomes tied up in the paddy soil, this phosphate is available to the rice plants as is the phosphate taken up by the *Azolla* when the latter is incorporated. Talley and Rains (68) noted that production of 93 kg $N \cdot ha^{-1}$ as *Azolla* N through the application of 30 kg $P \cdot ha^{-1}$ is encouraging in that 18–27 kg $P \cdot ha^{-1}$ is recommended for California rice (28) and much of the phosphorus added during the fallow period remains available for the summer rice crop.

Azolla is also subject to attack by pests, especially larvae of various lepidopterous and dipterous insects (26, 60, 61, 74) but also aphids (27) and other pests (26). While the insect larvae cause the *Azolla* leaves to roll around them, and can do serious damage in a very brief period of time, they can be controlled with the application of pesticides such as carbofuran (59). Here again it may well be possible to select *Azolla* strains which are more resistant than others to insect predation. Although it has been stated that *Azolla* species/strains do not carry any rice diseases or insect pests (62) the senior author (G.A. Peters) does not consider our knowledge adequate to make such a sweeping statement. Until this has been absolutely established by experts in these areas, caution is warranted. At the very least, when *Azolla* stocks are transferred from one geographical area to another the stocks should be fumigated and/or surface sterilized before conducting field studies to guard against the inadvertent introduction of a new pest or rice disease.

The water requirement of *Azolla* will necessarily limit its use as a green manure grown during the fallow season in some areas since the water is simply not available. Moreover, to the author's knowledge there is no definitive information on the extent to which *Azolla* may compete with rice for water in a dual culture situation. This clearly needs to be ascertained.

The fact that water turbulence causes fragmentation and eventual death of *Azolla* (2) minimizes the likelihood of an introduced strain running amuck and choking major waterways. It does not in itself exclude the possibility of *Azolla* creating a problem in other areas by taking over where it is not wanted. However, *Azolla* is quite sensitive to a number of herbicides. In fact, its disappearance from some areas to which it was indigenous may be at least partially attributed to herbicide usage. In a recent study of the effect of three insecticides and three herbicides on growth and physiological process in four *Azolla* species (70) we found the *Azolla* was quite sensitive to Butachlor (Machete EC) and Propanil (STAM F-34). Thus, undesirable *Azolla* growth can probably be readily controlled in those localized situations where it might pose a problem. It is worth noting that herbicides may also be found to have application in conjunction with *Azolla* in dual culture with rice. For example, rice is relatively tolerant to Propanil while *Azolla* is killed by Propanil at 1 ppm in solution (70). Thus, depending upon management practices and local restrictions, it may be possible to use a herbicide such as this to kill off the *Azolla* at the tillering stage of rice, timing the N input from the total *Azolla* biomass and decreasing the amount of labor required to get the *Azolla* N to the rice plants.

At present the use of *Azolla* is undisputably labor intensive. Maintenance ponds are required for multiplication of the inoculum for the field and the simple weight of the fresh material and its fragile nature necessarily preclude any long distance transport. In the future it may be possible to employ *Azolla* spores for an inoculum in much the same manner as a seed. This will be largely dependent upon our increased understanding of those factors which trigger sporulation, such that it can be controlled—i.e. caused or prevented. While this type of information is, in our opinion, quite central to the future utilization of *Azolla*, there are a number of ifs involved. However, *if* sporulation can be induced, and *if* a method can be developed for harvesting the spores, and *if* the spores can be stored for a prolonged period, a number of obstacles will be overcome and new vistas will be opened. The following are some examples. First, storage of the spores would provide a means of preserving germ plasm for existing species, especially strains with any specifically desired attributes. Second, spores could be readily transported and "sown" as an inoculum, thus overcoming the need for maintenance tanks or ponds and the problems associated with the fresh materials, provided the formation of the sporophyte is rapid enough. Third, spores provide the key to future attempts at the establishment of breeding programs and/or interspecific crosses which could provide associations with specifically desired attributes—as for example bringing together pest resistance from one strain, or possibly species, with temperature tolerance from another. Finally, if it is possible to determine factors such that sporulation can be induced at will it may be

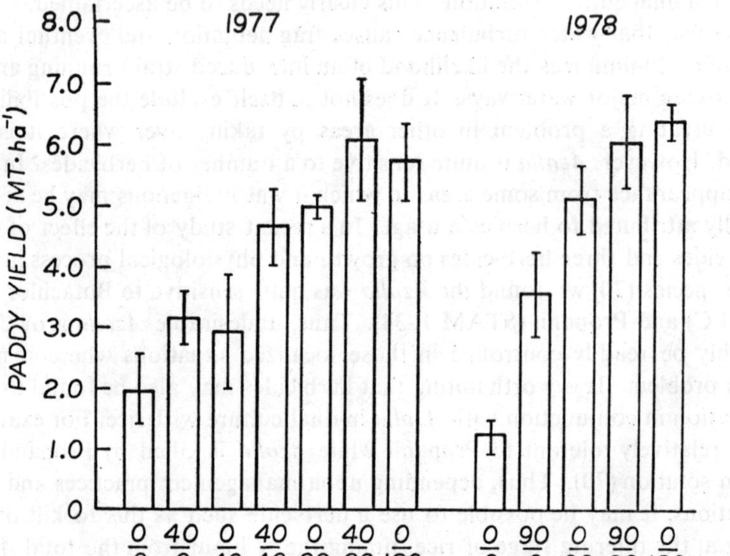

Fig. 6. Average and standard deviation of paddy yield for rice (ED7) as functions of *Azolla* and ammonium sulfate N fertilization during 1977 and 1978 at the U.C. Davis Rice Research Facility. Data for 1977 and 1978 respectively represent averages for 3 and 4 22.4 m^2 plots. Planting dates were 22 May, 1977 and 2 June, 1978 [after Talley and Rains (68)].

equally possible to prevent sporulation. Since sporulation results in diminished N_2 fixation and senescence of the *Azolla* population (68), its prevention could increase N input from a single crop in at least some areas. Thus, if a spore promoting factor were to be found, it might be possible to synthesize an antagonist which could prevent it having any effect.

In conclusion, the consideration of *Azolla* as an agronomic crop is taking hold. While there are problems associated with its widespread use these may be at least partially alleviated by future research. As with the improvement of other plant species this will necessitate an amalgamation of laboratory study, field testing, further refinement of management practices and possibly some mechanization through the design of mechanical equipment specifically adapted for use with *Azolla* in rice.

References

1. Anonymous. *Cultivation, Propagation and Utilization of Azolla*, Compiled by Institute of Soils and Fertilizers, Chekiang Academy of Agricultural Sciences, Agricultural Publishing House Peking (1975), English Translation for IRRI, 1977 by AD-EX Translations International, 129 pp.
2. Ashton, P.J. The effects of some environmental factors on the growth of *Azolla filiculoides* Lam., pp. 123–138 in *The Orange River Progress Report* (Edited by E.M.V. Zinderen Bakker, Sr.), Bloenfontein, South Africa, University of the Orange Free State (1974).
3. Ashton, P.J. and Walmsley, R.D. The aquatic fern *Azolla* and its *Anabaena* symbiont, *Endeavour*, *35*, 39–43 (1976).
4. Becking, J.H. Environmental requirements of *Azolla* for use in tropical rice production, pp. 345–374 in *Nitrogen and Rice*, International Rice Research Institute, Los Baños, Laguna, Philippines (1979).
5. Bonnet, A.L.M. Contribution l'etude des hydropteridees. III. Recherches sur *Azolla filiculoides*, *Revue de Cytologie et de Biologie Vegetales*, *18*, 1–86 (1957).
6. Bortels, H. Über die Bedeutung des Molydans für stickstoff bindende Nostocacean, *Archive für Mikrobiologie*, *11*, 155–186 (1940).
7. Calvert, H.E. and Peters, G.A. The *Azolla-Anabaena azollae* relationship. IV. Morphological analysis of leaf cavity hair populations, *New Phytologist* (in press) (1981).
8. Campbell, D.H. On the development of *Azolla filiculoides* Lam., *Annals of Botany* (*London*), *7*, 155–187 (1893).
9. Capaya, D.T. (Compiler). *International Bibliography of Azolla*, Library and Documentation Center, International Rice Research Institute, Los Baños, Laguna, Philippines (1979).
10. Cohn, J. and Renlund, R.N. Notes on *Azolla Caroliniana*, *American Fern Journal*, *43*, 7–11 (1953).
11. Dao, T.T. and Tran, Q.T. Use of *Azolla* in rice production in Vietnam, pp. 395–405 in *Nitrogen and Rice*, International Rice Research Institute, Los Baños, Laguna, Philippines (1979).
12. Demalsy, P. Novelles recherches sur le sporophyte d' *Azolla*, *Cellule*, *59*, 235–268 (1958).
13. Espinas, C.R., Berja, N.S., del Rosario, D.C. and Watanabe, I. Environmental conditions affecting *Azolla* growth, *Greenfields*, *9*, 14–19 (1979).

14. Fosberg, F.R. Uses of Hawaiian ferns, *American Fern Journal, 32,* 15–23 (1942).
15. Galston, A.W. The waterfern-rice connection, *Natural History, 84,* 10 (1975).
16. Gunning, B.E.S., Hughes, J.E. and Hardham, A.R. Formative and proliferative cell divisions, cell differentiation and developmental changes in the meristem of *Azolla* roots, *Planta, 143,* 121–144 (1978).
17. Haselkorn, R. Heterocysts, *Annual Review of Plant Physiology, 29,* 319–344 (1978).
18. Haselkorn, R., Mazur, B., Orr, J., Rice, D., Wood, N. and Rippka, R. Heterocyst differentiation and nitrogen fixation in cyanobacteria (blue-green algae), pp. 259–278 in *Nitrogen Fixation, Vol. II* (Edited by W.E. Newton and W.H. Orme-Johnson), University Park Press, Baltimore (1980).
19. Hill, D.J. The pattern of development of *Anabaena* in the *Azolla-Anabaena* symbiosis, *Planta, 133,* 237–242 (1975).
20. Hill, D.J. The role of *Anabaena* in the *Azolla-Anabaena* symbiosis, *New Phytologist, 78,* 611–616 (1977).
21. Holst, R.W. Anthocyanins of *Azolla*. *American Fern Journal, 67,* 99–100 (1977).
22. International Rice Research Institute—International Network on Soil Fertility and Fertilizer Efficiency in Rice. *Report on First Trials of Azolla Use to Rice,* pp. 1–20, IRRI, Los Baños, Laguna, Philippines (1980).
23. Johnson, G.V., Mayeux, P.A. and Evans, H.J. A cobalt requirement for symbiotic growth of *Azolla filiculoides* in the absence of combined nitrogen, *Plant Physiology, 41,* 852–855 (1966).
24. Kaplan, D. and Peters, G.A. The *Azolla-Anabaena azollae* relationship. X. $^{15}N_2$ fixation and transport in main stem axes, *New Phytologist* (in press) (1981).
25. Konar, R.N. and Kapoor, R.K. Embryology of *Azolla pinnata, Phytomorphology, 24,* 228–261 (1974).
26. Liu, C.C. Use of *Azolla* in rice production in China, pp. 375–394 in *Nitrogen and Rice,* International Rice Research Institute, Los Baños, Laguna, Philippines (1979).
27. Lumpkin, T.A. and Plucknett, D.L. *Azolla:* Botany, physiology and use as a green manure, *Economic Botany, 34,* 111–153 (1980).
28. Mikkelson, D.S. and Evat, N.S. Soils and fertilizers, pp. 76–87 in *Rice in the United States: Varieties and Production,* Agricultural Research Service, U.S. Department of Agriculture, Agricultural Handbook No. 289 (1973).
29. Moore, A.W. *Azolla:* Biology and agronomic significance, *Botanical Review, 35,* 17–34 (1969).
30. Newton, J.W. Photo production of molecular hydrogen by a plant algal

symbiotic system, *Science*, *191*, 559–561 (1976).
31. Newton, J.W. and Cavins, J.F. Altered nitrogenous pools induced by the *Azolla-Anabaena* symbiosis, *Plant Physiology*, *58*, 798–799 (1976).
32. Newton, J.W. and Herman, A.I. Isolation of cyanobacteria from the aquatic fern, *Azolla*, *Archives of Microbiology*, *120*, 161–165 (1979).
33. Olsen, C. On biological nitrogen fixation in nature, particularly in blue-green algae, *Comptes Rendus des Travaux du Laboratorie Carlsburg*, *37*, 269–283 (1972).
34. Pate, J.S. and Gunning, B.E.S. Transfer cells, *Annual Review of Plant Physiology*, *23*, 211–223 (1972).
35. Peters, G.A. The *Azolla-Anabaena azollae* relationship. III. Studies on metabolic capabilities and a further characterization of the symbiont, *Archives of Microbiology*, *103*, 113–122 (1975).
36. Peters, G.A. Studies on the *Azolla-Anabaena azollae* symbiosis, pp 592–610 in *Proceedings of the First International Symposium on Nitrogen Fixation, Vol. II* (Edited by W.E. Newton and C.J. Nyman), Washington State University Press, Pullman, Washington (1976).
37. Peters, G.A. The *Azolla-Anabaena azollae* symbiosis, pp. 231–258 in *Genetic Engineering for Nitrogen Fixation* (Edited by A. Hollaender), Plenum Press, New York (1977).
38. Peters, G.A. The *Azolla-Anabaena* symbiosis: Morphology and physiology, pp. 153–169 in *Proceedings, Second Review Meeting in I.N.P.U.T.S. Project* (Edited by S. Ahmed and H.P.M. Gunasena), East-West Center, Honolulu (1978).
39. Peters, G.A., Evans, W.R. and Toia, R.E., Jr. *Azolla-Anabaena azollae* relationship. IV. Photosynthetically-driven, nitrogenase-catalyzed H_2 production, *Plant Physiology*, *58*, 119–126 (1976).
40. Peters, G.A., Ito, O., Tyagi, V.V.S. and Kaplan, D. Physiological studies on N_2-fixing *Azolla*, in *Genetic Engineering of Symbiotic Nitrogen Fixation and Conservation of Fixed Nitrogen* (Edited by J.M. Lyons *et al.*), Plenum Press, New York (in press) (1981).
41. Peters, G.A., Ito, O., Tyagi, V.V.S., Mayne, B.C., Kaplan, D. and Calvert, H.E. Photosynthesis and N_2 fixation in the *Azolla-Anabaena* symbiosis, pp. 121–124 in *Current Perspectives in Nitrogen Fixation* (Edited by A.H. Gibson and W.E. Newton), Australian Academy of Science, Canberra (1981).
42. Peters, G.A. and Kaplan, D. Soluble carbohydrate pool in the *Azolla-Anabaena* symbiosis, *Plant Physiology*, *67*, S-37 (1981).
43. Peters, G.A. and Mayne, B.C. The *Azolla-Anabaena azollae* relationship. I. Initial characterization of the association, *Plant Physiology*, *53*, 813–819 (1974).
44. Peters, G.A. and Mayne, B.C. The *Azolla-Anabaena azollae* relationship. II. Localization of nitrogenase activity as assayed by acetylene reduction,

Plant Physiology, *53*, 820–824 (1974).

45. Peters, G.A., Mayne, B.C., Ray, T.B. and Toia, R.E., Jr. Physiology and biochemistry of the *Azolla-Anabaena* symbiosis, pp. 325–344 in *Nitrogen and Rice*, International Rice Research Institute, Los Baños, Laguna, Philippines (1979).
46. Peters, G.A., Ray, T.B., Mayne, B.C. and Toia, R.E., Jr. *Azolla-Anabaena* association: Morphological and physiological studies, pp. 293–309 in *Nitrogen Fixation, Vol. II* (Edited by W.E. Newton and W.H. Orme-Johnson), University Park Press, Baltimore (1980).
47. Peters, G.A., Toia, R.E., Jr., Evans, W.R., Crist, D.K., Mayne, B.C. and Poole, R.E. Characterization and comparisons of five N_2-fixing *Azolla-Anabaena* associations. I. Optimization of growth conditions for biomass increase and N content in a controlled environment, *Plant, Cell and Environment*, *3*, 261–269 (1980).
48. Peters, G.A., Toia, R.E., Jr. and Lough, S.M. *Azolla-Anabaena azollae* relationship. V. $^{15}N_2$ fixation, acetylene reduction and H_2 production, *Plant Physiology*, *59*, 1021–1025 (1977).
49. Peters, G.A., Toia, R.E., Jr., Raveed, D. and Levine, N.J. The *Azolla-Anabaena azollae* relationship. VI. Morphological aspects of the association, *New Phytologist*, *80*, 583–593 (1978).
50. Peterson, R.B., Dolan, E., Calvert, H.E. and Ke, B. Energy transfer from phycobiliproteins to photosystem I in vegetative cells and heterocyst of *Anabaena variabilis*, *Biochimica et Biophysica Acta*, *634*, 433–449 (1981).
51. Pieterse, A.H., DeLange, L. and Van Vliet, J.P. A comparative study of *Azolla* in the Netherlands, *Acta Botanica Neerlandica*, *36*, 433–439 (1977).
52. Rains, D.W. and Talley, S.N. Use of *Azolla* in rice production in North America, pp. 419–431 in *Nitrogen and Rice*, International Rice Research Institute, Los Baños, Laguna, Philippines (1979).
53. Rao, H.S. The structure and life history of *Azolla pinnata* R. Brown with remarks on the fossil history of the Hydropterideae, *Proceedings of the Indian Academy of Science*, *2B*, 175–200 (1936).
54. Rap, T.B., Mayne, B.C., Toia, R.E., Jr. and Peters, G.A. *Azolla-Anabaena* relationship. VIII. Photosynthetic characterization of the association and individual partners, *Plant Physiology*, *64*, 791–795 (1979).
55. Ray, T.B., Peters, G.A., Toia, R.E., Jr. and Mayne, B.C. *Azolla-Anabaena* relationship. VII. Distribution of ammonia-assimilating enzymes, protein and chlorophyll between host and symbiont, *Plant Physiology*, *62*, 463–467 (1978).
56. Reynaud, P.A. The use of *Azolla* in West Africa. Workshop on Biological Nitrogen Fixation Technology for Tropical Agriculture, Cali, Columbia (1981).
57. Saubert, G.G.P. Provisional communication on the fixation of elementary

nitrogen by a floating fern, *Annals of the Royal Botanical Garden* (Buitenzorg), *51*, 177–197 (1949).

58. Sculthorpe, C.D. *The Biology of Aquatic Vascular Plants*, Edward Arnold (Publishers) Limited, London (1967).
59. Singh, P.K. Multiplication and utilization of fern "Azolla" containing nitrogen-fixing algal symbiont as green manure in rice culture, *Riso*, *26*, 125–137 (1977).
60. Singh, P.K. Symbiotic algal N_2-fixation and crop productivity, pp. 37–65 in *Annual Reviews of Plant Science*, *Vol. I* (Edited by C.P. Malik), Kalyani Publishers, New Delhi (1979).
61. Singh, P.K. Use of *Azolla* in rice production in India, pp. 407–418 in *Nitrogen and Rice*, International Rice Research Institute, Los Baños, Laguna, Philippines (1979).
62. Singh, P.K. Introduction of "Green *Azolla*" biofertilizer in India, *Current Science*, *49*, 155–156 (1980).
63. Singh, P.K. Recent studies on *Azolla* cultivation and its effects on rice crop in India, in *Proceedings of Special Workshop on Nitrogen Fixation and Utilization in Rice Fields*, International Rice Research Institute, Los Baños, Laguna, Philippines (1980).
64. Stewart, W.D.P. A botanical ramble among the blue-green algae, *British Phycological Journal*, *12*, 89–115 (1977).
65. Subudhi, P.R. and Watanabe, I. Minimum level of phosphate in water for growth of *Azolla* determined by continuous flow culture, *Current Science*, *48*, 1065–1066 (1959).
66. Svenson, H.K. The new world species of *Azolla*, *American Fern Journal*, *34*, 69–84 (1944).
67. Talley, S.N. and Rains, D.W. *Azolla* as a nitrogen source for temperate rice, pp. 311–320 in *Nitrogen Fixation*, *Vol. II* (Edited by W.E. Newton and W.H. Orme-Johnson), University Park Press, Baltimore (1980).
68. Talley, S.N. and Rains, D.W. *Azolla filiculoides* Lam. as a fallow season green manure for rice in temperate climate, *Agronomy Journal*, *72*, 11–18 (1980).
69. Talley, S.N., Talley, B.J. and Rains, D.W. Nitrogen fixation by *Azolla* in rice fields, pp. 259–281 in *Genetic Engineering for Nitrogen Fixation* (Edited by A. Hollaender), Plenum Press, New York (1977).
70. Toia, R.E., Jr., Crist, D.K., Poole, R.E., Bent, P.E. and Peters, G.A. Effects of selected pesticides on physiology and composition of four *Azolla* species, *Plant Physiology*, *67*, S-81 (1981).
71. Tung, H.F. and Shen, T.C. Studies of the *Azolla pinnata—Anabaena azollae* symbiosis: Growth and nitrogen fixation, *New Phytologist*, *87*, 743–749 (1981).
72. Tyagi, V.V.S., Mayne, B.C. and Peters, G.A. Purification and initial characterization of phycobiliproteins from the endophytic cyanobacteri-

um of *Azolla*, *Archives of Microbiology*, *128*, 41–44 (1980).

73. Tyagi, V.V.S., Mayne, B.C. and Peters, G.A. Action spectra of acetylene reduction in the *Azolla-Anabaena* association and in the isolated symbiont, *Plant Physiology*, *65*, S-109 (1980).
74. Watanabe, I. *Azolla* and its use in lowland rice culture, *Soil and Microbe* (Japan), *20*, 1–10 (1978).
75. Watanabe, I., Espinas, C.R., Berja, N.S. and Alimagno, B.V. Utilization of the *Azolla-Anabaena* complex as a nitrogen fertilizer for rice, *International Rice Research Institute Research Paper Series*, *11*, 1–15 (1977).
76. Watanabe, I., Berja, N.S. and del Rosario, D.C. Growth of *Azolla* in paddy fields as affected by phosphorus fertilizer, *Soil Science and Plant Nutrition*, *26*, 301–307 (1980).
77. Yatazawa, M., Tomomatsu, N., Hosoda N. and Nunome, K. Nitrogen fixation in *Azolla-Anabaena* symbiosis as affected by mineral nutrient status, *Soil Science and Plant Nutrition*, *26*, 415–426 (1980).
78. Yoshida, S., Forno, D.A. and Cock, J.H. *Laboratory Manual for Physiological Studies of Rice*, pp. 61, International Rice Research Institute, Los Baños, Laguna, Philippines (1971).

N.S. SUBBA RAO

8. Biofertilizers

As the name would perhaps suggest, the term biofertilizers denotes all nutrient inputs for plant growth which are of biological origin. Essentially, such products can be more appropriately called 'microbial inoculants' and in a broad sense may include inputs in agriculture which have manurial value and which are aided by microbial processes. The first and foremost example of this kind is the *Rhizobium* or legume inoculant which was first marketed in the U.S.A. during the early part of this century (24).

The discovery of legume fixation of atmospheric nitrogen in root nodules mediated by nodule bacteria was perhaps the first realization that microorganisms play a part in the nutrient status of soil although composting was known since ancient times. The well-known monograph of Fred, Baldwin and McCoy (24) summarizes the events which laid the foundations for *Rhizobium* research. The Russian products 'Azotobakterin' and 'Phosphobacterin' containing cells of *Azotobacter chrococcum* and *Bacillus megatherium* var. *phosphaticum* respectively, were used in the U.S.S.R. and East European countries for the 'bacterization' of crops with a view to exploit the nitrogen-fixing and phosphate solubilizing properties of *Azotobacter* and *Bacillus.* Subsequent works on *Azotobacter* in Russia, Czechoslovakia, England and India have clearly demonstrated that these bacteria secrete growth substances and an antifungal antibiotic which may contribute to the improvement of seed germination and root growth (5, 30, 33, 45, 46, 48, 56, 57, 59, 63, 67, 78, 99).

During the last three decades, the work on the methodology of the mass cultivation of rhizobia, preparation of carrier-based inoculants and application of such inoculants to soil or seeds of legumes has been carried out in the U.S.A. (8, 9, 10) and Australia (3, 4, 11, 12, 13, 61, 62, 104, 105) which have been adopted by Asian (83–87), African and Latin American (50, 58) countries with modifications to suit their local needs and conditions.

A greater understanding of grass-bacteria associations and the re-discovery of the potentialities of associate symbiosis (*Azospirillum* associations) have recently led to their practical application as seed inoculants for cereals in developing countries where fertilizer nitrogen is scarce (2, 6, 7, 14, 16–19, 32, 34, 35, 37–39, 49, 51–53, 65, 69, 88, 89, 95, 96, 100, 107).

Waterlogged rice fields offer a great potentiality for both heterotrophic and photoautotrophic nitrogen-fixing microorganisms to function and contribute to the nitrogen nutrition of rice plants. The ability of algal species such as *Aulosira, plectonema, cylindrospermum, Nostoc, Anabaena, Tolypothrix* etc. to fix atmospheric nitrogen and secrete amino acids and growth-promoting substances has been well documented (1, 20, 21, 23, 25, 28, 29, 36, 42, 43, 60, 64, 68, 77, 79, 80–82, 92, 93, 101–103, 107, 108, 109). More recently the role of *Azolla* in the nitrogen nutrition of rice has been re-emphasized. Methods to harness it in practical agriculture (47, 54, 55, 66, 70–76, 94, 98, 110) have been tried in farmers, fields.

The purpose of this chapter is not to provide a detailed account of all the nitrogen-fixing microorganisms or associations but to highlight the production, use and benefits of microbial products designed to augment these microbial processes in soils.

Cultivation and mass production of *Rhizobium* inoculant

Rhizobia are maintained on yeast extract mannitol (YEM) agar medium either by subculturing at frequent intervals or by lyophilization procedures to be reconstituted into agar-based cultures whenever necessary. The primary culture is designated as the "Mother culture". One of the accepted liquid medium for large-scale cultivation is the YEM liquid medium with the following composition (g/1): yeast extract—1; mannitol—10 (replaced by sucrose or glucose); K_2HPO_4—0.5; $MgSO_4 \cdot 7H_2O$—0.2; NaC1—0.1; pH—6.5–7. The selection of suitable strains of a *Rhizobium* species is dependent upon many criteria—a single strain or more than one strain for a particular cultivar or a group of cultivars or crops. The selected strain is grown for three to four days on YEM agar slant depending on the fast or slow growing nature of the strain. The culture is tested for purity by well-known tests and transferred to large flasks containing sterile solid or liquid medium for four to nine days. This is called the 'starter culture' which is transferred to a seed tank fermentor and incubated for four to nine days. By about this time, a large quantity of liquid broth is formulated in the fermentor (the size depending upon the requirements), pH adjusted to 6.5 to 7 with KOH or H_2SO_4 and sterilized. After cooling to 30°C, inoculum from the seed tank fermentor is transferred aseptically to the production fermentor at the rate of 1 per cent by volume (11–13, 61, 62, 104–106). In some laboratories of the U.S.A., aeration is done through porous carborundum or stainless steel spargers in the bottom of the production fermentor. According to Burton, 0.15 atm O_2 partial pressure and 30–32°C temperature are optimum for rhizobia. He finds that a rhizobial count of 5×10^9 cells can be attained in 96 hours with a lag phase of 48 hours for *R. japonicum* and 24 hours for *R. meliloti* when the initial inoculum in the fermentor is 5 per cent of the volume of the medium and by increasing the inoculum level, the lag phase can also be considerably reduced. The factors influencing the output of cells are aeration, volume, initial inoculum level, bacterial strain, temperature and incubation time. The main objective is to attain high populations in the minimum time. Vincent mentions that viable counts up to 2000 × 10^6 or 4000 × 10^6 could be attained by manipulating these factors (104–106).

In the U.S.A. large automatic fermentors are used whereas in Australia it has been recommended that a container such as a drum of 10 to 100 litre capacity (60–80 litre capacity with 40–50 litre broth, with air-inlet, air-outlet, an inoculation point and a sampling tap) can be conveniently used with desirable results (11). The entire set-up can be sterilized in an autoclave (Fig. 1). In

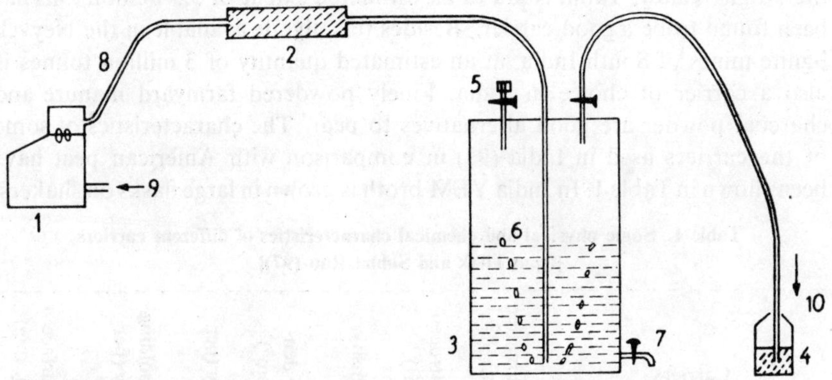

Fig. 1. Diagram of a fermentor used in Australia [after Date (11)].

India, most manufacturers use indigenously made shakers which can hold many conical flasks (750 ml capacity with 350 ml broth) to produce inoculants on a small scale to suit the needs of season-bound grain legume crops (84, 86). The resulting broth from fermentors is checked for purity according to methods outlined by Vincent and quality broths having cells more than 10^9/ml are used to prepare inoculants (104).

The product

The product is a free flowing carrier-based preparation containing live cells of specific rhizobia. In the U.S.A. sedge peat is ground to fine powder capable of passing through 100 mesh sieve and heat treated followed by neutralization with $CaCO_3$ to raise the pH of the peat carrier to 6.8. Diluted broth having rhizobial cell populations in excess of 10^9 cells/ml is blended with the peat carrier so as to bring the final moisture level of peat to 35–40 per cent on wet basis. The final product shall have at least 300 million (3 × 10^8) rhizobia per g of peat. In the process of blending, the broth is sprayed to powdered peat and left for curing in containers or trays for two to 10 days at 22–24°C. The final product is again milled and packed in polythene sheets where the numbers of rhizobia multiply. For example, *R. japonicum* increases more than 25×10^9 cells/g in four weeks while *R. meliloti* reaches the same figure in two weeks (8, 9, 10).

In Australia, milled dry peat having 15 per cent moisture is packed in polythene bags (0.089 mm gauge of medium density and 0.038–0.051 mm of the high density sheets) and sterilized by gamma radiation. The broth is injected into the polythene bags to bring the moisture level of peat to 50 per cent and incubated for five to 10 days, depending on the strain. By this way, the cell count from 10^9 to 10^{10} g peat can be attained at the time of manufacture (11).

In India, peat of high quality is not available. Peat-like material available in the Nilgiri Valley, Tamil Nadu to an estimated extent of 5.5 million tons has been found to be a good carrier. Besides this, lignite available in the Neyveli lignite mines of South India, at an estimated quantity of 3 million tonnes is also a carrier of choice in India. Finely powdered farmyard manure and charcoal powder are good alternatives to peat. The characteristics of some of the carriers used in India (97) in comparison with American peat have been shown in Table 1. In India YEM broth is grown in large flasks on shakers.

Table 1. Some physical and chemical characteristics of different carriers [from Tilak and Subba Rao (97)]

Carriers	Organic matter (per cent)	Total nitrogen (per cent)	Bulk density (g/cm³)	Particle density (g/cm³)	Porosity (per cent)	Water holding capacity (per cent)	Total surface area (sq. mig)
1. American peat	76.15	0.95	0.82	1.48	44.58	208.5	1063.3
2. Indian peat-soil	41.65	0.69	1.02	2.18	53.16	149.3	646.6
3. Indian peat soil + charcoal (1:1)	34.91	0.66	0.87	1.88	53.73	182.4	766.8
4. Farmyard manure (FYM)	79.05	0.93	0.79	1.77	55.39	153.4	911.1
5. FYM + charcoal (1:1)	58.19	0.78	0.64	1.72	62.70	169.4	885.8
6. Press mud	76.50	0.83	0.75	1.70	56.12	155.4	889.0
7. Press mud + charcoal (1:1)	55.25	0.85	0.72	1.80	65.50	165.4	870.1
8. Compost	55.06	0.55	0.75	1.82	58.84	171.3	940.8
9. Compost + charcoal (1:1)	42.71	0.57	0.61	1.77	65.52	177.2	888.5
10. Clay (vermiculite)	1.07	0.01	0.98	2.66	63.24	152.4	109.3
11. Teak leaf meal (TLM)	65.19	1.14	0.46	1.65	71.96	219.9	821.1
12. Clay + TLM (1:1)	40.00	0.55	0.81	1.95	58.91	196.7	563.6
13. Lignite	75.46	0.31	1.08	1.66	34.79	198.9	556.4
14. Coconut shell powder (CSP)	72.16	0.78	1.02	1.57	30.25	185.8	540.1
15. Lignite + CSP (1:1)	75.25	0.72	0.75	1.49	50.75	225.3	840.9
16. Lignite + TLM (1:1)	70.46	0.71	0.72	1.55	53.03	212.2	810.9
17. Soil	3.37	0.25	1.33	2.59	48.69	59.8	70.2
18. Charcoal	21.60	0.01	0.43	1.62	73.39	200.0	870.9
C.D. at 5 per cent	6.45	0.03	0.12	0.09	2.44	19.19	65.42

The powdered carrier (passing through 100 mesh sieve) is neutralized with $CaCO_3$ and autoclaved at 15 lb pressure for 4 hours. After cooling, a high count broth is mixed to bring the moisture of the carrier to 40 per cent, cured for two to five days in trays, and packed in polythene bags. The general procedure for the preparation of inoculants is systematically shown in Fig. 2. The

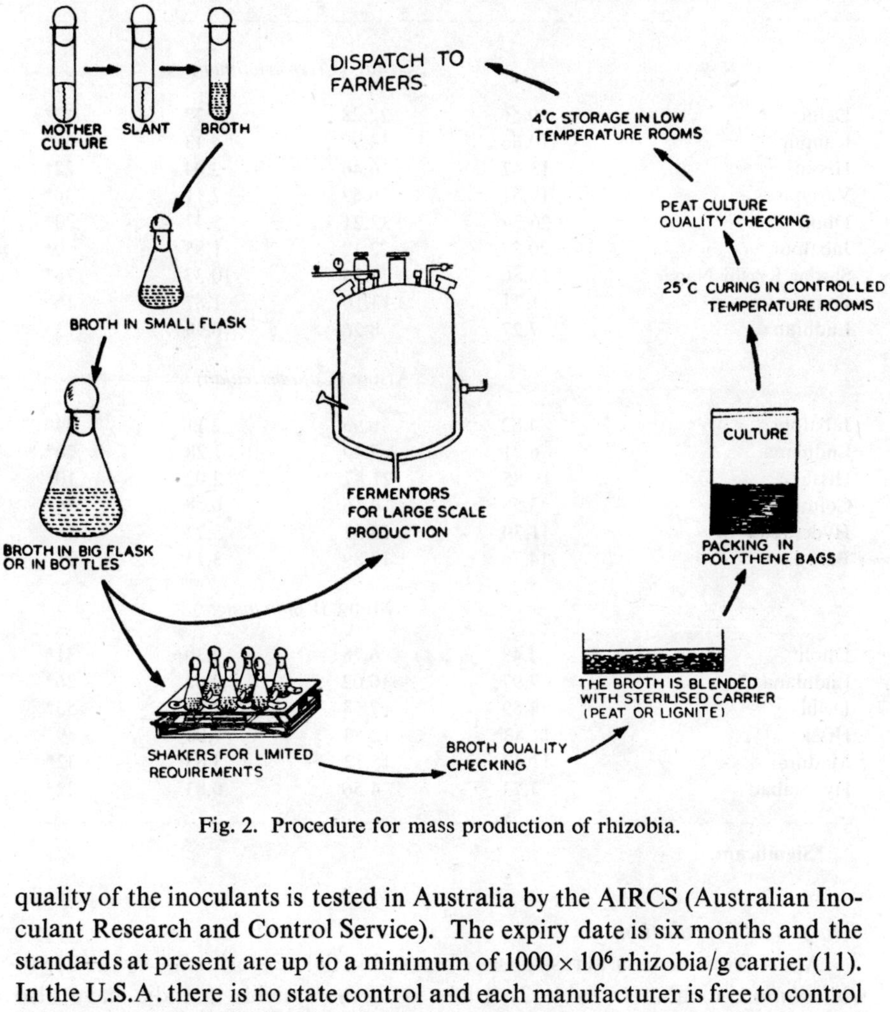

Fig. 2. Procedure for mass production of rhizobia.

quality of the inoculants is tested in Australia by the AIRCS (Australian Inoculant Research and Control Service). The expiry date is six months and the standards at present are up to a minimum of 1000×10^6 rhizobia/g carrier (11). In the U.S.A. there is no state control and each manufacturer is free to control his own product. In India, the Indian Standards Institution (I.S.I.) is entrusted with the task of quality control. In Czechoslovakia, the minimum requirement is 300×10^6 rhizobia/g carrier, whereas in Holland it is 4000×10^6/g carrier. In New Zealand the standard reaches up to 10^8 cells/g peat whereas in Russia, it ranges from 50–100 million rhizobia/g carrier (8).

Table 2. Effect of *Rhizobium* inoculation on the yield of different pulses at different locations, 1978. (Coordinated Research Project on pulses of the ICAR—Principal Investigator: R.B. Rewari)

Location	Yield/q/ha			
	Control	Inoculated	Increase	% Increase
		Gram (*Cicer arietinum*)		
Delhi	19.56	22.28	2.72	13*
Kanpur	10.86	14.99	4.13	38*
Hissar	13.42	16.46	3.04	22*
Varanasi	19.51	26.62	7.11	36*
Dholi	26.74	32.21	5.47	20*
Jabalpur	20.27	22.12	1.85	9*
Sardar Krishi Nagar	13.56	23.93	10.37	76*
Durgapura	11.23	13.10	1.87	16
Ludhiana	7.27	8.26	0.99	13
		Arhar (*Cajanus cajan*)		
Jabalpur	4.82	6.93	2.11	44*
Ludhiana	6.21	7.49	1.28	20*
Hissar	19.85	21.87	2.02	10
Coimbatore	3.58	4.16	0.58	16
Hyderabad	11.30	16.52	5.22	46
Baroda	14.76	17.89	3.13	21
		Mung (*Vigna mungo*)		
Dholi	4.48	6.78	2.30	51*
Ludhiana	7.97	10.02	2.05	26*
Delhi	4.89	7.58	2.69	55*
Hissar	11.83	12.83	1.00	9
Madurai	11.47	15.12	3.65	32*
Hyderabad	3.73	4.56	0.83	22*

*Significant.

Benefits of inoculation

Apart from successful responses observed in field experiments by the use of *Rhizobium* inoculants (Table 2), trials in farmers' fields in India have also proved to be a great success and serve to depict the importance of legume inoculation (Plate 1) in the improvement of grain legume yields (Table 3). There have been many advances in the method of inoculating seeds with rhizobia. Adhesives and seed pelleting agents such as gum arabic, lime, etc. have been advocated to improve the survival of bacteria on seed (4).

Table 3. Response to *Rhizobium* in farmers' fields (range of per cent increase in grain yield) (ICAR Coordinated Project on Pulses—Principal investigator: R.B. Rewari)

Crop	Location	1978	1979	1980
Chickpea	Hissar	7–22	5–11	11–13
	Durgapura	11–33	9–31	9–24
	Jabalpur			6–30
	Kanpur			7–15
	Dholi		38	
Lentil	Ludhiana		8–84	11–50
Pigeonpea	Hissar	20	10–16	
	Badnapur		5–6	
Moong	Hissar	15		
	Ludhiana	11–18	12–16	
	Madurai	33–60		
	Badnapur		11	
	Coimbatore		65	
Urid	Madurai	32–54		
	Durgapura	20	13	
Cowpea	Durgapura	30	26–30	

Mass cultivation of *Azotobacter* and phosphate solubilizers and the preparation of inoculants

Large-scale cultivation of *Azotobacter chroococcum*, *Bacillus megatherium* or *Pseudomonas striata* in India is done either by growing cultures in large flasks on rotary shakers or in batch fermentors depending on the quantity needed. All carrier materials used in the preparation of *Rhizobium* inoculants are also suitable for *Azotobacter* and phosphate-solubilizing bacteria. The medium used for *Azotobacter chroococcum* is as follows (g/1): Sucrose—20; K_2HPO_4—1; $MgSO_4 \cdot 7H_2O$—0.5; NaCl—0.5; $FeSO_4$—0.1; $CaCO_3$—2. The medium used for phosphate-solubilizing bacteria is as follows (g/l): Glucose—10; $Ca_3(PO_4)_2$—5; $(NH_4)_2SO_4$—0.5; KCl—0.2; $MgSO_4 \cdot 7H_2O$—0.1; $MnSO_4$—trace; $FeSO_4$—trace; yeast extract—0.5.

At the end of growth for the desired periods on the liquid substrate, the broth is poured into a powdered carrier and mixed uniformly to bring the level of moisture in the carrier to 40 per cent. After curing for two to five days, the carrier-based inoculant is packed in polythene bags for transport.

Seed inoculation is done in the way it is done for *Rhizobium*. However, in transplanted crops such as rice, cauliflower, cabbage etc., the roots of seedlings are dipped for 10 to 30 minutes in the water slurry of the inoculant before transplanting. For sugar cane and some millets, each manufacturer recommends multiple inoculations in the early stage of plant growth by advocating the pouring of the slurry of the inoculant near the root zone. Some of

Table 4. Effect of Azotobacterin on crops [from Mishustin and Shilnikova (46)]

Crop	No. of tests	Gain in yield by Azotobacterin (%)
Winter wheat	4	6.0
Spring wheat	15	7.6
Maize	13	5.1
Potato	14	10.4

Table 5. Summary of published results on field response of crops to *Azotobacter chroococcum* in India

Crop	Basal dressing	Increase (% over corresponding uninoculated control)	How the inoculation was done	Authors
Onion	0 kg N/ha 40 cartloads of FYM/ha 50 kg P_2O_5/ha 100 kg K_2O/ha	22	Roots dipped in a slurry of lignite based inoculant at transplanting	Joi and Shende (31)
Onion	100 kg N/ha 40 cartloads of FYM/ha 50 kg P_2O_5/ha 100 kg P_2O_5/ha	18	,, ,,	
Wheat Sonara 64	at varying levels of NPK	2–8	Seed inoculation with suspension of cells	Mehrotra and Lehri (44)
Rice	,, ,,	1–22	,, ,,	,,
Rice	at 120 kg N/ha 60 kg P_2O_5/ha 60 kg K_2O/ha	23*	,, ,,	,,
Brinjal or egg plant	at varying levels of NPK and FYM	1–42	Roots dipped in a slurry of lignite based inoculant at transplanting	,,
Tomato	,, ,,	2–29	,, ,,	,,
Cabbage	,, ,,	26–45*	,, ,,	,,
Brinjal or egg plant	,, ,,	15–62*	,, ,,	Lehri and Mehrotra (41)
Cabbage	,, ,,	25–50*		,,
Wheat	12,000 kg/ha FYM	10–20	Seed inoculation with cell suspension	Lehri and Mehrotra (40)
Wheat	Nil	10–20	,, ,,	,, Sundara Rao *et al.* (91)

*Indicates significant increase.

them even recommend mixing cultures in farmyard manure and broadcasting near the root zone.

Benefits of inoculation

Mishustin and Shilnikova (45, 46) have summarized the results of the numerous field experiments on *Azotobacter* inoculation in the U.S.S.R. (Table 4).

Table 6. The effect of *Azotobacter chroococcum* inoculation in India on the yield of three crops, 1978–79 (Dr. S.T. Shende, unpublished)

Crop	Location of field trials in India	Without *Azotobacter*	With *Azotobacter*	C.D. at 5%	% increase due to *Azotobacter*
Sorghum (q/ha)	Pali	12.8	14.0	1.22	9.3
	Dharwar	23.6	32.6	10.56	38.1
Maize (q/ha)	I.A.R.I.	7.8	13.4	4.80	71.7
	Dharwar	32.0	43.7	9.90	36.5
Cotton (q/ha)	Surat	1254	1339	241	6.7
	Indore	366	401	104	9.5
	Khandwa	559	708	768	20.6

Table 7. Effect of phosphate-solubilizing microorganisms on various crops in some field experiments in India

Name of the crop	Organism	No. of field experiments	No. of experiments showing significant increase	% yield increase over control	Reference
Berseem (*Trifolium alexandrinum*)	Phosphobacterin (*Bacillus megatherium* var. *phosphaticum*)	6	4	10–20	Sundara Rao (90)
Maize (*Zea mays*)		3	2	0–14	,,
Wheat (*Triticum aestivum*)		6	1	0–37	,,
Rice (*Oryza sativa*)		12	2	12–31	,,
Gram (*Cicer arietinum*)		1	0	—	,,
Arhar (*Cajanus cajan*)		1	0	—	,,
Urid (*Phaseolus aureus*)		3	1	16–19	,,
Soybean (*Glycine max*)		1	0	—	,,
Groundnut (*Arachis hypogea*)		1	0	—	,,
Potato (*Solanum tuberosa*)		1	0	—	,,
Wheat (*Triticum aestivum*)	*Pseudomonas striata*	1	1	10	Gaur *et al.* (27)
Rice (*Oryza sativa*)	*Bacillus polymyxa*	1	1	9.5	,,
Potato (*Solanum tuberosa*)	*Pseudomonas striata*	1	1	25.0	Gaur and Negi (26)

The results of work done in India (31, 40, 41, 44, 91) on *Azotobacter* field experiments have been summarized (85) by Subba Rao (Table 5). More recent work on the response of field crops such as maize and cotton to inoculation with new strains of *A. chroococcum* has shown the feasibility of using *Azotobacter* inoculants to minimize the use of fertilizer nitrogen (Table 6).

The results of work on phosphate-solubilizing bacteria to determine responses to inoculation in field conditions in India (26, 27, 90) are summarized in Table 7.

These results point out that while there are indications of response, (Plate 2), more experiments with a variety of crops in different conditions have to be carried out to arrive at definite conclusions.

Mass cultivation of *Azospirillum* and inoculant preparation

For large-scale cultivation, bottles or flasks with ammonium chloride containing liquid medium are being used at the Indian Agricultural Research Institute. These containers are incubated at 35°C on a rotary shaker and cells harvested after three days. The composition of the medium (51, 52) is as follows (g/l):

a)	K_2HPO_4	6.0
	KH_2PO_4	4.0
	Distilled water	500.0 ml
b)	$MgSO_4$	0.2
	NaCl	0.1
	$CaCl_2$	0.02
	NH_4Cl	1.0
	Malic acid	5.0
	NaOH	3.0
	Yeast extract	0.05
	Na_2MoO_4	0.002
	$MnSO_4$	0.001
	H_3BO_3	0.0014
	$Cu(NO_3)_2$	0.0004
	$ZnSO_4$	0.0021
	$FeCl_3$	0.002
	Distilled water	500.0 ml
	Bromothymol blue (0.5 per cent alcoholic solution).	2 ml

The phosphate buffer portion of the medium was made in half of the total volume required and also contained enough agar (1.5 to 2 per cent) for solidification. Part (a) and (b) were sterilized separately, mixed while hot and poured into plates and allowed to set.

The composition of an alternate medium (19) is as follows (g/l):

Malic acid	5.0
KOH	4.0
K_2HPO_4	0.5
$FeSO_4 \cdot 7H_2O$	0.05
$MnSO_4 \cdot H_2O$	0.01
$MgSO_4 \cdot 7H_2O$	0.10
NaCl	0.02
$CaCl_2$	0.01
Na_2MoO_4	0.002
Distilled water	1000.0 ml
Bromothymol blue (0.5% alcoholic solution)	2.0 ml
Agar	1.75
pH adjusted to 6.6–7	

The broth is incorporated into a carrier material consisting of powdered and sterilized farmyard manure and soil in the ratio of 1 : 1 and packed in polythene packets. This carrier has been found most suitable after testing three types of carriers (Fig. 3).

Benefits of inoculation

Several field experiments have been carried out in India (85, 89) by inoculating seeds of different crop plants with carrier-based cultures of efficient

Table 8. Response of sorghum to inoculation with *Azospirillum brasilense* at different locations in India (Field trials 1979–80) [from Subba Rao *et al.* (89)]

Treatment	Coimbatore	Dharwar	Hyderabad	Akola	Parbhani	Navasari	Indore	Udaipur	Pantnagar	All India mean
Control (uninoculated, no N)	25.9	5.2	34.7	3.7	10.7	21.6	16.0	19.6	24.3	18.0
Azospirillum brasilense	35.3* (36.6)	10.7 (105.8)	40.8 (17.2)	4.8 (29.7)	7.9	25.4 (17.6)	20.5 (28.1)	29.8* (22.6)	29.8 (22.6)	23.1 (28.3)
40 kg N/ha (urea at planting time)	36.0	21.8	44.2	11.5	23.3	24.0	28.2	29.5	32.5	27.9
40 kg N/ha + *A. brasilense*	41.7 (15.9)	21.4	47.0 (6.3)	9.4	22.7	20.6	16.8	30.0 (1.7)	30.9	27.3

The figures in parenthesis indicate per cent increase in yield over corresponding control.

Variety used—Udaipur—CSV 4; All other centres—CSH 5.

*Significant increase over corresponding control.

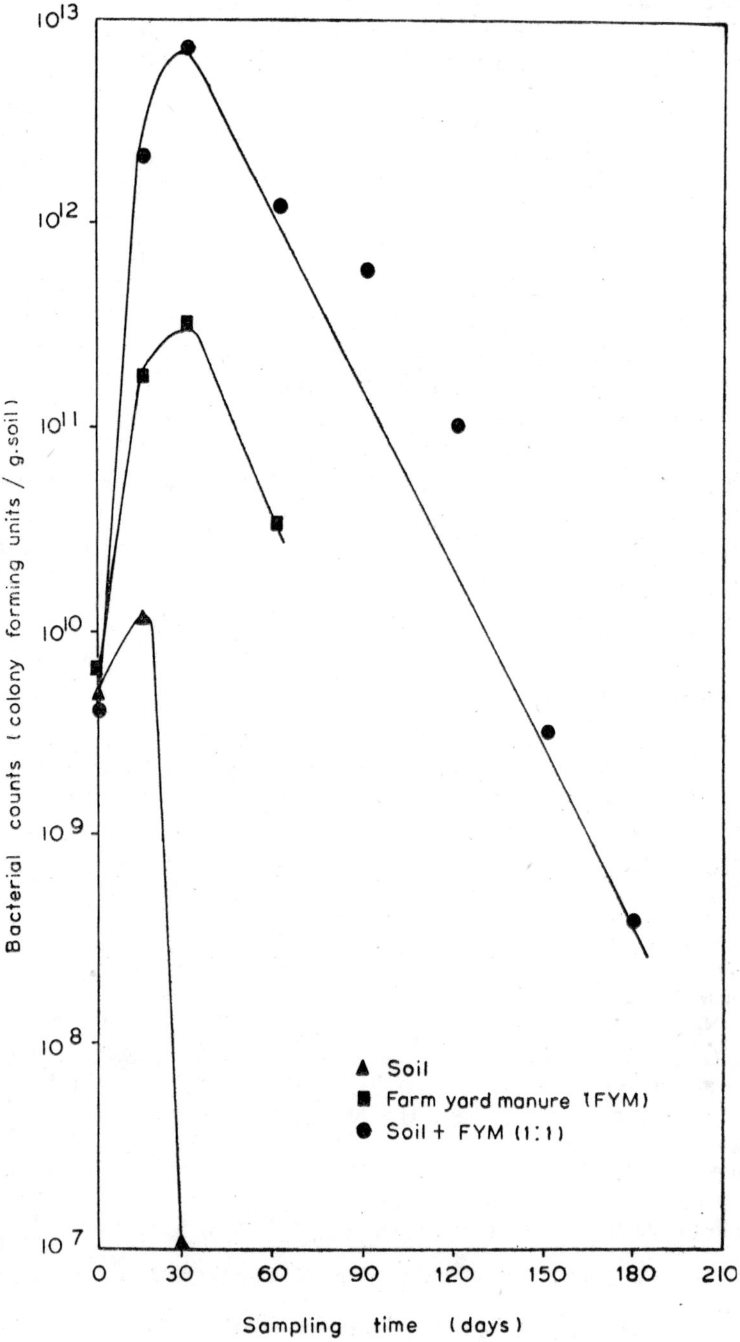

Fig. 3. Survival of *Azospirillum* in three carrier materials.

Table 9. Response of *Azospirillum brasilense* on the grain yield (kg/ha) of pearl millet (Bajra)—*Pennisetum americanum* var. BJ 104 (Field trial 1979–80) at different locations in India [from Subba Rao *et al.* (89)]

Treatment	Durgapura	Jodhpur	Aurangabad	Gwalior	Kanpur	Rajinder-nagar	Hissar	Jamnagar	All India mean
1. Control (uninoculated and no N)	640	357	1430	1217	1325	1267	2221	1065	1208
2. *Azospirillum brasilense*	1064**	556**	1789**	1267	1720**	1097	2279*	1164**	1396
3. 1/3rd N application	800	497	2053	1292	1610	1393	2194	1329	1406
4. 1/3rd N+*A. brasilense*	1224**	601	2150**	1399**	1845**	993	2408**	1195	1517
5. 1/2 N application	1012	581	2173	1683	1715	1212	2279	1349	1522
6. 1/2 N+*A. brasilense*	1212**	789**	2264**	1842**	1960**	1006	2408**	1263	1640
7. Full N application*	1092	751	2338	2350	1830	954	2725	1568	1720
8. Full N+*A. brasilense*	1372**	681	2421**	2280	2020**	915	2858**	1493	1792
S. Em ±	22	55	19	23	19	38	42	22	—
C.D. at 5%	67	161	58	69	38	115	127	66	—

*Jamnagar—Full N=80 kg N/ha (urea). All other places Full N=40 kg N/ha (urea).
**Significant increase over corresponding control.

strains of *Azospirillum brasilense* (Plate 2). The response of sorghum (*Sorghum bicolor*) and pearl millet (*Pennisetum americanum*) to *A. brasilense* under several agro-climatic conditions of India is shown in Tables 8 and 9. Quite recently, such responses to *Azospirillum* inoculations have been confirmed in Israel (53).

Mass cultivation of blue-green algae and their use

Watanabe and co-workers in Japan developed methods of mass culturing in illuminated tanks for blue-green algae by formulating a mineral-glucose liquid medium. Various carrier materials such as fine porous gravel and sponge rock were tried and it was observed that a combination of fine porous gravel and woolly fused magnesium phosphate was a suitable carrier material (109).

At the Indian Agricultural Research Institute, New Delhi, a simple farm-oriented open air method for bulk production of blue-green algae from a starter culture consisting of a soil-based mixture of *Aulosira*, *Tolypothrix*, *Nostoc*, *Anabaena* and *Plectonema* has been developed (29, 101, 102). Shallow trays of galvanized iron sheet (suggested dimension 6′ × 3′ × 9″) or brick and mortar or pits lined with polythene sheets are constructed into which 10 kg soil plus 200 g superphosphate are added followed by the filling of water to 2–6″ depth. The pH of the mixture must be neutral or made neutral by the addition of lime. When the soil settles down, sawdust is sprinkled along with the starter culture of algae. On hot sunny days, the algal growth in the open air tanks is quick and within a week a thick scum of algae is formed. The water is allowed to dry and the dried algal flakes from the surface are scraped and stored in bags (Plate 3). The production of dried algae on farmers' fields can be a continuous process (29).

It has been recommended that dried algae are broadcast at the rate of 10 kg/h over the standing water in rice fields, one week after the transplantation of rice seedlings. The cost of algal material has been calculated at Rs. 30 for 10 kg sufficient for one hectare of land. The mean increase in yield of grain due to algal inoculation works out to be 300 kg for Rs. 30 investment, which is indeed a worthwhile proposition (102). In terms of nitrogen, the benefit due to algal inoculation is about 25–30 kg N/ha per cropping season. The results of field experiments (79) are stated in Table 10.

Table 10. Blue-green algal trials in farmers' fields [from Srinivasan and Ponnayya (79)]

No.	Details			Mean yield data (kg/ha)		
	N	P (kg/ha)	K	Vellore (11 trials)	Trichy (8 trials)	Cuddalore (10 trials)
1.	50	25	25	4235	4557	3789
2.	50	25	25 + Algae	4455	4785	3884
3.	75	37.5	37.5	4400	4485	3953
4.	75	37.5	37.5 + Algae	4565	5117	4156

Plate 1. *Rhizobium* and legume root nodulation. (1) Colonies of *Rhizobium* growing on yeast extract mannitol agar plates; (2) Root hair of a clover seedling showing infection thread containing *Rhizobium* cells in it; (3) A good nodulated plant of gram or chickpea (*Cicer arietinum*); (4) Leonard jar assemblies containing chickpea plants showing the benefits of *Rhizobium* inoculation (courtesy: Dr. R.B. Rewari); (5) Soybean seedlings in a pot showing the benefits of inoculation; (6) The root system of *Rhizobium* inoculated and uninoculated soybean seedlings—note profuse root nodulation on roots of *Rhizobium* inoculated seedling on the right.

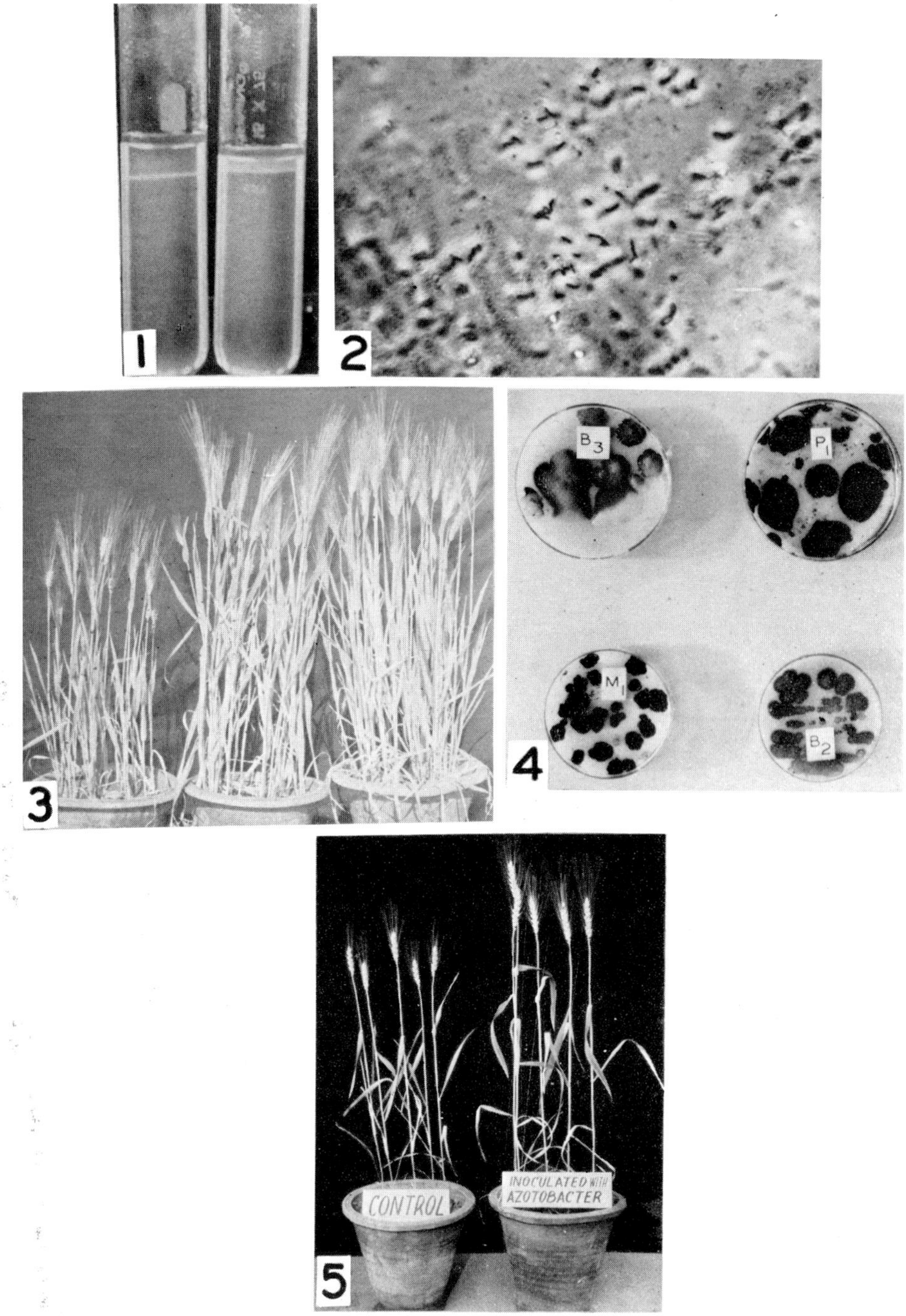

Plate 2. *Azospirillum* and *Azotobacter* effects on plants. (1) Test tubes containing malate nitrogen free medium showing characteristic pellicle formation below the surface of the semi-solid medium wherein the microaerophilic *Azospirillum* bacteria grow and fix atmospheric nitrogen (courtesy: Dr. S.K. Kavimandan); (2) Spiral or cork-screw shaped cells of *Azospirillum brasilense* as seen under a phase contrast microscope (courtesy: Dr. M. Lakshmi Kumari); (3) *Azospirillum* inoculation effects on oats (*Avena sativa*)—from left to right, pots showing control plants, *Azospirillum* inoculated plants and plants receiving 40 kg N/ha only (courtesy: Dr. K.V.B.R. Tilak); (4) *Azotobacter chroococcum* colonies on nitrogen free Jensen's media (courtesy: Dr. S.T. Shende); (5) *Azotobacter* inoculation effects on wheat (*Triticum aestivum*) in pots.

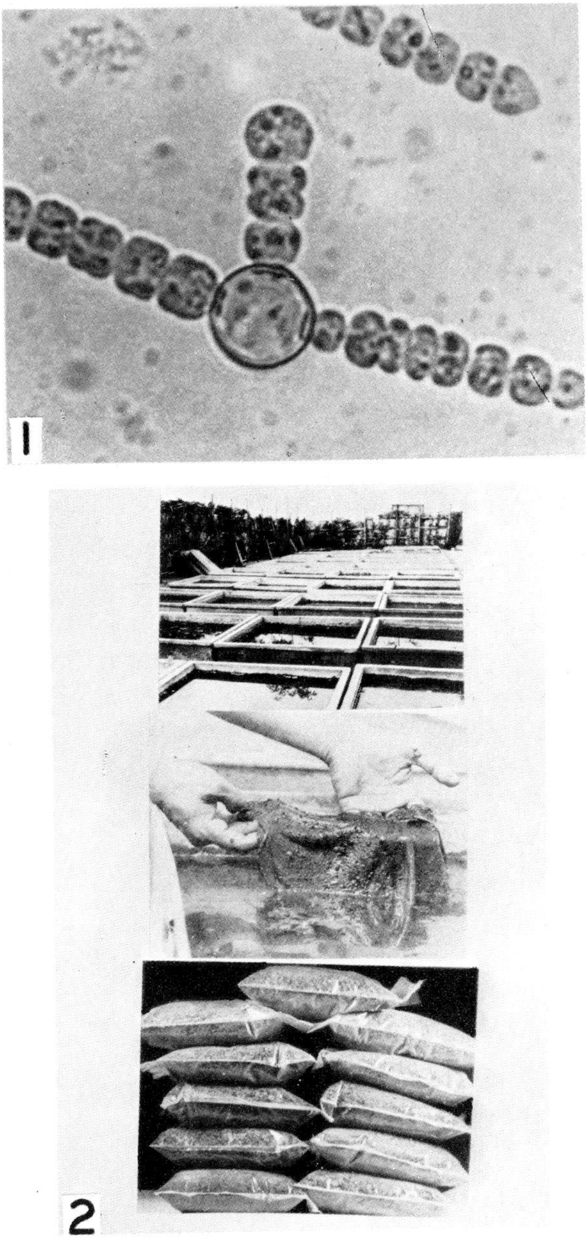

Plate 3. Blue-green algae in rice cultivation. (1) A filament of *Anabaena* sp. showing a large heterocyst (courtesy: Dr. Bansi Dhar); (2) Open air concrete tanks for cultivation of blue-green algae, also showing a mat of algal growth and plastic packets containing dried algal flakes for field application (courtesy: Dr. G.S. Venkataraman).

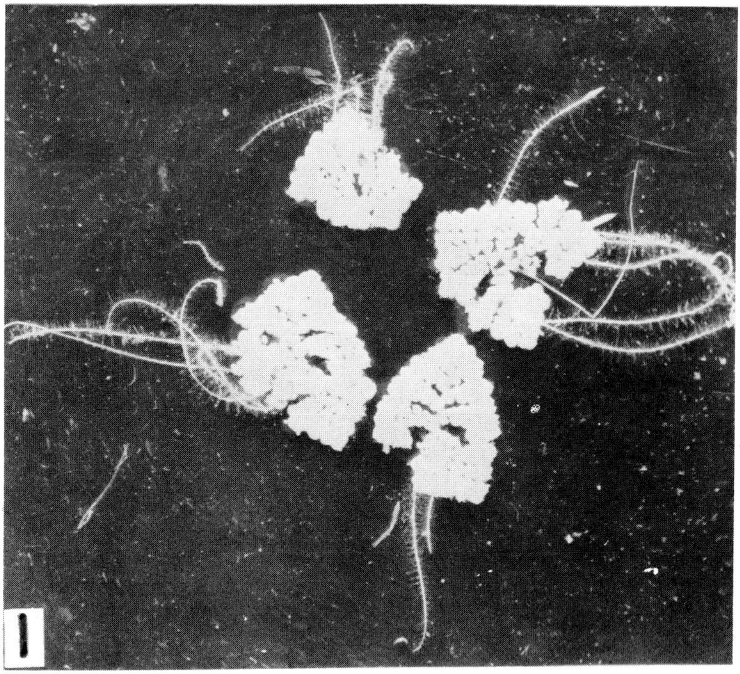

Plate 4. *Azolla* in rice cultivation. (1) Floating ferns of *Azolla pinnata* with aerial fronds and submerged rhizoids; (2) Mass multiplication method of *Azolla pinnata* in nurseries (courtesy: Dr. P.K. Singh).

Mass cultivation of *Azolla* and its use

The value of *Azolla* as a biofertilizer in rice cultivation was first demonstrated in North Vietnam, where *Azolla pinnata* is grown in 400,000 ha (98). In recent years, it has become a common input in rice cultivation in Thailand (47), Indonesia (66), China (43) and the Philippines (110).

There are six species of *Azolla*—*A. caroliniana, A. nilotica, A. filiculoides, A. mexicana, A. microphylla* and *A. pinnata.* Of these, the one commonly found in India is *A. pinnata.* The plant has a floating branched stem, bilobed leaves and true roots which penetrate water. The dorsal, fleshy chlorophyll-containing lobe has an algal symbiont, *Anabaena azollae*, within a cavity which fixes nitrogen (55).

The Chinese grow small nurseries where *Azolla* is initially grown for four weeks. At the time of rice cultivation, *Azolla* is seeded in the flooded rice fields at the rate of 7.5 t/ha. After five to ten days, the water in the field is drained and the *Azolla* is ploughed in by using tractors. This process of flooding the field, draining water and ploughing is repeated before transplanting the rice seedlings. However, *Azolla* can also be grown along with rice seedlings after transplantation when the ferns are buried with hands in the soil and the process repeated as often as necessary so that the *Azolla* mat may not cause choking of rice plants due to oxygen starvation. The Chinese methods are such that 50 per cent of the nitrogen need of rice is met by *Azolla* but this can be achieved by applying 150–225 kg/ha superphosphate (22).

In India, the Central Rice Research Institute offers the following recommendations (54, 70–76): Nurseries of *Azolla* are raised (Plate 4) in 5–10 cm standing water in microplots by inoculating fresh *Azolla* at the rate of 0.05–0.4 kg/sq m to the water body (pH 8.0; temp 14–35° C optimum) containing 4 to 20 kg P_2O_5/ha. Pesticides may be used (Furadan) to prevent insect attack; *Azolla* mat is formed in one to three weeks when it is harvested with bamboo sticks and a fresh inoculum introduced to repeat the cultivation of *Azolla* on the same plot. Harvested *Azolla* is liable to rapid decomposition (7–10 days) and hence fresh *Azolla* is inoculated at the rate of 10 t fresh mat/ha. This works out to an equivalent of 25–30 kg N/ha. Two methods of *Azolla* utilization are followed—as a green manure by incorporating in the fields prior to rice planting or by dual cropping with rice followed by the draining of water and the incorporation of *Azolla* in soil. The incorporation of *Azolla* in soil as a green manure appears to be a better method than the dual cropping of rice with *Azolla.*

Several experiments have been done at the Indian Central Rice Research Institute on the possible benefits of *Azolla* in rice cultivation using many varieties of rice (94). The incorporation of *Azolla* mat in soil as a green manure appears to be a better method than the dual cropping of rice and *Azolla.* The results of many field experiments with both high and low yielding cultivars of rice point out that *Azolla* can be used with corresponding benefits even with high doses of fertilizer nitrogen (Table 11).

Table 11. Dual cropping of *Azolla pinnata* ((Indian and Bangkok isolates) and rice variety Kalinga-2 during 1979 [from P.K. Singh (76)]

Treatment	*Azolla* growth	Rice crop response		
	Fresh wt. after 25 days of inoculation (t/ha).	Tillers/hill after 53 DAP	Effective tillers/ m^2	Grain yield (t/ha).
T_1—Control	—	5.32	224	2.1
T_2—*A. pinnata* (India) unincorporated	11.0	5.10	238	2.2 (4.2)
T_3—*A. pinnata* (India) incorporated after cover	8.3	5.43	224	2.5 (18.3)
T_4—*A. pinnata* (Bangkok) unincorporated	16.0	5.27	238	2.5 (17.2)
T_5—*A. pinnata* (Bangkok) incorporated after cover	12.9	6.30	280	2.9 (34.8)
T_6—30 kg N/ha as am. suph. (split)	—	6.99	280	2.7 (25.9)
T_7—30 kg N/ha as am. suph. split+ *A. pinnata* (India) incorporated after cover	11.0	6.79	280	3.4 (58.4)
T_8—30 kg N/ha as am. suph. (split) +*A. pinnata* (Bangkok) incorporated after cover	14.7	7.0	294	3.6 (71.3)
C.D. at 5%	4.3	1.106	46.8	0.86
C.D. at 1%	—	1.536	—	1.0

Plot size—20 m^2, *Azolla* inoculum—500 kg fresh wt/ha.
P_2O_5 dose—12 kg P_2O_5/ha (3 splits), Carbofuran—75 g/ha (3 splits).
DAP—Days after planting.
Figures in parentheses represent per cent increase over control.

References

1. Aiyar, R.S., Salahudeen, S. and Venkataraman, G.S. On a long term algalization field trial with high yielding rice varieties. Yield and economics, *Indian Journal of Agricultural Science*, *42*, 380–383 (1972).
2. Becking, J.H. Fixation of molecular nitrogen by an aerobic *Vibrio* or *Spirillum* sp., *Antonie van Leeuwenhoek Journal* of *Microbiology and Serology*, *29*, 326 (1963).

3. Brockwell, J. Studies on seed pelleting as an aid to legume seed inoculation I. Coating materials, adhesives and methods of inoculation, *Australian Journal of Agricultural Research, 13*, 638–649 (1962).
4. Brockwell, J. Application of legume seed inoculants, pp. 277–310 in *A Treatise on Dinitrogen Fixation, Section IV: Agronomy and Ecology* (Edited by R.W.F. Hardy and A.H. Gibson), John Wiley and Sons, New York (1977).
5. Brown, M. Population of *Azotobacter* in the rhizosphere and effect of artificial inoculation, *Plant and Soil, 17* (*3*), 15–25 (1962).
6. Bulow, J.F.W. von and Dobereiner, J. Potential for nitrogen fixation in maize genotypes in Brazil, *Proceedings of the National Academy of Sciences, U.S.A, 72*, 2389–2393 (1975).
7. Burris, R.H., Okon, Y. and Albrecht, S.L. Physiological studies of *Spirillum lipoferum*, pp. 445–450 in *Genetic Engineering for Nitrogen Fixation* (Edited by A. Hollaender), Plenum Press, New York (1977).
8. Burton, J.C. *Rhizobium* culture and use, in *Microbial Technology* (Edited by J. Peppler), Reinhold Publishing Corporation, New York (1967).
9. Burton, J.C. Methods of inoculating seeds and their effect on survival of rhizobia, pp. 175–189 in *Symbiotic Nitrogen Fixation in Plants* (Edited by P.S. Nutman), Cambridge University Press, Cambridge, England (1976).
10. Burton, J.C. New developments in inoculating legumes, in *Recent Advances in Biological Nitrogen Fixation* (Edited by N.S. Subba Rao), Oxford & IBH Publishing Co., New Delhi (1979).
11. Date, R.A. Legume inoculant production, *Proceedings of the Indian National Science Academy, 40* (B), 667–686 (1974).
12. Date, R.A. Principles in *Rhizobium* strain selection, pp. 137–150 in *Symbiotic Nitrogen Fixation in Plants* (Edited by P.S. Nutman), Cambridge University Press, Cambridge, England (1976).
13. Date, R.A. and Roughley, R.J. Preparation of legume seed inoculants, pp. 137–150 in *A Treatise on Dinitrogen Fixation, IV: Agronomy and Ecology*, Cambridge University Press, Cambridge, England (1977).
14. Day, J.M. and Dobereiner, J. Physiological aspects on N_2 Fixation by a *Spirillum* from *Digitaria* roots, *Soils Biology and Biochemistry, 8*, 45–50 (1976).
15. De, P.K. The role of blue-green algae in nitrogen fixation in rice fields, *Proceedings of the Royal Society, London, 1278*, 121–139 (1939).
16. Dewan, G. and Subba Rao, N.S. Seed inoculation with *Azospirillum brasilense* and *Azotobacter chroococcum* and the root biomass of rice (*Oryza sativa* L.), *Plant and Soil, 53*, 295–302 (1979).
17. Dobereiner, J. and Day, J.M. Nitrogen fixation in the rhizosphere of tropical grasses, pp. 39–56 in *Nitrogen Fixation by Free-living Microorganisms* (Edited by W.D.P. Stewart), Cambridge University Press (1975a).

18. Dobereiner, J. and Day, J.M. Associative symbioses in tropical grasses: Characterization of microorganisms and dinitrogen-fixing sites, International symposium on N_2 fixation—Interdisciplinary discussions 3–7 June, 1974, Washington State University, Washington State University Press, Pullman (1975b).
19. Dobereiner, J., Marriel, I.E. and Nergy, M. Ecological distribution of *Spirillum lipoferum* Beijerinck, *Canadian Journal of Microbiology, 22*, 1464–1473 (1976).
20. El-Nawawy, A.S., Lotfi, M. and Fahmy, H. Studies on the ability of some blue-green algae to fix atmosphere nitrogen and their effect on growth and yield of paddy, *Agricultural Research Reviews, Cairo, 36*, 308–319 (1958).
21. El-Nawawy A.S. and Hamdi, Y. Research on blue-green algae in Egypt, 1958–1972, pp. 219–228 in *Nitrogen Fixation by Free-living Microorganisms* (Edited by W.D.P. Stewart), Cambridge University Press, London (1975).
22. FAO. China: recycling of organic wastes in agriculture, Food and Agricultural Organization of the U.N., Rome (1977).
23. Fay, P. Nitrogen fixation in heterocysts, pp. 121–165 in *Recent Advances in Biological Nitrogen Fixation* (Edited by N.S. Subba Rao), Oxford & IBH Publishing Co., New Delhi (1979).
24. Fred, E.B., Baldwin, I.L. and McCoy, E. *Root Nodule Bacteria and Leguminous Plants*, University of Wisconsin, Madison, Wisconsin, U.S.A. (1932).
25. Fogg, G.E. Nitrogen fixation, pp. 161–170 in *Physiology and Biochemistry of Algae* (Edited by R.A. Lewin), Academic Press, New York (1939).
26. Gaur, A.C. and Negi, A.S. Bacterial solubilisation of phosphate as evidenced by yield of potato crop, *Plant and Soil* (in press) (1980).
27. Gaur, A.C., Ostwal, K.P. and Mathur, R.S. Save superphosphate by using phosphobacteria, *Kheti, 32*, 23–25 (1980).
28. Goyal, S.K. and Venkataraman, G.S. Response of high yielding rice varieties to algalization, II. Interaction of soil types to algal inoculation, *Phykos, 10*, 32–33 (1971).
29. IARI. Research bulletin No. 9, *Algal Technology for Rice*, Indian Agricultural Research Institute, New Delhi (1978).
30. Jensen, H.L. Notes on the biology of *Azotobacter*, *Proceedings of the Society of Applied Bacteriology, 74* (1), 89–94 (1951).
31. Joi, M.B. and Shinde, P.A. Response of onion crops to Azotobacterization, *Journal of Maharashtra Agricultural Universities, 1* (2–6), 151–162 (1976).
32. Kavimandan, S.K., Subba Rao, N.S. and Mohrir, A.V. Isolation of *Spirillum lipoferum* from different varieties of wheat and nitrogen fixation in enrichment cultures, *Current Science, 47*, 96 (1978).

33. Knowles, R. Free-living bacteria, pp. 25–40 in *Limitations and Potentials for Biological Nitrogen Fixation in the Tropics* (Edited by J. Dobereiner, R.H. Burris and A. Hollaender), Plenum Press, New York (1978).
34. Krieg, N.R. Taxonomic studies of *Spirillum lipoferum*, pp. 463–472 in *Genetic Engineering for Nitrogen Fixation* (Edited by A. Hollaender), Plenum Press, New York (1977).
35. Krieg, N.R. and Tarrand, J.J. Taxonomy of the root-associated nitrogen-fixing bacterium *Spirillum lipoferum*, pp. 317–333 in *Limitations and Potentials for Biological Nitrogen Fixation in the Tropics* (Edited by J. Dobereiner *et al.*), Plenum Press, New York (1978).
36. Kuksa, I.N. and Orleanskii, V. Development of scientific research on nitrogen fixing blue-green algae and their practical use in agriculture, *Mikrobiologiya, 34*, 743–747 (1965).
37. Lakshmi-Kumari, M., Kavimandan, S.K. and Subba Rao, N.S. Occurrence of nitrogen fixing *Spirillum* in roots of rice, sorghum, maize and other plants, *Indian Journal of Experimental Biology, 14*, 638–639 (1976).
38. Lakshmi, V., Satyanarayana Rao, A., Vijayalakshmi, K., Lakshmi-Kumari, M., Tilak, K.V.B.R. and Subba Rao, N.S. Establishment and survival of *Spirillum lipoferum, Proceedings of the Indian Academy of Science, 86* (B), 397–404 (1977).
39. Lakshmi-Kumari, M., Vijayalakshmi, K. and Subba Rao, N.S. Interaction between *Azotobacter* sp. and fungi. 1. *In vitro* studies with *Fusarium moniliforme* shield, *Phytopathology Zeitzschrift, 7, 75*, 27–30 (1972).
40. Lehri, L.K. and Mehrotra, C.L. Use of bacterial fertilizers in crop production in U.P., *Current Science, 37*, 494–495 (1968).
41. Lehri, L.K. and Mehrotra, C.L. Effect of *Azotobacter* inoculation on the yield of vegetable crops, *Indian Journal of Agricultural Research, 9* (3), 201–204 (1972).
42. Ley, S.H. The effect of nitrogen fixing blue-green algae on the yields of rice plant, *Acta Hydrobiologia, Sinica, 4*, 440–444 (1959).
43. Lin Pim-tung. Progress in the trial of an alternative to chemical fertilizer and its extension work. Sino-America Technical Cooperation Association, *21* (4), 1–3 (1976).
44. Mehrotra, C.L. and Lehri, L.K. Effect of *Azotobacter* inoculation on crop yields, *Journal of Indian Society of Soil Science, 19* (3), 243–248 (1971).
45. Mishustin, E.N. and Shilnikova, V.K. Free-living nitrogen-fixing bacteria of the genus *Azotobacter*, pp. 72–124 in *Soil Biology, Reviews of Research*, UNESCO Publication (1969).
46. Mishustin, E.N. and Shilnikova, V.K. *Biological Fixation of Atmospheric Nitrogen*, Macmillan, London (1971).
47. Moore, A.W. *Azolla*: Biology and agronomic significance, *Botanical Review, 35*, 17–34 (1969).

48. Mulder, E.G. and Brontonegoro. Free-living heterotrophic nitrogen-fixing bacteria, pp. 37–85 in *The Biology of Nitrogen Fixation* (Edited by A. Quispel), North-Holland Publishing Co., Amsterdam (1974).
49. Nayak, D.N. and Rao, V.R. Nitrogen fixation by *Azospirillum* sp. from rice roots, *Archives für Microbiology, 115*, 359 (1977).
50. Nutman, P.S. (Editor). *Symbiotic Nitrogen Fixation in Plants*, Cambridge University Press, Cambridge (1976).
51. Okon, Y., Albrecht, S.L. and Burris, R.H. Factors affecting growth and nitrogen fixation of *Spirillum lipoferum*, *Journal of Bacteriology, 127*, 1248–1254 (1976).
52. Okon, Y., Albrecht, S.L. and Burris, R.H. Methods for growing *Spirillum lipoferum* for counting it in pure cultures and in association with plants, *Applied Environmental Microbiology, 33*, 85 (1977).
53. Okon, Y., Kapulnik, Y., Sarig, S., Nur, I., Kigel, J. and Henis, Y. *Azospirillum* increases cereal crop yields in fields of Israel, *Proeeedings of the IV International Symposium on Nitrogen Fixation*, Canberra, Australia (1980).
54. Pande, H.K. Organic resource management. *Azolla:* Their potential role in developing Indian agriculture, Paper presented at the FAI Seminar, Fertilizer Association of India, New Delhi (1979).
55. Peters, G.A. The *Azolla-Anabaena azollae* symbiosis, pp. 231–258, in *Genetic Engineering for Nitrogen Fixation* (Edited by A. Hollaender), Plenum Press, New York (1977).
56. Postgate, J.R. (Editor). *The Chemistry and Biochemistry of Nitrogen Fixation*, Plenum Press, London-New York (1971).
57. Postgate, J.R. Prerequisites for biological nitrogen fixation in free-living heterotrophic bacteria, pp. 663–686 in *The Biology of Nitrogen Fixation* (Edited by A. Quispel), North-Holland Publishing Co., Amsterdam (1974).
58. Quispel, A. (Editor). *The Biology of Nitrogen Fixation*, North-Holland Publishing Co., Amsterdam (1974).
59. Rangaswami, G. and Sadasivan, K.V. Studies on the occurrence of *Azotobacter* in some soil types, *Journal of the Indian Society of Soil Science, 1* (12), 43–49 (1964).
60. Relwani, L.L. and Subramanyam, R. Role of blue-green algae, chemical nutrients and partial sterilization on paddy yield, *Current Science, 32*, 441–443 (1963).
61. Roughley, R.J. The preparation and use of legume seed inoculants, *Plant and Soil, 32*, 625–701 (1970).
62. Roughley, R.J. The production of high quality inoculants and their contribution to legume yield, pp. 125–136 in *Symbiotic Nitrogen Fixation in Plants* (Edited by P.S. Nutman), Cambridge University Press (1976).
63. Rovira, A.D. Effect of *Azotobacter*, *Bacillus* and *Clostridium* on the

growth of wheat, *Plant Microbes Relationships*, Prague, Czechoslovak Academy of Sciences, 193–200 (1965).

64. Sankaram, A. *Work done on Blue-green Algae in Relation to Agriculture*, Indian Council of Agricultural Research, New Delhi, pp. 20 (1971).
65. Sathyanarayana Rao, A. Studies on nitrogen fixing *Spirillum* species. M.Sc. thesis, P.G. School, Indian Agricultural Research Institute, New Delhi (1977).
66. Saubert, G.G.P. Provisional communication on the fixation of elementary nitrogen by a floating fern, *Annals of Royal Botanic Garden* (Buitenzorg), *51*, 177–197 (1949).
67. Shende, S.T., Apte, R.G. and Singh, T. Influence of *Azotobacter* on germination of rice and cotton seed, *Current Science*, *46*, 675 (1977).
68. Shitina, E.A. Fixation of free nitrogen in blue-green algae, pp. 66–79 in *The Ecology and Physiology of Blue-green Algae* (Edited by V.D. Federov and M.M. Tellichenko), Moscow University Press, USSR (in Russian) (1965).
69. Singh, C.S. and Subba Rao, N.S. Associative effect of *Azospirillum brasilense* and *Rhizobium Japonicum* on nodulation and yield of soybean (*Glycine max*), *Plant and Soil*, *53*, 387–392 (1979).
70. Singh, P.K. *Azolla* plants as fertilizer and feed, *Indian Farming*, *27*, 19–21 (1977a).
71. Singh, P.K. Multiplication and utilization of fern *Azolla* containing nitrogen fixing algal symbiont as a green manure in rice cultivation, *IL RISO*, *26* (2), 125–136 (1977b).
72. Singh, P.K. Effects of *Azolla* on the yield of paddy with and without application of N-fertilizer, *Current Science*, *46*, 642–644 (1977c).
73. Singh, P.K. Multiplication and utilization of fern *Azolla* containing nitrogen fixing algal symbiont as green manure in rice cultivation, *IL RISO*, *25*, 125–137 (1977d).
74. Singh, P.K. Use of *Azolla* in rice production in India, pp. 407–418 in *Nitrogen and Rice*, International Rice Research Institute, Manila, Philippines (1979).
75. Singh, P.K. Introduction of green *Azolla* biofertilizer in India, *Current Science*, *49*, 155 (1980).
76. Singh, P.K. Subject matter training-cum-discussion Seminar on the use of Biofertilizers with special reference to *Azolla* at the Central Rice Research Institute, Cuttack (1981).
77. Singh, R.N. *The Role of Blue-green Algae in Nitrogen Economy of Indian Agriculture*, Indian Council of Agricultural Research, New Delhi (1961).
78. Singh, T. Studies on interaction between *Azotobacter chroococcum* and some plant pathogens, Indian Agricultural Research Institute, Ph.D. thesis, New Delhi (1977).
79. Srinivasan, S. and Ponnayya, J.H.S. Blue-green algae as a biofertilizer—

Tamilnadu experience, Fertilizer Association of India Seminar, New Delhi (1978).

80. Stewart, W.D.P. Algal fixation of atmospheric nitrogen, *Plant and Soil*, *32*, 555–588 (1970).
81. Stewart, W.D.P. Physiological studies on nitrogen fixing blue-green algae, *Plant and Soil*, Special Volume, 377–391 (1971).
82. Stewart, W.D.P. Blue-green algae, pp. 202–287 in *The Biology of Nitrogen Fixation* (Edited by A. Quispel), North-Holland Publishing Co., Amsterdam (1974).
83. Subba Rao, N.S. Field response of legumes in India to inoculation and fertilizer application, pp. 255–268 in *Symbiotic Nitrogen Fixation in Plants* (Edited by P.S. Nutman), Cambridge University Press (1976).
84. Subba Rao, N.S. *Soil Microorganisms and Plant Growth*, Oxford & IBH Publishing Co., New Delhi (1977).
85. Subba Rao, N.S. (Editor). *Recent Advances in Biological Nitrogen Fixation*, Oxford & IBH Publishing Co., New Delhi (1979).
86. Subba Rao, N.S. and Balasundaram, V.R. *Rhizobium* inoculants for soybean, *Indian Farming*, *21*, 22–23 (1971).
87. Subba Rao, N.S. and Tilak, K.V.B.R. Rhizobial cultures—their role in pulse production, pp. 31–34, in *Souvenir Bulletin*, Directorate of Pulse Development, Government of India, Lucknow (1977).
88. Subba Rao, N.S., Tilak, K.V.B.R., Singh, C.S. and Lakshmi-Kumari, M. Response of a few economic species of graminaceous plants to inoculation with *Azospirillum brasilense*, *Current Science*, *48*, 133–134 (1979a).
89. Subba Rao, N.S., Tilak, K.V.B.R., Lakshmi-Kumari, M. and Singh, C.S. *Azospirillum*—a new bacterial fertilizer for tropical crops, *Science Reporter*, Council of Scientific and Industrial Research, India, *16* (10), 690–692 (1979b).
90. Sundara Rao, W.V.B. Phosphorus solubilization by microorganisms, pp. 21–29 in *Proceedings of the All India Symposium on Agricultural Microbiology*, University of Agricultural Sciences, Bangalore (1968).
91. Sundara Rao, W.V.B., Mann, H.S., Pal, N.S. and Mathur, R.S. Bacterial inoculation experiments with special reference to *Azotobacter*, *Indian Journal of Agricultural Science*, *33*, 279–290 (1963).
92. Subramanyan, R., Relwani, L.L. and Manna, G.B. Observations on the role of blue-green algae on rice yield compared with that of conventional fertilizers, *Current Science*, *33*, 485–486 (1964).
93. Subramanyan, R. and Sahay, M.N. Observations of nitrogen fixation by some blue-green algae and remarks on its potentialities in rice culture, *Proceedings of Indian Academy of Sciences*, *60* (B), 145–154 (1964).
94. Talley, S.N., Talley, B.J. and Rains, D.W. Nitrogen fixation by *Azolla* in rice fields, pp. 259–281 in *Genetic Engineering for Nitrogen Fixation* (Edited by A. Hollaender), Plenum Press, New York (1977).

95. Tarrand, J.J., Krieg, N.R. and Dobereiner, J. A taxonomic study of *Spirillum lipoferum* group with descriptions of a new genus *Azospirillum* gen. nov. and two species, *Azospirillum lipoferum* (Beijerinck) comb. nov. and *Azospirillum brasilense* sp. nov., *24*, 967–980 (1978).
96. Tilak, K.V.B.R., Lakshmi-Kumari, M. and Nautiyal, C. Survival of *Azospirillum brasilense* in different carriers, *Current Science*, *48*, 412–413 (1979).
97. Tilak, K.V.B.R. and Subba Rao, N.S. Carriers for legume (*Rhizobium*) inoculants, *Fertilizer News* (*India*), February issue (1978).
98. Tran, Quang Thuyet and Dao The Tuan. *Azolla:* A green manure. Agricultural Problems, 4, *Vietnamese Studies*, *No. 38*, 119–127 (1973).
99. Vancura, J. and Macura, J. The effect of root excretion on *Azotobacter*, *Folia Microbiology*, *6* (4), 250–259 (1961).
100. Vasil, I.K., Vasil, V. and Hubbell, D.H. Engineered plant cell or fungal association with bacteria that fix nitrogen, pp. 197–211 in *Genetic Engineering for Nitrogen Fixation* (Edited by A. Hollaender), Plenum Press, New York (1977).
101. Venkataraman, G.S. *Algal Biofertilizers and Rice Cultivation*, Today and Tomorrow Printers and Publishers, New Delhi (1972).
102. Venkataraman, G.S. and Kaushik, B.D. Save on 'N' fertilizers by the use of algae in rice fields, *Indian Farming*, October issue, pp. 27–30 (1980).
103. Venkataraman, G.S. and Neelakantan, S. Effect of the cellular constituents of the nitrogen fixing blue-green algae, *Cylindrospermum muscicola* on the root growth of rice seedlings, *Journal of General Applied Microbiology*, *13*, 53–61 (1967).
104. Vincent, J.M. *A Manual for the Practical Study of the Root Nodule Bacteria*, IBP Hand Book No. 15, Blackwell Scientific Publications, Oxford (1970).
105. Vincent, J.M. Root nodule symbioses with *Rhizobium*, pp. 265–341 in *The Biology of Nitrogen Fixation* (Edited by A. Quispel), North-Holland Publishing Co., Amsterdam and Oxford (1974).
106. Vincent, J.M. *Rhizhobium:* general microbiology, pp. 277–276 in *A Treatise on Dinitrogen Fixation: Section III: Biology* (Edited by R.W.F. Hardy and W.S. Silver), John Wiley and Sons, New York and London (1977).
107. Vlassak, K. and Reynolds, L. *Azospirillum* rhizocoenoses in agricultural practice, *Proceedings of the Fourth International Symposium on Nitrogen Fixation*, Canberra, Australia (1980).
108. Watanabe, A. The blue-green algae as nitrogen fixators, *Proceedings of the Ninth International Congress of Microbiology*, Moscow, 77–85 (1967).
109. Watanabe, A., Shirota, M., Endo, H. and Yamamoto, Y. An observation of the practical applications of nitrogen fixing blue-green algae for

rice cultivation, *Proceedings of the Third Global Impacts of Applied Microbiology Conference*, Bombay, pp. 53–64 (1971).

110. Watanabe, I., Espinas, C.R., Berja, N.S. and Alimagno, B.V. Utilization of the *Azolla-Anabaena* complex as a nitrogen fertilizer for rice, *Research Paper Series No. 11*, International Rice Research Institute Manila, Philippines (1977).

R. KNOWLES

9. Denitrification in Soils

Introduction

Denitrification is the process which is responsible for the release of gaseous nitrogen (N_2) which balances the global nitrogen cycle. It can also contribute nitrous oxide (N_2O) to the atmosphere and thus cause depletion of stratospheric ozone, and it is of great practical importance because it is a major mechanism by which fertilizer nitrogen is lost to the atmosphere. On the other hand, nitrogen removal in waste treatment processes aims to promote denitrification without at the same time increasing atmospheric N_2O concentrations. The process is therefore of first importance in the overall efficient conservation and utilization of nitrogen. Attempts to increase biological nitrogen fixation will be fruitless if much of the fixed nitrogen is subsequently lost to the atmosphere.

Nitrate (NO_3^-) and nitrite (NO_2^-) may undergo an assimilatory reduction to NH_4^+, a process which is repressed by NH_4^+ but not affected by O_2, and they may undergo a dissimilatory reduction to NH_4^+, the mechanism and regulation of which are not yet well understood. These oxides may also be used as electron acceptors in a dissimilatory reduction to the gaseous products NO, N_2O and N_2, a process which is repressed by O_2 but not affected by NH_4^+. The last process is referred to as denitrification and is the subject of this chapter.

The denitrification process

Denitrifying bacteria have the genetic potential to produce at least some of a series of nitrogen oxide reductases involved in the conversion of NO_3^- ultimately to N_2 (103, 108):

$$NO_3^- \xrightarrow{\text{NaR}} NO_2^- \xrightarrow{\text{NiR}} NO \xrightarrow{\text{NOR}} N_2O \xrightarrow{\text{N}_2\text{OR}} N_2$$

The interactions of these reductases with the electron transport components seems to depend on the species of bacterium. A model which was proposed for *Paracoccus denitrificans* is shown in Fig. 1, and it is seen that the nitrate reductase accepts electrons at the level of cytochrome *b*. The nitrite reductase, which in some denitrifiers itself incorporates a cytochrome *cd* but in others is a copper flavoprotein, accepts electrons from cytochrome *c* as does N_2O reductase. The participation of nitric oxide (NO) as an obligatory intermediate and of the NO reductase is still under debate (51, 138, 158).

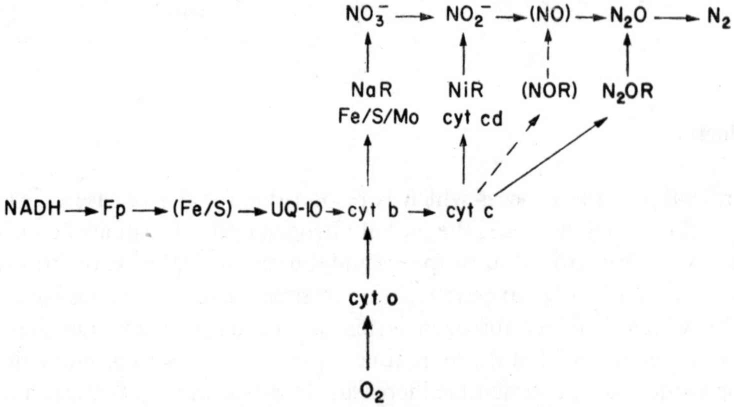

Fig. 1. Proposed scheme for electron transport from NADH to nitrogen oxides via their respective reductases in *Paracoccus denitrificans* [after Boogerd *et al.* (14)].

The nitrate reductase appears to be located on the inner face of the cytoplasmic membrane (64, 74, 86, 150) and participates in energy conservation (83). Nitrite reductases, though originally reported to be soluble, have been more recently reported to be periplasmic (91, 152) or on the inner membrane face (86, 123). This is consistent with the *in vivo* evidence of phosphorylation associated with NO_2^- reduction in *Pseudomonas denitrificans* (83). Utilization of N_2O as an electron acceptor also apparently involves energy conservation (83), consistent with a membrane location for N_2O reductase (103, 106). However, recent evidence suggests that this enzyme may be soluble (85).

Denitrifying bacteria

Many microorganisms use nitrogen oxides as electron acceptors with the evolution of gaseous products. These organisms, though apparently limited to the bacteria, are biochemically and taxonomically very diverse as is shown by the list in Table 1. Most utilize a wide variety of organic carbon compounds, some can utilize one-carbon compounds while others can grow autotrophically on H_2 and CO_2, or reduced sulphur compounds. One group is photosynthetic. While most possess all of the reductases necessary to reduce NO_3^- to N_2, some lack NO_3^- reductase (NaR) and are thus "NO_2^--dependent" (145, 157), others lack N_2O reductase (N_2OR) and thus yield N_2O as the terminal product (104). Yet others may possess the N_2OR but lack the ability to reduce NO_2^- to N_2O (154).

Two (*Azospirillum brasilense* and *Rhodopseudomonas sphaeroides*) have the ability to fix N_2 as well as to denitrify, but the ecological significance of such abilities in terms of their overall contribution to the nitrogen balance is not clear.

Factors in the regulation of denitrification

Environmental and other factors affect the overall rate of denitrification, but also, since they frequently exert differential effects on the successive reductases, they control the relative proportions of the intermediates and products which accumulate. Because of the toxicity and abiological reactions of NO_2^- (89), and the significance of N_2O in atmospheric chemistry (66), it is of great importance to understand the regulation of all of the denitrification reactions. For example, the N_2O reductase seems to be the most sensitive to low pH and the presence of O_2, NO_3^- and sulphide, and N_2O will therefore be expected as a significant product of denitrification under such conditions (52). The following discussion examines these regulating factors in more detail.

Oxygen

All of the nitrogen oxide reductases are repressed by oxygen which also inhibits preformed reductase, probably by competing for electrons. The reduction of NO_3^- to NO_2^- seems to be less sensitive to O_2 than are the later reductions (87, 122). Thus, an increasing proportion of N_2O is released as denitrification is inhibited by increasing exposure to O_2. This phenomenon has been incorporated in a model by Focht (54).

To describe the distribution of O_2 concentrations in such a complex system as soil is extremely difficult because of the existence of aggregates and the necessity to consider both intra-aggregate and inter-aggregate porosities (39). Such a treatment has been attempted mathematically in the hope that it will lead to a better understanding of anaerobic microsites in aggregated soils (125).

After the onset of anaerobiosis there are changes in the absolute and relative activities of the denitrifying reductases in soil (128). For up to three hours, activities apparently reflect the endogenous pre-existing conditions and N_2 is a dominant product. However, the NO_3^- and NO_2^- reductases then undergo sooner and further derepression than does the N_2O reductase such that N_2O becomes the major product. After about one to two days the activity of N_2O reductase increases and from this point N_2 again becomes the major product (53, 128). This points out the complexity of the study of soil systems in which O_2 availability is periodically changing.

The existence of anaerobic microenvironments in soil depends greatly on the moisture content of the soil since this is a major factor in limiting the intra- and inter-aggregate porosities.

Water content

Reports of denitrification occurring in soils of low water content (49) have usually been ascribed to the existence of anaerobic microsites. Most studies suggest that denitrification in air occurs only at water contents above 60 per cent of maximum water-holding capacity (19, 96). For a particular water content, denitrification increases with decreasing O_2 concentration, and for a particular O_2 level the denitrification increases with increasing water content. The absence of activity at low water contents and high O_2 concentrations is a feature of most reports in the literature (e.g. 19, 96, 110). An increase in activity is frequently observed between 100 and 200 percent of maximum water-holding capacity (19), probably due to the presence, at 100 percent water, of air-filled inter-aggregate pores providing high O_2 diffusion rates (39). These pores become water-filled at 200 percent water, restricting O_2 diffusion and increasing the volume of anaerobic microenvironments in which denitrification can occur (88).

Air drying and storage of dried soils increases the rate of denitrification when they are subsequently returned to optimal denitrifying conditions, but this is thought to be due to the release of available organic carbon on drying and wetting of soils (101), a soil phenomenon which has been recognized for many years (9).

Organic carbon availability

The most abundant denitrifiers (Table 1) are heterotrophs which require sources of electrons or reducing equivalents contained in soil organic matter, plant residues and root exudates. Soil microorganisms in general are usually limited by organic carbon and energy and it is not surprising that denitrification is usually stimulated by the addition of exogenous carbon (15, 57). Rates

Table 1. Genera of bacteria at least some strains of which have been reported to denitrify ionic nitrogen oxides to gaseous products*

Achromobacter	*Moraxella*
Agrobacterium	*Neisseria*
***Alcaligenes*	*Paracoccus*
Azospirillum	*Propionibacterium*
***Bacillus*	***Pseudomonas*
Corynebacterium	*Rhizobium*
Flavobacterium	*Rhodopseudomonas*
Halobacterium	*Thiobacillus*
Hyphomicrobium	*Zanthomonas*

*For further details of species and their characteristics, with references, see (81.)

**Probably numerically the most important denitrifiers in soils (56).

of denitrification are highly correlated with amounts of water-extractable soil organic carbon (19, 23) and of extractable reducing sugars (136). Interestingly, it is reported that different organic compounds which support equal rates of denitrification may, however, give different mole fractions of N_2O in the products (17), possibly because of differential effects on the reductases involved. On the other hand, it was reported that addition of carbon did not increase the rate of denitrification of $^{13}NO_3^-$ in very short-term experiments (141), suggesting that carbon availability was not rate limiting under the particular conditions employed.

The dependence on organic carbon results in higher denitrification potentials being found in surface soils. Such potentials, measured under anaerobic conditions, decrease greatly below the 20 cm depth (81). The effect of drying and wetting cycles on the release of available carbon and consequently on the denitrification potential was referred to above.

Nitrate concentration

When soils are exposed to relatively high concentrations of NO_3^--N (more than about 40 μg N g^{-1} or ml^{-1}) the denitrification reaction is frequently found to be zero-order (82, 124). This appears to be due to the low apparent K_M for NO_3^- reduction which has been reported to be below 2 μg NO_3^--Ng^{-1} and much lower for the reduction of N_2O (see 156 for references). Thus for many soils which contain endogenous NO_3^- at concentrations greater than 10 μg N g^{-1}, there is no great effect of a further NO_3^- supplement on the rate of denitrification.

Although NO_3^- can exert rate control at low concentrations, it can have other effects. Higher concentrations, for example, may (i) increase the rate of diffusion of NO_3^- to anaerobic microsites (25) and thus affect the apparent kinetics of NO_3^- reduction in unstirred water-saturated systems (107, 114), (ii) cause more prolonged poising of the E_h at about $+200$ mV which might otherwise drop to about -300 mV (5), (iii) inhibit the enzymatic reduction of NO and thus cause accumulation of intermediates (105), and (iv) inhibit reduction of N_2O and thus cause a greater mole fraction of N_2O in the products (10, 38, 88, 96, 140). Low concentrations of NO_3^- (5–10 μg N g^{-1}), on the other hand, promote derepression of the reductases and stimulate N_2O reduction in soil which has not been exposed previously to NO_3^- (11).

Some of the abovementioned effects may also be shown by NO_2^- and may be attributed to competition between different nitrogen oxides for electrons. For example, NO_2^- causes a lag in the reduction of $^{15}NO_3^-$ (96).

pH

The potential for denitrification in soil is positively correlated with pH (94), having an optimum between 7 and 8 (96, 144, 151). In an acid peat, the low pH (3.5) was the only factor which prevented denitrification from occurring (79). However, the N_2O reductase system seems to be more sensitive to low pH than are the other reductases, such that the mole fraction of N_2O increases as pH falls, and at about pH 4 N_2O is reported to be the major product (96, 151). Thus, both decreasing pH and increasing O_2 tend to decrease the overall denitrification rate while at the same time increasing the proportion of N_2O in the products evolved (54).

At low pH values a possible complication is that any NO_2^- which accumulates may undergo not only enzymatic reduction but several abiological reactions. These reactions (81) can yield a variety of products, including NO, N_2, N_2O, and CH_3NO_2.

Temperature

In the range 10 to 35°C the rate of denitrification in soil is very temperature-dependent (6, 96) with a Q_{10} close to 2 (134). The rate continues to increase at higher temperatures reaching a maximum at 60 to 75°C and then falling to zero above this temperature (19, 78). Whether this reaction at high temperature is due to thermophilic organisms or is an abiological reaction is not clear (78). The mole fraction of N_2O in the denitrification products at high temperatures tends to be very high (78, 81).

At lower temperatures the denitrification rate decreases further but is nevertheless measurable even in the range 0 to 5°C (6, 19, 124). Also, low temperatures are reported to result in the production of relatively more N_2O (96) and NO (4) but whether this is a general phenomenon is not certain.

The marked temperature dependence in soils is in contrast to observations of sediments where there is little variation with temperature, at least in the range 5 to 23°C (2, 30, 75).

Inhibitory substances

Acetylene strongly inhibits the reduction of N_2O but not the other reductases involved in denitrification (7, 50, 156). It has subsequently been shown that this provides a valid basis for measurements of denitrification in soils (79, 102, 120, 127, 155) and sediments (80, 130) in the laboratory, as well as in *in situ* studies in soil (100, 121) and sediment (31, 131). In using C_2H_2 as a measurement tool, however, it should be remembered that it has several side effects (Table 2) the implications of which, for the measurement of denitrification, are not always clear. Probably the most significant is the inhibition of nitrification (72) at the level of the oxidation of NH_4^+ to hydroxylamine (Hynes and Knowles, unpublished data). This invalidates the use of C_2H_2 for the measurement of nitrification-denitrification coupled reactions.

Sulphur compounds can also affect the denitrification process. Sulphate (100 to 500 μg SO_4^{2-}-S g^{-1}) was reported to delay the disappearance of NO_3^- in soil (84). Sulphide, in particular, was reported to inhibit soil denitrification (95) and more specifically the reduction of NO and N_2O to N_2 (133, 139).

Sulphide at about 8 μmol g^{-1} delays the reduction of N_2O by soil whereas thiosulphate has no such effect (139), the possibility therefore arising that relatively sulphidic denitrifying habitats could release NO and N_2O as significant products (132).

Table 2. Biological effects of acetylene

Effect	References
Inhibition of:	
N_2O reduction by denitrifiers	7, 156
N_2 fixation (can impose N limitation)	27
N_2ase catalyzed H_2 evolution	71
Proliferation of clostridia	22
Hydrogenases: conventional	126
uptake	32, 126
NH_4^+ oxidation: by *Nitrosomonas*	72
by soil	73, 147
Methanogenesis	112
Methane oxidation	44
Utilization of C_2H_2 *by:*	
Nocardia rhodochrous (aerobic)	77
Aerobic soils	60, 61
Anaerobic soil and sediment	43, 148

Sulphur compounds exert another effect of possible importance for the use of the C_2H_2 inhibition technique in the measurement of denitrification. In anaerobic systems, the presence of sulphide, or to a lesser extent thiosulphate from which sulphide may be readily produced, can cause the relief of the C_2H_2 inhibition of N_2O reduction (139). For example, in an anaerobically incubated laboratory soil system N_2O was reduced rapidly when added alone but not at all when added with C_2H_2. However, the presence of about 8 μmol sulphide or thiosulphate per g permitted the complete reduction of the N_2O within three or six days, respectively. This suggests that the accumulation of N_2O in C_2H_2-treated systems over relatively long time periods may not be a valid measure of denitrification if sulphide is present.

The nitrification inhibitor nitrapyrin (N-Serve) was reported to inhibit denitrification ia a paddy soil by 30 percent at a concentration of 14 ppm (93). However, in another study, 50 ppm of this compound inhibited denitrification in a pure culture of a pseudomonad but had no effect in a soil system (70). Pesticides which have been reported to inhibit soil denitrification are the dithiocarbamate VAPAM at 20 ppm (93), dalapon at 10 ppm (149) and certain toluidine derivatives (13) the latter causing accumulation of NO_2^- and N_2O. Some other pesticides tested did not affect denitrification (13, 93).

Methods of measurement

A detailed discussion of the problems inherent in many of the methods by

Table 3. Methods for the measurements of denitrification in soils

Method	Comments	References
N or ^{15}N balance	Must be relatively long-term. Low precision	1, 20, 33,
	Requires addition of N or ^{15}N	55, 115
Disappearance of NO_3^- or NO_2^-	Valid in some conditions	96, 151
	May be assimilated or reduced to NH_4^+, especially when C-rich and highly anaerobic	28, 29, 36, 135
Disappearance of added N_2O	Non-disturbing. Saturation of N_2O-reducing sites may overestimate rate. Nitrate or NO_2^- may compete for electrons	58, 59
Production of N_2O and (or) N_2	Requires closed system with air N_2 replaced by Ar or He. Changes O_2 concentrations	8
Production of $^{15}N_2O$ and (or) $^{15}N_2$ from $^{15}NO_3^-$, in lab. systems	Commonly used in closed systems	
	Precision relatively high but tedious and expensive	65, 69, 97, 98
in the field	Fluxes of ^{15}N gases determined in short-term enclosures. Insensitive	116
Reduction of $^{13}NO_3^-$ to $^{13}N_2O$ and (or) $^{13}N_2$	High sensitivity but very short (10 min) half-life. Used mainly in stirred slurries to promote mixing	62, 143
Production of N_2O in the presence and absence of C_2H_2	N_2O reductase is inhibited by C_2H_2 and N_2O accumulates	7, 156
in lab. systems	Appears to be reliable in 0.1 atm C_2H_2	79, 120, 127, 155
in the field	Flow-through or gas-tight enclosures	100, 120
	Incomplete inhibition sometimes observed	60, 61, 153
	C_2H_2 may be utilized, or inhibit nitrification and other processes	See Table 2

which denitrification has been measured is beyond the scope of this chapter, but may be found in recent articles (68, 81, 142). Table 3 summarizes the major methods which are available and comments on the more important advantages or disadvantages of each. Selected key references indicate sources of additional information. It is clear that balance studies and methods involving the determination of rates of disappearance of nitrogen oxides or production of N_2O or N_2 are all fraught with difficulty. If ^{13}N- or ^{15}N-labelled nitrogen oxide is added, certain difficulties are removed but such studies are expensive and relatively complex, such that adequate replication may be difficult. The measurement of the production of N_2O in the presence of C_2H_2 gives an estimate of the total amount of N_2O or N_2 which would be produced in the absence of C_2H_2. A parallel measure of N_2O production in the absence of C_2H_2 permits one to calculate the natural mole fraction of N_2O released in the products of denitrification:

$$\frac{N_2O}{N_2O+N_2}=\frac{N_2O \text{ in absence of } C_2H_2}{N_2O \text{ in presence of } C_2H_2}.$$

Since N_2O can be determined gas chromatographically by electron capture detection about 3,000 times more sensitively than by thermal conductivity detection, the potential exists for short-term measurements of one to four hours both in the laboratory and in the field.

Steady state, batch type, and *in situ* soil systems

In an attempt to develop models of nitrogen transformations in soils, steady state continuous flow columns have been used in the laboratory. The geometry and other conditions associated with such systems have varied widely (Table 4). Waterlogged conditions with NO_3^- in the influent showed denitrification exclusively, while an air space with influent NH_4^+ supported both nitrification and denitrification occurring simultaneously. Denitrification data were best fitted assuming first-order kinetics for some conditions (117), although others were best fitted with zero-order kinetics (48). Zones of nitrification were usually near the surface, with denitrification occurring in an underlying zone (137), however, a positive air flow through water-saturated aggregates not supplemented with organic carbon apparently inhibited denitrification, such that no N_2O was detected in the presence of inhibitory C_2H_2 (73).

Steady state columns permit mathematical treatments and the testing of assumptions. However, they do not mimic, very closely, agricultural systems which are usually of a batch type with pulsed inputs and fluctuating environmental conditions. The normal situation in agricultural systems is periodic flushes of denitrifying activity at times when conditions are most favourable for this process.

The wetting and drying events, or periodic flooding cycles, which occur in

Table 4. The conditions employed in some selected continuous-flow soil column studies of nitrogen transformations including denitrification

Column size (cm)	Influent solution (μg N ml^{-1})	Influent solution: Others	Water status (cm H_2O suction)	Pore vel. cm/h (flow rate ml h^{-1})	References
7.6 × 35	50–500 $(NH_4)_2SO_4$	+0.01 N $CaSO_4$	Saturated 85	(3)	92
13 × 15 × 100	50 NH_4Cl	+0.01 N $CaSO_4$	Saturated (10% air space)	0.25	137
6.0 × 10	100 ?NO_3^-	+0.01 N $CaSO_4$	Unsaturated + N_2	1.2–2.9	146
2.5 × (5–50)	10–100 KNO_3	+0.001 N $CaCl_2$	Waterlogged (bottom aerobic?)	0.2–1.34	47, 48
20 × 125	200 $Ca(NO_3)_2$	+0.6% glucose	Waterlogged	(15–30)	3
15.8 × 100	300 kgN/ha KNO_3	pulse	Saturated 24–75	0.05–0.1	117
3.3 × 60	10 KNO_3	none	Saturated (31% air space)	3.9–4.3	73
8.4 × 50	50 ?NO_3^-	+0.01 N $CaSO_4$	Saturated 50–100	(7)	34

many agricultural systems stimulate the occurrence of both nitrification and denitrification (21, 99, 113) and give rise to occasional short flushes of activity as well as longer-term seasonal variations. Fluxes of N_2O from soils in the field show marked variations, with peak rates in the range 0.01 to 0.1 kg N_2O-N $ha^{-1}d^{-1}$ (16, 25, 26, 119), but temperature-dependent diurnal peaks of 0.4 kg N_2O-N $ha^{-1}d^{-1}$ in the early afternoon have been reported from irrigated and fertilized vegetable plots (119). Direct-drilled soil showed peaks up to 0.7 kg N_2O-N $ha^{-1}d^{-1}$ (24). Fluxes of $^{15}N_2O$ and $^{15}N_2$ from $^{15}NO_3^-$ (300 kg N ha^{-1}) measured in temporary enclosures over one- to two-hour periods (116) also showed marked time-dependent variations with peak fluxes of N_2O and N_2 up to 0.4 and 60 kg N $ha^{-1}d^{-1}$, respectively. The N_2O flux represented 5 to 26 percent of the total denitrification, which in turn represented 1 to 73 per cent of the fertilizer added (116). Field measurements using the C_2H_2 inhibition method also showed sharp peaks up to 3.6 kg N denitrified $ha^{-1}d^{-1}$. The N_2O flux represented 13 to 30 percent of the total denitrification, which in turn represented 14 to 52 percent of the fertilizer added (118).

Fluxes of N_2O from soils are observed even at relatively low moisture contents which are considered too low to permit denitrification (35, 45, 46). Such losses are frequently greater from the nitrifiable fertilizers, NH_4^+ and urea, than from NO_3^- (16, 37) and may amount to as much as 0.03 kg N_2O-N ha^{-1} d^{-1} (37). It appears that this N_2O production in dry soils is due to nitrification (18) since NH_4^+-oxidizing *Nitrosomonas, Nitrosolobus, Nitrosospira* and *Nitrosococcus* can all produce small quantities of N_2O at high as well as at low O_2 concentrations (12, 63). Although this source of N_2O may be important for atmospheric chemistry it apparently represents only about a 0.1 per cent loss of applied fertilizer nitrogen (16).

Global aspects

The nitrogen cycle

The global nitrogen cycle is believed to be in balance, but the evidence in support of this assumption is not very strong because of the large uncertainties associated with the estimate of many global flux rates. These uncertainties are sometimes a factor of two or even an order of magnitude, and this is illustrated by the ranges given in Table 5 which are derived from four recent compilations (41, 67, 111, 129). Denitrification is clearly a very critical global process, and is the major factor balancing the nitrogen fixation occurring in biological, industrial, combustion, and atmospheric processes. The high maximum for denitrification in the sea is largely a result of an attempt to accommodate some high estimates of N_2O fluxes out of the oceans (67).

Fertilizers, nitrous oxide, and destruction of stratospheric ozone

Nitrous oxide which escapes from denitrifying systems or from other

Table 5. The global nitrogen balance (the units are Tg yr^{-1}. One Tg= 10^{12}g=10^6 metric tons)

Flux	Land	Sea	Global
Inputs:			
N_2 fixation: biological	149–180[a]	1–130	150–310
industrial	40–57	0	40–57
combustion	20	0	20
atmospheric	5	5	10
River run-off		10–40	10–40
Total	214–262	16–175	230–437
Outputs:			
Denitrification etc.: N_2+N_2O	108–185	25–270	133–455
River run-off	10–40		10–40
To sediments		10	10
Total	118–225	35–280	153–505

[a]The ranges given are minimum and maximum estimates derived from four sources (41, 67, 111, 129). Fluxes of NH_3, NH_4^+, organic-N and some NO_x in dust and precipitation etc. are not included since inputs and outputs are assumed to balance.

sources such as nitrification (18) or combustion (109) contributes to the weak gradient of N_2O in the troposphere and diffuses into the stratosphere above about 16 km. Here it undergoes the following photochemical reaction

$$N_2O+O(^1D) \rightarrow 2\,NO$$

and the NO released catalyzes the follwing reactions

$$NO+O_3 \rightarrow NO_2+O_2$$
$$NO_2+O \rightarrow NO+O_2$$

whose net result is

$$O_3+O \rightarrow 2O_2$$

Thus it was realized in the early 1970's (40, 76) that an increase in the use of inorganic fertilizers (from 40 Tg in 1974 to 150 Tg N by the year 2000) could lead to an increase in the N_2O flux to the stratosphere and bring about the destruction of stratospheric ozone by 5 to 20 percent (90).

On the other hand, stratospheric chemistry is not yet completely understood and hitherto undescribed reactions are occasionally reported. One such series of reactions is the following (42)

$$NO+HO_2 \rightarrow NO_2+HO$$
$$NO_2+h\nu \rightarrow NO+O$$
$$O+O_2+M \rightarrow O_3+M$$

The net result of these reactions is

$$HO_2+O_2 \rightarrow HO+O_3$$

which will tend to counteract the undesirable effects of the O_3-destructive reactions brought about by N_2O and NO above.

Conclusion

It is clear that before one can predict, with any precision, where and to what extent denitrification will occur, one must learn much more about the biochemical processes involved in denitrification, and the ways in which it is regulated. It is only during the last few years that relatively convenient assay methods have become available, and we may hope that as data accumulate, our understanding of the process of denitrification at the biochemical, *in vivo*, soil, farm, and global levels will improve. An ultimate objective must be to manage the nitrogen cycle in such a way that excessive losses of nitrogen from soil or excessive contributions of N_2O to the atmosphere do not occur.

References

1. Allison, F.E. The enigma of soil nitrogen balance sheets, *Advances in Agronomy*, *7*, 213–250 (1955).
2. Anderson, J.M. Rates of denitrification of undisturbed sediment from six lakes as a function of nitrate concentration, oxygen and temperature, *Archiv für Hydrobiologie*, *80*, 147–159 (1977).
3. Ardakani, M.S., Belser, L.W. and McLaren, A.D. Reduction of nitrate in a soil column during continuous flow, *Soil Science Society of America Proceedings*, *39*, 290–294 (1975).
4. Bailey, L.D. Effects of temperature and root on denitrification in a soil, *Canadian Journal of Soil Science*, *56*, 79–87 (1976).
5. Bailey, L.D. and Beauchamp, E.G. Nitrate reduction, and redox potentials measured with permanently and temporarily placed platinum electrodes in saturated soils, *Canadian Journal of Soil Science*, *51*, 51–58 (1971).
6. Bailey, L.D. and Beauchamp, E.G. Effects of temperature on NO_3^- and NO_2^- reduction, nitrogenous gas production, and redox potential in a saturated soil, *Canadian Journal of Soil Science*, *53*, 213–218 (1973).
7. Balderston, W.L., Sherr, B. and Payne, W.J. Blockage by acetylene of nitrous oxide reduction in *Pseudomonas perfectomarinus*, *Applied and Environmental Microbiology*, *31*, 504–508 (1976).
8. Barbarre, J.M. and Payne, W.J. Products of denitrification by a marine bacterium as revealed by gas chromatography, *Marine Biology*, *1*, 136–139 (1967).
9. Birch, H.F. The effect of soil drying on humus decomposition and nitrogen availability, *Plant and Soil*, *10*, 9–31 (1958).
10. Blackmer, A.M. and Bremner, J.M. Inhibitory effect of nitrate on reduction of N_2O to N_2 by soil microorganisms, *Soil Biology and Bioche-*

mistry, *10*, 187–191 (1978).

11. Blackmer, A.M. and Bremner, J.M. Stimulatory effect of nitrate on reduction of N_2O to N_2 by soil microorganisms, *Soil Biology and Biochemistry*, *11*, 313–315 (1979).
12. Blackmer, A.M., Bremner, J.M. and Schmidt, E.L. Production of nitrous oxide by ammonia-oxidizing chemoautotrophic microorganisms in soil, *Applied and Environmental Microbiology*, *40*, 1060–1066 (1980).
13. Bollag, J.M. and Kurek, E.J. Nitrite and nitrous oxide accumulation during denitrification in the presence of pesticide derivatives, *Applied and Environmental Microbiology*, *39*, 845–849 (1980).
14. Boogerd, F.C., van Verseveld, H.W. and Stouthamer, A.H. Electron transport to nitrous oxide in *Paracoccus denitrificans*, *FEBS Letters*, *113*, 279–284 (1980).
15. Bowman, R.A. and Focht, D.D. The influence of glucose and nitrate concentrations upon denitrification rates in a sandy soil, *Soil Biology and Biochemistry*, *6*, 297–301 (1974).
16. Breitenbeck, G.A., Blackmer, A.M. and Bremner, J.M. Effects of different nitrogen fertilizers on emission of nitrous oxide from soil, *Geophysical Research Letters*, *7*, 85–88 (1980).
17. Bremner, J.M. Role of organic matter in volatilization of sulphur and nitrogen from soils, pp. 229–240, in *Soil Organic Matter Studies, Volume II*, International Atomic Energy Agency, Vienna (1977).
18. Bremner, J.M. and Blackmer, A.M. Effects of acetylene and soil water content on emission of nitrous oxide from soils, *Nature*, *280*, 380–381 (1979).
19. Bremner, J.M. and Shaw, K. Denitrification in soil. II. Factors affecting denitrification, *Journal of Agricultural Science*, *51*, 40–52 (1958).
20. Broadbent, F.E. and Clark, F.E. Denitrification, pp. 344–359 in *Soil Nitrogen* (Edited by W.V. Bartholemew and F.E. Clark), American Society of Agronomy Inc., Madison (1965).
21. Broadbent, F.E. and Tusneem, M.E. Losses of nitrogen from some flooded soils in tracer experiments, *Soil Science Society of America Proceedings*, *35*, 922–926 (1971).
22. Brouzes, R. and Knowles, R. Inhibition of growth of *Clostridium pasteurianum* by acetylene: implication for nitrogen fixation assay, *Canadian Journal of Microbiology*, *17*, 1483–1489 (1971).
23. Burford, J.R. and Bremner, J.M. Relationships between denitrification capacities of soils and total water-soluble and readily decomposable soil organic matter, *Soil Biology and Biochemistry*, *7*, 389–394 (1975).
24. Burford, J.R., Dowdell, R.J. and Crees, R. Emission of nitrous oxide to the atmosphere from direct-drilled and ploughed clay soils, *Journal of the Science of Food and Agriculture*, *32*, 219–223 (1981).
25. Burford, J.R. and Greenland, D.J. Denitrification under an annual pas-

ture, in *International Grassland Conference*, Surfers Paradise, Queensland University Press, St. Lucia, Australia (1970).
26. Burford, J.R. and Millington, R.J. Nitrous oxide in the atmosphere of a red-brown earth, pp. 505–511 in *International Congress of Soil Science*, Adelaide, Vol. 2 (1968).
27. Burns, R.C. and Hardy, R.W.F. *Nitrogen Fixation in Bacteria and Higher Plants*, Springer-Verlag, Heidelberg (1975).
28. Caskey, W.H. and Tiedje, J.M. Evidence for clostridia as agents of dissimilatory reduction of nitrate to ammonium in soils, *Soil Science Society of America Journal*, *43*, 931–935 (1979).
29. Caskey, W.H. and Tiedje, J.M. The reduction of nitrate to ammonium by a *Clostridum* sp. isolated from soil, *Journal of General Microbiology*, *119*, 217–223 (1980).
30. Cavari, B.Z. and Phelps, G. Denitrification in Lake Kinneret in the presence of oxygen, *Freshwater Biology*, *7*, 385–391 (1977).
31. Chan, Y.K. and Knowles, R. Measurement of dentrification in two freshwater sediments by an *in situ* acetylene inhibition method, *Applied and Environmental Microbiology*, *37*, 1067–1072 (1979).
32. Chan, Y.K., Nelson, L.M. and Knowles, R. Hydrogen metabolism of *Azospirillum brasilense* in nitrogen-free medium, *Canadian Journal of Microbiology*, *26*, 1126–1131 (1980).
33. Chichester, F.W. and Smith, J.J. Disposition of ^{15}N-labeled fertilizer nitrate applied during corn culture in field lysimeters, *Journal of Environmental Quality*, *7*, 227–233 (1978).
34. Christensen, S. Percolation studies on denitrification, *Acta Agricultura Scandinavica*, *30*, 225–236 (1980).
35. Christianson, C.B., Hedlin, R.A. and Cho, C.M. Loss of nitrogen from soil during nitrification of urea, *Canadian Journal of Soil Science*, *59*, 147–154 (1979).
36. Cole, J.A. and Brown, C.M. Nitrite reduction to ammonia by fermentative bacteria: A short circuit in the biological nitrogen cycle, *FEMS Microbiology Letters*, *7*, 65–72 (1980).
37. Conard, R. and Seiler, W. Field measurements of the loss of fertilizer nitrogen into the atmosphere as nitrous oxide, *Atmospheric Environment*, *14*, 555–558 (1980).
38. Cooper, G.S. and Smith, R.L. Sequence of products formed during denitrification in some diverse western soils, *Soil Science Society of America Proceedings*, *27*, 659–662 (1963).
39. Craswell, E.T. and Martin, A.E. Effect of moisture content on denitrification in a clay soil, *Soil Biology and Biochemistry*, *6*, 127–129 (1974).
40. Crutzen, P.J. The influence of nitrogen oxides on atmospheric ozone content, *Quarterly Journal of the Royal Meteorological Society*, *96*, 320–325 (1970).

41. Crutzen, P.J. Upper limits on atmospheric ozone reductions following increased application of fixed nitrogen to the soil, *Geophysical Research Letters*, *3*, 169–172 (1976).
42. Crutzen, P.J. and Howard, C.J. The effect of the HO_2+NO reaction rate constant on one-dimensional model calculations of stratospheric ozone perturbations, *Pure and Applied Geophysics*, *116*, 497–510 (1978).
43. Culbertson, C.W., Zehnder, A.J.B. and Oremland, R.S. Anaerobic oxidation of acetylene by estuarine sediments and enrichment cultures, *Applied Environmental Microbiology*, *41*, 396–403 (1981).
44. De Bont, J.A.M. and Mulder, E.G. Invalidity of the acetylene reduction assay in alkane-utilizing, nitrogen-fixing bacteria, *Applied and Environmental Microbiology*, *31*, 640–647 (1976).
45. Denmead, O.T., Freney, J.R. and Simpson, J.R. Studies of nitrous oxide emission from a grass sward, *Soil Science Society of America Journal*, *43*, 726–728 (1979).
46. Denmead, O.T., Freney, J.R. and Simpson, J.R. Nitrous oxide emission during denitrification in a flooded field, *Soil Science Society of America Journal*, *43*, 716–718 (1979).
47. Doner, H.E., Volz, M.G., Belser, L.W. and Løken, J.P. Short term nitrate losses and associated microbial populations in soil columns, *Soil Biology and Biochemistry*, *7*, 261–263 (1975).
48. Doner, H.E., Volz, M.G. and McLaren, A.D. Column studies of denitrification in soil, *Soil Biology and Biochemistry*, *6*, 341–346 (1974).
49. Ekpete, D.M. and Cornfield, A.H. Losses, through denitrification, from soil of applied inorganic nitrogen even at low moisture contents, *Nature*, *201*, 322–323 (1964).
50. Fedorova, R.I., Milekhina, E.I. and Il'yukhina, N.I. Evaluation of the method of "gas metabolism" for detecting extraterrestrial life. Identification of nitrogen-fixing microorganisms, Izvestia Akademia Nauk, SSSR, *Ser. Biol.*, 1973 (6), 797–806 (1973).
51. Firestone, M.K., Firestone, R.B. and Tiedje, J.M. Nitric oxide as an intermediate in denitrification: evidence from nitrogen-13 isotope exchange, *Biochemical and Biophysical Research Communications*, *91*, 10–16 (1979).
52. Firestone, M.K., Firestone, R.B. and Tiedje, J.M. Nitrous oxide from soil denitrification: factors controlling its biological production, *Science*, *208*, 749–751 (1980).
53. Firestone, M.K. and Tiedje, J.M. Temporal changes in nitrous oxide and dinitrogen from denitrification following onset of anaerobiosis, *Applied and Environmental Microbiology*, *38*, 673–679 (1979).
54. Focht, D.D. The effect of temperature, pH, and aeration on the production of nitrous oxide and gaseous nitrogen—a zero-order kinetic model, *Soil Science*, *118*, 173–179 (1974).

55. Focht, D.D. and Stolzy, L.H. Long-term denitrification studies in soil fertilized with $(NH_4)_2$ SO_4-N15, *Soil Science Society of America Journal*, *42*, 894–898 (1978).
56. Gamble, T.N., Betlach, M.R. and Tiedje, J.M. Numerically dominant denitrifying bacteria from world soils, *Applied and Environmental Microbiology*, *33*, 926–939 (1977).
57. Garcia, J.L. Influence de la rhizosphère du riz sur l'activité dénitrifiante potentielle des sols de rizières du Sénégal, *Oecologia Plantarum*, *8*, 315–323 (1973).
58. Garcia, J.L. Réduction de l'oxyde nitreux dans les sols de rizières du Sénégal: mesure de l'activité dénitrifiante, *Soil Biology and Biochemistry*, *6*, 79–84 (1974).
59. Garcia, J.L. Évaluation de la dénitrification dans les rizières per la methode de réduction de N_2O, *Soil Biology and Biochemistry*, *7*, 251–256 (1975).
60. Germon, J.C. Étude quantitative de la dénitrification biologique dans le sol à l'aide de l'acétylene. I. Application à différents sols. *Annales de Microbiologie* (Institut Pasteur), *131B*, 69–80 (1980).
61. Germon, J.C. Étude quantitative de la dénitrification biologique dans le sol à l'aide de l'acétylène. II. Evolution de l'effet inhibiteur de l'acétylène sur la N_2O-réductase; incidence de l'acétylène sur la vitesse de dénitrification et sur la réorganisation de l'azote nitrique, *Annales de Microbiologie* (Institut Pasteur), *131B*, 81–90 (1980).
62. Gersberg, R., Krohn, K., Peek, N. and Goldman, C.R. Denitrification studies with ^{13}N-labeled nitrate, *Science*, *192*, 1229–1231 (1976).
63. Goreau, T.J., Kaplan, W.A., Wofsy, S.C., McElroy, M.B. Valois, F.W. and Watson, S.W. Production of NO_2^- and N_2O by nitrifying bacteria at reduced concentrations of oxygen, *Applied and Environmental Microbiology*, *40*, 526–532 (1980).
64. Graham, A. and Boxer, D.H. Agreement of respiratory nitrate reductase in the cytoplasmic membrane of *Escherichia coli*. Location of β subunit, *FEBS Letters*, *113*, 15–20 (1980).
65. Guiraud, G. and Berlier, Y. Détermination quantitative et isotopique, par spectrométrie de masse, des composés gazeux produits dans la dénitrification, *Chimie Analytique*, *52*, 53–56 (1970).
66. Hahn, J. The cycle of atmospheric nitrous oxide, *Philosophical Transactions of the Royal Society of London*, *A*, *290*, 495–504 (1979).
67. Hahn, J. and Junge, C. Atmospheric nitrous oxide—a critical review. *Zeitschrift für Naturforschung*, *A*, *32*, 190–214 (1977).
68. Hauck, R.D. and Bremner, J.M. Use of tracers for soil and fertilizer nitrogen research, *Advances in Agronomy*, *28*, 219–266 (1976).
69. Hauck, R.D., Melsted, S.W. and Yankwich, P.E. Use of N-isotope distribution in nitrogen gas in the study of denitrification, *Soil Science*, *86*, 287–291 (1958).

70. Henninger, N.M. and Bollag, J.M. Effect of chemicals used as nitrification inhibitors on the denitrification process, *Canadian Journal of Microbiology*, *22*, 668–672 (1976).
71. Hwang, J.C., Chen, C.H. and Burris, R.H. Inhibition of nitrogenase-catalyzed reductions, *Biochimica et Biophysica Acta*, *292*, 256–270 (1973).
72. Hynes, R.K. and Knowles, R. Inhibition by acetylene of ammonia oxidation in *Nitrosomonas europaea*, *FEMS Microbiology Letters*, *4*, 319–321 (1978).
73. Hynes, R.K. and Knowles, R. Denitrification, nitrogen fixation and nitrification in continuous flow laboratory soil columns, *Canadian Journal of Soil Science*, *60*, 355–363 (1980).
74. John, P. Aerobic and anaerobic bacterial respiration monitored by electrodes, *Journal of General Microbiology*, *98*, 231–238 (1977).
75. Johnston, D.W., Holding, A.J. and McCluskie, J.E. Preliminary comparative studies in denitrification and methane production in Loch Leven, Kinross, and other freshwater lakes, *Proceedings of the Royal Society of Edinburgh*, (*B*), *74*, 123–133 (1974).
76. Johnston, H. Newly recognized vital nitrogen cycle, *Proceedings of the National Academy of Sciences, U.S.A.*, *69*, 2369–2372 (1972).
77. Kanner, D. and Bartha, R. Growth of *Nocardia rhodochrous* on acetylene gas, *Journal of Bacteriology*, *139*, 225–230 (1979).
78. Keeney, D.R., Fillery, I.R. and Marx, G.P. Effect of temperature on the gaseous nitrogen products of denitrification in a silt loam soil, *Soil Science Society of America Journal*, *43*, 1124–1128 (1979).
79. Klemedtsson, L., Svensson, B.H., Lindberg, T. and Rosswall, T. The use of acetylene inhibition of nitrous oxide reductase in quantifying denitrification in soils, *Swedish Journal of Agricultural Research*, *7*, 179–185 (1977).
80. Knowles, R. Denitrification, acetylene reduction and methane metabolism in lake sediment exposed to acetylene, *Applied and Environmental Microbiology*, *38*, 486–493 (1979).
81. Knowles, R. Denitrification, pp. 323–369 in *Soil Biochemistry*, *Vol. 5* (Edited by E.A. Paul and J. Ladd), Marcel Dekker Inc., New York (1981).
82. Kohl, D.H., Vithayathil, F., Whitlow, P., Shearer, G. and Chien, S.H. Denitrification kinetics in soil systems: the significance of good fits of data to mathematical forms, *Soil Science Society of America Journal*, *40*, 249–253 (1976).
83. Koike, I. and Hattori, A. Energy yield of denitrification: an estimate from growth yield in continuous culture of *Pseudomonas denitrificans* under nitrate-, nitrite- and nitrous oxide-limited conditions, *Journal of General Microbiology*, *88*, 11–19 (1975).

84. Kowalenko, C.G. The influence of sulfur anions on denitrification, *Canadian Journal of Soil Science*, *59*, 221–223 (1979).
85. Kristjansson, J.K. and Hollocher, T.C. First practical assay for soluble nitrous oxide reductase of denitrifying bacteria and a partial kinetic characterization, *Journal of Biological Chemistry*, *255*, 704–707 (1980).
86. Kristjansson, J.K., Walter, B. and Hollocher, T.C. Respiration-dependent proton translocation and the transport of nitrate and nitrite in *Paracoccus denitrificans* and other denitrifying bacteria, *Biochemistry*, *17*, 5014–5019 (1978).
87. Krul, J.M. and Veeningen, R. The synthesis of the dissimilatory nitrate reductase under aerobic conditions in a number of denitrifying bacteria, isolated from activated-sludge and drinking water, *Water Research*, *11*, 39–43 (1977).
88. Letey, J., Valoras, N., Hadas, A. and Focht, D.D. Effect of air-filled porosity, nitrate concentration, and time on the ratio of N_2O/N_2 evolution during denitrification, *Journal of Environmental Quality*, *9*, 227–231 (1980).
89. Magee, P.N. Nitrogen as health hazard, *Ambio*, *6*, 23–125 (1977).
90. McElroy, M.B., Wofsy, S.C. and Yung, Y.L. Nitrogen cycle perturbations due to man and their impact on atmospheric N_2O and O_3, *Philosophical Transactions of the Royal Society of London*, *B*, *277*, 159–181 (1977).
91. Meijer, E.M., Van der Zwaan, J.W. and Stouthamer, A.H. Location of the proton-consuming site in nitrite reduction and stoichiometries for proton pumping in anaerobically grown *Paracoccus denitrificans*, *FEMS Microbiology Letters*, *5*, 369–372 (1979).
92. Misra, C., Nielsen, D.R. and Biggar, J.W. Nitrogen transformations in soil during leaching: III. Nitrate reduction in soil columns, *Soil Science Society of America Proceedings*, *38*, 300–304 (1974).
93. Mitsui, S., Watanabe, I., Honma, M. and Honda, S. The effect of pesticides on the denitrification in paddy soil, *Soil Science and Plant Nutrition*, *10*, 15–23 (1964).
94. Müller, M.M., Sundman, V. and Skujins, J. Denitrification in low pH spodosols and peats determined with the acetylene inhibition method, *Applied and Environmental Microbiology*, *40*, 235–239 (1980).
95. Myers, R.J.K. The effect of sulphide on nitrate reduction in soil, *Plant and Soil*, *37*, 431–433 (1972).
96. Nömmik, H. Investigations on denitrification in soil, *Acta Agricultura Scandinavica*, *6*, 195–228 (1956).
97. Nömmik, H. and Thorin, J. A mass spectrometric technique for studying the nitrogenous gases produced on the reaction of nitrite with raw humus, *Agrochimica*, *16*, 319–322 (1972).
98. Nömmik, H. and Thorin, J. Transformations of ^{15}N-labelled nitrite and

nitrate in forest raw humus during anaerobic incubation, pp. 369–382 in *Isotopes and radiation in soil-plant relationships including forestry*, International Atomic Energy Agency, Vienna (1972).

99. Patrick, W.H. and Wyatt, R. Soil nitrogen loss as a result of alternate submergence and drying, *Soil Science Society of America Proceedings, 28*, 647–653 (1964).
100. Patriquin, D.G., MacKinnon, J.C. and Wilkie, K.I. Seasonal patterns of denitrification and leaf nitrate reductase activity in a corn field, *Canadian Journal of Soil Science, 58*, 283–285 (1978).
101. Patten, D.K., Bremner, J.M. and Blackmer, A.M. Effects of drying and air-dry storage of soils on their capacity for denitrification of nitrate, *Soil Science Society of America Journal, 44*, 67–70 (1980).
102. Paul, E.A. and Victoria, R.L. Nitrogen transfer between the soil and atmosphere, pp. 525–541 in *Environmental Biogeochemistry and Geomicrobiology, Vol. 2: The Terrestrial Environment* (Edited by W.E. Krumbein), Ann Arbor Science (1978).
103. Payne, W.J. Reduction of nitrogenous oxides by microorganisms, *Bacteriological Reviews, 37*, 409–452 (1973).
104. Payne, W.J. and Balderston, W.L. Denitrification, pp. 339–345 in *Microbiology—1978* (Edited by D. Schlessinger), American Society of Microbiology, Washington (1978).
105. Payne, W.J. and Riley, P.S. Suppression by nitrate of enzymatic reduction of nitric oxide, *Proceedings of the Society for Experimental Biology and Medicine, 132*, 258–260 (1969).
106. Payne, W.J., Riley, P.S. and Cox, C.D. Separate nitrite, nitric oxide, and nitrous oxide reducing fractions from *Pseudomonas perfectomarinus, Journal of Bacteriology, 106*, 356–361 (1971).
107. Phillips, R.E., Reddy, K.R. and Patrick, W.H. The role of nitrate diffusion in determining the order and rate of denitrification in flooded soil: II. Theoretical analysis and interpretation, *Soil Science Society of America Journal, 42*, 272–278 (1978).
108. Pichinoty, F. La réduction bactérienne des composés oxygénés minéraux de l'azote, *Bulletin de l'Institut Pasteur, 71*, 317–395 (1973).
109. Pierotti, D. and Rasmussen, R.A. Combustion as a source of nitrous oxide in the atmosphere, *Geophysical Research Letters, 3*, 265 (1976).
110. Pilot, L. and Patrick, W.H. Nitrate reduction in soils: Effect of soil moisture tension, *Soil Science, 114*, 312–315 (1972).
111. Pratt, P.F., Barber, J.C., Corrin, M.L., Goering, J., Hauck, R.D., Johnston, H.S., Klute, A., Knowles, R., Nelson, D.W., Pickett, R.C. and Stephens, E.R. Effect of increased nitrogen fixation on stratospheric ozone, *Climatic Change, 1*, 109–135 (1977).
112. Raimbault, M. Étude de l'influence inhibitrice de l'acétylène sur la formation biologique du méthane dans un sol de rizière, *Annales Microbio-*

logique (Institut Pasteur), *126A*, 247–258 (1975).
113. Reddy, K.R. and Patrick, W.H. Effect of alternate aerobic and anaerobic conditions on redox potential, organic matter decomposition and nitrogen loss in a flooded soil, *Soil Biology and Biochemistry*, *7*, 87–94 (1975).
114. Reddy, K.R., Patrick, W.H. and Phillips, R.E. The role of nitrate diffusion in determining the order and rate of denitrification in flooded soil: I. Experimental results, *Soil Science Society of America Journal*, *42*, 268–272 (1978).
115. Rolston, D.E., Broadbent, F.E. and Goldhamer, D.A. Field measurement of denitrification: II. Mass balance and sampling uncertainty, *Soil Science Society of America Journal*, *43*, 703–708 (1979).
116. Rolston, D.E., Hoffman, D.L. and Toy, D.W. Field measurement of denitrification: I. Flux of N_2 and N_2O, *Soil Science Society of America Journal*, *42*, 863–869 (1978).
117. Rolston, D.E. and Marino, M.A. Simultaneous transport of nitrate and gaseous denitrification products in soil, *Soil Science Society of America Journal*, *40*, 860–865 (1976).
118. Ryden, J.C. and Lund, L.J. Nature and extent of directly measured denitrification losses from some irrigated vegetable crop production units, *Soil Science Society of America Journal*, *44*, 505–511 (1980).
119. Ryden, J.C., Lund, L.J. and Focht, D.D. Direct in-field measurement of nitrous oxide flux from soils, *Soil Science Society of America Journal*, *42*, 731–737 (1978).
120. Ryden, J.C., Lund, L.J. and Focht, D.D. Direct measurement of denitrification loss from soils: I. Laboratory evaluation of acetylene inhibition of nitrous oxide reduction, *Soil Science Society of America Journal*, *43*, 104–110 (1979).
121. Ryden, J.C., Lund, L.J., Letey, J. and Focht, D.D. Direct measurement of denitrification loss from soils: II. Development and application of field methods, *Soil Science Society of America Journal*, *43*, 110–118 (1979).
122. Sacks, L.E. and Barker, H.A. The influence of oxygen on nitrate and nitrite reduction, *Journal of Bacteriology*, *58*, 11–22 (1949).
123. Saraste, M. and Kuronen, T. Interaction of *Pseudomonas* cytochrome cd_1 with the cytoplasmic membrane, *Biochimica et Biophysica Acta*, *513*, 117–131 (1978).
124. Smid, A.E. and Beauchamp, E.G. Effects of temperature and organic matter on denitrification in soil, *Canadian Journal of Soil Science*, *56*, 385–391 (1976).
125. Smith, K.A. A model of the extent of anaerobic zones in aggregated soils and its potential application to estimates of denitrification, *Journal of Soil Science*, *31*, 263–277 (1980).

126. Smith, L.A., Hill, S. and Yates, M.G. Inhibition by acetylene of conventional hydrogenase in nitrogen-fixing bacteria, *Nature*, *262*, 209–210 (1976).
127. Smith, M.S., Firestone, M.K. and Tiedje, J.M. The acetylene inhibition method for short-term measurement of soil denitrification and its evaluation using nitrogen-13, *Soil Science Society of America Journal*, *42*, 611–615 (1978).
128. Smith, M.S. and Tiedje, J.M. Phases of denitrification following oxygen depletion in soil, *Soil Biology and Biochemistry*, *11*, 261–267 (1979).
129. Söderlund, R. and Svensson, B.H. The global nitrogen cycle, in *Nitrogen, Phosphorus and Sulphur—Global Cycles*. SCOPE Rep. 7 (Edited by B.H. Svensson and R. Söderlund), *Ecological Bulletins* (Stockholm), *22*, 23–73 (1976).
130. Sørensen, J. Capacity for denitrification and reduction of nitrate to ammonia in a coastal marine sediment, *Applied and Environmental Microbiology*, *35*, 301–305 (1978).
131. Sørensen, J. Denitrification rates in a marine sediment as measured by the acetylene inhibition technique, *Applied and Environmental Microbiology*, *36*, 139–143 (1978).
132. Sørensen, J. Occurrence of nitric and nitrous oxides in a coastal marine sediment, *Applied and Environmental Microbiology*, *36*, 809–813 (1978).
133. Sørensen, J., Tiedje, J.M. and Firestone, R.B. Inhibition by sulfide of nitric and nitrous oxide reduction by denitrifying *Pseudomonas fluorescens*, *Applied and Environmental Microbiology*, *39*, 105–108 (1980).
134. Stanford, G., Dzienia, S. and Vander Pol, R.A. Effect of temperature on denitrification rate in soils, *Soil Science Society of America Proceedings*, *39*, 867–870 (1975).
135. Stanford, G., Legg, J.O., Dzienia, S. and Simpson, E.C. Denitrification and associated nitrogen transformations in soils, *Soil Science*, *120*, 147–152 (1975).
136. Stanford, G., Vander Pol, R.A. and Dzienia, S. Denitrification rates in relation to total and extractable soil carbon, *Soil Science Society of America Proceedings*, *39*, 284–289 (1975).
137. Starr, J.L., Broadbent, F.E., Nielsen, D.R. Nitrogen transformations during continuous leaching, *Soil Science Society of America Proceedings*, *38*, 283–289 (1974).
138. St. John, R.T. and Hollocher, T.C. Nitrogen-15 tracer studies on the pathway of denitrification in *Pseudomonas aeruginosa*, *Journal of Biological Chemistry*, *252*, 212–218 (1977).
139. Tam, T.Y. and Knowles, R. Effects of sulfide and acetylene on nitrous oxide reduction by soil and by *Pseudomonas aeruginosa*, *Canadian Journal of Microbiology*, *25*, 1133–1138 (1979).
140. Terry, R.E. and Tate, R.L. The effect of nitrate on nitrous oxide reduc-

tion in organic soils and sediments, *Soil Science Society of America Journal*, *44*, 744–746 (1980).

141. Tiedje, J.M. Denitrification in soil, pp. 362–366 in *Microbiology—1978* (Edited by D. Schlessinger), American Society for Microbiology, Washington (1978).

142. Tiedje, J.M. Denitrification, in *Methods of Soil Analysis, Vol. 2* (Edited by R.H. Miller *et al.*), American Society of Agronomy, Madison (in press) (1982).

143. Tiedje, J.M., Firestone, R.B., Firestone, M.K., Betlach, M.R., Smith, M.S. and Caskey, W.H. Methods for the production and use of nitrogen-13 in studies of denitrification, *Soil Science Society of America Journal*, *43*, 709–719 (1979).

144. Van Cleemput, O. and Patrick, W.H. Nitrate and nitrite reduction in flooded gamma-irradiated soil under controlled pH and redox potential conditions, *Soil Biology and Biochemistry*, *6*, 85–88 (1974).

145. Vangnai, S. and Klein, D.A. A study of nitrite-dependent dissimilatory micro-organisms isolated from Oregon soils, *Soil Biology and Biochemistry*, *6*, 335–339 (1974).

146. Volz, M.G. and Starr, J.L. Nitrate dissimilation and population dynamics of denitrifying bacteria during short-term continuous flow, *Soil Science Society of America Journal*, *41*, 891–896 (1977).

147. Walter, H.M., Keeney, D.R. and Fillery, I.R. Inhibition of nitrification by acetylene, *Soil Science Society of America Journal*, *43*, 195–196 (1979).

148. Watanabe, I. and de Guzman, M.R. Effect of nitrate on acetylene disappearance from anaerobic soil, *Soil Biology and Biochemistry*, *12*, 193–194 (1980).

149. Weeraratna, C.S. Effect of dalapon-sodium on nitrification and denitrification in a tropical loam soil, *Weed Research*, *20*, 291–293 (1980).

150. Wientjes, F.B., van't Riet, J. and Nanninga, N. Immunoferritin labeling of respiratory nitrate reductase in membrane vesicles of *Bacillus licheniformis* and *Klebsiella aerogenes*, *Archives of Microbiology*, *127*, 39–46 (1980).

151. Wijler, J. and Delwiche, C.C. Investigations on the denitrifying process in soil, *Plant and Soil*, *5*, 155–169 (1954).

152. Wood, P.M. Periplasmic location of the terminal reductase in nitrite respiration, *FEBS Letters*, *92*, 214–218 (1978).

153. Yeomans, J.C. and Beauchamp, E.G. Limited inhibition of nitrous oxide reduction in soil in the presence of acetylene, *Soil Biology and Biochemistry*, *10*, 517–519 (1978).

154. Yoshinari, T. N_2O reduction by *Vibrio succinogenes*, *Applied and Environmental Microbiology*, *39*, 81–84 (1980).

155. Yoshinari, T., Hynes, R. and Knowles, R. Acetylene inhibition of nitrous oxide reduction and measurement of denitrification and nitrogen fixa-

tion in soil, *Soil Biology and Biochemistry*, *9*, 177–183 (1977).

156. Yoshinari, T. and Knowles, R. Acetylene inhibition of nitrous oxide reduction by denitrifying bacteria, *Biochemical and Biophysical Research Communications*, *69*, 705–710 (1976).
157. Youatt, J.B. Denitrification of nitrite by a species of *Achromobacter*, *Nature*, *173*, 826–827 (1954).
158. Zumft, W.G. and Cárdenas, J. The inorganic biochemistry of nitrogen bioenergetic processes, *Die Naturwissenschaften*, *66*, 81–88 (1979).

L. W. BELSER

10. Inhibition of Nitrification

Introduction

Nitrification is one of several microbially mediated processes associated with the terrestrial nitrogen cycle. During nitrification ammonium is oxidized to nitrate via the intermediate nitrite. The substrate and the endproduct of this process (i.e. ammonium and nitrate) form the major pools of biologically available nitrogen in soil. Of the two, however, nitrate is much more liable to loss, since it is susceptible to both denitrification and leaching. It has been proposed, therefore, that the biologically available nitrogen could be more efficiently retained in the soil, and thus more efficiently recycled to plants, if nitrate formation was reduced or prevented through the inhibition of nitrification (e.g. 30, 43, 44, 67, 68). Figure 1 shows how the nitrogen cycle would be modified by the inhibition of nitrification. It can be seen that the two pathways of denitrification and leaching are eliminated in the inhibition modified nitrogen cycle. In addition to improving the cycling of nitrogen to plants, it has also been argued that it is energetically better for plants to take up ammonium rather than nitrate, since nitrate nitrogen must be reduced to the oxidation state of ammonium before it can be utilized by a plant (67).

Two major questions are associated with inhibition of nitrification in terrestrial environments. First, do some plants or plant communities actively inhibit nitrification, as has been proposed by several investigators (43, 44, 67, 68)? Second, in agricultural systems where nitrate losses can often be significant, can inhibition be used to improve crop response to ammoniacal fertilizers (31, 40)? In this chapter these two questions will be considered. For more general considerations of nitrification see reviews by Schmidt (78), Belser (6), and Focht and Verstraete (28).

Background relating to inhibition proposals

Mechanisms for nitrogen loss

Due to differences in their chemical properties, ammonium is retained in most soils to a higher degree than nitrate. Ammonium, being in the most reduced state of nitrogen, cannot be lost via denitrification, since only oxidized

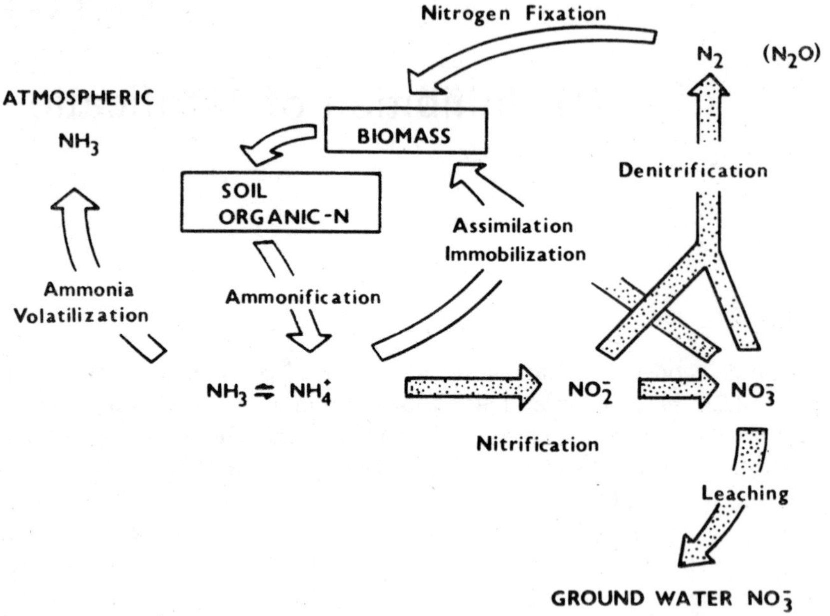

Fig. 1. The terrestrial nitrogen cycle showing the major microbiologically mediated pathways of nitrification, denitrification, ammonification, nitrogen fixation and immobilization. The shaded pathways are the ones affected by the inhibition of nitrification.

inorganic nitrogen compounds are susceptible to this reductive process. Nitrate is also more susceptible to leaching than ammonium. Ammonium, being a cation, is not readily leached through negatively charged soil matrixes, whereas nitrate, being an anion, is. The only major mechanism for ammonium loss is through the volatilization of ammonia, which occurs only under alkaline conditions.

Production of nitrate

To effectively inhibit nitrification by preventing the formation of nitrate in soil, the major pathway for nitrate formation must be nitrification. This appears to be the case. There is only one other biological process that could potentially produce nitrate in soil, and that is the process of heterotrophic nitrification (not shown in Fig. 1). Heterotrophic nitrification involves the mineralization of organic nitrogen to nitrite and nitrate. This process is considered negligible compared to the other mineralization process of ammonification. Therefore, most if not all soil nitrate which is derived from soil organic-N results from the autotrophic nitrification of ammonium formed by ammonification. And as long as only ammoniacal fertilizers are added to soil, autotrophic nitrification will limit the rate of nitrate production.

Microbiology of nitrification

Nitrification is carried out by the chemoautotrophic bacteria of the family Nitrobacteraceae. There are two steps associated with this process, the oxidation of ammonium to nitrite, and the oxidation of nitrite to nitrate. The nitrifiers have specialized in one or the other of these two processes, but no autotrophic nitrifier has been found to carry out both processes. All the energy required for growth and maintenance of nitrifier populations can be obtained from the oxidation of these inorganic nitrogen compounds. In addition, no organic carbon is required by these organisms, as they are able to fix carbon dioxide. Four genera of ammonium oxidizers have been isolated from terrestrial environments: *Nitrosomonas*, *Nitrosospira*, *Nitrosolobus*, and *Nitrosovibrio* (33, 95). It has been shown recently that at least the first three of these various genera can coexist in soil and are assumed to be common to many soils (7, 8). The distribution and abundance of *Nitrosovibrio* is unknown.

The only known genus of terrestrial nitrite oxidizer is *Nitrobacter* (95). The only other autotrophs known to oxidize nitrite are obligate marine bacteria (95).

The heterotrophic nitrifiers which can oxidize nitrogen from its most reduced state (organic-N or ammonium) to nitrate, do not obtain any energy from the oxidative processes. Since this process is not considered to produce significant amounts of nitrate in most soils, it will not be considered further here. More details on heterotrophic nitrification can be obtained in the review by Focht and Verstraete (28).

Little is known about the regulation of population sizes of nitrifiers. It seems likely that continuous oxidation of ammonium would be required to maintain populations in soil and that the size of a population would be governed by the rate at which ammonium became available (6). Since in many environments the process of ammonification controls the rate at which ammonium becomes available, it is possible that the ammonification rate may play an important role in limiting nitrifier population sizes. Of course additions of ammoniacal fertilizers would allow a population to exceed the sizes expected in unfertilized soils where natural ammonification was the main source of ammonium.

Additional factors could also limit the size of nitrifier populations. It has been suggested that space could limit population sizes, since populations of nitrifiers appear to reach saturation levels in some soils (47, 48). However such saturation population levels may not be due to space limitation, but due to growth inhibition induced by such factors as locally produced low hydrogen ion concentrations (3, 52).

Edaphic and climatic factors inhibitory to nitrification and ammonification

For a chemical inhibitor to be most effective in increasing available nitrogen, it must inhibit nitrification without inhibiting ammonification, since inhibition of ammonification would clearly defeat the objective of increasing ammonium in the soil. Factors, such as pH, moisture tension, and temperature, can differentially inhibit nitrification with respect to ammonification in a way similar to chemical inhibitors. Thus, when monitoring the retention of ammonium during inhibition studies it is particularly important to distinguish these edaphic or climatic effects from those induced by inhibitors.

Nitrification is much more sensitive to pH than is ammonification. At both low and high pH ammonium will tend to accumulate during soil incubations. However, as pH increases above 9, the accumulation of ammonium will tend to be offset by the volatilization of ammonia. In pure culture nitrifiers grow well only between pH 6 and 8.5. However, active nitrificacation has been reported in soils of pH 4–5. It is possible that such nitrification is associated with soil micro-environments of higher pH.

Nitrifiers are also more sensitive to moisture stress than are ammonifiers, with selective inhibition becoming significant at a pF (pF = −log (soil moisture potential)) of about 5 (25). Under very moist or waterlogged soil conditions, nitrification will also become inhibited due to restricted aeration. This can also be regarded as a differential inhibition with respect to ammonification, since ammonification proceeds readily under anaerobic as well as aerobic conditions (41).

Temperature also selectively inhibits nitrification with respect to ammonification. In soils from temperate climatic regions, Keeney and Bremner (41) found complete nitrification of ammonified nitrogen at temperatures below 35°C, but virtually no nitrification above this temperature. Ammonification had an optimum >40°C. In tropical soils, Meyers (51) has reported a higher tolerance to heat for the nitrifying bacteria. Mahendrappa *et al.* (45) have suggested that nitrifiers are adopted to a particular climate, showing corresponding different temperature tolerances. Interpretation of data suggesting adaption of nitrifiers to various climates is complicated by the fact that virtually nothing is known about the genera, species and strain composition of nitrifiers in soil.

Thus, stresses caused by extremes of temperature, moisture and pH have the potential for selective inhibition of nitrification with respect to the production of ammonium. Therefore, when assessing the possible role of biologically induced inhibition by higher plants on nitrification, it must be made clear that such environmental factors are not playing a role in the inhibitory process.

Evidence for plant-induced inhibition of nitrification

There have been two types of proposals in the literature concerning the possibility that certain plants excrete products for the specific purpose of inhibiting nitrification. One type suggests that plants associated with climax ecosystems are able to produce chemicals that effectively inhibit nitrification (67, 68). The other suggests that certain grassland species also have this ability (81, 86). These grasslands, however, are not necessary climax ecosystems.

Inhibition in climax ecosystems

The main body of evidence for the inhibition of nitrification by climax ecosystems comes from the work of Rice and Pancholy (67, 68). Their evidence is mainly field data consisting of observations that nitrate concentrations and nitrifier populations in soil (estimated by a viable counting technique) tend to decrease with ecological succession, while concentrations of

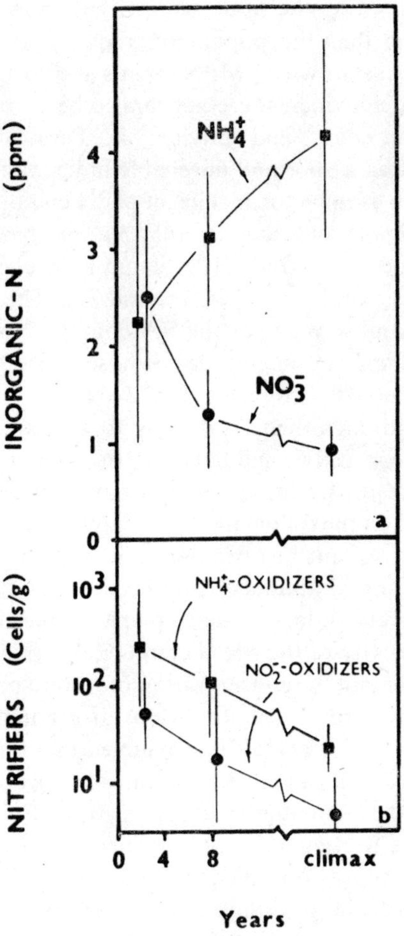

Fig. 2. Changes in MPN nitrifier counts, ammonium and nitrite during the succession of a Post Oak-Blackjack Oak ecosystem [data after Rice and Pancholy (67)].

ammonium tend to increase. Representative data from the work of Rice and Pancholy (67) are plotted in Fig. 2. All these data are consistent with the concept that climax ecosystems actively induce inhibition of nitrification. Other investigators observed that nitrate levels in streams increase after clearing mature forests (43, 93). This has been claimed to provide additional evidence for the inhibition hypothesis.

There is some question whether the conclusions drawn from such field data are valid. The viable count data are difficult to interpret since reliability of the counting methods used in these studies is not known. The counts were made by the most probable number (MPN) technique (1, 2), and there have been recent reports questioning the efficiency of this technique for estimating nitrifiers (6, 8, 46).

An evaluation of studies where both MPN counts and activity have been measured on the same sample, has shown that counts are often significantly smaller than the population required to produce the observed activity (6). Even in soils where MPN counts appear to be in a range compatible with observed activities, there appears to be a complete lack of correlation between nitrifier counts and activity (74). These observations show that MPN counts are often a poor and unreliable index of nitrifier activity.

An example of inefficient MPN counting is shown in Fig. 3. In this study by Rennie and Schmidt (65) *Nitrobacter* were counted by both fluorescent antibody (FA) and MPN counting techniques (for details of the FA technique see references 75, 76, and 77). The MPN counts were a factor of one thousand lower than the FA counts. Note that the MPN counts in Fig. 2 are similar in magnitude to those in Fig. 3 and there is a strong likelihood that the MPN counts in Fig. 2 are underestimating the populations present.

Although there is no way to assess accurately the counting efficiency in the work of Rice and Pancholy (67, 68), it appears that the counts significantly underestimate the populations at all stages of succession. This conclusion is based on the following two factors. First, since nitrate is continuously present in the various environments, it must be produced continuously. If it were not being continuously produced, its concentration would rapidly decrease to undetectable levels due to plant uptake, leaching or denitrification. Second, the activity of the MPN countable nitrifier populations would be insufficient to maintain the concentration of nitrate present. Ammonium oxidizers under optimal growth conditions oxidize ammonium at a maximum rate of 0.02 pmoles/cell/h (11). If it is assumed that this is similar to the maximum activity per cell that can be attained in soil, it would take three years for a population of 150 ammonium oxidizers/g to oxidize enough ammonium for 1 ppm of nitrate to accumulate. A population of 150 cells/g is larger than any of the MPN counts observed in the climax soils, while 1 ppm nitrate is typical of nitrate concentrations in these climax soils for much of the year. Clearly there must be more activity than this to maintain nitrate at the 1 ppm level. It seems likely that the populations are underestimated by a factor of between 100 and 1000.

There are other arguments against the climax inhibition theory. According to Rice and Pancholy, leaching of nitrate should be lowest under climax vegetation. However, Vitousek and Reiners (93) have presented a hypothesis that predicts that there would be more loss (leaching) of nitrate at the climax stage than during succession. In testing their hypothesis in New Hampshire forests, Vitousek and Reiners found higher nitrate outputs in streams draining the more mature ecosystem watersheds, than in less mature forests. Their findings directly contradict what is expected from the climax ecosystem

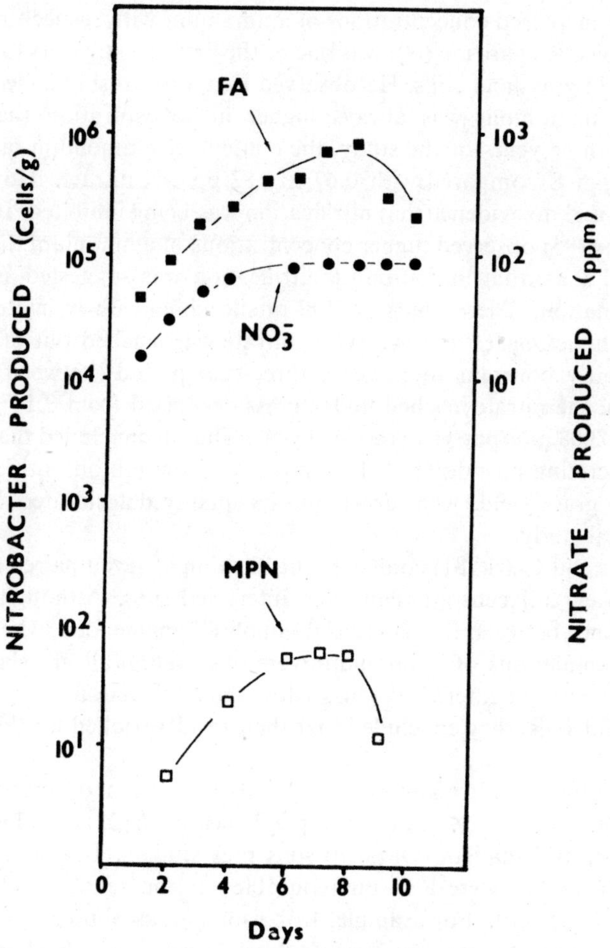

Fig. 3. Net production of nitrate and *Nitrobacter* populations (FA and MPN counts) as a function of time for Hubbard soil with a pH of 4.7.

inhibition hypothesis. Their theory also does not require any inhibition of nitrification.

Inhibition in grasslands

There have been several reviews which considered among other things the possibility that grassland ecosystems biologically inhibit nitrification. Clark and Paul (23) concluded that the balance of evidence favoured the hypothesis. But, according to Russell (72) there is no satisfactory explanation for the slow nitrification of ammonium in grassland soils. Harmsen and van Schreven (35) have also questioned the inhibition hypothesis.

One of the major lines of evidence said to indicate the occurrence of inhibition is the increased concentrations of ammonium with respect to nitrate in grassland soils. Richardson (69) was one of the first investigators to document this feature of grassland soils. He observed that with freshly collected grassland soils, ammonium was always higher in concentration than nitrate. During the three years of the study, the content of ammonium ranged from 4.6 to 6.6 ppm-N compared with 0.67 to 1.82 ppm-N nitrate. However, Richardson found no evidence that nitrification was being inhibited. It was later, when Theron (85) observed higher concentrations of ammonium than nitrates in grassland soils, that inhibition of nitrification was suggested as the most likely explanation. Theron measured nitrate leaching with lysimeters planted with grass, millet, or left fallow. While nitrate was leached out of the fallow soil at a fairly constant rate over a three-year period (between .3 and 4.4 g-N/pot/yr), the nitrate leached under grass decreased from 2.14 g-N/pot/yr initially to 0.008 g-N/pot/yr in the final year. Theron concluded that the grass must be excreting an inhibitor. It is not clear how inhibition benefited the grass since grass yields went down and its quality deteriorated during the experimental study.

Soulides and Clark (81) compared nitrification of seven paired soils representing closely adjacent grassland and intertilled sites. Although there was little difference between the paired soil samples if unamended, when urea was added, accumulations of ammonium were observed in all grassland soils in comparison with the intertilled soils. Although nitrification was occurring in the grassland soils, they concluded that their results backed up the theory of Theron.

On the other hand, many grassland soils have been observed to nitrify actively (e.g. 22, 69, 74, 82). For example, Chase *et al.* (22) found no evidence of inhibition in Canadian (Ontario) soils they studied. In other soils where nitrifying capacities were low, unfavourable environmental conditions have been blamed (15, 70). For example, Robinson (70) attributed poor nitrifying activity found in low fertility New Zealand Tussuck grassland soils to substrate limitation and a poor growth environment. Liming improved nitrification. Brar and Giddens (15) attempted to extract and exclude inhibitors from soil with low activity, but found no improvement in the nitrification rate. They concluded that conditions were not conducive to nitrification due to acidity. Soils from grass clover pasture soils have also been shown to nitrify actively (74, 82).

Thus, the possibility of the inhibition of nitrification in grassland soils is still open to question. There are some difficulties in interpreting experiments such as those by Theron (85) and Soulides and Clark (81), since the processes taking place in arable and grassland soils can be quite different. As pointed out by Richardson (69), one major consequence of cultivating a grassland is to decrease organic carbon and nitrogen. Converting an arable soil to grassland increases the organic matter. It is likely that many of the dynamic processes occurring in the two soils will be stimulated to a different extent. These processes will determine the rate of turnover of the inorganic nitrogen compounds, thereby greatly affecting their concentrations.

Inhibition of nitrification by plant extracts

Some investigators have chosen to investigate the role of plant and root extracts directly on the oxidative processes of nitrifiers. Rice (66), Munro (55) and Neal (56) found that extracts from macerated roots or macerated plant parts were inhibitory to nitrification. Rice and Pancholy (68) found that tannin like compounds extracted from soils could also be inhibitory. However, Bohlool *et al.* (13) have questioned the role of tannins as nitrification inhibitors, since they found little inhibition of nitrification with purified tannins from *Pinus radiata*.

All of the above extract studies have one thing in common. They tested the inhibitory properties of extracts by adding them to nitrifiers growing in liquid culture. Molina and Rovira (53) also used this approach, but expanded their studies to include work on rhizosphere effects. Instead of macerated root extracts, they used plant root exudates. They found during the first fifteen days of growth, that the rhizosphere stimulated nitrifiers, while exudates collected from plants during the same period were inhibitory. It is clear from this work that great care must be taken in extrapolating pure culture studies with extracts or exudates to the soil environment.

Inhibition of nitrification in agricultural ecosystems

The purpose of adding a nitrification inhibitor to an agricultural soil is to improve crop yield by increasing the efficiency of the nitrogen uptake. There are products available commercially that have been shown to inhibit nitrification specifically (18, 30, 62, 73). However, the problem exists as to whether an inhibitor can be used to improve crop yield consistently.

Studies on the inhibition of nitrification by chemicals can be placed in two categories. One category consists of studies dealing with the effectiveness of the inhibitor itself. These include studies with nitrifiers in soils and pure culture, as well as studies on the persistence of the inhibitor in the soil environment. The other category consists of studies of plant responses associated with the inhibition of nitrification. These include not only studies on plant yield in the presence or absence of an inhibitor but also studies associated with phytotoxicity and even studies on plants' preference for nitrate or ammonium.

Inhibitors of nitrification

There are many compounds that have been shown to be inhibitory to nitrification. Bundy and Bremner (18) have screened the inhibitory properties of 24 compounds (including the major commercial inhibitors) on nitrification in three soils. In these studies with inhibitors applied at a rate of 10 μg/g, they found that only six of the 24 inhibitors were at least 50 per cent effective (on average) in inhibiting the nitrification of 200 μg/g ammonium-N added to soil for 14 days at 30°C. The most effective inhibitor was 2-chloro-6-(trichloromethyl)-pyridine (nitrapyrin/N-Serve) followed by 4 amino-1, 2, 4, triazole (ATC), sodium or potassium azide, 2, 4-diamino-6-trichloromethyl-s-triazine (CL 1580), dicyandiamide, and phenyl acetate. Another eight compounds showed lesser inhibitory capabilities (these included 3-mercapto-1, 2, 4-triazole or 2-amino-4-chloro-6-methyl-pyrimidine (AM) and sulphathiazole (ST)). Ten of the compounds tested showed no significant inhibitory properties (i.e. <4 per cent). Acetylene has recently been shown to also inhibit ammonium oxidation (39, 94). Several of these compounds tested have been patented as nitrification inhibitors (i.e. N-Serve, ATC, CL 1580, AM and ST). Prasad *et al.* (62) have reviewed the properties of some of these commercial inhibitors (e.g. N-Serve, AM and ST). Of the compounds that are commercially available, only nitrapyrin, (N-Serve) has been widely used in agronomic studies. Therefore, most of the following discussion will deal with studies using this inhibitor.

Nitrapyrin has been shown to inhibit ammonium oxidation specifically leaving the nitrite oxidation step relatively unaffected (19, 20). This appears to be the case for most commercial nitrification inhibitors. It is essential that a nitrification inhibitor be most effective at the ammonium oxidation step of nitrification. Compounds such as chlorate which inhibit nitrite oxidation to a greater extent than they do ammonium oxidation, would allow nitrite to accumulate (7). Nitrite accumulation would be unwanted since nitrite is toxic to many plants and is also subject to the same loss process as nitrate.

While initial studies on the inhibition of nitrification by nitrapyrin in pure culture were done only with *Nitrosomonas* (19, 37), recent work has shown that the inhibitor is also effective against *Nitrosospira* and *Nitrosolobus* (12). However, this latter work has shown that various strains of ammonium oxidizer show varying degrees of sensitivity. This might have practical implications for long-term use of nitrapyrin in field soils. For example, it is possible that the continued use of the inhibitor could cause the selection of inhibition resistant strains.

Biochemical basis of inhibition

The biological oxidation of ammonium to nitrite by *Nitrosomonas* proceeds as follows (57):

$$NH_4^+ + 1/2O_2 \xrightarrow{-2e} NH_2OH + H^+ \qquad (1)$$
$$F = +4 \text{ kCal}$$

$$NH_2OH + O_2 \xrightarrow{-4e} NO_2^- + H_2O + H^+ \qquad (2)$$
$$F = -69 \text{ kCal}$$

Of the commercially available inhibitors apparently only nitrapyrin has been studied with regard to mode of action. Although the biochemical basis for its inhibition is still not understood, nitrapyrin is known to interfere with the initial biochemical step where ammonium (ammonia) is converted to hydroxylamine (a reaction with a positive free energy change). *Nitrosomonas* can oxidize both ammonium and hydroxylamine, but in the presence of nitrapyrin ammonium oxidation only is inhibited. The enzyme associated with the initial step is ammonia oxidase (26, 36). Hooper and Terry (37) have shown that this enzyme is selectively inhibited by over 40 other compounds. The original studies by Campbell and Aleem (19) on the nitrapyrin inhibitory process suggested that copper might in some way be chelated, since addition of copper reversed the inhibitory process. However, there is a problem with such "metal reversal" studies since the addition of copper may be simply binding up all the inhibitor (Hooper, personal communication).

The mechanisms by which the various inhibitors affect nitrification needs much more study. Research in this area, however, faces some difficulties. Since the reaction in which ammonium is converted to hydroxylamine has a positive free energy, it will not proceed spontaneously. In addition, ammonia oxidase appears to be membrane bound. This has hampered the purification of cell-free extracts necessary for biochemical studies on the mechanisms.

Effect of soil factors on inhibition

The effectiveness of the inhibitors of nitrification appears to be greatly influenced by soil organic matter. Goring (30) was the first to note this during his studies on the effectiveness of nitrapyrin in 87 soils of varying texture and organic matter content. He found a correlation between the amount of organic matter in the soil and a reduction in the effectiveness of the inhibitor. He attributed this to the sorption of nitrapyrin from the solution phase by the organic matter. He reasoned that only nitrapyrin in the solution phase was effective in inhibiting nitrification.

Bundy and Bremner (18) in their study of 24 potential inhibitors found that soil texture greatly influenced the effectiveness of the inhibitors. All inhibitors were more effective in light textured soils with low organic matter content but effectiveness decreased with decreasing sand content and increasing organic matter content. Nitrapyrin and 4 amino-1, 2, 4-triazole (ATC) were least affected by these changes, being greater than 66 per cent effective in the 3 soils tested. On the other hand, 3-mercapto-1, 2, 4-triazole and 2-amino-4-chloro-6-methyl-pyrimidine (AM) were between 64 and 69 per cent effective in the light textured soil, but were less than 6 per cent effective in the heavy textured soil.

Persistence of inhibition in soil

For an inhibitor to be effective in conserving nitrogen in soil its effect must persist over a prolonged period. The length of the inhibitory period will depend on how long the inhibitor remains in an active form in the soil, to what extent it decreases the viable nitrifier population, and how soon the nitrifier population reestablishes itself after the inhibitor has become ineffective.

The loss of an inhibitor from the soil environment is due to a combination of chemical breakdown and physical movement out of the soil environment (e.g. volatilization or leaching). There have been several measurements of the loss of nitrapyrin from soil. Redemann *et al.* (64) reported half lives of between four and 22 days for four soils incubated at 20°C. Studies of half lives as a function of temperature have shown that the rate of loss decreases rapidly with a decrease in temperature. Herlihy and Quirke (34) reported an increase in the half life for nitrapyrin from nine to 16 days at 20°C to 43–77 days at 10°C. Touchton *et al.* (89) studied the loss of nitrapyrin from two soils as a function of temperatuie (4, 13, and 21°C). In one soil (Ciscne silt loam; 2 per cent organic matter) nitrapyrin had a half life of 22 days at 4°C decreasing to less than seven days at the higher temperatures. In the other soil (Drummer silty clay loam; 5 per cent organic matter) the half lives were 92, 44, 22 days at 4, 13 and 21°C respectively. They also noted that nitrapyrin added to Drummer soil was much more effective in controlling nitrification than when it was added to the Ciscne soil. Briggs (17) also noted a longer half life in a high organic matter (peat treated) soil (50 days) than in a low organic matter soil (28 days). He also found slightly better inhibition in the soil of higher organic content.

Although, Briggs (17) results do not appear to be compatible with the results of Goring (30) and Bundy and Bremner (18) (the latter workers found inhibition more effective in low organic matter, lighter textured soils), this may not be the case, since there are two different effects occurring. One is the possible stabilization of nitrapyrin in soil by organic matter, while the other is the removal of the nitrapyrin from the aqueous phase by sorption to organic matter, thus reducing the effective inhibitory concentration in solution.

The loss of nitrapyrin from soil appears to be associated with two processes; breakdown and volatilization. Movement associated with leaching does not appear to be important. According to Redemann *et al.* (64) and Bremner *et al.* (16), the principal breakdown product of nitrapyrin is 6-chloropicolinic acid (6-CPA), the latter compound being ineffective as a nitrification inhibitor. In aqueous solution at 30°C it took three days to hydrolyze 40 per cent of the nitrapyrin and 14 days to hydrolyze 90 per cent. As for volatilization, Briggs (17) found that 80 per cent of nitrapyrin applied to fertilizer crystals could be volatilized overnight after a broadcast application of the fertilizer. However, when the fertilizer was incorporated 3 cm deep in the soil, the loss was negligible. Guthrie and Bomke (32) compared the effectiveness of nitrapyrin and 4-amino-1, 2, 4-triazole (ATC), a water soluble nonvolatile inhibitor. They found the latter more effective as an inhibitor in loamy sands while the two inhibitors were similar in silty soils. They observed that nitrapyrin effectiveness persisted four times as long in silt as in sandy loam, which they attributed to more rapid volatilization in the latter.

In studies by Briggs (17) and Touchton *et al.* (87) movement of nitrapyrin in soil was found not to be appreciable (e.g. < 7.5 cm from the application zone) in experiments lasting up to three months. Rudert and Locascio (71) concluded that in sandy soils nitrapyrin movement was effected more by volatilization than by soil solution leaching rate and that ammonium could be more motile at high leaching rates than nitrapyrin.

Nitrifier mortality and growth

After the loss of inhibitor from soil the recommencement of active nitrification will depend on how soon it takes the nitrifier population to reestablish itself. This depends on the extent of population mortality, and the suitability of the soil conditions for growth. There is little information on the mortality of populations due to the presence of an inhibitor. Goring (30) found evidence that ammonium oxidizer populations were significantly decreased in size by the addition of nitrapyrin. His studies showed that nitrification could be reestablished in nitrapyrin treated soils much faster if they were reinoculated. Inoculation was only effective after a minimum of eight weeks delay. The implication is that the inhibitor had decreased sufficiently in this time for nitrification to take place, but the indigenous population had been decreased to such an extent that little ammonium oxidation actually occurred. It is also clear from the work of Goring (30) that the populations were not completely killed, since nitrification could be reestablished in the uninoculated controls after 16 weeks incubation. Laskowski and Bidlack (42) using soil reperfusion techniques found that reestablishment of nitrification took between 50 and 60 days after the addition of nitrapyrin.

Since ammonium oxidizer populations may be significantly decreased by inhibitors, the reestablishment of nitrification will depend greatly on growth characteristics of these microorganisms. Most studies on the growth of nitrifiers have been done in pure culture. Generation time (i.e. time required for population to double) for ammonium oxidizers has been reported to range between 12 and 20 hours (11, 27). However, the growth of nitrifiers in soil is much slower. Generation times for nitrite oxidizers of between 8 and 17 hours were observed in pure culture by Belser (5) and Rennie and Schmidt (65), but in soils they found generation times of between 22 and 129 hours. Morrill and Dawson (54) also measured growth of nitrite oxidizers in soil and found generation times between 20 and 55 hours for six of the soils they studied.

Due to the slow growth of nitrifiers, after a significant die off, it could take an appreciable time for the population to recover. This time of recovery to normalcy would depend on the properties of the soil. Neutral soils would provide better conditions for growth than would acid soils. As noted by Goring (30) this might help explain the slightly better inhibitory characteristics of nitrapyrin for soils of lower pH. Decreased growth rates at lower temperatures could also help explain in part the increased effectiveness of all inhibitors at lower temperatures. Moisture stress will also affect the ability of nitrifier populations to recover to the original state.

Another factor that may affect growth rates is the presence of residual inhibitor. Belser and Schmidt (12) have shown that with low concentrations of nitrapyrin (0.05–0.2 ppm) ammonium oxidizers remained metabolically active, but their growth rates were greatly reduced.

It seems likely that the response of the ammonium oxidizers population has three phases after the application of inhibitor. The initial response is to a high concentration of inhibitor and results in complete inhibition of ammonium oxidation and cell death. As the inhibitor is lost from soil, a second phase is entered whereby the ammonium oxidizers are metabolically active but growth is partially inhibited. The final stage occurs when the inhibitor concentration is so low that both activity and growth are unaffected. There apparently has been no work done on this aspect of nitrifier reestablishment in soils.

Under laboratory conditions the inhibitor can be applied evenly throughout a soil sample. However in the field an inhibitor cannot be applied evenly. Due to this uneven distribution there will be localized zones in which nitrifier populations remain unaffected. These areas will act as inoculants to areas of depleted populations. This offers another mode for population reestablishment. This mode would be through recolonization and will depend on the motility of the ammonium oxidizers in soil.

Inhibition in the field

There have been a number of studies testing the effectiveness of nitrapyrin

Table 1. Effectiveness of nitrification inhibitors on the retention of ammonium in the field

Inhibitor	Inhibitor[a] application rate	Percent of added ammonium remaining in the soil		Time after application	Season	Reference
		inhibitor	control			
Nitrapyrin	1.0%[b]	30–35	< 2	14 days	spring	Huges and Welch (38)
Nitrapyrin	1.0%	15–20	< 10	41 days	early spring	Huges and Welch (38)
Nitrapyrin	0.8%[c]	53	15	59 days	over winter	Touchton *et al.* (86)
Nitrapyrin	0.8%[c]	71	17	66 days	spring	Touchton *et al.* (86)
Nitrapyrin	2.0%	40	11	15 weeks	over winter	Turner *et al.* (92)
Nitrapyrin	8.0%	20–40	< 2	4 weeks	—	Goring (31)
Nitrapyrin	1 ppm	100	< 2	4 months	over winter	Boswell and Anderson (14)
Nitrapyrin	1 ppm	55–72	2–17	5 months	over winter	Sabey (73)
AM	1 ppm	60	36	5 months	over winter	Sabey (73)

[a]Expressed either as a percent of fertilizer or ppm inhibitor on a soil basis.
[b]Based on 112 kg/ha N.
[c]Based on 134 kg/ha N.

in soil under field conditions. Results from some typical studies have been summarized in Table 1. All of these studies showed an improved retention of ammonium in the presence of inhibitor. Typically the delay of nitrification was longer for the studies where temperatures were lower (i.e. over winter incubation). For example, Touchton *et al.* (86) found best control when the temperature was below 10°C. The studies by Goring (31), Sabey (73) and Boswell and Anderson (14) were done under simulated field conditions.

Plant responses in the presence of inhibitors in the field

A number of experiments to determine the effect of the inhibitor on yield have been carried out by growing field crops with and without inhibitor. The results of such experiments have been mixed, as application of inhibitor with fertilizer increased yields in some studies while not in others. Results of some typical studies have been summarized in Tables 2 and 3. Studies where yields

Table 2. Studies where nitrification inhibitors improved crop yield

Inhibitor	Inhibitor[a] application rate (% of fertilizer-N)	Crop	Field conditions			Reference
			nitrification retarded[b]	denitrifying	leaching	
Nitrapyrin	3.3[c]	corn	Yes	–	+	Touchton *et al.* (88)[d]
Nitrapyrin	2.0	corn	NM	–	+	Prasad and Turkhede (63)
Nitrapyrin	5.0	corn	Yes	–	+	Kausta and Varsa (40)
Azide	0.6–1.2	corn	Yes	–	+	Kausta and Varsa (40)
Nitrapyrin	0.1–2.0	corn/cotton sugarbeet	NM	–	+	Swezey and Turner (83)
Nitrapyrin	1.0	rice	NM	+	–	Sharma and Prasad (80)
Nitrapyrin	0.7–2.0	rice	NM	+	–	Wells (96)
Nitrapyrin	5.0[c]	sugar cane	Yes	–	+	Prasad (61)

[a]Fertilizer application rates ranged between 50 and 268 kg/ha with typical rates being around 100 kg/ha.

[b]As measured by higher ammonium in soil with inhibitor than without. NM—soil ammonium not measured.

[c]Yield response greatest at lower fertilizer application rates.

[d]Yield response only when fertilizer applied in fall.

have been improved are given in Table 2. Studies where no crop response was observed are given in Table 3. Results from some of these studies are listed in both tables. These studies typically lasted two years, with good crop response in one year but not the other.

Theoretically, conditions must be conducive for either the denitrification or leaching of nitrate to occur so as to reflect any difference in yields with and without inhibitor. Therefore, these conditions are also tabulated in the tables. As can be seen from Table 2, in most studies where yields have increased such conditions appeared to have existed. Thus, Swezey and Turner (83)

Table 3. Studies where nitrification inhibitors did not improve crop yield

Inhibitor	Inhibitor application rate (% of fertilizer-N)	Crop	Field conditions: nitrification retarded[a]	Field conditions: denitri-fying	Field conditions: leach-ing	Reference
Nitrapyrin	0.6–1.2	corn	Yes	–	–	Kausta and Varsa (40)
Azide	5.6	corn	Yes	–	–	Kausta and Varsa (40)
Nitrapyrin	0.8–3.3	corn	NM[b]	–	–	Touchton *et al.* (90)
Nitrapyrin	1.6	rice (wetland)	NM[b]	+	–	Patrick *et al.* (60)
AM	1.6	rice (wetland)	NM[c]	+	–	Patrick *et al.* (60)
Nitrapyrin	0.5	corn	Yes[d]	–	–	Taber and Peterson (84)
Nitrapyrin	1.0	corn	Yes	–	–	Guthrie and Bomke (32)
ATC	1.0	corn	Yes	–	–	Guthrie and Bomke (32)
Nitrapyrin	1.6–3.3	corn	Yes	–	–	Touchton *et al.* (88)
Nitrapyrin	1.0	corn	No[e]	–	–	Townsend and McRae (91)

[a]As measured by higher ammonium in soil with inhibitor than without. NM—soil ammonium not measured.
[b]Laboratory tests showed that inhibitor was effective in soil used.
[c]Laboratory tests showed that inhibitor was ineffective in soil used.
[d]Indicated indirectly by decrease on nitrate in leaf tissue.
[e]Nitrification did not take place naturally during winter conditions.

found increases in yields with corn and sugarbeet (e.g. for corn a 25 per cent increase in dry matter yield) in soils which were sandy and under conditions which were conducive to leaching. Prasad and Turkhede (63) reported improved corn yields with a leachable soil under heavy rain conditions. Sharma and Prasad (80) and Wells (96) found increases in yield of wetland rice, in the presence of nitrapyrin. Conditions were believed to have been conducive to denitrification.

In most studies where inhibitors did not improve yields, conditions were not conducive for nitrate loss. Thus, Kausta and Varsa (40) did not find consistent improvement in corn yields in a leachable soil during a two-year study when drought conditions prevailed. Touchton *et al.* (90) found no increases of corn yield when nitrapyrin was added with anhydrous ammonia, but noted that soil moisture was not favourable during the growth season. In other studies with urea fertilization, Touchton *et al.* (88) found that corn yield increased when nitrapyrin was added to urea applied in fall (Table 2), but there was no improvement when applied in the spring (Table 3). While it appears that the over winter conditions provided conditions of nitrate loss, it was observed that moisture was not excessive during the growth period. Taber and Peterson (84) found no yield response with corn, but noted that low rainfall during both years of their study would have limited leaching losses. Guthrie and Bomke (32) observed no effect in corn yields by either the addition of nitrapyrin or 4-amino-

1, 2, 4-triazole (ATC) with banded urea. They noted that the conditions had not been favorable for denitrification or leaching. Townsend and McRae (91) found nitrapyrin to be ineffective in improving yields of corn in winter applied fertilizer conditions, but noted that anmmonium was neither nitrified nor lost under winter conditions at the study site (Nova Scotia, Canada). Only in the paddy rice study by Patrick *et al.* (60) was there no yield response under conditions that should have been conducive to denitrification. This is in contrast to the rice paddy studies of Prasad (80) and Wells (96) where yields were improved.

It is clear from the results in Table 2 that in cases where there was a significant improvement in yield the moisture regime in the soil was probably conducive to leaching or denitrification. It is also clear that when lack of yield response occurred (i.e. results in Table 3), it was not the result of lack of inhibition. In five out of seven studies summarized in Table 3, there was more ammonium retained in the soil in the presence of inhibitor than in controls. There is also evidence that inhibition was taking place during the study of Taber and Peterson (84). This is inferred from their observation that there were lower levels of nitrate in leaf tissue when plants were grown with nitrapyrin treated fertilizer.

Phytotoxicity of inhibitors

Not all responses of the plant to the presence of inhibitor may be beneficial. Parr *et al.* (59) concluded that root development in cotton and ryegrass was adversely affected by nitrapyrin (10 ppm). Geronimo *et al.* (29) also found nitrapyrin phytotoxic, but concluded that phytotoxic concentrations were higher than the highest recommended field application rates of inhibitor.

Nitrogen fixation

It has recently been shown that nitrapyrin can significantly reduce the nitrogen fixation rate by symbiotic rhizobia (21). Nothing is known of the effects of inhibitors on other nitrogen-fixing associations, or of their effects on free living nitrogen fixers.

There are several agricultural environments where both nitrification and nitrogen fixation do occur. For example, several investigators have observed high nitrification activities in New Zealand grass-clover (N-fixing) pastures (74, 82). Nitrification can also be very active in the rice paddy environment, where the association between *Azolla* and *Anabaena* can fix considerable amount of nitrogen (79). Needless to say, it is essential that nitrogen-fixing activities be measured in conjunction with inhibitor studies in this environment.

Plant growth on ammonium vs. nitrate

It is often implicitly assumed that plants can grow as well on ammonium as on nitrate. However it is not clear that this is always the case. For example,

Dibb and Welch (24) investigated the growth of corn as affected by ammonium or nitrate nutrition. They found that if certain cations (e.g. potassium) were not present in sufficient quantities, growth of corn on ammonium alone could have an adverse effect on yield.

It is often difficult to determine which form of nitrogen a plant is taking up in field studies. One way of investigating this problem is to analyse nitrate reductase activity in a crop. The nitrate reductase enzyme system is induced in response to nitrate in the soil and the activity of the enzyme system can be used as an index of nitrate assimilation. Using this procedure, Notton *et al.* (58) found a decrease in the nitrate reductase levels of plants (50–90%) in the presence of inhibitor.

Research needs

Although it is clear that certain compounds can specifically block nitrification allowing retention of ammonium in soil, there is still much research needed on the practical application of inhibitors to the field environment.

In research associated with inhibitor application to field crops, it is necessary to measure not only the yield response but also the extent of nitrifier inhibition and the environmental conditions during the study. Knowing the extent of inhibition and the environmental conditions are extremely important in the interpretation of studies where there is no improvement in yield in the presence of an inhibitor. Other research is needed on plant preference of nitrate or ammonium. Research is also needed on the inhibition of nitrification in agricultural environments where nitrogen fixation is an important source of nitrogen.

A better index of nitrifier inhibition than those presently used is also needed. Inhibition is most often inferred by relative increases or decreases in the concentrations of ammonium or nitrate in soils with and without inhibitor. To a lesser extent decreases in countable nitrifier populations have been used as an index of inhibition. While increased ammonium may infer that nitrification is being inhibited, decreases in viable nitrifier counts should be considered a potentially unreliable index of inhibition. A far better index of nitrifier inhibition would be to measure changes in nitrifier activity. There are several simple methods for assessing nitrifier activity (7, 74, 82) and these should be used in preference to concentration analysis alone, or viable nitrifier counts. Following nitrifier activity in soil not only would give a reliable measurement of inhibition, but would also give details of duration of inhibition and the dynamics of the recovery of the nitrifier populations.

Additional research is also needed on the biochemical basis of inhibition. Virtually no work has been done on inhibitors other than nitrapyrin, and even with nitrapyrin little is known of its mode of action.

One goal of inhibitor research should be to predict the effectiveness of an inhibitor in various environments. Due to the complex nature of the various interactions of inhibitors with soil and breakdown of inhibitors as a function of temperature, a modeling approach may be the best way to handle such a study. Meikle *et al.* (49, 50) have made a model for the retention of ammonium due to the inhibition of nitrapyrin in soil. Their model considers organic matter, temperature and the breakdown kinetics of an inhibitor. The one major aspect on the inhibition kinetics that is not considered in their model is the population dynamics of the nitrifiers. A model that combines the aspects of Meikle's model with a population dynamic model such as that developed by Beek and Frissel (4), could predict the longevity of inhibitor and the associated population depletion and recovery in soil. The population recovery dynamics would be essential to predict the duration of inhibition.

In order to investigate a possible role of the inhibition of nitrification in natural ecosystems, experiments have to be designed to study the relative amounts of nitrogen that pass through the various pathways. This would require that rates would have to be measured. The measurement of prime importance would be that the rate of nitrification relative to that of ammonification. If only a small amount of the ammonium that is produced via ammonification is utilized by the nitrification pathway, then the ecosystem would be a prime candidate for active inhibition. Such evidence alone would not be sufficient to prove biotic inhibition. Evidence would also have to be presented to show that low relative nitrification was not simply the result of the nitrifiers being out-competed for ammonium by plants or other microorganisms, or that the abiotic environment was not conducive to nitrification (e.g. low pH). Ultimately, proof may require that the inhibitory substance be characterized.

Conclusions

Two roles have been considered for nitrification inhibitors, both of which involve the improvement of nitrogen cycling by the retention of inorganic nitrogen in the ammonium form. In the case of the application of inhibitors to agricultural environments, there appears to be good prospects of improving yields by the addition of inhibitor. However it is clear that not all environments will respond favourably, and much work has to be done to determine which environments will respond favourably and under what conditions. Improving the assimilation of ammoniacal or ammonium producing fertilizers has clearly important economic and enviromental implications and is an extremely important area of research.

There is still some question whether natural plant compounds act as nitrification inhibitors in the natural environment. It is clear that plants can contain substances that are toxic to nitrifiers, at least in pure culture or at high concentrations. However, in nature due to dilution, adsorption, degradation and complexing with other soil components it is unclear whether these substances can be effective as nitrification inhibitors.

References

1. Alexander, M. Most-probable-number method for microbial populations, in *Methods of Soil Analysis* (Edited by C.A. Black), American Society of Agronomy, Madison (1965).
2. Alexander, M. and Clark, F.E. Nitrifying bacteria, in *Methods of Soil Analysis* (Edited by C.A. Black), American Society of Agronomy, Madison (1965).
3. Bazin, M.J., Saunders, P.T. and Prosser, J.I. Models of microbial reactions in soil, *CRC Critical Reviews in Microbiology*, *4*, 463–498 (1976).
4. Beek, J. and Frissel, M.J. *Simulation of Nitrogen Behaviour in Soils*, Center for Agricultural Publication and Documentation, Wageningen (1973).
5. Belser, L.W. Nitrate reduction to nitrite, a possible source of nitrite for growth of nitrite oxidizing bacteria, *Applied and Environmental Microbiology*, *34*, 403–410 (1977).
6. Belser, L.W. Population ecology of nitrifying bacteria, *Annual Review of Microbiology*, *33*, 309–333 (1979).
7. Belser, L.W. and Mays, E.L. Specific inhibition of nitrite oxidation by chlorate and its use in assessing nitrification in soils and sediments, *Applied and Environmental Microbiology*, *39*, 505–510 (1980).
8. Belser, L.W. and Schmidt, E.L. Diversity in the ammonia-oxidizing nitrifier population of a soil, *Applied and Environmental Microbiology*, *36*, 589–593 (1978).
9. Belser, L.W. and Schmidt, E.L. Serological diversity within a terrestrial ammonia-oxidizing population, *Applied and Environmental Microbiology*, *36*, 589–593 (1978).
10. Belser, L.W. and Schmidt, E.L. Nitrification in soils, in *Microbiology—1978* (Edited by D. Schlessinger), American Society for Microbiology, Washington, D.C. (1978).
11. Belser, L.W. and Schmidt, E.L. Growth and oxidation kinetics of three genera of ammonia oxidizers, *FEMS Microbiology Letters*, *7*, 213–216 (1980).
12. Belser, L.W. and Schmidt, E.L. Inhibitory effect of nitrapyrin on the growth of ammonium oxidizers, *Applied and Environmental Microbiology*, *41*, 819–821 (1981).
13. Bohlool, B.B., Schmidt, E.L. and Beasley, C. Nitrification in the intertidal zone: Influence of effluent type and effect of tannins on nitrifiers, *Applied and Environmental Microbiology*, *34*, 523–528 (1978).
14. Boswell, F.C. and Anderson, O.E. Nitrification inhibitor studies of soil in field-buried polyethylene bags, *Soil Science Society of America Proceedings*, *38*, 851–852 (1974).
15. Brar, S.S. and Giddens, J. Inhibition of nitrification in Bladen grassland soil, *Soil Science Society of America Proceedings*, *32*, 821–823 (1968).

16. Bremner, J.M., Blackmer, A.M. and Bundy, L.G. Problems in the use of nitrification to inhibit nitrification in soils, *Soil Biology and Biochemistry, 10*, 441–442 (1978).
17. Briggs, G.G. The behaviour of the nitrification inhibitor "N-serve" in broadcast and incorporated applications to soil, *Journal of the Science of Food and Agriculture, 26*, 1083–1092 (1975).
18. Bundy, L.G. and Bremner, J.M. Inhibition of nitrification in soils, *Soil Science Society of America Proceedings, 37*, 396–398 (1973).
19. Campbell, N.E.R. and Aleem, M.I.H. The effect of 2-chloro-6-(trichloromethyl) pyridine on the chemoautotrophic metabolism of nitrifying bacteria. I. Ammonia and hydroxylamine oxidation by *Nitrosomonas, Antonie van Leeuwenhoek Journal of Microbiology and Serology, 31*, 124–136 (1965).
20. Campbell, N.E.R. and Aleem, M.I.H. The effect of 2-chloro-6-(trichloromethyl) pyridine on the chemoautotrophic metabolism of nitrifying bacteria. I. Nitrite oxidation by *Nitrobacter, Antonie van Leeuwenhoek Journal of Microbiology and Serology, 31*, 137–141 (1965).
21. Chambers, C.A., Smith, S.E., Smith, F.A., Ramsey, M.D. and Nicholas, D.J.D. Symbiosis of *Trifolium subteranium* with mycorrhizal fungi and *Rhizobium trifolii* as affected by ammonium sulfate and nitrification inhibitors, *Soil Biology and Biochemistry, 12*, 93–100 (1980).
22. Chase, F.E., Corke, C.T. and Robinson, J.B. Nitrifying bacteria in soils, in *The Ecology of Soil Bacteria* (Edited by T.R.G. Gray and D. Parkinson), Liverpool University Press, Liverpool (1968).
23. Clark, F.E. and Paul, E.L. The Microflora of grassland, *Advances in Agronomy, 22*, 375–435 (1970).
24. Dibb, D.W. and Welch, L.F. Corn growth as affected by ammonium vs. nitrite absorbed from soil, *Agronomy Journal, 68*, 89–94 (1976).
25. Dommergues, J.R. *Biologie du Sol*, Presses Universitaire de France, Paris (1977).
26. Dua, R.D., Bhandari, B. and Nicholas, D.J.D. Stable isotope studies on the oxidation of ammonia to hydroxylamine by *Nitrosomonas europaea, FEBS Letters, 106*, 401–404 (1979).
27. Engel, M.S. and Alexander, M. Growth and autotrophic metabolism of *Nitrosomonas europaea, Journal of Bacteriology, 76*, 217–222 (1958).
28. Focht, D.D. and Verstraete, W. Biochemical ecology of nitrification and denitrification, *Advances in Microbial Ecology, 1*, 135–214 (1977).
29. Geronimo, L.L., Smith, G.M., Stockdale, G.D. and Goring, C.A.I. Comparative phytotoxicity of nitrapyrin and its principal metabolite 6-chloropicolinic acid, *Agronomy Journal, 65*, 689–692 (1973).
30. Goring, C.I.A. The control of nitrification by 2-chloro-6-(trichloromethyl) pyridine, *Soil Science, 93*, 211–218 (1962).
31. Goring, C.I.A. Control of nitrification of ammonium fertilizers and urea

by 2-chloro-6-(trichloromethyl) pyridine, *Soil Science, 93*, 431–439 (1962).
32. Guthrie, T.F. and Bomke, A.A. Nitrification inhibition by N-serve and ATC in soils of varying texture, *Soil Science Society of America Journal, 44*, 314-320 (1980).
33. Harms, H., Koops, H.P. and Wehrmann, H. An ammonia-oxidizing bacterium: *Nitrosovibrio tenuis* nov. gen. nov. sp., *Archive for Microbiology, 108*, 105–111 (1976).
34. Herlihy, M. and Quirke, W. The persistence of 2-chloro-6-(trichloromethyl) pyridine in soil, *Communications in Soil Science and Plant Analysis, 6*, 513–520 (1975).
35. Harmsen, G.W. and van Schreven, D.A. Mineralization of organic nitrogen in soil, *Advances in Agronomy, 7*, 299–385 (1955).
36. Hooper, A.B. Nitrogen oxidation and electron transport in ammonia-oxidizing bacteria, in *Microbiology—1978* (Edited by D. Schlessinger), American Society for Microbiology, Washington, D.C. (1978).
37. Hooper, A.B. and Terry, K.R. Specific inhibitors of ammonia oxidation in *Nitrosomonas, Journal of Bacteriology, 115*, 480–485 (1973).
38. Hughes, T.D. and Welch, L.F. 2-chloro-6-(trichloromethyl) pyridine as a nitrification inhibitor for anhydrous ammonia applied in different seasons, *Agronomy Journal, 62*, 821–824 (1970).
39. Hynes, R.K. and Knowles, R. Inhibition by acetylene of ammonia oxidation in *Nitrosomonas europaea, FEMS Microbiology Letters, 4*, 319–322 (1978).
40. Kausta, G. and Varsa, E.C. Nitrification inhibitors—do they work? *Down to Earth, 28*, 21–23 (1972).
41. Keeney, D.R. and Bremner, J.M. Determination and isotope-ratio analysis of different forms of nitrogen in soils: 6. mineralizable nitrogen, *Soil Science Society of America Proceedings, 31*, 34–39 (1967).
42. Laskowski, D.A. and Bidlack, H.D. Nitrification recovery in soil after inhibition by nitrapyrin, *Down to Earth, 33*, 12–17 (1977).
43. Likens, G.E., Bormann, F.H. and Johnson, N.M. Nitrification: importance to nutrient losses from a cutover forested ecosystem, *Science, 163*, 1205–1206 (1969).
44. Lodhi, M.A.K. Comparative inhibition of nitrifiers and nitrification in a forest community as a result of the allelopathic nature of various tree species, *American Journal of Botany, 65*, 1135–1137 (1978).
45. Mahendrappa, M.K., Smith, R.L. and Christianson, A.T. Nitrifying organisms affected by climatic region in western United States, *Soil Science Society of America Proceedings, 30*, 60–62 (1966).
46. Matulewich, V.A., Strom, P.F. and Finstein, M.S. Length of incubation for enumerating nitrifying bacteria present in various environments, *Applied Microbiology, 29*, 265–268 (1975).
47. McLaren, A.D. Nitrification in soil: Systems approaching a steady state,

Soil Science Society of America Proceedings, *33*, 551–556 (1969).
48. McLaren, A.D. Kinetics of nitrification in soil: Growth of the nitrifiers, *Soil Science Society of America Proceedings*, *35*, 91–95 (1971).
49. Meikle, R.W. Prediction of ammonium nitrogen fertilizer disappearance from soils in the presence and absence of N-serve nitrogen stabilizers, *Down to Earth*, *34*, 6–10 (1978).
50. Meikle, R.W. Prediction of ammonium nitrogen fertilizer disappearance from soils in the presence and absence of N-serve nitrogen stabilizers, *Soil Science*, *127*, 292–298 (1979).
51. Meyers, R.J.K. Temperature effects on ammonification and nitrification in a tropical soil, *Soil Biology and Biochemistry*, *7*, 83–86 (1975).
52. Molina, J.A.E., Gerard, G. and Mignolet, R. Asynchronous activity of ammonium oxidizer clusters in soil, *Soil Science Society of America Journal*, *43*, 728–731 (1979).
53. Molina, J.A.E. and Rovira, A.D. The influence of plant roots on autotrophic nitrifying bacteria, *Canadian Journal Microbiology*, *10*, 249–257 (1968).
54. Morrill, L.G. and Dawson, J.E. Growth rates of nitrifying chemoautotrophs in soil, *Journal of Bacteriology*, *83*, 205–206 (1962).
55. Munro, P.E. Inhibition of nitrifiers by grass root extracts, *Journal of Applied Ecology*, *3*, 231–238 (1966).
56. Neal, J.L. Inhibition of nitrifying bacteria by grass and forb root extracts, *Canadian Journal of Microbiology*, *15*, 633–635 (1969).
57. Nicholas, D.J.D. Intermediary metabolism of nitrifying bacteria, with particular reference to nitrogen, carbon and sulfur compounds, in *Microbiology—1978* (Edited D. Schlessinger), American Society for Microbiology, Washington, D.C. (1978).
58. Notton, B.A., Watson, E.F. and Hewitt, E.J. Effects of N-serve (2-chloro-6-(trichloromethyl) pyridine) formulations on nitrification and on loss of nitrate in sand culture experiments, *Plant and Soil*, *51*, 1–12 (1979).
59. Parr, J.F., Carroll, B.R. and Smith, S. Nitrification inhibition in soil: I. a comparison of 2-chloro-6-(trichloromethyl) pyridine and potassium azide formulated with anhydrous ammonia, *Soil Science Society of America Proceedings*, *35*, 469–473 (1971).
60. Patrick, W.H., Peterson, F.J. and Turner, F.T. Nitrification inhibitors for lowland rice, *Soil Science*, *105*, 103–105 (1968).
61. Prasad, R. Nitrogen nutrition and yield of sugar cane as affected by N-serve, *Agronomy Journal*, *68*, 343–346 (1976).
62. Prasad, R., Rajale, G.B. and Lakhive, B.A. Nitrification retarders and slow-release nitrogen fertilizers, *Advances in Agronomy*, *23*, 337–383 (1971).
63. Prasad, R. and Turkhede, B.B. Relative efficiency of nitrogen fertilizer for 'Ganga 101' maize (*Zea mays* L.) as influenced by rainfall, *Indian*

Journal of Agricultural Science, *41*, 485–489 (1971).
64. Redemann, C.T.R., Meikle, R.W. and Widofsky, J.G. The loss of 2-chloro-6-(trichloromethyl) pyridine from soil, *Journal of Agricultural and Food Chemistry*, *12*, 201–207 (1964).
65. Rennie, R.J. and Schmidt, E.L. Autecological and kinetic analysis of competition between strains of *Nitrobacter* in soils, *Ecological Bulletins* (Stockholm), *25*, 431–441 (1977).
66. Rice, E.L. Inhibition of nitrogen-fixing and nitrifying bacteria by seed plants, *Ecology*, *45*, 824–837 (1964).
67. Rice, E.L. and Pancholy, S.K. Inhibition of nitrification by climax vegetation, *American Journal of Botany*, *59*, 1033–1040 (1972).
68. Rice, E.L. and Pancholy, S.K. Inhibition of nitrification by climax ecosystems. II. Additional evidence and possible role of tannins, *American Journal of Botany*, *60*, 691–702 (1973).
69. Richardson, H.L. The nitrogen cycle in grassland soils: with especial reference to the Rothamsted Park grass experiment, *Journal of Agricultural Science*, *28*, 73–121 (1938).
70. Robinson, J.B. Nitrification in a New Zealand grassland soil, *Plant and Soil*, *19*, 173–183 (1963).
71. Rudert, B.D. and Locascio, S.J. Differential mobility of nitrapyrin and ammonium in a sandy soil and its effect on nitrapyrin efficiency, *Agronomy Journal*, *71*, 487–489 (1979).
72. Russell, E.W. *Soil Conditions and Plant Growth*, 10th edition, Longman, London (1973).
73. Sabey, B.R. The influence of nitrification suppressants on the rate of ammonium oxidation in Midwestern USA field soils, *Soil Science Society of America Proceedings*, *32*, 675–679 (1968).
74. Sarathchandra, S.V. Nitrification activities of some New Zealand soils and the effect of some clay types on nitrification, *New Zealand Journal of Agricultural Research*, *21*, 615–621 (1978).
75. Schmidt, E.L. Nitrification in soil, in *Soil Nitrogen* (Edited by F.J. Stevenson, J.M. Bremner, R.D. Hauk and D.R. Keeney), American Society of Agronomy, Madison (1981).
76. Schmidt, E.L. Nitrifying microorganisms and their methodology, in *Microbiology—1978* (Edited D. Schlessinger), American Society for Microbiology, Washington, D.C. (1978).
77. Schmidt, E.L. Quantitative autecological study of microorganisms in soil by immunofluorescence, *Soil Science*, *118*, 141–149 (1974).
78. Schmidt, E.L. Fluorescent antibody techniques for the study of microbial ecology, *Bulletins from the Ecological Research Committee* (Stockholm), *17*, 67–76 (1973).
79. Shanmugan, K.T., O'Gara, F.O., Anderson, K., Morandi, C. and Valentine, R.C. Control of biological nitrogen fixation, in *Nitrogen in the*

Environment, Vol. 2 (Edited by D.R. Nielson and J.G. MacDonald), Academic Press, New York (1978).

80. Sharma, S.N. and Prasad, R. Effects of rates of nitrogen and relative efficiency of sulfur-coated urea and nitrapyrin-treated urea in dry matter production and nitrogen uptake by rice, *Plant and Soil*, *55*, 389–396 (1980).
81. Soulides, D.A. and Clark, F.E. Nitrification in grassland soils, *Soil Science Society of America Proceedings*, *22*, 308–311 (1958).
82. Steele, K.W., Wilson, A.T. and Saunders, W.M.H. Nitrification activities in New Zealand grassland soils, *New Zealand Journal of Agricultural Research*, *23*, 249–256 (1980).
83. Swezey, A.W. and Turner, G.O. Crop experiments on the effect of 2-chloro-6-(trichloromethyl) pyridine for the control of nitrification of ammonium and urea fertilizers, *Agronomy Journal*, *54*, 532–535 (1962).
84. Taber, H.G. and Peterson, L.E. Effect of nitrogen source and nitrapyrin on sweet corn, *Horticultural Science*, *14*, 34–36 (1979).
85. Theron, J.J. The influence of plants on the mineralization of nitrogen and the maintenance of organic matter in the soil, *Journal of Agricultural Science*, *41*, 289–296 (1951).
86. Touchton, J.T., Hoeft, R.G. and Welch, L.F. Effect of nitrapyrin on nitrification of fall and spring-applied anhydrous ammonia, *Agronomy Journal*, *70*, 805–810 (1978).
87. Touchton, J.T., Hoeft, R.G. and Welch, L.F. Nitrapyrin degradation and movement in soil, *Agronomy Journal*, *70*, 811–816 (1978).
88. Touchton, J.T., Hoeft, R.G. and Welch, L.F. Effect of nitrapyrin on nitrification of broadcast-applied urea plant nutrient concentrations and corn yield, *Agronomy Journal*, *71*, 787–791 (1979).
89. Touchton, J.T., Hoeft, R.G., Welch, L.F. and Argyilan, W.L. Loss of nitrapyrin from soils as affected by pH and temperature, *Agronomy Journal*, *71*, 865–869 (1979).
90. Touchton, J.T., Hoeft, R.G., Welch, L.F., Mulvaney, D.L., Oldham, M.G. and Zajiek, F.E. N uptake and corn yield as affected by application of nitrapyrin with anhydrous ammonia, *Agronomy Journal*, *71*, 238–242 (1979).
91. Townsend, L.R. and McRae, K.B. The effect of the nitrification inhibitor nitrapyrin on yield and nitrogen fraction in soil and tissue of corn, *Canadian Journal of Plant Science*, *60*, 337–347 (1980).
92. Turner, G.O., Warren, L.E. and Andriessen, F.G. Effect of 2-chloro-6-(trichloromethyl) pyridine on the nitrification of ammonium fertilizers in field soils, *Soil Science*, *94*, 270–273 (1962).
93. Vitousek, P.M. and Reiners, W.A. Ecosystem succession and nutrient retention: A hypothesis, *Bioscience*, *25*, 376–381 (1975).
94. Walter, H.M., Keeney, D.R. and Fillery, I.R. Inhibition of nitrification

by acetylene, *Soil Science Society of America Journal*, *43*, 195–196 (1979).
95. Watson, S.W. Gram-negative chemolithotrophic bacteria Family I, in *Bergey's Manual of Determinative Bacteriology*, 8th edition (Edited by R.E. Buchanan and N.E. Gibbons), The Williams and Wilkins Company, Baltimore (1974).
96. Wells, B.R. Nitrapyrin (2-chloro-6-(trichloromethyl) pyridine) as a nitrification inhibitor for paddy rice, *Down to Earth*, *32*, 28–32 (1977).

N.S. SUBBA RAO

11. Phosphate Solubilization by Soil Microorganisms

Next only to nitrogen, phosphorus is another important key element not only in plant metabolism but also in soil microbiological processes. The energy cycle in cell protoplasm is intimately related to this element. Soil microorganisms are involved in many activities in soil such as the solubilization of bound phosphates by acid secretion, the mineralization of organic phosphatic compounds by which organic phosphorus is converted to inorganic forms and the immobilization of phosphorus by which inorganic ions are incorporated into microbial cell material. These and other related reactions constitute the phosphorus cycle (Fig. 1).

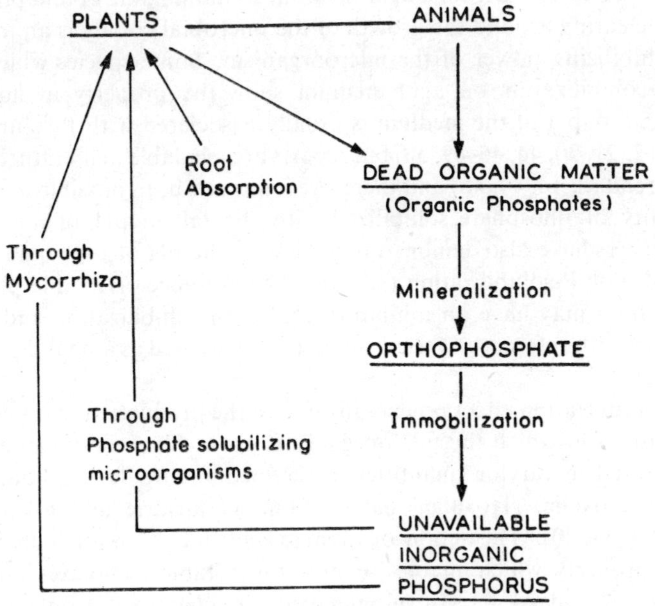

Fig. 1. The phosphorus cycle.

There are both inorganic as well as organic sources of phosphorus in soil. The inorganic sources include mono-, di-, tri- and tetra-phosphates of K^+, Na^+, Ca^{++}, Mg^{++}, Al^{+++} and Fe^{+++}. The organic sources comprise phytin, phospholipids, nucleic acids, phosphorylated sugars and co-enzymes. Organic matter rich soils are also rich in organic phosphorus derived from vegetation and protoplasm of microorganisms which undergo decomposition.

Phosphorus is added to soil in the form of phosphatic fertilizer, part of which is utilized by the plant and the remainder converted into fixed and insoluble forms of phosphorus. Phosphatic fertilizers are in short supply in developing countries and hence they have to be imported to meet indigenous needs. Large quantities of low grade rock phosphate are, however, available in developing countries like India but they are unfit as raw material for the production of superphosphate. There is, nevertheless, enough scope to use finely ground rock phosphate directly on the farm, especially in acidic soils.

Solubilization of phosphates by microorganisms

From time to time, the ability of different microorganisms to solubilize bound phosphates (Table 1) incorporated in agar, soil or liquid media have been demonstrated (1–57). Mycorrhizal fungi are also endowed with the property of phosphate mobilization and this aspect is covered elsewhere in the book (Chapters 12 and 13). On solid agar medium containing the bound phosphate, a zone of clearing around the growth of the microbial colony is an indication of the solubilizing power of the microorganism. Some species which do not exhibit P-solubilization on agar medium show the property in liquid medium. A fall in pH of the medium is usually associated with P-solubilization (2, 9, 10, 12, 21, 30, 44, 46–49, 56) but reports are available in literature to show no such trend (2, 14, 44, 48) and therefore it has not been possible to correlate the quantity of phosphate solubilized with the fall in pH of the medium. Many workers have also demonstrated a rise in the pH of the liquid medium associated with P-solubilization (12, 44). The presence of soluble phosphate in the medium may have an inhibitory action on solubilization and this has been shown with reference to tricalcium phosphate used as a bound phosphate source (12).

The solubilization effect is generally due to the production of organic acids in the medium in which the microorganisms grow. The non-volatile organic acids liberated in varying quantities in the medium are citric, glutamic, succinic, lactic, oxalic, glyoxalic, maleic, fumaric, tartaric and α-ketobutyric acids (8, 20, 38, 40). The action of organic acids has been attributed to their chelation property which enables them to form stable complexes with Ca^{++}, Mg^{++}, Fe^{+++} and Al^{+++}. In sulphur oxidizing chemoautotrophs (ex. *Thiobacillus*) and ammonium oxidizing chemoautotrophs (ex. *Nitrosomonas*), sulphuric acid and nitric acid are formed, respectively. Both the organic as well as

Table 1. Microorganisms and sources of phosphate which have been reported to be involved in phosphate solubilization (from several published reports cited in the references)

Microorganisms	Phosphate sources
Bacteria	*Mineral*
Bacillus sp., *B. pulvifaciens, B. megaterium, B. circulans, B. subtilis, B. mycoides, B. mesentericus, B. fluorescence, B. circulans*	Tricalcium phosphate
	Calcium phosphate
	Iron phosphate
Pseudomonas sp., *P. putida, P. liquifaciens, P. calcis, P. rathonia*	Hydroxyapatite
	Fluorapatite
Escherichia freundii, E. intermedia	Rock phosphate
Xanthomonas spp.	*Organic*
Flavobacterium spp.	Calcium phytate
Brevibacterium spp.	Calcium glycerophosphate
Serratia spp.	Phytin
Alcaligenes spp.	Lecithin
Achromobacter spp.	Hexose monophosphatic ester
Aerobacter aerogenes	
Erwinia spp.	Phenyl phosphate
Nitrosomonas spp.	Other organic phosphates
Thiobacillus thiooxidans	
Fungi	
Aspergillus sp., *A. niger, A. flavus, A. fumigatus, A. terreus, A. awamori*	
Penicillium sp., *P. lilacinum, P. digitatum*	
Fusarium sp., *F. oxysporum*	
Curvularia lunata	
Humicola sp.	
Sclerotium rolfsii	
Pythium sp.	
Acrothecium sp.	
Phoma sp.	
Mortierella sp.	
Paecilomyces sp.	
Cladosporium sp.	
Rhizoctonia sp.	
Cunninghamella sp.	
Rhodotorula sp.	
Candida sp.	
Schwanniomyces occidentalis	
Oideodendron sp.	
Pseudogymnoascus sp.	
Actinomycetes	
Streptomyces sp.	

Table 2. Some published reports on beneficial effects of phosphate-solubilizing microorganisms

Microorganisms	Details of study	Reference
Rhizosphere microorganisms capable of phosphate dissolution	Improved growth and phosphate uptake in oat, mustard etc.	Gerretsen (21) Pikovskaya (41)
Lecithin hydrolysing microorganisms	Improved growth of potato and sugar beets	Fialova (15)
'Phosphobacterin'	Improved growth of cabbage and tomato—better germination	Mishustin and Naumova (37)
Phosphobacteria	Significant increase in tomato by 7.5 per cent over control but not on wheat	Smith *et al.* (45)
'Phosphobacterin'	In association with 'Azotobacterin' suppressed fungi in the root zone and stimulated plant growth	Mishustin and Naumova (38)
B. megaterium var. *phosphaticum*	Reduces disease in agricultural plants caused by pathogenic fungi	Menkina (36)
Phosphate dissolving bacteria	Increase P uptake and plant growth of barley	Taha *et al.* (56)
'Phosphobacterin' from the U.S.S.R. 'Fosfo 24' from Czechoslovakia	10–20 per cent increase in P uptake in berseem (*Trifolium alexandrinum*) and 7–33 per cent in wheat	Sundara Rao and Paul (54)
'Phosphaterin' from the U.S.S.R. and Indian culture	The Russian culture was better than Indian culture in significantly increasing yields and P uptake in tomato in pots and wheat in field trials	Sundara Rao *et al.* (55)
Bacillus circulans	Inhibited the growth of *Alternaria solani*, *Helminthosporium oryzae*, *Rhizoctonia solani* and *Fusarium udum*	Vasantharajan (57)
'Phosphobacterin'	37 field trials on wheat, berseem, maize, chickpea, soybean, groundnut, arhar or red gram and rice were done out of which 10 experiments showed beneficial effects	Sundara Rao (51, 52)
Bacillus polymyxa and *B. pulvifaciens*	*B. polymyxa* in the presence of rock phosphate significantly increased grain and straw yield and P uptake of wheat whereas *P. pulvifaciens* did not do so	Gaur and Ostwal (18)
Pseudomonas striata	With superphosphate and rock phosphate significantly increased yield of wheat	Gaur *et al.* (19)
Bacillus polymyxa	With rock phosphate significantly increased the yields of rice	Gaur and Negi (17)
Pseudomonas striata	With rock phosphate and superphosphate significantly increased potato yield	

the inorganic acids convert tricalcium phosphate into monobasic phosphates, depending on the extent of carbohydrate utilization in the production of these acids. These reactions take place in the rhizosphere and because P-solubilizing microorganisms render more phosphates into solution than is required for their growth and metabolism, the surplus gets absorbed by plants (3).

Use of ^{32}P in studies on P-solubilization

Several studies have been done by using tagged phosphate sources to quantify the amount of phosphorus made available to plants by the activity of P-solubilizing microorganisms. Such studies have been done in sterile as well as non-sterile soils with and without inoculation of P-solubilizing microorganisms (53). In one of the studies (7), tagged superphosphate and hydroxy apatite were used as phosphate sources and *Bacillus megaterium* var. *phosphaticum* and *B. circulans* were used as inoculants. *Bacillus megaterium* was found to increase the availability of phosphorus in sterile soil in the presence of farmyard manure and ammonium sulphate. In another investigation using *Bacillus pumilis*, *B. polymyxa* and *B. pulvifaciens*, it was observed that the microorganisms no doubt solubilized radioactive phosphorus but inoculation did not improve solubilization significantly in hydroxyapatite added sterile and non-sterile soils (54). Similarly, tagged hydroxyapatite was used to assess the effect of seed inoculation with *Pseudomonas* sp. wherein it was found that the amount of hydroxyapatite dissolved by the culture did not significantly differ between inoculated and control series (28). Studies have been done wherein tagged superphosphate and hydroxyapatite were used as phosphate sources for cowpea (*Vigna* sp.) with inoculation of phosphate solubilizing microorganisms. The per cent phosphorus in the plant increased due to inoculation whereas in the uninoculated controls superphosphate was found to be a better source of the element than hydroxy apatite.

Benefits due to inoculation with phosphate dissolving microorganisms

Several investigators have repeatedly attempted to test the value of seed inoculation with phosphate-dissolving microorganisms on growth and yield of plants (Table 2). These plants include oats, mustard, sugar beets, cabbage, tomato, barley, Egyptian clover, maize, potato, red gram, rice, bengal gram or chickpea, soybean and groundnut. Vegetables appear to respond more favourably to phosphobacteri inoculation than grain crops. Significant increases in yield in field experiments due to inoculation have not been consistently obtained. Of the 37 field trials carried out in India, only 10 trials showed significant increases in yields with various crops (52, 54). More recently, studies on this aspect are being renewed to find out the usefulness of such microbial inoculation with rock phosphate application in acidic as well as neutral soils (17, 19).

References

1. Agnihotri, V.P. Solubilization of insoluble phosphates by some soil fungi isolated from nursery seed-beds, *Canadian Journal of Microbiology, 16*, 887–880 (1970).
2. Ahmad, N. and Jha, K.K. Solubilization of rock phosphate by micro-organisms isolated from Bihar soils, *Journal of General and Applied Microbiology, 14*, 89–95 (1968).
3. Alexander, M. *Introduction to Soil Microbiology*, John Wiley and Sons, New York (1977).
4. Aurora, D. Studies of rock phosphate solubilizing microorganisms in relation to crop yield and phosphate uptake, Ph.D. thesis, Post Graduate School, Indian Agricultural Research Institute, New Delhi (1975).
5. Bajpai, P.D. and Sundara Rao, W.V.B. Phosphate solubilizing bacteria. I. Solubilization of phosphate in liquid culture by selected bacteria as affected by different pH values, *Soil Science and Plant Nutrition, 17* (2), 41–43 (1971a).
6. Bajpai, P.D. and Sundara Rao, W.V.B. Phosphate solubilizing bacteria. II. Extracellular production of organic acids by selected bacteria solubilizing insoluble phosphate, *Soil Science and Plant Nutrition, 17* (2), 44–45 (1971b).
7. Bajpai, P.D. and Sundara Rao, W.V.B. Phosphate solubilizing bacteria. III. Soil inoculation with phosphorus solubilizing bacteria, *Soil Science and Plant Nutrition, 17* (2), 46–53 (1971c).
8. Bardiya, M.C. Studies on microbiological mobilization of rock phosphate, Ph.D. thesis, Post Graduate School, Indian Agricultural Research Institute, New Delhi (1970).
9. Bardiya, M.C. and Gaur, A.C. Rock phosphate dissolution by bacteria, *Indian Journal of Microbiology, 12*, 269–271 (1972).
10. Bardiya, M.C. and Gaur, A.C. Isolation and screening of micro-organisms dissolving low grade rock phosphate, *Folia Microbiology, 19*, 386–389 (1974).
11. Casida, L.E. Phosphatase activity of some common soil fungi, *Soil Science, 87*, 305–310 (1959).
12. Chhonkar, P.K. and Subba Rao, N.S. Phosphate solubilization by fungi associated with legume root nodules, *Canadian Journal of Microbiology, 13*, 749–753 (1967).
13. Cooper, R. Bacterial fertilizers in Soviet Union, *Soils and Fertilizers, 22* (5), 327–333 (1959).
14. Das, A.C. Utilization of insoluble phosphate by soil fungi, *Journal of Indian Society of Soil Science, 11*, 203–207 (1963).
15. Fialova, J. Experiments with bacteria decomposing organic phosphorus compounds, *Soils and Fertilizers, 19*, 832 (1955).

16. Gaur, A.C., Madan, M. and Ostwal, K.P. Solubilization of phosphatic compounds by native microflora of rock phosphates, *Indian Journal of Experimental Biology*, *11*, 427–429 (1973).
17. Gaur, A.C. and Negi, A.S. Bacterial solubilization of phosphate as evidenced by yield of potato crop, *Plant and Soil* (in press).
18. Gaur, A.C. and Ostwal, K.P. Influence of phosphate dissolving bacilli on yield and phosphate uptake by wheat crop, *Indian Journal of Experimental Biology*, *10*, 393–394 (1972).
19. Gaur, A.C., Ostwal, K.P. and Mathur, R.S. Save superphosphate by using phosphobacteria, *Kheti*, *32*, 23–25 (1980).
20. Gaur, A.C. and Pareek, K.P. Organic acids in soil during degradation of organic residues, *Proceedings of Indian National Science Academy*, *40*, 68 (1972).
21. Gerretsen, F.C. The influence of microorganisms on the phosphorus uptake by the plant, *Plant and Soil*, *1*, 51–81 (1948).
22. Goswami, K.P. and Sen, A. Solubilization of calcium phosphate by three strains of phosphobacterin, *Indian Journal of Agricultural Science*, *32*, 96–101 (1962).
23. Greaves, M.P., Anderson, G. and Webley, D.M. A study of breakdown of organic phosphates by microorganisms from the root region of certain pasture grasses, *Journal of Applied Bacteriology*, *28*, 454–464 (1965).
24. Greaves, M.P. and Webley, D.M. The hydrolysis of myo-inositol hexaphosphate by soil microorganisms, *Soil Biology and Biochemistry*, *1*, 37–43 (1969).
25. Johnson, D.D. and Broadbent, F.E. Microbial turnover of phosphorus in soil, *Proceedings of Soil Sicence Society of America*, *16*, 56–59 (1952).
26. Katznelson, H. and Bose, B. Metabolic activity and phosphate dissolving ability of bacterial isolates from wheat rhizosphere and non-rhizosphere soil, *Canadian Journal of Microbiology*, *5*, 79–85 (1959).
27. Katznelson, H., Peterson, E.A. and Rouatt, J.W. Phosphate dissolving microorganisms on seeds and in the root zone of plants, *Canadian Journal of Biology*, *40*, 1181–1186 (1962).
28. Kavimandan, S.K. and Gaur, A.C. (1971). Effect of seed inoculation with *Pseudomonas* sp. on phosphate uptake and yield of maize, *Current Science*, *40*, 439–440 (1971).
29. Kelley, W.P. Effects of nitrifying bacteria on the solubility of tricalcium phosphate, *Journal of Agricultural Research*, *12*, 671–683 (1918).
30. Louw, H.A. and Webley, D.M. A study of soil bacteria dissolving certain mineral phosphate fertilizers and related compounds, *Journal of Applied Bacteriology*, *22*, 227–233 (1959).
31. Mahmoud, S.A.Z., Abdel-Hafez, El-Sawy, M. and Hanafy, E.A. Ribonucleic acid and lecithin hydrolyzing bacteria in soil and rhizosphere of

wheat and broad bean cultivated in Egypt, *Zentral Bakteriologie* Abt. II, *128*, 196–202 (1973a).

32. Mahmoud, S.A.Z., Abdel-Hafez, A.M., El-Sawy, M. and Hanafy, E.A. Phenolphthalein diphosphate splitting bacteria in soil and rhizosphere of wheat and broad bean in soils of Egypt, *Zentral Bakteriologie* Abt. II, *128*, 524–527 (1973b).
33. Mahmoud, S.A.Z., Abdel-Hafez, A.M., El-Sawy, M. and Hanefy, E.A. Phytin-hydrolysing bacteria in soils of Egypt, *Zentral Bakteriologie* Abt. II, *128*, 528–531 (1973c).
34. Mehta, Y.R. and Bhide, V.P. Solubilization of tricalcium phosphate by some soil fungi, *Indian Journal of Experimental Biology*, *8* (3), 229 (1970).
35. Menkina, R.A. Bacteria which mineralise organic phosphorus compounds, *Mikrobiologiya*, *19*, 308–316 (1950).
36. Menkina, R.A. Bacterial fertilizers and their importance for agricultural plants, *Mikrobiologiya*, *33*, 352–358 (1963).
37. Mishustin, E.N. and Naumova, A.N. The use of bacterial fertilizers in sowing vegetable seeds into peat manure nutrient cubes, *Mikrobiologiya*, *25*, 41–48 (1956).
38. Mishustin, E.N. and Naumova, A.N. Bacterial fertilizers, their effectiveness and mode of action, *Mikrobiologiya*, *31*, 543–555 (1962).
39. Ostwal, K.P. and Bhide, V.P. Solubilization of tricalcium phosphate by soil *Pseudomonas*, *Indian Journal of Experimental Biology*, *10*, 153–154 (1972).
40. Pareek, R.P. and Gaur, A.C. Release of phosphate from tricalcium phosphates by organic acids, *Current Science*, *42*, 278–279 (1973).
41. Pikovskaya, R.I. Mobilization of phosphorus in soil in connection with vital activity of some microbial species, *Mikrobiologiya*, *17*, 362–370 (1948).
42. Rose, R.E. Techniques of determining the effect of microorganisms on insoluble inorganic phosphates, *New Zealand Journal of Science and Technology*, *38*, 773–780 (1957).
43. Sen, A. and Paul, N.B. Solubilization of phosphates by some common soil bacteria, *Current Science*, *26*, 222 (1957).
44. Sethi, R.P. and Subba Rao, N.S. Solubilization of tricalcium phosphate and calcium phytate by soil fungi, *Journal of General Applied Microbiology*, *14*, 329–331 (1968).
45. Smith, J.H., Allison, F.E. and Soulides, D.A. Evaluation of phosphobacterin as a soil inoculant, *Soil Science Society of America Proceedings*, *25*, 109–111 (1961).
46. Sperber, J.I. Solution of mineral phosphates by soil bacteria, *Nature*, London, *180*, 994–995 (1957).
47. Sperber, J.I. The incidence of apatite dissolving organisms in the rhizo-

sphere and soil, *Australian Journal of Agricultural Research*, *9*, 778–781 (1958a).
48. Sperber, J.I. Solution of apatite by soil microorganisms producing organic acids, *Australian Journal of Agricultural Research*, *9*, 782–787 (1958b).
49. Sperber, J.I. Release of phosphate from soil minerals by hydrogen sulphide, *Nature*, London, *181*, 934 (1958c).
50. Subba Rao, N.S. and Bajpai, P.D. Fungi on the surface of legume root nodules and phosphate solubilization, *Experentia*, *21*, 386–387 (1965).
51. Sundara Rao, W.V.B. Bacterial fertilizers, in *Handbook of Manures and Fertilizers*, Indian Council of Agricultural Research, New Delhi (1965).
52. Sundara Rao, W.V.B. Phosphorus solubilization by microorganisms, pp. 210–29, in *Proceedings of All India Symposium on Agricultural Microbiology*, University of Agricultural Sciences, Bangalore (1968).
53. Sundara Rao, W.V.B., Bajpai, P.D., Sharma, J.P. and Subbaiah, B.V. Solubilization of phosphorus solubilizing organisms using ^{32}P as tracer and the influence of seed bacterization on the uptake by the crop, *Journal of Indian Society of Soil Science*, *11*, 209–219 (1963).
54. Sundara Rao, W.V.B. and Paul, N.B. Bacterization with phosphobacterin, in *Radioisotopes, Fertilizer and Cowdung Gas Plant*, Indian Council of Agricultural Research Symposium Proceedings, 322–326 (1959).
55. Sundara Rao, W.V.B. and Sinha, M.K. Phosphate dissolving organisms in the soil and the rhizosphere, *Indian Journal of Agricultural Sciences*, *33*, 272–278 (1963).
56. Taha, S.M., Mahmoud, S.A.Z., El-Damaty, A.H. and El-Hafez. Activity of phosphate dissolving bacteria in Egyptian soils, *Plant and Soil*, *31* (1), 149–160 (1969).
57. Vasantharajan, V.N. Inhibition of plant pathogenic fungi by *B. circulans*, solubilizing phosphorus, *Indian Journal of Microbiology*, *4*, 27 (1964).

R. MOLINA and J.M. TRAPPE

12. Applied Aspects of Ectomycorrhizae

Introduction

Mycorrhizae, or fungus-root associations, are the norm for most vascular plants (15). Many plants depend on their mycorrhizal structures for adequate uptake of nutrients and survival in natural ecosystems (11, 48, 81). Various types of mycorrhizae are known, but this paper will address only ectomycorrhizae and their applied aspects. Ectomycorrhizal hosts include species primarily in the Pinaceae, Fagaceae, Betulaceae, and Saliceae and a few genera in other families, e.g., *Eucalyptus*, *Tilia*, and *Arbutus* (47); several members of the Caesalpiniaceae and Dipterocarpaceae are prevalent ectomycorrhizal hosts in the tropics (61). Numerous fungi, mostly higher Basidiomycetes and Ascomycetes, form ectomycorrhizae (70, 79); a few Zygomycetes in the Endogonaceae are also involved (12).

Three major features characterize the ectomycorrhiza structure: 1) fungus colonization or sheathing of the short feeder roots, the fungus mantle; 2) intercellular fungus penetration between cortical cells, the Hartig net; and 3) morphological differentiation of colonized feeder roots via increased branching and elongation (Plate 1, A–F). This final characteristic together with the often dense mycelial connections with the soil allow for a highly active absorptive organ which colonizes large volumes of soil. Thus, a major benefit to the host is increased water and nutrient uptake, particularly of immobile ions such as phosphates. Other benefits include increased tolerance to drought, high soil temperatures, soil toxins, and extreme pH, as well as protection against root pathogens (25, 28, 36, 81). In return, the fungi depend on their hosts for carbohydrates, mostly in the form of simple sugars and vitamins. The physiology and ecology of ectomycorrhizae have been extensively reviewed (22) and will not be attempted here.

Much of our understanding on the functions of ectomycorrhizae has come from research directed towards practical application in forestry. Early in this century, for example, repeated failures in establishing exotic pine plantations in the tropics and other areas where ectomycorrhizal hosts do not naturally occur clearly demonstrated the dependence of these trees on their fungal symbionts. Only after inoculation with forest soil containing ectomycorrhizal fungus propagules could these trees survive and function properly (48, 49). Intensive mycorrhiza research in the past 30 years has increased our understanding of the complex physiology and ecology of ectomycorrhizae and our appreciation of the role of symbiosis (22). Most importantly, this information provides many necessary tools and concepts for strengthening forestry programs around the world.

Today, widescale inoculation of forest nurseries with selected ectomycorrhizal fungi appears imminent. Commercial interest in producing pure cultures of ectomycorrhizal fungus inoculum expands the possibilities of worldwide application. The success of these inoculation programs hinges on selection of effective and beneficial fungal symbionts. Little data exist on which of the thousands of possible ectomycorrhizal fungi may be the best candidates for inoculating a particular host species. Other considerations, such as nursery management practices and location of outplanting sites, complicate the selection process. New inoculation programs must be strongly research oriented from the outset (80).

In this paper we briefly review the considerations involved in determining the need for ectomycorrhizal inoculation, the techniques available for inoculation, and some relevant criteria for selecting specific fungi for inoculation. The readers are referred to the comprehensive reviews of Mikola (48, 49) for historical perspectives of worldwide development and applications, Trappe (80) for selection of fungi for nursery inoculation, and Marx (31) for detailed information on current inoculation techniques.

Forestry uses of ectomycorrhiza inoculation

Ectomycorrhizal host trees must be accompanied by their mycorrhizal fungi to survive when planted in areas lacking suitable fungi. This has been experienced many times in the introduction of exotic pines into the Southern Hemisphere and tropical islands (6, 48). Afforestation attempts in the treeless grasslands of the U.S.A. and the steppes of Russia have also required inoculation for success (18, 45). Schramm (66) and Marx (26, 30) have shown the absolute requirement for ectomycorrhizal planting stock for tree establishment on stripmined lands and other severely perturbed sites.

Although successful inoculation of tree seedlings already planted in the field have been reported (6), nursery inoculation is more common. Seedlings inoculated in the nursery can establish a healthy ectomycorrhiza system before outplanting. The increasing use of soil fumigation to eliminate pests in nurseries makes mycorrhizal inoculation of nurseries especially critical. Complete soil fumigants, such as the commonly used methyl bromide-chloropicrin* mixes, can thoroughly eradicate ectomycorrhizal fungus populations (4). Tree seedlings lacking ectomycorrhizae suffer severe nutrient deficiencies early in their first growing season; the deficiencies persist until mycorrhizae are formed (43, 82). Although deleterious to resident fungal populations, nursery fumigation is necessary for present inoculation techniques; competition from resident ectomycorrhizal fungi is eliminated or reduced as a prerequisite for successful establishment of the selected fungal inoculum.

New nurseries often show mycorrhizal deficiency symptoms, particularly when established on heavily fumigated former agricultural land. Trappe and Strand (82) report a striking example of this in the Willamette Valley of Oregon. The first crop of Douglas-fir seedlings exhibited severe phosphorus deficiency, unexpected because soil analyses had indicated no such deficiency. Subsequent heavy fertilization with phosphates did not alleviate the deficiency. Only after natural inoculation by windblown spores did the seedlings recover and begin to grow (Plate 2, A). A similar example was described by Marx *et al.* (43) for an Oklahoma nursery established on former pasture with few nearby ectomycorrhizal host trees. Significant improvement in seedling quality occurred after inoculation with pure fungus cultures. Substantial economic losses and disruption of forestation programs can result from failure to recognize and correct mycorrhiza deficiencies in nurseries.

The millions of containerized seedlings produced around the world offer another situation in which ectomycorrhiza inoculation may be important. Many cultural practices used in raising containerized seedlings—e.g., artificial potting mixes, frequent applications of concentrated soluble fertilizers, and greenhouse rearing—minimize or severely retard ectomycorrhiza formation. Most containerized seedlings we have examined appear vigorous and healthy but routinely lack mycorrhizae. We suspect that the mycorrhizal root system will increase survival and initial growth of containerized seedlings after outplanting. Ruehle (62) recently reported better survival and growth of containerized loblolly pine seedlings inoculated with *Pisolithus tinctorius* than of nonmycorrhizal seedlings planted on severely disturbed soil. Interactions between inoculation and the cultural practices mentioned above are currently under intensive investigation.

Ectomycorrhiza inoculation of tree seedlings offers promise for increasing reforestation success of suboptimal sites. Inoculation of seedlings destined for cutover lands has sometimes been thought unnecessary due to the abundance of mycorrhizal fungus propagules still present. We have found this to be true for high quality sites in the Oregon Coast Ranges (Molina and Trappe, unpublished data). Reforestation of sites stressed by drought, heat,

*This paper reports work involving pesticides. It does not include recommendations for operational use nor does it imply that uses discussed here are registered unless specifically stated. All uses of pesticides in the U.S.A. must be registered by appropriate State and Federal Agencies before they can be recommended.

or cold, however, will be improved by planting seedlings inoculated with fungi specifically adapted to those habitats. In droughty sites, for example, the density of fungal propagules may be low due to the frequent failure of a fungal fruiting season. Marx (31) also emphasizes that temporarily adverse conditions are common to many forest sites following timber harvest and site preparation. Harvey *et al.* (16, 17), recently found that both partial and total cutting of timber, followed by broadcast burning, reduced the ectomycorrhizal activity. The resulting delay in ectomycorrhiza formation on non-mycorrhizal seedlings or seedlings mycorrhizal with fungi not adapted to such sites may well decrease seedling survival or growth. Marx *et al.* (39) recently reported that inoculation with the fungal symbiont *Pisolithus tinctorius*, significantly increased growth and survival of five southern pine species planted in various sites in North Carolina and Florida. Much research is still needed on selection of ectomycorrhizal fungi for specific reforestation sites.

Techniques of inoculation

Once the need for inoculation is recognized, several methods are available. Spontaneous inoculation from windblown spores may suffice for nurseries established near stands of ectomycorrhizal host trees. Usually, however, spontaneous inoculation is erratic. When sowing follows spring fumigation, the chances for adequate natural inoculation during the first growing season is slight in regions of summer drought. In the Pacific Northwest of the U.S.A. and many other regions, for example, the height of the fungal fruiting season comes with the onset of fall rains and tapers off over winter and into spring; inoculum density of wind disseminated spores is relatively low in the spring.

Relatively few species of ectomycorrhizal fungi commonly inhabit nurseries. *Thelephora terrestris*, *Laccaria laccata*, and *Inocybe lacera* are common in Douglas-fir nurseries in the Pacific Northwest; *Thelephora* spp. frequently dominate in conifer nurseries around the world. These fungi are aggressive and well adapted to the cultivated, highly fertile, irrigated nursery conditions. Such nursery fungi may not be well suited to many planting sites (31, 80). Other common nursery fungi in temperate regions are the Discomycetes that form ectendomycorrhizae (19). These mycorrhizae are infrequent in natural forests, so their effectiveness is questionable for forest sites.

Four primary sources of inoculum for use in nurseries are listed below with their primary disadvantages (31, 47, 80). The advantages of each are then elaborated, but the promising technique of inoculation with pure cultures is particularly emphasized.

I. Soil inoculum

1. Large amounts are needed (10 per cent volume).
2. Pests and pathogens may be introduced, although in past experience this has not proved to be a problem.

3. Fungal symbionts are unknown.

II. Mycorrhizal nurse seedlings interplanted in seedbeds
1. Pests and pathogens may be introduced.
2. Mycorrhizal colonization is uneven.
3. Fungal symbionts are unknown.

III. Spores and sporocaps
1. Collection is seasonally limited and yearly availability unpredictable.
2. Adequate quantities of sporocarps may be difficult to obtain.
3. Much hand labor is needed in sporocarp collection.
4. Spore viability is difficult to determine; long-term storage requirements are unknown.
5. Several weeks are needed for mycorrhiza formation after inoculation.

IV. Pure cultures
1. Isolation of some ectomycorrhizal fungi is difficult.
2. Most of the fungi grow slowly in pure culture.
3. Production of sufficient inoculum is time consuming and expensive.
4. Conditions for survival of fungal inoculum in the soil are poorly known.
5. Effective and beneficial fungal symbionts must be selected.

The most commonly used and probably the most reliable ectomycorrhizal inoculum is soil taken from beneath ectomycorrhizal hosts (49). About 10 percent by volume of soil inoculum is incorporated into the top ± 10 cm of nursery beds prior to sowing or transplanting. Soil inoculum can also be added to the planting hole when seedlings are outplanted. Soil inoculation has been instrumental in establishment of exotic pine plantations in the Southern Hemisphere and continues as a regular practice there today (48).

A major disadvantage of soil inoculation is the relatively large amount of soil needed and its transportation over long distances. Inoculation of new or fumigated beds with soil from established beds is often feasible within individual nurseries. A possible danger with soil inoculum is the introduction of pathogens and other pests, although this has not generally been a problem (80). Nursery managers are nonetheless often reluctant to incorporate nonfumigated soil into fumigated beds. Soil inoculation remains a reliable method of eliminating mycorrhizal deficiencies and deserves further research attention.

Planting mycorrhizal "nurse" seedlings or incorporating chopped roots of ectomycorrhizal hosts into nursery beds as a source of the fungi for neighboring young seedlings has been successful (48, 49) (Plate 2, B and C). Mycorrhizal colonization, however, may spread slowly and unevenly. Chevalier and Grente (7) were able to inoculate seedlings with the highly prized truffle fungus *Tuber melanosporum* by use of nurse seedlings already mycorrhizal with this fungus. The major disadvantages of this method parallel those of soil inoculation—the possibility of introducing unwanted pathogens and other pests, and, the identities and effectiveness of the introduced fungi are unknown. In modern, mechanized nurseries, the presence of large nurse seedlings may interfere with cultural practices.

Basidio- and ascospores or crushed sporocarps have been used occasionally as inoculum, usually in small experiments. Some investigators have reported good success with this technique (10, 27, 34, 42, 63, 65, 73, 75). Asexual spores and sclerotia are further sources of inoculum (21, 76). Theoretically, the use of spores would most closely imitate natural inoculation. Practical application is limited, however, by the generally short season for collecting sporocarps in quantity. In some regions, adverse weather may even eliminate the collecting season. Also, in many regions the fungi fruit in fall whereas the nurseries fumigate and sow seeds in the spring. Spore inoculum then must be stored over winter, but little is known of the storage requirements of ectomycorrhizal fungal spores. This is complicated further by the difficulty of germinating sexual spores of most ectomycorrhizal fungi.

The Gasteromycetes (puffballs and related fungi), with abundant spore masses, offer better sources of large numbers of spores than the gilled fungi. Most recent research has been with *Pisolithus tinctorius* (27, 34, 42, 63, 65). Billions of basidiospores can be easily collected from this large puffball, which often fruits in abundance. Inoculum rates of $5.5 \times 10^8 - 1.3 \times 10^{10}$ spores/m^2 of soil surface have been used to successfully inoculate several bareroot nurseries in southern United States (31). The powdery spore mass is easily manipulated, so several application techniques work well. Mixing of spores in hydromulch (wood pulp suspended in water) and broadcasting with a tractor-drawn applicator has been particularly effective (31, 42). Container-grown seedlings are also easily inoculated with *Pisolithus* spores (34, 63, 65), with best results when they are dusted onto germinants or young seedlings before competition arises from natural inoculum sources (63). Marx (31) also notes that *Pisolithus* spores have been stored dry for five years at 5°C without apparent loss of viability.

Inoculations with spores of *Rhizopogon* species also appear promising. In Australia, Theodorou (72) and Theodorou and Bowen (74) coated seed of *Pinus radiata* with basidiospores of *R. luteolus*; abundant *Rhizopogon* mycorrhizae formed on seedlings produced from the coated seed. In addition to fresh spores, they also found that freeze-dried and air-dried spores produced mycorrhizae but only at higher inoculum rates. Similarly, Donald (10) used air dried, pulverized sporocarps of *R. luteolus* to successfully inoculate *Pinus radiata* in South Africa nurseries. Castellano and Trappe (unpublished data) found fresh and dried spore suspensions of *R. vinicolor* and *R. colossus* to be effective in inoculating bareroot and containerized Douglas-fir; also, *R. ochraceorubens* spores worked well with containerized *Pinus contorta*, *P. ponderosa*, and *P. radiata*. Their similar experiments with *Gautieria* and

Hysterangium species failed, however, as have attempts by others with various other fungi (31, 80). More needs to be learned about the basic biology of spores before their use in nursery inoculation can be operationally dependable.

The final type of inoculum is pure cultures of ectomycorrhizal fungi. Although many difficulties remain in using this source, techniques for wide-scale application are now being developed (31).

A pure culture of a specific fungus must first be isolated either from the fruiting body or the ectomycorrhiza itself; occasionally, isolates can be obtained from spores, surface sterilized rhizomorphs or sclerotia. The ubiquitous ectomycorrhizal fungus *Cenococcum geophilum* (=*C. graniforme*) is easily isolated from its hard black sclerotia (78). Isolation from sporocarps is, however, easiest for most fungi. Isolations from ectomycorrhizae or rhizomorphs are more difficult, and species often cannot be identified (84). Unfortunately, many ectomycorrhizal fungi grow extremely slowly or not at all in culture. Still, many do grow well in culture, e.g. most species of *Suillus*, *Hebeloma*, *Laccaria*, *Amanita*, *Rhizopogon*, and *Pisolithus*. Consequently, much research attention has been devoted to these species. As improved isolation and culturing procedures are developed, many other fungi can be considered.

Most pure culture inoculation has been restricted to small-scale experiments, although Moser (57, 59) successfully inoculated nursery beds of *Pinus cembra* in Austria with pure cultures of *Suillus plorans* more than 20 years ago. Vozzo and Hacskaylo (83) later used Moser's methods to inoculate pine seedlings in Puerto Rico. Similarly, Theodorou (72) and Theodorou and Bowen (74) inoculated *Pinus radiata* with isolates of *Rhizopogon luteolus*, *Suillus granulatus*, *S. luteolus* and *Cenococcum geophilum*. Marx and Bryan (37) further refined Moser's technique and report excellent results in inoculating nursery beds with *Pisolithus tinctorius* (see Marx (31), for a complete description of these techniques). They grew the fungus three to four months in 2-liter jars containing a sterilized peatmoss-vermiculite substrate moistened with modified Melin-Norkrans nutrient solution (24). After the fungus completely penetrates the substrate, the inoculum is removed from the jars and thoroughly leached with cool, running tap water to remove unassimilated nutrients. The inoculum is spread on the soil at the rate of about 1 l/m^2 and cultivated 8 to 10 cm into the soil. The bed surface is then smoothed and sown. Thorough fumigation of the nursery soil, soon before inoculation preferably with a methyl bromide-chloropicrin mix, is critically needed to reduce competition and antagonism by other soil organisms. Fumigation in autumn before sowing the following spring is often ineffective, because antagonist populations can build up overwinter, especially in areas with cool but mild winters (Trappe and Hung, unpublished data).

Marx (31) recently discussed other pertinent data on development and refinement of pure culture inoculation techniques with *P. tinctorius*. Leached vermiculite inoculum is heavy from water saturation and is difficult to transport, spread and mix evenly into the soil. After drying leached inoculum to 12 percent moisture at 28–30°C for 56 hours in a forced-air oven, he reports that the dried inoculum performed as well if not better than wet inoculum at several application rates. He attributes this to better mixing of the dried inoculum into the soil. In another nursery study, *Pisolithus* inoculum added at rates of 2.8, 2.16, 1.62, 1.08 and 0.5 $1/m^2$ of soil surface all produced *Pisolithus* mycorrhizae, with 1.08 $1/m^2$ the least amount that was most effective. *Pisolithus* inoculum can also be stored with little loss of viability for seven to nine weeks at 5 and 23°C and for five to seven weeks at 30°C. Marx (31) further notes that *Pisolithus* inoculum can survive in soil for 30 days without a host and maintain inoculum effectiveness. *P. tinctorius* has also been shown to overwinter in nursery plots after seedling removal; seedlings grown the following year in these previously inoculated, then fallow, plots formed abundant *Pisolithus* mycorrhizae. Data of this nature are critically needed to evaluate potential use of other ectomycorrhizal fungi for nursery inoculation.

Inoculation of containerized seedlings with pure cultures also holds great promise (1, 8, 9, 23, 34, 44, 51, 53, 55, 60, 62, 63, 65, 67, 68). A peatmoss-vermiculite mix is used both as an inoculum substrate and container potting mix, so the inoculum is easily incorporated when containers are filled. *Pisolithus tinctorius* has again received the most research attention and has been successfully inoculated on container-grown seedlings in the genera *Pinus* (9, 34, 44, 51, 60, 62, 63), *Psuedotsuga* (44, 51), *Tsuga* (44), and *Quercus* (1, 9, 23, 44, 64). We have effectively inoculated western U.S. species of *Pinus*, *Tsuga*, *Picea*, *Larix*, and *Pseudotsuga* with cultures of *Laccaria laccata* (53, 55, 67, 68), *Cenococcum geophillum* (53, 67, 68), and *Hebeloma crustruliniforme* (68). *L. laccata* and *H. crustuliniforme* appear particularly promising in that the entire root system and substrate are completely colonized by these fast growing fungi. *Thelephora* spp., common natural colonizers of containerized seedlings, can likewise be introduced via pure cultures (62, Trappe and Molina, unpublished data).

High fertility rates, particularly of soluble fertilizers commonly used in rearing containerized seedlings, often restrict or impede mycorrhizal development following inoculation (23, 34, 44, 64, 65, 68); most experience indicates that reduced fertility levels significantly improve mycorrhizal development. Successful mycorrhiza inoculation, however, rarely improves growth of containerized seedlings. Inoculated seedlings often have smaller tops than non-inoculated controls (34, 44, 51, 53, 67). Active fungus utilization of limited host photosynthates impinges on seedling top growth but improves the size and surface of the root system. In addition, some fungi such as *Laccaria laccata* use host photosynthate to form mushrooms in the containers during the first growing season in addition to totally colonizing the root system and substrate (53, 55, 67). Not all fungi respond the same to fertility levels. *Piso-*

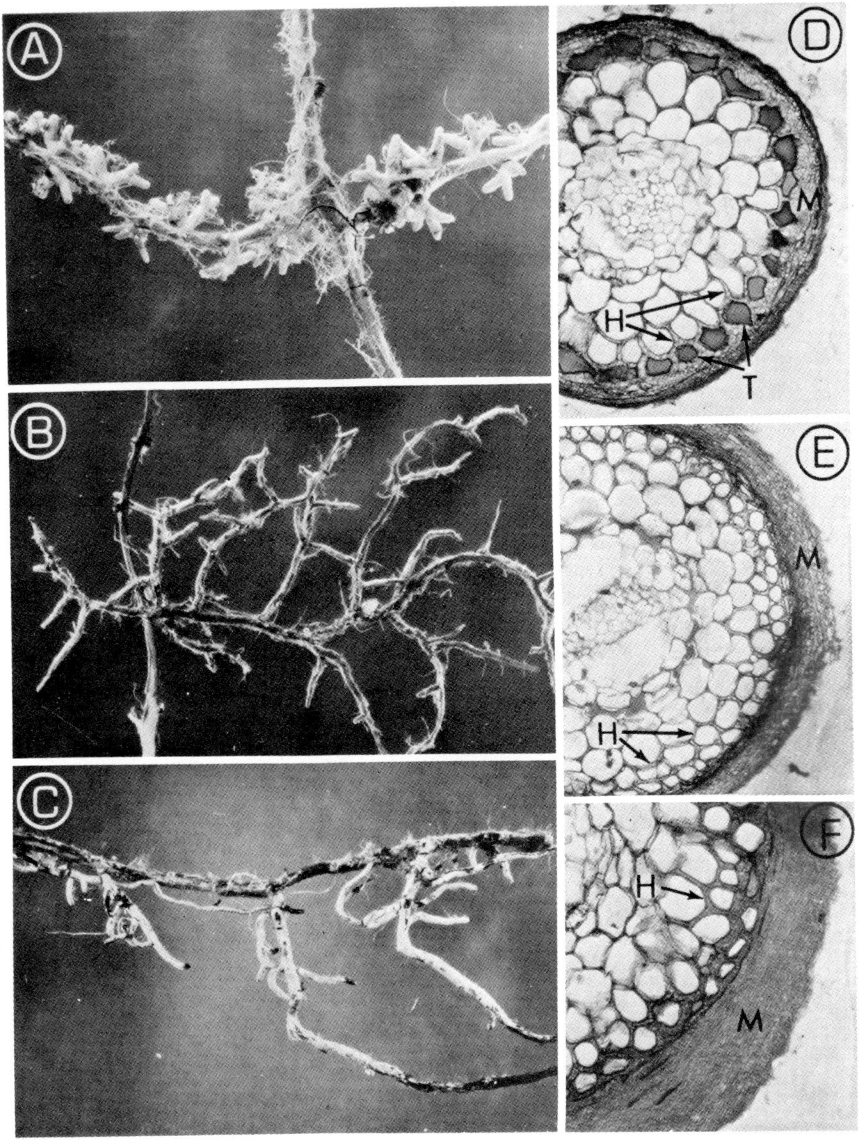

Plate 1. Examples of ectomycorrhizae, form and structure.

A-C: Branching forms and elongation. (A) *Rhizopogon subcaerulescens* + *Pinus contorta* ectomycorrhizae, × 3.3. Note the typical bifurcate (forked) branching typical of *Pinus* ectomycorrhizae. (B) *Rhizopogon ellenae* + *Tsuga heterophylla* ectomycorrhizae, × 3.3. (C) *Paxillus involutus* + *Larix occidentalis* ectomycorrhizae, × 3.3.

D-F: Cross sections of typical ectomycorrhizae (M = Mantle, H = Hartig net, T = Tannin layer). All figures × 175. (D) *Suillus brevipes* + *Pinus ponderosa*. (E) *Astraeus pteridis* + *Tsuga heterophylla*. (F) *Amanita muscaria* + *Larix occidentalis*.

Plate 2. (A) Two-year-old *Pseudotsuga menziesii* seedlings in the Canby Nursery, Willamette Valley of Oregon. Scattered clumps of vigorously growing seedlings (arrows) are mycorrhizal from natural (wind disseminated) inoculum sources [see Trappe and Strand (82)]. (B) First year growth of *Pseudotsuga menziesii* seedlings inoculated with chopped mycorrhizal roots (predominantly *Laccaria laccata* mycorrhizae) at the Humbolt Nursery in northern California. (C) First year seedlings receiving no mycorrhizal inoculation show severe stunting (Figures B and C courtesy of Mike Shrago, U.S. Forest Service).

lithus tinctorius appears especially sensitive to high fertility levels in the rooting substrate (8, 34, 44, 64); high nitrogen levels, however, applied as a foliar mist significantly stimulated *P. tinctorius* development on containerized pine seedlings (8). On the other hand, Molina (unpublished data) found that *Laccaria laccata* inoculum performed equally well regardless of high or low fertility levels. Further research on fertility × mycorrhization interactions is needed to optimize mycorrhiza development and size of containerized nursery stock.

The logistics of producing massive quantities of inoculum presently limits wide-scale use of pure culture inoculum. Large scale production methods are now being developed, however, by Abbott Laboratories*, Long Grove, Illinois. Large volumes of dried *Pisolithus tinctorius* inoculum in a vermiculite carrier are produced quickly in industrial fermentors. A U.S.-wide evaluation of this inoculum has demonstrated it to be effective in producing *Pisolithus* mycorrhizae on several tree species in both bareroot and container nurseries (31, 44). Other firms are also experimentally producing pure culture inoculum of ectomycorrhizal fungi. Industry representatives and mycorrhiza researchers are optimistic that effective commercial inoculum will soon be available on the market.

Selection of fungi for inoculation

The promising outlook for pure culture inoculation raises still another important question: which fungus is best for a particular host or habitat? The need for information on effectiveness of the various mycorrhizal fungi on different host species has been repeatedly emphasized in the literature, yet little data exist. Thousands of ectomycorrhizal fungi and numerous hosts have been reported (77), so careful selection of the best fungi for particular hosts is critical.

Many important criteria must be considered when selecting fungus candidates for nursery inoculation. The major criteria are:

1. Ease of isolation.
2. Growth rate in pure culture.
3. Effectiveness as inoculum.
4. Effects on growth and vigor of host seedlings tops and roots.
5. Ecological adaptability and ecotypic variation.
6. Interactions with soil microorganisms.
7. Host specificity.

Other criteria may be added for special circumstances. It must be stressed

*The use of trade, firm, or corporation names in this paper is for the information and convenience of the reader. Such use does not constitute endorsement by the U.S. Department of Agriculture of any product or service to the exclusion of others which may be suitable.

that many species and ecotypes of fungi are closely adapted to particular habitats, so each isolate must be tested on its own merits. Careful experimentation and good record keeping are essential throughout evaluations of each isolate to document how well it meets the criteria as compared to alternative isolates.

Criteria 1 and 2 are basic to all the others: one must first be able to isolate the particular fungus and grow it reasonably well in culture. Following the experience of Moser (58), we routinely isolate in the field using a small portable hood to reduce air movement. For isolation, we have had best results with small agar slants containing either Melin-Norkrans agar as modified by Marx (24) or potato-dextrose agar.

Relatively fast growing fungi are generally preferred for inoculation because of their short incubation period. Unfortunately, many otherwise desirable ectomycorrhizal fungi grow slowly. As the physiological growth requirements of mycorrhizal fungi become better understood, growth of the slow-growers may be improved for use in inoculation. Fungi that do not grow or grow only slowly in culture may be highly specialized in their symbiotic relationship to the host and may benefit their host greatly. Clearly, further study of these recalcitrant fungi is needed.

Marx (31) emphasizes two additional points about culture characteristics. First, fresh cultures are preferred to cultures repeatedly transferred and stored for several years. Some ectomycorrhizal fungi lose their mycorrhiza-forming capacity after longterm storage, presumably due to adaptive enzyme shifts in utilizing artificial substrates. This can be overcome for many fungi by storing them under refrigeration in sterile water for several years without transferring (41). Marx (31, 32) further suggests passing important fungus cultures through a host, i.e. host inoculation and mycorrhiza formation followed by reisolation, every few years to maintain mycorrhiza-forming capacity. *Pisolithus tinctorius* isolates used for wide-scale inoculations have significantly improved in mycorrhiza-forming capacity and enhancement of seedling growth following repeated reisolations (32). Secondly, fungi which produce large hyphal strands or rhizomorphs in culture or soil may be superior in soil exploration and mineral uptake to those which lack rhizomorphic growth (3, 4, 31). For example, Bowen (3, 4) suggests that the extensive rhizomorph network of *Rhizopogon luteolus* is largely responsible for enhanced nutrient uptake and seedling growth. Marx (31) believes that hyphal strands of *P. tinctorius* not only enhance nutrient uptake but also increase seedling survival potential under adverse conditions. Schramm (66) and Marx (28) have traced mycelial strands of *P. tinctorius* through coal spoils up to 4 m from seedlings to sporocarps. Comparable data and observations are needed for the many fungi which do not form such hyphal structures.

Criteria 3 and 4 are next considered in the selection process. After the fungal inoculum has been prepared, its effectiveness must be determined.

Feeder roots susceptible to mycorrhizal colonization do not form on seedlings until six to eight weeks after seed germination. During this period the vermiculite particles are believed to provide a protective niche for the naked mycelium. Survival and effectiveness of the inoculum is determined by examining roots for ectomycorrhizal formation. Roots should be sampled periodically during the first growing season. The numbers and kinds of other native mycorrhizal types should also be noted to assess effectiveness of the soil fumigation and the competitive ability of the inoculated fungus. Marx (31) emphasizes that the candidate fungi must be aggressive in forming mycorrhizae as soon as feeder roots are produced and in maintaining superiority over native fungi in the nursery. With *Pisolithus tinctorius*, maximum benefit for pine seedlings results only after at least two-thirds of the feeder roots are colonized by *P. tinctorius*. All nursery cultural practices should also be carefully monitored and recorded for future reference.

The most crucial criterion in the selection process deals with the benefits the host derives from inoculation with a specific symbiont. Differences between inoculated and noninoculated seedlings in height, top and root weights, and stem caliper must be compared. Marked improvement in bareroot nursery seedlings can be expected by effective inoculation with a highly beneficial fungus. Because the seedlings are raised for forestation purposes, however, the critical test is survival and growth after outplanting as compared to normally produced nursery stock. Survival and growth data must be collected over at least the first three years. Nurserymen will not want to inoculate seedlings unless it significantly improves nursery production or field performance. Readers are referred to Marx (31) for a detailed review on current performance of outplanted inoculated nursery stock.

Criterion 5 deals with fungal physiology with special reference to ecological adaptability. Field observations as well as laboratory tests are important. Data should be recorded on the ecological range of the fungus as well as specific habitat types in which it is found. Environmental conditions of outplanting sites also need consideration. Planting sites characterized by drought or temperature extremes are commonly difficult to reforest. Trappe (unpublished data) has found that conifer seedlings inoculated with *Pisolithus tinctorius* in Oregon survive better than "nursery run" seedlings on hot, droughty sites. *Pisolithus* did not improve performance of seedlings on cool, high elevation sites, however. In mine spoils soil toxicity is a major problem. Temperature and moisture requirements of the fungus can easily be estimated from simple laboratory tests (80). Our working hypothesis is that fungi already adapted to conditions similar to the planting sites should be the primary candidates for inoculation. Trappe (80) states: "In essence, the provenance of the fungus should be considered along with the provenance of tree seed."

Special emphasis should be placed on the ecotypic variation displayed within fungal species (80). For example, Molina (51) found that six isolates of *Pisolithus tinctorius* differed significantly in ability to colonize feeder roots of container-grown Douglas-fir and lodgepole pine seedlings. Marx (32) provided further evidence for worldwide variation among *P. tinctorius* isolates and made several suggestions for selecting the best strain. On the other hand, Molina (55) recently noted that several isolates of *Laccaria laccata* lacked significant differences on mycorrhiza colonization following inoculation of four conifer species; all isolates completely colonized the root systems and sporocarp production was prolific. Thus, each fungus isolate must be tested on its own merit and with several performance criteria in mind.

Criterion 6 has received scant attention in selection of fungi for nursery inoculation, but it has potential. Ectomycorrhizal fungi protect host roots to varying degrees against certain pathogens (25). Nurseries with continuing root pathogen problems may wish to introduce mycorrhizal fungi selected for the ability to protect seedling roots from disease. For example, *Laccaria laccata* protected Douglas-fir seedlings against *Fusarium oxysporum* even in the absence of mycorrhiza formation (69, 71). Recent work by Bowen and Theodorou (5) on interactions of ectomycorrhizal fungi and bacteria emphasizes the need to assess compatibility and potential antagonisms of the resident microflora with the introduced fungi.

The final criterion is host specificity. Many ectomycorrhizal fungi exhibit wide host ranges: *Amanita muscaria*, *Boletus edulis*, *Laccaria laccata*, *Pisolithus tinctorius* and *Cenococcum geophilum*, to mention a few. Others are more restricted. Some are known only to fruit in association with a single host or genus of hosts. The association of *Suillus grevillei* with *Larix* species and *Leccinum scabrum* with *Betula* species are two commonly cited examples. The genus *Pinus* appears to have its select group of "pine" mycorrhizal fungi. *Pseudotsuga menziesii* has many mycorrhizal fungi common only to its distribution. Precise data of this nature is very important if we plan to inoculate a wide range of tree species, especially in regions where many different host genera are raised commercially. The Pacific Northwest of the U.S.A., for example, contains at least 16 native genera of ectomycorrhizal hosts, including over 60 species. At least a third of these are raised in bareroot and container nurseries. A single nursery may raise 10 species. Should many different fungi be inoculated to satisfy the different hosts, or is it better to inoculate with one fungus capable of colonizing them all? Mikola (49) believes that, due to its more specialized relationship with a particular host, a host-specific fungus would benefit its host more than would a broad-ranging fungus. This hypothesis warrants further research, especially with the development of wide scale, pure culture inoculation.

Pure culture synthesis of mycorrhizae as developed by Melin (46) and modified by Hacskaylo (13) and others provides good evidence on host specificity. Seedlings are raised aseptically in two-membered culture with an introduced mycorrhizal fungus. With large glass test tubes for the chambers, numerous combinations of host species and fungi can be readily assessed in three to six months when the seedlings are harvested (52, 54). With this technique we have found that fungus-host specificity is more complex than simply a constant association of sporocarps with particular hosts in the field (52, 54, 56). Some fungi that fruit with specific hosts form typical ectomycorrhizae with other, unassociated hosts.

The *Pisolithus* story: A case in point

The intensive research on *Pisolithus tinctorius* conducted by Dr. Donald Marx and co-workers at the U.S. Forest Service Institute for Mycorrhizal Research and Development, Athens, Georgia, has gathered the support of both the research community and industry and has brought mycorrhizal applications in forestry to the forefront. Although focusing on one fungal symbiont, *P. tinctorius*, their underlying hypothesis is that growth and survival of seedlings can be significantly improved by inoculations with specific mycorrhizal fungi. A brief summary of their work follows with special emphasis on their integrated use of many of the concepts presented in this paper (see Marx (31), for a more detailed review).

Schramm (66) reported the extensive development of *P. tinctorius* ectomycorrhizae and sporocarps associated with pine roots growing on anthracite mining wastes in Pennsylvania. *P. tinctorius* was often the pioneering mycorrhizal symbiont on the young, most vigorous pine seedlings. Realizing that the extremely high soil temperatures reported by Schramm might limit fungal symbionts to a few adapted species, Marx *et al.* (40) explored the temperature-growth interactions of *P. tinctorius*. They found that it formed more ectomycorrhizae with *Pinus taeda* seedlings at 34°C than at lower temperatures; mycelial cultures grew at temperatures as high as 40°C. Marx and Bryan (35) later found that aseptically grown *Pinus taeda* seedlings colonized with *P. tinctorius* survived and grew as well at 40° as at 24°C; comparative nonmycorrhizal seedlings and those colonized with the fungal symbiont *Thelephora terrestris* had less survival and did not grow at 40°C. Clearly, the adaption to higher temperatures was a major factor in allowing *P. tinctorius* to invade the coal wastes.

Realizing the practical significance of these results, Marx (26) and co-workers surveyed strip-mined lands for the presence of *P. tinctorius*. They found it to be the dominant and often only ectomycorrhizal fungus of pine roots growing on coal wastes in Indiana, Pennsylvania, Ohio, West Virginia, Virginia, Kentucky, Tennessee, and Alabama and on kaolin soils in Georgia.

These results prompted extensive investigation of ways to inoculate and establish *P. tinctorius* ectomycorrhizae on roots of pine seedlings destined for outplanting on mine spoils. Marx and Bryan (37) developed techniques as previously described for preparing pure culture inoculum and inoculating nursery soil with *P. tinctorius*. They report excellent success in establishing *P. tinctorius* in the nursery with doubled growth of inoculated seedlings over uninoculated controls (38). Inoculation with *P. tinctorius* basidiospores has also succeeded (27, 38, 42). More importantly, *P. tinctorius* inoculation has significantly increased survival and growth of seedlings on mine spoils (2, 26, 30, 33, 62). Many of these sites had a history of repeated failures of pine plantations. *P. tinctorius* ectomycorrhizal colonization can also increase survival and growth of southern pines on routine reforestation sites (39).

Development of the *P. tinctorius* inoculation program included many of the selective criteria discussed previously. *P. tinctorius* is easily isolated from sporocarps and grows rapidly in culture. Inoculation with pure cultures of vegetative mycelium as well as basidiospores often results in complete colonization of the entire feeder root system. Overall host response and performance in both nursery and plantation are excellent. Both field and laboratory studies emphasize the ecological adaptiveness of *P. tinctorius* to stressful sites, including tolerance to high soil temperatures, moisture stress, and soil toxicity. Feasibility of utilizing commercially produced *P. tinctorius* inoculum is being studied and appears promising. Finally, *P. tinctorius* is distributed worldwide and forms ectomycorrhizae with over 48 species of trees (29). Worldwide use of this highly beneficial fungus is quite possible and experimentation in many regions is underway (50).

These impressive results and the concepts they represent have stimulated inoculation research programs around the world. Clearly, with the increased demand for and dwindling supply of wood resources, the regeneration of cut-over lands and establishment of man-made forests is of highest priority. Inoculation with highly beneficial mycorrhizal fungi specifically selected for certain traits can enormously increase the chances of meeting this priority.

References

1. Beckjord, P.R., Adams, R.E. and Smith, D.W. Effects of nitrogen fertilization on growth and ectomycorrhizal formation of red oak, *Forest Science, 26*, 529–536 (1980).
2. Berry, C.R. and Marx, D.H. Effects of *Pisolithus tinctorius* ectomycorrhizae on growth of loblolly and Virginia pines in The Tennessee Copper Basin. United States Department of Agriculture, Forest Service, Research Note SE-264, Southeastern Forest Experiment Station, Ashville, North Carolina, 6 pp. (1978).
3. Bowen, G.D. Phosphate uptake by mycorrhizas and uninfected roots of *Pinus radiata* in relation to root distribution, *Transactions of the 9th International Congress of Soil Science, 2*, 219–228 (1968).

4. Bowen, G.D. Mineral nutrition of ectomycorrhizae, in Reference 22, 151–205 (1973).
5. Bowen, G.D. and Theodorou, C. Interactions between bacteria and ectomycorrhizal fungi, *Soil Biology and Biochemistry*, *11*, 119–126 (1979).
6. Briscoe, C.B. Early results of mycorrhizal inoculation of pine in Puerto Rico, *Caribbean Forester*, *20*, 73–77 (1959).
7. Chevalier, G. and Grente, J. Propagation de la mycorrhization par la truffe a partir de racines excisées et de plantules inséminatrices, *Annales de Phytopathologie*, *4*, 317–318 (1973).
8. Dixon, R.K., Garnett, R.K. and Cox, G.S. Containerized shortleaf pine seedlings show superior growth and ectomycorrhizal development with mist foliar fertilization, *Southern Journal of Applied Forestry*, *3*, 154–157 (1979).
9. Dixon, R.K., Wright, G.M., Behrms, G.T., Teskey, R.O. and Hinckcley, T.M. Water deficits and root growth of ectomycorrhizal white oak seedlings, *Canadian Journal of Forest Research*, *10*, 545–548 (1980).
10. Donald, D.G.M. Mycorrhizal inoculation of pines, *South African Forestry Journal*, *92*, 27–29 (1975).
11. Gerdemann, J.W. Vesicular-arbuscular mycorrhiza and plant growth, *Annual Review of Phytopathology*, *6*, 397–418.
12. Gerdemann, J.W. and Trappe, J.M. The Endogonaceae in the Pacific Northwest, *Mycologia Memoir*, *5*, 1–76 (1974).
13. Hacskaylo, E. Pure culture synthesis of pine mycorrhizae in terra-lite, *Mycologia*, *45*, 971–975 (1953).
14. Hacskaylo, E. and Palmer, J.G. Effects of several biocides on growth of seedling pines and incidence of mycorrhizae in field plots, *Plant Disease Reporter*, *41*, 354–358 (1957).
15. Harley, J.L. *The Biology of Mycorrhiza*, Leonard Hill, London, 334 pp. (1969).
16. Harvey, A.E., Jurgensen, M.F. and Larsen, M.J. Clearcut harvesting and ectomycorrhizae: Survival activity on residual roots and influence on a bordering forest stand in Montana, *Canadian Journal of Forest Research*, *10*, 300–303 (1980).
17. Harvey, A.E., Larsen, M.J. and Jurgensen, M.F. Partial cut harvesting and ectomycorrhizae: Early effects in Douglas-fir-larch forest of western Montana, *Canadian Journal of Forest Research*, *10*, 436–440 (1980).
18. Imshenetskii, A.A. (Editor). *Mycotrophy in Plants*, United States Department of Commerce Translations TT 67-51290 (1955).
19. Laiho, O. Further studies on the ectendotrophic mycorrhiza, *Acta Forestalia Fennica*, *79*, 1–35 (1965).
20. Laiho, O. and Mikola, P. Studies on the effect of some eradicants on mycorrhizal development in forest nurseries, *Acta Forestalia Fennica*, *77*, 1–34 (1964).

21. Lamb, R.J. and Richards, B.N. Survival potential of sexual and asexual spores of ectomycorrhizal fungi, *Transactions of the British Mycological Society*, *62*, 181–191 (1974).
22. Marks, C.G. and Kozlowski, T.T. (Editors). *Ectomycorrhizae—Their Ecology and Physiology*, Academic Press, New York, 444 pp. (1973).
23. Maronek, D.M. and Hendrix, J.W. Growth acceleration of pine-oak seedlings with a mycorrhizal fungus, *Horticultural Science*, *14*, 627–628 (1979).
24. Marx, D.H. The influence of ectotrophic mycorrhizal fungi on the resistance of pine roots to pathogenic infections. I. Antagonism of mycorrhizal fungi to root pathogenic fungi and soil bacteria, *Phytopathology*, *59*, 153–163 (1969).
25. Marx, D.H. Mycorrhizae and feeder root diseases, in Reference 22, 351–382 (1973).
26. Marx, D.H. Use of specific mycorrhizal fungi on tree roots for forestation of disturbed surface areas, pp. 47–65 in *Proceedings of the Conference on Forestation of Disturbed Areas*, Birmingham, Alabama (Edited by K.A. Utz), United States Department of Agriculture, Atlanta (1976).
27. Marx, D.H. Synthesis of ectomycorrhizae on loblolly pine seedling with basidiospores of *Pisolithus tinctorius*, *Forest Science*, *22*, 13–20 (1976).
28. Marx, D.H. The role of mycorrhizae in forest production, *TAPPI Conference Papers*, Annual Meeting, Atlanta, Georgia, 151–161 (1977).
29. Marx, D.H. Tree host range and world distribution of the ectomycorrhizal fungus *Pisolithus tinctorius*, *Canadian Journal of Microbiology*, *23*, 217–223 (1977).
30. Marx, D.H. Role of mycorrhizae in forestation of surface mines, pp. 109–116 in *Symposium on Trees for Reclamation in the Eastern United States*, Lexington, Kentucky (1980).
31. Marx, D.H. Ectomycorrhiza fungus inoculations: a tool for improving forestation practices, pp. 13–71 in *Tropical Mycorrhiza Research* (Edited by P. Mikola), Oxford University Press, Oxford (1981).
32. Marx, D.H. Variability in ectomycorrhizal development and growth among isolates of *Pisolithus tinctorius* as affected by source, age, and reisolation, *Canadian Journal of Forest Research*, *11*, 168–174 (1981).
33. Marx, D.H. and Artman, J.D. *Pisolithus tinctorius* ectomycorrhizae improve survival and growth of pine seedlings on acid coal spoils in Kentucky and Virginia, *Reclamation Review*, *2*, 23–31 (1979).
34. Marx, D.H. and Barnett, J.P. Mycorrhizae and containerized forest tree seedlings pp. 85–92 in *Proceedings of The North American Containerized Forest Tree Seedling Symposium* (Edited by R.W. Tinus, W.I. Stein, and W.E. Balmer), Great Plains Agricultural Council Publication No. 68 (1974).
35. Marx, D.H. and Bryan, W.C. Influence of ectomycorrhizae on survival and growth of aseptic seedlings of loblolly pine at high temperature, *Forest Science*, *17*, 37–41 (1971).

36. Marx, D.H. and Bryan, W.C. The significance of mycorrhizae to forest trees, pp. 107–117 in *Forest Soils and Forest Land Management* (Edited by B. Bernier and C.H. Winget), Laval University Press (1975).
37. Marx, D.H. and Bryan, W.C. Growth and ectomycorrhizal development of loblolly pine seedlings in fumigated soil infested with the fungal symbiont *Pisolithus tinctorius, Forest Science, 21*, 245–254 (1975).
38. Marx, D.H., Bryan, W.C. and Cordell, C.E. Growth and ectomycorrhizal development of pine seedlings in nursery soils infested with the fungal symbiont *Pisolithus tinctorius, Forest Science, 22*, 91–100 (1976).
39. Marx, D.H., Bryan, W.C. and Cordell, C.E. Survival and growth of pine seedlings with *Pisolithus* ectomycorrhizae after two years on reforestation sites in North Carolina and Florida, *Forest Science, 23*, 363–373 (1977).
40. Marx, D.H., Bryan, W.C. and Davey, C.B. Influence of temperature on aseptic synthesis of ectomycorrhizae by *Thelephora terrestris* and *Pisolithus tinctorius* on loblolly pine, *Forest Science, 16*, 424–431 (1970).
41. Marx, D.H. and Daniel, W.J. Maintaining cultures of ectomycorrhizae and plant pathogenic fungi in sterile water cold storage, *Canadian Journal of Microbiology, 22*, 338–341 (1976).
42. Marx, D.H., Mexal, J.G. and Morris, W.G. Inoculation of nursery seedbeds with *Pisolithus tinctorius* spores mixed with hydromulch increases ectomycorrhizae and growth of loblolly pines, *Southern Journal of Applied Forestry, 3*, 175–178 (1979).
43. Marx, D.H., Morris, W.G. and Mexal, J.G. Growth and ectomycorrhizal development of loblolly pine seedlings in fumigated and nonfumigated soil infested with different fungal symbionts, *Forest Science, 24*, 193–203 (1978).
44. Marx, D.H., Ruehle, J.L., Kenny, D.S., Cordell, C.E., Riffle, J.W., Molina, R.J., Pawuk, W.H., Navratil, S., Tinus, R.W. and Goodwin, O.C. Development of *Pisolithus tinctorius* ectomycorrhizae on containerized tree seedlings with vegetative inocula produced by commercial and research procedures, *Forest Science* (in press).
45. McComb, A.L. The relations between mycorrhizae and the development and nutrient absorption of pine seedlings in a prairie nursery, *Journal of Forestry, 36*, 1148–1154 (1938).
46. Melin, E. Über die Mycorrhizenpilze von *Pinus silvestris* L. und *Picea abies* (L.) Karst., *Svensk Botanisk Tidskrift, 15*, 192–203 (1921).
47. Meyer, F.H. Distribution of ectomycorrhizae in native and man-made forests, in Reference 22, 79–105 (1973).
48. Mikola, P. Mycorrhizal inoculation in afforestation, *International Review of Forestry Research, 3*, 123–196 (1970).
49. Mikola, P. Application of mycorrhizal symbiosis in forest practices, in Reference 22, 383–411 (1973).

50. Mikola, P. (Editor). *Tropical Mycorrhiza Research,* Oxford University Press, Oxford, 270 pp. (1981).
51. Molina, R. Ectomycorrhizal inoculation of containerized Douglas-fir and lodgepole pine seedlings with six isolates of *Pisolithus tinctorius, Forest Science, 25,* 585–590 (1979).
52. Molina, R. Pure culture synthesis and host specificity of red alder mycorrhizae, *Canadian Journal of Botany, 57,* 1223–1228 (1979).
53. Molina, R. Ectomycorrhizal inoculation of containerized western conifer seedlings, United States Department of Agriculture, Forest Service Research Note PNW-357, Pacific Northwest Forest and Range Experiment Station, Portland, Oregon, 10 pp. (1980).
54. Molina, R. Ectomycorrhizal specificity in the genus *Alnus, Canadian Journal of Botany, 59,* 325–334 (1981).
55. Molina, R. Use of the ectomycorrhizal fungus *Laccaria laccata* in forestry. I. Consistency between isolates in effective inoculation of containerized western conifer seedlings (in review).
56. Molina, R. Patterns of ectomycorrhizal host specificity and potential among Pacific Northwest conifers and fungi, *Forest Science* (in review).
57. Moser, M. Die Künstliche Mykorrhizaimpfung von Forstpflanzen. II. Die Torfstreukultur von Mykorrizapilzen, *Forstwissenschaftliches Centralblatt, 77,* 257–320 (1958).
58. Moser, M. Der Einfluss tiefer Temperaturen auf das Wachstum und die Lebenstätigkeit höherer Pilze mit spezieller Berücksichtigung von Mykorrhizapilzen, *Sydowia, 12,* 386–399 (1958).
59. Moser, M. Die Künstliche Mykorrhizaimpfung von Forstpflanzen. III. Die Impfmethodik im Forstgarten, *Forstwissenschaftliches Centralblatt, 78,* 193–202 (1959).
60. Pawuk, W.H., Ruehle, J.L. and Marx, D.H. Fungicide drenches affect ectomycorrhizal development of container-grown *Pinus palustris* seedlings, *Canadian Journal of Forest Research, 10,* 61–64 (1980).
61. Redhead, J.F. Mycorrhiza in natural tropical forests, in Reference *50,* 127–142 (1981).
62. Ruehle, J.L. Growth of containerized loblolly pine with specific ectomycorrhizae after two years on an amended borrow pit, *Reclamation Review, 3,* 95–101 (1980).
63. Ruehle, J.L. Inoculation of containerized loblolly pine seedlings with basidospores of *Pisolithus tinctorius,* United States Department of Agriculture, Forest Service Research Note SE-291, Southeastern Forest Experiment Station, Asheville, North Carolina, 4 pp. (1980).
64. Ruehle, J.L. Ectomycorrhizal colonization of container-grown Northern red oak as affected by fertility, United States Department of Agriculture, Forest Service Research Note SE-297, Southeastern Forest Experiment Station, Asheville, North Carolina, 5 pp. (1980).

65. Ruehle, J.L. and Marx, D.H. Developing ectomycorrhizae on containerized pine seedlings, United States Department of Agriculture, Forest Service Research Note SE-292, Southeastern Forest Experiment Station, Asheville, North Carolina, 8 pp. (1977).
66. Schramm, J.R. Plant colonization studies on black wastes from anthracite mining in Pennsylvania, *Transactions of the American Philosophical Society*, *56*, 1–194 (1966).
67. Shaw, C.G., III, and Molina, R. Formation of ectomycorrhizae following inoculation of containerized Sitka spruce seedlings, United States Department of Agriculture, Forest Service Research Note PNW-351, Pacific Northwest Forest and Range Experiment Station, Portland, Oregon, 8 pp. (1980).
68. Shaw, C.G., III, Molina, R. and Walden, J. Development of ectomycorrhizae following inoculation of containerized Sitka and white spruce seedlings, *Canadian Journal of Forest Research* (in review).
69. Sinclair, W.A., Cowles, D.P. and Hee, S.M. *Fusarium* root rot of Douglas-fir seedlings: Suppression by soil fumigation, fertility management, and inoculation with spores of the fungal symbiont *Laccarialaccata*, *Forest Science*, *21*, 390–399 (1975).
70. Smith, A.H. Taxonomy of ectomycorrhiza-forming fungi, pp. 1-8 in *Mycorrhizae* (Edited by E. Hacskaylo), United States Department of Agriculture, Forest Service Miscellaneous Publication No. 1189 (1971).
71. Stack, R.W. and Sinclair, W.A. Protection of Douglas-fir seedlings against *Fusarium* root rot by a mycorrhizal fungus in the absence of mycorrhiza formation, *Phytopathology*, *65*, 468–472 (1975).
72. Theodorou, C. Inoculation with pure cultures of mycorrhizal fungi of radiata pine growing in partially sterilized soil, *Australian Forestry*, *31*, 303–309 (1967).
73. Theodorou, C. Introduction of mycorrhizal fungi into soil by spore inoculation of seed, *Australian Forestry*, *35*, 17–22 (1971).
74. Theodorou, C. and Bowen, G.D. Mycorrhizal responses of radiata pine in experiments with different fungi, *Australian Forestry*, *34*, 183–191 (1970).
75. Theodorou, C. and Bowen, G.D. Inoculation of seeds and soil with basidospores of mycorrhizal fungi, *Soil Biology and Biochemistry*, *5*, 765–771 (1973).
76. Trappe, J.M. *Cenococcum graniforme*—Its distribution, morphology, mycorrhiza formation, and inherent variation. Ph. D. thesis, University of Washington, Seattle, 148 pp. (1962).
77. Trappe, J.M. Fungus associates of ectotrophic mycorrhizae, *Botanical Review*, *28*, 538–606 (1962).
78. Trappe, J.M. Studies on *Cenococcum graniforme*. I. An efficient method

for isolation from sclerotia, *Canadian Journal of Botany*, *47*, 1389–1390 (1969).

79. Trappe, J.M. Mycorrhiza-forming Ascomycetes, in *Mycorrhizae* (Edited by E. Hacskaylo), United States Department of Agriculture, Forest Service Miscellaneous Publication No. 1189, 19–37 (1971).
80. Trappe, J.M. Selection of fungi for ectomycorrhizal inoculation in nurseries, *Annual Review of Phytopathology*, *15*, 203–222 (1977).
81. Trappe, J.M. and Fogel, R. Ecosystematic functions of mycorrhizae, pp. 205–214, in *The Belowground Ecosystem: A Synthesis of Plant Associated Processes* (Edited by J. Marshall), Colorado State University, Range Science Department, Science Series 26 (1977).
82. Trappe, J.M. and Strand, R.F. Mycorrhizal deficiency in a Douglas-fir region nursery, *Forest Science*, *15*, 381–389 (1969).
83. Vozzo, J.A. and Hacskaylo, E. Inoculation of *Pinus caribaea* with ectomycorrhizal fungi in Puerto Rico, *Forest Science*, *17*, 239–245 (1971).
84. Zak, B. Classification of ectomycorrhizae, in Reference 22, 43–78 (1973).

D.S. HAYMAN

13. Practical Aspects of Vesicular-Arbuscular Mycorrhiza

Introduction

World-wide interest in vesicular-arbuscular mycorrhiza is increasing at a phenomenal rate. At one conference alone in 1979 (4th North American Conference on Mycorrhiza, Fort Collins, Colorado) about one half of the more than 150 papers presented were on vesicular-arbuscular (VA) mycorrhiza. Several hundred papers have been published since the year 1973 when Mosse (116) estimated that just over a hundred had been published during the five-year period between her and Gerdemann's (46) review, following 96 and 60 in the two preceding ten-year periods and 26 in the 1930's. Yet the subject has been scarcely mentioned in all but the most recent biology textbooks.

Why this earlier neglect? The answer can be found in two of the main features of VA mycorrhiza (VAM). Firstly, this fungus-root association is a symbiotic one in which young feeder roots show no sign of damage even when densely infected. Secondly the VAM fungi can neither be cultured in the absence of a living root nor isolated on agar plates by standard microbiology techniques.

We were reminded by Gerdemann (46) that more plant tissue may be infected by VAM fungi than by any other fungi including the major plant pathogens. In addition it is now well established that many plants cannot grow adequately without VAM fungi, especially in phosphate-deficient soils (47, 65, 116, 156, 160). Recognition of these two factors has been a major stimulus to recent research on the function of VAM, in contrast to the mainly descriptive work prior to the late 1960's.

In this chapter we are concerned with practical aspects of VA mycorrhizas, i.e. studying them with the objective of harnessing them to our own advantage, especially in relation to improved phosphate nutrition of major crops. Probably their impact in tropical agriculture will be greater than in temperate regions because in the tropics phosphate-deficient, phosphate-fixing soils are widespread, superphosphate fertilizer is often in short supply, and the high temperatures enhance microbial activity (109). As many workers in the tropics may lack ready access to much of the relevant mycorrhizal literature, this review will include basic methodology in addition to conceptual principles and current trends in applied research on the subject.

Where to look for VA mycorrhizas

The answer to this question is in association with most plant species growing almost anywhere. However they are especially abundant in certain plants and habitats and rare in others.

Distribution in the plant kingdom

VA mycorrhizas occur in most angiosperms as well as in some gymnosperms, pteridophytes and bryophytes. Meyer (108) estimated that most flowering plants have endomycorrhiza (almost entirely of the VA type) in contrast to only 3 per cent with ectomycorrhiza. As far as we know VA mycorrhiza is absent only from a few plant families, mainly those which form only ectomycorrhiza (Pinaceae, Betulaceae) or the two other specific types of endomycorrhiza (Ericales, Orchidaceae). Biological boundaries are never rigid, however, and some plant families and even species can form both ecto- and VA mycorrhizal associations, e.g. oak, hazel, juniper, sweetgum, poplar, *Leptospermum*, the tulip tree and some eucalypts.

There are some plant families where the situation is confusing. Gerdemann (46) listed 14 which are believed to have little or no mycorrhiza, including the Cruciferae, Chenopodiaceae, Caryophyllaceae, Polygonaceae, Juncaceae and Cyperaceae. Although there are occasional reports of VAM infections in some of these (94, 195), their appearance is atypical because arbuscules are lacking. Experimental evidence that these are not truly VAM families comes from studies with brassicas and chenopods grown with and without a mycorrhizal "nurse" plant such as citrus, onions, maize, lettuce, barley and potato (77, 135). Small amounts of infection (hyphae, vesicles, but no arbuscules) developed in the "non-host" when a host "nurse" plant was present, but far less or even none developed in "non-hosts" grown alone.

Most plant species in the two families of major economic importance, the Leguminosae and Gramineae, are normally mycorrhizal. Yet even here some species are more prone to VAM infection than others. The root systems of many fodder legumes, e.g. *Stylosanthes* and *Trifolium* spp. usually have dense VAM infections whereas lupins have little or none (113, 189). Rye may be less mycorrhizal than other cereals (178), and maize is usually heavily infected (46). Wheat (9), *Phaseolus* beans (181) and alfalfa/lucerne (95) vary according to cultivar. Other crops where considerable VAM infections have been reported include sorghum, barley, upland rice, cassava, grapevine, olive, citrus, cacao, tobacco, cotton, sugarcane, pineapple, lettuce, onions, cowpea, soybean, strawberries, apple, rubber, coffee, tea, papaya, oil palm and various ornamental bulbs.

Occurrence in the major plant habitats

On a global scale VA mycorrhizas are virtually ubiquitous, being present in tropical, temperate and arctic regions. It is remarkable that even a single VAM fungal species can have a worldwide distribution. For example, spores of *Acaulospora laevis* (49), also called "honey-coloured sessile" (118), have been reported in Australia, Brazil, England, New Zealand, Pakistan, Scotland, South Africa and the U.S.A. This is difficult to explain as VAM spores are not readily airborne like conidia or rust uredospores, for example, and therefore unlikely to be carried in air currents over vast distances. Limited dispersal can occur in water wash, with soil animals and in wind-blown soil dust, but broader dispersal is usually accounted for by spread in association with the spread of host plants. This would be facilitated by their exceptionally wide host range compared to other plant-infecting fungi. In addition fossil plants from some 300 million years ago contain fungal structures interpreted by many as VA mycorrhizal (24). If this is so, then perhaps the VAM fungi had a very ancient origin and spread around the globe before separation of the continents and evolution of the phanerogams.

Within the different global regions VA mycorrhizas have a broad ecological range. They are found in most ecosystems including dense rain forests, open woodlands, scrub, savanna, grasslands, heaths, sand dunes and semi-deserts. The iroccurrence within these systems varies according to localized environmental conditions and plant cover. This is readily seen in the relatively simple sand dune system, where Nicolson (130) demonstrated that both the percentage VAM root infection in dune grasses and the amount of external mycelium were related to the ecological succession stage. Thus there was most VAM in the recently fixed yellow dunes and least in the open embryo and mobile dunes and in the older fixed closed community. The importance of the VAM mycelium external to the root in binding the sand particles and hence contributing to dune stability was shown by Koske, Sutton and Sheppard (92) who found bound aggregates accounted for 5 to 9 per cent of the weight of sand. More recently VAM hyphae have been shown to stabilize soil aggregates (188), which may be valuable in agricultural land.

Redhead (145) found VAM in all 15 exotic and 44 out of 51 indigenous plant species that he examined in a lowland tropical rain forest in Nigeria. Of the remaining seven, three (Caesalpiniaceae) had ectomycorrhiza. St. John (176) reported an abundance of VAM in species with magnoleoid root systems (16) in a Brazilian rain forest. Most woody plants from the 'cerrado' regions of Brazil were also reported to have endomycorrhiza (187). In a deciduous woodland in England every plant examined, except the ectomycorrhizal trees, had VAM (121), although there were few VAM spores in the soil. This survey included 50 individuals of seven different species, and an average of two out of three fine rootlets and approximately one-third of their total length were mycorrhizal. It is fairly typical to find most VAM infection in the fine feeder roots, a point which should be remembered when looking for VAM in plants. By contrast VAM can be rare in those temperate forests which are dominated by ectomycorrhizal conifers. Very wet areas such as rice paddy fields generally lack them completely, although Mejstrik (104) found fairly extensive endomycorrhiza in a marshy habitat. Recently disturbed areas may have little or no VAM infection (110, 147), but reclaimed lands such as coal tips may have more (34).

VA mycorrhiza is also abundant in many temperate grasslands. Crush (28) found both VAM infections and resting spores to be widespread in the native tussock hill grasslands of New Zealand. Records were complicated by the presence of two morphologically distinct infections, termed "coarse endophyte" and "fine endophyte" (55). The coarse type was predominant, especially on woody hosts, but the fine endophyte was proportionally more common at higher altitudes. Both types are now known to occur widely and represent numerous VAM species. Read, Koucheki and Hodgson (144) found all the most important plant species in the semi-natural hill grasslands in England that they examined were mycorrhizal, with most individual plants heavily infected, especially members of the Gramineae. They considered spore populations to be low and that infection arose mainly from root to root contact rather than from spores, resulting in a network of mycelium linking different plants within the sward. Much VAM infection but few spores were reported in acid hill grasslands in northern England (174), mid-Wales (73) and the western United States and Canada (112).

The studies cited so far and those of Mosse and Bowen (119) in Australia indicate that spore populations of VAM fungi in natural ecosystems are not regularly correlated with the abundance of VAM root infections. Some habitats nevertheless still yield appreciable spore numbers. Large species of *Gigaspora*, which seem to be most abundant in hot climates (132), have been successfully recovered from Australian heaths (184) and sand dunes (91), and *Glomus* and *Acaulospora* species from New Zealand grasslands (28, 30) and forests (56, 57, 64). Thapar and Khan (185) isolated 11 spore types from forest soils in India. Grasslands, scrub and forests in southern New Zealand showed an enormous diversity in spore densities, ranging from 6 to 1590 per 100 g soil (64). All these studies suggest that it is well worthwhile surveying a wide diversity of habitats in the search for a representative range of local VAM endophytes, and using a variety of isolation techniques (described later) in addition to collecting individual spores.

The frequent difficulties in predicting levels of indigenous VAM populations arise from the large number of factors that can affect their distribution.

These include soil fertility, soil moisture, plant susceptibility, soil depth, light intensity, altitude, soil organic matter, soil disturbance, physical movement and random variation. This last factor necessitates bulking several subsamples from a wide area in quantitative studies. Redhead (146) found it was important to sample for spores in the top 15 cm of soil because below that depth numbers decreased abruptly. Some spores have been found in the stomachs of rodents and may be spread in their faeces (39). Soil disturbance due to logging drastically reduced the number of VAM sporocarps in sugar maple stands (84). Despite this last example, it is generally assumed that perennial ecosystems contain fewer VAM spores than cultivated fields subject to annual disturbance because they exert no evolutionary pressure to select species or strains of VAM fungi able to produce long-lived propagules such as resting spores to survive periods without plant cover. In the natural system the endophytes can survive from year to year inside the living roots.

Occurrence in cultivated areas

VAM populations of cultivated lands are affected by the various soil, plant and environmental factors that affect them in natural ecosystems plus various agricultural and horticultural practices. Important among the latter are fertilizer amendments, pesticide applications and crop rotations.

As mentioned earlier, most crop species can become mycorrhizal. Some are heavily infected, some only moderately so, whereas others, such as swedes and sugar beet, have almost no infection. In southern Spain, for example, maize, *Phaseolus* beans and grapevine were consistently heavily mycorrhizal, olives were variable, and tomatoes consistently fairly lightly infected even when present at the same sites as the first three (66). In mixed cropping infection in the host plant wheat was reduced by the non-host mustard (80).

According to Butler's review (24), VAM root infections may be more abundant in orchard and plantation crops than in annual field crops, although recent studies (65) indicate more VAM spores in the latter. This may be explained, as already suggested, by selection pressures on a mixed VAM population that favour those endophytes able to survive as spores the fallow periods between crops and at least a year with a non-host crop (69). VAM populations are believed to be very low in intensively cultivated garden soils, probably because of their high fertility.

Differences in crop susceptibility to VAM presumably account in part for changes in VAM populations with different rotations. A lightly infected crop will obviously leave behind less infected root material than a heavily infected one. However the effects of different rotations of annual crops are not altogether consistent. In one study, for example, Kruckelman (94) observed most spores in wheat monoculture, but fewer in wheat after oats than in oat monoculture. There were fewest spores with potatoes. Probably the volume of soil occupied by the root systems of a crop as well as the per cent root length infected influence the number of spores produced. These factors could also explain the higher spore numbers at intermediate than at high or low levels of phosphate fertilizer in a Rothamsted field (69)—presumably there was abundant total infected root material at the intermediate level because high phosphate decreased per cent root infection and low phosphate decreased root

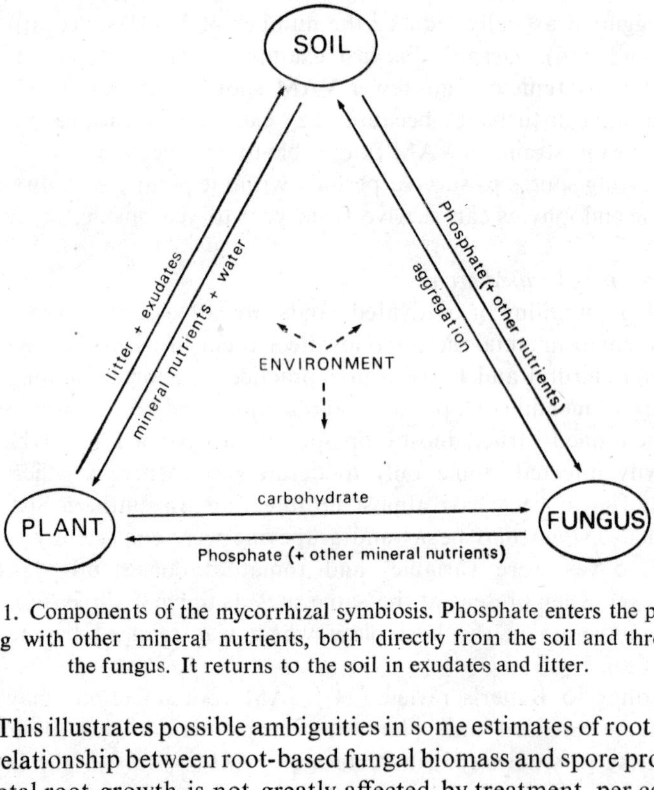

Fig. 1. Components of the mycorrhizal symbiosis. Phosphate enters the plant, along with other mineral nutrients, both directly from the soil and through the fungus. It returns to the soil in exudates and litter.

growth. This illustrates possible ambiguities in some estimates of root infection and the relationship between root-based fungal biomass and spore production. Where total root growth is not greatly affected by treatment, per cent infection and spore numbers can be closely related, e.g. in their negative responses to nitrogen fertilizer (62). Thus the considerable variation in field populations of VAM fungi to be found within a single site can be partly explained by the inhibitory effects of large applications of nitrogen and phosphate fertilizers, also reported by Strzemska (179). However, fertilizers may have a positive effect on VAM if the initial soil fertility is very low (94).

Fungicides such as benomyl applied to soil depress VAM populations, although paradoxically the insecticide/nematicide aldicarb increased it except that high nitrogen levels overrode this effect (133). Some other nematicides may increase VAM (20), but insecticides generally decrease it (94). VAM species can differ in their susceptibility to fungicides (175).

At a single site in Florida, Schenck and Kinloch (165) found marked differences in populations of VAM fungi between different crops grown in monoculture for seven years on a newly cleared woodland site. There were more spores with soybean than with the other crops and fewest in the woodland. Three species of *Gigaspora* were most numerous around soybean roots, whereas two *Glomus* species were most prevalent with Bahia grass and *Acaulospora* spp. with cotton and peanut. Sorghum had the largest number of VAM species. This is one of the most detailed reports on the selective effects of host species under field conditions.

Variation in VAM populations between sites, unlike that within a site, cannot necessarily be accounted for by differences in soil nutrients. The lack of any correlation between infection and fertility when a range of sites were compared in southern Spain, for example, points to the overriding importance of natural variability in propagule distribution (66).

The species composition of a VAM population in a field can be changed by soil amendments. Nitrogen fertilizer, for example, depressed numbers of 'white reticulate' spore types more than 'laminate' types in a sandy loam (65). Other factors, as yet unknown, are also involved as in adjacent fields at Rothamsted with the same crop and soil type and similar fertility but very different endophyte composition (62). In Welsh hill grasslands populations of indigenous endophytes were reduced more than introduced *Glomus mosseae* and *G. fasciculatus* by additions of lime and phosphate fertilizer (73). The distribution of 'honey-coloured sessile' and 'yellow vacuolate' spore types in Western Australia was related to soil pH (1). In Pakistan reticulate spore types seemed most common in sandy soils and vacuolate types in clay loams (85). Clearly the ecology of VAM species is a complicated subject needing considerable further study.

Usually more spores are found towards the middle or end of the growing season than at the beginning (62, 101). This is attributed to increased spore production as root growth slows down or ceases. VAM root infections usually reach a maximum towards the end of the growing season (62), following a three-phase pattern of growth (158, 181). With field-grown *Phaseolus* beans and soybeans, Sutton (181) noted an initial lag phase of 20 to 25 days, attributed to rapid root growth of the seedlings and the time required for spore germination, germ-tube growth and penetration of the host plant root. In the second phase, lasting 30 to 35 days, extensive mycorrhizal development coincided with most shoot growth and copious spread of external mycelium leading to multiple infections. During the third phase, from host fruiting to senescence, the proportion of mycorrhizal to non-mycorrhizal roots remained constant.

The foregoing information suggests that VA mycorrhizas are best looked for in late summer in agricultural land not intensively cultivated. The latter may contain most spores where a heavily mycorrhizal crop such as maize is growing. Natural ecosystems, especially those rich in plant species, should be examined most thoroughly at the time when young feeder roots are abundant.

How to collect VA mycorrhizas

Having decided where to look, the next step is to collect root and soil samples from which satisfactory VAM infections and spores can be recovered. A small amount of soil, preferably close to a plant, should be dug out with a trowel to a depth of 10 to 15 cm after scraping away the top cm or two. Alternatively a metal tube or soil auger can be used where a more exact volume is required. It is essential that plants are not just pulled out of the soil, otherwise the root cortex may be stripped off, taking the VAM infection with it. For quantitative studies, because of the variability of VAM populations in soil, bulked samples of up to 15 sub-samples should be replicated at least three times per site. The samples should be placed in polythene bags, labelled, dated and stored at about 2°C until they can be processed.

The most common method of processing a sample, after it has been thoroughly mixed, is the wet sieving and decanting technique. Originally developed by nematologists, this technique was adapted for VAM studies by Gerdemann and Nicolson (48). One version is as follows: (1) mix a measured portion of soil (50 g is suitable for quantitative studies) by hand in luke-warm water (200 ml is a convenient volume) in a large beaker until all soil aggregates have dispersed to leave a uniform suspension; (2) decant most of the suspension through a 710 μm sieve (supported in a funnel) into a tall 1 litre graduated cylinder; (3) resuspend the residue in more water and decant, repeating this four or five times to give about 700 ml in the cylinder and leaving only grit, sand and heavy organic particles in the beaker; (4) wash roots and other organic matter on the sieve with a fine jet of water, e.g. from a squeeze bottle, and collect washings in the cylinder; (5) resuspend material in cylinder by stirring several times and decant through a 250 μm sieve into a second 1 litre cylinder, retaining a small volume which is resuspended in a further 300 ml of water and poured through the sieve to almost fill the second cylinder; (6) wash material on sieve and add washings to second cylinder; (7) resuspend material in second cylinder and pour most through a 105 μm sieve; (8) resuspend the residue in another litre of water, pour through same sieve, wash material on sieve; (9) resuspend these last two litres and pour through a 53 μm sieve; (10) wash each sieving into separate small beakers and examine each in turn; (11) discard residues in large beaker and cylinders (but check in the first attempts with the technique that no spores are left behind).

The root pieces retained on the 710 μm sieve can be examined under a dissecting microscope for attached hyphae, spores and sporocarps, then stained for VAM infections (see later). It may be necessary to process more soil to obtain an adequate root sample. The organic matter from the 250 μm sieve is examined for sporocarps and large spores. The 105 μm sieving should yield most spores since their commonest size range is between 100 and 250 μm and spores smaller than 100 μm often occur in clusters trapped on the 106 μm sieve. Small detached spores may be found on the 53 μm sieve. This sieving technique works well with clay and especially sandy soils but can be tedious with soils containing much organic matter. The fastest, yet simplest, way of examining sievings is in a petri dish with several partitions, e.g. the Doncaster nematode-counting dish (75).

Root samples alone can of course be obtained simply by washing without wet sieving. For spores several other techniques have been described and each has its own disciples. One of the first was a floatation-centrifugation technique with sucrose density gradients (136), but this could damage the spores by osmosis. Subsequently gelatin (124) and percoll (Macdonald, pers. comm.) columns have been used. Other methods include a floatation-adhesion technique where spores in suspension stick to the sides of a separatory funnel as the water is drained away (182), and bubbling air through a suspension of spores in 50 per cent glycerol to maintain them at a specific level in a column (44).

With any method care must be taken to differentiate between VAM spores and other objects of similar size and shape, to distinguish apparently viable (with contents) from dead (empty) ones, and to enumerate separately spores so small as to be partly overlooked or uncertainly identified. These problems may account for the occasional large discrepancies in published reports of spore numbers in soil. To reduce confusion viable spores should be categorized according to genus or species whenever possible and dubious or dead ones listed separately.

Identification of VAM infections

The anatomical features characteristic of VA mycorrhizas cannot be seen unless the infected roots are suitably stained. The only other evidence of a VAM infection is a yellow colouration in some plants, e.g. maize, onions, clover (which disappears on exposure to light), or a granular/milky appearance in the cortex of relatively transparent young roots, e.g. *Coprosma robusta.* There is no obvious effect of VAM on root morphology and no distortion of the host tissue.

Standard staining methods involve treating the roots with a clearing agent, e.g. chloral hydrate, potassium hydroxide or phenol, staining the fungus with, for example, trypan blue, cotton blue or acid fuchsin, then removing excess stain from the host tissues with lactophenol, lactic acid, etc. A commonly used method (139) is as follows: (1) wash roots (either fresh or fixed) in tap water, but not vigorously enough to detach all the external mycelium; (2) simmer at about 90°C for one or two hours in 10 per cent KOH; (3) rinse four times in tap water; (4) acidify by immersing for five minutes in 2 per cent (approx.)

HCl; (5) pour off the acid and add the stain, viz. 0.05 per cent trypan blue in lactophenol; (6) boil roots in stain for three minutes; (7) pour off stain, add lactophenol and stand overnight to destain the host tissues; (8) examine under a microscope. Operations (2) to (7) must be done in a fume hood. Roots can be examined in latic acid or 1:1 lactic acid : glycerol instead of lactophenol which is a health hazard. Mounts of roots on slides can be made semi-permanent by sealing the edges of the coverslip with nail varnish or by mixing PVA (polyvinyl alcohol) with the lactophenol.

The presence of vesicles and arbuscules is, by definition, the diagnostic criterion for identifying a vesicular-arbuscular mycorrhizal fungus in a root. Vesicles (Plate 1, C and D; Plate 2, D) are usually oval, sometimes round and occasionally (e.g. *Acaulospora laevis*) irregularly lobed. They form between the cortical cells or occasionally inside them and are attached to the internal hyphae in a terminal, or more rarely intercalary, position. They contain oil, sometimes as a large single globule (e.g. *Glomus fasciculatus* 'E3') and are believed to function as storage organs. In older roots they can develop thick walls and presumably function as resting spores when the roots decay. Some rootlets may contain many more vesicles than other rootlets in the same root system, possibly because of different physiological ages.

Arbuscules (Plate 2, A, B, C and D) are more or less equivalent to the haustoria of obligate parasitic fungi such as rusts and mildews but are believed to function in bidirectional transfer of nutrients. Essentially this transfer involves carbohydrates from plant to fungus and minerals, especially phosphate, from fungus to plant. One arbuscule develops in a single host cell. Arbuscules (literally "little trees") have been seen within a few days of the fungus penetrating the root and at this young stage they show dichotomous branching from a thick trunk where the fungus entered the cell. The branches become finer and finer until the tips are barely visible (around 1 μm). Electron microscope pictures reveal the arbuscule to be outside the host plasmalemma. They last two to three weeks and then develop a granular appearance as they degenerate. In the older literature this is referred to as "digestion", but there is no evidence of the root cell positively destroying them in the many recently published fine structure studies (e.g. 27, 88). A new arbuscule can replace the degenerated one in the same cell. Young infected roots are often seen to be almost full of arbuscules.

The arbuscules are borne on side branches of the distributive hyphae. These hyphae grow parallel to the root axis and can occupy much of the primary cortex. They do not penetrate the endodermis or stele. Their branching pattern can help in distinguishing between some endophytes (3). Often these internal hyphae are most prevalent in the inner cortex and are connected by characteristic zig-zagging or coiling hyphae to the entry point at the root surface (see Fig. 2). The entry points extend from the appressoria (infection cushions) formed on the epidermis or, more rarely, on a root hair. The

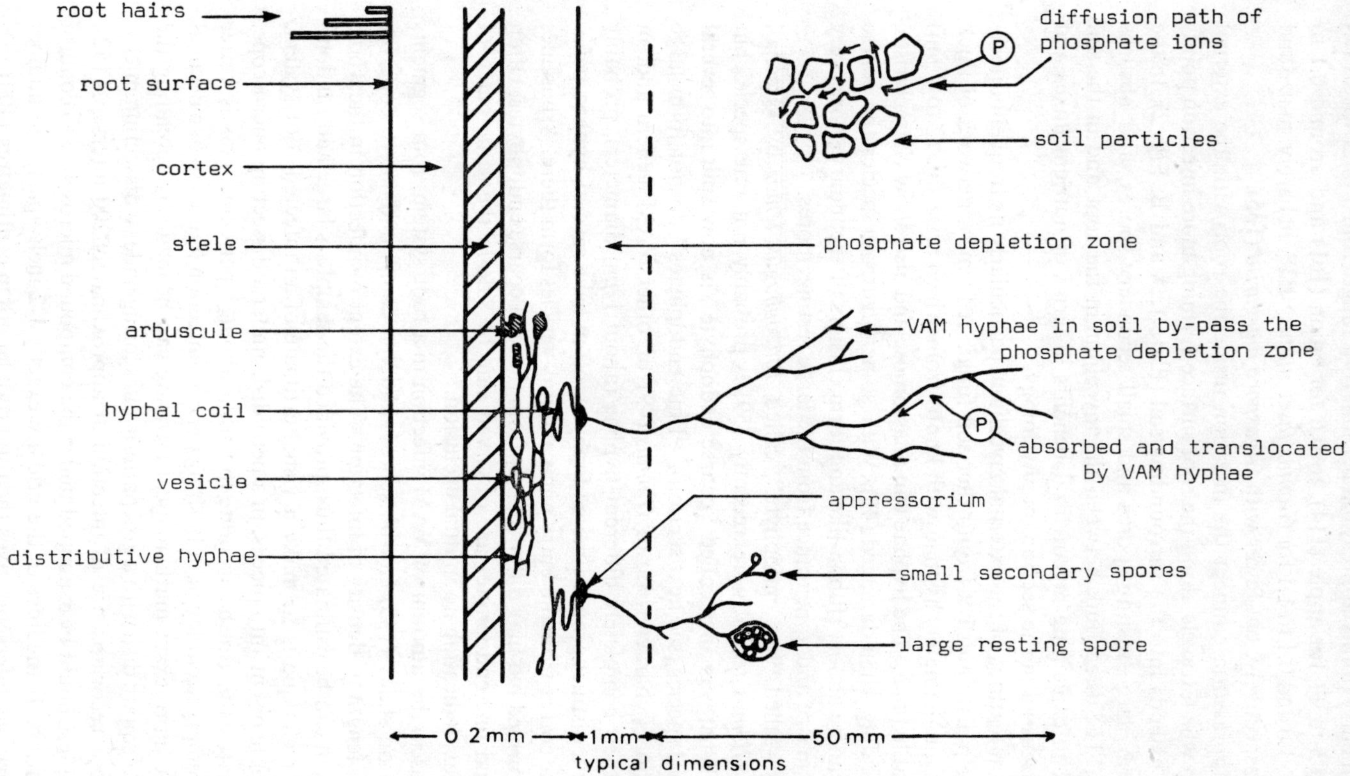

Fig. 2. Diagram (not to scale) illustrating the major features of a **VA** mycorrhiza and the chief mechanism whereby it is believed to enhance the uptake of phosphate from soil. The slow diffusion path of phosphate ions in soil, resulting from tortuous pathways and reversible adsorption to the soil particles, is short-circuited by direct transfer of phosphate to the root through the fungal hyphae which extend well beyond the root hair zone. P = primarily $H_2PO_4^-$.

number of entry points/appressoria per mm root range from 3 to 21 for strawberry and 5 to 11 for apple (114), to 0.6 for onion (161) and around 0.1 to 0.2, 0.1 to 1.0 and 0.1 to 1.1 for *Glomus fasciculatus* "E3" on barely, maize and onions, respectively, but fewer with *Gigaspora margarita* (135).

The appressoria connect the infection inside the root with the external mycelium which travels along the root surface forming more infection points and also extends into the surrounding soil (Plate 1, A and B; Fig. 2). In the soil it forms large resting spores and small secondary spores, and absorbs nutrients. The direct link between the mycelium in the root and in the soil is critical in explaining mycorrhizal benefits in terms of nutrient fluxes and will be discussed in the section on physiology.

Certain features of the external mycelium, particularly its frequent dimorphism, help distinguish it from other soil fungi. It forms a network of thick (20 to 30 μm diameter) hyphae with knobs from which arise fine (2 to 7 μm) ephemeral hyphae. The hyphae can anastomose and also show characteristic wound healing when damaged (45). With some *Gigaspora* species the coarse hyphae can be almost ribbon-like and form clusters of secondary spores very different in size and appearance from the large resting spores.

The so-called "fine endophyte", called *Rhizophagus tenuis* by Greenall (55) and *Glomus tenuis* more recently (59) is probably not one species but several. What was once called "coarse endophyte" is now split into several genera and species (see next section). "Fine endophytes" do not form large resting spores characteristic of *Glomus* (49) or other VAM genera and, in view of their completely different hyphae (about 1 μm diameter), probably represent a distinct group. Fine endophytes often have several hyphae penetrating the cortex from the entry point and tiny vesicles form there. Arbuscules are also formed but may differ slightly from those of coarse endophytes. Often coarse and fine endophytes are observed side by side in the same patch of root cortex, indicating no mutual exclusion.

To assess the amount of VAM infection in a root system, root segments mounted on slides (135) can be scored under a compound microscope for presence, length or density of infection. The extent of infection in the whole root system can be estimated thus, provided the sample is large and random. A faster technique is to make a visual estimate of an infected root system, either whole or cut into pieces, in a petri dish under a dissecting microscope. With practice this can be accurate to within about 5 per cent. This is better than the empty accuracy of publishing per cent infection figures to two decimal places. A more exact mathematical estimate can be made by spreading the roots on a square dish with a grid pattern and scoring under a dissecting microscope presence or absence of infection at approximately 200 intersepts (52). Sub-samples should be examined under the compound microscope to confirm identification. If the lines of the grid are exactly 1/2 inch apart, the number of root/gridline intersepts gives the total root length in centimetres (100).

Identification of VAM spores

Species and genera of VAM fungi are identified by the morphology of their large resting spores. Anatomical differences between different endophyte infections are sufficient to permit only tentative identification of a few types or groups.

The spore types able to form VA mycorrhiza were divided into four genera by Gerdemann and Trappe (49), viz. *Glomus*, *Gigaspora*, *Acaulospora* and *Sclerocystis*. These are grouped in the family Endogonaceae, order Mucorales, class Zygomycetes. This is part of the old grouping Phycomycetes which accounts for the term "phycomycetous mycorrhiza" sometimes seen in earlier literature. A fifth genus, *Endogone*, formerly the name used for all VAM fungi, is now restricted to species that form ecto- or no mycorrhiza. It is the only genus believed to form true zygospores resulting from fusion of gametangia and nuclei. The resting spores of the genus *Gigaspora* are called azygospores because they have attached to them a structure interpreted as a vestigial gametangium or suspensor. This is the bulbous swelling at the base of the spore where it is attached to the hyphal stalk (Plate 3, F). Sometimes there is a second, smaller suspensor-like organ attached (Plate 3, E) but fusion is doubtful. Spores with straight or angular (but not bulbous) stalks are grouped together as *Glomus* (Plate 3, A, C and G). These are vegetative chlamydospores where no fusion is involved. They are borne free or in sporocarps, usually on a single subtending hypha. Spore types without stalks are literally called *Acaulospora*. This type is unique because the resting spore is borne sessile on the side of the stalk of a mother spore (Plate 3, B). *Sclerocystis* forms chlamydospores which resemble *Glomus* except for being arranged regularly on a central core in a sporocarp (Plate 3, D)—in those *Glomus* species forming sporocarps the spores are borne at random.

There are several papers on the taxonomy of the Endogonaceae. Thaxter's (186) classic monograph emphasizes types with large sporocarps and describes species forming zygospores, chlamydospores and sporangiospores. Gerdemann paid particular attention to individual spores found free in the soil or on hyphae attached to the root. In 1963 he and Nicolson (48) distinguished six spore types which they classified numerically. Five years later (131) they described more spores, using latin binomials (*Endogone* spp.). At about the same time Mosse and Bowen (118) discovered nine distinct spore types and gave them each a descriptive name. Currently the 1974 scheme of Gerdemann and Trappe (49) is used as a basis for identifying known species and naming new ones. In addition several descriptions have been published since then of single or groups of species, e.g. a key to species of *Gigaspora* (17), details of several Florida endophytes (132) and a key to all Endogonaceae species known up until recently (59). Two new genera, *Complexipes* (192) and *Entrophospora* (6) have also been described but neither has yet been proved to form

VA mycorrhiza. *Modicella* and *Glaziella* (49) are the remaining non-VAM genera in the Endogonaceae.

It is clear that the taxonomy of the Endogonaceae is in a state of flux. The traditional arguments between "lumpers" and "splitters" are well underway. At this stage it is probably best to try and match any new isolate with a published description and use the appropriate Latin binomial if possible. Where there are doubts, however, descriptive names are safer and avoid inaccuracies such as where isolates given the same Latin names by different workers have been shown to be different species when examined side by side. *Glomus fasciculatus* exemplifies some of the problems because there is disagreement as to whether it is one or several species. Other difficulties include the existence of spore types which do not "fit", e.g. the "white reticulate" type of Mosse and Bowen (118) is a *Glomus* according to its shape and stalk, whereas its cytoplasmic structure is more like some *Gigaspora* types. For experimental purposes the main thing is to be able to distinguish between one's own isolates, and a number or descriptive name will suffice for some time. However, correct naming is important in ecological studies and in communicating with other research workers.

The following diagnostic criteria are used in spore identification:

(1) Size and shape: Spores are usually spherical or oval but their sizes range widely, e.g. around 50 μm diameter or less for *Glomus microcarpus*, 100 to 200 μm or more for *Glomus mosseae*, *Gigaspora calospora* and *Glomus caleaonius*, 200 to 400 μm for *Gigaspora margarita* and nearly 1 mm for a few *Gigaspora* isolates.

(2) Colour: This should be determined under a dissecting microscope with incident light because under transmitted light the dense contents of some pale spores, e.g. white *Gigaspora margarita* spores, make them appear dark. Other colours include pale greenish yellow, yellow brown, reddish brown, greyish brown, honey and amber.

(3) Stalk attachment: Simple (*Glomus* and *Sclerocystis*), bulbous (*Gigaspora*) or absent (*Acaulospora*).

(4) Cytoplasmic structure: Two categories (118), reticulate (network of cytoplasmic strands around small vacuoles) and vacuolate (cytoplasm contains large vacuoles which ultimately coalesce to form a few or even one large one).

(5) Wall thickness and the number or types of layers present: Generally there is a fine white outer wall and a thick, chitinous, brittle, coloured inner wall in the spores with vacuolate cytoplasm, e.g. *Glomus mosseae* (49)/yellow vacuolate (118). *Glomus fasciculatus* "E3" has simple walls. The reticulate types generally have the thin, white membranous wall inside the thick, chitinous outer coloured wall e.g. *Gigaspora calospora* (49)/bulbous reticulate (118) and *Acaulospora laevis* (49)/honey-coloured sessile (118). *Gigaspora margarita* and *G. heterogama* seem to have thick single walls.

(6) Method of germination: With vacuolate types one germ tube is usually

produced which emerges through the main stalk, whereas reticulate types can form several germ tubes which emerge through the side of the spore and the cytoplasm sometimes becomes polarized (118).

(7) Secondary spores. These are distinct from the main thick-walled resting spore. They are smaller, smooth or spiny, borne singly or in clusters or arranged spirally, and are sometimes called external vesicles. In *Glomus* spp. they tend to look more like vesicles whereas with some *Gigaspora* spp. they can be spiny and arranged in characteristic spirals.

(8) Sporocarps—formed by some species. In *Glomus mosseae* they are up to 1 mm in diameter with an outer wall of loosely woven hyphae enclosing some three to 12 irregularly arranged chlamydospores. In some temperate and especially tropical species they may be one or more cm in diameter and contain hundreds of spores embedded in a dense matrix. *Sclerocystis rubiformis* sporocarps are small (around 300 to 500 μm), have no outer wall and their spore arrangement resembles raspberry fruits. *Glomus fasciculatus* forms loose naked clusters of up to several hundred spores. Species of *Gigaspora* and *Acaulospora* are not known to form sporocarps.

Within a morphological species there are likely to be physiological strains which differ in their interaction with both plant and soil. This is a difficult topic to unravel while we cannot culture the fungi.

Techniques for isolating VAM endophytes

Essentially the starting point for practical studies on VA mycorrhiza is the isolation of local endophytes and their establishment in pot cultures. Inoculum obtained from collections elsewhere may be useful for comparative purposes, but is no substitute for indigenous endophytes which are likely to be more effective because they are adapted to local conditions. This point is illustrated by the work of Schenck, Graham and Green (162) who showed that spores of species from hot climates did not germinate as well as the temperate species, at low temperatures.

Since VAM fungi are obligate symbionts they must be established on host plants. These are grown in sterilized soil in pots and their purity is checked by wet-sieving aliquots to check if spores are all of the same morphological type or species. This does not, of course, reveal the possible presence of different physiological strains. There are various ways of obtaining infective VAM propagules from soil and establishing pot cultures.

(1) Individual spores. These are obtained by the techniques of wet-sieving, density gradients, etc. described earlier. The wet-sieving method is convenient because it yields undamaged spores in water which can be removed with a fine capillary pipette under a dissecting microscope, after using fine needles to separate them from the surrounding organic matter. Spores of different types are transferred to separate watch glasses and any remaining

organic matter discarded. A few (10 to 20 is a suitable number), identified as one type, are then transferred to 1 cm squares of filter paper. One or more of these squares can then be wrapped around the roots of a small seedling while transplanting it from a seed tray of sterilized sand : soil mix to a pot of similar growth medium.

After three or four months (this can vary according to host plant and growing conditions), a portion of soil should be removed, wet-sieved and the spore population checked for uniformity, i.e. that a single species is present. By this time there should be hundreds or even thousands of spores present which makes it easier to select 10 to 20 for further inoculation purposes should the presence of more than one species make it necessary to repeat the isolation procedure.

(2) Soil sievings. If there are few spores but a reasonable level of root infection, then clearly other propagules are present. These are likely to be in the organic matter of a wet-sieving, so a pad of the material from the sieve (see previous description of wet-sieving technique) should be placed against the roots of a transplanted seedling and the procedure described for individual spores followed. Thus from the many spores produced in pot culture, morphologically distinct species can then be separated.

(3) Infected roots. This is an alternative to soil sievings if few spores are present, particularly if the presence of non-sporing types or fine endophytes is suspected. Pieces of root are cut from plants shown to be infected by staining a portion of their root system. One or more cm of root can then be placed against the roots of a transplanted seedling and the procedure described for spores and sievings followed. As there may be more than one endophyte infecting 1 cm or less of root, several inoculations should be made, and smaller root pieces used if mixed infections persist.

(4) Infected plants. Young seedlings where natural infections are likely to have just started can be dug up, the roots thoroughly washed, and the seedlings replanted in sterilized soil in pots. Infection should then become well established and spores formed later. Mixed infections can be dealt with as described previously.

(5) Baiting. This method (50) involves growing seedlings in unsterile field soil just long enough for infection to start. They are then removed, washed, replanted in sterilized soil and subsequently checked by the procedures described for spores and sievings. In addition this method can be modified to give an estimate of soil infectivity by measuring the rate of infection at different dilutions of the test soil.

Stock plants

Once a pot culture of a single species is achieved, inoculum from it can be used to establish a large number of stock plants to provide inoculum for future experiments. To eliminate carryover of contaminating micro-organisms to new plants, the spores can be surface-sterilized by immersing in a solution containing 2 per cent (w/v) chloramine T and 200 ppm streptomycin for 15 minutes, followed by washing three times in sterile water (125). This treatment will not kill parasites inside the spores so any suspect ones, usually those with unclear contents, should be rejected and only healthy-looking spores retained. Larger spores such as *Glomus mosseae* are readily separated visually in this way. Subsequent inoculation of stock plants can be done with small amounts of soil from the purified cultures. This form of inoculum rapidly produces new infections.

The choice of stock plant species depends on availability, longevity and to a lesser extent endophyte preference. It should be perennial and not be too large for glasshouse conditions. *Nardus stricta*, *Coprosma robusta*, onions, black pepper, citrus, sorghum and *Stylosanthes* are commonly used. Maize is useful for building up large quantities of inoculum quickly although it dies after one growing season. Regarding endophyte preference, *Gigaspora margarita* infects onions better than lettuce, for example, and *Nardus stricta* is good for sporocarp production by *Glomus mosseae*.

Several soils should be tested before selecting the one or two most suitable ones. They should be of moderately, but not excessively, low fertility and have a good physical structure. Soil-endophyte specificity (see later) must also be taken into account. The soil must be partially sterilized, sufficient to kill the indigenous VAM fungi. Gamma-irradiation (0.8 or 1 Mrad) is preferred, otherwise steaming (about 1 hr for pots containing 1 kg soil, longer for larger volumes of soil) or chemical treatment (e.g. methyl bromide) can be used. Often mixing the soil 1 : 1 with sand helps maintain satisfactory drainage and aeration.

Maintaining stock plants on sand trays watered by upward capillarity avoids problems of cross-contamination by splashing. Adjacent pots should be adequately spaced. Pots with different endophyte species should not be kept on the same sand tray because of possible root contact, although root growth into the sand can be inhibited by adding chemicals such as Algamine to the sand (4 ml per 70 × 10 cm sand tray). Dilute nutrient solution should be applied to the pots at intervals, enough to maintain satisfactory plant growth without inhibiting infection.

Spore numbers reach a maximum during the first year, but may decline after two or three years. If many spores are needed for experiments, stock plants need to be renewed annually.

Initial pot trials

Since we are concerned with practical aspects of VA mycorrhiza, our initial pot experiments should be planned to give basic information relevant to subsequent field trials. Thus we need to screen local VAM endophytes selected from our newly established culture collection on crop species of economic importance locally in arable soils representative of the district, possibly amended with different phosphate fertilizers. Optimum combinations of plant, soil and endophyte should be noted. The first screenings should be done in sterilized soil, then the competitive ability of selected endophytes tested in unsterile soil.

A useful first experiment would be with one host plant in two soils with four inoculation treatments (none and three different local endophytes), three phosphate treatments (none, superphosphate and rock phosphate) and at least three replicates. This gives a minimum of 72 pots so obviously experiments must be designed carefully within the limits of available time and space.

In these experiments inoculum can be conveniently added as sievings placed against the roots of uninfected seedling transplants or placed as a thin layer about 2 cm below the seeds sown directly in the test pots. Non-inoculated controls must be given filtered washings from the inoculum (containing contaminating micro-organisms which might affect plant growth, but lacking VAM propagules) and possibly an equivalent volume of sterilized sievings to compensate for any nutrient carryover in them.

Published pot experiments on VAM are far too numerous to cite here (see reviews 47, 65, 116, 156), although a few will be referred to in later sections on the principles of VAM symbiosis. In most growth experiments with VAM in pots done by various investigators in several countries, it has become almost commonplace to show a dramatic enhancement of phosphate uptake by VAM inoculation of plants in P-deficient soil.

Return to the field

The ultimate test on the usefulness of VAM inoculation is whether it can increase crop yields under standard field conditions. Unfortunately field trials are not without their tribulations. Often results from pot experiments do not match closely those from field experiments (73, 93, 143, 167), although this could be accounted for by obvious differences in environmental factors, e.g. temperature, light, soil moisture and constrictions on root growth. The last factor is illustrated in reports of some plants responding to VAM inoculation in large but not in small pots. Nevertheless, although pot experiments cannot predict absolute growth effects in the field, they can be used as a guide to soil-plant-endophyte compatibility to reduce the amount of screening necessary in the field.

There are four major factors that determine potential benefits from mycorrhizal inoculation in the field:

(1) THE NATIVE MYCORRHIZAL POPULATION. This varies greatly in size and

effectiveness. If the natural level of VAM propagules in a field soil is low, infection may be slow to build up in the early stages of plant growth when demand for phosphate is high. This could check growth such that the plants never catch up. Correct inoculation should speed up infection in such soils and even more so in soils where VAM fungi have been eliminated by fumigation, etc. Furthermore some indigenous endophytes, although abundant, may be relatively ineffective symbionts. If these infect rapidly but ineffectively, an inoculum must be introduced which is highly competitive as well as symbiotically efficient so as to establish well against the native ones. Some assessment should therefore be attempted of the native VAM population before field plans are made, although this is hindered by our incomplete knowledge of propagule forms; for example spore populations are not always good indicators of soil infectivity (75).

(2) SOIL FERTILITY. Growth responses to mycorrhiza tend to be greatest in soil deficient in plant-available phosphate. Therefore a soil phosphate analysis, e.g. $NaHCO_3$-extractable P, also called Olsen P, (137), should be done, although this cannot always predict responses to inoculation because of complications with the native VAM (177). If the soil is a "phosphate-fixing" one, responses to inoculation are possible even where appreciable amounts of fertilizer P are added. To avoid exhaustion of the labile P reserves (see section on mechanisms), some P should be added. Recent experiments suggest that plants may more efficiently use small dressings of soluble superphosphate and less soluble forms such as rock phosphate when they are mycorrhizal. Indeed the chief benefit of mycorrhizal inoculation in agriculture may in future prove to lie in its more efficient utilization of applied phosphate fertilizers.

(3) CROP SPECIES. Some plant species and even cultivars of the same species are more dependent on mycorrhiza than others. This is governed by their demand for and ability to take up phosphate from soil and will be reviewed in detail later. As a broad generalization, crops such as wheat, with extensive fine root systems, would not be expected to respond to VAM except in rather P-deficient soils. Others with coarser, less hairy roots such as onions, citrus and *Stylosanthes*, are very responsive to VAM even in soils with moderate P levels.

Often the greatest impact of early mycorrhizal infection in nature is on seedling establishment. This principle may also apply to crops transplanted as seedlings. In some grasslands a seedling's roots may need to become incorporated into the VAM mycelial network in the soil if it is to survive (93).

(4) CULTIVATION PRACTICES. Some rather drastic practices such as soil fumigation to kill plant pathogenic fungi and nematodes can have their usefulness reduced if they kill essential VAM fungi, e.g. in citrus nurseries (89). However, local recommendations for pest control should not be ignored, otherwise there could be no experiment to harvest! High levels of nitrogen and phosphate fertilizer may also depress VAM infection. Moreover, the

native population may decline in the absence of suitable host plants during long fallow periods, necessitating the introduction of fresh VAM inoculum.

Bearing these factors in mind, field experiments can be set up to test various plant-soil-endophyte-phosphate combinations that look promising from the initial pot experiments. Plots should be as uniform as possible, of reasonable size, and the experimental design approved statistically. They must be maintained carefully to avoid cross-contamination, e.g. from inoculum carried on shoes. Apart from using single species inoculum, mixed inoculum might also be considered, based on the premise that the crop will select the endophyte(s) most adapted to the site. The infectivity and longevity of different endophytes can also be monitored.

In summary, benefits from inoculation with VAM are most likely in infertile soils containing few or ineffective endophytes and with crops that have coarse root systems and a high demand for phosphate. Specific sites should be tested empirically and endophytes from those sites compared with ones already selected from stock plants. For the practical use of VAM on a large-scale inoculum must be produceable in bulk, readily transportable to the field, and easily introduced into the seedling root zone. These aspects will be discussed next.

Methods for field inoculation and inoculum production

Inoculation techniques

Various techniques have been devised for introducing VAM inoculum into a field-grown crop.

(1) PRE-INOCULATED TRANSPLANTS. A seedling can be inoculated with a selected VAM endophyte, grown in a suitable compost up to the time of outplanting, and the pre-transplant mycorrhizal status determined. Thus a strong, selected VAM infection can be established before exposure of the plant to indigenous endophytes of unknown effectiveness. On a practical scale this could be useful for many agronomically important tropical trees, e.g. coffee, tea, cacao, papaya and oil palm. Some other tree seedlings e.g. *Liquidambar styraciflua*, seem to be obligately mycorrhizal in tree nurseries (170), and vegetatively propagated cuttings of *Liriodendron tulipifera* mostly stayed dormant unless made mycorrhizal (90).

Preliminary trials with pre-inoculated transplants of plants normally sown direct in the field may be useful in situations where there is a risk of direct inoculation in the field failing and hence precluding observations of mycorrhizal potential.

(2) DIRECT INCORPORATION INTO SEED FURROWS OF SOIL FROM POT CULTURES. This soil inoculum contains spores, hyphae and infected root pieces. It has been especially successful with experiments in fumigated soils where there was no competition from indigenous endophytes, but has also worked

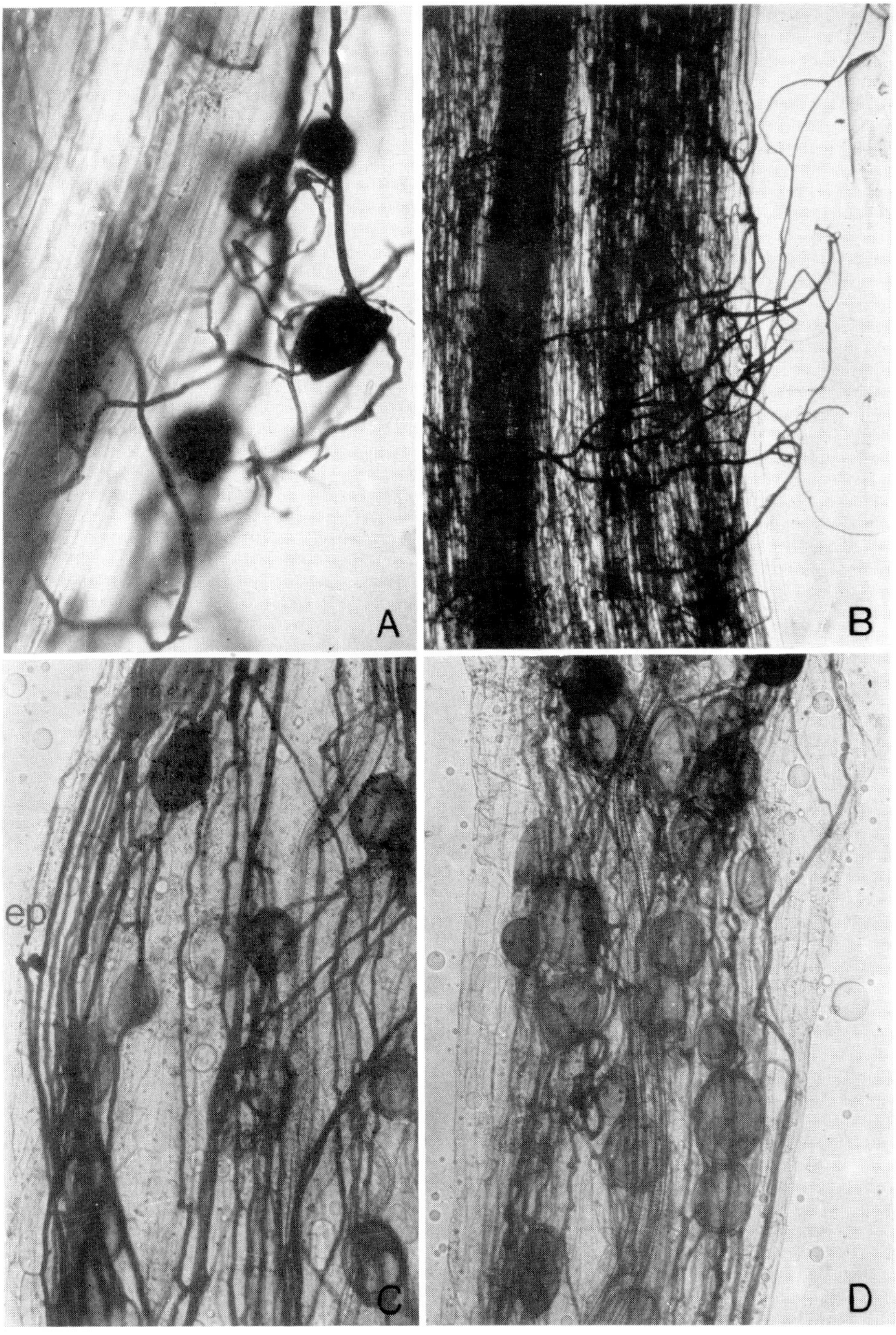

Plate 1. VA mycorrhizal infections. (A) Spores and external mycelium attached to root ($\times$ 220); (B) Part of root with clumps of external mycelium ($\times$ 85); (C) Entry point (ep), distributive hyphae and vesicles ($\times$ 220); (D) Root with extensive vesicular infection ($\times$ 220).

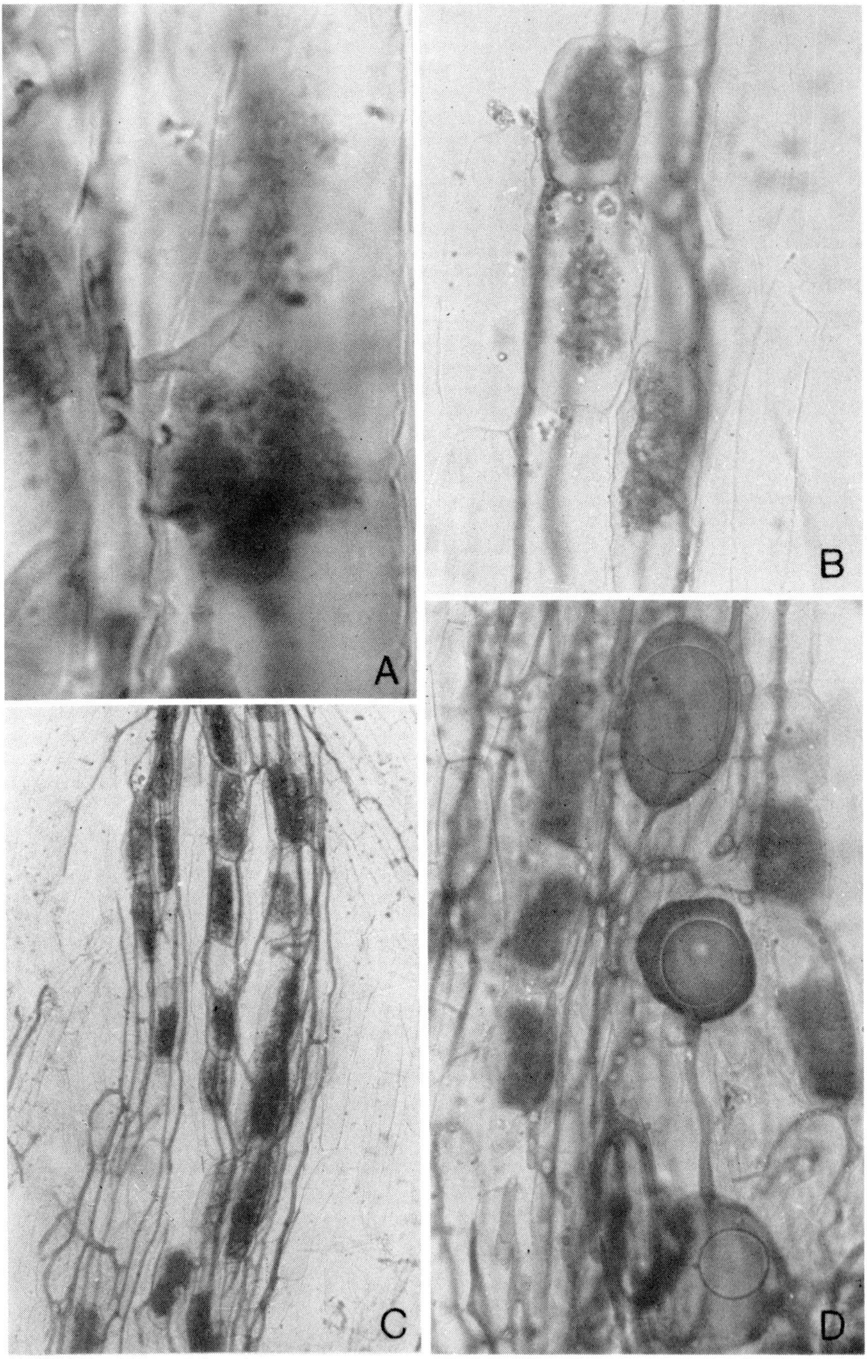

Plate 2. VA mycorrhizal infections. (A) Arbuscule showing main trunk (× 1320); (B) Arbuscules showing branching pattern (× 590); (C) Arbuscules and distributive hyphae (× 230); (D) Arbuscules, showing granular appearance, and vesicles with oil globules (× 590).

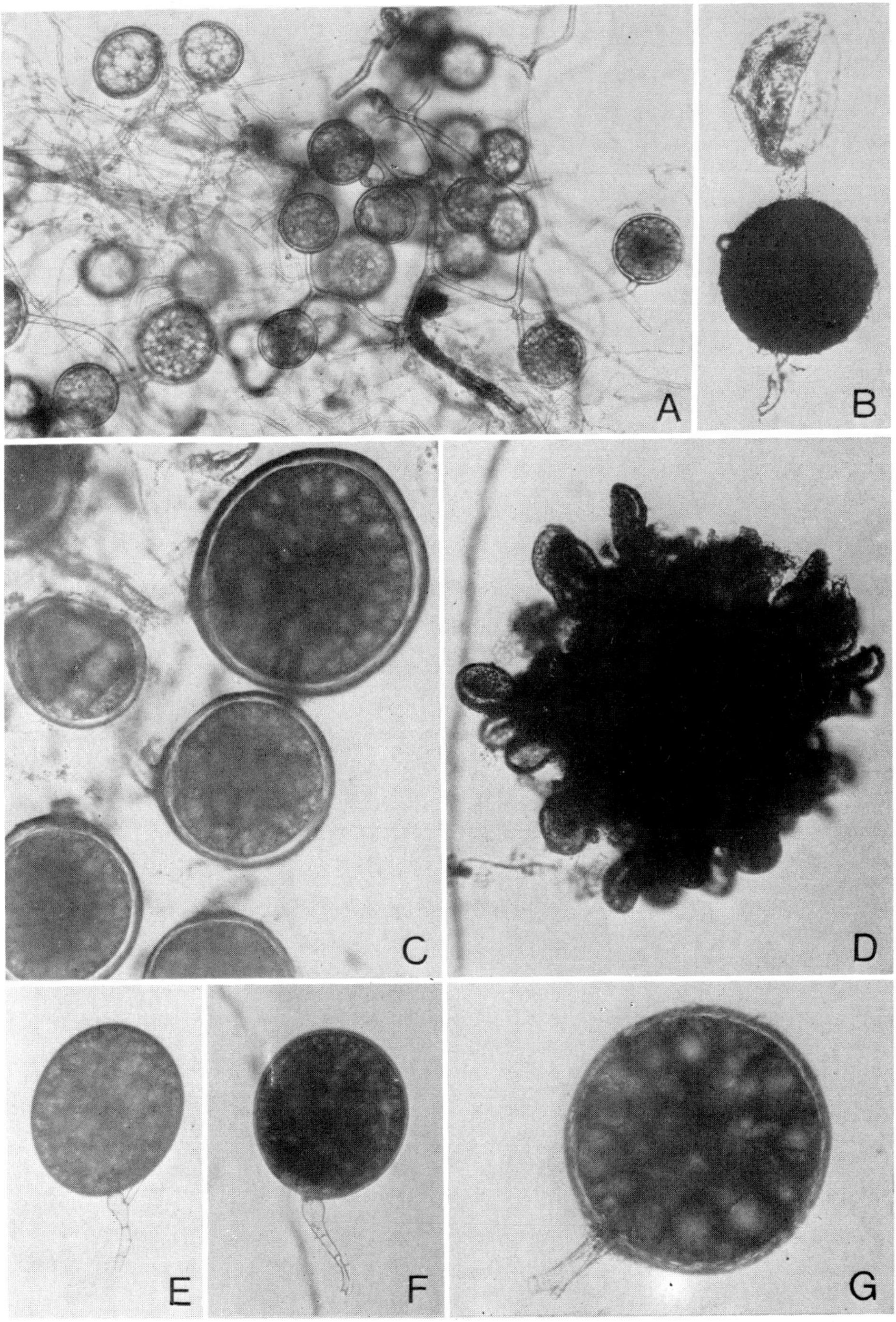

Plate 3. Spores of VA mycorrhizal fungi. (A) *Glomus fasciculatus* 'E3' (× 210); (B) *Acaulospora* sp. (× 230); (C) *Glomus clarus* (×230) (courtesy of C. Clarke); (D) *Sclerocystis rubiformis* (sporocarp) (× 230); (E & F) *Gigaspora* sp. "bulbous reticulate" (× 210); (G) *Glomus caledonius* "laminate" (× 230).

well in unsterile field plots (25, 70, 138). Unfortunately pot culture inoculum is crude and bulky and it has been calculated that 2 or 3 tonnes per hectare would be needed (138). Such a large amount is obviously difficult to produce in pots and impractical to apply on an agricultural scale.

(3) FLUID DRILLING. Germinated seeds and inoculum are suspended in a viscous medium such as 4 per cent methyl cellulose and added to the seed furrows as a slurry. This has proved very successful with field-grown red clover in soil containing various indigenous endophytes (70). The advantage of this technique is that the inoculum is less bulky (it is primarily the organic fraction concentrated to about one-seventh by wet-sieving the original soil inoculum) and does not appreciably increase the total volume of material being drilled. Suitable strains of *Rhizobium* can also be incorporated in the fluid.

(4) SEED PELLETING. Inoculum is attached to individual seeds by an adhesive such as methyl cellulose. This has given fairly good results with plants in pots (70) and with citrus in field nurseries (60). A major difficulty is that VAM propagules, unlike bacterial inoculants, are rather large to stick onto small seeds—the spores are mostly 100 to 200 μm in diameter and the infected root pieces 1 to 2 mm long.

(5) MULTI-SEEDED PELLETS. Several seeds can be incorporated in a protective carrier material such as lignite or soil stabilized with clay, mixed with sievings or soil from pot cultures. Seeds need to be on or near the surface to receive adequate aeration for germination. Results can be variable (70) but this is a technique with large-scale potential (143).

(6) HIGHLY INFECTIVE SOIL. The VAM population can be increased in a field plot by growing a crop there which will become heavily mycorrhizal. In addition the top soil can be removed and used as a crude inoculum elsewhere (21).

Inoculum production

The first five inoculation techniques described require large quantities of inoculum raised in pot cultures of mycorrhizal stock plants. Inoculum can be supplied in the form of infective soil, infected roots, soil sievings and perhaps one day as axenic cultures. Infected roots grown in NFT culture (nutrient film technique) would be cleaner and less bulky than soil inoculum, but at present create problems of storage, viability and longevity (40). Methods of storing inoculum, such as refrigerated sievings and lyophilized roots, need further investigation. A special case is that of *Glomus epigaeus* which produces sporocarps on the soil surface. These can be scraped off and used as concentrated fungal inoculum (36). Finally infectivity can be raised *in situ* by appropriate pre-cropping.

Infective soil from pot cultures open to the atmosphere could become contaminated with harmful nematodes and fungi. To reduce this risk specific pesticides should be developed. Another way of avoiding possible carryover of specific plant pathogens is to raise inoculum on a different host plant to the crop to be inoculated in the field. Menge, Lembright and Johnson (107) describe a practical scheme for this with Sudan grass as the stock plant and citrus as the test plant.

Current status of field inoculation studies world-wide

Most experiments on the effects of VA mycorrhiza on plant growth have been done in pots, but the impetus for more realistic growth studies done under field conditions is gaining momentum. Mosse and Hayman (120) cite 20 field experiments in their 1980 review and more have been published since then or are in progress. It is simplest to discuss field experiments in two categories, those done in unsterile soil and those done in fumigated soil.

Field experiments in unsterile soil

These have mostly been done in agriculturally marginal lands where the soil was deficient in both phosphate and effective VAM endophytes, but some encouraging results have also been obtained in arable fields. VAM may also be harnessed in land reclamation schemes involving plant establishment on coal tips, semi-deserts, etc.

(1) Cereals. Khan (86, 87) tested pre-inoculation of maize and wheat with a type of *Glomus mosseae* before transplanting to some infertile soils in Pakistan deficient in phosphate and native VAM. *G. mosseae* more than doubled shoot dry weight, a bigger effect than adding phosphate fertilizer at 50 kg P/ha. Later Saif and Khan (159) reported very similar results with barley.

In a Rothamsted soil containing only 10 ppm $NaHCO_3$-soluble P, Owusu-Bennoah and Mosse (138) obtained a 30 per cent increase in barley shoot growth with VAM inoculation. Clarke and Mosse (25) in the same field increased ear growth of barley two-fold by inoculation; adding P at 82.6 kg/ha increased it $5\frac{1}{2}$ times without inoculation, four times in the presence of *Glomus mosseae* and *G. fasciculatus* 'E3' inoculum and $6\frac{1}{2}$ times with *G. caledonius* (adding fertilizer raised the soluble P to approximately 40 ppm). *G. caledonius* was also the best endophyte in previous experiments at this site (138) yet it produced the least infection of all three endophytes. This shows growth responses are not necessarily directly related to the amount of infection.

(2) Potatoes. Black and Tinker (21) increased potato yields by 20 per cent in an infertile arable field containing few VAM spores since it had been kept fallow for two years. For inoculum they used highly infective soil from an adjacent area growing barley and containing many VAM spores. Adding superphosphate at 82 kg P/ha also increased growth over the controls but cancelled the effect of mycorrhiza.

(3) White clover. This forage legume is used extensively in temperate

regions to supply fixed nitrogen to pasture grasses and protein and minerals to grazing animals. Recent experiments in the agriculturally marginal hill grasslands of Britain and New Zealand have shown that better establishment and growth of white clover in these areas can be achieved by inoculation with certain VAM endophytes. In a series of tests by Powell (142, 143) in New Zealand inoculation with *Glomus fasciculatus* 'E3' produced 50 per cent more shoot growth on a cold south-facing slope but 42 per cent less on a warm north-facing slope. *Glomus tenuis* and *Gigaspora margarita* increased growth by 80 per cent in other tests, but certain introduced endophytes were less effective than some of the indigenous ones.

Comparable results were obtained by Hayman and Mosse (73) who found pre-inoculation with a mixture of *Glomus mosseae* and *Glomus fasciculatus* 'E3' doubled seedling growth in plots given standard dressings of phosphate fertilizer (90 kg P/ha) as basic slag. Nodulation was also much better on the VAM-inoculated plants, probably due to their enhanced P status. However small uninoculated plants sometimes had high P contents, suggesting an effect of mycorrhiza beyond a straightforward P effect. At other sites growth responses to inoculation varied, partly due to the effectiveness of the indigenous VAM population (68).

(4) RED CLOVER. In an experiment at Rothamsted (74) plants showed an early response to superphosphate but by the end of the second year yields were high in all plots, viz. 15 t/ha dry matter production, despite only 10 ppm $NaHCO_3$-soluble P in the soil. This overall good growth was attributed to one of the introduced endophytes, *Glomus caledonius*, because it had grown and sporulated profusely in all plots, including those inoculated with two other endophytes, and in previous experiments had greatly stimulated lucerne which is similar to red clover in its mycorrhiza response. This experiment shows that ability to spread and infect in the field may be as important as symbiotic efficiency and should be considered when choosing inoculum. It also suggests that instead of comprehensive inoculation of perennial crops, it may be possible to space out inoculum at infection centres from which a suitable endophyte will spread rapidly.

(5) LUCERNE (alfalfa). A large response to inoculation with *Glomus mosseae* was obtained in southern Spain in a field subject to standard cultivation practices (10). Inoculation with *Rhizobium* increased growth of *Glomus*-inoculated plants but not of the controls without *Glomus*, suggesting an interaction between symbionts.

Lucerne and other forage legumes are particularly responsive to mycorrhiza compared to many graminaceous plants. For example a four-fold increase in growth from inoculation with *Glomus caledonius* was reported in a low phosphate field where barley was much less mycorrhiza-responsive (138).

(6) COWPEA. Islam, Ayanaba and Sanders (81) transplanted seedlings with and without *Glomus fasciculatus* to the field and observed most growth in

crops given both *Glomus* and rock phosphate. This was a sequel to earlier experiments at IITA, Nigeria, where Islam had concluded that rock phosphate might be a better long-term source of phosphate than triple superphosphate because the indigenous VAM fungi were actively involved in P uptake, and superphosphate, but not rock phosphate, suppressed them. This suppression would create greater crop dependency on superphosphate fertilizer in subsequent years, making it less economical than rock phosphate quite apart from manufacturing and transport expenses.

Other field experiments with cowpea in the tropics are those of La Torraca (97) in the Amazon region of Brazil. She found mycorrhizal inoculation increased yield, especially when superphosphate was also added.

(7) SOYBEAN. This plant did not respond to field inoculation with *Glomus fasciculatus* in India (11), although it has shown responses in pots and in fumigated soil.

(8) LOTUS PEDUNCULATUS. This plant established slightly better when sown with mycorrhizal compared to non-mycorrhizal soil pellets on an eroded hill site in New Zealand (58).

Field experiments in fumigated soils

Most studies have been done in the United States where fumigation of soil with methyl bromide and chloropicrin is used to control root diseases in certain crops. Similar studies are likely in countries where two or three successive crops are raised in one year and soil fumigation to prevent pathogen build-up will keep indigenous VAM populations low or absent.

(1) CITRUS. The severe stunting of citrus seedlings in Californian nurseries after soil fumigation, normally corrected by adding much phosphate fertilizer, was overcome by mycorrhizal inoculation to replace the killed native VAM fungi. *Glomus mosseae* increased growth by 0 to 150 per cent, depending on citrus species (89) and *G. fasciculatus* increased it by 70 to 300 per cent (60). In Florida *Glomus macrocarpus* var. *geosporus* increased citrus growth in fumigated soil in the field, whereas *Gigaspora calospora* did not although it doubled growth of plants grown in pots (167). In was concluded that VAM inoculation was too inconsistent to correct citrus stunt in those particular commercial nurseries. Many other reports of VAM effects on citrus have been published in the last five years, showing some endophyte-cultivar combinations to be much more effective than others.

(2) PEACH. Seedlings of peach in a commercial nursery in California were more vigorous, showed no nutrient deficiency, and contained more per cent Zn in their tissues when they were inoculated with VAM fungi (96). These results agree with those of Gilmore (51) for similar zinc-deficient soils in pots.

(3) SOYBEAN. VAM inoculation increased growth and yield by 34 to 40 per cent in small plots consisting of rectangular bins buried in the ground, and by 29 per cent and zero at two sites with larger field plots (154). Ross (151)

continued these studies in bins and found that with no added phosphate inoculation almost doubled dry matter production, seed yield and per cent P. Inoculation had less effect with P at 44 kg/ha and very little effect at 176 kg P/ha. Without mycorrhiza yields were directly related to the levels of P added.

Schenck and Hinson (163) grew two soybean isolines in small field plots. VAM increased growth of the nodulating line by 53 per cent but had no significant effect on the non-nodulating line. This suggests that VAM did not significantly affect nitrogen uptake but may have indirectly enhanced N_2-fixation in the nodulating line because phosphate was not limiting growth in this soil. These dual effects of VAM may be of great practical significance and will be discussed in the special section on legumes.

(4) CASSAVA. This plant did not respond to phosphate fertilizer in unsterile soil but did so in fumigated plots (190). This was attributed to VAM fungi in the unsterile soil because fumigation was beneficial to Chinese cabbage, which is not infected by VAM fungi, whereas cassava is very responsive to VAM and has a high demand for P in nutrient solution. A similar conclusion was reached by Kang *et al.* (83) who found field-grown cassava plants, which were heavily mycorrhizal, had a low P requirement, whereas in sterilized soil in pots inoculation with VAM fungi increased growth and lowered the P response.

(5) SWEETGUM. Seedlings of this plant grown in a fumigated soil mix in large boxes in a tree nursery grew bigger with VA mycorrhiza but not when fertilizer alone was added, although mineral concentration was not affected (170). It appeared that the stimulus from mycorrhiza was physiological rather than nutritional.

Components of the VAM symbiosis

Vesicular-arbuscular mycorrhiza is a balanced mutualistic symbiosis in which both partners can benefit. The system as a whole incorporates a three-way interaction between soil, plant and fungus in which the movement of phosphate is of prime importance (Fig. 1). Phosphate moves in three stages: uptake by the soil-based hyphae, transfer into the root-based mycelium via the connecting links (entry points), and release into the plant. The interactions are influenced by environmental conditions. Of greatest practical interest are the large effects VAM can have on plant growth.

The soil

Soil fertility is a key factor governing the extent to which a plant can benefit from mycorrhiza, as illustrated in data from the field experiments cited in the preceding section. Often a growth response to VAM can be predicted on the basis of soil analysis, especially estimates of plant-available phosphate for which the Olsen test (137) for P soluble in $NaHCO_3$ is commonly used. The results of Hayman and Mosse (71) in a range of soils show a gross inverse correlation between $NaHCO_3$-soluble P ("Olsen P") and growth of onion plants, but more exact correlations within a narrower P range were not observed. The effects of adding VAM or P fertilizer were similar in four of the soils; VAM was better than added P in one soil and P better than VAM in the other three. In one of these last three the external mycelium remained close to the root surface without exploring much soil. In later pot experiments with lucerne (Hayman, unpublished) VAM increased growth and shoot per cent P in soil containing 8 ppm Olsen P, increased per cent P only in soil with 26 ppm Olsen P, and increased neither in soil with 40 ppm Olsen P. These and other studies show that the effects of VAM on P-uptake and growth are not always consistent for different soils and broad generalizations on mycorrhizal function based on results from one soil should be cautioned.

Most pot experiments have been done in sterilized soils in order to have non-mycorrhizal controls. These can be criticized to the extent that soil sterilization itself can resullt in better plant growth from the release of nutrients and elimination of pathogens and microbial competition for nutrients. However, growth responses to VAM in unsterile soils similar to those in sterilized soils have been obtained where the controls developed little or ineffective infection (121, 123).

The plant

The different degree to which different plant species depend on mycorrhiza ("mycotrophy") is well illustrated by the doubling of growth of *Stylosanthes* by VAM in the soil with 40 ppm Olsen P where the normally mycotrophic lucerne did not respond (Hayman, unpublished). It is surprising therefore that *Stylosanthes* spp. can grow so successfully in tropical regions in soils containing less than 3 ppm Olsen P. This suggests it may be an obligately mycorrhizal species under field conditions. Cassava too seems to be obligately mycorrhizal, which probably explains why it can grow in very poor low-P soils yet needs a high P concentration in solution culture (190).

Usually more than one plant species are found growing together and mycorrhiza may affect their relative competitive ability. Where two-species systems have been tested experimentally, it seems that although each species separately can benefit considerably from VAM, one can benefit much more than the other when they are grown together. For example, Fitter (41) grew the two grasses *Holcus lanatus* and *Lolium perenne* in the same pot. Uptake of P was similar in non-mycorrhizal *Lolium* and *Holcus*, but with mycorrhiza it was reduced to about two-thirds in *Lolium* and increased by about one-third in *Holcus* when the plants were growing in competition with their roots intermingled. Crush (29) tested two species used in pasture improvement, viz. *Lolium perenne* and *Trifolium repens*. He found that when the plants were grown together in pots of sterilized, moderately low-P soil, *Trifolium* could compete much more successfully with *Lolium* when mycorrhizal inoculum was added. The effects of VAM on inter-species competition for scarce nutrients is likely to be important in marginal or eroded lands where the extraction of P from soil may be more critical than competition for light, for example. Another aspect is that in some mixed plant communities there may be a network of VAM hyphae linking plants of different species. Thus interspecific transfer of phosphate via hyphal links is an exciting possibility. This has been tested experimentally (76) but the evidence is not yet conclusive because of difficulties in distinguishing whether transfer from donor to receiver plant is via interconnecting hyphae or via the soil.

Can we predict which plant species are most likely to benefit from mycorrhiza? One general answer is based on root geometry. In broad terms plants with coarse roots and few root hairs often respond to VAM because they cannot tap sufficient soil P, e.g. citrus and onion, whereas plants with fine, hairy roots are frequently unresponsive, e.g. ryegrass, although some tropical grasses are more responsive, e.g. *Paspalum* and *Brachiaria*. Baylis (15) pursued this concept in a P-deficient (8 ppm Truog P) sterilized soil where he found an inverse relationship between root hair production of a species and the amount of phosphate that had to be added to produce a significant growth increase. In the same soil, kept unsterile, all plants became mycorrhizal and did not respond to added P. Recently St. John (176) obtained data for a range of species in a Brazilian rain forest which support Baylis's "root-hair" hypothesis. Crush (29) found an inverse correlation between root hair length and response to VAM in four legumes.

Two other factors that affect plant mycotrophy and which are interconnected are its growth rate and its minimum P content for healthy foliage. Comparing two indigenous New Zealand tree species, for example, Hall (56) found *Weinmannia racemosa* was much more heavily mycorrhizal in nature and responded more rapidly to VAM in pots than *Metrosideros umbellata*, yet the latter has few root hairs. Hall concluded that, despite its root geometry, the slow growth rate of *Metrosideros* was the key factor in making it independent of VAM because it resulted in a low demand for P.

The fungus

It should be clear from examples given already that VAM fungi vary greatly in their symbiotic effectiveness. Factors affecting this are their preferences for particular soils or host-plants ("specificity"), effectiveness in stimulating plant growth, competitive ability and tolerance to added fertilizer or pesticides.

Establishment of the fungus may be restricted by the unsuitability of a host plant, e.g. lettuce for *Gigaspora margarita* compared to *Glomus fasciculatus* 'E3' (135). However virtually any VA mycorrhizal plant species can be infected to some extent by any VAM fungal species. More important in "specificity" is the suitability of the soil. For example *Acaulospora laevis* and

Glomus fasciculatus 'E3' are best in acid soils, *Glomus mosseae* in alkaline-neutral soils. Endophytes also differ in their adaptation to temperature and minor element toxicity. Endophytes native to infertile soils often tolerate less fertilizer than introduced ones. Populations change when virgin land is brought into cultivation (64, 165). By contrast endophytes adapted to old, well fertilized pastures were not affected by applications of superphosphate (140).

Where several endophytes are present they may compete for sites and nutrients, although they do not necessarily exclude one another from the same piece of root cortex. Ross and Ruttencutter (155) found more VAM infection in peanuts and soybean inoculated with *Glomus* plus *Gigaspora* than with *Glomus* alone. However numbers of chlamydospores of *Glomus macrocarpus* var. *geosporus* increased rapidly following field inoculation by this fungus alone, but not when *Gigaspora gigantea* was included. This was attributed to either its poor competitive ability compared to the *Gigaspora* or to specific hyperparasites.

From Powell's (141) experiments in ^{32}P-labelled soils it seems that different endophytes use the same source of soil P. In view of the importance of root-soil surface contact the best endophytes would be expected to form the most external mycelium, but this correlation is too simple. Possibly endophytes differ in arbuscule formation and breakdown which would affect nutrient exchange. One important principle is that the best endophytes for stimulating plant growth are not necessarily the most infective. A corollary of this is that per cent infection is not closely related to growth response, levels of say 50 and 80 per cent, for example, being unlikely to produce different growth effects in practice. The speed with which an endophyte infects and starts functioning may be critical, but this would need to be monitored by serial harvests during an experiment. Finally external conditions may affect endophyte function, e.g. in soils of pH greater than 7 *Glomus fasciculatus* 'E3' may infect a root as much as *Glomus mosseae* but be less active outside because it prefers acid soils.

The environment

Light may affect mycorrhizal development, seemingly changing the appearance of an infection, e.g. better arbuscule formation under high than under low light intensities (63), rather more than the total infection. The concomitant effect on plant growth may have been related to carbohydrate supply from plant to fungus.

Low temperatures may cause the endophyte to change from beneficial to mildly parasitic (43). Nyabenda (111) showed that the ability of the endophyte in *Eupatorium odoratum* to absorb P from insoluble forms was affected by temperature. Plants with VAM showed a steep rise in their growth response from 20 to 30°C whereas non-mycorrhizal plants were unaffected by temperature. Two other plant species did not show this effect, however.

Mycorrhiza is also affected by pesticides, especially fungicides (129). Benomyl and thiophanate methyl, apparently acting on the fungal component of the symbiosis, reduced P uptake by mycorrhizal onion and strawberry plants much more than by non-mycorrhizal ones (22). Thus in some soils the net benefit of applying fungicides may be the increased yield from disease control minus the possible decreased yield due to inhibition of VAM.

Physiology of the host-endophyte interaction

It will be obvious by now that most research on VAM effects on plant nutrition has been concerned with phosphate because it is a major plant nutrient and VAM can produce dramatic benefits to plants growing in P-deficient soils. However VAM can also enhance the uptake of other nutrients. It can improve zinc uptake by peach from zinc-deficient soils in California (51, 96), and by maize, wheat and potatoes from some zinc-deficient soils in India (183). Sulphur uptake by red clover and maize grown in sand and given nutrient solution is also increased by VAM (54). Like phosphate zinc and sulphur are translocated to VAM roots through the external hyphae (26, 149). VAM may also increase the uptake of heavy metals such as cadmium, which could be deleterious on reclaimed coal tips and soils treated with sewage sludge by increasing plant toxicity to mammals. Conversely it is speculated that VAM may increase plant tolerance of aluminium and manganese which can reach toxic levels in some tropical soils. The whole question of nutrient translocation in VA mycorrhizas has been thoroughly reviewed by Rhodes and Gerdemann (150). VAM endophytes do not appear to increase the uptake of nitrate and there is no evidence of them fixing atmospheric nitrogen. It should be stressed that when mineral concentrations are compared between plants with and without mycorrhiza, the plants should be of similar size and not showing abnormalities due to P-deficiency.

There is some speculation that water uptake can be increased by VAM. Certainly recovery from water stress has been shown to be better in rough lemon seedlings when mycorrhizal, apparently due to effects on stomatal regulation (98). The ability of avocado seedlings to survive transplanting injury better when mycorrhizal is attributed to improved water absorption (106). Resistance to water transport was decreased by VAM infection in soybeans, probably because the fungus improved their nutritional status (157).

Occasionally the host-endophyte balance tilts from mutualistic symbiosis to parasitism. For example Crush showed an unfavourable effect of some endophytes on certain pasture legumes at high levels of added superphosphate (31) and of *Rhizophagus tenuis* on some grasses in a soil containing 8 ppm Truog P compared to its enhancement of their growth in very infertile soil containing 4 ppm Truog P (28, 30). The practical implication of this in pasture production is that any VAM inoculum introduced must be carefully selected on its performance at realistic P levels. This does not appear to be the case with barley and wheat, however, since Jensen and Jakobsen (82) found no differences in their shoot P content under field conditions in Denmark. There a low level of soil P seemed to be offset by higher VAM infection and vice versa. Thus the symbiosis seemed to be self-regulating and suggests that VAM may be significant in the phosphate nutrition of cereals, rather than deleterious at high P levels. At very high levels of P Mosse (115) found that onions grew worse with mycorrhiza than without, but observed that the fungus had changed morphologically and eventually died out at the highest P levels. This correlated with a high per cent P in the plant tissues. Again this suggests a self-regulating mechanism, the plant discarding the fungus when its P requirement is more than satisfied. In some systems one endophyte may be beneficial, another harmful (105). Finally the VAM fungus may be slightly detrimental to the host plant under poor light and temperature conditions (43, 63). Whether this is due to a carbohydrate drain on the host when photosynthesis is restricted is currently a matter of some controversy.

There is recent evidence that VAM affects the hormone balance in plants, e.g. cytokinins (5). Also the overcoming of first-year dormancy of cuttings, e.g. yellow poplar, by VAM inoculation may be a hormonal effect, as is perhaps the slowness of many woody plants to start growing again after transplanting which can be partly countered by VAM inoculation. Related to this may be the fact that some plants, e.g. sweetgum (*Liquidambar styraciflua*) and hoop pine (*Araucaria cunninghamii*) are better with VA mycorrhiza than with any amount of fertilizer.

Other physiological effects of VAM infection include an increase in plant chlorophyll content (38). Reactivation of cortical cells when infected with VAM endophytes is suggested by an increase in mitochondria, endoplasmic reticulum, respiration, and size of the nucleus, and a decrease in starch content.

Nutrients are believed to be released from the fungus to the plant by secretion or leakage from hyphae and intact arbuscules and during arbuscule breakdown. Electron microscope studies indicate transfer of material in oil globules, exocytotic vesicles from intact arbuscules and pieces of fungal membrane from disintegrating arbuscules (27, 88). Much of the movement of phosphate along the hyphae is now believed to be by mass flow of polyphosphate granules in the cytoplasm.

The fungus receives carbohydrate from the host plant. This is illustrated by radioactivity found in the fungus attached to plants growing in air containing $^{14}CO_2$ (78), especially in fungal proteins and organic acids (18). However the quantities of these materials and the amount of mycelium relative to root tissue suggests there is no carbon sink equivalent to that in the ectomycorrhizal sheath. There is no sign of the fungal carbohydrates trehalose and mannitol in VAM roots (18, 63).

The main benefit to the fungal component of VAM is the provision of an ecological niche because it cannot grow independently. Inside the root it receives nutrients and is protected from microbial antagonisms in the soil. Nevertheless, there is speculation that the fungus can obtain some carbohydrate externally, and evidence that it may be capable of some saprophytic growth in soil (134, 193).

Not all the fungal mycelium that stains with trypan blue or acid fuchsin is active. Cytochemical techniques (99) can be used to reveal the enzymatically active hyphae and the ratio of active to inactive infection calculated after normal staining of the same roots.

Mechanisms of increased phosphate uptake

Physiological mechanisms

The large-scale occupation of a root cortex by a fungus must influence its physiology and metabolism. Indeed root segments are more active in P uptake when mycorrhizal, as shown by increased short-term uptake of ^{32}P from solution and a nullification of this by adding fungicide (53). This ^{32}P concentrates in the fungal structures (4, 23). However these effects will not account for most of the increased phosphate uptake in infertile soils where there are too few ions at the root surface. They could account for some uptake where a phosphate threshold is reached below which non-mycorrhizal roots cannot absorb phosphate whereas mycorrhizal roots or hyphae can, e.g. < 3 ppm Olsen P for *Centrosema pubescens* and *Paspalum notatum* (122), where roots stay functional longer when mycorrhizal (23), and because phosphate uptake requires much metabolic energy to absorb against a gradient of around 10^{-6} to 10^{-7} M in the soil solution to 10^{-2} to 10^{-3} M inside the root (19).

Chemical mechanisms

Some 95 to 99 per cent of the total P in soil exists in insoluble forms unavailable to plants. The remaining soluble P is mostly on exchangeable sites in equilibrium with the small amount of P in the soil solution (10^{-6} M in the soil solution is equivalent to about 0.03 ppm P and soils often contain around 500 ppm total P). Therefore any solubilization of insoluble P by VAM fungi could significantly add to the available pool. This possibility was tested experimentally by labelling the soluble soil P with ^{32}P and measuring the specific activity (^{32}P/total P) taken up by plants with and without mycorrhiza. No differences were found, showing that both used the same phosphate pool and that mycorrhizal plants did not tap additional (unlabelled or weakly labelled) sources. This happened in eight soils including some where mycorrhiza had increased P uptake manifold (72) and in one soil at different stages of plant growth (161). It has been concluded from these and other (14, 153) experiments that mycorrhizal plants do not dissolve insoluble native soil P more efficiently than non-mycorrhizal plants.

Physical mechanisms

The morphology of VA mycorrhizas and the kinetics of phosphate movement in soil now support a physical interpretation of the main way in which VAM functions (see Fig. 2). Ions such as phosphate are rather immobile in soil and a zone of phosphate-depletion develops around roots in phosphate-deficient soils because the root absorbs phosphate ions much faster than they can be replenished at the root surface by diffusion. The VAM hyphae attached to the root extend beyond this zone and directly translocate nutrients from the soil to the hyphae and arbuscules inside the root cortex where transfer to the plant occurs. Translocation across 1.9 cm (61) and even 7 cm (148) of soil has been demonstrated experimentally. In addition the number of connections between the fungus inside and outside the root is calculated to be sufficient for hyphal transfer to account for the increased flux of phosphate into mycorrhizal roots (19, 161).

The improved growth of mycorrhizal plants given sparingly soluble forms of phosphate, e.g. rock phosphate (116, 128), is now attributed to greater physical contact between phosphate particles and hyphae than between roots and these particles. The hyphae are believed to mop up more efficiently than roots phosphate ions dissociating chemically at the particle surface.

The physical effect of VAM can explain the importance of root geometry and the reason why plants with least root-soil contact benefit most from the additional absorbing surface provided by the mycelial network around their roots. Thus VAM external hyphae are broadly analogous to extra root hairs, a concept supporting Baylis's root hair theory. Furthermore, when several plants are grown together in a pot such that depletion zones overlap micorrhizal benefits often disappear.

In conclusion physical factors largely, but not entirely, override physiological factors. Exceptions include the different P-thresholds of plants with equally hairy roots (122), fertile soils where P-depletion zones do not arise, and stunted non-mycorrhizal seedlings which contain as much P per cent as mycorrhizal ones but need a "trigger" to start growing again.

Legumes, a special case

The occurrence of two symbionts in legume roots, *Rhizobium* bacteria and VA mycorrhizal fungi which can sometimes act synergistically, makes legumes a special case. This is well illustrated in *Stylosanthes guyanensis* grown in very low phosphate (2 ppm Olsen P) soil from the Brazilian cerrado (126) where the combination of *Glomus fasciculatus* 'E3' and rock phosphate greatly increased nodulation and N_2-fixation in addition to P-uptake and growth. Thus the increased P uptake by VAM stimulated the activity of *Rhizobium* which is well known to depend on an adequate supply of phosphate. Serial harvests of lucerne showed that the effects of mycorrhiza on nodulation and nitrogenase activity preceded those on growth (172, 173). This suggests nodules had first call on the phosphate. For this reason, in future studies on *Rhizobium* × VAM interactions, measurement of acetylene reduction during an experiment and total plant nitrogen at harvest is recommended. Utilization of added P fertilizer in the *Rhizobium* × VAM interaction is a vital area for practical studies. The importance of N × P interactions in legumes is underlined by data showing VAM improved growth of a nodulating but not of a non-nodulating soybean isoline (163).

Legumes generally have less extensive root systems than graminaceous plants, for example, and many are poor foragers for soil phosphate. In addition many strains of *Rhizobium* found in culture collections are adapted to media containing 100 times more P than is usual in the soil solution (127), and can require at least 0.1 per cent P in the shoot tissues to produce nodules (126). Some indigenous rhizobia, on the other hand, can nodulate at lower concentrations of P to which they are adapted (117), an important consideration in dual inoculation of legumes in P-deficient tropical soils.

Usually the effects of VAM on legume growth, nodulation and N_2-fixation are similar to those of adding phosphate. The enhanced nodulation of legumes reported by Asai (7) was probably due to VAM. This effect has since been shown experimentally with various legumes including *Centrosema pubescens* (29), *Medicago sativa* (172), *Phaseolus* sp. (32, 67), *Glycine max* (151, 163), *Arachis hypogea* (33), *Vigna unguiculata* (81), *Pueraria* sp. (191), *Trifolium repens* (73, 126), *Trifolium subterraneum* (2), and *Stylosanthes* sp. (117, 126, 191).

Mycorrhizal effects on the uptake of other elements besides phosphorus may be important in legume nodulation, e.g. zinc [cf. peach (51)] and copper, but not apparently molybdenum (67). However too much zinc inhibited nodulation and mycorrhiza in Pinto beans (103).

The degree of mycorrhizal infection varies considerably with legume species (180). Also legumes vary in their mycorrhiza-dependence. In general forage legumes are more responsive than grain legumes; some lupins lack mycorrhiza completely despite being able to obtain phosphate from very infertile soil. Islam *et al.* (81) reported that in the same soil inoculation with VAM doubled growth (dry weights) of cowpeas but did not affect soybeans. *Leucaena leucocephala*, with virtually no root hairs, is strongly mycorrhiza-dependent (127).

Evidence that mycorrhizal infection can lead to increased nitrogen contents of shoots and seeds (151, 163) is of fundamental significance for human nutrition and underlines the need for further studies on mycorrhizal effects on protein levels in legumes.

Interactions with other microorganisms

Roots are obviously changed by mycorrhizal infection. Possible VAM-induced changes in root exudates and the rhizosphere population, as well as possible physical barriers and chemical inhibitors from VAM fungi, may have practical implications in the biological control of some plant disease organisms. Much of the relevant literature was comprehensively reviewed by Schenck and Kellam (164). The changed nutritional status of a plant when mycorrhizal may also affect above-ground pathogens. Schönbeck (168) recently presented a general picture where VAM often, but not always, decreases root diseases but increases leaf diseases.

Fungi

VAM inhibited chlamydospore formation by *Thielaviopsis basicola* on tobacco roots, probably because it increased root concentrations of arginine and citrulline (12). Tomato wilt, caused by *Fusarium oxysporum* f. sp. *lycopersici*, was also reduced by mycorrhizal infection which decreased leakage from host cells and increased their chlorophyll content (38). In addition to these protective physiological effects, the internal mycelium may present a physical barrier. This is referred to as "territorial occupation of living root tissues" by Wilhelm (194) who speculates that one reason why juvenile roots may be more susceptible to disease than older roots is because they have not had time to develop mycorrhiza.

Ross (152), by contrast, found mycorrhizal infection increased root-rot of soybeans by *Phytophthora megasperma* var. *sojae*. Possibly entry of the pathogen was facilitated where large VAM vesicles had split the roots. Also *Phytophthora cinnamomi* was more severe on mycorrhizal than on non-mycorrhizal avocado, but other *Phytophthora* species pathogenic to citrus and alfalfa were not affected by mycorrhiza (37). In other studies there was no definite effect of VAM on fungal pathogens (102, 164). Unlike with ectomycorrhizas, no chemical (antibiotic) barriers have yet been shown in VA mycorrhizas.

Bacteria

In addition to stimulating *Rhizobium*, VA mycorrhiza can also influence rhizosphere bacteria beneficial to the plant. For example *Azotobacter paspali* survived longer around roots of *Paspalum notatum* when they were mycorrhizal (13). Phosphate-solubilizing bacteria survived longer around mycorrhizal than non-mycorrhizal roots of maize and lavender and sometimes acted synergistically with the mycorrhiza to increase plant growth, especially where rock phosphate was added to the soil (8). *Azotobacter chroococcum* may act synergistically with *Glomus fasciculatus* (11).

Viruses

Inoculation of tobacco with *Glomus mosseae* increased leaf infection by

TMV (169). In another study *Glomus macrocarpus* var. *geosporus* infection increased the titres of other viruses inoculated onto tomato, strawberry and petunia (35). The results in both studies were attributed to the improved phosphate status of the host plants when mycorrhizal.

Nematodes

There is a rapidly growing literature on mycorrhiza-nematode interactions. In many instances the nematodes and VAM fungi adversely affect one another, probably by mutual suppression of reproduction processes and competition for the same root niches, e.g. *Heterodera solanacearum* and *Gigaspora gigantea* on tobacco (42) and *Meloidogyne* spp. and *Glomus mosseae* on several crops (171). Sometimes there is no mutual suppression (79). The nematode-fungus interaction is more complex on soybeans, varying with VA endophyte species and plant cultivar (166).

Conclusions

It seems that the most practical areas for VA mycorrhizal research, particularly in tropical and sub-tropical countries, are: (1) more efficient use of applied fertilizer, especially rock phosphate and sub-optimal levels of superphosphate; (2) enhanced nodulation and N_2-fixation in grain legumes; (3) better establishment of forage legumes in pastures; (4) re-establishment of vegetation in eroded or degraded sites and in semi-deserts; (5) re-introduction of VAM after soil sterilization; (6) better establishment of plantation crops by pre-inoculation. Indigenous VAM endophytes should be isolated and screened for these purposes. It would be valuable to establish reference collections within a country, perhaps at one university and one large research institute, to help streamline the choice of endophytes for field trials.

The significance of VA mycorrhiza in food production is becoming more widely appreciated, and the number of studies on VAM still seems to be increasing almost exponentially. It is not an agricultural panacea, however, but must be studied in specific situations where it is most likely to give practical results. With certain crops inoculated with highly efficient endophytes in regions of phosphate-deficient soils its impact on productivity could be very large and more than justify the research efforts applied to harnessing it.

References

1. Abbott, L.K. and Robson, A.D. The distribution and abundance of vesicular arbuscular endophytes in some Western Australian soils, *Australian Journal of Botany*, *25*, 515–522 (1977).

2. Abbott, L.K. and Robson, A.D. Growth stimulation of subterranean clover with vesicular-arbuscular mycorrhizas, *Australian Journal of Agricultural Research*, *28*, 639–649 (1977).
3. Abbott, L.K. and Robson, A.D. A quantitative study of the spores and anatomy of mycorrhizas formed by a species of *Glomus*, with reference to its taxonomy, *Australian Journal of Botany*, *27*, 363–375 (1979).
4. Ali, B. Cytochemical and autoradiographic studies of mycorrhizal roots of *Nardus*, *Archiv für Mikrobiologie*, *68*, 236–245 (1969).
5. Allen, M.F., Moore, T.S. and Christensen, M. Phytohormone changes in *Bouteloua gracilis* infected by vesicular-arbuscular mycorrhizae. I. Cytokinin increases in the host plant, *Canadian Journal of Botany*, *58*, 371–374 (1980).
6. Ames, R.N. and Schneider, R.W. *Entrophospora*, a new genus in the Endogonaceae, *Mycotaxon*, *8*, 347–352 (1979).
7. Asai, T. Über die Mykorrhizenbildung der leguminosen Pflanzen, *Japanese Journal of Botany*, *13*, 463–485 (1944).
8. Azcón, R., Barea, J.M. and Hayman, D.S. Utilisation of rock phosphate in alkaline soils by plants inoculated with mycorrhizal fungi and phosphate-solubilizing bacteria, *Soil Biology and Biochemistry*, *8*, 135–138 (1976).
9. Azcón, R. and Ocampo, J.A. Factors affecting the vesicular-arbuscular infection and mycorrhizal dependency of thirteen wheat cultivars, *New Phytologist* (in press) (1981).
10. Azcón-G de Aguilar, C., Azcón, R. and Barea, J.M. Endomycorrhizal fungi and *Rhizobium* as biological fertilizers for *Medicago sativa* in normal cultivation, *Nature*, *279*, 325–327 (1979).
11. Bagyaraj, D.J., Manjunath, A. and Patil, R.B. Interaction between a vesicular-arbuscular mycorrhiza and *Rhizobium* and their effects on soybean in the field, *New Phytologist*, *82*, 141–145 (1979).
12. Baltruschat, H. and Schönbeck, F. Untersuchungen über den Einfluss der endotrophen Mycorrhiza auf den Befall von Tabak mit *Thielaviopsis basicola*, *Phytopathologische Zeitschrift*, *84*, 172–188 (1975).
13. Barea, J.M., Brown, M.E. and Mosse, B. Association between VA mycorrhiza and *Azotobacter*, *Rothamsted Report for 1972, Part 1*, pp. 81–82 (1973).
14. Barrow, N.J., Malajczuk, N. and Shaw, T.C. A direct test of the ability of vesicular-arbuscular mycorrhiza to help plants take up fixed soil phosphate, *New Phytologist*, *78*, 269–276 (1977).
15. Baylis, G.T.S. Root hairs and phycomycetous mycorrhizas in phosphorus-deficient soil, *Plant and Soil*, *33*, 713–716 (1970).
16. Baylis, G.T.S. The magnolioid mycorrhiza and mycotrophy in root systems derived from it, pp. 373–389 in *Endomycorrhizas* (Edited by F.E. Sanders, B. Mosse and P.B. Tinker), Academic Press, London (1975).

17. Becker, W.N. and Hall, I.R. *Gigaspora margarita*, a new species in the Endogonaceae, *Mycotaxon*, *4*, 155–160 (1976).
18. Bevege, D.I., Bowen, G.D. and Skinner, M.F. Comparative carbohydrate physiology of ecto- and endomycorrhizas, pp. 149–174 in *Endomycorrhizas* (Edited by F.E. Sanders, B. Mosse and P.B. Tinker), Academic Press, London (1975).
19. Bieleski, R.L. Phosphate pools, phosphate transport, and phosphate availability, *Annual Review of Plant Physiology*, *24*, 225–252 (1973).
20. Bird, G.W., Rich, J.R. and Glover, S.U. Increased endomycorrhizae of cotton roots in soil treated with nematicides, *Phytopathology*, *64*, 48–51 (1974).
21. Black, R.L.B. and Tinker, P.B. Interaction between effects of vesicular-arbuscular mycorrhiza and fertilizer phosphorus on yields of potatoes in the field, *Nature*, *267*, 510–511 (1977).
22. Boatman, N., Paget, D., Hayman, D.S. and Mosse, B. Effects of systemic fungicides on vesicular-arbuscular mycorrhizal infection and plant phosphate uptake, *Transactions of the British Mycological Society*, *70*, 443–450 (1978).
23. Bowen, G.D., Bevege, D.I. and Mosse, B. Phosphate physiology of vesicular-arbuscular mycorrhizas, pp. 241–260 in *Endomycorrhizas* (Edited by F.E. Sanders, B. Mosse and P.B. Tinker), Academic Press, London (1975).
24. Butler, E.J. The occurrences and systematic position of the vesicular-arbuscular type of mycorrhizal fungi, *Transactions of the British Mycological Society*, *22*, 274–301 (1939).
25. Clarke, C. and Mosse, B. Plant growth responses to vesicular-arbuscular mycorrhiza. XII. Field inoculation responses of barley at two soil P levels, *New Phytologist*, *87* (in press) (1981).
26. Cooper, K.M. and Tinker, P.B. Translocation and transfer of nutrients in vesicular-arbuscular mycorrhizas. II. Uptake and translocation of phosphorus, zinc and sulphur, *New Phytologist*, *81*, 43–52 (1978).
27. Cox, G. and Sanders, F. Ultrastructure of the host-fungus interface in a vesicular-arbuscular mycorrhiza, *New Phytologist*, *73*, 901–912 (1974).
28. Crush, J.R. Significance of endomycorrhizas in tussock grassland in Otago, New Zealând, *New Zealand Journal of Botany*, *11*, 645–660 (1973).
29. Crush, J.R. Plant growth responses to vesicular-arbuscular mycorrhiza. VII. Growth and nodulation of some herbage legumes, *New Phytologist*, *73*, 743–749 (1974).
30. Crush, J.R. Occurrence of endomycorrhizas in soils of the Mackenzie Basin, Canterbury, New Zealand, *New Zealand Journal of Agricultural Research*, *18*, 361–364 (1975).
31. Crush, J.R. Endomycorrhizas and legume growth in some soils of the

Mackenzie Basin, Canterbury, New Zealand, *New Zealand Journal of Agricultural Research*, *19*, 473–476 (1976).

32. Daft, M.J. and El-Giahmi, A.A. Effect of *Endogone* mycorrhiza on plant growth. VII. Influence of infection on the growth and nodulation in french bean (*Phaseolus vulgaris*), *New Phytologist*, *73*, 1139–1147 (1974).
33. Daft, M.J. and El-Giahmi, A.A. Studies on nodulated and mycorrhizal peanuts, *Annals of Applied Biology*, *83*, 273–276 (1976).
34. Daft, M.J. and Hacskaylo, E. Arbuscular mycorrhizas in the anthracite and bituminous coal wastes of Pennsylvania, *Journal of Applied Ecology*, *13*, 523–531 (1976).
35. Daft, M.J. and Okusanya, B.O. Effect of *Endogone* mycorrhiza on plant growth. V. Influence of infection on the multiplication of viruses in tomato, petunia and strawberry, *New Phytologist*, *72*, 975–983.
36. Daniels, B.A. and Menge, J.A. Evaluation of the commercial potential of the vesicular-arbuscular mycorrhizal fungus, *Glomus epigaeus*, *New Phytologist*, *87*, 345–354 (1981).
37. Davis, R.M., Menge, J.A. and Zentmyer, G.A. Influence of vesicular-arbuscular mycorrhizae on *Phytophthora* root rot of three crop plants, *Phytopathology*, *68*, 1614–1617 (1978).
38. Dehne, H.W. and Schönbeck, F. Untersuchungen über den Einfluss der endotrophen Mykorrhiza auf die *Fusarium*-Welke der Tomate, *Zeitschrift für Pflanzenkrankheiten und Pflanzenschutz*, *82*, 630–632 (1975).
39. Dowding, E.S. Ecology of *Endogone*, *Transactions of the British Mycological Society*, *42*, 449–457 (1959).
40. Elmes, R. and Mosse, B. Vesicular-arbuscular mycorrhiza: Nutrient film technique, *Rothamsted Report for 1979, Part 1*, p. 188 (1980).
41. Fitter, A.H. Influence of mycorrhizal infection on competition for phosphorus and potassium by two grasses, *New Phytologist*, *79*, 119–125 (1977).
42. Fox, J.A. and Spasoff, L. Interaction of *Heterodera solanacearum* and *Endogone gigantea* on tobacco, *Journal of Nematology*, *4*, 224–225 (Abstr.) (1972).
43. Furlan, V. and Fortin, J.A. Formation of endomycorrhizae by *Endogone calospora* on *Allium cepa* under three temperature regimes, *Naturaliste Canadien*, *100*, 467–477 (1973).
44. Furlan, V. and Fortin, J.A. A flotation-bubbling system for collecting Endogonaceous spores from sieved soil, *Naturaliste Canadien*, *102*, 663–667 (1975).
45. Gerdemann, J.W. Wound-healing of hyphae in a phycomycetous mycorrhizal fungus, *Mycologia*, *47*, 916–918 (1955).
46. Gerdemann, J.W. Vesicular-arbuscular mycorrhiza and plant growth, *Annual Review of Phytopathology*, *6*, 397–418 (1968).
47. Gerdemann, J.W. Vesicular-arbuscular mycorrhizae, pp. 575–591 in

The Development and Function of Roots: 3rd Cabot Symposium, Harvard Forest (Edited by J.G. Torrey and D.T. Clarkson), Academic Press, London (1975).

48. Gerdemann, J.W. and Nicolson, T.H. Spores of mycorrhizal *Endogone* species extracted from soil by wet-sieving and decanting, *Transactions of the British Mycological Society*, *46*, 235–244 (1963).
49. Gerdemann, J.W. and Trappe, J.M. The Endogonaceae in the Pacific Northwest, *Mycologia Memoir*, No. 5, 76 pp. (1974).
50. Gilmore, A.E. Phycomycetous mycorrhizal organisms collected by open-pot culture methods, *Hilgardia*, *39*, 87–105 (1968).
51. Gilmore, A.E. The influence of endotrophic mycorrhizae on the growth of peach seedlings, *Journal of the American Society for Horticultural Science*, *96*, 35–38 (1971).
52. Giovannetti, M. and Mosse, B. An evaluation of techniques for measuring vesicular-arbuscular mycorrhizal infection in roots, *New Phytologist*, *84*, 489–500 (1980).
53. Gray, L.E. and Gerdemann, J.W. Uptake of phosphorus-32 by vesicular-arbuscular mycorrhizae, *Plant and Soil*, *30*, 415–422 (1969).
54. Gray, L.E. and Gerdemann, J.W. Uptake of sulphur-35 by vesicular-arbuscular mycorrhizae, *Plant and Soil*, *39*, 687–689 (1973).
55. Greenall, J.M. The mycorrhizal endophytes of *Griselinia littoralis* (Cornaceae), *New Zealand Journal of Botany*, *1*, 389–400 (1963).
56. Hall, I.R. Endomycorrhizas of *Metrosideros umbellata* and *Weinmannia racemosa*, *New Zealand Journal of Botany*, *13*, 463–472 (1975).
57. Hall, I.R. Species and mycorrhizal infections of New Zealand Endogonaceae, *Transactions of the British Mycological Society*, *68*, 341–356 (1977).
58. Hall, I.R. Growth of *Lotus pedunculatus* Cav. in an eroded soil containing soil pellets infested with endomycorrhizal fungi, *New Zealand Journal of Agricultural Research*, *23*, 103–105 (1980).
59. Hall, I.R. and Fish, B.J. A key to the Endogonaceae, *Transactions of the British Mycological Society*, *73*, 261–270 (1979).
60. Hattingh, M.J. and Gerdemann, J.W. Inoculation of Brazilian sour orange seed with an endomycorrhizal fungus, *Phytopathology*, *65*, 1013–1016 (1975).
61. Hattingh, M.J., Gray, L.E. and Gerdemann, J.W. Uptake and translocation of ^{32}P-labelled phosphate to onion roots by endomycorrhizal fungi, *Soil Science*, *116*, 383–387 (1973).
62. Hayman, D.S. *Endogone* spore numbers in soil and vesicular-arbuscular mycorrhiza in wheat as influenced by season and soil treatment, *Transactions of the British Mycological Society*, *54*, 53–63 (1970).
63. Hayman, D.S. Plant growth responses to vesicular-arbuscular mycorrhiza. VI. Effect of light and temperature, *New Phytologist*, *73*, 71–80 (1974).

64. Hayman, D.S. Mycorrhizal populations of sown pastures and native vegetation in Otago, New Zealand, *New Zealand Journal of Agricultural Research, 21*, 271–276 (1978).
65. Hayman, D.S. Endomycorrhizae, pp. 401–442 in *Interactions between Non-pathogenic Soil Microorganisms and Plants* (Edited by Y.R. Dommergues and S.V. Krupa), Elsevier, Amsterdam (1978).
66. Hayman, D.S., Barea, J.M. and Azcón, R. Vesicular-arbuscular mycorrhiza in Southern Spain: its distribution in crops growing in soil of different fertility, *Phytopathologia Mediterranea, 15*, 1–6 (1976).
67. Hayman, D.S. and Day, J.M. Influence of VA mycorrhiza on plant growth: Uptake of molybdenum, *Rothamsted Report for 1977, Part 1*, p. 240 (1978).
68. Hayman, D.S. and Hampson, K.A. VA Mycorrhiza. Field inoculation trial (white clover in Welsh upland soil), *Rothamsted Report for 1978, Part 1*, pp. 238–239 (1979).
69. Hayman, D.S., Johnson, A.M. and Ruddlesdin, I. The influence of phosphate and crop species on *Endogone* spores and vesicular-arbuscular mycorrhiza under field conditions, *Plant and Soil, 43*, 489–495 (1975).
70. Hayman, D.S., Morris, E.J. and Page, R.J. Methods for inoculating field crops with mycorrhizal fungi, *Annals of Applied Biology* (submitted) (1981).
71. Hayman, D.S. and Mosse, B. Plant growth responses to vesicular-arbuscular mycorrhiza. I. Growth of *Endogone*-inoculated plants in phosphate-deficient soils, *New Phytologist, 70*, 19–27 (1971).
72. Hayman, D.S. and Mosse, B. Plant growth responses to vesicular-arbuscular mycorrhiza. III. Increased uptake of labile P from soil, *New Phytologist, 71*, 41–47 (1972).
73. Hayman, D.S. and Mosse, B. Improved growth of white clover in hill grasslands by mycorrhizal inoculation, *Annals of Applied Biology, 93*, 141–148 (1979).
74. Hayman, D.S., Page, R.J. and Clarke, C.A. Vesicular-arbuscular mycorrhiza: Field inoculation studies with red clover, Sawyers I, *Rothamsted Report for 1980, Part I* (in press) (1981).
75. Hayman, D.S. and Stovold, G.E. Spore populations and infectivity of vesicular-arbuscular mycorrhizal fungi in New South Wales, *Australian Journal of Botany, 27*, 227–233 (1979).
76. Heap, A.J. and Newman, E.I. The influence of vesicular-arbuscular mycorrhizas on phosphorus transfer between plants, *New Phytologist, 85*, 173–179 (1980).
77. Hirrel, M.C., Mehravaran, H. and Gerdemann, J.W. Vesicular-arbuscular mycorrhizae in the Chenopodiaceae and Cruciferae: do they occur? *Canadian Journal of Botany, 56*, 2813–2817 (1978).
78. Ho, I. and Trappe, J.M. Translocation of ^{14}C from *Festuca* plants to

their endomycorrhizal fungi, *Nature New Biology*, *244*, 30–31 (1973).

79. Hussey, R.S. and Roncadori, R.W. Interaction of *Pratylenchus brachyurus* and *Gigaspora margarita* on cotton, *Journal of Nematology*, *10*, 16–20 (1978).
80. Iqbal, S.H. and Qureshi, K.S. The influence of mixed sowing (cereals and crucifers) and crop rotation on the development of mycorrhiza and subsequent growth of crops under field conditions, *Biologia* (*Pakistan*), *22*, 287–298 (1976).
81. Islam, R., Ayanaba, A. and Sanders, F.E. Response of cowpea (*Vigna unguiculata*) to inoculation with VA mycorrhizal fungi and to rock phosphate fertilization in some unsterilized Nigerian soils, *Plant and Soil*, *54*, 107–117 (1980).
82. Jensen, A. and Jakobsen, I. The occurrence of vesicular-arbuscular mycorrhiza in barley and wheat grown in some Danish soils with different fertilizer treatments, *Plant and Soil*, *55*, 403–414 (1980).
83. Kang, B.T., Islam, R., Sanders, E.F. and Ayanaba, A. Effect of phosphate fertilization and inoculation with VA-mycorrhizal fungi on performance of cassava (*Manihot esculenta* Crantz) grown on an alfisol, *Field Crops Research*, *3*, 83–94 (1980).
84. Kessler, K.J. and Blank, R.W. *Endogone* sporocarps associated with sugar maple, *Mycologia*, *64*, 634–638 (1972).
85. Khan, A.G. Occurrence of *Endogone* spores in West Pakistan soils, *Transactions of the British Mycological Society*, *56*, 217–224 (1971).
86. Khan, A.G. The effect of vesicular-arbuscular mycorrhizal associations on growth of cereals. I. Effects on maize growth, *New Phytologist*, *71*, 613–619 (1972).
87. Khan, A.G. The effect of vesicular-arbuscular mycorrhizal associations on growth of cereals. II. Effects on wheat growth, *Annals of Applied Biology*, *80*, 27–36 (1975).
88. Kinden, D.A. and Brown, M.F. Electron microscopy of vesicular-arbuscular mycorrhizae of yellow poplar. IV. Host-endophyte interactions during arbuscular deterioration, *Canadian Journal of Microbiology*, *22*, 64–75 (1976).
89. Kleinschmidt, G.D. and Gerdemann, J.W. Stunting of citrus seedlings in fumigated nursery soils related to the absence of endomycorrhizae, *Phytopathology*, *62*, 1447–1453 (1972).
90. Kormanik, P.P., Bryan, W.C. and Schultz, R.C. Endomycorrhizal inoculation during transplanting improves growth of vegetatively propagated yellow poplar, *The Plant Propagator*, *23*, 4–5 (1979).
91. Koske, R.E. *Endogone* spores in Australian sand dunes, *Canadian Journal of Botany*, *53*, 668–672 (1975).
92. Koske, R.E., Sutton, J.C. and Sheppard, B.R. Ecology of *Endogone* in Lake Huron sand dunes, *Canadian Journal of Botany*, *53*, 87–93 (1975).

93. Koucheki, H.K. and Read, D.J. Vesicular-arbuscular mycorrhiza in natural vegetation systems. II. The relationship between infection and growth in *Festuca ovina* L., *New Phytologist, 77*, 655–666 (1976).
94. Kruckelmann, H.W. Effects of fertilizers, soils, soil tillage, and plant species on the frequency of *Endogone* chlamydospores and mycorrhizal infection in arable soils, pp. 511–525 in *Endomycorrhizas* (Edited by F.E. Sanders, B. Mosse and P.B. Tinker), Academic Press, London (1975).
95. Lambert, D.H., Cole, H. and Baker, D.E. Variation in the response of alfalfa clones and cultivars to mycorrhizae and phosphorus, *Crop Science, 20*, 615–618 (1980).
96. La Rue, J.H., McClellan, W.D. and Peacock, W.L. Mycorrhizal fungi and peach nursery nutrition, *California Agriculture, 29*, 6–7 (1975).
97. La Torraca, S. Effects of inoculation with VA mycorrhiza on growth and nodulation of *Vigna unguiculata* (L.) Walp in three 'terra firme' soils, Dissertation, INPA, Manaus, Brazil (1979).
98. Levy, Y. and Krikun, J. Effect of vesicular-arbuscular mycorrhiza on *Citrus jambhiri* water relations, *New Phytologist, 85*, 25–31 (1980).
99. Macdonald, R.M. and Lewis, M. The occurrence of some acid phosphatases and dehydrogenases in the vesicular-arbuscular mycorrhizal fungus *Glomus mosseae, New Phytologist, 80*, 135–141 (1978).
100. Marsh, B.a' B. Measurement of length in random arrangements of lines, *Journal of Applied Ecology, 8*, 265–267 (1971).
101. Mason, D.T. A survey of numbers of *Endogone* spores in soil cropped with barley, raspberry and strawberry, *Horticultural Research, 4*, 98–103 (1964).
102. Mataré, R. and Hattingh, M.J. Effect of mycorrhizal status of avocado seedlings on root rot caused by *Phytophthora cinnamomi, Plant and Soil, 49*, 433–435 (1978).
103. McIlveen, W.D., Spotts, R.A. and Davis, D.D. The influence of soil zinc on nodulation, mycorrhizae, and ozone-sensitivity of Pinto bean, *Phytopathology, 65*, 645–647 (1975).
104. Mejstrik, V.K. Vesicular-arbuscular mycorrhizas of the species of a Molinietum coeruleae L.I. association: the ecology, *New Phytologist, 71*, 883–890 (1972).
105. Meloh, K.A. Untersuchungen zur Biologie der endotrophen Mycorrhiza bei *Zea mays* L. und *Avena sativa* L., *Archiv für Mikrobiologie, 46*, 369–381 (1963).
106. Menge, J.A., Davis, R.M., Johnson, E.L.V. and Zentmyer, G.A. Mycorrhizal fungi increase growth and reduce transplant injury in avocado, *California Agriculture, 32*, 6–7 (1978).
107. Menge, J.A., Lembright, H. and Johnson, E.L.V. Utilization of mycor-

rhizal fungi in citrus nurseries, *Proceedings of the International Society of Citriculture*, *1*, 129–132 (1977).

108. Meyer, F.H. Distribution of ectomycorrhizae in native and man-made forests, pp. 79–105 in *Ectomycorrhizae: Their Ecology and Physiology* (Edited by G.C. Marks and T.T. Kozlowski), Academic Press, New York (1973).
109. Mikola, P. (Editor). *Tropical Mycorrhiza Research*, University Press, Oxford (1980).
110. Miller, R.M. Some occurrences of vesicular-arbuscular mycorrhiza in natural and disturbed ecosystems of the Red Desert, *Canadian Journal of Botany*, *57*, 619–623 (1979).
111. Moawad, M. Ecophysiology of vesicular-arbuscular mycorrhiza in the tropics, pp. 197–209 in *The Soil-Root Interface* (Edited by J.L. Harley and R. Scott Russell), Academic Press, London (1979).
112. Molina, R.J., Trappe, J.M. and Strickler, G.S. Mycorrhizal fungi associated with *Festuca* in the western United States and Canada, *Canadian Journal of Botany*, *56*, 1691–1695 (1978).
113. Morley, C.D. and Mosse, B. Abnormal vesicular-arbuscular mycorrhizal infections in white clover induced by lupin, *Transactions of the British Mycological Society*, *67*, 510–513 (1976).
114. Mosse, B. Observations on the extra-matrical mycelium of a vesicular-arbuscular endophyte, *Transactions of the British Mycological Society*, *42*, 439–448 (1959).
115. Mosse, B. Plant growth responses to vesicular-arbuscular mycorrhiza. IV. In soil given additional phosphate, *New Phytologist*, *72*, 127–136 (1973).
116. Mosse, B. Advances in the study of vesicular-arbuscular mycorrhiza, *Annual Review of Phytopathology*, *11*, 171–196 (1973).
117. Mosse, B. Plant growth responses to vesicular-arbuscular mycorrhiza. X. Responses of *Stylosanthes* and maize to inoculation in unsterile soils, *New Phytologist*, *78*, 277–288 (1977).
118. Mosse, B. and Bowen, G.D. A key to the recognition of some *Endogone* spore types, *Transactions of the British Mycological Society*, *51*, 469–483 (1968).
119. Mosse, B. and Bowen, G.D. The distribution of *Endogone* spores in some Australian and New Zealand soils, and in an experimental field soil at Rothamsted, *Transactions of the British Mycological Society*, *51*, 485–492 (1968).
120. Mosse, B. and Hayman, D.S. Mycorrhiza in agricultural plants, pp. 213–230 in *Tropical Mycorrhiza Research* (Edited by P. Mikola), University Press, Oxford (1980).
121. Mosse, B. and Hayman, D.S. Mycorrhiza, in *The Ecology of Meathop Wood, an Integrated Study* (Edited by J.E. Satchell) (in press) (1982).

122. Mosse, B., Hayman, D.S. and Arnold, D.J. Plant growth responses to vesicular-arbuscular mycorrhiza. V. Phosphate uptake by three plant species from P-deficient soils labelled with ^{32}P, *New Phytologist*, *72*, 809–815 (1973).
123. Mosse, B., Hayman, D.S. and Ide, G.J. Growth responses of plants in unsterilized soil to inoculation with vesicular-arbuscular mycorrhiza, *Nature*, *224*, 1031–1032 (1969).
124. Mosse, B. and Jones, G.W. Separation of *Endogone* spores from organic soil debris by differential sedimentation on gelatin columns, *Transactions of the British Mycological Society*, *51*, 604–608 (1968).
125. Mosse, B. and Phillips, J.M. The influence of phosphate and other nutrients on the development of vesicular-arbuscular mycorrhiza in culture, *Journal of General Microbiology*, *69*, 157–166 (1971).
126. Mosse, B., Powell, C.Ll. and Hayman, D.S. Plant growth responses to vesicular-arbuscular mycorrhiza. IX. Interactions between VA mycorrhiza, rock phosphate and symbiotic nitrogen fixation, *New Phytologist*, *76*, 331–342 (1976).
127. Munns, D.N. and Mosse, B. Mineral nutrition of legume crops, pp. 115–125 in *Advances in Legume Science* (Edited by R.J. Summerfield and A.H. Bunting), HMSO, London (1980).
128. Murdoch, C.L., Jackobs, J.A. and Gerdemann, J.W. Utilization of phosphorus sources of different availability by mycorrhizal and non-mycorrhizal maize, *Plant and Soil*, *27*, 329–334 (1967).
129. Nesheim, O.N. and Linn, M.B. Deleterious effects of certain fungitoxicants on the formation of mycorrhiza on corn by *Endogone fasciculata* and on corn root development, *Phytopathology*, *59*, 297–300 (1969).
130. Nicolson, T.H. Mycorrhiza in the Gramineae. II. Development in different habitats, particularly sand dunes, *Transactions of the British Mycological Society*, *43*, 132–145 (1960).
131. Nicolson, T.H. and Gerdemann, J.W. Mycorrhizal *Endogone* species, *Mycologia*, *60*, 313–325 (1968).
132. Nicolson, T.H. and Schenck, N.C. Endogonaceous mycorrhizal endophytes in Florida, *Mycologia*, *71*, 178–198 (1979).
133. Ocampo, J.A. and Hayman, D.S. Effects of pesticides on mycorrhiza in field-grown barley, maize and potatoes, *Transactions of the British Mycological Society*, *74*, 413–416 (1980).
134. Ocampo, J.A. and Hayman, D.S. Influence of plant interactions on vesicular-arbuscular mycorrhizal infections. II. Crop rotations and residual effects of non-host plants, *New Phytologist*, *87*, 333–343 (1981).
135. Ocampo, J.A., Martin, J. and Hayman, D.S. Influence of plant interactions on vesicular-arbuscular mycorrhizal infections. I. Host and non-host plants grown together, *New Phytologist*, *84*, 27–35 (1980).
136. Ohms, R.E. A flotation method for collecting spores of a phycomycet-

ous mycorrhizal parasite from soil, *Phytopathology*, *47*, 751–752 (1957).
137. Olsen, S.R., Cole, C.V., Watanabe, F.S. and Dean, L.A. Estimation of available phosphorus in soils by extraction with sodium bicarbonate, United States Department of Agriculture, Circular No. 939, pp. 19 (1954).
138. Owusu-Bennoah, E. and Mosse, B. Plant growth responses to vesicular-arbuscular mycorrhiza. XI. Field inoculation responses in barley, lucerne and onion, *New Phytologist*, *83*, 671–679 (1979).
139. Phillips, J.M. and Hayman, D.S. Improved procedures for clearing roots and staining parasitic and vesicular-arbuscular mycorrhizal fungi for rapid assessment of infection, *Transactions of the British Mycological Society*, *55*, 158–161 (1970).
140. Porter, W.M., Abbott, L.K. and Robson, A.D. Effect of rate of application of superphosphate on populations of vesicular-arbuscular endophytes, *Australian Journal of Experimental Agriculture and Animal Husbandry*, *18*, 573–578 (1978).
141. Powell, C.Ll. Plant growth responses to vesicular-arbuscular mycorrhiza. VIII. Uptake of P by onion and clover infected with different *Endogone* spore types in ^{32}P labelled soils, *New Phytologist*, *75*, 563–566 (1975).
142. Powell, C.Ll. Mycorrhizas in hill country soils. III. Effect of inoculation on clover growth in unsterile soils, *New Zealand Journal of Agricultural Research*, *20*, 343–348 (1977).
143. Powell, C.Ll. Inoculation of white clover and ryegrass seed with mycorrhizal fungi, *New Phytologist*, *83*, 81–85 (1979).
144. Read, D.J., Koucheki, H.K. and Hodgson, J. Vesicular-arbuscular mycorrhiza in natural vegetation systems. I. The occurrence of infection, *New Phytologist*, *77*, 641–653 (1976).
145. Redhead, J.F. Mycorrhizal associations in some Nigerian forest trees, *Transactions of the British Mycological Society*, *51*, 377–387 (1968).
146. Redhead, J.F. Endotrophic mycorrhizas in Nigeria: species of the Endogonaceae and their distribution, *Transactions of the British Mycological Society*, *69*, 275–280 (1977).
147. Reeves, F.B., Wagner, D., Moorman, T. and Kiel, J. The role of endomycorrhizae in revegetation practices in the semi-arid west. I. A comparison of incidence of mycorrhizae in severely disturbed vs. natural environments, *American Journal of Botany*, *66*, 6–13 (1979).
148. Rhodes, L.H. and Gerdemann, J.W. Phosphate uptake zones of mycorrhizal and non-mycorrhizal onions, *New Phytologist*, *75*, 555–561 (1975).
149. Rhodes, L.H. and Gerdemann, J.W. Hyphal translocation and uptake of sulphur by vesicular-arbuscular mycorrhizae of onion, *Soil Biology and Biochemistry*, *10*, 355–360 (1978).
150. Rhodes, L.H. and Gerdemann, J.W. Nutrient translocation in vesicular-

arbuscular mycorrhizae, pp. 173–195 in *Cellular Interactions in Symbiosis and Parasitism* (Edited by C.B. Cook, P.W. Pappas and E.D. Rudolph), The Ohio State University Press, Columbus, Ohio (1978).

151. Ross, J.P. Effect of phosphate fertilization on yield of mycorrhizal and non-mycorrhizal soybeans, *Phytopathology*, *61*, 1400–1403 (1971).
152. Ross, J.P. Influence of *Endogone* mycorrhiza on *Phytophthora* rot of soybean, *Phytopathology*, *62*, 896–897 (1972).
153. Ross, J.P. and Gilliam, J.W. Effect of *Endogone* mycorrhiza on phosphorus uptake by soybeans from inorganic phosphates, *Soil Science Society of America Proceedings*, *37*, 237–239 (1973).
154. Ross, J.P. and Harper, J.A. Effect of *Endogone* mycorrhiza on soybean yields, *Phytopathology*, *60*, 1552–1556 (1970).
155. Ross, J.P. and Ruttencutter, R. Population dynamics of two vesicular-arbuscular endomycorrhizal fungi and the role of hyperparasitic fungi, *Phytopathology*, *67*, 490–496 (1977).
156. Safir, G.R. Vesicular-arbuscular mycorrhizae and crop productivity, pp. 231–252 in *The Biology of Crop Productivity* (Edited by P.S. Carlson), Academic Press, New York (1980).
157. Safir, G.R., Boyer, J.S. and Gerdemann, J.W. Nutrient status and mycorrhizal enhancement of water transport in soybean, *Plant Physiology*, *49*, 700–703 (1972).
158. Saif, S.R. The influence of stage of host development on vesicular-arbuscular mycorrhizae and Endogonaceous spore population in field-grown vegetable crops. I. Summer-grown crops, *New Phytologist*, *79*, 341–348 (1977).
159. Saif, S.R. and Khan, A.G. The effect of vesicular-arbuscular mycorrhizal associations on growth of cereals. III. Effects on barley growth, *Plant and Soil*, *47*, 17–26 (1977).
160. Sanders, F.E., Mosse, B. and Tinker, P.B. (Editors). *Endomycorrhizas:* Proceedings of Symposium, University of Leeds, Academic Press, London (1975).
161. Sanders, F.E. and Tinker, P.B. Phosphate flow into mycorrhizal roots, *Pesticide Science*, *4*, 385–395 (1973).
162. Schenck, N.C., Graham, S.O. and Green, N.E. Temperature and light effect on contamination and spore germination of vesicular-arbuscular mycorrhizal fungi, *Mycologia*, *67*, 1189–1192 (1975).
163. Schenck, N.C. and Hinson, K. Response of nodulating and non-nodulating soybeans to a species of *Endogone* mycorrhiza, *Agronomy Journal*, *65*, 849–850 (1973).
164. Schenck, N.C. and Kellam, M.K. The influence of vesicular-arbuscular mycorrhizae on disease development, Bulletin 798, University of Florida Agricultural Experiment Station (1978).
165. Schenck, N.C. and Kinloch, R.A. Incidence of mycorrhizal fungi on six

field crops in monoculture on a newly cleared woodland site, *Mycologia, 72*, 445–456 (1980).
166. Schenck, N.C., Kinloch, R.A. and Dickson, D.W. Interaction of endomycorrhizal fungi and root-knot nematode on soybean, pp. 606–617 in *Endomycorrhizas* (Edited by F.E. Sanders, B. Mosse and P.B. Tinker), Academic Press, London (1975).
167. Schenck, N.C. and Tucker, D.P.H. Endomycorrhizal fungi and the development of citrus seedlings in Florida fumigated soils, *Journal of the American Society for Horticultural Science, 99*, 284–287 (1974).
168. Schönbeck, F. Endomycorrhiza in relation to plant diseases, pp. 271–280 in *Soil-borne Plant Pathogens* (Edited by B. Schippers and W. Gams), Academic Press, London (1979).
169. Schönbeck, F. and Schinzer, U. Untersuchungen über den Einfluss der endotrophen Mycorrhiza auf die TMV-Läsionenbildung in *Nicotiana tabacum* L. var. *Xanthi*-nc., *Phytopathologische Zeitschrift, 73*, 78–80 (1972).
170. Schultz, R.C., Kormanik, P.P., Bryan, W.C. and Brister, G.H. Vesicular-arbuscular mycorrhiza influence growth but not mineral concentrations in seedlings of eight sweetgum families, *Canadian Journal of Forest Research, 9*, 218–223 (1979).
171. Sikora, R.A. and Schönbeck, F. Effect of vesicular-arbuscular mycorrhiza (*Endogone mosseae*) on the population dynamics of the root-knot nematodes *Meloidogyne incognita* and *Meloidogyne hapla*, *VIII International Plant Protection Congress, Moscow*, pp. 158–164 (1975).
172. Smith, S.E. and Daft, M.J. Interactions between growth, phosphate content and nitrogen fixation in mycorrhizal and non-mycorrhizal *Medicago sativa*, *Australian Journal of Plant Physiology, 4*, 403–413 (1977).
173. Smith, S.E., Nicholas, D.J.D. and Smith, F.A. Effect of early mycorrhizal infection on nodulation and nitrogen fixation in *Trifolium subterraneum* L. *Australian Journal of Plant Physiology, 6*, 305–316 (1979).
174. Sparling, G.P. and Tinker, P.B. Mycorrhizal infection in pennine grassland. I. Levels of infection in the field, *Journal of Applied Ecology, 15*, 943–950 (1978).
175. Spokes, J.R., Macdonald, R.M. and Hayman, D.S. Effects of plant protection chemicals on vesicular-arbuscular mycorrhizas, *Pesticide Science, 12* (in press) (1981).
176. St. John, T.V. Root size, root hairs and mycorrhizal infection: a re-examination of Baylis's hypothesis with tropical trees, *New Phytologist, 84*, 483–487 (1980).
177. Stribley, D.P., Tinker, P.B. and Snellgrove, R.C. Effect of vesicular-arbuscular mycorrhizal fungi on the relations of plant growth, internal phosphorus concentration and soil phosphate analyses, *Journal of Soil*

Science, *31*, 655–672 (1980).

178. Strzemska, J. Investigations on mycorrhiza in cereals (in Polish), *Acta Microbiologica Polonica*, *4*, 191–204 (1955).
179. Strzemska, J. Mycorrhiza in farm crops grown in monoculture, pp. 527–535 in *Endomycorrhizas* (Edited by F.E. Sanders, B. Mosse and P.B. Tinker), Academic Press, London (1975).
180. Strzemska, J. Occurrence and intensity of mycorrhiza and deformation of roots without mycorrhiza in cultivated plants, pp. 537–543 in *Endomycorrhizas* (Edited by F.E. Sanders, B. Mosse and P.B. Tinker) (1975).
181. Sutton, J.C. Development of vesicular-arbuscular mycorrhizae in crop plants, *Canadian Journal of Botany*, *51*, 2487–2493 (1973).
182. Sutton, J.C. and Barron, G.L. Population dynamics of *Endogone* spores in soil, *Canadian Journal of Botany*, *50*, 1909–1914 (1972).
183. Swaminathan, K. and Verma, B.C. Responses of three crop species to vesicular-arbuscular mycorrhizal infection on zinc-deficient Indian soils, *New Phytologist*, *82*, 481–487 (1979).
184. Sward, R.J., Hallam, N.D. and Holland, A.A. *Endogone* spores in a heathland area of south-eastern Australia, *Australian Journal of Botany*, *26*, 29–43 (1978).
185. Thapar, H.S. and Khan, S.N. Studies on endomycorrhiza in some forest species, *Proceedings of the Indian National Science Academy*, *39*, 687–694 (1973).
186. Thaxter, R. A revision of the Endogonaceae, *Proceedings of the American Academy of Arts and Sciences*, *57*, 291–350 (1922).
187. Thomazini, L.I. Mycorrhiza in plants of the 'Cerrado', *Plant and Soil*, *41*, 707–711 (1974).
188. Tisdall, J.M. and Oades, J.M. Stabilization of soil aggregates by the root systems of ryegrass, *Australian Journal of Soil Research*, *17*, 429–441 (1979).
189. Trinick, M.J. Vesicular-arbuscular infection and soil phosphorus utilization in *Lupinus* spp., *New Phytologist*, *78*, 297–304 (1977).
190. Vander Zaag, P., Fox, R.L. De la Pena, R.S. and Yost, R.S. P nutrition of cassava, including mycorrhizal effects on P, K, S, Zn and Ca uptake, *Field Crops Research*, *2*, 253–263 (1979).
191. Waidyanatha, U.P. de S., Yogaratnam, N. and Ariyaratne, W.A. Mycorrhizal infection on growth and nitrogen fixation of *Pueraria* and *Stylosanthes* and uptake of phosphorus from two rock phosphates, *New Phytologist*, *82*, 147–152 (1979).
192. Walker, C. *Complexipes moniliformis*: a new genus and species tentatively placed in the Endogonaceae, *Mycotaxon*, *10*, 99–104 (1979).
193. Warner, A. and Mosse, B. Independent spread of vesicular-arbuscular mycorrhizal fungi in soil, *Transactions of the British Mycological Society*, *74*, 407–410 (1980).

194. Wilhelm, S. Principles of biological control of soil-borne plant diseases, *Soil Biology and Biochemistry*, *5*, 729–737 (1973).
195. Williams, S.E., Wollum, A.G. and Aldon, E.F. Growth of *Atriplex canescens* (Pursh) Nutt. improved by formation of vesicular-arbuscular mycorrhizae, *Soil Science Society of America Proceedings*, *38*, 962–965 (1974).

SECTION B

Management of Pathogens, Pests and Weeds through Microorganisms

N. WALKER

14. Interactions of Pesticides with Soil Microorganisms

Introduction

Within the last thirty years a virtual revolution has taken place in modern agriculture in that numerous synthetic organic compounds have been prepared and used to combat pests and diseases that affect crop plants. Prior to 1940 with the exception of fertilizers, only a few chemicals were used as auxiliaries in agriculture. A few natural products were used as insecticides, for example, nicotine, derris, pyrethrum and some inorganic substances such as sulphur, copper and arsenic salts; certain coal tar products, for example, naphthalene or cresols were used as disinfectants and as insect repellants. None of these substances was very specific in its effects and the mode of action was not well understood. It is well known that crop plants may be affected by diseases caused by fungi, bacteria or viruses and they are especially liable to attack by animal pests, including insects, beetles and other invertebrates, nematodes and others; in the field, crops constantly face competition from weeds. The successful control of these varied pests and diseases by chemical means has resulted in greatly improved crop yields.

The discovery, in 1939, of the insecticidal activity of the chlorinated hydrocarbon, DDT, and of the selective herbicidal activity of 2,4-D (2–4 dichlorophenoxyacetic acid) between 1940 and 1945 signalled the beginning of the present era of chemical plant protection. There are now 200 synthetic chemicals in widespread use in modern agriculture throughout the world. The two main groups of plant protection chemicals are herbicides and insecticides, but, in addition, there are fungicidal compounds, nematicides and other substances. While the term "pesticide" is often used as a general word for all plant protection chemicals, strictly speaking it should refer only to chemicals used against animal pests. We shall be concerned here with pesticides as so defined, herbicides and fungicides. The scientific literature on all aspects of this subject is now so vast that it is impossible to survey it, and within the scope of this chapter, we will only touch briefly on a few typical aspects.

Many monographs have been published (refer to listing at the end of the chapter) and a whole series of journals are now devoted to studies on the use and behaviour of pesticides in the environment.

Ideally, chemicals used in plant protection should effectively kill or inhibit the pests or weeds in question and cause little or no harm to nontarget plants and animals. Unfortunately this is seldom the case; consequently public concern has increasingly focussed on the harmful side effects of pesticides, their mitigation and more generally, environmental pollution of all kinds. Public concern was greatly stimulated by the publication of Rachel Carson's book *Silent Spring* (8); since then many detailed studies have appeared on the possible harmful consequences of pesticides or their residues to man, birds, mammals, other nontarget plants and animals and the environment.

The great majority of plant protection chemicals (although not all) are subject to transformation or degradation in the soil mainly through the activities of soil microorganisms. Some chemicals decompose or dissipate in other ways, such as by volatilization, chemical or photochemical reactions, leaching and so on but these nonbiological actions will not be considered here. The interactions, therefore, that are of importance to the safe and efficient use of chemicals in agriculture include the degradation of pesticides by microorganisms and the effects of chemicals on useful soil microorganisms as well as the influence of soil and climatic factors on these interactions.

Soil biology and microbiology

Soils harbour a vast array of living organisms, including fungi, bacteria, actinomycetes, algae and other microorganisms, as well as soil animals of many species from microscopic protozoa to beetles and earthworms. Population of these multifarious organisms is dependent on the nutritive conditions of the soil and other factors such as soil texture, pH, water content, aeration and so on. Some organisms prey on others, some complete for available nutrients and oxygen; the supply of water, oxygen or carbon dioxide fluctuates and both space and soil moisture can be limiting. Thus physical, chemical and climatic factors affect the growth and survival of soil fauna and soil microorganisms and these variable conditions mean that, even in undisturbed soil, any quasi equilibrium between the soil biological populations is seldom long lasting. For example, temperature in the soil varies seasonally and within a day, water content changes according to the climate, soils may dry out and later become wet, they may freeze and then thaw; on drying, gases are lost from the soil and many nutrients and inorganic salts are subject to movement up and down the soil profile in relation to water changes. All these changes affect the activities of living soil organisms.

We will not discuss soil characters and composition in greater detail but it is necessary to emphasize the very complex nature of soils and their biology because this invariably complicates the investigation of the behaviour of pesticidal chemicals in the soil. This account will be restricted to considering degradation of chemicals by microorganisms but first some general comments must be made on the chemicals themselves.

Plant protection chemicals

This miscellaneous collection of substances comprises fairly complicated synthetic organic compounds often possessing one or more aromatic carbocyclic rings, one or more halogen atoms, sometimes sulphur or phosphorus atoms and frequently the substances are only sparingly soluble in water. They are a very diverse group of compounds and systematic classification is scarcely feasible. Some broad grouping, however, is possible, for example, there are chlorinated hydrocarbons, organophosphorus esters, substituted triazines, phenoxyacetic acid derivatives, substituted phenylureas and substituted phenylcarbamates; but within these groupings there can be marked differences between individual compounds in structure, properties and susceptibility to microbial attack. Generalisations regarding the probable stability in the soil of compounds even belonging to the same broad group may frequently be only tentative. Briggs (5) has attempted with some success however, to predict the behaviour and persistence in soil of compounds from a knowledge of their octanol/water distribution coefficients and adsorption characteristics. Indications of possible susceptibility to microbial decomposition based on substituting atoms or groups may be feasible although, in view of the complexity of soil situations, such deductions again may not be absolutely dependable. Moreover, the enzymes that may be induced in microbial cells in response to the presence of a particular compound are often highly specific, reacting on occasion with only one particular optical or geometrical isomer of the substance in question. Adaptation to one compound by a degrading microorganism does not imply that reaction with similar compounds will occur. Thus, Raymond and Alexander (40) found that a p-nitrophenol-grown soil bacterium would not grow with m-nitrophenol as substrate although the latter isomer was co-oxidized to nitrohydroquinone in the presence of the para isomer.

Characteristics of microbial decomposition processes

To some extent the decomposition of unusual synthetic compounds by microorganisms in the soil reflects the stages and characteristics of a microbial culture but this process may be complicated by the soil properties and the presence of other living organisms. However, when a foreign synthetic compound is added to the soil certain reactions may be triggered and recognition of them often provides evidence that the process is indeed microbiological.

Firstly, depending on its particular properties the compound may be adsorbed to some extent on the clay or the organic matter fraction of the soil. If a solution of the substance were percolating through the soil, the result would be a rapid initial fall in concentration. Then, assuming that the compound is reasonably stable, a long time would elapse without further change in concentration until some microorganisms (19) acquired the necessary enzyme systems to react with the substance. We still do not fully understand how enzymes are induced and whether genetic changes or mutations are involved. It seems clear that adapted organisms can metabolize the foreign compounds and so obtain carbon and energy for growth and a population of such organisms then develops with a resulting advantage over other organisms. If all conditions for microbial growth are favourable then growth will occur with consequent decline in the concentration of the foreign molecule.

Audus, who first studied the kinetic behaviour of the herbicide 2,4-D in a soil percolator system between 1948 and 1951 noted all the above phases in the development of the 2,4-D-decomposing microbial population and isolated an *Arthrobacter* sp. as the responsible microbe. The herbicide is not particularly toxic and high concentrations were successfully metabolized; successive doses decomposed without further lag. Characteristic features of the process are the long lag before degradation begins (two or three weeks in the case of 2, 4-D), the fairly rapid time (two to four days) for utilization of the compound, absence of any lag in the degradation of subsequent doses, the transfer of degrading activity to fresh soil by means of small inocula and the suspension of degradation by sterilization treatments. These features and the sequence of events have been demonstrated repeatedly by later researchers in investigations of many other pesticide chemicals. If the degrading microorganism is an aerobe then oxygen is necessary for degradation; degradation then can be inhibited by excluding air or by adding respiratory poisons such as cyanide, azide or fluoride. Favourable temperatures, hydrogen ion concentrations, perhaps the presence of trace elements may all be necessary for successful degradations. Some degradations occur only under anaerobic conditions (49, 33, 29) and there are some where the course of degradation differs depending on whether aerobic or anaerobic conditions prevail.

It is possible for a pesticide to be partially degraded by a microorganism that grows on and derives its carbon and energy from the usual substrate without recourse to enzyme induction to degrade the pesticide by the organism. This phenomenon is called cometabolism or co-oxidation and is now believed to play a significant role in decomposition of many chemicals that are not readily metabolized. The process was first reported by Leadbetter and Foster (31) who observed the partial oxidation of gaseous hydrocarbons by a methane organism. Aromatic hydrocarbons can also be partially oxidized in *Nocardia* cultures growing on paraffin (41). Many workers (21, 13, 31, 36, 22, 42, 46, 47) have reported cometabolism by various microorganisms of numerous compounds including a variety of stable pesticides, such as DDT, triazine herbicides and others. Insecticides are generally used in much smaller quantities than herbicides and except for a few soil-applied insecticides, they reach the soil only accidentally through spray dripping from foliage and so on. Consequently the amounts in the soil are small and it has been suggested that these may be degraded incidentally by non-induced, constitutive enzymes of the microbial cell. Matsumura and Benezet (35) have called this process incidental degradation. Finally, decomposition of various subtances may also take place by mixed cultures of microorganisms that is 'consortia' of organisms in which the several members grow permanently in the culture, even after long period under continuous culture conditions. It is not always clear whether every member of the consortium is essential. The phenomenon has been discussed by Bull (7) who rightly points out that in nature there is always a very mixed population of organisms in the soil and it would be rare for a single organism to dominate the entire course of a decomposition. Occasionally it has been noted that two different microorganisms have jointly caused the degradation of a pesticide, for example, diazinon (17), which was not metabolized by one organism alone.

The general level of microbial activity in soils can be influenced in various ways and it seems possible that certain pesticide degradation can be promoted by increasing microbial activity in general. Such general microbial activity, which can be assessed (2) by counting viable organisms, measuring soil respiration or by determining biomass, is mainly dependent on the nutrient supply. This can be supplemented by exudation of organic substances from plant roots, sloughing of root cells and by the fall of leaves and other plant parts onto the soil. When soil fauna or other organisms die, the dead protoplasm becomes an easily available nutrient for the surviving microorganisms. This also results from partial sterilization treatments and even the drying of soils may kill sensitive organisms such as protozoa thus promoting the growth of other microorganisms when the soil is remoisturised. It is difficult to assess the extent to which the degradation of foreign synthetic substances like pesticides may be stimulated by these general nutritional effects through perhaps cometabolism but we are quite certain this process occurs (24, 36).

Effects of microorganisms on pesticides

1. Herbicides

Published information on the degradation of plant protection chemicals by soil microorganisms is now very extensive. It is impossible to survey even briefly and in the present short chapter only a few typical examples of such studies will be mentioned. During the past decade, several monographs on this subject have appeared. We now present a brief selection of some investigations on the breakdown of representative plant protection chemicals. Two general types of microbial degradation have been mentioned; first, decomposition in which the pesticide is more or less completely metabolized providing carbon and energy for the growth of the microorganisms in question and, second, degradations which are usually only partial, do not provide carbon and energy for growth and in which the degrading activity may be slow.

A classic example of the first type of decomposition is the breakdown of the selective 'hormone' herbicides, 2,4-dichlorophenoxyacetic acid (2,4-D) and 2-methyl-4-chlorophenoxyacetic acid (MCPA). More than a dozen different species of soil bacteria and some fungi have been identified as causing the degradation of 2,4-D and MCPA in various soils and this problem has been repeatedly studied by numerous works (Fig. 1). At least two metabolic pathways for the degradation have been proposed and some of the reactions have been examined using cell-free enzyme systems. More interesting now, however, is the fate of 2,4,5-trichlorophenoxyacetic acid (2,4,5-T) which is much more persistent than either 2,4-D or MCPA and for which no pure culture of an organism has yet been isolated to completely metabolize it. There is evidence of the cometabolism of 2,4,5-T by benzoate-grown microorganisms (21) resulting in the formation of 3,5-dichlorocatechol and the release of a chlorine atom. We cannot confirm (unpublished) earlier claim by Audus that successive pretreatment of soil with 2,4-D followed by MCPA produces a soil adapted to metabolize 2,4,5-T. The long persistence in the soil of 2,4,5-T and particularly the presence of the toxic, teratogenic impurity 2,3,7,8-tetrachlorodibenzo-p-dioxin in commercial grade 2,4,5-T have led some countries, including the USA, to ban this herbicide. The stability of 2,4,5-T against microbial attack may be due to the presence of the chlorine atom at the meta position since *m*-chlorophenoxyacetic acid is also very resistant to microbial oxidation unlike the paraisomer which is readily degraded.

Certain herbicidal chlorinated benzoic acids are stable compounds in contrast to benzoic acid itself which is a readily available substrate to many microorganisms. For example, there is little evidence that 2,3,6-trichlorobenzoic acid can act as a carbon source for microorganisms although Horvath (22) isolated a *Brevibacterium* sp. which, after growth on benzoate, could partially oxidize the tri-chloro compound to form 3,5-dichlorocatechol and release CO_2 and one chlorine atom. Spokes and Walker (47) have reported the cometabolism of monochlorobenzoates by various soil bacteria including *Bacillus* sp. Dicamba, 3,6-dichloro-2-methoxybenzoic acid, at field rates of application (1 kg/ha) is degraded fairly rapidly in Canadian prairie soils with the formation of 3,6-dichlorosalicylic acid and liberation of carbon dioxide (45). Nevertheless, repeated attempts to isolate organisms able to degrade Dicamba from soil enrichments have failed (N. Walker, unpublished).

Bromoxynil (3,5-dibromo-4-hydroxybenzonitrile) and Ioxynil (3,5-diiodo-4-hydrozybenzon trile) are more easily degraded in soils. Smith and Cullimore (46) found that 3,5-dibromo-4-hydroxybenzamide and 3-5-dibro-

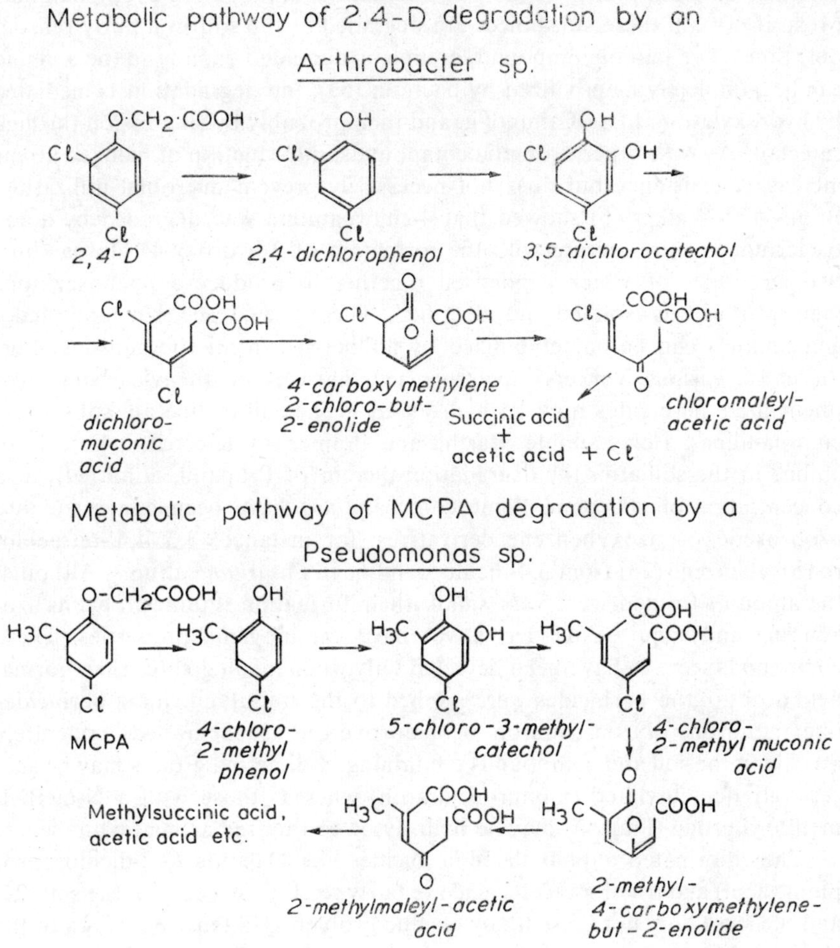

Fig. 1.

mo-4-hydroxybenzoic acid accumulated in *Flexibacterium* cultures containing bromoxynil. Hsu and Camper (23) reported the degradation of Ioxynil by pure cultures of the soil fungus, *Fusarium solani*, using ^{14}C ring labelled or ^{14}C cyano labelled Ioxynil. They detected at least eight intermediate products including 3,5-diiodo-4-hydroxybenzamide and 3-5-diiodo-4-hydroxybenzoic acid. With both these benzonitrile herbicides, the –CN group is converted to a carboxyl group by hydrolysis and oxidation.

Many important herbicides belong to the substituted phenylurea and phenylcarbamate groups of compounds and all are derived from appropriate aromatic amines, for example, chloroanilines. Some other herbicides are acyl derivatives of amines; for example, Propanil is 3,4-dichloro-propionanilide. Most, if not all, these substances are degraded in the soil in a fairly reasonable time. The parent compound, aniline, is degraded rapidly in the soil and can be completely metabolized by bacteria (53); the degradation is mediated by hydroxylation, loss of ammonia and then probably by ring fission through catechol. As with other aromatic compounds, introduction of halogen atoms increases persistence but does not necessarily prevent microbial utilization. Briggs and Walker (6) showed that 4-chloroaniline was degraded by a soil bacterium in pure culture with the formation of 2-hydroxy-4-chloroaniline, two molecules of which condensed together to produce a phenoxazinone pigment that was isolated and identified. There is evidence that mono-halogenoanilines can be cometabolized by aniline-grown microorganisms. Furthermore, various workers have reported that certain phenylcarbamate or phenylurea herbicides may be hydrolysed in the soil to liberate substituted chloroanilines. For example, Bartha and Pramer (4) detected 3,4-dichloroaniline in the soil after the degradation therein of Propanil, admittedly at a concentration of 500 ppm. Some chloroanilines have been converted into azobenzene or azoxybenzene derivatives; for instance, 3,3′-4,4′-tetrachloroazoxybenzene (28) from 3,4-dichloroaniline in *Fusarium* cultures. Although the amounts formed were very small, their formation is undesirable as azobenzene and some of its derivatives have carcinogenic properties. These compounds seem to have been detected only when much greater than normal field doses of the herbicides were applied to the soil. Phenylurea herbicides like monuron, diuron, linuron or chloroxuron are degraded moderately quickly in the soil and compounds containing N-dimethyl groups may be successively demethylated (monuron, diuron) whereas those with a N-methyl-methoxyl group (linuron) may be hydrolysed to the free aromatic amine.

The chlorinated aliphatic acid herbicides like Dalapon (2,2-dichloropropionic acid) and trichloroacetic acid are fairly readily degraded in the soil (25) and their use does not present any residue problems. Extensive reviews of the work on these and the phenylurea and phenylcarbamate herbicides are given in the monographs listed at the end of the chapter.

In contrast, the various triazine herbicides such as simazine, atrazine, prometryne and others are outstanding for their great persistence in the soil with half lives of many months. In most cases, these herbicides are degraded slowly and are not completely metabolized. In partial degradations, usually the side chain alkyl groups are utilized with some other substance providing either the carbon or the nitrogen for the attacking microorganisms; this is often a fungus, *Aspergillus, Penicillium, Trichoderma* or other species. A supplementary source of either nitrogen or carbon is therefore required for the fungus to partially degrade the triazine compound. Most of the investigations on this class of herbicide have been done using ^{14}C-labelled material; this is a very sensitive method and indicates the small amount of material that may be degraded in a reasonable time (30, 34).

The bipyridyl herbicides, diquat and paraquat, are unusual in having rapid defoliant action on plants and in being quickly inactivated by their strong adsorption to clay or to soil organic matter. They can be used therefore to clear weeds immediately before sowing a crop. These herbicides are best degraded anaerobically and, apart from certain anaerobic bacteria, yeasts of the genus *Lypomyces* can adapt to degrade both paraquat and diquat.

2. Insecticides

Although insecticides are used in smaller quantities than are herbicides, they are generally much toxic to mammals and their degradation or detoxification is consequently important.

Chlorinated hydrocarbon insecticides have very low water solubility, are adsorbed strongly to soil constituents and consequently their availability to microbial action is limited. The best known compound is DDT [1,1,1-trichloro-2,2-bis (p-chlorophenyl) ethane]; it has a long half-life in soil but suffers partial degradation by a variety of microorganisms and is decomposed more readily under anaerobic than aerobic conditions. Most research so far has shown cometabolism to be the main mode of degradation either of DDT itself or of a few related compounds. For example, Focht and Alexander (13) demonstrated the cometabolism of dichlorodiphenylmethane by diphenylmethane-grown *Hydrogenomonas* sp. and identified some metabolites produced from the cometabolism of 1,1-diphenyl-2,2,2-trichloroethane, but no chloride ions were released. Pfaender and Alexander (38) found various products including 1-chloro-2,2-bis (p-chlorophenyl) ethane and 4, 4′-dichlorobenzophenone when DDT reacted with *Hydrogenomonas* cell extracts anaerobically and, on adding whole cells and oxygen, p-chlorophenylacetic acid was detected. This demonstrated that ring-cleavage of one of the phenyl groups in DDT had occurred. An *Arthrobacter* strain metabolized p-chlorophenylacetic acid. Pure cultures of *Mucor alternans* (1) could degrade DDT to water soluble metabolites, but adding other fungi to the *Mucor* culture repressed the DDT metabolism. There is no doubt that DDT can be slowly degraded by microorganisms but the action seems to be incidental and incomplete.

Hexachlorocyclohexane (γ-BHC, Lindane) is less persistent than DDT and is also subject to very slow degradation by microorganisms especially under anaerobic conditions (18). Many metabolites have been detected by investigators and include isomers of BHC and products formed by loss of one or more chlorine atoms but again the degradation is only partial and incidental (29, 33).

An important group of insecticides includes various organophosphorus esters, the development of which can be traced to earlier work on "nerve gases" and almost all of these compounds have marked activity as inhibitors of cholineesterase which largely explains their insecticidal action. Organophosphorus insecticides were important at first as effective alternatives to chlorinated hydrocarbon compounds which were too persistent. The former insecticides are much more readily degraded in the soil. Typical products are parathion (Fig. 2), malathion and diazinon; parathion unfortunately is highly toxic to mammals but all three insecticides are degraded microbiologically. Parathion (diethyl-4-nitro-phenyl phosphorothionate) can be decomposed in various ways: by reduction and hydrolysis forming aminoparathion, p-nitrophenol and p-aminophenol (16, 24), by hydrolysis to yield diethyl thiophosphoric acid and p-nitrophenol; the latter is rapidly metabolized by several bacterial species (16, 40). Sethunathan and co-workers isolated from rice fields, among other organisms, a *Flavobacterium* that converted parathion to p-nitrophenol, presumably by a constitutive enzyme. Forrest *et al.* (14) also isolated a *Flavobacterium* strain from arable soil in Britain that hydrolysed parathion to p-nitrophenol. We isolated (unpublished) from Brazilian soils a bacterium

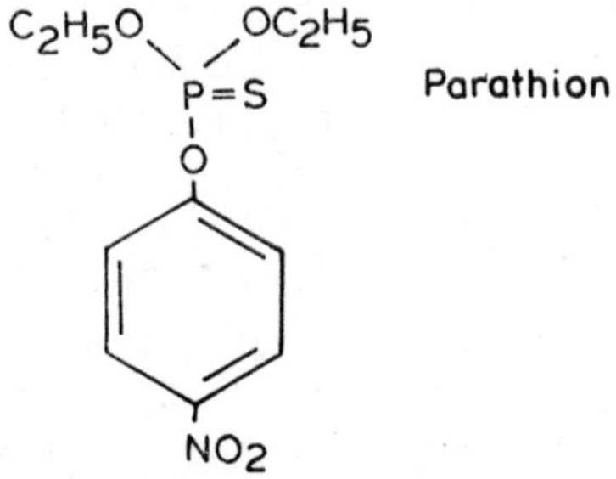

Fig. 2.

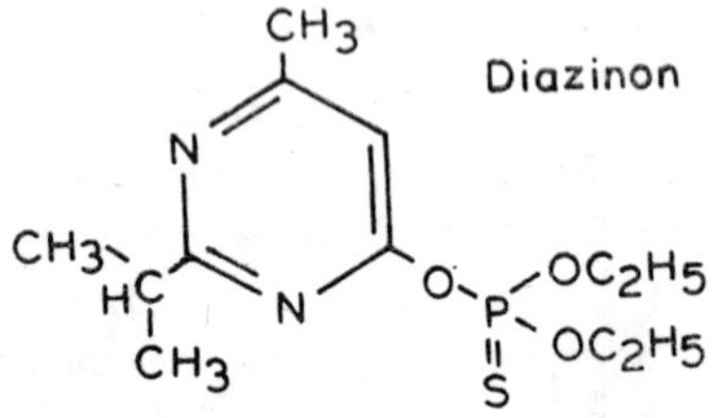

Fig. 3.

that grew in pure culture with parathion as the sole organic carbon source, forming p-nitrophenol which was subsequently further metabolized. Diazinon (diethyl-2-isopropyl-6-methyl-4-pyrimidinyl phosphorothionate) (Fig. 3) is also readily degraded in soil by a variety of microorganisms. Gunner and Zuckerman (17) observed the synergistic effect of a streptomycete and an *Arthrobacter* sp. which together degraded diazinon but diazinon was not attacked by either organism alone. Interestingly repeated applications of diazinon to rice eventually led to its failure to control brown plant-hoppers and simultaneously to the increase in diazinon-decomposing activity in rice soil (44). Sethunathan and his co-workers isolated several microorganisms that degraded diazinon by hydrolysis to the constituent 2-isopropyl-6-methyl-4-hydroxypyrimidine, which in some cases was metabolized further. Under anaerobic conditions, this metabolite accumulated. From soil in Britain on which a commercial diazinon preparation was applied 14 times over four years, Forrest *et al.* (14) isolated a *Flavobacterium* sp., which although differing in minor ways from Sethunathan's *Flavobacterium*, also degraded diazinon in pure culture. Washed cells of peptone-grown *Flavobacterium* quickly hydrolysed diazinon to 2-isopropyl-6-methyl-4-hydroxypyrimidine, parathion to p-nitrophenol and chlorpyriphos (Dursban) probably to its constituent trichloropyridinel. Thus, the enzyme hydrolysing these organophosphorus esters was constitutive and not an induced enzyme. In the soil, however, only diazinon was decomposed and under anaerobic conditions the substituted hydroxypyrimidine compound accumulated. Failure to degrade parathion or other organophosphorus insecticides in moist soil could be due to their adsorption on the soil colloids. There are many reports of microbial degradations of organophosphorus insecticides (24, 37) from different countries and even proposals to use microorganisms or their enzymes to dispose of accidental accumulations of these pesticides. Some of these substances are subject to cometabolism and the breakdown of malathion by this means can be promoted by enrichment of the co-substrate (36, 37). Repeated use of the carbamate ester insecticide, carbofuran, in rice soils has also led to enrichment of microorganism capable of degrading it and hence to the loss of effectiveness of the insecticide (49). Carbaryl, a rare example of an aromatic carbamate insecticide, is derived from 1-naphthol, which is readily metabolized by a variety of bacteria and fungi and so residue problems do not usually arise. Carbaryl itself is decomposed by a variety of microorganisms; the carbamate radicle is hydrolysed and certain fungal species produce the hydroxylated intermediates, 1-naphthyl-(hydroxymethyl) carbamate, 4- and 5-hydroxy-naphthyl methylcarbamate. The metabolic pathways for the bacterial degradation of the parent compound, naphthalene, is now well known and has been studied extensively.

Finally the recent development of synthetic pyrethrin analogues as highly potent insecticides with very low mammalian toxicity must be mentioned. These are mainly highly water-insoluble esters of substituted cyclopropanecarboxylic acids; permethrin (Fig. 4), for example, is the 3-phenoxybenzyl ester of *cis* or *trans* 3-(2,2-dichlorovinyl)-2,2-dimethyl-cyclopropane carboxylic acid (see Fig. 4). It is more photostable than some earlier compounds and therefore more effective and persistent. Kaneko *et al.* (27) reported a half-life of 6 to 12 days for *cis*- or *trans*-permethrin in an upland soil at 25°C; in a study of ^{14}C labelled permethrin, they detected hydroxylated derivatives and

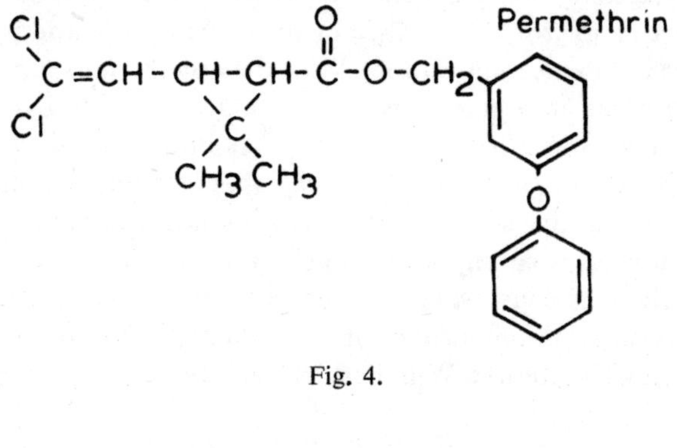

Fig. 4.

3-phenoxybenzyl alcohol and 3-phenoxybenzoic acid as breakdown products. Organisms responsible for this degradation have not been isolated and it seems likely that persistence in soil will vary with soil type, temperature and other factors. Degradation seems to be an aerobic process (N. Walker, unpublished).

Effects of plant protection chemicals on soil microorganisms

When pesticides were first used on a large scale, their poisonous effects on mammals or birds were observed and became a matter of public concern. As information and publicity increased, other aspects of the safe use of sometimes very toxic chemical compounds received attention and now potential pesticide compounds must be carefully examined for possible harmful effects on nontarget animals, plants and microorganisms. The effects on useful soil microorganisms therefore need to be considered.

We have already mentioned the diversity of living organisms in soils and the varied factors that control the ecological balance of the manifold populations of different species. Clearly, nutrient availability is an important factor. The soil ecology is very resilient and it would take treatments that are little short of complete or partial sterilization to induce serious, long-term disturbances in the balance of populations. Moreover, there is some evidence that many soil microorganisms survive or grow slowly in the soil until there is a sudden influx of fresh nutrients. Nevertheless, microorganisms are often inhibited by toxic substances and it may be some time before their populations recover. It is difficult to decide what are essential, useful organisms because some functions may be performed by a variety of different organisms. For example, the breakdown and mineralization of natural organic matter, although a vital activity, is not the prerogative of any narrow group of microorganisms, but many varied organisms participate. There is little evidence to suggest that any pesticide, even at rates of application higher than normal, seriously disturbs such general functions. We must look, therefore, to specialized groups of microorganisms to find some useful or essential activity the inhibition of which, by, say, a toxic chemical could serve as an indicator of harmful effects. Possible indicator microbial activities include nitrogen fixation, nitrification and perhaps cellulose decomposition. With the exception of nitrogen fixation by blue-green algae in aquatic environments and possibly by organisms in the rhizosphere of some tropical grasses, it is doubtful whether free-living nitrogen-fixing bacteria, such as *Azotobacter* sp. contribute significantly to the nitrogen supply of arable soils. On the contrary, nitrogen fixation by legume-*Rhizobium* symbiosis is so important that any harm caused to this system by pesticides would be serious. A lot of work has been done in testing pesticides against *Rhizobium* spp., but at normal field rates of application there is little evidence of serious inhibition. With herbicides, if there is any harm it is always greater to the legume host plant than to the nodule bacteria. Occasionally rhizobial strains are found that can degrade certain pesticides for example, malathion (37).

Autotrophic nitrifying bacteria are a special group of microbes because they obtain energy for growth by oxidizing ammonia to nitrite or nitrite to nitrate, they use carbon dioxide as a carbon source and in general grow very slowly, at best one or two divisions per day. A lot of attention has been given to the effects of pesticides on these bacteria. Conversion of ammonia to nitrate has some disadvantages in agriculture in that nitrate is more easily washed out of the soil than is ammonia, but the accumulation of nitrite is undesirable should nitrite-oxidizers be inhibited because nitrite is not only toxic but under some circumstances it may react with secondary amines to form toxic, carcinogenic nitrosamines (3). Nitrification also flourishes under soil conditions which favour plant growth (neutral reaction, moist, well-aerated soil, presence of ammonia) and can thus be an indication of fertile soil. A great deal of work has been done on testing pesticides of all kinds for their effects on nitrifying bacteria (2, 9, 10, 11, 15, 26, 39, 48, 50, 52). Debona and Audus (9) reported that, of the ten herbicides studied, only Propanil was sufficiently inhibitory to autotrophic nitrifiers in the soil to cause any serious depression of nitrification at field rates of application. Wainwright and Pugh (50) found that very low concentrations of the fungicides Captan, Thiram and Verdasan stimulated nitrifiers but that only high doses were inhibitory and ammonia oxidation was more susceptible than nitrite oxidation. The fungicide, Benomyl, was reported (39) to inhibit nitrification but its degradation product, methyl 2-benzimidazole carbamate, was not inhibitory; in practice, therefore, the breakdown of Benomyl would eliminate any inhibition of nitrification. *Nitrosomonas europaea* was inhibited to some extent by 3,4- and 2,5-dichloroanilines at five parts per million, a concentration not likely to be found in the soil (48). Monochloro anilines were much less inhibitory. Several insecticides, for example, Aldrin, Lindane and Parathion were tested by Garretson and San Clemente (15) against pure cultures of either *Nitrosomonas europaea* or *Nitrobacter agilis* and found inhibitory at 1 to 10 parts per million, but even these concentrations are more than would normally occur in the soil.

A general scheme for testing pesticides for side effects on soil microorganisms, including nitrifying bacteria, has been presented by Johnen and Drew (26).

Other interactions

The effect of pesticides on microorganisms and the converse become complicated if two or more chemicals are applied together to the soil, for example, when a fungicide is present where some herbicide or insecticide is subject to degradation by fungi. So far few studies of interactions of this kind have been made. Soil factors or treatments directly affecting soil ecology have been investigated fairly often. It has been shown by Walker (51) that the degradation of Methazole (2-3,4-dichlorophenyl)-4-methyl-1,2,4-oxidiazolidine-3, 5-dione) was influenced by temperature and moisture; at field capacity, methazole had a half life of 3.5 days at 25°C, but 31 days at 5°C, whereas with moisture at 50 per cent field capacity at 25°C the half life was five days and at 25 per cent capacity, it was 9.6 days. There are other such examples in the literature. Forrest *et al.* (14) observed that in a soil adapted to degrade diazinon successive drying and remoisturising the soil accelerated degradation when additions of diazinon were also made. Similarly, repeated freezing and thawing of the soil did not adversely affect the diazinon-degrading activity but the latter was enhanced only when diazinon was also present. Anderson and Lichtenstein (1) noted that some soil fungi and certain insecticides could affect to some extent the ability of *Mucor alternans* to degrade DDT. The synergism of insecticides by herbicides has been noted by Liang and Lichtenstein (32). Hsu and Bartha (24) studied the mineralization of parathion and diazinon measured by the evolution of CO_2 in the soil with and without the presence of growing bean plant roots. More insecticide was mineralized in the presence of roots and they interpreted this as being due to better cometabolism by the rhizosphere microorganisms. The effect of pesticides disulfoton, permethrin or prometryne on the soil microflora, although slight may be modified by the growing crop because differences in numbers of bacteria, actinomycetes and fungi were seen in soils under lettuce compared with those under carrots (34). In flooded soils, herbicides behave differently than in upland or arable soils where conditions are more aerobic. Kuwatsuka and Niki (30) discussed these differences and listed various herbicides such as phenoxyphenyl compounds possessing nitro groups and highly chlorinated compounds pentachlorophenol, which are degraded preferentially in anaerobic flooded soils. Often conditions in the soil are quite different from those in laboratory cultures and it is impossible to correlate them. For example, the herbicide Zytron (*o*-2,4-dichlorophenyl *o*-methyl isopropyl phosphoramidothioate), which is an ester of 2,4-dichlorophenol, liberates the latter compound on degradation. Although more than 10 ppm of 2,4-dichlorophenol is inhibitory to fungi in cultures, there is no evidence of the accumulation of dichlorophenol in the soil and both this and Zytron are degraded by *Aspergillus clavatus* (11). Rosenzweig and Stotzky (43) observed that although *Aspergillus niger* was inhibited in cultures at a suitable pH by the bacterial pigment prodigiosin (formed by *Serratia marcescens*), this inhibition was not noted in the soil. The absence of this phenomenon in the soil was attributed either to adsorption of the pigment to clay or soil particles or to being unavailable because of low solubility. Helweg (20) showed that the fungicide Carbendazim and its degradation product 2-aminobenzimidazole were slowly degraded in the soil and relatively more of the compounds were decomposed as the dose went higher. This indicated that a limited amount of adsorption to the soil components protected a proportion of the fungicide from degradation and the slow breakdown process suggested that the degradation was due to cometabolism. As Carbendazim by itself is a degradation product of another fungicide, Benomyl, this study suggests that neither product would cause residue problems even though the rate of degradation was slow.

Conclusions

While we have made no attempt to summarize the vast literature now available on the behaviour and fate of pesticides and their interactions with the soil microflora, we have a few examples to show the role played by microorganisms in decomposing a wide variety of synthetic chemical plant protection substances in soils. Some of these products can be used as carbon and energy substrates by various bacteria, fungi and other microorganisms and this results in disappearance of the pesticide and the parallel development of a population of the degrading organism. Other pesticides are more difficult to metabolize and often only a partial degradation is effected by microorganisms that grow on more normal nutritive substrates; this is cometabolism. Some substances, like insecticides which reach the soil in only very small quantities may be slowly but incidentally degraded. Finally there are a few compounds which are recalcitrant and are not at all susceptible to microbial attack. Such products, like DDT, Dieldrin and a few others may be gradually replaced in practice by less persistent but equally effective agents. The aim of future research will be to search for more specific chemicals with sufficient stability but causing minimal side reactions. Interactions with soil microorganisms are complex because of the intricate nature of soil properties and its complex fauna and microflora. Harmful effects are usually related to quantities or concentrations of chemicals used and correct dosage is often the key to successful use without harmful side effects. In any case life in the soil is a resilient ecosystem.

References

1. Anderson, J.P.E. and Lichtenstein, E.P. Effects of various soil fungi and insecticides on the capacity of *Mucor alternans* to degrade DDT, *Canadian Journal of Microbiology*, *18*, 553–560 (1972).
2. Anderson, J.R. and Drew, E. Effects of pure Paraquat dichloride, Gramoxone W and formulation additives on soil microbiological activities. 3. Estimation of soil microflora and enzyme activity in field-treated soil, *Zentralblatt für Bakteriologie, Parasitenkunde, Infektionskrankheiten und Hygiene, IIte Abteilung*, *131*, 247–258 (1976).

3. Ayanaba, A. and Alexander, M. Microbial formation of nitrosamines *in vitro*, *Applied Microbiology, 25*, 862–868 (1973).
4. Bartha, R. and Pramer, D. Pesticide transformation to aniline and azo compounds in soil, *Science, 156*, 1617–1618 (1967).
5. Briggs, G.G. Degradation in soils, *Proceedings of the British Crop Protection Council Symposium on "Persistence of Insecticides and Herbicides"*, pp. 41–54 (1976).
6. Briggs, G.G. and Walker, N. Microbial metabolism of 4-chloroaniline, *Soil Biology and Biochemistry, 5*, 695–697 (1973).
7. Bull, A.T. Biodegradation: some attitudes and strategies of microorganisms and microbiologists, pp. 107–136 in *Contemporary Microbial Ecology* (Edited by D.C. Ellwood, J.N. Hedger, M.J. Latham, J.M. Lynch and J.H. Slater), Academic Press, New York (1980).
8. Carson, R. *Silent Spring,* Hamish Hamilton, London (1963).
9. Debona, A.C. and Audus, L.J. Studies on the effects of herbicides on soil nitrification, *Weed Research, 10*, 250–263 (1970).
10. Domsch, K.H. and Paul, W. Simulation and experimental analysis of the influence of herbicides on soil nitrification, *Archiv für Mikrobiologie, 97*, 283–301 (1974).
11. Fields, M.L. and Hemphill, D.D. Effect of Zytron and its degradation products on soil microorganisms, *Applied Microbiology, 14*, 724–731 (1966).
12. Fisher, A.J., Appleton, J. and Pemberton, J.M. Isolation and characterization of the pesticide-degrading plasmids pJP_1 from *Alcaligenes paradoxus, Journal of Bacteriology, 135*, 798–804 (1978).
13. Focht, D.D. and Alexander, M. Aerobic cometabolism of DDT analogues by *Hydrogenomonas* sp., *Journal of Agricultural and Food Chemistry, 19*, 20–22 (1971).
14. Forrest, Margaret, Lord, K.A., Walker, N. and Woodville, H.C. The influence of soil treatments on the bacterial degradation of diazinon and other organophosphorus insecticides, *Environmental Pollution, Series A* (1981).
15. Garretson, A.L. and San Clemente, C.L. Inhibition of nitrifying chemolithotrophic bacteria by several insecticides, *Journal of Economic Entomology, 61*, 285–288 (1968).
16. Griffiths, D.C. and Walker, N. Microbiological degradation of Parathion, *Mededelingen Faculteit Landbouw Wetenschappen Gent, 35*, 805–810 (1970).
17. Gunner, H.B. and Zuckerman, B.M. Degradation of diazinon by synergistic microbial action, *Nature*, London, *217*, 1183–1184 (1968).
18. Haider, K. and Jagnow, G. Abbau von ^{14}C-, ^{3}H- and ^{36}Cl-markiertem γ-Hexachlorocyclohexan durch anaerobe Bodenmikroorganismen, *Archives of Microbiology, 104*, 113 (1975).

19. Hankin, L. and Hill, D.E. Proportion of bacteria in agricultural soils able to produce degradative enzymes, *Soil Science, 126*, 40–43 (1978).
20. Helweg, A. Degradation and adsorption of Carbendazim and 2-aminobenzimidazole in soil, *Pesticide Science, 8*, 71–78 (1977).
21. Horvath, R.S. Microbial cometabolism of 2,4,5-trichlorophenoxyacetic acid, *Bulletin of Environmental Contamination and Toxicology, 5*, 537–541 (1970).
22. Horvath, R.S. Cometabolism of the herbicide 2,3,6-trichlorobenzoate, *Journal of Agricultural and Food Chemistry, 19*, 291–293 (1971).
23. Hsu, J.C. and Camper, N.D. Degradation of Ioxynil by a soil fungus, *Fusarium solani, Soil Biology and Biochemistry, 11*, 19–22 (1979).
24. Hsu, T.S. and Bartha, R. Accelerated mineralization of two organophosphorus insecticides in the rhizosphere, *Journal of Applied and Environmental Microbiology, 37*, 36–41 (1979).
25. Jensen, H.L. Decomposition of chloro substituted aliphatic acids by soil bacteria, *Canadian Journal of Microbiology, 3*, 151–164 (1957).
26. Johnen, B.G. and Drew, E.A. Ecological effects of pesticides on soil microorganisms, *Soil Science, 123*, 319–324 (1977).
27. Kaneko, H., Ohkawa, H. and Miyamoto, J. Degradation and movement of permethrin isomers in soil, *Journal of Pesticide Science, 3*, 43–51 (1978).
28. Kaufman, D.D., Plimmer, J.R., Iwan, J. and Klingebiel, U.I. 3,3′-4,4′-Tetrachloroazoxybenzene from 3,4-dichloroaniline in microbial culture, *Journal of Agricultural and Food Chemistry, 20*, 916–919 (1972).
29. Kohnen, R., Haider, K. and Jagnow, G. Investigations on the microbial degradation of Lindane in submerged and aerated moist soil, in *Environmental Quality and Safety, Vol. III* (Edited by F. Coulston and F. Korte), IUPAC 3rd International Congress of Pesticide Chemistry, Helsinki (1974). Thieme Verlag, Stuttgart (1975).
30. Kuwatsuka, S. and Niki, Y. Fate and behaviour of herbicides in soil environments with special emphasis on the fate of principal paddy herbicides in flooded soils, *Review of Plant Protection Research* of the Pesticide Science Society of Japan, Vol. 9, 143–163 (1976).
31. Leadbetter, E.R. and Foster, J.W. Oxidation products formed from gaseous alkanes by the bacterium *Pseudomonas methanica, Archives of Biochemistry and Biophysics, 82*, 491–492 (1959).
32. Liang, T.T. and Lichtenstein, E.P. Synergism of insecticides by herbicides, effect of environmental factors, *Science, 186*, 1128–1130 (1974).
33. MacRae, I.C., Raghu, K. and Bautista, E.M. Anaerobic degradation of the insecticide lindane by *Clostridium* sp., *Nature*, London, *221*, 859–860 (1969).
34. Mathur, S.P., Belanger, A., Hamilton, H.A. and Khan, S.U. Influence on microflora and persistence of field-applied disulfoton, permethrin and

prometryne in an organic soil, *Pedobiologia*, *20*, 237–242 (1980).

35. Matsumura, F. and Benezet, H.J. Microbial degradation of insecticides, pp. 627–628 in *Pesticide Microbiology* (Edited by I.R. Hill and S.J.L. Wright), Academic Press, London, New York, San Francisco (1978).
36. Merkel, G.J. and Perry, J.J. Increased co-oxidative biodegradation of malathion in soil via co-substrate enrichment, *Journal of Agricultural and Food Chemistry*, *25*, 1011–1012 (1977).
37. Mostafa, I.Y., Fakhr, I.M.I., Bahig, M.R.E. and El-Zawahry, Y.A. Metabolism of organophosphorus insecticides. XIII. Degradation of Malathion by *Rhizobium* spp., *Archiv für Mikrobiologie*, *86*, 221–224 (1972).
38. Pfaender, F.K. and Alexander, M. Extensive microbial degradation of DDT *in vitro* and DDT metabolism by natural communities, *Journal of Agricultural and Food Chemistry*, *20*, 842–846 (1972).
39. Ramakrishna, C., Gowda, T.K.S. and Sethunathan, N. Effect of Benomyl and its hydrolysis products, MBC and AB, on nitrification in a flooded soil, *Bulletin of Environmental Contamination and Toxicology*, *21*, 328–333 (1979).
40. Raymond, D.G.M. and Alexander, M. Microbial metabolism and cometabolism of nitrophenols, *Pesticide Biochemistry and Physiology*, *1*, 123–130 (1971).
41. Raymond, R.L., Jamison, V.W. and Hudson, J.O. Microbial hydrocarbon co-oxidation. 1. Oxidation of mono- and di-cyclic hydrocarbons by soil isolates of the genus *Nocardia*, *Applied Microbiology*, *15*, 857 (1967).
42. Rosenberg, A. and Alexander, M. Microbial metabolism of 2,4,5-T in soil, soil suspensions and axenic culture, *Journal of Agricultural Food Chemistry*, *28*, 297–302 (1980).
43. Rosenzweig, W.D. and Stotzky, G. Prodigiosin and the inhibition of *Aspergillus niger* by *Sarratia marcescens* in soil, *Soil Biology and Biochemistry*, *12*, 295–296 (1980).
44. Sethunathan, N. and Pathak, M.D. Development of a diazinon-degrading bacterium in paddy water after repeated applications of diazinon, *Canadian Journal of Microbiology*, *17*, 699–702 (1971).
45. Smith, A.E. Breakdown of the herbicide, Dicamba, and its degradation product, 3,6-dichloro-salicyclic acid, in prairie soils, *Journal of Agricultural and Food Chemistry*, *22*, 601–605 (1974).
46. Smith, A.E. and Cullimore, D.R. The *in vitro* degradation of the herbicide bromoxynil, *Canadian Journal of Microbiology*, *20*, 773–776 (1974).
47. Spokes, J.R. and Walker, N. Chlorophenol and chlorobenzoic acid cometabolism by different genera of soil bacteria, *Archiv für Mikrobiologie*, *96*, 125–134 (1974).
48. Thompson, F.R. and Corke, C.T. Persistence and effects of some chlorinated anilines on nitrification in soil, *Canadian Journal of Microbiology*, *15*, 791–796 (1969).

49. Venkateswarlu, K. and Sethunathan, N. Degradation of Carbofuran in rice soils as influenced by repeated applications and exposure to aerobic conditions following anaerobiosis, *Journal of Agricultural and Food Chemistry, 26*, 1148–1151 (1978).
50. Wainwright, M. and Pugh, G.J.F. The effect of three fungicides on nitrification and ammonification in soil, *Soil Biology and Biochemistry, 5*, 577–584 (1973).
51. Walker, A. The degradation of methazole in soil. 1. Effects of soil type, temperature and soil moisture content, *Pesticide Science, 9*, 326–332 (1978).
52. Walker, N. A soil *Flavobacterium* sp. that degrades sulphanilamide and Asulam, *Journal of Applied Bacteriology, 45*, 125–129 (1978).
53. Walker, N. and Harris, D. Aniline utilization by a soil pseudomonad, *Journal of Applied Bacteriology, 32*, 457–462 (1969).

Some monographs and reference works for further study

Pesticide Microbiology, edited by I.R. Hill and S.J.L. Wright, Academic Press, London, New York, San Francisco, xx + 844 (1978).

The Bipyridinium Herbicides, L.A. Summers, Academic Press, London (1980).

Pesticides in Soil and Water, edited by W.D. Guenzi *et al.*, Soil Science Society of America, Inc., Madison, Wisconsin, USA, xviii + 562 (1974).

Microbial Aspects of Pollution, edited by G. Sykes and F.A. Skinner, Society for Applied Bacteriology Symposium Series No. 1, Academic Press, London and New York, xiii + 289 (1971).

Organic Chemicals in the Soil Environment, edited by C.A.I. Goring and J.W. Hamaker, Marcel Dekker, New York (1972).

Silent Spring, Rachel Carson, Hamish Hamilton, London (1963).

K. AIZAWA

15. Microbial Control of Insect Pests

Although chemical insecticides have been valuable in the control of insect pests of crop plants, their use has posed certain problems. They tend to harm non-target organisms such as man, domestic animals, beneficial insects and wildlife, and their residues not only remain on the crops, but also in the air, water, and soil. Sometimes, insect pests become resistant to chemical insecticides. Due to such complexities arising from the use of chemical insecticides, the use of microorganisms as agents to control insect pests of crops has become very significant in recent years, as one of the promising alternatives.

Quite a number of pathogenic microorganisms on insects, such as viruses, rickettsiae, mycoplasma, bacteria, fungi, and protozoa have been recorded. Insect pathology and insect microbiology have the twin objectives of controlling diseases of beneficial insects such as the silkworm, *Bombyx mori* and the honey bee, *Apis mellifera* and to control insect pests of crops through microorganisms. Microbial preparations and formulations in the insect pest control programme are known as 'microbial insecticides'.

Since microorganisms pathogenic to insects are occurring naturally, microbial insecticides are usually non-toxic to humans, domestic animals, and plants. They are non-pathogenic to non-target insects owing to their relatively high degree of specificity. The selective use of microorganisms towards target insects is advantageous in the conservation of ecosystems. Theoretically, resistance of insects to microbial insecticides is possible but there has been no record of any authentic instances of such resistance.

The idea of the microbial control of insect pests emerged 100 years ago. In fact Metchnikoff and Pasteur were the originators of this idea (55). During the 1940s, the investigations on the microbial control of insect pests began to advance rapidly and the mass production and formulation of microbial preparations as microbial insecticides were initiated in the 1950s. During the past 30 years, new microbial insecticides have been developed and some commercial microbial insecticides have been marketed.

There are some disadvantages in the use of pathogenic microorganisms in the control of insect pests. Firstly, the immediate effectiveness is not seen owing to the incubation period of the microorganism in the body of insect. Secondly, the specificity or the narrow host range of microbial insecticides is disadvantageous in practical insect pest control. Thirdly, the contact infection occurs only in the case of fungal infections and in the case of others, insects should necessarily ingest leaves coated with microbial insecticides before being infected.

In this review, recent advances in microbial insecticides are described with particular reference to the production and use of *Bacillus thuringiensis* as a bacterial insecticide (6, 33, 36, 42, 54).

I. Insect pathogenic microorganisms and microbial control of insect pests

A. VIRUSES

Virus infections have been found in approximately 600 insect species.

Representative insect viruses and their cryptograms (20) in the classification of viruses are shown in Table 1. Among these insect viruses, nuclear-

Table 1. Cryptogram of insect viruses

1. DNA viruses
   a) Baculovirus
      i) Nuclear polyhedrosis: Polyhedra which contain virions formed in the nucleus. D/2: 80/8–15: U_o/(E): I/0
      ii) Granulosis: Granular inclusions are formed mainly in the cytoplasm and one inclusion usually contains one virion. D/2: */10–15: U_o/(E): I/0
      iii) *Oryctes rhinoceros* virus: Inclusion bodies are not formed. D/2: 92/14: U/(E): I/0
   b) Entomopoxvirus: Spheroid and spindle inclusions are formed in the cytoplasm. Virions are contained in the former. A member of poxvirus group. D/2: 140–240/5–6: X/*: I/0
   c) Iridovirus (Iridescent virus): Icosahedral virions are formed in the cytoplasm. D/2: 130/15–20: S/S: I/I
   d) Densonucleosis: A member of parvovirus group. D/1: 1.5–2.2/19–32: S/S: 1/0

2. RNA viruses
   a) Cytoplasmic polyhedrosis: Polyhedra which contain icosahedral virions are formed in the cytoplasm of midgut. A member of reovirus group. R/2: Σ13–Σ18/25–30: S_o/S: I/0
   b) Cricket paralysis virus: A member of picorna virus group. R/1: 2.8–3/30: S/S: I/0
   c) Sigma virus of *Drosophila*: Possible member of rhabdovirus group. */*: */*: U_e/E: I/C

1st pair: Type of nucleic acid/strandness of nucleic acid
2nd pair: Molecular weight of nucleic acid (in millions)/percentage of nucleic acid in infective particles
3rd pair: Outline of particle/outline of nucleocapsid
S=essentially spherical
U=elongated with parallel sides, end(s) rounded
o in first term=virion occluded in viral protein matrix

e in first term=presence of viral envelope
E=elongated with parallel sides, ends not rounded
X=complex or none of above
4th pair: Kinds of host infected/kinds of vector
I=invertebrate
O=spreads without a vector via a contaminated environment
C=congenital
In all instances
*=property of the virus is not known
( )=enclosed information is doubtful or unconfirmed

---

polyhedrosis viruses, cytoplasmic-polyhedrosis viruses, and granulosis viruses have been mainly used for controlling insect pests. Some examples of the successful application of viral insecticides are as follows: In the United States of America, 200 gallons of suspension of nuclear polyhedra (5×10^6/ml) were sprayed in 1949 by aeroplane to control the alfalfa caterpillar, *Colias eurytheme*, over 40 acres and good control was obtained. In Canada, a nuclear-polyhedrosis virus was imported from Sweden and mass-produced polyhedra were sprayed in 1952 (22 gallons over 50 acres, 5×10^6/ml) to control the European sawfly, *Neodiprion sertifer*. Twenty-one days after dissemination of the virus, more than 94 per cent of larvae were infected with the virus. In France, a cytoplasmic-polyhedrosis virus was used in 1958 in the control of the pine processionary moth, *Thaumetopoea pityocampa*. Nine tons of a dust preparation were sprayed over 320 ha by helicopter (1.2×10^{12} polyhedra/ha) and two months later 80 per cent of the larvae were infected with the virus.

B. BACTERIA

There are many bacteria which cause septicaemia in insects by injection, but bacteria which cause infections by feeding are relatively few.

1) *Bacillus thuringiensis*

Berliner (6) proposed the name *Bacillus thuringiensis* for the bacillus which he isolated from the diseased larvae of the Mediterranean flour moth, *Anagasta* (*Ephestia*) *kühniella* in 1911. But in reality, *B. thuringiensis* had earlier been discovered by Ishiwata in 1901 from a severe flacherie of the silkworm, *Bombyx mori*. Ishiwata (33) named it "sotto" (fainting or sudden-collapse) bacillus although his taxonomical description of the causative bacillus of the "sotto" disease was incomplete. Mattes (42) reisolated the same bacillus from the same insect host which Berliner had found earlier. This strain is now maintained as *B. thuringiensis* subsp. *thuringiensis* (serotype 1).

B. thuringiensis is a sporeformer and is closely related to *B. cereus*. The important characteristic of *B. thuringiensis* is the production of a toxic crystalline inclusion (parasporal body, toxic crystal, δ-endotoxin) while *B. cereus* does not produce any such inclusion.

The classification of *B. thuringiensis* has been done by the flagellar agglutination which demonstrates the existence of flagellar antigen (H-antigen). Previously, the classification was done by the combination of H-antigens (serotypes) (12) and biotypes, particularly the esterase types (45). However, recently the name of a new subspecies of *B. thuringiensis* has been proposed based on the H-antigens. The classification of *B. thuringiensis* is shown in Table 2.

Table 2. Classification of *Bacillus thuringiensis*

Serotype (H-antigen)		Esterase type	Subspecies	Production of β-exotoxin in reference strain
H1		Berliner	subsp. *thuringiensis*	+
H2		Finitimus	subsp. *finitimus*	−
H3	3a	Alesti	subsp. *alesti*	−
	3a : 3b	Galleriae	subsp. *kurstaki*	−
H4	4a : 4b	Sotto	subsp. *sotto*	−
	4a : 4b	Dendrolimus	subsp. *dendrolimus*	−
	4a : 4c	Kenya	subsp. *kenyae*	−
H5	5a : 5b	Galleriae	subsp. *galleriae*	−
	5a : 5c	*	subsp. *canadensis*	−
H6		Entomocidus	subsp. *entomocidus*	−
			subsp. *subtoxicus*	−
H7		Galleriae	subsp. *aizawai*	−
H8	8a : 8b	Morrison	subsp. *morrisoni*	+
	8a : 8c	Ostrinia	subsp. *ostriniae*	−
H9		Tolworth	subsp. *tolworthi*	+
H10		Darmstadt	subsp. *darmstadiensis*	+
H11	11a : 11b	Toumanoff	subsp. *toumanoffii*	+
	11a : 11c	*	subsp. *kyushuensis*	−
H12		Thompson	subsp. *thompsoni*	−
H13		Pakistan	subsp. *pakistani*	−
H14		Israel	subsp. *israelensis*	−
H15		*	subsp. *dakota*	*
H16		*	subsp. *indiana*	*
H17		*	subsp. *tohokuensis*	−
H18		*	subsp. *kumamotoensis*	+
H19		*	subsp. *tochigiensis*	−
—		Galleriae	subsp. *wuhanensis*	−

*Undecided

a) INSECTICIDAL TOXINS PRODUCED BY *B. thuringiensis*. *B. thuringiensis* produces several toxins such as α-, β-, γ-exotoxins, δ-endotoxin (27), and louse-factor (24); however, β-exotoxin and δ-endotoxin are important from the viewpoint of microbial insecticides.

The β-exotoxin was first reported by McConnell and Richards (43) and is composed of adenine, ribose, glucose, and allaric acid with a phosphate group

(17, 35). The molecular structure has been proposed by Farkaš *et al.* (17). The β-exotoxin is produced by some but not all strains of *B. thuringiensis* and the production is a strain-specific property rather than a serotype (subspecies)-specific property (46). The β-exotoxin is known as "thermostable exotoxin" or "fly toxin" and is toxic not only to insects but also to mammals and plants. The presence of β-exotoxin in any commercial formulation of *B. thuringiensis* is forbidden in the United States of America.

Usually one toxic crystal is formed in one sporangium and the shape is bipyramidal, although the size and shape of toxic crystals vary depending on the bacterial strains. Spindle-shaped inclusions are frequently observed and either cubic or irregular-shaped inclusions have been reported.

Although the δ-endotoxin is composed of a glycoprotein subunit (8), the toxic crystal is not a single entity and toxic polypeptides are considered as δ-endotoxin. The proteinaceous toxin shows toxicity by feeding and not by injection. However, toxic polypeptides which show toxicity both by injection and by feeding have been reported. The toxicity of δ-endotoxin to a given insect varies among *B. thuringiensis* strains.

Lepidopterous insects show different responses to δ-endotoxin (27, 28). Type I insects exhibit general paralysis after ingestion of the δ-endotoxin accompanied by an increase in the alkalinity of the hemolymph, a fall in the intestinal pH, and destruction of the epithelium. The silkworm, *B. mori* is a typical insect of this type. Type II insects suffer gut paralysis shortly after feeding on the δ-endotoxin but there is no leakage from the gut into the hemolymph. Accordingly, changes in blood pH and general paralysis are not observed. The insects starve for want of food and quickly succumb to the bacteria invading their bodies and die in two to four days. Many lepidopterous insects belong to this type. Type III insects die due to the ingestion of both spores and δ-endotoxins. The Mediterranean flour moth, *A. kühniella* and the gypsy moth, *Porthetria dispar* belong to this type.

A number of workers have reported on the nature of the subunit protein from toxic crystals. The molecular weight of the subunit protein varies depending on the method of the dissolution of crystals (alkaline solution and the addition of reducing agents or enzymes) and on different *B. thuringiensis* strains. The protein subunits having molecular weights of 230,000, 120,000 to 140,000, and 40,000 to 80,000 have been reported (8, 31, 38, 40). Toxic fragments of quite low molecular weight have also been reported (18, 19).

A polypeptide of molecular weight 40,000 has been synthesized in a cell-free system and this has reacted with the rabbit antiserum against toxic crystal proteins (50). Antigen analysis of crystals has been done and the relationship between the crystal serotype and insecticidal activity has been reported (37).

The number of reports on the correlation between the presence of plasmid DNA and the formation of toxic crystals is increasing (5, 9, 41, 51, 53). Recently, the formation of crystals in the protoplast of *B. subtilis* by

B. thuringiensis plasmid has been reported. Furthermore, a toxic protein was formed in *E. coli* by recombinant DNA techniques (51). Partially digested fragments of subsp. *kurstaki* (HD-1) plasmid DNA by *Sau* 3Al were ligated into the *Bam* Hl site of pBR 322 and were transformed into *E. coli* strain HB 101. A protein which reacted with the anti-rabbit serum against the crystal proteins had the same electrophoretic mobility as the dissolved crystals. Protein extracts from the transformant were toxic to larvae of the tobacco hornworm, *Manduca sexta*. On the other hand, transduction experiments in *B. thuringiensis* using bacteriophage have been reported (39, 49, 56).

b) DISTRIBUTION OF *B. thuringiensis* IN NATURE. *B. thuringiensis* has been frequently isolated from dead insects, litters of sericultural farms, and soils. This indicates the wide distribution of the bacterium in nature. For example, in Japan serotype 4a : 4b (subsp. *sotto* and subsp. *dendrolimus*) and serotype 8 (subsp. *morrisoni*) are distributed widely in the sericultural farms. Serotype 3a (subsp. *alesti*) and serotype 7 (subsp. *aizawai*) have a more restricted distribution. Serotype 1 (subsp. *thuringiensis*), serotype 3a : 3b (subsp. *kurstaki*), serotype 4a : 4c (subsp. *kenyae*), and serotype 10 (subsp. *darmstadiensis*) have been occasionally isolated (1). However, there are only a few reported cases of epizootics in lepidopteran forest insects and in stored product insects.

c) MULTIPLICATION OF *B. thuringiensis* IN THE SOIL AND DRIFT EXPERIMENTS IN MULBERRY PLANTATION. If the reproduction of spores and toxic crystals will occur when *B. thuringiensis* multiplies in the soil, damage will be caused. Multiplication of *B. thuringiensis* in the soil was investigated when bacterial preparations were applied singly. The cell number decreased gradually depending on the nature of the soil and *B. thuringiensis* strains. No accumulation of the bacterium was observed by repeated applications of preparations (1). Drift experiments with *B. thuringiensis* preparation were carried out. The cell numbers of *B. thuringiensis* in the soil of mulberry plantation and that in mulberry leaves were estimated. The results showed that mulberry leaves taken at a distance of 70–100 m from the sprayer used for spraying *B. thuringiensis* did not affect silkworm rearing (1).

d) INSECT PESTS IN SOUTHEAST ASIA SUSCEPTIBLE TO *B. thuringiensis*. Preparations of *B. thuringiensis* showed effectiveness on the following insect pests in Southeast Asia: common cabbageworm, *Pieris rapae crucivora*, diamondback moth, *Plutella xylostella*, fall webworm, *Hyphantria cunea*, oriental tobacco budworm, *Helicoverpa assulta*, persimmon fruit moth, *Stathmopoda masinissa*, summer fruit tortrix, *Adoxophyes orana fasciata*, rice skipper, *Parnara guttata*, beet semi-looper, *Plusia nigrisigna*, Japanese tussock moth, *Orgyia thyellina*, sweetpotato leaf worm, *Aedia leucomelas*, sweetpotato leaf folder, *Brachmia triannulella*, persimmon cochlid, *Scopelodes contracta*, tea tussock moth, *Euproctis pseudoconspersa*, pine caterpillar, *Dendrolimus spectabilis*, tobacco cutworm, *Spodoptera litura*, cabbage armyworm, *Mamestra brassicae*, apple tortrix, *Archips fuscocupreanus*, pear leafminer, *Bucculatrix pyrivorella*, gypsy moth, *Lymantria dispar japonica*, corn earworm, *Heliothis armigera*, oriental corn borer, *Ostrinia furnacalis*, orange dog butterfly, *Papilio demoleus malayanus*, rice leafroller, *Cnaphalocrocis medinalis*, green rice caterpillar, *Naranga aenescens*, soybean pod borer, *Leguminivora glycinivorella*, cabbage webworm, *Hellula undalis*, brown tail moth, *Euproctis similis*, cone pyralid, *Dioryctria abietella*, cherry hornworm, *Smerinthus planus*, and black riceworm, *Aglossa dimidiata*.

2) *Bacillus popilliae* and *Bacillus lentimorbus*

The Japanese beetle, *Popillia japonica*, causes extensive damage in the eastern United States to more than 275 kinds of trees, shrubs, and turf grass. The adult beetle feeds on fruit, flowers, and foliage. The grub feeds on roots and underground stems of plants and causes serious damage to grasses and lawns. The grubs are subject to several diseases and the most important one is the milky disease (16) caused by two types of bacterial species, *Bacillus popilliae* and *Bacillus lentimorbus*.

B. popilliae is the causative bacterium of type A milky disease and forms a parasporal (refractive) body in the sporangium. *B. lentimorbus* is the causative bacterium of type B milky disease and does not produce the parasporal body and the sporangium is more spindle-shaped than that of *B. popilliae*. About 70 insect species are susceptible to *B. popilliae* by injection but some resist oral infection.

Spores in the infested larva germinate in the gut and the vegetative cells invade into the hemocoel where they multiply. The sporulation of *B. popilliae* occurs easily in insect hemocoel, although no satisfactory method has been established to mass produce spores *in vitro*. Accordingly, the spores of *B. popilliae* are mass produced by the injection of spores into the hemocoel of larvae. The parasporal body of *B. popilliae* is not readily separated from the sporangium. The parasporal bodies are proteinaceous and it was shown that the solubilized parasporal bodies are toxic to insects by injection.

3) *Bacillus sphaericus*

B. sphaericus is an acrystalliferous bacterium and at least 10 species of mosquitoes belonging to the genera *Culex*, *Anopheles*, and *Psorophora* are susceptible to this bacillus. *B. sphaericus* is less active than *B. thuringiensis* subsp. *israelensis* (serotype 14) on *Aedes* but the former is more active on *Anopheles albimanus*. Recently a highly pathogenic strain of *B. sphaericus* against *Culex* has been isolated. *B. sphaericus* cells are digested in the gut of mosquito larva and an insecticidal component associated with the bacterial cell wall is released which is toxic to the larva. Death of the larva occurs within eight to twelve hours and non-digested *B. sphaericus* cells grow in the dead larva. The extraction of the *B. sphaericus* toxin was attempted and LC_{50} of purified cell-wall fraction to *Culex fatigans* was 6 ng/ml. Toxicity was annulled by boiling and by 0.01 N NaOH treatment. Mammalian toxicity of *B. sphaericus* has not been shown (52).

4) *Bacillus moritai*

B. moritai closely resembles *B. cereus* but differs from the latter in some bacteriological features. The results of screening showed that this sporeforming bacterium was pathogenic to the housefly, *Musca domestica* (4, 23). A preparation from this bacterium was exempted from the requirement of tolerance by the Japanese Government in 1973.

5) *Bacillus larvae*

B. larvae is the causative agent of the American foulbrood of the honey bee.

6) *Rickettsiae*

Insect pathogenic rickettsiae have been found from Coleoptera, Orthoptera, and Lepidoptera (*Rickettsiella popilliae*, *Rickettsiella melolonthae*, *Rickettsiella grylli*, *Rickettsiella tipulae*, etc.).

C. FUNGI

Insects are infected by fungi through the body surface and this property is different from the infections caused by viruses, bacteria, and protozoa. Conidia attached on the insect integument germinate and the germination tubes penetrate the insect body under optimum temperature and humidity. The fungus proliferates in the insect body which is covered with mycelia and conidia. The newly formed conidia are dispersed and cause subsequent infection. This property proves to be very advantageous in the microbial control of insect pests.

The generic names of insect pathogenic fungi are shown in Table 3 and successful control of insect pests has been obtained with the following entomogenous fungi (21).

Beauveria bassiana: Colorado potato beetle, *Leptinotarsa decemlineata* 1–1.5 kg of Beauverin (or Boverin) (fungus preparation; 3×10^{10} conidia/g)/ha [Soviet Union]

B. bassiana: Codling moth, *Carpocapsa pomonella* 1 kg of Boverin (2.4×10^{10} conidia/g)/ha [Soviet Union]

Nomuraea rileyi: Tobacco budworm, *Heliothis virescens* 1.1×10^{13} conidia/acre (1.2×10^{11} conidia/g) [U.S.A.]

Aschersonia placenta: Citrus whitefly, *Dialeurodes citri* 10^6–10^8 spores/ml; 900–1,000 l/ha [Soviet Union]

Hirsutella thompsoni: Citrus rust mite, *Phyllocoptruta oleivora* [U.S.A.]

Metarhizium anisopliae: *Zulia entreriana*, *Aeneolomia selecta selecta* (pasture), *Mahanarva posticata* (sugar cane) [Brazil]

M. anisopliae: Rhinoceros beetle, *Oryctes rhinoceros* [Tonga]

Aschersonia sp.: Greenhouse whitefly, *Trialeurodes vaporariolum* 5×10^6–8×10^7 spores/ml; 20–30 l/a [Soviet Union]

Verticillium lecanii: Brown soft scale, *Coccus hesperidum* [Czechoslovakia]

Beauveria tenella (=*B. brongniartii*): *Melolontha melolontha* 2×10^{10} conidia/m^2 [France]

The following fungi have been successfully used to control insect pests in Japan (1).

B. tenella: Soybean beetle, *Anomala rufocuprea*, Larger striated chafer, *Anomala costata*

M. anisopliae: Black rice bug, *Scotinophara lurida*

Paecilomyces fumosoroseus: Peach fruit moth, *Carposina niponensis*

Paecilomyces farinosus: Citrus ground mealy bug, *Rhizoecus kondonis*

Aschersonia aleyrodis: Citrus whitefly, *Dialeurodes citri*

Paecilomyces lilacinus: Potato tuberworm, *Phthorimaea opeculella*

Table 3. Fungi pathogenic to insects

Mastigomycotina
- Chytridiomycetes
  - Blastocladiales
    - *Coelomomyces*

Zygomycotina
- Zygomycetes
  - Entomophthorales
    - *Entomophthora*
    - *Massospora*

Ascomycotina
- Pyrenomycetes
  - Sphaeriales
    - *Cordyceps*
    - *Podonectria*
    - *Torrubiella*

Deuteromycotina
- Coelomycetes
  - *Aschersonia*
- Hyphomycetes
  - *Aspergillus*
  - *Beauveria*
  - *Fusarium*
  - *Hirsutella*
  - *Metarhizium*
  - *Nomuraea*
  - *Paecilomyces*

There are several disadvantages in the utilization of entomogenous fungi.

Firstly, it is not easy to estimate the potency of the fungal preparation not only by bioassay but also by conidia counts. Secondly, insects are not attacked by the fungus at higher temperature, sometimes even at 30°C. Thirdly, the stability of the fungal preparation is not good. Fourthly, the production of mycotoxins is not well known in some cases and the safety of fungal preparations should be investigated in greater detail.

The problems of infection at higher temperature and the stability of fungal preparation can be solved by the selection and improvement of strains.

D. PROTOZOA

Approximately 210 species of protozoa pathogenic on insects have been recorded and most potential candidates in microbial control belong to the family Nosematidae. Among them, one protozoan, *Nosema locustae* has been experimentally used in the control of grasshoppers (29, 30) and several species of crickets. *Nosema bombycis* is well known as a causative agent of pébrine of the silkworm.

E. NEMATODES

Nematodes parasitic on insects have been recorded in the case of black flies, mosquitoes, chironomids, rootworms, grubs, caterpillars, grasshoppers, and bark beetles. The two nematodes which have potential for possible commercial formulations are *Neoaplectana carpocapsae* (DD-136) which kills over 1,000 species of insects and *Romanomermis culicivorax* (= *Reesimermis nielseni*) which kills over 60 species of mosquitoes.

II. Microbial insecticides

A. BACTERIAL PREPARATIONS

1) *Bacillus thuringiensis preparation*

Representative *B. thuringiensis* preparations have been produced in the United States of America, France, and the Soviet Union. In the U.S.A., *B. thuringiensis* preparations are exempted from tolerance requirements, and are applied at recommended doses even up to the day of plant harvest.

a) FORMULATION. *B. thuringiensis* preparations are available as wettable powder, dusts, and water-dispersible emulsions. The active ingredient is *B. thuringiensis* and the potency of preparation is expressed as an international unit (IU)/mg based on insect bioassay data.

b) DETERMINATION OF POTENCY. Standardized bioassay for the formulation of *B. thuringiensis* preparation has been established and the potency of the product is calculated by the following formula (10, 15):

$$\text{Potency of sample} = \frac{LD_{50}\ \text{standard}}{LD_{50}\ \text{sample}} \times \text{IU/mg standard}$$

International Standard E-61	1,000 IU/mg
U.S. Standard (HD-1-S-1971)	18,000 IU/mg
New U.S. Standard (HD-1-S-1980)	16,000 IU/mg

c) BIOASSAY OF FORMULATION OF *B. thuringiensis* IN JAPAN. In Japan, at least four *B. thuringiensis* serotypes, which vary from low to high toxicity to the silkworm, are used in the formulation of preparations and there exists a mixture of serotypes. This situation has caused much difficulty in the standardization of *B. thuringiensis* preparations. The Study Committee on *Bacillus thuringiensis* Preparations of the Plant Protection Association of Japan has investigated the methods of evaluation of the potency of *B. thuringiensis* preparations and adopted the following procedures for the regulation of the potency of preparations (1, 2, 3).

i) Test insect: Silkworm, *B. mori*, strain Shunrei × strain Shogetsu, 3rd instar, 2nd day. Silkworm larvae are reared on an artificial diet at 24–25°C.

ii) Procedure for bioassay: One g of *B. thuringiensis* preparation is mixed with 40 ml of distilled water and blended in a waring blender at 12,000 rpm for one minute. Dilution will depend on the nature of the test sample or standard preparation, but is usually 1⅓–2 times. Ten g of artificial diet and 0.5 ml of diluted sample are mixed thoroughly with a spoon. The diet incorporated sample is spread 2–3 mm thick, in a 9–12 cm Petri dish. Thirty larvae (10 larvae × three dishes) are used for each dilution of the sample or standard preparation. Larvae are reared for three days on a diet inoculated with the *B. thuringiensis* preparation and for an additional two days on a normal diet. Mortalities are recorded and plotted on a log-probability paper. If more than two out of 30 control larvae die, the results are discarded.

iii) Calculation of results: Each formulator should prepare a "self reference" and should deposit it with the Agricultural Chemicals Inspection Station, Ministry of Agriculture, Forestry and Fisheries. Each "self reference" has 1,000 BmU (*Bombyx mori* Unit)/mg, irrespective of toxicity to the silkworm. The potency of the sample is calculated by the following formula:

$$\text{Potency of sample} = \frac{LC_{50}\ (\mu g/ml)\ \text{"self reference"}}{LC_{50}\ (\mu g/mg)\ \text{sample}} \times 1{,}000\ \text{BmU/mg.}$$

The potency should remain between 85 and 200 per cent of the indicated BmU of the sample. Each "self reference" will be expressed as IU/mg using E-61 or U.S. Standard, if necessary. This procedure is used for the regulation of the potency of preparations only.

d) SELECTION AND STRAIN IMPROVEMENT OF *B. thuringiensis*. In the U.S.A., previously subsp. *thuringiensis* was used for mass production. However, a strain which showed higher toxicity to insects other than subsp. *thuringiensis* was selected. This strain belonged to subsp. *kurstaki* and was designated as HD-1 (13). The strain HD-1 replaced the subsp. *thuringiensis* for *B. thuringiensis* preparations. Furthermore, an effective strain (HD-187) was selected (14). Recently a strain which belongs to subsp. *aizawai* is being mass produced to control the greater wax moth, *Galleria mellonella*. The potency is expressed as GMU (*Galleria mellonella*) unit/mg.

A strain which showed high toxicity against mosquito larva was isolated from Israel (25). This strain was found to belong to a new serotype and was named as subsp. *israelensis* (serotype 14) (11). Subsp. *israelensis* is quite effective for the control of larvae of mosquitoes (*Anopheles*, *Culex*, *Aedes*), the blackfly (26) and the other aquatic dipteran insects. Experimental applications in the field have been successful. The method for the bioassay of the bacterial preparation has been established. A standard preparation (IPS-78) has 1,000 toxic units per mg using *Aedes aegypti*.

Subsp. *israelensis* is highly toxic to mosquito larvae but not toxic to lepidopterous larvae. In this connection, based on the preferential toxicity, it was shown that strains belonging to subsp. *darmstadiensis* (serotype 10) were divided into three groups from the viewpoint of toxicity against lepidopterous and mosquito larvae. The reference strain (from Institut Pasteur) was toxic to the lepidopterous insects but nontoxic to mosquito larvae. The second group shows no toxicity on lepidopterous insects but is toxic to mosquito larvae. The third group produced large irregular-shaped parasporal inclusions nontoxic to larvae of lepidopterous insects and mosquitoes (48).

In sericultural countries, particularly in Japan, it is desirable to have microorganisms with low pathogenicity to the silkworm, and at the same time being highly pathogenic to insect pests. Along this line of thinking, the selection and breeding of *B. thuringiensis* strains with low toxicity for the silkworm have been tried. One strain was designated as AF 101 which belongs to serotype 4a : 4b (subsp. *sotto*) and a product of this strain is now utilized for experimental use in the control of the common cabbageworm, *Pieris rapae crucivora*, the fall webworm, *Hyphantria cunea*, and the diamondback moth, *Plutella xylostella*.

The tobacco cutworm, *Spodoptera litura*, is an important injurious insect for vegetables and pastures in Japan and it is resistant to ordinary *B. thuringiensis* strains. However, subsp. *aizawai* (serotype 7) and subsp. *kurstaki* (serotype 3a : 3b) showed toxicity on the tobacco cutworm. The selection of subsp. *aizawai* strains was done and a strain designated as AY was obtained. A product of this strain is now used for experimental use.

Table 4 shows the result of the estimation of potency which is expressed in IU/mg of self references of AF 101 and AY strains on the silkworm, the fall webworm, and the tobacco cutworm (3).

The development and success of microbial insecticides have been dependent upon the selection of effective pathogenic microorganisms in the control of insect pests. However, at present, not only the selection but also the

Table 4. Potencies of *Bacillus thuringiensis* preparations

Preparation	Subspecies	Spore count/g	IU*/mg Silkworm (*Bombyx mori*)	Fall webworm (*Hyphantria cunea*)	Tobacco cutworm (*Spodoptera litura*)
US Standard (HD-1-S-1971)	subsp. *kurstaki*	3.00×10^{10}	18,000	18,000	18,000
AY Reference (1,000 BmU**/mg)	subsp. *aizawai*	2.08×10^{10}	28,000	6,800	38,500
AF 101 Reference (1,000 BmU**/mg)	subsp. *sotto*	2.63×10^{10}	95	12,600	0

*International unit.
***Bombyx mori* (silkworm) unit.

breeding of effective strains should be emphasized and encouraged. One investigation, the breeding of a *B. thuringiensis* strain which does not form spore (S^-) and forms toxic crystal (C^+) is noteworthy. S^-C^+ strains were obtained by treatment with N-methyl-N'-nitro-N-nitrosoguanidine (NTG). Furthermore, in addition to NTG treatment, the selection was carried out by the combination of resistance to rifampicin and *ts* mutant. This strain ($S^{ts}C^+$) forms spores and toxic crystals at 27°C and forms only toxic crystals at 37°C. In Japan, one producer of microbial insecticides is producing spore-killed preparation by chemical treatment taking into consideration the effect of living spores contained in *B. thuringiensis* preparations on silkworm rearing. The preparation and use of $S^{ts}C^+$ strain as a microbial insecticide should be further improved by investigation which will contribute eventually to the development of breeding of effective strains (1).

2) *Bacillus popilliae preparation*

A commercial preparation of *B. popilliae* known as 'Doom' contains not less than 10^8 viable spores of either or both *B. popilliae* or *B. lentimorbus* per gram of preparation. One pound of powder will treat 4,000 square feet of lawn grass and one level teaspoon of Doom will have to be applied on the grass in spots which are four feet apart and in rows which are also four feet apart.

B. VIRUS PREPARATIONS

In the U.S.A., three nuclear-polyhedrosis virus (NPV) preparations have been registered. These NPVs are *Heliothis* NPV (trade name of the preparation: Elcar), *Lymantria dispar* NPV (trade name of the preparation: Gypcheck) and *Orgyia pseudotsugata* NPV (trade name of the preparation: Biocontrol-1).

Elcar is a commercial name of a NPV preparation in the control of the tobacco budworm, *Heliothis virescens*, the cotton bollworm, *Heliothis zea*, and other *Heliothis* species (*H. armigera*, *H. phloxiphaga*, *H. punctigera*, and *H. obtectus*). At least 4×10^9 PIB (polyhedral inclusion bodies = nuclear polyhedra) as active ingredient are contained per gram of the preparation (wettable powder). Elcar should be applied first at the time of substantial egg deposition or on the occurrence of newly hatched larvae and repeated as long as egg deposition continues at three to seven-day intervals. Elcar is applied at the rate of 150 g/ha on the ground or by aerial equipment.

Gypcheck is used for the control of the gypsy moth and Biocontrol-1 is used in the control of the Douglas-fir tussock moth, *Orgyia pseudotsugata*.

In North America, the following NPVs are currently under development by industry and institutions: alfalfa looper, *Autographa californica* NPV for *Spodoptera*, *Trichoplusia*, *Heliothis*, *Estigmene*, and *Plutella*; cabbage looper, *Tricoplusiani* NPV for cabbage looper; red-headed pine sawfly, *Neodiprion leconti* NPV for red-headed pine sawfly; European pine sawfly, *Neodiprion sertifer* NPV for European pine sawfly; spruce budworm, *Choristoneura fumiferana* NPV for spruce budworm; codling moth, *Laspeyresia pomonella* granulosis virus for codling moth.

In Japan, the utilization of a cytoplasmic-polyhedrosis virus of the pine caterpillar, *Dendrolimus spectabilis*, has been investigated since 1960 by the Government Forest Experiment Station, Ministry of Agriculture, Forestry and Fisheries. Polyhedron suspension (10^6 polyhedra/ml) was sprayed on the branches of pine trees and were fed with seventh instar larvae. About two weeks after the virus inoculation, dead and virus-infected larvae were collected and homogenized. Crude polyhedra were obtained by filtration and centrifugation. The number of polyhedra formed in a larva was about 5×10^8. Clay and white carbon were added to crude polyhedra and pulverized. Good success in the control of the pine caterpillar was obtained by the application of 10^{11} polyhedra per hectare (34).

Safety of the cytoplasmic-polyhedrosis virus was tested upon mice, rabbits, and hamsters by feeding and intracerebral, intravenous, intramuscular, and subcutaneous injections of polyhedra. A product (commercial name: Matsukemin) was registered in April, 1974.

The tobacco cutworm, *Spodoptera litura* was successfully controlled in the field using a nuclear-polyhedrosis virus preparation (47). The artificial diet containing nuclear polyhedra (5×10^7 PIB/g) was fed to the middle fifth instar larvae and all larvae died in the sixth instar. 8×10^8 polyhedra on an average were formed in one sixth instar larva. More than 50,000 infected larvae could be produced by two persons in five weeks. The larval mortality by the ultra low volume spraying (2×10^8 PIB/ml, 0.5 1/10a) was higher than those by ordinary spraying (100 1/10a) or mist spraying (20 1/10a) using the same amount of polyhedra.

The procedure for the estimation of potency of the nuclear-polyhedrosis virus preparation was investigated and the method to calculate the potency of a sample by the following formula was established

$$\text{Potency of test sample} = \frac{\text{LD}_{50}\text{ value of standard}}{\text{LD}_{50}\text{ value of test sample}}\ (\text{SLU*}).$$

C. FUNGAL PREPARATION

In the Soviet Union, a preparation of *Beauveria bassiana* (name of product: Beauverin or Boverin) is used for the control of the codling moth, *Carpocapsa pomonella*, and the Colorado potato beetle, *Leptinotarsa decemlineata*, at a rate of 1–2 kg/ha.

A commercial product of *Hirsutella thompsoni* as a wettable powder is being marketed in the U.S.A. For citrus, 2–4 lb of the product per acre have been recommended to be applied in the control of the citrus rust mite, *Phyllocoptruta oleivora*.

D. PROTOZOAN PREPARATION

In the U.S.A., based on the lack of pathogenicity of *Nosema locustae*, its preparation has been temporarily exempted from tolerance and is experimentally used for the control of grasshoppers. The technique of mass production of *N. locustae* has been improved. Each infected grasshopper contains approximately 3×10^9 spores which are enough to treat 1.2 ha (7).

E. NEMATODE PREPARATION

Romanomerimis culicivorax has been mass produced and sold (commercial name of preparation: Skeeter Doom) in the U.S.A. for mosquito control. A 700 gram package can treat 1,000 square feet of pond (44).

III. Safety of microbial insecticides

Commercial preparations of *B. thuringiensis* and *Heliothis* nuclear-polyhedrosis virus are exempted from residue tolerance in the U.S.A. (22, 32) and the following safety tests on microbial insecticides have been carried out:

Acute toxicity-pathogenicity (oral, intravenous, and intraperitoneal toxicities) to mammals and birds

Virulence and persistence in blood

Inhalation

**Spodoptera litura* unit.

Dermal, skin, and eye toxicity
Chronic toxicity-pathogenicity
Human volunteer experiment
Effect on fishes and wildlife
Mutation and selection for pathogenicity
Phytotoxicity

IV. Microbial control of insect pests of rice

Rice is the major agricultural product in the Asian countries and the importance of rice as a world food crop is widely recognized. It is necessary to develop the methods of the pest control of rice and the present status of investigations on the microbial control of insect pests of rice is included in this review.

Major insect pests of rice are as follows: planthoppers (brown planthopper, *Nilaparvata lugens*, white-backed planthopper, *Sogatella furcifera*, smaller brown planthopper, *Laodelphax striatellus*), green rice leafhopper, *Nephotettix cincticeps*, yellow rice borer, *Tryporyza incertulas*, rice gall midge, *Orseolia oryzae*, rice caseworm, *Nymphula depunctalis*, rice leaf folder, *Cnaphalocrosis medinalis*, *rice bugs* (*Leptocorisa acuta*, *Leptocorisa oratorius*), rice water weevil, *Lissorhoptrus oryzophilus*. At present, the brown planthopper is either the most or one of the most important insect pests to rice production in Asia.

Several entomopathogenic fungi have been isolated from planthoppers, leafhoppers, and rice bugs. *Conidiobolus coronatus* infections in the brown planthopper and the green rice leafhopper are usual in the rearing cages in the tropical Asian countries and an epizootic by *Entomophthora delphacis* in the brown planthopper was observed in Japan. Some insect pests of rice are highly susceptible to intrahaemocoelic inoculation of *Chilo* iridescent virus originally found in the rice stem borer, *Chilo suppressalis*. Some nematodes have been found in the brown planthopper, the white-backed planthopper, and the rice stem borer.

A few experimental disseminations of microbial control agents have been done and the black rice stink bug, *Scotinophara lurida*, was successfully controlled by *Metarhizium anisopliae* and *Paecilomyces lilacinus*. The rice-plant skipper, *Parnara guttata* can be controlled by *B. thuringiensis* preparation in the paddy field.

Although the present knowledge on the microbial diseases of major insect pest of rice and of field trials with microbial control agents is not enough, the use of fungi as microbial insecticides is promising.

Among entomopathogenic fungi such as *Beauveria bassiana*, *Beauveria tenella*, *Paecilomyces farinosus*, *Paecilomyces fumosoroseus*, *Nomuraea riley*, it was found that the most pathogenic fungus to the green rice leafhopper was

B. bassiana. The existence of three serotypes in *B. bassiana* has been demonstrated by the agglutination of blastospores. Generally, serotypes 1 and 2 are more virulent than serotype 3 on the silkworm, *Bombyx mori*. On the contrary, serotype 3 is more virulent than serotype 2 on the green rice leafhopper. Serotype 2 has been isolated from the silkworm, although serotype 3 was isolated from the brown planthopper and the green rice leafhopper but not from the silkworm (Aizawa *et al*., unpublished).

V. Future development and research requirements

For the development of microbial insecticides, the importance and the necessity for the selection of effective strains, microbial genetics including genetic manipulation in insect microbiology, and the inspection of microbial insecticides from the point of view of fermentation technology should be emphasized. The success in these areas will definitely contribute to the development of microbial insecticides.

Research topics for the development of microbial insecticides are as follows:

1. Target insects
   Decision of target insects
   Mass rearing of target insects
2. Selection and strain improvement of microorganisms
   Selection of effective microorganisms
   Screening of effective strains
   Microbial breeding and genetic manipulation for strain improvement
3. Effectiveness
   Laboratory test of microorganisms
   Field test for insect pest control
4. Safety of microorganisms
5. Effects of microorganisms on environment
   Effect of microorganisms on non-target organisms
   Microbial ecology
   Check of development of resistance to microorganisms in insects
6. Mass production of microorganisms
7. Formulation and quality control of microbial insecticides
   Formulation
   Bioassay for the measurement of potency
   Stability
8. Application technology
   Rationalization of application
   Countermeasures for any problems caused by microbial insecticides

References

1. Aizawa, K. Microbial insecticides, pp. 382–387 in *Plant Protection in Japan, 1976* (*Agriculture Asia, Special Issue No. 10*) (1976).
2. Aizawa, K. Recent development in the production and utilization of microbial insecticides in Japan, *Proceedings of the First International Colloquium on Invertebrate Pathology and IXth Annual Meeting*, Society for Invertebrate Pathology, Queen's University, Kingston, pp. 59–63 (1976).
3. Aizawa, K., Fujiyoshi, N., Ohba, M. and Yoshikawa, N. Selection and utilization of *Bacillus thuringiensis* strains for microbial control, *Proceedings of the First International Congress of IAMS, Vol. 2* (Edited by T. Hasegawa), pp. 597–606 (1975).
4. Aizawa, K., Kawamura, A., Fujiyoshi, N. and Maebashi, H. Human feeding test using the bacterial insecticide—*Bacillus moritai*, *Japanese Journal of Hygiene*, *29*, 275–280 (1974).
5. Alikhanian, S.I., Ryabchenko, N.F., Bukaonov, N.O. and Sankanyan, V.A. Transformation of *Bacillus thuringiensis* subsp. *galleria* protoplasts by plasmid pBC16, *Journal of Bacteriology*, *146*, 7–9 (1981).
6. Berliner, E. Über die Schlaffsucht der Mehlmottenraupe (*Ephestia kühniella* Zell.) und ihren Erreger *Bacillus thuringiensis*, n. sp., *Zeitschrift für Angewandte Entomologie*, *2*, 29–56 (1915).
7. Brooks, W.M. Production and efficacy of protozoa, *Biotechnology and Bioengineering*, *22*, 1415–1440 (1980).
8. Bulla, L.A., Jr., Kramer, K.J. and Davidson, L.I. Characterization of the entomocidal parasporal crystal of *Bacillus thuringiensis*, *Journal of Bacteriology*, *130*, 375–383 (1977).
9. Bulla, L.A., Jr., Davidson, L.I., Kramer, K.J. and Jones, B.L. Purification of the insecticidal toxin from the parasporal crystal of *Bacillus thuringiensis* subsp. *kurstaki*, *Biochemical and Biophysical Research Communications*, *91*, 1123–1130 (1979).
10. Burges, H.D. The standardization of products based on *Bacillus thuringiensis*, pp. 306–314 in *Insect Pathology and Microbial Control:* Proceedings of the International Colloquium on Insect Pathology and Microbial Control, Wageningen, The Netherlands, September 5–10, 1966 (Edited by P.A. van der Laan), North-Holland Publishing Company, Amsterdam (1967).
11. de Barjac, H. Une nouvelle variété de *Bacillus thuringiensis* très toxique pour les Moustiques. *B. thuringiensis* var. *israelensis* sérotype 14, *Comptes Rendus Hebdomadaires des Séances de l'Académie des Sciences* (*Paris*), *286D*, 797–800 (1978).
12. de Barjac, H. and Bonnefoi, A. Mise au point sur la classification des *Bacillus thuringiensis*, *Entomophaga*, *18*, 5–17 (1973).
13. Dulmage, H.T. Insecticidal activity of HD-1, a new isolate of *Bacillus*

thuringiensis var. *alesti, Journal of Invertebrate Pathology, 15*, 232–239 (1970).
14. Dulmage, H.T. and de Barjac, H. HD-187, a new isolate of *Bacillus thuringiensis* that produces high yields of δ-endotoxin, *Journal of Invertebrate Pathology, 22*, 273–277 (1973).
15. Dulmage, H.T., Boening, O.P., Rehnborg, C.S. and Hansen, G.D. A proposed standardized bioassay for formulations of *Bacillus thuringiensis* based on the international unit, *Journal of Invertebrate Pathology, 18*, 240–245 (1971).
16. Dutky, S.R. The milky diseases, pp. 75–115 in *Insect Pathology: An Advanced Treatise* (Edited by E.A. Steinhaus), Vol. 2, Academic Press, New York and London (1963).
17. Farkaš, J., Šebesta, K., Horská, K., Samek, Z., Dolejš, L. and Šorm, F. The structure of exotoxin, *Collection of Czechoslovak Chemical Communications, 34*, 1118–1120 (1969).
18. Fast, P.G. and Martin, W.G. *Bacillus thuringiensis* parasporal crystal toxin: Dissociation into toxic low molecular weight peptides, *Biochemical and Biophysical Research Communications, 95*, 1314–1320 (1980).
19. Faust, R.M., Hallam, G.M. and Travers, R.S. Degradation of the parasporal crystal produced by *Bacillus thuringiensis* var. *kurstaki, Journal of Invertebrate Pathology, 24*, 365–373 (1974).
20. Fenner, F. Classification and nomenclature of viruses (Second Report of the International Committee on Taxonomy of Viruses), *Intervirology, 7*, 1–115 (1976).
21. Ferron, P. Biological control of insect pests by entomogenous fungi, *Annual Review of Entomology, 23*, 409–442 (1978).
22. Fisher, R. and Rosner, L. Toxicology of the microbial insecticide, Thuricide, *Agricultural and Food Chemistry, 7*, 686–688 (1959).
23. Fujiyoshi, N. Studies on the utilization of sporeforming bacteria for the control of house flies and mosquitoes, *Research Report of the Seibu Chemical Industry Co. Ltd.*, No. 1, 1–37 (1973) (in Japanese with English summary).
24. Gingrich, R.E., Allan, N. and Hopkins, D.E. *Bacillus thuringiensis:* Laboratory tests against four species of biting lice (Mallophaga: Trichodectidae), *Journal of Invertebrate Pathology, 23*, 232–236 (1974).
25. Goldberg, L.J. and Margalit, J. A bacterial spore demonstrating rapid larvicidal activity against *Anopheles sergentii, Uranotaenia unguiculata, Culex univitattus, Aedes aegypti* and *Culex pipiens, Mosquito News, 37*, 355–358 (1977).
26. Guillet, P. and de Barjac, H. Toxicité de *Bacillus thuringiensis* var. *israelensis* pour les larves de Simulies vectrices de l'Onchocercose, *Comptes Rendus Hebdomadaires des Séances de l'Académie des Sciences (Paris), 289D*, 549–552 (1979).

27. Heimpel, A.M. A critical review of *Bacillus thuringiensis* var. *thuringiensis* Berliner and other crystalliferous bacteria, *Annual Review of Entomology*, *12*, 282–322 (1967).
28. Heimpel, A.M. and Angus, T.A. The site of action of crystalliferous bacteria in Lepidoptera larvae, *Journal of Insect Pathology*, *1*, 152–170 (1959).
29. Henry, J.E. Experimental application of *Nosema locustae* for control of grasshoppers, *Journal of Invertebrate Pathology*, *18*, 389–394 (1971).
30. Henry, J.E., Oma, E.A. and Onsager, J.A. Relative effectiveness of ULV spray applications of spores of *Nosema locustae* against grasshoppers, *Journal of Economic Entomology*, *71*, 629–632 (1978).
31. Herbert, B.N., Gould, H.J. and Chain, E.B. Crystal protein of *Bacillus thuringiensis* var. *tolworth:* Subunit structure and toxicity to *Pieris brassicae*, *European Journal of Biochemistry*, *24*, 366–375 (1971).
32. Ignoffo, C.M. Effects of entomopathogens on vertebrates, *Annals of the New York Academy of Sciences*, *217*, 141–164 (1973).
33. Ishiwata, S. On a severe flacherie (sotto disease), *Dainihon Sanshi Kaiho*, No. 114, 1–5 (1901) (in Japanese).
34. Katagiri, K. Review on microbial control of insect pests in forests in Japan, *Entomophaga*, *14*, 203–214 (I969).
35. Kim, Y.T. and Huang, H.T. The *β*-exotoxins of *Bacillus thuringiensis*. I. Isolation and characterization, *Journal of Invertebrate Pathology*, *15*, 100–108 (1970).
36. Krieg, A. *Bacillus thuringiensis* Berliner: Über seine biologie, Pathogenie und Anwendung in der biologischen Schädlingsbekämpfung, Mitteilungen aus der Biologischen Bundesanstalt für Land- und Forstwirtschaft, Heft 103, 79 pp. Berlin-Dahlem (1961).
37. Krywienczyk, J., Dulmage, H.T. and Fast, P.G. Occurrence of two serologically distinct groups within *Bacillus thuringiensis* serotype 3ab var. *kurstaki*, *Journal of Inuertebrate Pathology*, *31*, 372–375 (1978).
38. Lecadet, M.-M. and Martouret, D. Enzymatic hydrolysis of the crystals of *Bacillus thuringiensis* by the proteases of *Pieris brassicae*. II. Toxicity of the diffeaent fractions of the hydrolysate for larvae of *Pieris brassicae*, *Journal of Invertebrate Pathology*, *9*, 322–330 (1967).
39. Lecadet, M.-M., Blondel, M.-O. and Ribier, J. Generalized transduction in *Bacillus thuringiensis* var. *berliner* 1715 using bacteriophage CP-54 Ber, *Journal of General Microbiology*, *121*, 203–112 (1980).
40. Lüthy, P. Insecticidal toxins of *Bacillus thuringiensis*, *FEMS Microbiology Letters*, *8*, 1–7 (1980).
41. Martin, P.A.W., Lohr, J.R. and Dean, D.H. Transformation of *Bacillus thuringiensis* protoplasts by plasmid deoxyribonucleic acid, *Journal of Bacteriology*, *145*, 980–983 (1981).
42. Mattes, O. Parasitäre Krankheiten der Mehlmotten Larven und Versuche über ihre Verwendbarkeit als biologische Bekämpfungsmittel (Zugleich

ein Beitrag zur Zytologie der Bakterien), *Sitzungsberichte der Gesellschaft zur Beförderung der gesamten Naturwissenschaften zu Marburg*, *62*, 381–417 (1927).

43. McConnell, E. and Richards, A.G. The production by *Bacillus thuringiensis* Berliner of a heat-stable substance toxic for insects, *Canadian Journal of Microbiology*, *5*, 161–168 (1959).
44. Nickle, W.R. Possible commercial formulations of insect-parasitic nematodes, *Biotechnology and Bioengineering*, *22*, 1407–1414 (1980).
45. Norris, J.R. The classification of *Bacillus thuringiensis*, *Journal of Applied Bacteriology*, *27*, 439–447 (1964).
46. Ohba, M. Tantichodok, A. and Aizawa, K. Production of heat-stable exotoxin by *Bacillus thuringiensis* and related bacteria, *Journal of Invertebrate Pathology*, *38*, 26–32 (1981).
47. Okada, M. Studies on the utilization and mass production of *Spodoptera litura* nuclear polyhedrosis virus for control of the tobacco cutworm, *Spodoptera litura* Fabricius, *Review of Plant Protection Research*, *10*, 102–128 (1977).
48. Padua, L.E., Ohba, M. and Aizawa, K. The isolates of *Bacillus thuringiensis* serotype 10 with a highly preferential toxicity to mosquito larvae, *Journal of Invertebrate Pathology*, *36*, 180–186 (1980).
49. Perlak, F.J., Mendelsohn, C.L. and Thorne, C.B. Converting bacteriophage for sporulation and crystal formation in *Bacillus thuringiensis*, *Journal of Bacteriology*, *140*, 699–706 (1979).
50. Petit-Glatron, M.-F. and Rapoport, G. *In vivo* and *in vitro* evidence for existence of stable messenger ribonucleic acids in sporulating cells of *Bacillus thuringiensis*, *Spores*, *6*, 255–264 (1975).
51. Schnepf, H.E. and Whiteley, H.R. Cloning and expression of the *Bacillus thuringiensis* crystal protein gene in *Escherichia coli*, *Proceedings of the National Academy of Sciences of the United States of America*, *78*, 2893–2897 (1981).
52. Singer, S. *Bacillus sphaericus* for the control of mosquitoes, *Biotechnology and Bioengineering*, *22*, 1335–1355 (1980).
53. Stahly, D.P., Dingman, D.W., Bulla, L.A., Jr. and Aronson, A.I. Possible origin and function of the parasporal crystal in *Bacillus thuringiensis*, *Biochemical and Biophysical Research Communications*, *84*, 581–588 (1978).
54. Steinhaus, E.A. Possible use of *Bacillus thuringiensis* Berliner as an aid in the biological control of the alfalfa caterpillar, *Hilgardia*, *20*, 359–381 (1951).
55. Steinhaus, E.A. Microbial control—The emergence of an idea: A brief history of insect pathology through the nineteenth century, *Hilgardia*, *26*, 107–160 (1956).
56. Thorne, C.B. Transduction in *Bacillus thuringiensis*, *Applied and Environmental Microbiology*, *35*, 1109–1115 (1978).

T.E. FREEMAN

16. Microbial Herbicides

Interest in the use of plant pathogenic microorganisms as microbial herbicides has increased dramatically since the early 1970's. Prior to this time, little effort had been focused on their use for biological control of noxious plants (31, 32). The increased interest no doubt relates to the increasing cost and decreasing availability of fossil fuel and to a heightened awareness of environmental deterioration. Pesticides are related to these factors. Biological control, as visualized in integrated pest management (IPM) programs, offers an attractive alternative to the use of chemical pesticides. Since herbicides are the most widely used of the chemical pesticides, it was natural that an interest developed in alternate methods of weed control.

The success of biological control of weeds with insects is an accepted fact that is well documented (7, 9, 15). It is now realized that plant pathogens hold the same potential and the methodology for their use is developing rapidly. This methodology extends beyond the classical approach to biological control used with insects to the microbial herbicide approach (11, 24, 25). However, plant pathogens are amenable to use in both approaches. In addition, they have several advantages that have been pointed out by Zettler and Freeman (31), Freeman (10) and Freeman *et al.* (13). Plant pathogens are numerous and diverse. It has been estimated that there are over 100,000 plant diseases. They are caused by pathogens belonging to such diverse groups as algae, bacteria, fungi, flowering plants, mycoplasma, nematodes, protozoa, viroids and viruses. Plant pathogens are equally as diverse in their ecological niches. They survive in soils ranging from the arctic tundra to tropical jungles, in fresh and salt water, in the air and in plant and animal vectors. They vary in their parasitic activity from obligate parasites to facultative parasites. Plant pathogens are easily disseminated being air-, water- and soil-borne. They propagate in numbers that stagger the imagination. Many plant pathogens are host specific, even beyond the species to the variety or biotype level. Most important, few, if any, have been shown to adversely affect either man or other animals. Certainly their ability to destroy plants is well documented; however, throughout recorded history they have never completely eliminated a plant species. This in itself is an advantage in the overall scheme of natural constraints.

Efforts by plant pathologists to utilize plant pathogens to control weeds have developed along two lines (10, 13):

1) The search for and introduction of pathogens exotic to the area in which the weed is a problem. This is the classical approach used so successfully with insects. It is predicted on the fact that problem plants are frequently introduced species in the area where they are of concern. Presumably, they were introduced, either deliberately or accidentally, into the area without their constraining natural pests. Thus, the rationale behind this approach is to search for pathogens that occur in the native range of the pest plant but not in the problem area. These pathogens are then introduced into the latter area. Once introduced, the pathogen is generally allowed to become established and spread naturally. If successful, the pest plant is reduced to a manageable population level.

2) The use of plant pathogens as microbial herbicides. The rationale behind this approach is the augmentation of a disease that is already present by massive inoculations of the pest plant with the pathogen. Although endemic pathogens have been most frequently used, this approach could also be followed with exotic pathogens once they have become established. This is a more viable method for use in controlling weeds in crop production.

Both approaches have been used successfully in weed control programs. These successes have served to stimulate work on a large number of important weeds. Those weeds targeted for biological control with plant pathogens were compiled by Freeman *et al.* (13) in 1976 and Charudattan in 1978 (3). Their compilations are summarized in Table 1. Since publication of these compilations, several other weed species have been targeted for biological control with plant pathogens.

Notable among the successes with exotic pathogens have been: (a) the use of *Puccinia chrondrillina* Bubak & Syd from southern Europe for skeleton weed (*Chondrilla juncea* L.) control in Australia (8, 16); (b) control of Hamakera pamakani (*Ageratina riparia* L.) in Hawaii with *Cercosporella riparia* Trujillo introduced from Jamaica (28); and (c) control of wild blackberry (*Rubus* spp.) in Chile with the introduced rust *Phragmidium violaceum* (Schulz) Winter (18).

Notable among the successes with endemic pathogens using the microbial herbicide strategy are:

(a) The control of waterhyacinth (*Eichhornia crassipes* (Mort.) Solm) in lakes and waterways of the states of Florida and Louisiana, U.S.A., using *Cercospora rodmanii* Conway (5, 12, 27).

(b) The utilization of *Colletotrichum gloeosporioides* (Penz) Sacc. as a biocontrol of northern jointvetch (*Aeschynomene virginica* (L.) B.S.P.) in fields of rice (*Oryza sativa* L.) in the state of Arkansas, U.S.A. (23).

(c) The use of a host-specific pathotype of *Phytophthora citrophthora* for control of milkweed vine (*Morrenia odorata* Lindl.) in *Citrus* spp. in the state of Florida, USA (19).

Table 1. Compilation of aquatic and terrestrial weeds targeted for biological control with plant pathogens [after Charudattan (3) and Freeman *et al.* (13)]

Weeds	
Charudattan	*Freeman*
Aeschynomene virginica (L.) B.S.P.	*Acroplilon repens* (L.) DC
Albrizzia julibraissin Durazzini	*Aeschynomene virginica* (L.) B.S.P.
Alternanthera philoxeriodes (Mart.) Griseb	Algae (Blue-green)
Avena fatua L.	*Alternanthera philoxeroides* (Mart.) Griseb
Cannabis sativa L.	*Ambrosia artemisiifolia* L.
Cassia surrattensis Lamarck.	*Brassenia schreberi* Gmel
Chrondrilla juncea L.	*Cassia surrattensis* Lamarck.
Cyperus esculentus L.	*Chronrilla juncea* L.
Cyperus rotundus L.	*Convolvulus arvensis* L.
Eichhornia crassipes (Mart.) Solm	*Eichhornia crassipes* (Mart.) Solm
Hydrilla verticillata (L. fil.) Royle	*Hydrilla verticillata* (L. fil.) Royle
Hydrocotyl umbellata L.	*Lantana camara* L.
Jussiae decurrens (Walt) DC.	*Morrenia odorata* Lindl.
Limmobium spongia (Bosc) Steyd.	*Myriophyllum spicatum* L.
Morrenia odorata Lindl.	*Nuphar luteum* Engelm.
Nuphar luteum Engelm.	*Nymphaea odorata* Ait.
Panicum dicotomiflorum Michx.	*N. tuberosa* Paine
Populus spp.	*Nymphoides obiculata* L.
Quercus spp.	*Oxalis* spp.
Sida spinosa L.	*Pistia statiotes* L.
Solanum dulcambra L.	*Rubus constrictus* L. et M.
S. elaeagnifolium Cav.	*R. umbifolius* Shott
Xanthium spinosa L.	*Rumex crispus* L.
X. strumarium L.	*Salvinia* sp.
	Sida spinosa L.
	Solanum elaeagnifolium Cav.
	Xanthium spp.

Because of the need for international cooperation, exotic plant pathogenic biocontrols have been researched and their use developed by governmental agencies or other publicly funded agencies such as universities in various countries of the world. Foremost among these agencies have been the Australian Commonwealth Scientific and Industrial Research Organization (CSIRO), Canadian Department of Agriculture, various stations of the Commonwealth Institute of Biological Control (CIBC), the United States Department of Agriculture (USDA), University of Florida and the University of Hawaii.

In the case of endemic plant pathogens using the microbial herbicide approach, both public and private research organizations have become involved in research and development (26). The latter being motivated to enter the picture because of the potential for production of a marketable product for distribution and sale to public and private agencies and to individuals. Foremost among the public agencies following this approach have been the Division of Plant Industry (DPI) of the Florida Department of Agriculture and Consumer Affairs, University of Arkansas and the University of Florida. Abbott Laboratories of North Chicago, Illinois, USA and Upjohn Company of Kalamazoo, Michigan, U.S.A., have been the leading private organizations in the development of microbial herbicides. Thus far, the research and development of microbial herbicides has been a cooperative effort between these public and private agencies. The more fundamental work, such as efficacy evaluation, host range tests, epidemiology studies, determination of host-pathogen interaction, etc., has been conducted by the public agency. After completion of these foundation studies, private industry has done the developmental type of work, such as mass inoculum production methods, distribution, satisfying the many government regulations, etc.

There is usually a formal agreement drawn up between the public and private agency in order to protect the rights of each. This has been the procedure used with the microbial herbicide examples cited. For example, in the case of *Cercospora rodmanii*, the University of Florida owns the patent on the use of the fungus but has assigned the market rights to Abbott Laboratories (6, 14). This latter firm has also worked with the DPI in the development of *Phytophthora citrophthora* for milkweed vine control (20). Development of *Colletotrichum gloeosporioides* as a microbial herbicide for control of jointvetch has been a cooperative effort between the University of Arkansas and the Upjohn Company (26). Indications are that the same procedure will continue to be the trend in future development. Templeton *et al.* (25) point out that this procedure differs from the way in which chemical pesticides are researched and developed. Initial work with chemicals has usually been done by private industry prior to turning them over to the public agency for developmental evaluation. This process is the reverse of that used with microbial herbicides where the initial work is conducted by the public sector.

Despite about equal numbers of successes, the emphasis in recent work has been slanted toward the utilization of endemic pathogens using the microbial herbicide strategy (25). A major part of the reason for the popularity of this approach was noted by Freeman *et al.* (13); namely, regulations preventing the importation and use of exotic pathogens prompted, in part, by "pathophobia." In general, it is much easier to import macroorganisms such as insects than microorganisms such as plant pathogens. However, of probably more importance in prompting the use of endemic plant pathogens as microbial herbicides is the fact that they can be produced and disseminated in large quantities. This is one of the factors that made them viable candidates as biocontrol agents to begin with. Commercially produced inoculum of *C. rodmanii* contained in excess of 10^6 propagules per gram of dry weight (14). Five to ten grams of this inoculum were then used to inoculate one square meter area of waterhyacinths. Templeton *et al.* (25) reported that they had obtained 95 to 100 percent kill of jointvetch in 17 rice fields, totaling 240 hectares, by dispersing *C. gloeosporioides* on the fields by airplane, at the rate of 1.5×10^6 spores per ml in 96 liters of water per hectare. That comes to a staggering 144 billion potentially infective units per hectare. In addition, most of the organisms now being researched incite compound interest diseases (30), i.e., disease increase progresses at an exponential rate due to subsequent inoculum production. This subsequent inoculum production can be exceedingly high, e.g., at the height of a *C. rodmanii* epidemic in a waterhyacinth population. Freeman *et al.* (14) reported trapping over 900 spores per cubic meter of air in a 24-hour period. Only with microorganisms could such tremendous inoculum potentials be realized. In addition, pathogens are easily disseminated using conventional spray equipment (5, 20, 25). They have the added advantage of being capable of spreading from the initial point of inoculation; thus, the 100 per cent coverage needed with chemical herbicides is not required with microbial ones (5). Based on these factors, the microbial herbicide tactic, which permits the utilization of endemic plant pathogens, is a workable alternative to the classical approach using exotic biocontrol agents. It also offers a less costly and environmentally safer approach to weed control than that offered by the use of chemical pesticides.

These latter facts, coupled with the successes noted, led to the initiation, in 1978, of a massive research program in the southern United States on the use of fungal plant pathogens as weed control agents (23). Plant pathologists and weed scientists from the southern region of the U.S.A. which includes the states of Alabama, Arizona, Arkansas, Florida, Georgia, Kentucky, Louisiana, Mississippi, North Carolina, Oklahoma, Tennessee, Texas, South Carolina and Virginia have agreed to work cooperatively on the Southern Regional Project 136 (SR-136) entitled "Biological Control of Weeds with Fungal Plant Pathogens." In this region of the U.S.A. weeds cause an estimated loss of \$2.5 billion annually (23). A large portion of these losses is caused by the weeds targeted for research efforts under this project. They are: cocklebur (*Xanthium pennsylvanicum* Wallr.), morning glory (*Ipomoea purpurea* (L.) Roth and *I. herderacea* (L.) Jacq.), prickly sida (*Sida spinosa* L.), sicklepod (*Cassia obtusifolia* L.), hemp sesbania (*Sesbania exaltata* (Raf.) Corry), Johnsongrass (*Sorghum halepense* (L.) Pers.), nutsedge (*Cyperus esculentus* L. and *C. rotundus* L.), alligator weed (*Alternanthera philoxe-*

roides (Mart.) Griseb.), and waterhyacinth (*Eichhornia crassipes* (Mart.) Solms). The objectives of SR-136 are twofold: (1) to identify and determine the distribution of fungi infecting weeds in the southern United States; and (2) to evaluate the potential of fungal pathogens as biocontrol agents for reducing weed problems. The research efforts are heavily weighed in favor of the use of endemic fungal pathogens utilizing the microbial herbicide strategy. The original technical committee designated to conduct the research consisted of 22 scientists from the region. The interest in the program has been high and since its inception, several scientists have requested to be placed on the project. Many of these have been from outside the region. The project is an ambitious one that hopefully will finally establish the fact that plant pathogens are viable control agents for many noxious plant species.

Templeton *et al.* (25) have pointed out some of the constraints to the use of fungi as microbial herbicides. These include such factors as host resistance, environment and spatial isolation of the host. In addition, narrow environmental requirements for infection, spore dormancy and long incubation periods are factors that may be operative. All of these constraints would be especially important using microbial herbicides to control weeds in annual crops, such as control of jointvetch in rice fields. They may be less of a factor in control of weeds in situations that permit a longer range time table, such as weeds in rangelands, perennial crops and aquatic environments (10). However, Baker and Cook (1) as well as Snyder *et al.* (21) have expressed pessimism about the effectiveness of fungi, which is the major group of pathogens thus far researched as microbial herbicides, as biocontrol agents. In fact, Baker and Cook (1) consider them to be third in potential importance behind actinomycetes and bacteria. It should be noted, however, that these authors are generally referring to their use as biological control agents for other pathogens. Nevertheless, many of the constraints they noted, such as competition, predation and parasitism, would be the same for their use in weed control. Freeman (11) considered overcoming these constraints as not an insurmountable task. He notes, however, that their may be other disadvantages that are not definable in the light of present knowledge. These are the questions that haunt progenitors of biocontrol programs. Namely, what are the long-term effects on nontarget species and what will be the eventual influence on the total agro- and/or natural ecosystems?

In future efforts, plant pathologists should plan to expand their search to include other potential groups of plant pathogenic organisms. To date, most research has been done on fungal pathogens, with only a small amount of effort directed toward nematodes (19, 30) and viral pathogens (4). Presently, no one is considering the biocontrol potential of bacterial pathogens. Certainly members of this group would appear to have great potential, especially when considering the importance of secondary soft-rotting bacteria in the successful control of *Opuntia* in Australia and Hawaii (9, 15) using the insect

Coctoblastis sp. as the primary invader (9, 15). In addition, there are entire groups of pathogens, such as mycoplasma, viroids, protozoa, etc. that have only recently come to light. All of these other groups of plant pathogens need to be researched and exploited.

There are many additional weeds for which plant pathogenic biocontrols should be sought. Few of the seventy-six world's worst weeds as listed by Holm *et al.* (17) are included in biocontrol programs with plant pathogens. In fact, until the advent of the SR-136 project, waterhyacinth was the only weed listed in Holm *et al.*'s top twenty weeds that had been targeted for biocontrol efforts. This project added three others, *C. rotundus, C. esculentus* and *S. halapensi.* Many other weeds should be added, because they are all affected by pathogens with microbial herbicide potential.

There is also a need for more basic studies in the weed host-pathogen interaction, especially as it relates to epidemiology. Berger (2) outlined the epidemiological strategies needed to achieve economic plant disease control. They were: 1) reduce initial inoculum, 2) slow the rate of disease increase, and 3) shorten the time of host exposure to the pathogen. He advocates the use of modeling techniques and computer simulation to simplify the assessment of specific epidemiological events. The practicing artisans of biocontrol can use the same techniques to determine the underlying basis for disease occurrence. He can then use this knowledge in his efforts to induce epidemics by using the reciprocal of Berger's strategies for controlling them, i.e. 1) increase initial inoculum potential, 2) increase the rate of disease spread, and 3) extend the time of host exposure to the pathogen. Some work has been conducted on epidemiological aspects of biocontrol with pathogens (14, 22), but more is needed. In fact, it is imperative that this aspect be thoroughly researched. It is especially important in the application of the microbial herbicide tactic. It is fruitless to make massive inoculations with microbial herbicides if conditions are not optimum for infection and subsequent disease development and spread.

More futuristically, researchers should begin to think in terms of conventional, mutation, and recombinant DNA genetics. Using such techniques, it is theoretically possible to "tailor-make" a bio-control agent to fit any set of conditions, i.e., host specificity, toxin production, increased virulence, etc. Admittedly, such research would be painstakingly slow, expensive and subject to controversy. However, the rewards could be great.

The use of plant pathogens in IPM programs should be explored to the fullest. Plant pathogens are usually considered enemies in such programs, but they can also be allies if used properly. This is an especially fruitful approach for use in developing countries.

References

1. Baker, K.E. and Cook, R.J. *Biological Control of Plant Pathogens*, W.H. Freeman & Co., San Francisco (1974).
2. Berger, R.D. Application of epidemiological principles to achieve plant disease control, *Annual Review of Phytopathology*, *15*, 165–183 (1977).
3. Charudattan, R. *Biological Control Projects in Plant Pathology. A Directory*, Plant Pathology Miscellaneous Publications, University of Florida, Gainesville, USA (1978).
4. Charudattan, R., Zettler, F.W., Cordo, H.A. and Christie, R.G. Susceptibility of the Florida milkweed vine *Morrenia odorata* to a potyvirus from *Araujia angustifolia*, *Proceedings of American Phytopathological Society*, *3*, 272 (Abstr.) (1976).
5. Conway, K.E. Evaluation of *Cercospora rodmanii* as a biological control of waterhyacinths, *Phytopathology*, *66*, 914–971 (1976).
6. Conway, K.E., Freeman, T.E. and Charudattan, R. Method and compositions of controlling waterhyacinths, United States Patent 4,097,261, June 27 (1978).
7. Coulson, J.R. Biological control of alligatorweed, 1959–1972: A review and evaluation, *United States Department of Agriculture Technology Bulletin* 1547 (1977).
8. Cullen, J.M., Kable, P.F. and Colt, M. Epidemic spread of rust imported for biological control, *Nature*, *244*, 462–464 (1973).
9. Dodd, A.P. The biological control of prickly pear in Australia, *Australian Commonwealth Council of Scientific Industrial Research Bulletin*, *34*, 1–44 (1927).
10. Freeman, T.E. Biological control of aquatic plants with plant pathogens, *Aquatic Botany*, *3*, 145–184 (1977).
11. Freeman, T.E. Use of Conidial fungi in biological control, in *Biology of Conidial Fungi* (Edited by G.T. Cole and B.R. Kenrick), Academic Press, New York (in press) (1980).
12. Freeman, T.E., Zettler, F.W. and Charudattan, R. Phytopathogens as biocontrols for aquatic weeds, PANS, *20*, 181–184 (1974).
13. Freeman, T.E., Charudattan, R. and Conway, K.E. Status of the use of plant pathogens in the biological control of weeds, *Proceedings of IV International Symposium on Biological Control of Weeds*, University of Florida, pp. 201–206 (1976).
14. Freeman, T.E., Charudattan, R. and Conway, K.E. Biological control of aquatic weeds with plant pathogens, *Contract Completion Report*, United States Army Waterways, Experiment Station Vicksburg, Massachusetts, U.S.A. (1981).
15. Fullway, D.T. Biological control of cactus in Hawaii, *Journal of Economic Entomology*, *47*, 696–700 (1956).

16. Hasan, S. First introduction of rust fungus in Australia for biological control of skeleton weed, *Phytopathology*, *64*, 253–254 (1974).
17. Holm, L.G., Plucknett, D.L., Pancho, J.V. and Herberger, J.P. *The World's Worst Weeds*, University of Hawaii Press, Honolulu (1977).
18. Dehrens, E. Biological control of the blackberry through the introduction of rust, *Phragmidium violaceum* in Chile, Food and Agriculture Organization, *Plant Protection Bulletin*, *25* (1), 26–28 (1977).
19. Orr, C.C., Abernathy, J.R. and Hudspeth, E.B. *Nothanguina phyllobia*, a nematode parasite of nightshade, *Plant Disease Reporter*, *59*, 416–418 (1975).
20. Ridings, W.H., Mitchell, D.R., Schoulties, C.L. and El-Gholl, N.E. Biological control of milkweed vine in Florida citrus groves with a pathotype of *Phytophthora citrophthora*, *Proceedings of IV International Symposium on Biological Control of Weeds*, University of Florida, pp. 224–240 (1976).
21. Snyder, W.C., Wallis, G.W. and Smith, S.N. Biological control of plant pathogens, pp. 521–539 in *Theory and Practice of Biological Control* (Edited by C.B. Haffaker and P.S. Messenger), Academic Press, New York (1976).
22. Te Beest, D.O., Templeton, G.E., and Smith R.J. Temperature and moisture requirements for development of anthracnose and northern jointvetch, *Phytopathology*, *68*, 389–393 (1978).
23. Technical Committee. Biological control of weeds with fungal plant pathogens, *Southern Regional Research Project*, S-136 (1978).
24. Templeton, G.E., Te Beest, D.O. and Smith, R.J. Development of an endemic fungal pathogen as a mycoherbicide for biocontrol of northern jointvetch in rice, *International Symposium on Biology of Weeds*, pp. 214–216, University of Florida (1976).
25. Templeton, G.E., Te Beest, D.O. and Smith, R.J. Biological weed control with mycoherbicides, *Annual Review of Phytopathology*, *17*, 301–310 (1979).
26. Templeton, G.E., Smith, R.J., Klomparens, Jr. W. Commercialization of fungi and bacteria for biological control, *Biological Control News and Information*, *1*, 291–294 (1980).
27. Theriot, E. and Sanders, D.R. Large-scale operational management test (LSOMT) using *Cercospora rodmanii* for control of waterhyacinth, *Proceedings of Research Planning Conference*, *Miscellaneous Paper A-80-3*, U.S. Army Engineers Waterways, Experiment Station Vicksburg, Massachusets, U.S.A. (1980).
28. Trujillo, E.E. Biological control of *Hamakera pamakani* with plant pathogens, *Proceedings of American Phytopathology Society*, *3*, 298 (Abstr.) (1976).

29. Van Der Plank, J.R. *Principles of Plant Infection*, Academic Press, New York (1975).
30. Watson, A.K. The biological control of Russian knapweed with a nematode, pp. 221–223 in *Proceedings of IV International Symposium on Biological Control of Weeds*, University of Florida (1976).
31. Wilson, C.L. Use of plant pathogens in weed control, *Annual Review of Phytopathology*, *7*, 411–434 (1969).
32. Zettler, F.W. and Freeman, T.E. Plant pathogens as biocontrols of aquatic weeds, *Annual Review of Phytopathology*, *10*, 455–470 (1972).

A. KERR

17. Biological Control of Soil-borne Microbial Pathogens and Nematodes

"A truly extraordinary variety of alternatives to the chemical control of insects is available. Some are already in use and have achieved brilliant success. Others are in the stage of laboratory testing. Still others are little more than ideas in the minds of imaginative scientists, waiting for the opportunity to put them to test. All have this in common: they are *biological* solutions."

Rachel Carson: *Silent Spring* (12)

Introduction

Rachel Carson exposed the dangers of excesses of chemical methods of pest and disease control. Since 1962, when *Silent Spring* was first published, biological control has been a very popular phrase. One can almost imagine a chant from *Animal Farm* (41), "Biological control good, chemical control bad." Has this faith in biological control been justified? There have certainly been some spectacular successes, mainly in the control of animal pests and weeds but even here, the new catch phrase "integrated control" has replaced "biological control," presumably because it has been found that biological methods of control by themselves rarely give adequate protection. What about the biological control of plant diseases? Before we can answer this question, we have to define what we mean by biological control. There have been several definitions. A very wide definition might be *any method of control that involves biological rather than physical or chemical methods.* This would include disease resistance and the control of disease by cultural practices such as crop rotation. Such a definition is much too broad to be treated adequately in this chapter. Readers who are interested in a broad treatment of biological control should read Baker and Cook (8). In this chapter the much more restricted definition of biological control as *the use of one specific organism to control another specific organism* is used.

One can now return to the question raised earlier, "What about the biological control of plant diseases?" The answer is quite simple—there are very few (only five or six) plant diseases where biological control is practised commercially. Three of these cannot be included with soil-borne diseases because they are caused by viruses. So that leaves only two or three soil-borne diseases that are controlled biologically. An important aim of this chapter is to try to understand this situation and to consider ways of remedying it.

At first sight this lack of success apears surprising. Thousands of antibiotics are now known and most of these are produced by soil-inhabiting organisms. Many of the antibiotics are active against fungi, including plant pathogens. Why cannot these antibiotic-producing organisms be used to control disease? Many persons have tried and many successes have been achieved when the soil was sterilized prior to inoculation but when natural soil was used, little or no control was achieved. As Garrett (25) pointed out "Such attempts to boost the population of an antagonistic microorganism by inoculation alone have been doomed to failure from their inception, because they are in flagrant contradiction to the ecological axiom that the population is a reflection of the habitat, and that any change due to plant introduction without change of the habitat must be a transient one." Perhaps he was unduly pessimistic but it is certainly true that microorganisms introduced into natural soil usually dwindle very rapidly. One says that the soil is "biologically buffered" by the resident microbial population.

Before Garrett's pessimism affects us unduly, it might be appropriate to emphasize that there is strong evidence that biological control of soil-brone plant pathogens is an important natural phenomenon. The evidence comes from situations where biological control has been destroyed. When soil is fumigated with chemicals or treated with steam, a biological vacuum is created—the resident organisms are killed and a substrate is left uncolonized. If a fast-growing pathogen is introduced into such treated soil, it usually becomes rampant and causes severe damage to susceptible crops growing in the treated, contaminated soil. Most root disease fungi are much more destructive in sterilized than in unsterilized soil. Further evidence comes from hydroponics. Tomato plants cultivated by this method suffer severe damage from the fungus *Didymella lycopersici*, apparently because the natural antibiosis associated with soil culture is lost. One of the most intriguing examples of biological control in nature is the phenomenon of suppressive soils, sometimes called long-life soils in contrast to short-life or conducive soils. Important early work was with bananas growing in Central America (55). In short-life soils, bananas were quickly attacked by the soil-borne wilt fungus *Fusarium oxysporum* f. sp. *cubense* and the average life expectancy of a plantation on these soils was six years. In contrast, bananas grown in long-life soils rarely developed wilt and a plantation would be economical for 25 years or more. A similar situation has been observed with many diseases caused by soil-borne pathogens. The phenomenon has a biological basis because the effect is lost if long-life soil is heated. Possible explanations will be discussed in the section on Fungal Pathogens.

It is one thing to demonstrate that biological control of plant diseases is operating in nature but it is another thing entirely to be able to manipulate the phenomenon to control disease to an economically satisfactory level. The task is formidable because before a pathogen can be described as "soil-borne," it must have evolved a system of living in, or at least surviving in soil which is a very harsh and competitive environment. One gram of a normal fertile soil may contain 10^9 bacteria, 7×10^5 actinomycetes, 4×10^5 fungi, 10^5 protozoa, plus nematodes, mites and other soil fauna. A soil-borne pathogen has to cope with these competing organisms. It would seem unlikely that one biological control organism could be found and manipulated in such a way that the pathogen could no longer cope. Nevertheless, two soil-borne pathogens have been successfully controlled in this way and there are several others where the prospects of biological control are good.

There is clearly a need for biological control. Soil-borne pathogens are notoriously difficult to control except by crop rotation, but because of economic factors, there is an increasing trend towards monoculture and in many situations, crop rotation is no longer an acceptable method of control. The use of chemicals is restricted because the target organs, the roots, are embedded in soil and therefore protected from most chemicals. Systemic chemicals can be used in some situations but are expensive and their effectiveness may be limited because many pathogens can mutate to resistance. Other limitations on chemicals are that they must be environmentally acceptable and pose no health hazards either in their application or in their manufacture. Several widely used soil nematicides have recently been banned because they are dangerous to health. Although breeding of resistant varieties is, perhaps, the ideal method of disease control, there is no satisfactory source of resistance to many of our most damaging soil-borne pathogens. It is important therefore, that the possibilities of biological control be fully investigated.

This chapter deals first with the basic strategies of biological control—what sorts of organisms are available, how can they be found, how and when should they be applied and so on. Secondly, the biological control of diseases caused by bacteria, fungi and nematodes are considered separately (I know of no work on the biological control of soil-borne viruses). Finally, a short overall assessment of the current situation is attempted.

This chapter is not a review. It is a personal assessment of recent work on the biological control of soil-borne plant pathogens with particular emphasis on the two proven procedures for the biological control of *Agrobacterium tumefaciens* and *Heterobasidium annosum* (*Fomes annosus*) and on prospects for developing new economic biological controls for other pathogens.

Basic Strategies

What organisms?

There would seem to be no shortage of organisms with potential for biological control of plant pathogens. Apart from the antibiotic-producers already mentioned, there is a wide variety of parasites and predators of all the major groups of pathogens. There are viruses that attack bacteria and fungi, bacteria that attack other bacteria and also fungi and nematodes; there are fungi that parasitize other fungi, catch and devour nematodes and also parasitize them; there are nematodes that eat bacteria, fungi and other nematodes and there are amoebae and other protozoa that devour most soil-borne plant pathogens. Where do we start and what do we look for?

Perhaps it would be useful to try to define the characteristics of an ideal biological control organism: (a) It has to be able to survive for an extended period in soil either in an active or an inactive form. (b) There must be a high probability that it contacts the pathogen either directly or indirectly by diffusion of chemicals. Perhaps this is one of the most serious deficiencies in biological control agents for plant pathogens. Few potentially useful organisms have the ability to seek out the target, one of the prime requirements for an insect predator used as an agent of biological control (27). With soil-borne diseases, contact between agent and pathogen is severely inhibited because soil is such an impenetrable environment. Procedures which aid contact e.g. application of inoculum to roots, help overcome this constraint. (c) Multiplication in the laboratory should be simple and inexpensive; this usually means that it should grow well in culture and, if it is a fungus, sporulate prolifically. (d) It should be amenable to a simple, efficient and inexpensive process of packaging, distribution and application. (e) If possible, it should be specific for the target organism; the more specific it is, the less environment upset will it cause. (f) It should not be a health hazard in its preparation, distribution or application. (g) It should be active under the appropriate environmental conditions. For example, there is no point in trying to control a pathogen in soil at temperatures of 5–10°C, if the controlling organism does not grow at these temperatures. (This is one of the dangers of glasshouse experiments.) The same situation could apply to soil moisture. (h) Finally and of greatest importance, it has to control the target pathogen efficiently and economically.

It should be borne in mind that some of these conditions are automatically satisfied if the controlling organism is a non-pathogenic variant of the pathogen. The conditions that favour growth of the pathogen will also favour growth of the control organism. If a biological control agent is being sought, first consider the possibility of using an organism closely related to the pathogen. In theory, at least, it would seem unwise to use an organism that is different from the pathogen in its response to environmental factors. For example, bacteria and fungi respond differently to soil moisture (26). If the pathogen is a fungus and the control organism a bacterium, under certain soil moisture regimes, the fungus will be active when bacteria are inactive and this would considerably reduce the chance of control. However, such generalizations should never be accepted dogmatically. It will be seen later that one of the most promising developments in biological control is the use of *Pseudomonas fluorescens* to control two important fungal diseases. We shall return to the choice of biological control agents when we have considered other basic strategies.

Reduce inoculum or reduce infection?

The final objective is to reduce the amount of disease but this can be achieved directly by altering conditions in the infection court or indirectly by reducing the amount of inoculum. All soil-borne pathogens have an inactive phase and most survive this phase by producing resistant structures such as oospores, chlamydospores, sclerotia, and cysts. These structures are adapted to resist attack by other soil organisms and it might seem futile to direct biological control measures against them. However, as is seen in later sections, this is one of the most promising developments in biological control. Although the survival structures are resistant to attack by most soil organisms, there are some parasites that are specialized to attack them. A great advantage of attempting to reduce inoculum is that the survival phase usually lasts for several weeks, months or even years and this gives greater opportunity for a biological control agent to be effective. Where control is directed against infection, there is only a very short vulnerable phase—between germination or hatching of the survival structures and penetration of the host.

When testing a biological control agent for its effectiveness in reducing the level of inoculum, the relationship between inoculum density and disease incidence should be considered (Fig. 1). When the level of inoculum is low, there is a simple direct relationship between inoculum density and disease incidence but at high levels of inoculum, the curve flattens out. In the example given in Fig. 1, it can be seen that reducing the amount of inoculum from C propagules to B propagules would have little effect on disease incidence. However, reducing it from B to A would be significant. Inoculum is reduced by the same amount in both cases. In any experiment, using the wrong level of inoculum can produce very misleading results. The most common error is to use a very high level of inoculum, a level that does not occur naturally in soil. It cannot be over emphasized that a natural inoculum level should be used in all experiments. Natural levels of infestation may be less than one propagule per gram soil for some pathogens to about 4000 for others. Where nematode cysts or fungal sclerotia are used, these levels can be achieved simply by weighing the soil and adding the propagules to give the required density. With fungi such as *Pythium*, *Phytophthora* or *Fusarium* it is usually not so easy. Oospores or chlamydospores could be used as inoculum but these are not always easy to separate from the fungal thallus or substrate. One solution is to inoculate soil with mycelium, conidia, etc., and measure the level of inoculum over time. When it has become stable, the inoculated soil can be mixed with uninoculated soil to adjust the level of inoculum. For other than exceptional circumstances, untreated soil should always be used.

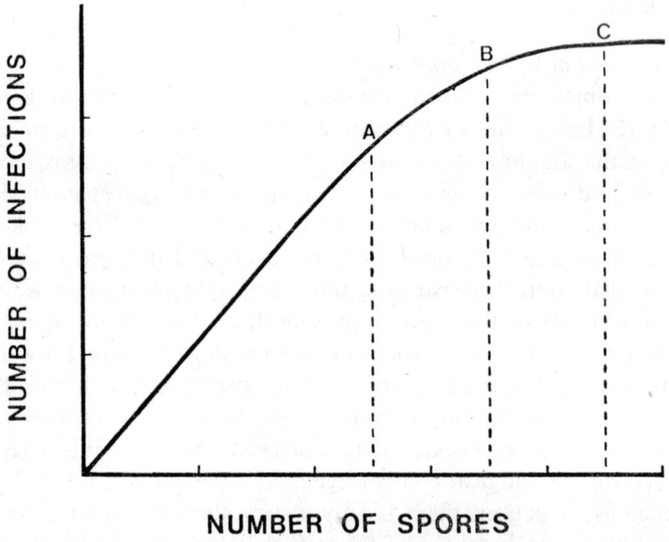

Fig. 1. Relationship between inoculum density and disease incidence (diagrammatic).

Another approach can be taken with bedding plants which are grown in "partially sterilized" soil. Following treatment, the soil can be inoculated with biological control agents to prevent re-colonization by pathogens. This approach has been tried (10) but does not show much promise; it is certainly not practised commercially. The main problem appears to be the specificity of the control organisms. An organism will control some soil pathogens but not others and this is not satisfactory for bedding plants which are usually subject to damping-off by several pathogens.

When considering organisms that might directly reduce the level of infection, there are three ways this can be achieved: (a) by competition, (b) by antibiotic production at the infection sites, or (c) by induced resistance of the host.

Competition can be for infection sites (hypothetical in many cases) or for a resource. Competition for sites without antibiotic production, is unlikely to be successful. The chances are slight that all possible infection sites will be occupied and for complete protection, the control agent would have to be present in very high numbers. Of course, the nature of the disease is also important. With some diseases, such as *Pythium* root rot, a 50 per cent reduction in infection might lead to a 50 per cent reduction in disease loss. On the other hand, with a vascular wilt disease caused either by fungus or bacterium, one infection can cause as much damage as 100 infections—the plant still wilts. It would appear that the biological control of wilt diseases might be impossible to achieve by competition for infection sites.

With one notable exception, competition for resources did not, until recently, show great promise as a mechanism of biological control. The notable exception is the control of *Heterobasidium annosum* by *Peniophora gigantea* which will be described in detail in the section on Fungal Pathogens. It is one of the two biological control procedures practised commercially. The recent development that has improved the prospect of using organisms that compete for resources concerns *Pseudomonas fluorescens*. This organism produces siderophores that bind strongly to iron, making it unavailable to other soil microorganisms which cannot grow for lack of it (33). This also will be described in more detail under the section on Fungal Pathogens.

Probably more work has been done on studying the possible uses of antibiotic-producing organisms as biological control agents than on any other aspect of biological control. There are still persons who argue that antibiotic production does not occur in soil, that antibiotics are just chance products of intermediary metabolism and have no function in nature. However, some antibiotics have been detected in soil (66) and there is strong circumstantial evidence that others are produced there (29, 32). One of the reasons for the popularity of antibiotic-producing organisms is that they are easy to detect and isolate. A soil suspension is diluted and plated on agar to produce isolated colonies and then the plate is sprayed with a propagule suspension of the pathogen. A zone of inhibition indicates antibiotic production. However, it cannot be assumed that because an antibiotic is produced in an agar medium, it is also produced in soil. Substrate can have a dramatic effect on antibiotic production (Plate 1, B). It is probably better to use both a rich and a poor substrate when screening for antibiotics. However, there is no hard and fast rule that can be followed, unless of course, one is looking for siderophores in which case a medium with a low iron content is essential. Theoretically, antibiotic-producing organisms should be much more effective than non-antibiotic-producers at preventing infection because, even if the whole root surface is not colonized, the antibiotic will diffuse over unoccupied areas.

The final mechanism of reducing infection is by inducing resistance in the host. This is usually achieved by using hypovirulent or avirulent strains of the pathogen or of a closely related species. In some fungal pathogens where hypovirulent strains have been discovered, hypovirulence appears to be caused by an infectious agent which can be transmitted to the pathogen rendering it hypovirulent also. This would seem to be an ideal method of controlling disease but at present, such agents can be transmitted only by anastomosis; this means that only strains in the same anastomosis group can be controlled. Perhaps a study of the genetics of incompatibility would be profitable. Mutants with wide compatibility and harbouring an infectious agent which confers hypovirulence would seem to have great potential.

Although biological control of plant pathogens has considerable theoretical and intellectual interest, practical application will depend on economic factors such as cost-benefit ratio, on ecological factors such as environmental impact and on health hazards. There would seem to be little point in developing a biological control system that is less efficient or more expensive than a chemical treatment unless the chemical poses environmental or health problems. It is interesting that although the purpose of research into biological control was largely to reduce our dependence on chemicals, legislation to protect the environment and the community from dangerous chemicals has been applied to agents of biological control and is hampering their introduction and use.

Bacterial pathogens

Agrobacterium tumefaciens

Most bacterial pathogens are not soil borne but one of the few exceptions is *A. tumefaciens.* It is closely related to *Rhizobium meliloti* and *R. leguminosarum* (63) and like these bacteria, is most abundant on and around root surfaces. In other words, it is a rhizosphere organism. It causes a disease called crown gall (Plate 1, C). Most dicotyledonous plants are susceptible but economic damage is confined to relatively few plants, the most important being stone fruit, pome fruit and other rosaceous crops, walnuts, grapevine and some floricultural crops such as chrysanthemum and dahlia.

CROWN GALL INDUCTION. It is necessary to understand how the disease is caused before appreciating the subtleties of the biological control system. It was recently established that the genes controlling pathogenicity in *Agrobacterium* are located on a large plasmid, the Ti (for tumour-inducing) plasmid (58, 60). It consists of circular DNA with a molecular weight of around 120×10^6 or about 5 per cent of total cellular DNA. If the plasmid is eliminated, the bacteria appear unaltered but they can no longer induce crown gall. The Ti plasmid is a conjugative plasmid which means that it can promote its own transfer from one cell (the donor) to another cell (the recipient). When this happens, the recipient is converted from a non-pathogen to a pathogen.

Great progress has been made in understanding how crown gall is caused (46). The plant tissue must be wounded and there is some evidence for an "illegitimate" conjugation between bacterial cell and plant cell but the details of this stage of the process are not known. It is clearly established, however, that part of the Ti plasmid, the T-DNA, enters a plant cell and causes it to divide out of control. This results in galling. Once cell division has started, living bacteria are no longer necessary and can be eliminated by antibiotics without affecting gall development. Another interesting

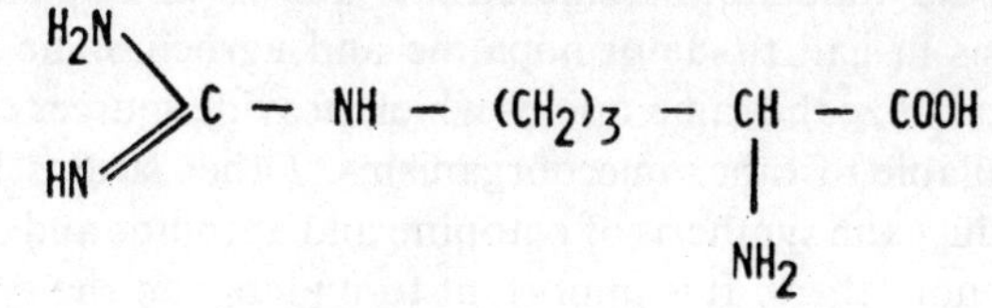

ARGININE

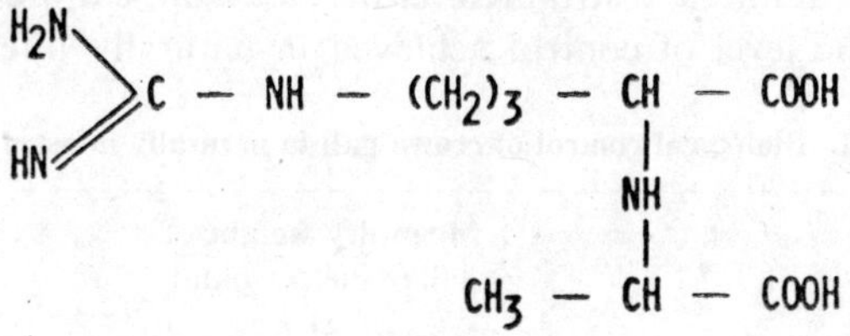

OCTOPINE

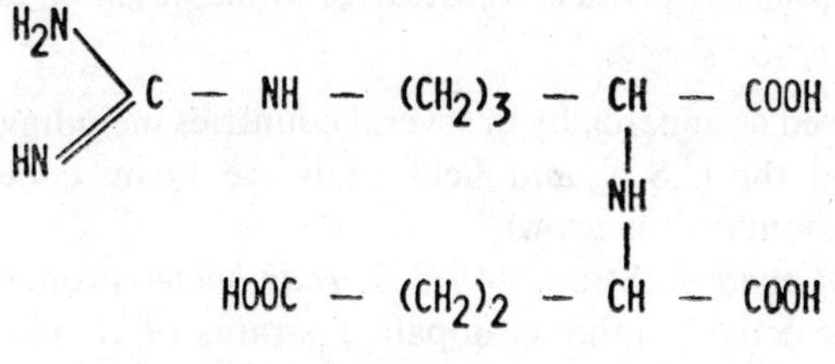

NOPALINE

Fig. 2. Structure of arginine and the opines, octopine and nopaline.

feature of the disease is that unusual chemicals called "opines" are synthesized in crown gall tissue, even in sterile tissue cultures. The T-DNA directs the plant cell to synthesize these compounds which are found nowhere else in the plant kingdom. The chemical structures of two opines are shown in Fig. 2. Which opine is synthesized is determined by the strain of bacterium responsible for tumour induction. Some strains known as nopaline strains induce the synthesis in gall tissue of nopaline and agrocinopine A (see later); they can also catabolize the same opines which provide sources of carbon and nitrogen not available to other microorganisms. Other strains known as octopine strains induce the synthesis of octopine and agropine and can catabolize these opines but not others. It is important to understand the system because only nopaline strains are subject to biological control.

BIOLOGICAL CONTROL. Crown gall is controlled by using strain 84 of the nonpathogenic *A. radiobacter* as a biological control agent. Since 1973, commercial stonefruit and rose growers in Australia have protected their crops by dipping their planting material in a suspension of *A. radiobacter* cells (c. 10^7 cells ml^{-1}), achieving nearly complete control of the disease. The bacteria colonize the root system and can be isolated from roots several months after inoculation. Inoculum is distributed either as agar cultures, or in peat. Table 1 shows the level of control achieved in naturally infested soil. The

Table 1. Biological control of crown gall in naturally infested soil

Treatment	Mean dry weight of gall tissue per plant[1] (g)	Control[2] (%)
None	11.64	—
Seed inoculation with strain 84	2.50	78.5
Root inoculation with strain 84	0.59	94.9
Seed and root inoculation with strain 84	0.14	98.8

[1]Least significant difference of mean dry weight 0.97 (P=0.05).

[2]Percentage control is the difference between the weight of gall tissue on inoculated and noninoculated plants expressed as a percentage of the weight of gall tissue on non-inoculated plants.

method is now used commercially in several countries including New Zealand, South Africa and the U.S.A. and field trials are being conducted in most countries where stonefruit is grown.

Mechanism of control: Strain 84 of *A. radiobacter* produces an unusual antibiotic that selectively inhibits nopaline strains of *A. tumefaciens* (Plate 1, A). The antibiotic has been called agrocin 84 and belongs to a new group of antibiotics known as nucleotide bacteriocins (56). A bacteriocin is an antibiotic substance produced by certain strains of bacteria and active against other strains of the same or closely related species. Most are proteins and bacteriocins are sometimes defined as proteins but this is no longer valid. Agrocin 84 is a fraudulent adenine nucleotide with two substitutions (Fig. 3). Evidence for the involvement of agrocin 84 in biological control is based on correlations and on genetic data.

AGROCIN 84

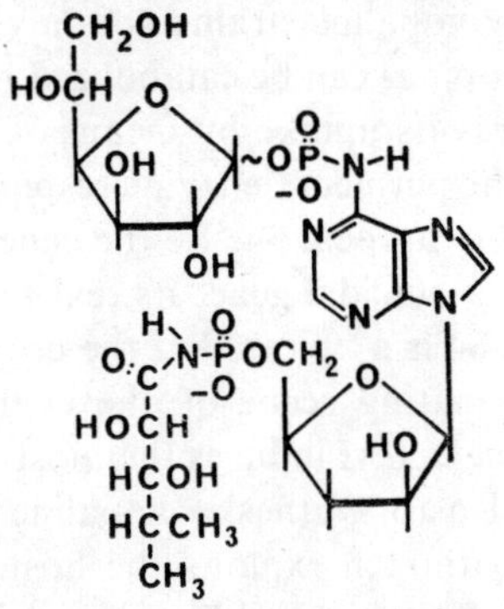

Fig. 3. The structure of agrocin 84.

Only strains of *A. radiobacter* that produce agrocin 84 control crown gall; non-producers (Plate 1, A) are ineffective. Only strains of *A. tumefaciens* sensitive to agrocin 84 in a laboratory test can be controlled; strains resistant to agrocin 84 (e.g. strain A6 in Plate 1, A) are not subject to control. All octopine strains such as those isolated from grapevine cannot be controlled by strain 84. Fortunately crown gall on stonefruit is caused by nopaline strains and can be controlled.

The production of agrocin 84 is coded for by a plasmid with a molecular weight of 30×10^6 (23). It is quite different from the Ti plasmid. As far as is known, the agrocin plasmid is not a conjugative plasmid—it cannot promote its own transfer from one bacterial cell to another. However, strain 84 has a second plasmid that codes for the catabolism of nopaline. This plasmid is conjugative. It not only promotes conjugation and its own transfer to another cell, it appears able to mobilize the agrocin 84 plasmid which can then be transferred to another nonpathogenic recipient bacterium. The recipient can now produce agrocin 84 and has also acquired the ability to control crown gall. Again, control is dependent on agrocin 84 production. The evidence seems overwhelming that biological control of crown gall operates through production of agrocin 84.

The genetic manipulation of agrocin 84 production could have important practical applications. There are probably many hosts, the roots of which are not efficiently colonized by strain 84 which would, therefore, be unsuitable as a biological control agent on these hosts even if the strains of *A. tumefaciens* inducing crown gall were sensitive to agrocin 84. However, if the agrocin 84 plasmid were transferred to a nonpathogenic recipient that was an efficient root colonizer, biological control might be achieved.

Why are only nopaline strains sensitive to agrocin 84? It has been shown that the genes coding for sensitivity to agrocin 84 are located on the nopaline Ti plasmid. They are not present in octopine strains. A gene for sensitivity to agrocin 84 is really a "suicidal" gene. Why should such a gene evolve in nopaline strains? Recent evidence (24) points to the explanation. Galls induced by nopaline strains not only synthesize nopaline but also agrocinopine A. Both opines can be catabolized by nopaline strains. Catabolism of a substrate requires its uptake by means of a permease enzyme. It has been established that the permease enzyme responsible for the uptake of agrocinopine A also takes up agrocin 84. So the gene responsible for agrocin 84 sensitivity is not really a suicidal gene; its real function is opine uptake. It would appear that strain 84 is a "pirate" of the crown gall system. The overall picture presents a fascinating scene of the exploiter being exploited. A nopaline strain of *A. tumefaciens* induces the host plant cells to grow and divide out of control and also to synthesize nopaline which can be utilized by the inducing strain. The pathogen exploits the host. However, nopaline can also be utilized by strain 84 which, although a nonpathogen, possesses a nopaline catabolic plasmid. So it can compete with the pathogen for this specialized substrate. Not only can it compete on equal terms for the substrate, it can also kill its opponent. It produces a toxic substance, agrocin 84, which is taken up by strains possessing a nopaline Ti plasmid, by means of a permease designed for another opine. Some quantitative ecology to define the importance of the interactions in nature is badly needed.

Possible breakdown in control: Although there have been no reports from Australia of a breakdown in the biological control of crown gall on stonefruit, breakdowns have been reported from Greece and the U.S.A. (39). It has not always been realized that only nopaline strains are subject to control. Some of the reports might reflect this ignorance. Another possible explanation for reported failures is that strain 84 prevents crown gall, it does not cure it. In some cases crown gall induction could have occurred before strain 84 was applied. However, all reported failures cannot be explained in these ways. In Greece, three strains of *A. tumefaciens* isolated from peach trees could not be controlled because they were resistant to agrocin 84. It is possible that such strains will become more common when inoculation with strain 84 is widely practised.

An even more dangerous situation has been reported from Greece (42). It has been shown that the agrocin 84 plasmid can be transferred, probably by mobilization, to pathogenic recipients. As a result, the transconjugant pathogens produce agrocin 84 and are resistant to it; they are no longer subject to biological control (Fig. 4). It seems that this is not a common occurrence in commercial nurseries where strain 84 is used (22) but there can be

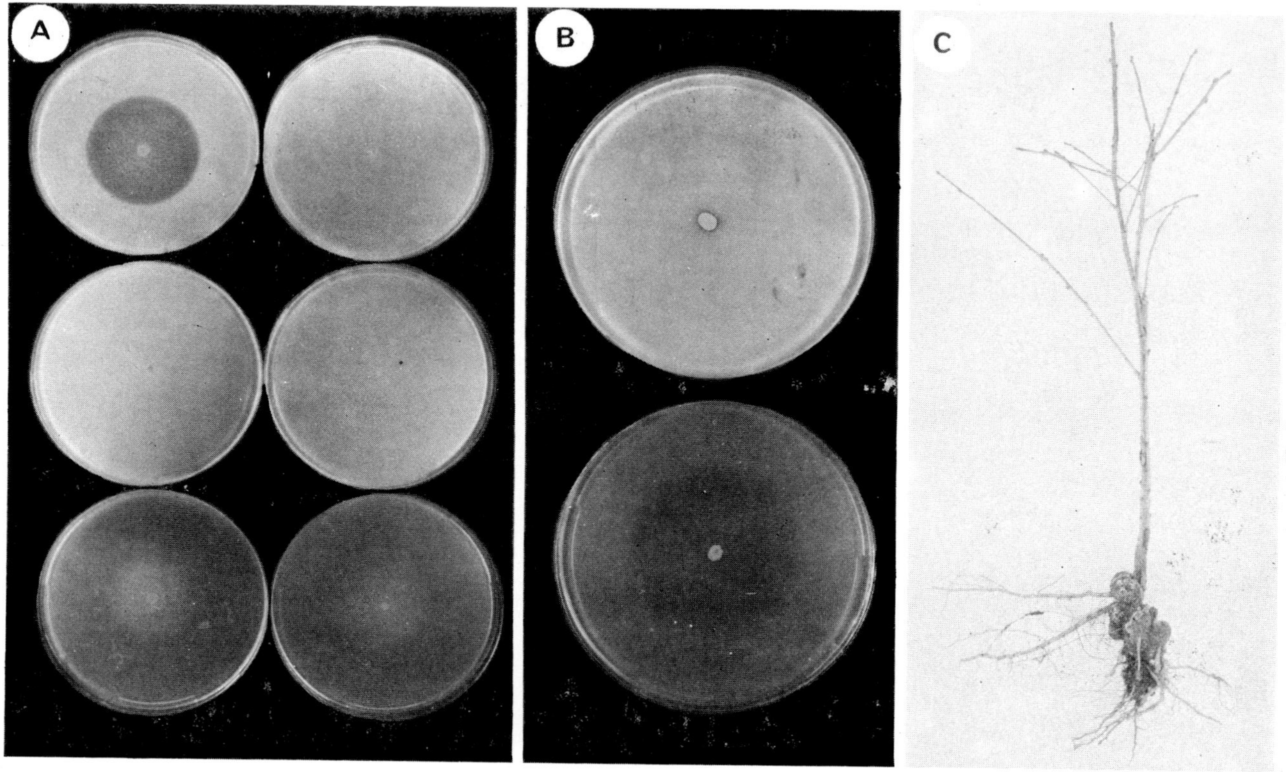

Plate 1. (A) Agrocin production and specificity against strains of *Agrobacterium*. Left: Strain K84 which produces agrocin 84; Right: A strain K84 mutant which does not produce agrocin 84. Both K84 and mutant were inoculated in centre of plates, incubated for 48 h and then killed with chloroform. Top: A nopaline strain K198; Middle: A non-pathogenic strain K57; Bottom: An octopine strain A6. The three indicator strains were added to cool molten agar which was then poured over the plates. (B) Effect of substrate on antibiotic production. The antibiotic producing strain is *Agrobacterium tumefaciens* T37; the indicator strain is *A. tumefaciens* K198. Top plate: rich mannitol-yeast medium; Bottom plate: weak glutamic acid-citric acid medium. (C) Crown gall caused by *A. tumefaciens* on a young peach tree.

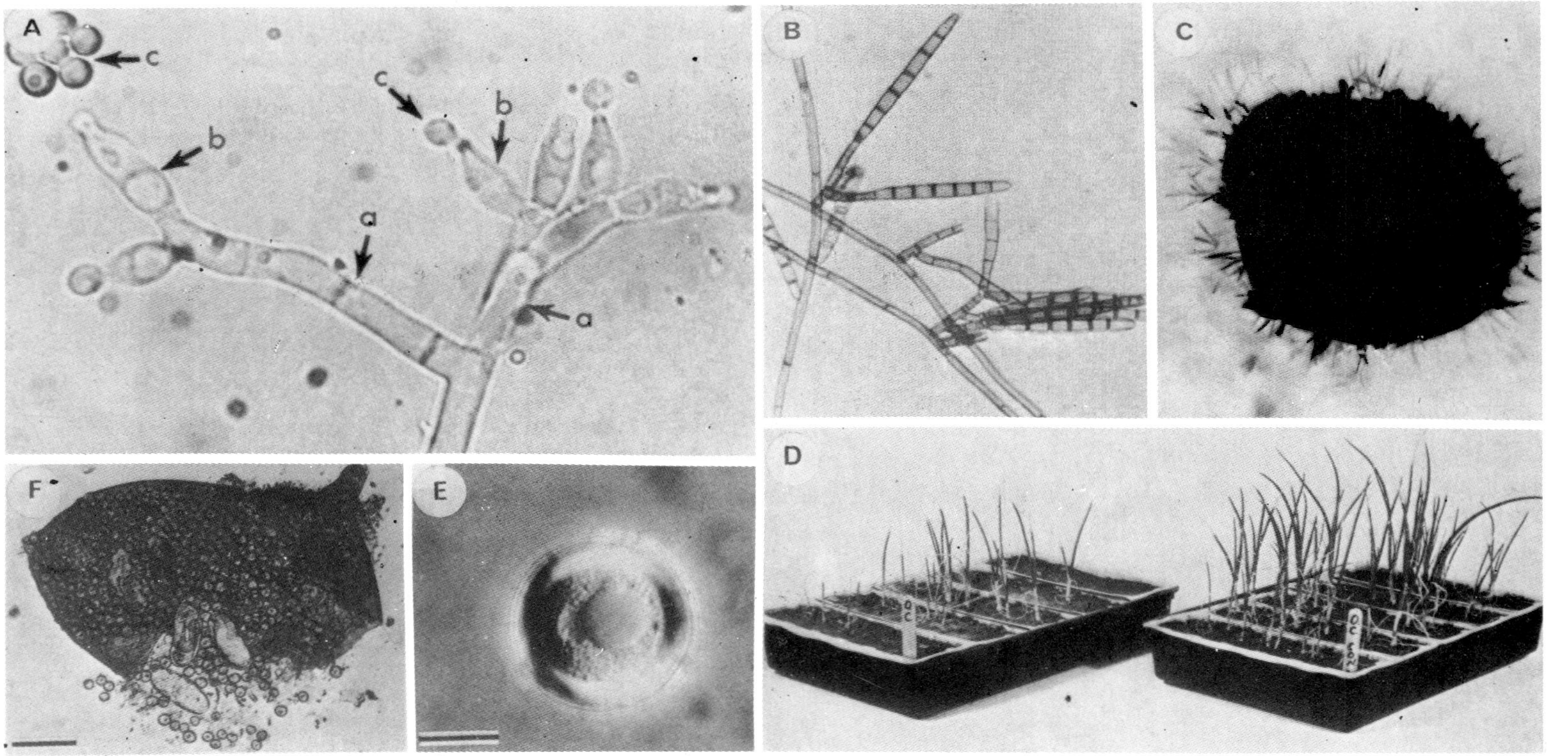

Plate 2. (A) *Trichoderma harzianum* [Wells *et al.* (62)]. (B) *Sporidesmium sclerotivorum* macroconidia [Uecker *et al.* (57)]. (C) *Sporidesmium sclerotivorum* on a sclerotium of *Sclerotinia sclerotiorum* [Uecker *et al.* (57)]. (D) White rot of onions caused by *Sclerotium cepivorum*. Biological control by *Coniothyrium minitans* [Ahmed and Tribe (2)]. Right: treated; Left: untreated. (E) *Nematophthora gynophila* oospore [Kerry and Crump (30)]. (F) *N. gynophila*: nematode cyst containing oospores [Kerry and Crump (30)].

little doubt that pathogenic, agrocin 84-resistant strains will eventually arise in this way. Mutants with a defective transfer or mobilization system are being sought in an attempt to prevent this happening.

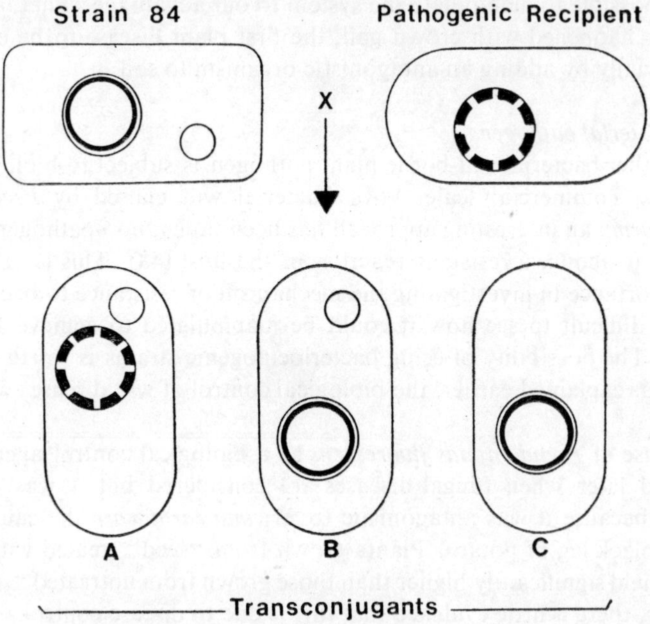

Fig. 4. Diagrammatic representation of a cross between strain 84 and a pathogenic recipient of *Agrobacterium tumefaciens.* Chromosomes are not shown. Strain 84 contains two plasmids, one (single line) coding for agrocin 84 production and for resistance to agrocin 84 and the other (double line) coding for nopaline catabolism and for conjugation. The pathogen has one plasmid (broken line) that codes for pathogenicity and for agrocin 84 sensitivity. The most common transconjugants are shown. Strain A combines pathogenicity with resistance to agrocin 84.

General principles. When considering how the principles behind the biological control of crown gall could be applied to other soil-borne pathogens, it is important to realize that the biological *raison d'être* of strain 84 is not to control crown gall but to exploit it. In nature, the biological success of strain 84 depends on the formation of crown galls which synthesize nopaline. By killing the pathogens which induced the galls, strain 84 has virtually sole access to a carbon and nitrogen source which forms about 2 per cent of the dry weight of gall tissue. This has an important bearing on where one looks for biological control agents. Baker and Cook (8) say that one should "seek antagonists... where the disease does not occur, declines or cannot develop." They are almost certainly correct for some diseases but, if seeking a biological control agent for crown gall, they would be wrong. Strain 84 was isolated from soil surrounding a crown gall and we now know that this is the logical place to find such an organism. Perhaps new biological control agents will be found in similar locations. The fact that disease still exists in a crop does not mean that a successful biological control agent will not be found there. It may be possible to manipulate the system to our advantage. That, at least, is what has happened with crown gall, the first plant disease to be controlled commercially by adding an antagonistic organism to soil.

Other bacterial pathogens

No other bacterial soil-borne plant pathogen is subject to biological control on a commercial scale. With bacterial wilt caused by *Pseudomonas solanacearum* an interesting approach has been taken; non-pathogenic strains are used to induce a resistant reaction in the host (48). This is of considerable importance in investigating the mechanism of resistance to bacterial wilt but it is difficult to see how it could be manipulated to achieve biological control. The possibility of using bacteriocinogenic strains is worth considering but, as explained earlier, the biological control of wilt diseases will not be easy.

The use of *Pseudomonas fluorescens* as a biological control agent will be described later when fungal diseases are considered but it was originally selected because it was antagonistic to *Erwinia carotovora* the cause of soft rot and black leg of potato. Plants grown from "seed" treated with *P. fluorescens* yield significantly higher than those grown from untreated "seed" (33). However, there is little evidence that this is due to disease control.

Another important soil-borne bacterial plant pathogen is *Streptomyces scabies*. It causes common scab of potato and other hosts. No work with specific antagonists is being undertaken but it does exhibit monoculture decline (37). When potatoes are grown in virgin soil, the level of scab increases rapidly but after several consecutive crops, scab incidence declines. This phenomenon will be discussed in more detail when take-all of cereals is described but the fact that it occurs with *S. scabies* implies that it might be subject to biological control, if the phenomenon of monoculture decline were understood.

Fungal pathogens

Heterobasidium annosum

H. annosum causes root decay and heart rot of conifers. In the U.K., it is considered to be the most important disease of pine trees. Pine forests are planted at a high density, partly for weed control and partly to reduce the development of branches on young trees. As the forest grows, the plantings have to be thinned to prevent overcrowding. Thinning is an important factor in the development of root decay and heart rot. After felling, roots attached to a stump start to die and they become much more susceptible to *H. annosum.* Any incipient infections that were present before the tree was felled now develop rapidly and *H. annosum* colonizes the tree stump from the roots. It also colonizes the cut surface of the stump. This is because *H. annosum* has an airborne phase as well as a soil-borne phase. Polypore fructifications are produced at or near soil level; spores are released and transported by wind. Pine stumps provide highly selective substrates for the spores because few other fungi can colonize them initially. *H. annosum* spores germinate and the fungus grows into the massive body of the stump and thence out along stump roots. So the stump becomes colonized by *H. annosum* from two directions and acts as a source of inoculum for neighbouring trees whose roots become infected when they contact the stump.

The method of control is to prevent the stump from being colonized by *H. annosum.* In the forests, it was observed that when a stump became colonized by *Peniophora gigantea,* it did not also become colonized by *H. annosum.* Perhaps if stumps were inoculated with *P. gigantea,* colonization of stumps by *H. annosum* would be prevented. This proved to be so (Table 2). Inoculation of stumps was introduced into British forests in 1962 and now, all suitable

Table 2. Natural colonization of pine stumps after felling [data from Rishbeth (44)]

Treatment	Area of stump section (%) colonized after 8 months by	
	Heterobasidium annosum[1]	*Peniophora gigantea*[1]
None	20	55
Inoculated with *P. gigantea*	1	87

[1]Mean of 12 replicates.

plantations, covering 62,000 hectares are treated. A stump 16 cm in diameter is inoculated with 10^4 spores immediately after felling. A convenient method has been developed for the distribution and application of the spores. Dehydrated tablets containing 10^7 viable spores are prepared. They can be stored for two months at 22°C. When required for use, they are suspended in liquid to which a non-toxic dye has been added for marking treated stumps; the spore suspension is applied to the stump surface and then distributed by brush.

The method is also used in America where *Peniophora gigantea* is registered for use as a plant protectant. Unfortunately, not all conifers that are susceptible to *H. annosum* can be protected by *P. gigantea.* It is effective on *Pinus sylvestris* and *P. nigra* but ineffective on *Picea sitchensis* and *Larix decidua.* Nevertheless, it is an impressive achievement and for many years was the only example of biological control of a plant disease applied commercially. The research was carried out at Cambridge, England and this description is based largely on an article by Rishbeth (44).

Fusarium oxysporum

This pathogen was chosen because of the phenomenon of suppressive soils mentioned in the introduction. Other pathogens could have been chosen but by far the most work has been done on *F. oxysporum*, the cause of *Fusarium* wilt.

F. oxysporum occurs as several forms which are specialized for particular hosts. *F. oxysporum* f. sp. *cubense* causes wilt of banana but of no other crop; similarly *F. oxysporum* f. sp. *pisi* affects peas, *F. oxysporum* f. sp. *melonis* affects muskmelon and so on. If a soil is suppressive for one form of *F. oxysporum*, it will be suppressive for other forms but may be quite conducive for different pathogens including other species of *Fusarium*, e.g., *F. solani*. On the other hand, some soils appear to be suppressive for many different pathogens. When a soil is suppressive, high numbers of *F. oxysporum* propagules can be added but very few susceptible plants growing in that soil will contract the disease in marked contrast to those growing in infested conducive soils (Table 3).

Table 3. Incidence of *Fusarium* wilt of muskmelon in two soils inoculated with different propagule numbers of the pathogen, *Fusarium oxysporum* f. sp. *melonis* [data from Alabouvette *et al.* (3)]

Number of propagules added per gram soil	Percentage of healthy plants	
	Soil 1	Soil 2
0	100	100
400	67	100
1600	8	100
8100	0	100
24000	0	92

A great deal of work has been done to compare suppressive and conducive soils. One major difference has been revealed; suppressive soils contain montmorillenoid clays whereas conducive soils do not. Montmorillenoid clays have a high buffering capacity and it is believed that they allow soil microorganisms to metabolise actively over a much longer period than in soils lacking this type of clay. The high microbial activity, it is argued, inhibits pathogens and could explain the suppressive nature of these soils (54). If this be the true explanation, it is not very promising for biological control. It would be impossible to alter, on a large scale, the clay content of soil to prevent disease.

Other workers believe that the main difference between suppressive and conducive soils is microbiological and is quite independent of clay content.

A recent development has been a study of certain strains of *Pseudomonas fluorescens* present in suppressive soils (32). Strain B10 has received most attention. Adding this bacterium to conducive soil to give 10^5 cells per gram soil made the soil suppressive. Strain B10 is a rhizosphere organism and readily colonizes the root system following inoculation of seed or young plants. It produces a siderophore which has a strong specific affinity for iron with which it forms a stable complex. The affinity constant is 10^{32} and Fe''' is bound so tightly that it becomes unavailable to soil microorganisms which do not produce siderophores or produce those with a lower affinity for iron. It is argued that suppression of *F. oxysporum* (and other pathogens) operates through this mechanism; the pathogen is deprived of Fe''' and cannot grow. If 50 μM $(Fe'''EDTA)^-$ is added at the same time as strain B10, the soil does not become suppressive (Table 4). Presumably, there is now enough iron available for the pathogen. If this is a correct interpretation of the data, it should be possible to change a suppressive soil into a conducive soil by addition of $(Fe'''EDTA)^-$. Indeed, this has been achieved (Table 5). Further

Table 4. Treatment of conducive soils with *Pseudomonas fluorescens* and with $(Fe'''EDTA)^-$ [data from Kloepper *et al.* (32)]

Soil inoculated with	Treatment	Surviving seedlings (%)
Fusarium oxysporum f. sp. *lini*	H_2O	45
F. oxysporum f. sp. *lini*	B10[1]	83*
F. oxysporum f. sp. *lini*	B10 + 50μM $(Fe'''EDTA)^-$	25
Gaeumannomyces graminis var. *tritici*	H_2O	25
G. graminis var. *tritici*	B10[2]	88*
G. graminis var. *tritici*	B10 + 50μM $(Fe'''EDTA)^-$	25

[1]Transplants dipped in bacterial suspension prior to planting.
[2]Seeds dipped in bacterial suspension prior to planting.
*Significantly different from water controls (P = 0.05).

Table 5. Treatment of suppressive soils with $(Fe'''EDTA)^-$ [data from Kloepper *et al.* (32)]

Soil inoculated with	Treatment	Surviving seedlings (%)
F. oxysporum f. sp. *lini*	H_2O	82
F. oxysporum f. sp. *lini*	50μM $(Fe'''EDTA)^-$	47*
Not inoculated	50μM $(Fe'''EDTA)^-$	90
Gaeumannomyces graminis var. *tritici*	H_2O	83
G. graminis var. *tritici*	50μM $(Fe'''EDTA)^-$	38*
Not inoculated	50μM $(Fe'''EDTA)^-$	85

*Significantly different from water controls (P=0.05).

evidence supports the interpretation that suppression of pathogens results from the complexing of Fe''' by siderophores produced by *P. fluorescens*. The siderophore produced by strain B10 has been isolated and partially characterized. It is a low molecular weight peptide and has been given the trivial name, pseudobactin. If 50 μM pseudobactin is added to a conducive soil on alternate days, the soil becomes suppressive.

There are several unresolved problems in the effectiveness of *P. fluorescens* in controlling *F. oxysporum*, not least being the explanation of how a bacterium which requires a relatively high soil moisture content for activity can control fungi which can grow at much lower moisture regimes. Nevertheless, this is one of the most exciting developments in biological control of plant disease in recent years. According to the Californians who did the work, "...seed treatment could easily become a standard commercial practice." However, the work has still to be confirmed by independent research. Perhaps partial confirmation has come from workers in Colorado (47) who have also been able to convert a conducive soil to a suppressive soil by the addition of a species of *Pseudomonas*; in this case, the mechanism of control was not established. No antibiosis was observed in culture but the effect of siderophores would not be detected if adequate Fe''' were available. If the same mechanism operates in both Colorado and in California, the prospects for biological control are good, especially as strain B10 has also been reported to control *Gaeumannomyces graminis* var. *tritici*.

Gaeumannomyces graminis var. *tritici*

G. graminis var. *tritici* causes take-all of wheat and barley, a serious root disease in most countries where these crops are grown. An excellent review has been published by Walker (59). The main reason for including *G. graminis* var. *tritici* in this chapter is to examine the phenomenon of take-all decline. It is really a particular example of monoculture decline but more work has been done on take-all than on any other disease that declines following continuous monoculture. It should not be assumed, however, that all decline phenomena have the same underlying cause.

When land has not been used for cereal culture for a few years, the first wheat crop to be planted is usually free from take-all. With continuous cropping to wheat, the disease builds up and reaches a high level, usually after four years. Thereafter, disease incidence falls and after a few more years, reaches a relatively stable, low level (Fig. 5). This is known as take-all decline and has been reported from most countries where wheat monoculture is practised. If, after decline has occurred, a crop not susceptible to *G. graminis* var. *tritici* is grown for one or more years and then wheat monoculture resumed, the whole process starts again. The factor which caused the decline has been lost and only reappears after several years of continuous wheat. If the cause of take-all decline could be identified, it might be possible to introduce

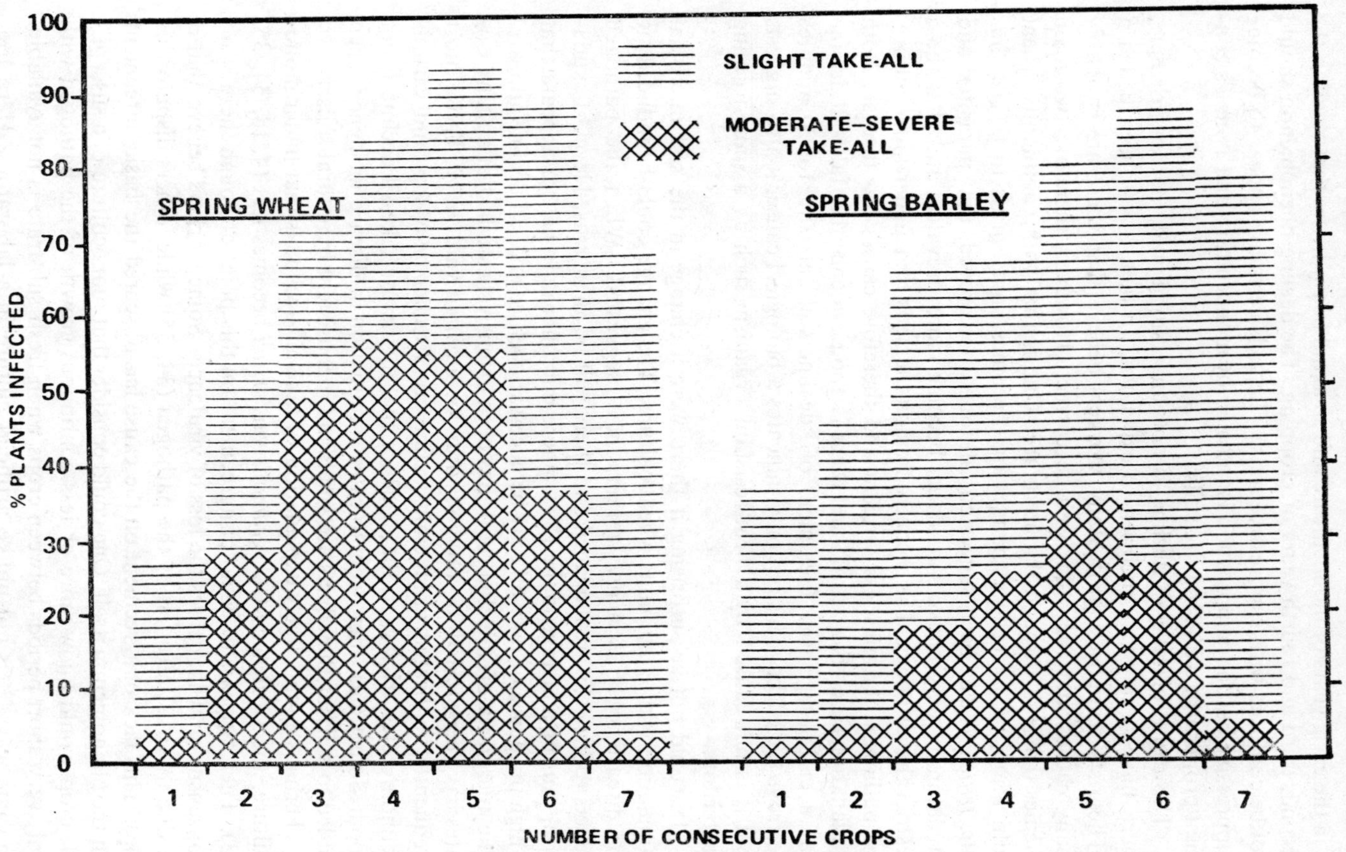

Fig. 5. Take-all decline [after Shipton (49)].

it in the early years of monoculture and so avoid heavy losses that occur in the second, third and fourth years of cropping. The decline phenomenon should not be confused with suppressive soils. The soils certainly appear to become suppressive but the suppression is induced by monoculture and is not a constant property as with long-life soils.

The cause of take-all decline is not understood and, not surprisingly, there have been many theories to explain it. In fact, it has been called a "Theorists' Paradise" (28). Many workers believe that there are probably many causes and they may well be correct. However, there are two main themes that run through the literature. One is that with continuous wheat cultivation, the microflora changes and becomes antagonistic to *G. graminis* var. *tritici.* It has been suggested, although not proved, that parasites and predators of *G. graminis* var. *tritici* become more prevalent and that these restrict the pathogen. The second common theme is that under monoculture, the pathogen becomes less virulent and therefore causes less disease. Both explanations are "biological" and there is good evidence that take-all decline has a biological basis. Treatment of "decline soil" at 60°C for 30 minutes destroys the effect. This not only indicates a biological cause, it also suggests that it is not caused by a heat-resistant organism such as a spore-forming bacterium.

It would not be surprising if there were a change in the soil microbial population with continuous wheat culture. The change could be induced by the disease itself, as many workers have suggested. When the pathogen infects roots, cells are damaged and there will be an increased level of exudation, causing a change in the rhizosphere microflora. In fact, it appears that a high incidence of take-all may be necessary before take-all decline appears. The microflora of infected roots has been extensively studied and many isolates have been tested for antagonism to *G. graminis* var. *tritici.* Those most frequently implicated in take-all decline are *Pseudomonas* spp. and various actinomycetes. However, the evidence that they cause take-all decline is not good. Sometimes, they seem to cause a reduction in disease incidence when added to soil but the effects are small and inconsistent in natural soils.

Perhaps a more likely explanation of take-all decline is that under monoculture, the virulence of *G. graminis* var. *tritici* becomes less (4, 18, 34, 35, 43). There appears to be no disagreement on this point although there is no consensus on what causes the loss of virulence. Some workers believe that it is due to virus infection of the pathogen (34, 35) while others dispute this (43). This lack of agreement on the cause has obscured the basic agreement on the phenomenon itself. One could consider that monoculture of a susceptible crop would allow more or less continuous growth of the pathogen with only very short periods between crops when fresh substrate is not available. In some ways, it is not unlike artificial culture in a laboratory. Here too, *G. graminis* var. *tritici* loses virulence (15). This work has also shown that avirulent cultures do not develop the black pigment melanin when buried in soil and this means that they cannot survive in soil for long periods. In contrast, virulent strains become melanized and can survive much better in soil (16). These two characteristics, loss of virulence under continuous growing conditions and poor survival in soil of avirulent forms could explain take-all decline. If true, it might be possible to introduce the avirulent form into the first or second year of a monoculture and so prevent the high losses caused by take-all in two to four years. Such work is being done in France (35). Finally, it should be re-emphasized that the cause of take-all decline has not been unequivocally established.

A closely related topic is the use of avirulent, or hypovirulent forms of *G. graminis* and the closely related *Phialophora radicicola* to control take-all (9, 19, 20, 50, 64). The work derives more from studies on these fungi in pastures than from work on take-all decline. The fungi infect wheat roots but cause little or no damage. It is believed that they induce a defence reaction in the host which, as a result, can resist infection by *G. graminis* var. *tritici*. Presumably it is a form of cross protection but there is no evidence for the production of phytoalexins (21). This method of biological control seems reasonably promising, especially when hypovirulent strains are used with the same temperature responses as the pathogen (65).

Rhizoctonia solani

R. solani causes root rot and damping-off in many crops. There are numerous studies on the biological control of *R. solani*. The most interesting and probably the most promising are those dealing with the use of diseased, hypovirulent strains of the pathogen. Table 6 shows that both soil and furrow inoculation achieved very efficient control (13).

Hypovirulence is associated with the presence of three double stranded RNA elements (dsRNA) with molecular weights of 2.2, 1.5 and 1.1×10^6 (14).

Table 6. Treatment of sugar beet seed furrows with hypovirulent mycelium to control damping-off caused by *Rhizoctonia solani* [data from Castanho and Butler (13)]

Treatment[1]	Pre-emergence damping-off (%)	Post-emergence damping-off (%)
Control (noninfested)	0	0
Control (infested)	48*	80*
Hypovirulent mycelium	3	5
Autoclaved hypovirulent mycelium	24*	95*

[1]UC mix soil was previously infested with a virulent strain of *R. solani*. One gram dried mycelium of hypovirulent strain added to each furrow sown with 50 seeds.
*Significantly different from noninfested control (P=0.01).

The disease agent is either dsRNA *per se* or a mycovirus but virus particles have not been detected. Infected strains are very slow growing, produce few or no sclerotia and have limited survival in soil. When an infected strain is mixed with a healthy strain, the dsRNA elements are transmitted following anastomosis and the disease spreads. The dsRNA elements are also transmitted to basidiospores, by cytoplasmic inheritance.

Considerably more work is required on this phenomenon. Presumably, biological control operates both by reducing virulence of *R. solani* and by reducing survival in soil. At present, however, the method has severe limitations because the dsRNA appears to be strain specific. Other strains, even in the same anastomosis group, are not susceptible. Many other diseased strains of *R. solani* have been isolated. They all contain dsRNA elements but of different sizes from those of the original strain. It may be possible to select or engineer a dsRNA element with a wide host range. This would have great potential as an agent of biological control.

Another promising organism for the control of *R. solani* is *Laetisaria* sp. (= *Corticium* sp.?) (40); it is parasitic on *R. solani* in culture but this may not be the method of control in the field because it also controls *Pythium* ultimum on which it is not parasitic. Methods of application, an important consideration for all biological control agents, have been studied. The seeds can be treated by dipping them in 1 per cent (w/v) methyl cellulose and then rolling them in *Laetisaria* inoculum. Coated seeds are dried for at least 12 hours. Field application also seems feasible. The control organism can be mixed either with diatomaceous earth or with sugar beet pulp and applied directly to the field. Early protection from *R. solani* is achieved but there is still doubt as to whether the method gives protection throughout the entire growing season of the crop. Promising features of the method are that *Laetisaria* is easy to grow in culture and can be stored for more than three years as an air-dry preparation of mycelium and sclerotia under non-sterile conditions.

Sclerotinia and *Sclerotium*

Perhaps the immediate prospect for introducing routine measures for the biological control of *Sclerotinia* and *Sclerotium* is brighter than for any other plant pathogens not currently being controlled biologically. Several organisms have been extensively studied, including *Trichoderma harzianum* (Plate 2, A), *Coniothyrium minitans* and *Sporidesmium sclerotivorum* (Plate 2, B and 2, C). All three fungi reduce disease incidence by parasitizing sclerotia, the resting structures of these pathogens.

Trichoderma harzianum. In 1932, Weindling (61) suggested that a species of *Trichoderma* could be used for the biological control of soil-borne plant pathogens. It still looks quite promising but has not yet been used commercially against soilborne pathogens although it is the genus most frequently cited in publications on biological control (17). Field control of *Sclerotium rolfsii* by *T. harzianum* was first reported in 1972 (62) but required the addition of 4,200 kg ha^{-1} of inoculum for effective control, a somewhat daunting amount. A more economical system has been developed (7). A diatomaceous earth granule impregnated with molasses was found suitable for growth of *T. harzianum* and for delivery to peanut fields. Only 140 kg ha^{-1} were required and control was equivalent to that achieved by 112 kg ha^{-1} of 10 per cent PCNB granules. Results are shown in Table 7. The method is in experimental use (17). As *S. rolfsii* is reported to cause heavy damage each year in Georgia alone, a successful biological control treatment would have considerable economic impact.

Table 7. Disease incidence and yield of peanut plants growing in field soil treated with *Trichoderma harzianum* and with PCNB [data from Backman and Rodriguez-Kabana (7)]

Treatment	Dead plants (number per 20 m row)	Yield (kg per ha)
Trichoderma granules[1] 140 kg/ha	4.53*	3336*
PCNB (10% granules) 112 kg/ha	4.39*	3360*
Untreated control	7.85	2928

[1]*T. harzianum* introduced on granules of diatomaceous earth impregnated with 10% molasses.

*Significantly different from control (P=0. 01).

Coniothyrium minitans. C. minitans is being tested as a biological control agent in Australia, Canada, the U.K. and the U.S.A. The original work was done in the U.K. (2). The fungus was isolated from naturally infected sclerotia of *Sclerotium cepivorum*, the causal agent of white rot of onions. It appears to be specific for sclerotial fungi. Inoculum of the fungus is prepared by growing it on sterilized, milled rice or bran for one week, followed by drying, grinding and sieving. It is known as pycnidial dust because many pycnidia are formed in culture; 10 g dust are added to 50 ml 1 per cent methyl cellulose. Treatment of seeds with this inoculum gives effective control (Plate 2, D); as effective as the standard treatment with calomel, a hazardous mercury compound (Table 8). *C. minitans* has many of the attributes of the ideal biological control agent. It grows well in pure culture and sporulates prolifically. The conidia are thick-walled and pigmented and appear to survive collection, storage and application procedures. Melanized pycnidia are also produced and harvested and give added protection to conidia. The fungus has been shown to survive in soil for at least 18 months; a truly promising biocontrol agent.

Sporidesmium sclerotivorum. The specific epithet is descriptive (it means sclerotium eater) but unfortunate. The fungus is used to control *Sclerotinia*

Table 8. Biological control of *Sclerotium cepivorum* by *Coniothyrium minitans* [data from Ahmed and Tribe (2)]

Seed treatment	Healthy plants out of 200 at 12 weeks in: Soil infested with *S. cepivorum*	Uninfested soil
Untreated	45 ± 6	200
1% methyl cellulose	41 ± 4	200
1% methyl cellulose plus *C. minitans*	141 ± 7	N.T.*
Calomel	139 ± 13	N.T.

*N.T. = Not tested.

sclerotiorum. So readers will have to distinguish *S. sclerotivorum*, the control agent from *S. sclerotiorum*, the target organism.

S. sclerotivorum was first described in 1978 (57) and, like *C. minitans* is a specific parasite of sclerotia. Unlike *C. minitans*, it is nearly an obligate parasite. It will not grow on many media and grows only slowly on cornmeal agar. The most satisfactory medium for growth contains a concoction of sclerotia (6).

S. sclerotivorum can be easily detected in and isolated from soil. Sclerotia of *S. sclerotiorum* are used as bait; they are added to soil and examined every five weeks for macroconidia of *Sporidesmium* (Plate 2, B and C). When the mycoparasite is present, 80 to 90 per cent of the sclerotia are destroyed after 15 weeks (5). *S. sclerotivorum* can be successfully introduced into soils; it is grown in large batches on a sand-sclerotia mixture and added to soil at the rate of 100 macroconidia per gram. Following inoculation of natural soil, more than 95 per cent of sclerotia of *S. sclerotiorum* are infected and destroyed in 10 weeks or less at 25°C. It is also very effective against *S. minor* and appears to be a major factor in the natural decline of sclerotia. Artificial infestation has a similar effect (Fig. 6). It is less effective against *Sclerotium cepivorum* and ineffective against *Macrophomina phaseolina*. One of its most useful features is that it can spread through soil by growth from one sclerotium to another. Another is that the macroconidia are converted to chlamydospores which can survive for long periods in soil. The biological control of *S. sclerotiorum* and other species of *Sclerotinia* by *Sporidesmium sclerotivorum* is an attractive possibility.

Other fungal pathogens

The genera *Bacillus* and *Streptomyces* are frequently mentioned in papers on the biological control of soil-borne plant pathogens (17). There is no doubt that isolates of these organisms produce antibiotics which are active against many plant pathogens including *Fusarium oxysporum*, *F. roseum*, *Helminthosporium* spp., *Phymatotrichum omnivorum*, *Phomopsis sclerotioides*, *Phytoph-*

thora cinnamomi, *Pythium* spp. and *Thielaviopsis paradoxa* (17). There are also frequent references to an increase in yield resulting from inoculation of planting material with either *Bacillus* and *Streptomyces*. Sometimes the yield increase is spectacular. For example, a 40 per cent increase in yield of wheat and carrots has been reported to result from seed inoculation with these

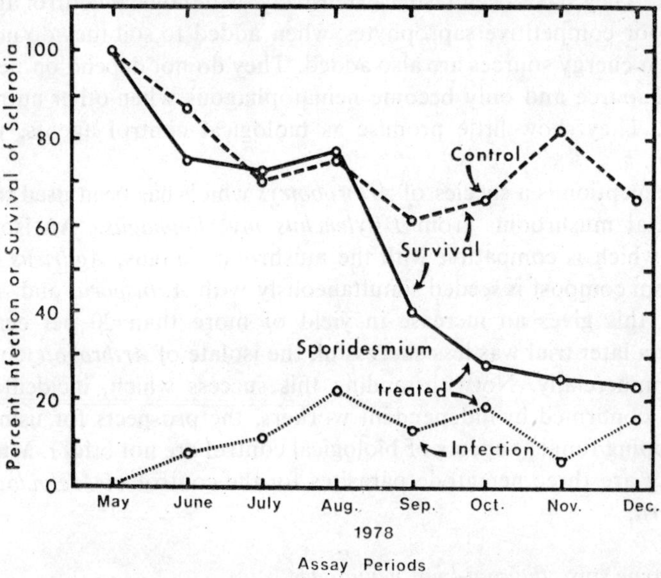

Fig. 6. The effect of *Sporidesmium sclerotivorum* added to soil at the rate of 100 macroconidia per gram of soil on infection and survival of sclerotia of *Sclerotinia minor* in the field [Adams and Ayers (1)].

organisms (38). However, there is no evidence that the yield response is due to control of disease. Production of growth substances seems to be a more likely explanation. Even so, such spectacular responses would be worthwhile, whatever the cause. Unfortunately the effect is erratic and unpredictable and this prevents widespread adoption of the method, although in the Soviet Union, seed bacterization of non-leguminous crops is widely practised (11).

Nematode pathogens

Two excellent reviews on the biological control of nematodes have recently been published (36, 45) and this Section will be based largely on them.

The biological control of nematodes is entirely dependent on predators and parasites. Predators include turbellarians, tardigrades, enchytriads, insects, mites, fungi, amoebae and other nematodes. The parasites include viruses, bacteria, protozoa and fungi. Most publicity has been given to the nematode-trapping fungi. They are fascinating organisms, easy to isolate from soil and easy to work with. Many have very sophisticated mechanisms for capture and penetration of nematodes. One, *Haptoglossa heterospora*, even shoots a minute projectile through the cuticle of passing nematodes in response to mechanical stimulation. The major genera are *Arthrobotrys*, *Dactylaria*, *Dactylella* and *Monacrosporium*. All are predators rather than parasites. They have been tested extensively as biological control agents but all are poor competitive saprophytes; when added to soil they do not persist even when energy sources are also added. They do not depend on nematodes as a food source and only become nematophagous when other nutrients are deficient. They show little promise as biological control agents, with one exception.

The exception is a species of *Arthrobotrys* which has been used to protect commercial mushrooms from *Ditylenchus myceliophagus*. An isolate was selected which is compatible with the mushroom fungus, *Agaricus bisporus*. Mushroom compost is seeded simultaneously with *A. bisporus* and *Arthrobotrys* and this gives an increase in yield of more than 20 per cent. Even although a later trial was less successful, the isolate of *Arthrobotrys* was marketed commercially. Notwithstanding this success which, incidentally, has not been confirmed by independent workers, the prospects for using nematode-trapping fungi as agents of biological control are not bright. Much more promising are three nematode parasites for the control of *Meloidogyne* and *Heterodera*.

Meloidogyne spp., *the root-knot nematodes*

Perhaps the most exciting prospect for the biological control of an important plant parasitic nematode is the use of *Bacillus penetrans* against *Meloidogyne* spp. The life-cycles of both organisms are shown in Fig. 7. *B. penetrans* was originally described as a protozoan but is now known to be a spore-forming bacterium. The spores enable it to survive in soil for long periods, in marked contrast to the nematode-trapping fungi. The spores are inactive but adhere to the surface of infectious second-stage female larvae. There is marked specificity in this process. The spores do not adhere to larvae of other nematodes, nor even to male larvae of *Meloidogyne* spp. This is an interesting phenomenon in its own right and deserves further study. Following adherence of spores to second-stage female larvae, infection occurs but is not apparent until the adult stage when the female is literally transformed into a bag of spores. The effectiveness of *B. penetrans* in controlling root-knot of tomato is shown in Table 9. A useful feature of *B. penetrans* is that it is resistant to most nematicides and could, therefore, be used in an integrated control programme. There is good evidence that *B. penetrans* is an important factor in reducing natural populations of *Meloidogyne* spp. It is widespread in medium-aged and old vineyards in South Australia and sufficient females are parasitized to implicate *B. penetrans* in the natural biological

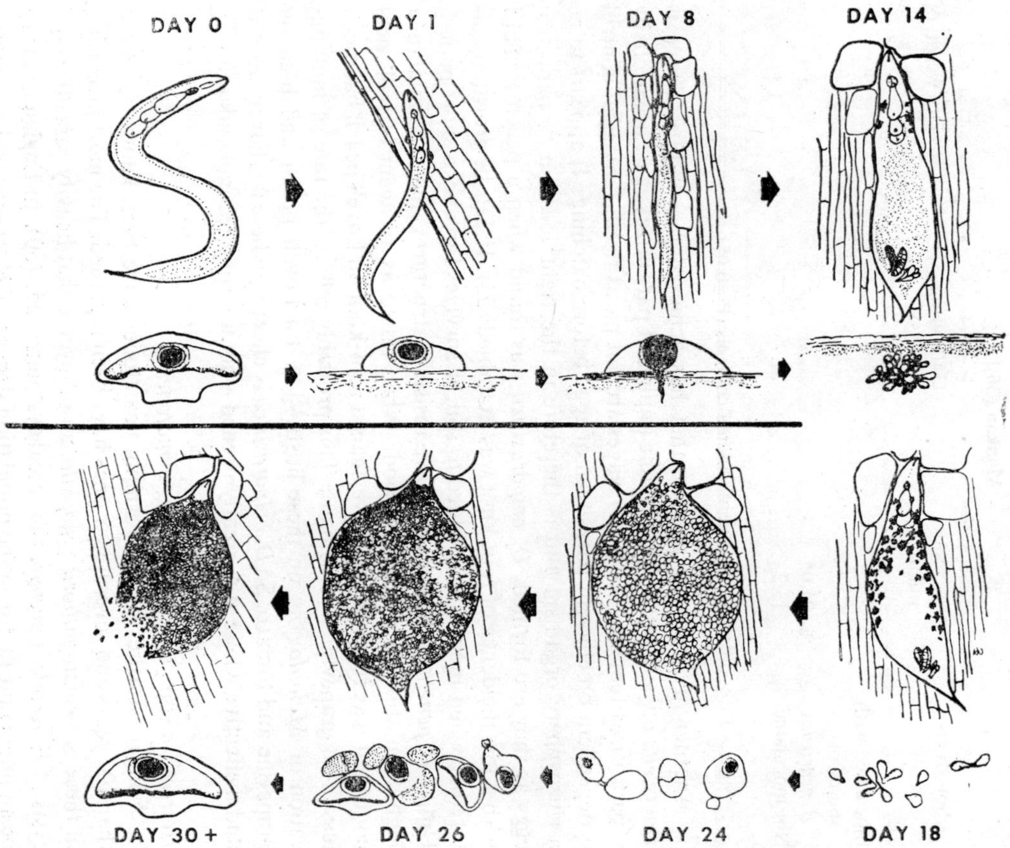

Fig. 7. Life-cycles of the root-knot nematode and *Bacillus penetrans*; enlarged life stage of *B. penetrans* is shown below each developmental stage of the nematode [Sayre (45)].

control of the nematode (51).

The most serious disadvantage of *B. penetrans* is that it appears to be an obligate parasite and cannot be grown in axenic culture. Although this is a serious disadvantage, it is not insuperable. In fact, a system has been developed to build up large numbers of spores in *Meloidogyne* females on tomato roots (52). *B. penetrans* presents one of the best opportunities for biological control of a major crop pathogen.

Table 9. Biological control of root-knot of tomato by *Bacillus penetrans*. Assessment 70 days after inoculation of plants with 10,000 larvae of *M. incognito* [data from Mankau (36)]

Soil treatment	Leaves/ plant	Gall rating (0–25)	Dry weight of tops (g)
Inoculated with spores of *B. penetrans*	21.0*	6.5*	5.32*
Not inoculated	13.2	22.7	3.42

*Significantly different from uninoculated controls (P=0.05).

Another *Meloidogyne* parasite that has achieved recent prominence is *Dactylella oviparasitica*. As the name implies, it parasitizes eggs; occasionally it parasitizes females. As larval stages are not parasitized, biological control is dependent on efficient parasitism of eggs before hatching. It appears to be a rhizosphere organism and is therefore in the right location to parasitize eggs as they are formed. *D. oviparasitica* was found when a peach orchard with a low incidence of root knot was examined (53). It can be grown in culture but when inoculum is added to soil, it survives for only a short period. Unlike *B. penetrans*, it does not have resistant structures that can maintain a high population over a long period. When tested as an agent of biological control, it was relatively effective against root-knot on Lovell peach but much less so on grapevine and tomato. This probably reflects the rate of multiplication of *Meloidogyne* on these hosts—low on Lovell peach and high on grapevine and tomato. So *D. oviparasitica* is effective where it is hardly needed and ineffective where it is greatly needed—not a strong recommendation.

Heterodera spp., *the cyst-forming nematodes*

Fungal parasites of cyst-forming nematodes have been known for more than 100 years and several new ones have been recorded. The most promising of these is *Nematophthora gynophila*, an oomycete that heavily parasitizes the cysts of *Heterodera avenae*, the cereal cyst nematode (30). In England, it may be an important factor in the population decline of *H. avenae* under continuous cereal culture where it is the most prevalent parasite. The fungus produces biflagellate zoospores that penetrate the developing females of *H. avenae* as they erupt through the cortex. Both females and eggs are killed. Masses of thick-walled oospores are produced in infected cysts (Plate 2, E and F). The fungus appears to be an obligate parasite and a special technique for multiplying it will have to be developed if it is to be used commercially as an agent of biological control. *N. gynophila* has been reported in cysts of *H. avenae* only from Britain although what appears to be the same fungus was found in *H. schachtii* females from Sweden. In glasshouse tests, *H. carotae*, *H. cruciferae*, *H. goettingiana*, *H. schachtii* and *H. trifolii* were infected but not the potato cyst nematode, *Globodera rostochiensis*. Cysts of *G. rostochiensis*, unlike those of *H. avenae* do not decline in numbers under continuous monoculture of a susceptible host. Instead, the soil becomes "sick" and economic crops of potato can no longer be grown. It is interesting to speculate that this may be because it is not susceptible to *N. gynophila*. Perhaps an effective parasite might be found in Central or South America, the centre of origin of the potato.

The world distribution of *N. gynophila* is not known. Where *H. avenae* is a serious pest, as in Australia, it may be because *N. gynophila* is absent. Alternatively, it may be because soil moisture conditions are not suitable for *N. gynophila*. Infection is by motile zoospores and therefore adequate moisture will be necessary when *Meloidogyne* females are emerging from roots (31). Even allowing for this important constraint, the possibility exists of extending the biological control of *H. avenae* and perhaps of other *Heterodera* spp. by *N. gynophila*.

Conclusions

Only two biological control agents are presently being used routinely to control soil-borne pathogens. *Peniophora gigantea* was first used commercially to control *Heterobasidium annosum* in 1962 and is now widely used in the U.K. and the U.S.A. The method depends on competition for substrate viz., the butts of felled pine trees. The biological control procedure is to ensure that all stumps become colonized by *P. gigantea* rather than by *H. annosum*. As with many other situations, possession is nine-tenths of the law. It could be argued that an airborne pathogen is being controlled, not a soil-borne pathogen. Certainly control is directed against the airborne phase of *H. annosum* and it is therefore difficult to see how the method could be extended or developed to control other root pathogens.

Biological control of *Agrobacterium tumefaciens* was introduced to commercial nurseries eleven years later, in 1973. Its great significance is that it is the first successful example of a disease control organism operating in soil. The organism, strain 84 of *A. radiobacter*, colonizes the roots of inoculated plants and the circumstantial evidence is overwhelming that it produces on the roots an antibiotic called agrocin 84 that inhibits the crown gall-inducing pathogen. The procedure is not "in flagrant contradiction" to basic ecological principles. Inoculation may not even increase the total number of cells of *Agrobacterium* on the root surface; it is just that one particular strain is given an advantage and it predominates for long enough to prevent infection by the pathogen. If it can be done with one organism, it should be possible to do it with others. Another important feature of the biological control of crown gall is that the control organism was found on the surface of diseased tissue and we now know that that is where we would expect to find it. So we should not confine our search for biological control agents to situations where there is no disease or where disease incidence has declined. There is one main problem in the biological control of crown gall—the effectiveness of the treatment could break down because the genes controlling agrocin 84 production can be transferred to pathogenic strains which, as a result, become resistant to agrocin 84. As agrocin 84 production is the basis of control, such strains are no longer subject to control.

There have been several promising developments in the biological control of other soil-borne pathogens. Perhaps the most promising is the control of *Sclerotinia* spp. by *Sporidesmium sclerotivorum* followed by the control of *Sclerotium* spp. by *Coniothyrium minitans.* Both control organisms are parasitic on sclerotia. It is interesting that both methods involve the destruction of resting structures which are normally considered highly resistant to attack by other microorganisms. The vulnerable stage has always been thought to be the period between germination and infection. Perhaps we should revise our ideas.

Nematode parasites, as opposed to predators, seem to be the most promising organisms for nematode control. Of these, the bacterium *Bacillus penetrans* to control *Meloidogyne* spp. has greatest potential, followed by the fungus *Nematophthora gynophila* to control *Heterodera* spp. It is surprising that parasites of pathogens have not been more successful as biological control agents. Wherever bacteria or fungi are grown commercially such as in cheese making, antibiotic production and mushroom culture, they become infected by parasites and great precautions have to be taken to protect them. Why cannot parasites be used to control unwanted bacteria and fungi? The reason is not known. Perhaps it is because few of the parasites have the ability to seek their host; infection is at random and may not give adequate control. Alternatively, parasites have been overlooked until recently and successful control measures using them may yet be developed. Only time will tell.

Work on *Pseudomonas fluorescens* strains which produce siderophores that deprive pathogens of iron is too recent to assess accurately the potential of this organism for biological control. It looks promising but much more work is required. However, as the mechanism is understood, the prospects for manipulation are greatly improved. Perhaps bacterization of non-leguminous seed will become a common feature of world agriculture and will not be confined to the Soviet Union. Experimentally, *P. fluorescens* has been shown to control the wilt pathogen, *Fusarium oxysporum* and also *Gaeumannomyces graminis* var. *tritici* the cause of take-all of wheat and barley. Other prospects for the control of the latter disease are hypovirulent fungi closely related to the pathogen. An understanding of take-all decline might be of considerable importance in this regard. Hypovirulence associated with dsRNA elements is an intriguing phenomenon and may be implemented in the control of *Rhizoctonia solani* and other soil-borne pathogens.

The reader may have gained the impression that the widespread use of biological methods in plant pathology is just around the corner. If so, this impression was not intended. If all the organisms mentioned in this chapter were developed to a stage where they were used commercially to control disease, it would be a major achievement but it would still mean that relatively few plant diseases are subject to biological control, in the restricted sense of the definition used in the introduction. Breeding for disease resistance, chemical sprays and dusts, cultural practices and hygiene in all its aspects, especially the provision of disease-free planting material will remain the most important methods of disease control. Rachel Carson was too optimistic, it seems.

References

1. Adams, P.B. and Ayers, W.A. *Sporidesmium sclerotivorum*: distribution and function in natural biological control of sclerotial fungi, *Phytopathology*, *71*, 96–93 (1981).
2. Ahmed, A.H.M. and Tribe, H.T. Biological control of white rot of onion (*Sclerotium cepivorum*) by *Coniothyrum minitans*, *Plant Pathology*, *26*, 75–78 (1977).
3. Alabouvette, C., Rouxel, F. and Louvet, J. Characteristics of *Fusarium* wilt-suppressive soils and prospects for their utilization in biological control, in *Soil-borne Plant Pathogens* (Edited by B. Schippers and W. Gams), Academic Press, London, New York, San Francisco (1979).
4. Asher, M.J.C. Interactions between isolates of *Gaeumannomyces graminis* var. *tritici*, *Transactions of the British Mycological Society*, *71*, 367–373 (1978).
5. Ayers, W.A. and Adams, P.B. Mycoparasitism of sclerotia of *Sclerotinia* and *Sclerotium* species by *Sporidesmium sclerotivorum*, *Canadian Journal of Microbiology*, *25*, 17–23.
6. Ayers, W.A. and Adams, P.B. Factors affecting germination, mycoparasitism and survival of *Sporidesmium sclerotivorum*, *Canadian Journal of*

Microbiology, *25*, 1021–1026 (1979).

7. Backman, P.A. and Rodriguez-Kabana, R. A system for the growth and delivery of biological control agents to the soil, *Phytopathology*, *65*, 819–821 (1975).
8. Baker, K.F. and Cook, R.J. *Biological Control of Plant Pathogens*, W.H. Freeman and Co., San Francisco (1974).
9. Balis, C. A comparative study of *Phialophora radicicola*, an avirulent root parasite of grasses and cereals, *Annals of Applied Biology*, *66*, 59–73 (1970).
10. Broadbent, Pat, Baker, K.F. and Waterworth, Yvonne. Bacteria and actinomycetes antagonistic to fungal root pathogens in Australian soils, *Australian Journal of Biological Sciences*, *24*, 925 (1971).
11. Brown, M.E. Seed and root bacterization, *Annual Review of Phytopathology*, *12*, 181–197 (1974).
12. Carson, R. *Silent Spring*, Hamilton, London (1963).
13. Castanho, B. and Butler, E.E. *Rhizoctonia* decline: studies on hypovirulence and potential use in biological control, *Phytopathology*, *68*, 1511–1514 (1978).
14. Castanho, B., Butler, E.E. and Shepherd, R.J. The association of double-stranded RNA with *Rhizoctonia* decline, *Phytopathology*, *68*, 1515–1519 (1978).
15. Chambers, S.C. Pathogenic variation in *Ophiobolus graminis*, *Australian Journal of Biological Science*, *23*, 1099–1103 (1970).
16. Chambers, S.C. and Flentje, N.T. Studies on variation with *Ophiobolus graminis*, *Australian Journal of Biological Sciences*, *20*, 941–951 (1967).
17. Charudattan, R. *Biological Control Projects in Plant Pathology—A Directory*, Institute of Food and Agricultural Sciences, University of Florida (1978).
18. Cunningham, P.C. Some consequences of cereal monoculture on *Gaeumannomyces graminis* (Sacc.) Arx & Olivier and the take-all disease, *European and Mediterranean Plant Protection Organisation*, *5*, 297–317 (1975).
19. Deacon, J.W. *Phialophora radicicola* and *Gaeumannomyces graminis* on roots of grasses and cereals, *Transactions of the British Mycological Society*, *61*, 471–485.
20. Deacon, J.W. Biological control of the take-all fungus *Gaeumannomyces graminis* by *Phialophora radicicola* and similar fungi, *Soil Biology and Biochemistry*, *8*, 275–283 (1976).
21. Deverall, B.J., Wong, P.T.W. and McLeod, S. Failure to implicate antifungal substances in cross-protection of wheat against take-all, *Transactions of the British Mycological Society*, *72*, 1233–1236 (1979).
22. Ellis, J.G. and Kerr, A. Transfer of agrocin 84 production from strain 84 to pathogenic recipients: a comment on the previous paper, in *Soil-*

borne Plant Pathogens (Edited by B. Schippers and W. Gams), Academic Press, London, New York, San Francisco (1979).

23. Ellis, J.G., Kerr, A., Van Montagu, M. and Schell, J. *Agrobacterium*: genetic studies on agrocin 84 production and the biological control of crown gall, *Physiological Plant Pathology*, *15*, 311–319 (1979).
24. Ellis, J.G. and Murphy, P.J. Four new opines from crown gall tumours—their detection and properties, *Molecular and General Genetics*, *181*, 36–43 (1981).
25. Garrett, S.D. *Biology of Root-infecting Fungi*, Cambridge University Press, London (1956).
26. Griffin, D.M. *Ecology of Soil Fungi*, Chapman and Hall, London (1972).
27. Hagen, K.S., Borbosch, S. and McMurtry, J.A. The biology and impact of predators, in *Theory and Practice of Biological Control* (Edited by C.B. Huffaker and P.S. Messenger), Academic Press, London, New York, San Francisco (1976).
28. Hornby, D. Take-all decline: a theorist's paradise, in *Soil-borne Plant Pathogens* (Edited by B. Schippers and W. Gams), Academic Press, London, New York, San Francisco (1979).
29. Kerr, A. Biological control of crown gall through production of agrocin 84, *Plant Disease*, *64*, 25–30 (1980).
30. Kerry, B.R. and Crump, D.H. Two fungi parasitic on females of cyst-nematodes (*Heterodera* spp.), *Transactions of the British Mycological Society*, *74*, 119–125 (1980).
31. Kerry, B.R., Crump, D.H. and Mullen, L.A. Parasitic fungi, soil moisture and multiplication of the cereal cyst nematode, *Heterodera avenae*, *Nematologica*, *26*, 57–68 (1980).
32. Kloepper, J.W., Long, J., Teintze, M. and Schroth, M.N. *Pseudomonas* siderophores: a mechanism explaining disease-suppressive soils, *Current Microbiology*, *4*, 317–320 (1980).
33. Kloepper, J.W., Leong, J., Teintze, M. and Schroth, M.N. Enhanced plant growth by siderophores produced by plant growth-prompting rhizobacteria, *Nature*, *286*, 885–886 (1980).
34. Lapierre, H., Lemaire, J.-M., Jouan, B. and Molin, G. Mise en évidence de particules virales associeés à une perte de pathogénicité chez le piétin-échaudage des céréales, *Ophiobolus graminis* Sacc., *Comptes rendus hebdomedaire des seances de l'Academie des Sciences, Paris D*, *271*, 1833–1836 (1970).
35. Lemaire, J.-M., Jouan, B., Capperet, M., Perraton, B. and Lecarre, L. Lutte biologique contre le piétin-échaudage des céréalis par l'utilisation de souches hypoagressives d'*Ophiobolus graminis*. Le caractère hypoagressif est il contagieux? *Sciences Agronomiques Rennes*, 63–65 (1976).
36. Mankau, R. Biological control of nematode pests by natural enemies, *Annual Review of Phytopathology*, *18*, 415–440 (1980).

37. Menzies, J.D. Occurrence and transfer of a biological factor in soil that suppresses potato scab, *Phytopathology*, *49*, 648–652 (1959).
38. Merriman, P.R., Price, R.D., Kollmorgen, J.F., Piggott, T. and Ridge, E.H. The effect of seed inoculation with *Bacillus subtilis* and *Streptomyces griseus* on the growth of cereals and carrots, *Australian Journal of Agricultural Research*, *25*, 219–226 (1974).
39. Moore, L.W. and Warren, G. *Agrobacterium radiobacter* strain 84 and biological control of crown gall, *Review of Phytopathology*, *17*, 163–179 (1979).
40. Odvody, G.N., Boosalis, M.G. and Kerr, C.D. Biological control of *Rhizoctonia solani* with a soil-inhabiting basidiomycete, *Phytopathology*, *70*, 655–658 (1980).
41. Orwell, G. *Animal Farm*, Secker and Warburg, London (1959).
42. Panagopoulos, C.G., Psallidas, P.G. and Alivizatos, A.S. Evidence of a breakdown in the effectiveness of biological control of crown gall, in *Soil-borne Plant Pathogens* (Edited by B. Schippers and W. Gams), Academic Press, London, New York, San Francisco (1979).
43. Rawlinson, C.J., Hornby, D., Pearson, V. and Carpenter, J.M. Virus-like particles in the take-all fungus, *Gaeumannomyces graminis*, *Annals of Applied Biology*, *74*, 197–209 (1973).
44. Rishbeth, J. Stump inoculation: a biological control of *Fomes annosus*, in *Biology and Control of Soil-borne Plant Pathogens* (Edited by G.W. Bruehl), American Phytopathological Society, St. Paul, Minnesota (1975).
45. Sayre, R.M. Promising organisms for biocontrol of nematodes, *Plant Disease*, *64*, 526–532 (1980).
46. Schell, J., Van Montagu, M., De Beuckeleer, M., De Block, M., Depicker, A., De Wilde, M., Engler, G., Genetello, C., Hernalsteens, J.P., Holsters, M., Seurinck, J., Silva, A., Van Vliet, F. and Villarroell, R. Interactions and DNA transfer between *Agrobacterium tumefaciens*, the Ti plasmid and the plant host, *Proceedings of the Royal Society, London B204*, 251–266 (1979).
47. Scher, F.M. and Baker, R. Mechanism of biological control in *Fusarium*-suppressive soil, *Phytopathology*, *70*, 412–417 (1980).
48. Sequeira, L., Gaard, G. and De Zoeten, G.A. Interaction of bacteria and host cell walls: its relation to mechanisms of induced resistance, *Physiological Plant Pathology*, *10*, 43–50 (1977).
49. Shipton, P.J. Take-all decline during cereal monoculture, in *Biology and Control of Soil-borne Plant Pathogens* (Edited by G.W. Bruehl), The American Phytopathological Society, St. Paul, Minnesota (1975).
50. Sivasithamparam, K. and Parker, C.A. Effect of certain isolates of soil fungi on take-all of wheat, *Australian Journal of Botany*, *28*, 421–427 (1980).
51. Stirling, G.R. and White, A.M. The distribution of a parasite of root-

knot nematodes in South Australian vineyards, *Plant Disease*, in press.

52. Stirling, G.R. and Wachtel, M.F. Mass production of *Bacillus penetrans* for the biological control of root-knot nematodes, *Nematologica*, *26*, 308–312 (1980).
53. Stirling, G.R., McKenry, M.V. and Mankau, R. Biological control of root-knot nematodes (*Meloidogyne* spp.) on peach, *Phytopathology*, *69*, 806–809 (1979).
54. Stotsky, G. and Rem, L.T. Influence of clay minerals on micro-organisms I–IV, *Canadian Journal of Microbiology*, *12*, 547–563, 831–848, 1235–1246, *13*, 1535–1550 (1966–67).
55. Stover, R.H. *Fusarium* wilt (Panama disease) of bananas and other *Musa* species, *Commonwealth Mycological Institute Phytopathological Paper*, *4*, 1–117 (1962).
56. Tate, M.E., Murphy, P.J., Roberts, W.P. and Kerr, A. Adenine N^6-substituent of agrocin 84 determines its bacteriocin-like specificity, *Nature, London*, *280*, 697–699 (1979).
57. Uecker, F.A., Ayers, W.A. and Adams, P.B. A new hyphomycete on sclerotia of *Sclerotinia sclerotiorum*, *Mycotaxon*, *7*, 275–282 (1978).
58. Van Larebeke, N., Genetello, C., Schell, J., Schilperoort, R.A., Hermans, A.K., Hernalsteens, J.P. and Van Montagu, M. Acquisition of tumour-inducing ability by non-oncogenic bacteria as a result of plasmid transfer, *Nature, London*, *255*, 742–743 (1975).
59. Walker, J. Take-all diseases of Gramineae: a review of recent work, *Review of Plant Pathology*, *54*, 113–144 (1975).
60. Watson, B., Currier, T.C., Gordon, M.-P., Chilton, M-D. and Nester, E.W. Plasmid required for virulence of *Agrobacterium tumefaciens*, *Journal of Bacteriology*, *123*, 255–264 (1975).
61. Weindling, R. *Trichoderma lignorum* as a parasite of other soil fungi, *Phytopathology*, *22*, 837–845 (1932).
62. Wells, H.D., Bell, D.K. and Jawenski, C.A. Efficacy of *Trichoderma harzianum* as a biocontrol for *Sclerotium rolfsii*, *Phytopathology*, *62*, 442–447 (1972).
63. White, L.O. The taxonomy of the crown gall organism *Agrobacterium tumefaciens* and its relationship to rhizobia and other agrobacteria, *Journal of General Microbiology*, *72*, 565–574 (1972).
64. Wong, P.T.W. Cross-protection against the wheat and oat take-all fungi by *Gaeumannomyces graminis* var. *graminis*, *Soil Biology and Biochemistry*, *7*, 189–194 (1975).
65. Wong, P.T.W. Effect of temperature on growth of some avirulent fungi and cross-protection against the wheat take-all fungus, *Annals of Applied Biology*, *95*, 291–299.
66. Wright, J.M. Biological control of a soil-borne *Pythium* infection by seed inoculation, *Plant and Soil*, *8*, *2*, 132–140 (1957).

T. MISATO and K. YONEYAMA

18. Agricultural Antibiotics

Introduction

One of the main obstacles in crop cultivation is the damage caused by plant diseases and insect pests. Although the use of pesticides in pest control seems to be indispensable to attain high agricultural productivity, there is growing concern about pesticides being harmful to mammals or wildlife, and a possible source of environmental pollution. It has become important to search for pesticides free from such shortcomings.

Among the pesticides with more desirable characteristics, antibiotics are more valued as they are generally selective between target organisms and non-target ones, and are decomposable more easily in the environment. However, their selective action does become a demerit often causing resistance in the target microorganisms or insect pests after repeated application.

The antibiotics first introduced in agriculture were those originally developed for medical purposes. The first success was the use of streptomycin for the control of pear fire blight in the U.S.A. Medical antibiotics, however, have limitations when used for plant protection because of the different requirements to be met in the agricultural field; for example: they must not have phytotoxicity; must not be rapidly inactivated by sunlight or washed away by rain; and the manufacturing cost must be reasonably low.

For this reason antibiotics usable for agricultural purposes have been specifically developed. The first was blasticidin S developed in Japan and still widely used for rice blast control by farmers. Since then many have been developed and have been put to practical use in agricultural fields. Table 1 lists those being used in Japan where antibiotics in agricultural production are used more frequently than in other countries.

In the following pages, the chemistry, biological activity, toxicity and utilization as plant disease control agents of major agricultural antibiotics used presently, as well as those used in the past, are outlined. Those showing promise for practical application are also mentioned.

Table 1. Practical use of agricultural antibiotics in Japan

Antibiotics	Formulation types	Plant diseases to be applied	Antibiotic concentrations for application
Antifungal antibiotics			
Cycloheximide	Wettable powder cycloheximide 0.5 or 1%	Onion downy mildew Shoot blight of Japanese larch	1.3 ppm 5 ppm
Blasticidin S	Dust blasticidin S 0.08%		3–4 kg/10 a
	Wettable powder blasticidin S 2%	Rice blast	20 ppm
	Emulsion blasticidin S 1%		7–10 ppm
Kasugamycin	Dust kasugamycin 0.2 or 0.3%		2–4 kg/10 a
	Wettable powder kasugamycin 2%	Rice blast	20 ppm
	Emulsion kasugamycin 2%		20 ppm
Polyoxins	Dust polyoxin D 0.04 or 25%	Rice sheath blight	3–4 kg/10 a
	Wettable powder polyoxin B 10% Emulsion polyoxin AL 10% in polyoxin B	Fungal diseases of fruits and vegetables	100–200 ppm
	polyoxin Z 2.2% in polyoxin D	Rice sheath blight	22 ppm
Validamycin A	Dust validamycin A 0.3%	Rice sheath blight	4 kg/10 a
	Liquid validamycin A 3%		30–60 ppm
Antibacterial antibiotics			
Streptomycin	Wettable powder streptomycin 5 or 20% Liquid streptomycin 20%	Bacterial diseases of fruits and vegetables	100–200 ppm
Oxytetracycline	Wettable powder streptomycin 15% + oxytetracycline 1.5%	Citrus canker Peach bacterial leaf spot	10–15 ppm
Insecticidal antibiotic			
Tetranactin	Emulsion tetranactin 16% + CPCBS 20% or tetranactin 12% + BPMC 30%	Carmine mite of fruits and tea	120–160 ppm

Agricultural antibiotics and the pollution problem

Environmental hazard caused by conventional agricultural chemicals is classified into two categories: a) non-selective toxicity (parathion), and b) concentration and accumulation of toxic compounds in the environment (DDT and BHC). Pollution-free pesticides, therefore, should have selective toxicity to target organisms and be sensitive to the photolysis and degradation by soil microorganisms. From these points of view, antibiotics may be presumed to be useful biodegradable pesticides. As is true for every scientific technique, the use of agricultural antibiotics has its advantages and limitations.

Advantages

1) SELECTIVE TOXICITY TO TARGET ORGANISMS. Since most antibiotics have selective toxicity to target organisms and low mammalian toxicity, they can be safely used without harming man, livestock, fish and crops (Table 2).

Table 2. Mode of action of antibiotics

Antibiotic	Primary action site
Polyoxins	Chitin synthesis of cell wall
Tetranactin	Cation leakage from mitochondria
Validamycin A	Biosynthesis of inositol
Blasticidin S Kasugamycin Cycloheximide Streptomycin Tetracyclines Chloramphenicol	Protein synthesis
Cellocidin Novobiocin	DNA synthesis

2) DEGRADATION BY SOIL MICROORGANISMS. After application to the crop, antibiotics are liable to be rapidly broken down in the environment, so that there is no likelihood of the danger of environmental pollution and food contamination.

3) SMALL AMOUNT OF A COMPOUND USED IN A UNIT AREA. Since agricultural antibiotics are sprayed at a very low concentration, the amount of a compound sprayed in a unit area is far less (1/10–1/100) than those of other conventional pesticidal chemicals.

4) MANUFACTURING OF BIOACTIVE COMPOUNDS WITH COMPLEX CHEMICAL STRUCTURES. Novel bioactive compounds with very complex chemical structures which are outside the domain of organic synthesis, can be isolated and manufactured on a commercial basis.

5) FAVOURABLE INVESTMENT IN EQUIPMENT. Various antibiotics can be produced by using a single set of equipment and facilities. This is an advantage towards the low initial cost of antibiotic production.

6) UTILIZATION OF SOLAR ENERGY: Antibiotics are produced by utilizing agricultural products which are obtained from the photosynthetic conversion of solar energy. The production of antibiotics, therefore, does not consume as much of non-renewable energy such as oil and coal.

Limitations

1) DIFFICULTY FOR ANALYSIS IN MICRO-SCALE. Antibiotics are generally mixtures of various structurally related components like polyoxins. This complexity creates a difficulty in micro-scale analysis and in the safety evaluation of compounds.

2) RESISTANCE OF PLANT PATHOGENS TO ANTIBIOTICS. Tolerance or resistance of pathogenic microorganisms to antibiotics has often occurred shortly after their application to control plant diseases. To reduce or avoid the emergence of tolerant fungi and bacteria the alternate or combined application of chemicals which have different mechanisms of action is recommended.

Public health aspects

A limited number and relatively small quantities of medical antibiotics have been introduced to agricultural use but most agricultural antibiotics are those which have been used only for plant protection purposes. Agricultural antibiotics are not concerned primarily with the health of human beings and their use has macro-environmental consequences. Most human infectious diseases are caused by bacteria and viruses, while plant pathogens are mostly classified as fungi. Accordingly, whereas most medical antibiotics are effective against bacteria, agricultural antibiotics are generally fungicidal. Selective agricultural antibiotic action can, therefore, be toxic to fungi responsible for the target plant disease without damaging other microorganisms such as bacteria which are parasitic on humans. The application of antibiotics to control plant pests will never result in the development of microorganisms resistant to medical antibiotics. Some antibiotics can be synthesized chemically. But in this respect there is no difference between antibiotics and synthetic chemicals.

Antifungal antibiotics

Cycloheximide

Cycloheximide, also known as Actidione, is obtained from a byproduct of streptomycin in the culture filtrate of *Streptomyces griseus* (101). It is markedly active against a wide range of fungi and yeasts, but remains inactive against bacteria. Its use against plant diseases is rather limited by its phytotoxicity.

CHEMISTRY. Cycloheximide belongs to glutarimide group of antibiotics which have 1-hydroxy-2-(3-glutarimidyl)-ethyl moiety in their molecules. The physicochemical properties of cycloheximide are as follows: crystalline needle; melting point 118–119°C; $[\alpha]_D^{25}$–6.8° in H_2O; soluble in water and most organic solvents except petroleum ether; stable in aqueous solution at pH 3 to 5, but unstable in alkaline solution.

The chemical structure of cycloheximide is β-[2-(3,5-dimethyl-2-oxocyclohexyl)-2-hydroxyethyl]-glutarimide (96), as shown in Fig. 1. It has four asymmetric centres at C-2, C-4, C-6, and C-2′, and therefore a number of stereoisomers are possible. Naramycin B, a stereoisomer of cycloheximide, is produced from *Streptomyces naraensis* (123), and similarly isocycloheximide is found in aged filtrate from culture of *S. griseus* (103). Isocycloheximide is also obtained from the isomerization of cycloheximide with acid deactivated alumina. Both naramycin B and isocycloheximide appear to be less active than cycloheximide against yeasts; the former is only 32% as active to *Saccharomyces sake* (122), and the latter about 30% as active to *S. pastorianus* (103).

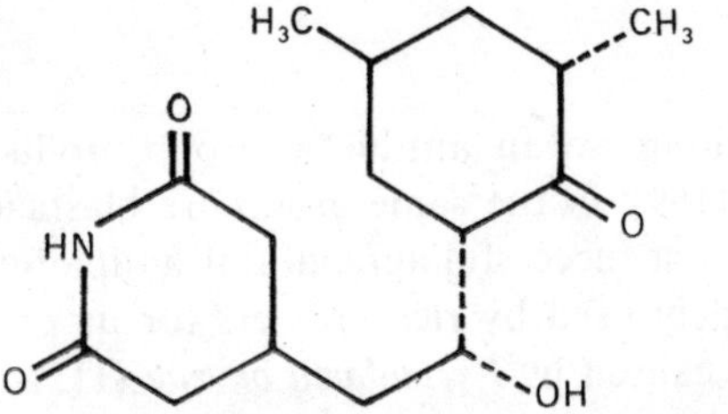

Fig. 1. Structure of cycloheximide.

BIOLOGICAL ACTIVITY. Cycloheximide is active on a wide spectrum of organisms, including plant pathogenic fungi, yeasts, algae, and protozoa, but it is not effective against bacteria (178). The antibiotic has antiviral and antitumour activities (34), although it is apparently too toxic for practical use. It is a highly effective rat repellent (177).

Even though cycloheximide is active to a wide range of organisms, it shows marked specificity for closely related organisms. For example, growth of some species of yeasts is completely suppressed at concentrations less than 0.2 μg/ml, while other species grow readily in the presence of 1000 μg/ml (178). Concentrations of the antibiotic required to inhibit growth of 33 species of phytopathogenic fungi range from 0.125–1000 μg/ml (179).

Cycloheximide is a potent inhibitor of protein synthesis in eucaryotic cells such as fungi and animals (180, 181). It acts on the 60S subunit of 80S ribosome, and inhibits the function of translocation of peptidyl-tRNA from A to P ribosomal sites, release of deacylated tRNA from the ribosome, and reinitiation of new peptide chains (16). On the other hand, cycloheximide markedly inhibits DNA synthesis in fungal cells (95). Whether inhibition of DNA synthesis results from an action independent of that of protein synthesis is not clear (148).

ANIMAL AND PLANT TOXICITIES. The toxicity of cycloheximide is high; the oral LD_{50}'s for mice and guinea pigs are 133 mg/kg and 65 mg/kg respectively, and the LD_{50}'s for intravenous are 190 μg/kg in mice and 2.5 μg/kg in rats. The TLm of the antibiotic to carp is 2.3 ppm. Cycloheximide is phytotoxic to many crops at more than 2 ppm. Cycloheximide-injured leaves also show a decrease in the amount of chlorophyll and an increase in the content of RNA. It also causes chromosome aberrations and an abnormal cell division in root-tip cells of wheat. Semicarbazone derivatives of cycloheximide show a reduced phytotoxicity without decreasing the original antifungal activity.

USE AGAINST PLANT DISEASES. Application of cycloheximide to plants is limited because of its high phytotoxicity. It has been used to control downy mildew of onions by spraying 2 ppm solution. The antibiotic is also used as a fungicide in forests, and it is effective against shoot blight of Japanese larch at 3 ppm.

Blasticidin S

Blasticidin S belongs to an antibiotic group produced by *Streptomyces griseochromogenes* (169); in the same group are blasticidin A, B and C (40). Blasticidin S is the first successful agricultural antibiotic developed in Japan, and it has been widely used by rice growers for its preventive and curative effects on rice blast caused by *Pyricularia oryzae* (112). Blasticidin S benzylaminobenzene sulphonate is least phytotoxic to the rice plant without reducing antifungal activity against *P. oryzae* (6).

CHEMISTRY. Blasticidin S is obtained as a white needle crystal, and the molecular formula is $C_{17}H_{26}O_5N_8$. Its physical and chemical properties are: melting point 235–236°C for free base; specific rotation $[\alpha]_D^{11^\circ} = +108.4°$ in 1 per cent aqueous solution; maximum UV absorption 275 nm $[E^{1\%}_{1cm}\ 349]$ in N/10 HCl and 266 nm $[E^{1\%}_{1cm}\ 266]$ in N/10 NaOH; very soluble in water and acetic acid, insoluble in organic solvents such as methanol, ethanol, acetone, benzene, chloroform and ether; stable in solution at pH 5–6, but unstable at pH 4 (by heating for 10 minutes).

The chemical structure is 1-(1'N-cytocinyl)-4-[L-3'-amino-5' (1''-N-methylguanidino) valerylamino]-1,2,3,4,-tetradeoxy-β-D-erythro-hex-2-eneuronic acid, as shown in Fig. 2. The molecule of blasticidin S consists of a novel nucleoside designated cytosinine and a new β-amino acid named blasticidic acid (ε-N-Methyl-β-arginine) (124, 125, 186).

In the biosynthesis of blasticidin S, the pyrimidine ring is derived from cytosine, the sugar moiety from glucose, the N-methyl group of blasticidic acid from methionine, and arginine is used as the precursor of blasticidic acid (145, 146). A metabolic intermediate on its biosynthetic pathway is known to be leucylblasticidin S (144).

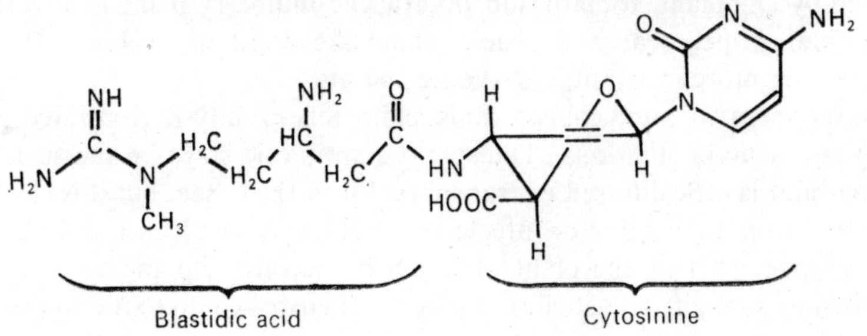

Fig. 2. Structure of blasticidin S.

Biological activity. Blasticidin S inhibits the growth of a wide range of bacterial and fungal cells (169), and it also has antiviral (57) and antitumour (164) activities. It has inhibitory effects on the spore germination, mycelial growth and sporulation of *P. oryzae* at low concentrations of less than 1 μg/ml. Its excellent curative effect on rice blast may be due to its perfect inhibitory activity on the mycelial growth even at 0.01 μg/ml (112).

Blasticidin S is an inhibitor of protein synthesis. The antibiotic markedly blocks the incorporation of ^{14}C-amino acid into protein in the intact cells or cell-free systems of *P. oryzae*, it does not affect glycolysis, the succinate dehydrogenase system, the electron transport system, oxidative phosphorylation, or the incorporation of ^{32}p into nucleic acids (69, 113, 114). Although the action of blasticidin S is still unknown in detail at the molecular level, the site of inhibition may be involved in certain processes of a peptidyl transferase activity in protein synthesis (20, 186, 191).

Animal and plant toxicities. Blasticidin S is rather toxic to mammals with the LD_{50} oral of 53.3 mg/kg for mice. Fortunately, it can be used in the paddy field since its toxicity is low to fish. Blasticidin S causes severe irritation to eyes and inflammation of skin on direct contact. A simple method to alleviate eye irritation is by the addition of calcium acetate to blasticidin S dust formulation to reduce irritation without impeding its antiblast effect. This improved dust is now used in practice for agriculture. Detoxin isolated from the culture broth of *Streptomyces caespitosum*, also reduces the phytotoxicity and irritant effect on eyes without reducing antifungal activity (188), although there are still some difficulties for it to be put into practical use.

Use against plant diseases. In the field application to control rice blast, the effective concentration of blasticidin S is usually 10–20 ppm (1–3 g blasticidin S/10a). If the concentration is increased it occasionally causes chemical injury to rice leaves when sprayed. Such phytotoxicity of blasticidin S on rice leaves is also closely related to the frequency and time of application, rice variety grown, or atmospheric conditions such as temperature and moisture. Among other crops, the tobacco plant is most susceptible to blasticidin S, followed by egg plant, tomato and potato; the mulberry plant is also rather susceptible; grape, pear and peach plants are resistant; water melon and cucumber are more resistant, like the rice plants.

ENVIRONMENTAL METABOLISM. Blasticidin S is rapidly decomposed under the environmental conditions. The sprayed antibiotic stays on the surface of the plant and is little diffused or transported into the tissue, but it is easily incorporated from the wound or infected part of leaves and translocated mainly into the upper part of the plant. The antibiotic on the plant surface is decomposed by sunlight. Blasticidin S fallen and adsorbed on to the soil surface is easily decomposed to carbon dioxide by soil microorganisms (185).

Kasugamycin

Kasugamycin is an aminoglycoside antibiotic, which is produced by *Streptomyces kasugaensis* (171). Kasugamycin has been widely used as an agricultural antibiotic for rice blast control, and it can be safely used on crop plants without any phytotoxicity. It has very low toxicity to mammals and fish.

CHEMISTRY. Kasugamycin is a water-soluble and basic antibiotic. Kasugamycin hydrochloride is water-soluble, but insoluble in methanol, ethanol, acetone, ethyl acetate, chloroform, benzene and other solvents, and it has a melting point of 202–204° C (decomposition), specific rotation $[\alpha]_D^{25} = +120°$. The chemical structure of kasugamycin is shown in Fig. 3. The molecule consists of three moieties, namely, D-inositol, kasugamine (2,3,4,6-tetradeoxy-2,4-diamino-hexopyranose) and an iminoacetic acid side chain (72, 156, 157).

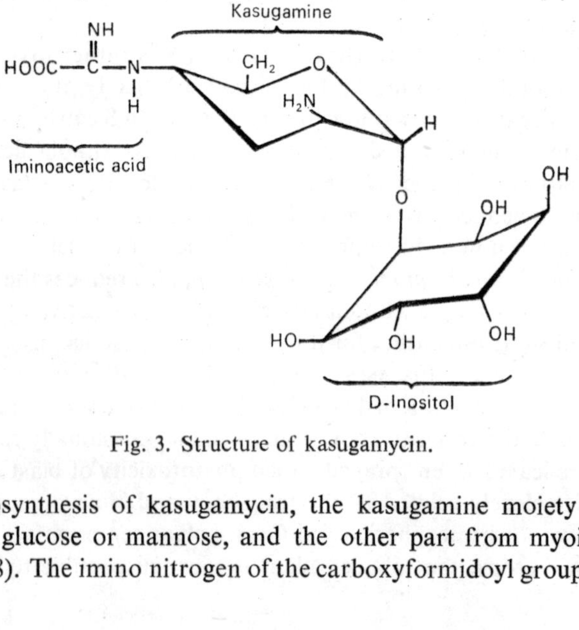

Fig. 3. Structure of kasugamycin.

In the biosynthesis of kasugamycin, the kasugamine moiety is mainly derived from glucose or mannose, and the other part from myoinositol or glycine (37, 38). The imino nitrogen of the carboxyformidoyl group is derived from the nitrogen atom of glycine (39).

BIOLOGICAL ACTIVITY. Kasugamycin selectively inhibits the growth of *Pyricularia oryzae* in acidic media (pH 5.0), but hardly inhibits it in neutral media (pH 7) (47). The antibiotic also inhibits some bacteria including *Pseudomonas* species, and unlike fungi it shows stronger inhibition against *Pseudomonas* at pH 7 than at pH 5 or 6. Its effect on *P. oryzae* is expressed in the plant and *in vitro* only under acidic conditions (74).

Kasugamycin inhibits protein synthesis in a cell-free system of *Escherichia coli*, but it does not affect the synthesis of nucleic acids (163, 167). In the *E. coli* system, it inhibits protein synthesis by interfering with the binding of aminoacyl-tRNA to the mRNA-30S ribosomal subunit complex but does not cause miscoding (166, 162). Kasugamycin sensitivity involves the 16S RNA but not the proteins of the 30S ribosomal subunit. The resistant mutant of *E. coli* cannot methylate two adjacent adenine residues near 3' end of the 16S RNA (53). The sensitive strain contains an RNA methylase which is capable of methylating the 16S RNA from the resistant mutant, but this methylase is lacking or inactive in the kasugamycin-resistant mutant (54).

ANIMAL AND PLANT TOXICITIES. Kasugamycin has no acute or chronic toxicity to mice, rats, rabbits, dogs, monkeys, and human beings. The oral LD_{50} for mice is 2 g/kg, and the TLM to carp is 1 mg/ml. There is no phytotoxicity to crops.

USE AGAINST PLANT DISEASES. Kasugamycin controls rice blast disease at a concentration as low as 20 ppm (74). For practical purposes, it is mainly applied as a dust containing only 0.3 per cent of the active ingredient. Since kasugamycin-resistant strains had been developed in the fields, the mixtures of kasugamycin and chemicals with different modes of action are usable in practice. When the rice seed is coated with 2 per cent kasugamycin wettable powder, the plants raised from such treated seeds are protected in the fields from rice blast for a month. Therefore, the antibiotic may be used as a seed disinfectant.

DEVELOPMENT OF RESISTANCE. The development of resistance in fungi to kasugamycin was not observed in the field for some years after its first application in 1965. However, since 1972, the development of a kasugamycin-resistant strain of *P. oryzae* in fields has become a serious problem. Thus, successive applications of kasugamycin in the same field should be avoided. In laboratory experiments, kasugamycin-resistant strains of *P. oryzae* are obtained from colonies growing on a kasugamycin-containing medium, and the resistant strains prove different in infectivity from the sensitive parent strains (119). At the same time, the ribosomes from the resistant strains are insensitive to kasugamycin.

Polyoxins

Polyoxins are nucleoside antibiotics consisting of at least 13 closely related compounds produced by *Streptomyces cacaoi* var. *asoensis* (77–79, 158, 159). They are active against some filamentous fungi but inactive against bacteria and yeasts. Polyoxins can be safely used for agricultural purposes with no toxicity to mammals, fish and plants. Such excellent characteristics of polyoxins come possibly from their selectivity in inhibiting the synthesis of cell wall chitin of sensitive fungi.

CHEMISTRY. Polyoxins have the chemical structures shown in Fig. 4 (75, 81–83). Polyoxins A, C and H are colourless needle crystals while B, D, E, F, G and I are amorphous powder. They are water soluble, and amphoteric. Their main physico-chemical properties are shown in Table 2. Polyoxin C component is the smallest molecule among polyoxins and its structure is 1-β-(5'-amino-5'-deoxy-D-allofuranuronosyl)-5-hydroxymethyluracil (82). Also it is a key compound to elucidate the structure of polyoxins, since all hydrolytic degradation products of polyoxins give polyoxin C or its analogues (75). Polyoxin J can be chemically synthesized (98).

In the biosynthesis of polyoxins, the 5-substituted uracil base is synthesized from both uracil and C-3 of serine by a new enzyme system which differs from thymidylate synthetase (80), and L-isoleucine utilized as a precursor for polyoximic acid (3-ethylidene-L-azetidine-2-carboxylic acid) (76, 84).

BIOLOGICAL ACTIVITY. Polyoxins, except inactive C and I, have selective antifungal activity against some plant pathogenic fungi (75, 159). Polyoxin D is most effective against rice sheath blight pathogen, whereas polyoxins B and L are most active on pear black spot fungus and apple cork spot fungus (77, 79). Polyoxins do not affect the rate of spore germination, but cause an abnormal swelling phenomenon on germ tubes of spores and hyphal tips of *Alternaria kikuchiana* at low concentration, and finally inhibit the growth of germ tubes or mycelia of the pathogen (27).

Polyoxins are inhibitors of cell wall synthesis. Polyoxin D markedly inhibits the incorporation of ^{14}C-glucosamine into cell wall chitin of *Cochliobolus miyabeanus*, without any inhibitory effect on the respiration and synthesis of macromolecules such as protein or nucleic acids (137). In the cell-free systems of *Neurospora crassa*, it also blocks competitively the incorporation of N-acetylglucosamine from UDP-N-acetylglucosamine into chitin (30). There is a close relation between polyoxine structure and the inhibitory activity on chitin synthetase—the carbamoylpolyoxamic acid moiety polyoxins serves to stabilize the polyoxin-enzyme complex, and the pyrimidine nucleoside moiety of the antibiotics blocks the binding site of chitin synthetase (59).

ANIMAL AND PLANT TOXICITIES. Polyoxins are nontoxic in intravenous injection of 800 mg/kg or oral administration of 15 g/kg to mice. When polyoxin solutions of 40 mg/ml were instilled into the conjunctival sac of rabbits, there was no irritation. No dermal toxicity was detected. They were nontoxic to fish during a 72-hour period of exposure at 10 ppm. Foliar sprays of 200 ppm polyoxins produced no phytotoxicity to most crops (78, 79).

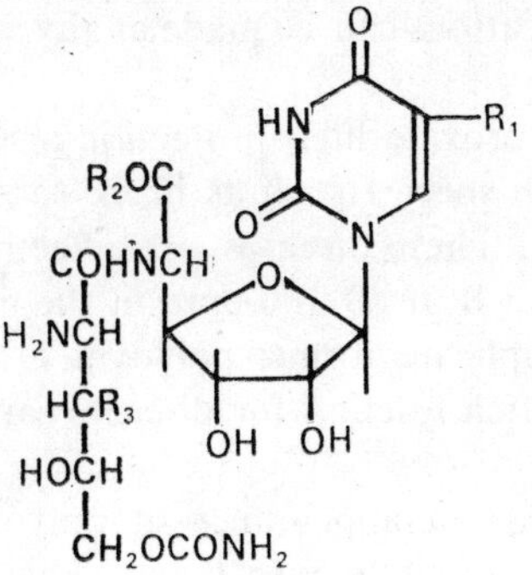

Polyoxin	R_1	R_2	R_3
A	CH_2OH	*	OH
B	CH_2OH	HO	OH
D	COOH	HO	OH
E	COOH	HO	H
F	COOH	*	OH
G	CH_2OH	HO	H
H	CH_3	*	OH
J	CH_3	HO	OH
K	H	*	OH
L	H	HO	OH
M	H	HO	H

* 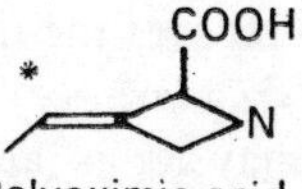

Polyoximic acid

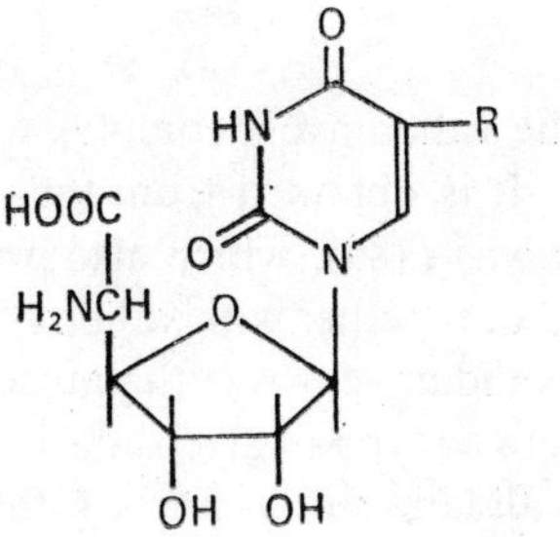

	R
C	HO
I	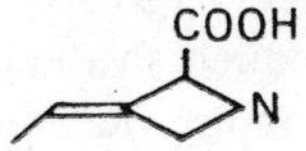

Fig. 4. Structure of polyoxins.

USE AGAINST PLANT DISEASES. Polyoxins have been widely used for controlling plant diseases caused by some pathogenic fungi such as *A. kikuchiana, Rhizoctonia solani C. miyabeanus,* etc.

In pot tests to prevent infection of rice sheath blight caused by *R. solani,* polyoxin complex had better efficacy than organoarsenate, methanearsonic acid-ferric ammonium salt complex (136). When polyoxin B was sprayed on rice plants at various concentrations, it persisted on the plants for at least 12 days at 100 ppm, and for at least nine days at 50 ppm. In field tests two applications of polyoxin B at 50 ppm to control the sheath blight, showed the same efficacy as two applications of organoarsenate at 32.5 ppm. Moreover, the larger the number of applications or the higher the concentration, the greater the tendency to increase yield. Applications can be made at any stage of plant growth without causing phytotoxicity.

Besides sheath blight control, polyoxins provide high protection against plant diseases of fruits caused by *Alternaria* species such as black spot of pear and *Alternaria* leaf spot of apple (27). These diseases are effectively controlled by several applications of polyoxin B at 50–100 ppm in the field. Polyoxin complex is practically used in duplicate forms; polyoxin D-rich fraction for the sheath blight control and B-rich fraction for diseases caused by *Alternaria* species.

DEVELOPMENT OF RESISTANCE. Since 1973, an appearance of polyoxin-resistant strains of fungi has been observed in some fields in Japan. This has possibly resulted from the chemical selection of the insensitive fungi to actually occurring polyoxins. A possible mechanism of the resistance in *A. kikuchiana* is a decreased permeability of polyoxins through the cell membrane into the site of chitin synthesis (58).

The inhibition of mycelial growth of *R. solani* by polyoxins is protected glycyl-L-alanine, glycyl-DL-valine and DL-alanyl-glycine (115). These dipeptides also recover the inhibition of chitin synthetase by polyoxins in the intact cells of *Pyricularia oryzae*, but not in the cell-free system. Therefore, the dipeptides may act as antagonists to the transport of polyoxins into the fungal cells (60).

Validamycin

Validamycin A is a main component of the validamycin complex which is used to control sheath blight in rice plants. It is obtained from the culture filtrate of *Streptomyces hygroscopicus* var. *limoneus* (89), which also produces five additional components validamycins B to F, together with validoxylamine A and B (64, 87). Among the validamycins, validamycin A is the most effective against *Rhizoctonia solani* (64), and the site or type of glycoside linkage is closely related to antifungal activity (174). Validamycin A can be safely used due to its non-phytotoxicity and very low toxicity to mammals. Also it is nontoxic to birds, fish and insects (68, 86).

CHEMISTRY. Validamycin A is a unique aminoglycoside antibiotic, which has two kinds of new hydroxymethyl-branched cyclitols in its molecule (62). Validamycins A to F and validoxylamine B are amorphous powders, and validoxylamine A is a colourless crystal. They are soluble in water, N,N-dimethylformamide and dimethylsulphoxide, but slightly soluble or insoluble in ethanol, acetone, ether, benzene, chloroform or ethylacetate. Their physico-chemical properties (64) are shown in Tables 3 and 4.

The chemical structure of validamycin A is N-[(1S)-1,4,6/5)-3-hydroxymethyl-4,5,6-trihydroxy-2-cyclohexenyl] [0-β-D-glucopyranosyl-(1–3)-(1S)-(1,2,4/3,5)-2,3,4-trihydroxy-5-hydroxymethyl-cyclohexyl] amine, as shown in Fig. 5. Two cyclitol moieties are linked to each other by an imino linkage,

Table 3. Physicochemical properties of polyoxins

Polyoxin	Formula	$[\alpha]^{20°}_{D}$ (c=1, water)	$\lambda^{0.05N HCl}_{max}$ ($E^{1\%}_{1cm}$) mμ	$\lambda^{0.05N NaOH}_{max}$ ($E^{1\%}_{1cm}$) mμ
A	$C_{23}H_{32}N_6O_{14}$	−30°	262 (143)	264 (103)
B	$C_{17}H_{25}N_5O_{13}$	+34°	262 (172)	264 (130)
D	$C_{17}H_{23}N_5O_{14}$	+30°	218 (217), 276 (127)	271 (137)
E	$C_{17}H_{23}N_5O_{13}$	+19°	218 (200), 276 (200)	271 (128)
F	$C_{23}H_{30}N_6O_{15}$	−18°	215sh (257), 276 (181)	271 (118)
G	$C_{17}H_{25}N_5O_{12}$	+37°	262 (170)	264 (134)
H	$C_{23}H_{32}N_6O_{13}$	−38°	265 (127)	266 (103)

Table 4. Physicochemical properties of validamycins A～F and validoxylamine A, B

	Molecular formula	$[\alpha]$ D (H_2O)	pK'a value
Validamycin A	$C_{20}H_{35}NO_{13}\cdot H_2O$	+112.5°	6.0±0.2
Validamycin B	$C_{20}H_{35}NO_{14}\cdot H_2O$	+102.3°	5.0±0.2
Validamycin C	$C_{26}H_{45}NO_{18}\cdot H_2O$	+132.9°	6.0±0.2
Validamycin D	$C_{20}H_{35}NO_{13}\cdot H_2O$	+169.3°	6.0±0.2
Validamycin E	$C_{26}H_{45}NO_{18}\cdot H_2O$	+148.2°	6.1±0.2
Validamycin F	$C_{26}H_{45}NO_{18}\cdot H_2O$	+130.7°	6.1±0.2
Validoxylamine A	$C_{14}H_{25}NO_8\cdot H_2O$	+170.0°	6.2±0.2
Validoxylamine B	$C_{14}H_{25}NO_9\cdot H_2O$	+172.4°	5.0±0.2

and these cyclitols are quite different from known aminocyclitol (63, 64, 94). The aglycone part of validamycin A or B is called validoxylamine A or B, respectively (61). Validamycins A, C, D, E and F have validoxylamine A as a common moiety in their molecules (63, 64), which are different from each other in at least one of the characteristics such as the configuration of anomeric centre of glucoside, the position of glucoside linkage, or the number of D-glucose molecules.

Biological activity. Validamycin A is effective against certain plant diseases caused by *Rhizoctonia* species such as sheath blight, and rot, damping-off, seed decay, black scurf, etc. (174). It shows, however, no antimicrobial activity against about 3000 species of bacteria and fungi when tested by using ordinary assay methods (85, 86). Although the antibiotic does not remarkably suppress the growth of *R. solani* in a nutritionally rich medium, it causes abnormal branching at the tips of hyphae of the pathogen on a poor medium, followed by the cessation of further growth (86, 117). The effect of validamycin A on *R. solani* is not fungicidal but suppressive to mycelial growth under a poor nutrient condition, since validamycin A-suppressed mycelia recover normal growth after transferring them into validamycin A-free medium (174).

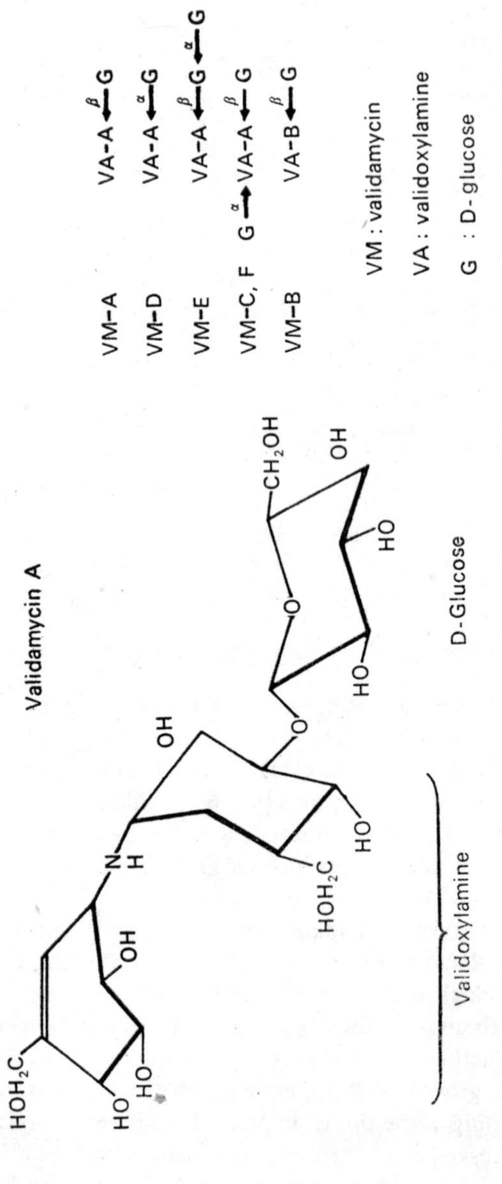

Fig. 5. Structure of validamycins.

As for the mode of action of validamycin A, it inhibits the biosynthesis of myo-inositol in *R. solani* (175), and therefore myo-inositol may be indispensable for the normal growth and pathogenic activity of the fungus. Also, validamycin activity on the mycelial growth or pathogenicity of the fungus is remarkably blocked by the addition of inositol (174).

ANIMAL AND PLANT TOXICITIES. The acute and subacute toxicities of validamycin A to mammals are very low; in oral administration at a dose of 10 g/kg to mice and rats, or in subcutaneous and intravenous administration at a dose of 2 g/kg to mice, all the animals tested survived without any effect for seven days. When the ^{14}C-validamycin is orally administered to rat, it is decomposed by enteric bacteria and gives rise to D-glucose and validoxylamine, in which the former is utilized as a nutrient source and the latter is excreted in the faeces without being absorbed into the body through the intestine. In the case of intravenous injections, it is rapidly excreted in the urine without decomposing. There is no irritating effect on the skin or the cornea of the rabbit (86). Validamycin A is slightly phytotoxic only to mulberry, but not to other 150 species of plants tested (174, 175).

USE AGAINST PLANT DISEASES. Validamycin A has been commercially used for rice sheath blight control. When applied in the early logarithmic phase of lesion expansion on a rice plant, sufficient control is achieved by a single application of 30 ppm solution. Its curative effect on rice sheath blight is more dramatic than its preventive effect, although the antibiotic has both the effects (174). Moreover, validamycin A may be applicable to control black scurf of potato, seed decay of cucumber, and damping-off of cotton at 60–100 ppm (174).

ENVIRONMENTAL METABOLISM. Validamycins are easily decomposed by soil microbes such as *Endomyces decipiens, Rhizopus nigricans, Erwinia carotovora, Pseudomonas denitrificans,* and result in complete loss of biological activity. Their half-life in soil is less than five hours, and they are mostly decomposed in 24 hours. Microbial degradation of validamycin A by *P. denitrificans* produces D-glucose and validoxylamine, which is further decomposed into validamine and valienamine (93). Microflora on plants and in soil are not affected by validamycin A (174).

Griseofulvin

Griseofulvin was first isolated from *Penicillium griseofulvum* (126), subsequently re-isolated as a curling factor of fungal hyphae from *P. raistrickki* (9), and more recently recognized as a metabolic product of many *Penicillium* species. It is a natural antifungal antibiotic containing a chlorine atom. This antibiotic is used both in the medical and agricultural fields.

CHEMISTRY. Griseofulvin is obtained as a stout octahedra or rhombic crystal from benzene. The physicochemical properties (111) are as follows: melting point 220°C; specific rotation $[\alpha]_D^{17}$ +370° in chloroform; maximum

UV absorption 286 and 325 nm; practically insoluble in water and ether, soluble in N,N-dimethylformamide, slightly soluble in ethanol, methanol, acetone, benzene, chloroform, acetic acid or ethyl acetate.

The chemical structure of griseofulvin (44) is shown in Fig. 6. It has asymmetric centres at positions 2 and 6 in the molecule, and the stereochemistry also is critical for antifungal activity; its racemic form has only half of the activity of griseofulvin itself, and its diastereoisomer known as a epi-griseofulvin is inactive (44).

When griseofulvin-producing organisms are grown by the addition of potassium bromide to the chloride-deficient medium, the dechloro analogue, 7-bromo-7-dechlorogriseofulvin is produced (104, 105). However, the analogue is less active than griseofulvin itself.

BIOLOGICAL ACTIVITY. Griseofulvin is active against some plant and animal pathogenic fungi, but inactive against bacteria and yeasts. Although the antibiotic inhibits the growth of fungi with chitinous cell walls, it has no effect on fungi with cellulose cell walls (10).

Griseofulvin does not inhibit the germination of fungal spores, but it has some remarkable influences on fungal development; stunting of hyphae, excessive branching, hyphal distortion, and loss of apical dominance on germinating conidia of *Botrytis alli* (7). It also interferes with mitosis of cell division in the metaphase, causes multipolar mitosis, and produces abnormal nuclei such as unusually large and irregularly shaped ones, though it does not inhibit the formation of spindle fibres in the nuclear division (128, 129, 170). Furthermore, it has mutagenic effect on *Microsporum gypseum* and *Nannizzia incurvate* (99). The antibiotic does not inhibit respiration, glycolysis and the synthesis of protein, chitin and total lipids in fungal cells, but it stimulates or reduces the synthesis of RNA and DNA. Although the action of griseofulvin is not completely elucidated, its site of action may be involved in the replicatory system of fungal cells (70).

ANIMAL AND PLANT TOXICITIES. Griseofulvin is one of the safest antibiotics, and produces essentially no side effect by oral administration to humans. The oral LD_{50} of the antibiotic to rats is 400 mg/kg, and when given intraperitoneally the rats can tolerate a daily dose of 2000 mg/kg. The antibiotic also is relatively non-phytotoxic, and slowly degraded in plant tissues but more quickly in soil (21).

USE AGAINST PLANT DISEASES. Utilization of griseofulvin for plant protection was first introduced in the control of early blight of tomato and *Botrytis* disease of lettuce (11). The antibiotic is effective against a number of mildews such as powdery mildews of roses and chrysanthemums, or downy mildew of cucumber (8), and has a curative effect on *Fusarium* wilt of melons when a dilute paste of the antibiotic is painted on the diseased parts of the plant. It also protects apple plants from the blossom blight by dusting a mixture of apple pollen grains and lycopodium containing 25 per cent griseofulvin dust at the time of artificial pollination. This antibiotic, however, seems to be too costly for agricultural use.

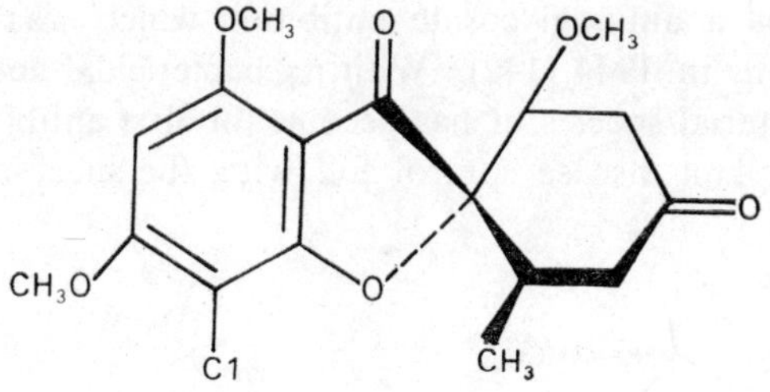

Fig. 6. Structure of griseofulvin.

Ezomycin

Ezomycins are antifungal antibiotics produced by a strain very similar to *Streptomyces kitazawaensis* (140). They show specific biological activity in suppressing the growth of very limited species of plant pathogenic fungi such as *Sclerotinia* and *Botrytis* species.

CHEMISTRY. Ezomycins are pyrimidine nucleoside antibiotics consisting of eight closely related compounds, which contain a new 3-aminouronic acid named ezoaminuroic acid (138–141). The chemical structure of ezomycin A (138) is shown in Fig. 7.

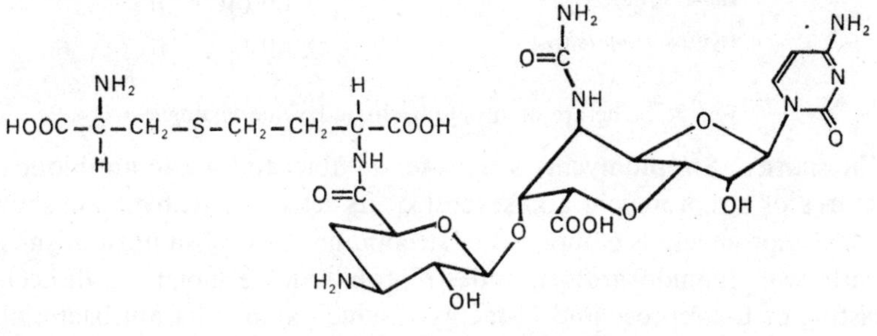

Fig. 7. Structure of ezomycin A.

BIOLOGICAL PROPERTIES. Ezomycins are active against *Sclerotinia* species such as *Sclerotinia sclerotiorum*, and *Botrytis* species such as *Botrytis cinerea*, but inactive against other fungal species, yeast and bacteria (140). The L-crystathionine in ezomycin molecule is responsible for specific antifungal activity [140]. The toxicity to mammals and fish is low.

Ezomycin complex was introduced in 1970 as an agricultural antibiotic in the control of stem rot of kidney bean, but it has scarcely been on the market because of limited consumption.

Antibacterial antibiotics

Streptomycin

Streptomycin is a aminoglycoside antibiotic which was isolated from *Streptomyces griseus* in 1944 (142). With its bactericidal activity against a wide range of bacterial species, it has become the first antibiotic introduced in agriculture for plant disease control following the successful use against human diseases.

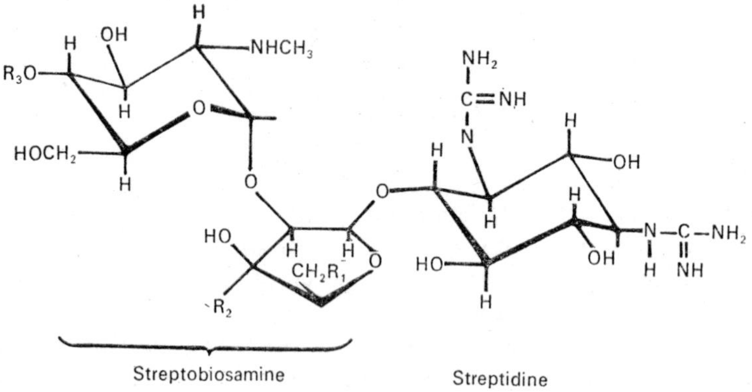

	R_1	R_2	R_3
Streptomycin	H	CHO	H
Dihydrostreptomycin	H	CH_2OH	H
Hydroxystreptomycin	OH	CHO	H

Fig. 8. Structure of streptomycin and some analogues.

CHEMISTRY. Streptomycin is a water-soluble and basic antibiotic. The structures of streptomycin and several of its active derivatives are shown in Fig. 8. Streptomycin is composed of streptidine moiety, an inositol substituted with two guanido groups, and streptobiosamine moiety, a disaccharide consisting of L-streptose and N-methyl-L-glucosamine. Its antibacterial activity is destroyed by cleavage of the glycosidic bond between streptidine and streptobiosamine by replacing the guanido groups with amino groups, or by carbobenzyloxylation of the secondary amine (132).

BIOLOGICAL ACTIVITY. Streptomycin is effective against both Gram-positive and Gram-negative bacteria, but inactive to fungi, rickettsiae, and viruses. The amount of streptomycin required for inhibition varies over a considerable range within the same bacterial species and is influenced by various factors such as the pH or ionic strength of the medium (90).

Streptomycin exerts a wide variety of metabolic effects *in vivo:* these include inhibition of protein synthesis (33), stimulation of RNA synthesis (154), inhibition of cellular respiration, loss of low molecular materials from intracellular pools (3), or leakage of ions from cells (26). Such effects imply that streptomycin may act on nucleic acids and cell membranes in addition to inhibiting protein synthesis (41). Also, in bacterial cells streptomycin binds only to the 30S ribosomal subunit without the 50S subunit (18, 92, 173), and causes misreading of the genetic code in protein synthesis (23, 102). The mutants of *Escherichia coli* highly resistant to streptomycin are known to involve modification of a protein of the bacterial ribosome (the P10 protein of the 30S subunit) (17, 108, 127). However, the primary site of the lethal action of streptomycin has been not yet clarified.

ANIMAL AND PLANT TOXICITIES. Although ototoxicity, nephrotoxicity, and hypersensitivity reactions are recognized in clinical use, animal cells are relatively insensitive to streptomycin. With regard to phytotoxicity of streptomycin, it causes chemical injuries to a range of vegetables and rice plant at a high concentration. When the antibiotic is applied to the rice plant at 500 ppm or more, especially at near-heading stage, it causes not only chlorotic spots on the rice leaves, but also reduces rice yield. In the injured rice plants the functions of starch translocation and manganium absorption decrease. A mixture of streptomycin sulphate and iron chloride or citrate is effective in reducing the phytotoxicity of the antibiotic.

USE AGAINST PLANT DISEASES. Streptomycin, the first antibiotic introduced in agriculture, was used in the United States of America for the control of apple and pear fire blights by foliar application at a concentration of 200 ppm. It is effective against wild fire of tobacco at 200 ppm, and against bacterial leaf blight of rice at 200–500 ppm. A mixture of streptomycin and oxytetracycline is effective to control bacterial canker of peach, citrus canker, soft rot of vegetables, and various other bacterial diseases.

Streptomycin absorbed by potato seedling when dipped in a 100 ppm solution translocates to the top of the seedling after three hours. Also, the antibiotic absorbed from tobacco stems translocates to upper leaves, and then reaches the lower leaves, but it does not translocate when absorbed from the leaves (55, 65).

DEVELOPMENT OF RESISTANCE. Streptomycin-resistant strains are distributed in a wide range of plant pathogenic bacteria such as *Xanthomonas oryzae* (176), *X. citri*, *Pseudomonas tabaci*, *P. lachrymans*, etc., and they may show cross-resistance to several other antibiotics. In agricultural use, the alternate or combined application of streptomycin and other chemicals with different mechanisms of action is recommended in order to reduce or avoid the development of streptomycin-resistant strains in the field.

Tetracyclines

Tetracycline, the prototype of a group of tetracycline antibiotics, is produced by *Streptomyces viridifaciens* (52). Some other members include chlortetracycline (aureomycin) produced by *S. aureofaciens* (12), oxytetracycline (5-hydroxytetracycline) by *S. rimosus* (32), and demethylchlortetracycline by a variant of *S. aureofaciens*. Tetracyclines are active against a wide range of microorganisms, including Gram-positive and Gram-negative bacteria, rickettsiae, mycoplasma, and protozoa (100). In the agricultural field, oxytetracycline has been used in the control of plant bacterial diseases by mixing with streptomycin.

CHEMISTRY. The chemical structures of tetracycline, oxytetracycline (168) and chlortetracycline (25) are shown in Fig. 9. Oxytetracycline is an amphoteric antibiotic which produces readily acidic and basic salts such as oxytetracycline hydrochloride, and oxytetracycline disodium salt dihydrate. These salts of oxytetracycline are soluble in water, methanol, ethanol, acetone, or butanol. Aqueous solutions of the hydrochloride at 3 to 9 show no loss in potency on storage at 5°C for at least 30 days, but aqueous solutions of the disodium salt dihydrate lose its potency rapidly.

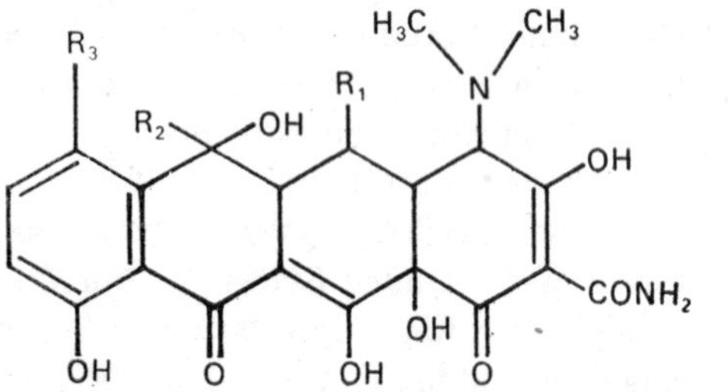

	R_1	R_2	R_3
Tetracycline	H	CH_3	H
Oxytetracycline	OH	CH_3	H
Chlortetracycline	H	CH_3	Cl

Fig. 9. Structure of tetracyclines.

BIOLOGICAL ACTIVITY. Tetracyclines are effective against some species of plant pathogenic bacteria, including *Pseudomonas, Xanthomonas,* and *Erwinia* species. In addition, they show protective effects on plant diseases caused by mycoplasma-like organisms (73).

Tetracycline is a potent inhibitor of bacterial protein biosynthesis (51, 155) and has only weak effects on mammalian cell extracts. It binds both the 30S and 50S bacterial ribosomal subunits, although the 30S subunit binds twice the amount of the 50S subunit (24, 109). It also has little or no effect on translocation or transpeptidation (131), and does not induce miscoding.

Many studies indicate that tetracycline inhibits greatly the binding of aminoacyl-tRNA, and the termination factors RF1 and RF2 to the A site of bacterial ribosome (16).

ANIMAL AND PLANT TOXICITIES. The toxicity of tetracyclines is low to mammals and fish. Tetracyclines have low phytotoxicity. A mixture of oxytetracycline and streptomycin occasionally causes to express chlorosis on the leaves of some plants when sprayed in the conditions of high temperature and high humidity, but the plant injury is practically recovered in 10 days.

USE AGAINST PLANT DISEASES. Oxytetracycline has been used to control plant bacterial diseases caused by *Pseudomonas* and *Erwinia* species, although it is used as a mixture of streptomycin in order to avoid the development of streptomycin-resistant strains in plant bacteria. Oxytetracycline is easily absorbed from plant leaves, especially from stomata, and is rapidly translocated into plant tissues.

Furthermore, tetracycline and chlortetracycline antibiotics show remarkable effects in suppressing mulberry dwarf disease caused by mycoplasma-like organism at a concentration of 100 ppm (73). Plants infected with clover phyllody mycoplasma recover from the disease after dipping the root in oxytetracycline solution of 100 ppm (149). When the application of these antibiotics was discontinued, however, the disease-recovered plants tend to develop their symptoms again (73).

Chloramphenicol

Chloramphenicol is a fermentation product of *Streptomyces venezuelae* (28). It is the first broad-spectrum antibiotic introduced into medical use, and also the first antibiotic to be chemically synthesized (19). In agriculture, the antibiotic has been used to control a bacterial leaf blight of rice and bacterial canker of tomato.

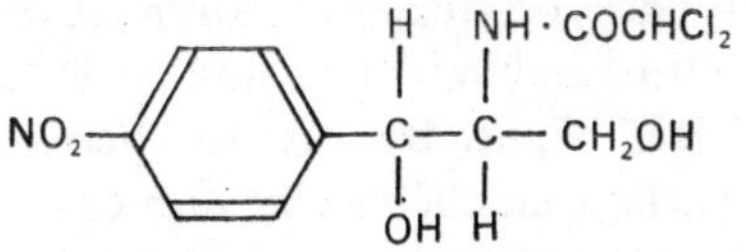

Fig. 10. Structure of chloramphenicol.

CHEMISTRY. Chloramphenicol is a colourless crystal, and soluble in ethanol, and acetone. The chemical structure is D-(-)-threo-1-p-nitrophenyl-2-dichloroacetamide-1,3-propanediol (133), as shown in Fig. 10. Chloramphenicol possesses two asymmetric carbon atoms, which give rise to four stereoisomers by chemical synthesis. Only the natural antibiotic which is the D-(-)-threo isomer has significant antimicrobial activity (110). The configuration of the propanediol moiety of chloramphenicol is essential for the activity

(46). When either of the two hydroxy groups is esterified, or replaced with the hydrogen atom, its antimicrobial activity is almost lost (46).

BIOLOGICAL ACTIVITY. Chloramphenicol is chiefly a bacteriostatic agent. It is a broad spectrum antibiotic inhibiting the growth of Gram-positive and Gram-negative bacteria. Eucaryotic cells are resistant to the antibiotic with a few exceptions (45).

Chloramphenicol inhibits protein synthesis in bacteria (42, 181), but it does not inhibit the synthesis of nucleic acids (42), cell wall peptidoglycan (48, 106) or polysaccharides (67). It selectively binds the 50S ribosomal subunit of bacterial cells (31, 172), and interferes with the function of mRNA (182). The selectivity of chloramphenicol between bacterial cells and eucaryotic cells is closely related to their ribosomal structures. The antibiotic inhibits incorporation of amino acids into protein in cell-free extracts from *Escherichia coli* (116, 118), but not in those from eucaryotic cells (2, 134, 153). It also inhibits protein synthesis in intact mitochondria (150) and in their extracts (97); similarly in chloroplasts and their extracts (29, 107), in which proteins are synthesized on 70S ribosomes similar to bacteria.

ANIMAL AND PLANT TOXICITIES. The oral LD_{50} of chloramphenicol for mice is 2,640 mg/kg and the LD_{50} by intravenous injection to mice is 245 mg/kg. The TLm to carp is more than 40 ppm. Chloramphenicol is phytotoxic to plants at high concentrations.

USE AGAINST PLANT DISEASES. Chloramphenicol is effective against bacterial leaf blight of rice by foliar application at a concentration of 100–200 ppm, but it occasionally causes phytotoxicity on rice leaves when sprayed beyond the concentration described above. A mixture of chloramphenicol and copperhydroxide sulphate is used to control bacterial canker of tomato caused by *Corynebacterium michiganense*.

Cellocidin

Cellocidin is an antibiotic produced by *Streptomyces chibaensis* (160). It has antibacterial and antitumour activities, and shows an excellent protective effect on rice bacterial leaf blight. Because its chemical structure is simple, the antibiotic can be synthesized chemically with ease.

CHEMISTRY. Cellocidin is obtained as a white and needle crystal, and it is slightly soluble in water, methanol, ethanol, or acetone, but insoluble in other organic solvents. It is relatively stable in the pH range of 2–7 by heating for 10 minutes at 100°C, but rather unstable in alkaline solution. Its maximal UV absorption is 299 nm [$E^{1\%}_{1\,cm}$ 290] in 0.1 N sodium hydroxide solution.

The chemical structure of cellocidin is an acetylene-dicarbosyamide composed of only four carbon atoms (161), as shown in Fig. 11. Technical grade cellocidin for commercial formulations is chemically synthesized from fumaric acid or butynediol.

BIOLOGICAL ACTIVITY. Cellocidin is effective against Gram-positive and Gram-negative bacteria, and it has a strong bactericidal activity on *Xanthomonas oryzae* and *Pseudomonas solanacearum* at 10 ppm (162). It also has anticancer action, but its use should be avoided because of its high toxicity (160).

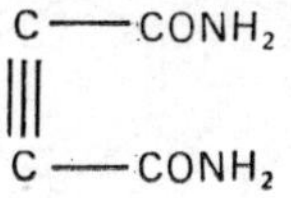

Fig. 11. Structure of cellocidin.

The antibacterial activity of cellocidin is antagonized by sulphydryl compounds such as cystein or glutathione, and the antagonistic mechanism is due to the addition of sulphydryl compounds to triple bond carbon of cellocidin molecule (189). Cellocidin inhibits the metabolism of α-keto-glutarate to succinate in *X. oryzae* (120, 121), and also inhibits the incorporation of radioactive thymidine into DNA in *Bacillus subtilis* (165).

ANIMAL AND PLANT TOXICITIES. Cellocidin exhibits high mammalian toxicity when injected intravenously, and the LD_{50} for mice is 11 mg/kg. However, it is not so highly toxic in oral administration and skin application, and the LD_{50}'s for mice are 89.2–125 mg/kg and 667 mg/kg, respectively. Its toxicity to fish is lower than that of DDT. Cellocidin sometimes causes plant injuries to rice, radish, cucumber, tomato at above 200 ppm.

USE AGAINST PLANT DISEASES. Cellocidin shows excellent preventive effects against bacterial leaf blight by a foliar application of 100–200 ppm. It has been in practical use since 1964, but its consumption has been remarkably reduced because of its phytotoxicity.

Novobiocin

Novobiocin, also known by the names cathomycin, streptonivicin, cardelmycin, albamycin and vulcamycin, is produced by *Streptomyces spheroides* and other actinomycetes (66, 91). It is active against Gram-positive bacteria and certain kinds of Gram-negative bacteria, but weakly or negligibly active against fungi (15). The antibiotic is used both in medical and agricultural fields.

CHEMISTRY. The chemical structure of novobiocin is shown in Fig. 12. The antibiotic has two acidic groups, a weakly acidic phenol, and a more strongly acidic enol on the coumarin ring (43, 147). The solubility of novobiocin is remarkably affected by pH; the free acid is insoluble in water, while the sodium salt is highly water-soluble. Novobiocin produces insoluble salts with a variety of divalent and trivalent cations, including the alkaline earth metals, zinc, aluminium, manganese, and iron.

The molecule of novobiocin consists of novobiocic acid and 3-0-carbamyl noviose. The existence and position of the 0-carbamyl group on the noviose sugar are essential for antibacterial activity (15).

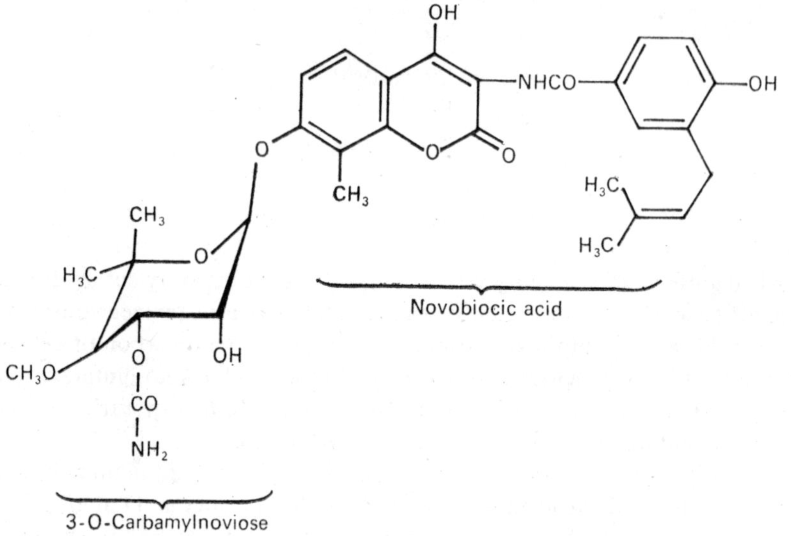

Fig. 12. Structure of novobiocin.

Biological activity. Novobiocin has a strong activity against Gram-positive bacteria, except mycobacteria and corynebacteria which are slightly less active. It also acts on certain Gram-negative bacteria, especially certain *Klebsiella* and *Proteus* species (180). Novobiocin causes filament formation in Gram-negative rods (152), chain formation in *Streptococcus faecium* (15), and permeability effects in *Escherichia coli* EL 35 (14), and its growth inhibiting activity is readily reversed by magnesium ions, or to a lesser extent by other alkaline earth metals (13).

Novobiocin inhibits respiration, fermentation, electron transport, oxidative phosphorylation, ATPase activity, amino acid transport and the incorporation of amino acids into tRNA and of radioactive uracil into RNA (15). The antibiotic also inhibits luminescence, and the activity of DNA and RNA polymerases (150). These processes affected by novobiocin are related to magnesium requirements of cells (15). Another evidence for the primary effect of novobiocin is the inhibition of DNA synthesis in *E. coli*. The antibiotic inhibits primarily DNA synthesis, later decreases RNA synthesis, then protein and cell wall synthesis. The inhibition of nucleic acid synthesis by the antibiotic is on the template-polymerase complexes (150, 151).

Animal and plant toxicities. The oral LD_{50}'s of novobiocin for mice is

100 mg/kg. The TLm to carp is more than 100 ppm. The antibiotic is phytotoxic to tomato seedlings at a high concentration.

USE AGAINST PLANT DISEASES. Novobiocin is effective against bacterial canker of tomato. The disease can be controlled by dipping tomato seedlings in 100 ppm of novobiocin aqueous solution overnight before transplanting. The antibiotic occasionally causes plant injury to tomato when used beyond the concentration described above.

Insecticidal antibiotics

Tetranactin

Tetranactin is a macrotetrolide antibiotic produced by *Streptomyces aureus* strain S-3466 (4). It is the first potential miticidal antibiotic which was used as miticide for plants. It shows low toxicity to warm-blooded animals and no phytotoxicity to plants (4).

CHEMISTRY. Tetranactin is crystallized as rhombic prisms from the filter cake of fermented broth of the producing organism. The physico-chemical properties of tetranactin are as follows: melting point 105–106°C; specific rotation $[\alpha]_D^{23.5}$ 0° (C–1, in chloroform); readily soluble in most organic solvents such as n-hexane, benzene, chloroform, acetone or alcohol, but insoluble in water. Tetranactin is relatively stable to pH, heating and sunlight, since no loss of activity is observed at pH 2-13 for five hours at room temperature, at 60°C for 15 days, and on exposure to sunlight for several days. The chemical structure is a cyclic polyester composed of four units of homononactic acid (71), as shown in Fig. 13.

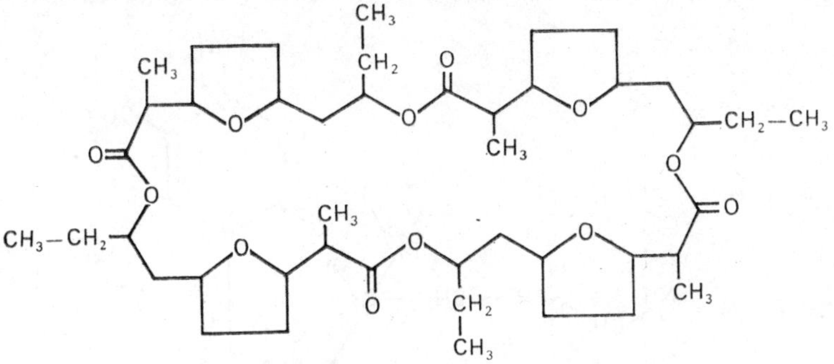

Fig. 13. Structure of tetranactin.

BIOLOGICAL ACTIVITY. Tetranactin shows a unique property in biological activity. Although the antibiotic shows no activity by direct contact to mites in a dry state, it however, exerts mitocidal activity by direct contact to mites in aqueous suspension which means water is an essential factor for the exertion of the miticidal activity. Tetranactin is effective against the adults of the carmine spider mite, two-spotted spider mite, or European red mite, but its ovicidal effect is not so strong. Adzukibean weevil and larva of mosquito are moderately sensitive to the antibiotic, while other pests, such as house fly and cockroach, are insensitive (135).

As for the mode of action of tetranactin, it is an uncoupler in cockroach mitochondria, and causes the leakage of alkali cation such as K^+ through the lipid layer of the biomembrane in mitochondria (5).

ANIMAL AND PLANT TOXICITIES. Acute toxicity of tetranactin is very low; the oral LD_{50}'s for mice and rats is more than 25 g/kg and 2.5 g/kg, respectively. Its toxicity to fish is relatively high; the TLm to carp is 0.003 ppm. It occasionally causes irritation to the human eyes. Tetranactin shows no phytotoxicity to apple, mandarin orange, and tea even at high concentration (56).

USE AGAINST INSECT PESTS. When tetranactin suspensions were applied to an apple tree at 100 ppm in a field, proliferation of both the Kanzawa spider and the European red mites under field conditions was completely retarded during 32 days (135). Tetranactin shows no systemic miticidal activity by translocating from plant roots (56). In order to avoid the development of resistant mites, the following mixtures of tetranactin with other chemicals are used in practice—tetranactin-CPCBS (p-chlorophenyl p-chlorobenzene sulphonate) mixture, and tetranactin-BPMC (bufencarb) mixture.

Other promising antibiotics

Mildiomycin

Mildiomycin, a new nucleoside antibiotic, is isolated from the culture filtrate of *Streptoverticillium rimofaciens* B-98891 (88). Mildiomycin is a water-

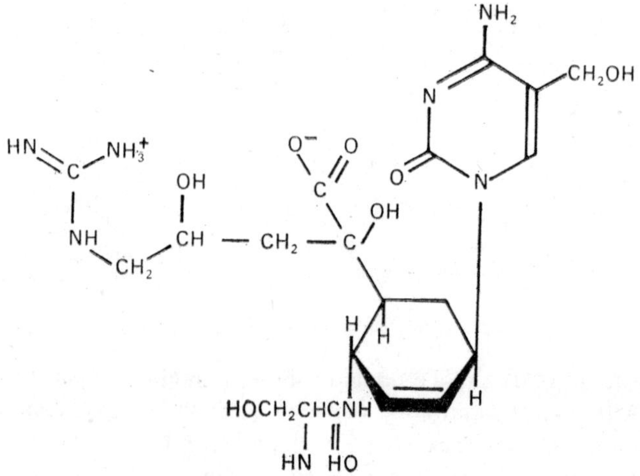

Fig. 14. Structure of mildiomycin.

soluble and basic antibiotic (49). The chemical structure (50) is shown in Fig. 14.

Mildiomycin is highly active against the growth of various powdery mildews, but weakly active against Gram-positive and Gram-negative bacteria, most fungi, and some yeast (88). It also shows excellent activity for the control of powdery mildew of various plants.

The toxicity of mildiomycin is very low (88). Preliminary acute toxicity in rats and mice is the LD_{50} 500–1000 mg/kg by intravenous and subcutaneous injections, and 2.5–5.0 g/kg by oral administration. At a concentration of 1000 ppm there is no irritation to the cornea and skin of rabbits during 10 days. Its toxicity to killifish is not observed at a concentration of 20 ppm for seven days.

Aabomycin A

Aabomycin A is obtained from culture broth of *Streptomyces hygroscopicus* var. *aabomyceticus* (143). Aabomycin A is easily soluble in many organic solvents, but insoluble in water (1). The chemical structure is not still confirmed. Aabomycin A is active against many fungi, especially against *Pyricularia oryzae* at a concentration of less than 0.01 μg/ml (1). It inhibits spore germination, mycelial growth and sporulation of *P. oryzae*, and suppresses markedly the rice blast disease on rice plants at a concentration of 20 ppm (184). Another activity of Aabomycin A is to inhibit the multiplication of tobacco mosaic virus at a concentration of 1–100 μg/ml. The antibiotic is also found effective in inhibiting disease development of cucumber mosaic virus and alfalfa mosaic virus in at pot test (184).

The toxicity of Aabomycin A to animals and fish is very low (1). The antibiotic is also non-phytotoxic even at a concentration of 1000 μg/ml (184).

Nikkomycin

Nikkomycin is obtained from the fermentation broth of *Streptomyces tendae* Tü 901 (22). It is a nucleoside-peptide antibiotic consisting of uracil, and amino hexuronic acid and a new amino acid containing pyridine ring (22). The chemical structure of nikkomycin is shown in Fig. 15.

Nikkomycin is active against a limited species of fungi, but inactive against Gram-negative and Gram-positive bacteria. It is an inhibitor of chitin biosynthesis (22). These observations show that nikkomycin is very similar to polyoxins in its structure and mode of action.

Myomycin

Myomycin is a water-soluble and basic antibiotic which is produced by a species of genus *Nocardia* (35). Myomycin culture broth contains an antibiotic complex consisting of one major component (myomycin B) and two minor

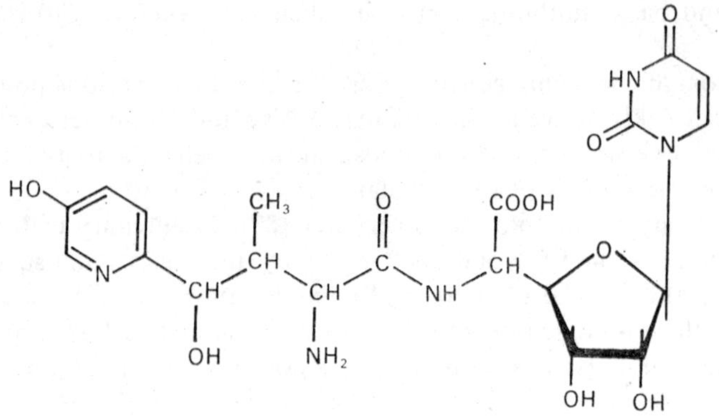

Fig. 15. Structure of nikkomycin.

components (myomycin A and C). The chemical structure of myomycin B (36) is shown in Fig. 16.

Myomycin is active against Gram-positive and Gram-negative bacteria, but inactive against fungi (35). The antibiotic also inhibits the growth of some plant pathogenic bacteria at a concentration of 2–100 μg/ml when measured by a turbidometric method. In a pot test *in vivo*, myomycin protects Chinese cabbage from soft rot caused by *Erwinia carotovora* at 45 μg/ml, and cucumber from angular leaf spot caused by *Pseudomonas lachrymans* (189).

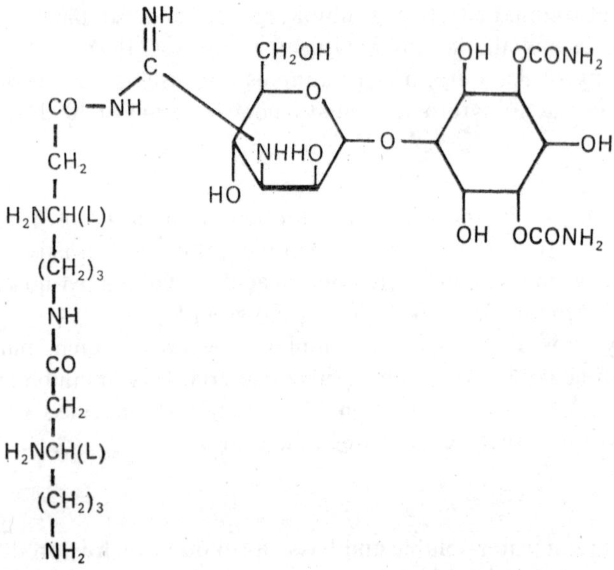

Fig. 16. Structure of myomycin.

References

1. Aizawa, S., Nakamura, Y., Shirato, S., Taguchi, R., Yamaguchi, I. and Misato, T. Aabomycin A, a new antifungal antibiotic. I. Production, isolation and properties of aabomycin A, *Journal of Antibiotics*, *22*, 457–462 (1969).
2. Allen, E.H. and Schweet, R.S. Synthesis of hemoglobin in a cell-free system. I. Properties of the complete system, *Journal of Biological Chemistry*, *237*, 760–767 (1962).
3. Anand, N. and Davis, B.D. Damage by streptomycin to the cell membrane of *Escherichia coli*, *Nature*, *185*, 22–23 (1960).
4. Ando, K., Oishi, H., Hirano, S., Okutomi, T., Suzuki, K., Okazaki, H., Sawada, M. and Sagawa, T. Tetranactin, a new miticidal antibiotic. I. Isolation, characterization and properties of tetranactin, *Journal of Antibiotics*, *24*, 347–352 (1971).
5. Ando, K., Sagawa, T., Oishi, H., Suzuki, K. and Nawata, Y. Tetranactin, a pesticidal antibiotic. *Proceedings of the First International Congress of IAMS*, *3*, Science Council of Japan (1974).
6. Asakawa, M., Misato, T. and Fukunaga, K. Studies on the prevention of the phytotoxicity of blasticidin *S*, *Pesticide and Technique*, *8*, 24–29 (1963).
7. Brian, P.W. Studies on the biological activity of griseofulvin, *Annals of Botany*, *13*, 59–77 (1949).
8. Brian, P.W. Griseofulvin. *The British Mycological Society Transactions*, *43*, 1–13 (1960).
9. Brian, P.W., Curtis, P.J. and Hemming, H.G. A substance causing abnormal development of fungal hyphae produced by *penicillium janczewskii* ZAL. I. Biological assay, production and isolation of "curling factor", *The British Mycological Society Transactions*, *29*, 173–176 (1946).
10. Brian, P.W., Curtis, P.J. and Hemming, H.G. A substance causing abnormal development of fungal hyphae produced by *Penicillium janczewskii* ZAL. III. Identity of "curling factor" with griseofulvin, *The British Mycological Society Transactions*, *32*, 30–33 (1949).
11. Brian, P.W., Wright, J.M., Stubbs, J. and Way, A.M. Uptake of antibiotic metabolites of soil microorganisms by plants, *Nature*, *167*, 347–349 (1951).
12. Broschard, R.W., Dornbush, A.C., Gordon, S., Hutchings, B.L., Kohler, A.R., Krupka, G., Kushner, S., Lefemine, D.V. and Pidacks, C. Aureomycin, a new antibiotic, *Science*, *109*, 199–200 (1949).
13. Brock, T.D. Studies on the mode of action of novobiocin, *Journal of Bacteriology*, *72*, 320–323 (1956).
14. Brock, T.D. Effects of magnesium ion deficiency on *Escherichia coli* and

possible relation to the mode of action of novobiocin, *Journal of Bacteriology, 84*, 679–682 (1962).
15. Brock, T.D. Novobiocin, in *Antibiotics I. Mechanism of Action* (Edited by D. Gottlieb and P.D. Shaw), Springer-Verlag, Berlin, Heidelberg, New York, (1967).
16. Caskey, C.T. Inhibitors of protein synthesis, in *Metabolic Inhibitors*, Vol. IV (Edited by R.M. Hochster, M. Kates and J.H. Quastel), Academic Press, New York (1973).
17. Chang, F.N. and Flaks, J.G. Topography of the *Escherichia coli* 30S ribosomal subunit and streptomycin binding, *Proceedings of National Academy of Sciences U.S.A., 67*, 1321–1328 (1970).
18. Chang, F.N. and Flaks, J.G. The binding of dihydrostreptomycin to *E. coli* ribosomes. Characteristics and equilibrium of the reaction, *Antimicrobial Agents and Chemotherapy, 2*, 294–307 (1972).
19. Controulis, J., Rebstock, M.C. and Crooks, H.M. Chloramphenicol (chloromycetin). V. Synthesis, *Journal of American Chemical Society, 71*, 2463–2468 (1949).
20. Coutsogeorgopoulos, C. Formation of olygophenylalanine in the presence of certain inhibitors of protein synthesis, *Federation Proceedings, 28*, 844 (1969).
21. Crowdy, S.H., Grove, J.F., Hemming, H.G. and Robinson, K.C. The translocation of antibiotics in higher plants. II. The movement of griseofulvin in broad bean and tomato, *Journal of Experimental Botany, 7*, 42–64 (1956).
22. Dähn, U., Hagenmaier, H., Höhne, H., König, W.A., Wolf, G. and Zähner, H. Stoffwechsel produkte von Mikroorganismen. 154. Mitteilung. Nikkomycin, eine neuer Hemmstoff der Chitinsynthese bei Pilzen, *Archives of Microbiology, 107*, 143–160 (1976)
23. Davies, J., Gilbert, W. and Gorini, L. Streptomycin, suppression and the code, *Proceedings of National Academy of Sciences U.S.A., 51*, 883–890 (1964).
24. Day, L.E. Tetracycline inhibition of cell-free protein synthesis. II. Effect of the binding of tetracycline to the components of the system, *Journal of Bacteriology, 92*, 197–203 (1966).
25. Donohue, J., Dunitx, J.D., Trueblood, K.N. and Webster, M.S. The crystal structure of aureomycin (chlortetracycline) hydrochloride. Configuration, bond distances and conformation, *Journal of American Chemical Society, 85*, 851–856 (1963).
26. Dubin, D.T., Hancock, R. and Davis, B.D. The sequence of some effects of streptomycin in *Escherichia coli*, *Biochimica et Biophysica Acta, 74*, 476–489 (1963).
27. Eguchi, J., Sasaki, S., Ohta, N., Akashiba, T., Tsuchiyama, T. and Suzuki, S. Studies on polyoxins, antifungal antibiotics. Mechanism

of action on the diseases caused by *Alternaria* spp. *Annals of the Phytopathological Society of Japan, 34*, 280–288 (1968).

28. Ehrlich, J., Bartz, Q.R., Smith, R.M., Joslyn, D.A. and Burkholder, P.R. Chloromycetin, a new antibiotic from a soil actinomycete, *Science, 106*, 417 (1947).
29. Ellis, R.J. Chloroplast ribosomes. Stereospecificity of inhibition by chloramphenicol, *Science, 163*, 477–478 (1969).
30. Endo, A. and Misato, T. Polyoxin D, a competitive inhibitor of UDP-N-acetylglucosamine. Chitin N-acetylglucosaminyl-transferase in *Neurospora crassa, Biochemical and Biophysical Research Communications, 37*, 718–722 (1969).
31. Fernandez-Munoz, R., Monro, R.E., Torres-pinedo, R. and Vazquez, D. Substrate- and antibiotic-binding sites at the peptidyl-transferase centre of *E. coli* ribosomes, *European Journal of Biochemistry, 23*, 185–193 (1971).
32. Finlay, A.C., Hobby, G.L., Pan, S.Y., Regna, P.P., Routien, J.B., Seeley, D.B., Shull, G.M., Sobin, B.A., Solomons, I.A., Vinson, J.W. and Kane, J.H. Terramycin, a new antibiotic, *Science, 111*, 85 (1950).
33. Fitzgerald, R.J., Bernheim, F. and Fitzgerald, D.B. The inhibition by streptomycin of adaptive enzyme formation in *Mycobacteria, Journal of Biological Chemistry, 175*, 195–200 (1948).
34. Ford, J.H. and Klomparens, W. Cycloheximide (actidione) and its non-agricultural uses, *Antibiotics and Chemotherapy, 10*, 682–687 (1960).
35. French, J.C., Bartz, Q.R. and Dion, H.W. Myomycin, a new antibiotic, *Journal of Antibiotics, 26*, 272–283 (1973).
36. French, J.C., Shores, S.C., Anderson, L.E., Woods, H., Burge, R.H., Clemens, M. and Howells, J.D. Polyamine compounds and methods for their production, U.S. patent 3,795,668 (1974).
37. Fukagawa, Y., Sawa, T., Takeuchi, T. and Umezawa, H., Studies on biosynthesis of kasugamycin. I. Biosynthesis of kasugamycin and the kasugamine moiety, *Journal of Antibiotics, 21*, 50–54 (1968a).
38. Fukagawa, Y., Sawa, T., Takeuchi, T. and Umezawa, H. Studies on biosynthesis of kasugamycin. III. Biosynthesis of the D-inositol, *Journal of Antibiotics, 21*, 185–188 (1968b).
39. Fukagawa, Y., Sawa, T., Takeuchi, T. and Umezawa, H. Studies on biosynthesis of kasugamycin. V. Biosynthesis of the amidine group, *Journal of Antibiotics, 21*, 410–412 (1968c).
40. Fukunaga, K., Misato, T. Ishii, I. and Asakawa, M. Blasticidin. A new anti-phytopathogenic fungal substance, *Bulletin of the Agricultural Chemical Society of Japan, 19*, 181–188 (1955).
41. Gale, E.F., Cundlife, E., Reynolds, P.E., Richmond, M.H. and Waring, M.J. (Editors). *The Molecular Basis of Antibiotic Action*, John Wiley and Sons, London, New York, Sydney, Toronto (1972).

42. Gale, E.F. and Folkes, J.P. The assimilation of amino acids by bacteria. 15. Actions of antibiotics on nucleic acid and protein synthesis in *Staphylococcus aureus*, *Biochemical Journal*, *53*, 493–498 (1953).
43. Golding, B.T. and Rickards, R.W. Conformation of some noviose glycosides and of novobiocin, *Chemistry and Industry*, 1081–1083 (1963).
44. Grove, J.F., MacMillan, J., Mulholland, T.P.C. and Rogers, M.A.T. Griseofulvin. Part IV. Structure, *Journal of the Chemical Society*, 3977–3987 (1952).
45. Hahn, F.E. Chloramphenicol, in *Antibiotics I. Mechanism of Action*, (Edited by D. Gottlieb and P.D. Shaw), Springer-Verlag, Berlin, Heidelberg, New York (1967).
46. Hahn, F.E., Hayes, J.E., Wisseman, C.L., Hopps, H.E. and Smadel, J.E. Mode of action of chloramphenicol. VI. Relation between structure and activity in the chloramphenicol series, *Antibiotics and Chemotheraphy*, *6*, 531–543 (1956).
47. Hamada, M., Hashimoto, T., Takahashi, S., Yoneyama, M., Miyake, T., Takeuchi, Y., Okami, Y. and Umezawa, H. Antimicrobial activity of kasugamycin, *Journal of Antibiotics*, Ser. *A*, *18*, 104–106 (1965).
48. Hancock, R. and Park, J.T. Cell-wall synthesis by *Staphylococcus aureus* in the presence of chloramphenicol, *Nature*, *181*, 1050–1052 (1958).
49. Harada, S., and Kishi, T. Isolation and characterization of mildiomycin, a new nucleoside antibiotic, *Journal of Antibiotics*, *31*, 519–524 (1978).
50. Harada, S., Mizuta, E. and Kishi, T. Structure of mildiomycin, a new antifungal nucleoside antibiotic, *Journal of American Chemical Society*, *100*, 4895–4897 (1978).
51. Hash, J.H., Wishnick, M. and Miller, P.A. On the mode of action of the tetracycline antibiotics in *Staphylococcus aureus*, *Journal of Biological Chemistry*, *239*, 2070–2078 (1964).
52. Heineman, B. and Hooper, I.R. Tetracycline recovery from fermentation broths, U.S. patent 2,886,595, (1959).
53. Helser, T.L., Davies, J.E. and Dahlberg, J.E. Change in methylation of 168 ribosomal RNA associated with mutation to kasugamycin resistance in *Escherichia coli*, *Nature*, *233*, 12–14 (1971).
54. Helser, T.L., Davies, J.E. and Dahlberg, J.E. Mechanism of kasugamycin resistance in *Escherichia coli*, *Nature*, *235*, 6–9, (1972).
55. Hidaka, J. and Murano, H. The absorption and translocation of ^{35}S labelled streptomycin sulfate in the tobacco plant, *Annals of the Phytopathological Society of Japan*, *24*, 161–174 (1959).
56. Hirano, S., Sagawa, T., Takahashi, H., Tanaka N., Oishi, H., Ando, K. and Togashi, K. Tetranactin, a new miticidal antibiotic. IV. Some properties of Tetranactin, *Journal of Economic Entomology*, *66*, 349–351 (1973).

57. Hirai, T. and Shimomura, T. Blasticidin S, an effective antibiotic against plant virus multiplication, *Phytopathology*, *55*, 291–295 (1965).
58. Hori, M., Eguchi, J., Kakiki, K. and Misato, T. Studies on the mode of action of polyoxins. VI. Effect of polyoxin B on chitin synthesis in polyoxin-sensitive and resistant strains of *Alternaria kikuchiana*, *Journal of Antibiotics*, *27*, 260–266 (1974).
59. Hori, M., Kakiki, K. and Misato, T. Studies on the mode of action of polyoxins. Part IV. Further study on the relation of polyoxin structure to chitin synthetase inhibition, *Agricultural and Biological Chemistry*, *38*, 691–698 (1974).
60. Hori, M., Kakiki, K. and Misato, T. Antagonistic effect of dipeptides on the uptake of polyoxin A by *Alternaria kikuchiana*, *Journal of Pesticide Science*, *2*, 139–149 (1977).
61. Horii, S., Iwasa, T. and Kameda, Y. Studies on validamycins, new antibiotics. V. Degradation studies, *Journal of Antibiotics*, *24*, 57–58 (1971).
62. Horii, S., Iwasa, T., Mizuta, E. and Kameda, Y. Studies on validamycins, new antibiotics. VI. Validamine, hydroxy-validamine and validatol, new cyclitols, *Journal of Antibiotics*, *24*, 59–63 (1971).
63. Horii, S. and Kameda, Y. Structure of validamycin A, *Journal of the Chemical Society*, 747–748 (1972).
64. Horii, S., Kameda, Y. and Kawahara, K. Studies on validamycins, new antibiotics. VIII. Validamycins C, D, E and F, *Journal of Antibiotics*, *25*, 48–53 (1972).
65. Hidaka, J. and Murano, H. Studies on streptomycin for plants. II. Translocation of streptomycin in plant body and control of bacterial diseases by the surface absorption, *Annals of the Phytopathological Society of Japan*, *21*, 49–53 (1956).
66. Hoeksema, H., Bergy, M.E., Jackson, W.G., Shell, J.W., Hinman, J.H., DeVries, W.H. and Crum, G.F. Streptonivicin, a new antibiotic. II. Isolation and characterization, *Antibiotics and Chemotherapy*, *6*, 143–148 (1956).
67. Hopps, H.E., Wisseman, C.L. and Hahn, F.E. Mode of action of chloramphenicol. V. Effect of chloramphenicol on polysaccharide synthesis by *Neisseria perflava*, *Antibiotics and Chemotherapy*, *4*, 857–858 (1954).
68. Hosokawa, S., Ogiwara, S. and Murata, Y. Oral subacute toxicity of validamycin for 4 months in beagle dogs, *Journal of the Takeda Research Laboratories*, *33*, 119–128 (1974).
69. Huang, K.T., Misato, T. and Asuyama, H. Effect of blasticidin S on protein synthesis of *Piricularia oryzae*, *Journal of Antibiotics*, Ser. *A*, *17*, 65–70 (1964).
70. Huber, F.M. Griseofulvin, in *Antibiotics I. Mechanism of Action* (Edited by D. Gottlieb and P.D. Shaw), Springer-Verlag, Berlin, Heidelberg, New York (1967).

71. Iitaka, Y., Sakamaki, T. and Nawata, Y. The molecular structures of tetranactin and its alkali metal ion complexes, *Chemistry Letters*, 1225–1230 (1972).
72. Ikekawa, T., Umezawa, H. and Iitaka, Y. The structure of kasugamycin hydrobromide by Z-ray crystallographic analysis, *Journal of Antibiotics*, Ser. *A*, *19*, 49–50 (1966).
73. Ishiie, T., Doi, Y., Yora, K. and Asuyama, H. Suppressive effects of antibiotics of tetracycline group on symptom development of mulberry dwarf disease, *Annals of the Phytopathological Society of Japan*, *33*, 267–275 (1967).
74. Ishiyama, T., Hara, I., Matsuoka, M., Sato, K., Shimada, S., Izawa, R., Hashimoto, T., Hamada, M., Okami, Y., Takeuchi, T. and Umezawa, H. Studies on the preventive effect of kasugamycin on rice blast, *Journal of Antibiotics*, Ser. *A*, *18*, 115–119 (1965).
75. Isono, K., Asahi K. and Suzuki, S. Studies on polyoxins, antifungal antibiotics. XIII. The structures of polyoxins, *Journal of American Chemical Society*, *91*, 7490–7505 (1969).
76. Isono, K., Crain, R.F., Odiorne, T.F., McCloskey, J.A. and Suhadolnik, R.J. Biosynthesis of the 5-fluoropolyxins, aberrant nucleoside antibiotics, *Journal of American Chemical Society*, *95*, 5788–5789 (1973).
77. Isono, K., Kobinata, K. and Suzuki, S. Isolation and of polyoxins J, K and L, new components of polyoxin complex, *Agricultural and Biological Chemistry*, *32*, 792–793 (1968).
78. Isono, K., Nagatsu, J. Kawashima, Y. and Suzuki, S. Studies on polyoxins, antifungal antibiotics. Part I. Isolation and characterization of polyoxins A and B, *Agricultural and Biological Chemistry*, *29*, 848–854 (1965).
79. Isono, K., Nagatsu, J., Kobinata, K., Sasaki, K. and Suzuki, S. Studies on polyoxins, antifungal antibiotics. Part V. Isolation and characterization of polyoxins C, D, E, F, G, H and I, *Agricultural and Biological Chemistry*, *31*, 190–199 (1967).
80. Isono, K. and Suhadolnik, R.J. The Biosynthesis of Natural and Unnatural polyoxins by *Streptomyces cacaoi*, *Archives of Biochemistry and Biophysics*, *173*, 141–153 (1976).
81. Isono, K. and Suzuki, S. The structures of polyoxins D, E, F, G, H, I, J, K and L, *Agricultural and Biological Chemistry*, *32*, 1193–1197 (1968).
82. Isono, K. and Suzuki, S. The structure of polyoxin C, *Tetrahedron Letters*, 203–208 (1968).
83. Isono, K. and Suzuki, S. The structures of polyoxins A and B, *Tetrahedron Letters*, 1133–1137 (1968).
84. Isono, K., Funayama, S. and Suhadolnik, R.J. Biosynthesis of the polyoxins, nucleoside peptide antibiotics. A new metabolic role for

L-isoleucine as a precursor for 3-ethylidene-ethylidene-L-azetidine-2-carboxylic acid (polyoxamic acid), *Biochemistry*, *14*, 2992–2996 (1975).
85. Iwasa, T., Higashide, E. and Shibata, M. Studies on validamycins, new antibiotics. III. Bioassay methods for the determination of validamycin, *Journal of Antibiotics*, *24*, 114–118 (1971).
86. Iwasa, T., Higashide, E., Yamamoto, H. and Shibata, M. Studies on validamycins, new antibiotics. II. Production and biological properties of VM-A and B, *Journal of Antibiotics*, *24*, 107–113 (1971).
87. Iwasa, T., Kameda, Y., Asai, M., Horii, S. and Mizuno, K. Studies on validamycins, new antibiotics. IV. Isolation and characterization of VM-A and B, *Journal of Antibiotics*, *24*, 119–123 (1971).
88. Iwasa, T., Suetomi, K. and Kusaka, T. Taxonomic study and fermentation of producing organism and antimicrobial activity of mildiomycin, *Journal of Antibiotics*, *31*, 511–518 (1978).
89. Iwasa, T., Yamamoto, H. and Shibata, M. Studies on validamycins, new antibiotics. I. *Streptomyces hygroscopicus* var. *limoneus* No. Var., validamycin-producing organism, *Journal of Antibiotics*, *23*, 595–602 (1970).
90. Jacoby, G.A. and Gorini, L. The effect of streptomycin and other aminoglycoside antibiotics on protein synthesis, in *Antibiotics I. Mechanism of Action* (Edited by D. Gottlieb and P.D. Shaw), Springer-Verlag, Berlin, Heidelberg, New York (1967).
91. Kaczka, E.A., Wolf, F.J., Pathe, F.P. and Folkers, K. Cathomycin. I. Isolation and characterization, *Journal of American Chemical Society*, *77*, 6404–6405 (1955).
92. Kaji, H., and Tanaka, Y. Binding of Dihydrostreptomycin to ribosomal subunit, *Journal of Molecular Biology*, *32*, 221–230 (1968).
93. Kameda, Y., Horii, S. and Yamano, T. Microbial transformation of validamycins, *Journal of Antibiotics*, *28*, 298–306 (1975).
94. Kamiya, K., Wada, Y., Horii, S. and Nishikawa, M. Studies on validamycins, new antibiotics. VII. The X-ray analysis of validamine hydrobromide, *Journal of Antibiotics*, *24*, 317–318 (1971).
95. Kerridge, D. The effect of actidione and other antifungal agents on nucleic acid and protein synthesis in *Saccharomyces carlsbergensis*, *Journal of General Microbiology*, *19*, 497–506 (1958).
96. Kornfeld, E.C., Jones, R.G. and Parke, T.V. The structure and chemistry of actidione, an antibiotic from *Streptomyces griseus*, *Journal of American Chemical Society*, *71*, 150–159 (1949).
97. Kroon, A.M. Protein synthesis in heart mitochondria. 1. Amino acid incorporation into the protein of isolated beef-heart mitochondria and fractions derived from them by sonic oscillation, *Biochemica et Biophysica Acta*, *72*, 391–402 (1963).

98. Kuzuhara, H., Ohrui, H. and Emoto, S. Total synthesis of polyoxin J, *Tetrahedron Letters*, 5055–5058 (1973).
99. Lanhart, K. Mutagenic effect of griseofulvin, *Mykosen*, *12*, 687–693 (1969).
100. Laskin, A.I. Tetracyclines, in *Antibiotics, I. Mechanism of Action*, (Edited by D. Gottlieb and P.D. Shaw), Springer-Verlag, Berlin, Heidelberg, New York (1967).
101. Leach, B.E., Ford, J.H. and Whiffen, A.J. Actidione, an antibiotic of *Streptomyces griseus*, *Journal of American Chemical Society*, *69*, 474 (1947).
102. Likover, T.E., Kurland, C.G. The contribution of DNA to translation errors induced by streptomycin *in vitro*, *Proceedings of National Academy of Sciences U.S.A.*, *58*, 2385–2392 (1967).
103. Lemin, A.J. and Ford, J.H. Isocycloheximide, *Journal of Organic Chemistry*, *25*, 344–346 (1960).
104. MacMillan, J. Dechlorogriseofulvin—a metabolic product of *Penicillium griseofulvum* DIECK and *Penicillium janczewskii* ZAL, *Chemistry and Industry*, 719 (1951).
105. MacMillan, J. Griseofulvin. Part 9. Isolation of the bromoanalogue from *Penicillium griseofulvum* and *Penicillium nigricans*, *Journal of the Chemical Society*, 2585–2587 (1954).
106. Mandelstam, J. and Rogers, H.J. Chloramphenicol-resistant incorporation of amino acids into *Staphylococci* and cell-wall synthesis, *Nature*, *181*, 956–957 (1958).
107. Margulies, M.M. and Brubaker, C. Effect of chloramphenicol on amino acid incorporation by chloroplasts and comparison *in vivo*, *Plant Physiology*, *45*, 632–633 (1970).
108. Masukawa, H. Localization of sensitivity to kanamycin and streptomycin in 30S ribosomal proteins of *Escherichia coli*, *Journal of Antibiotics*, *22*, 612–622 (1969).
109. Maxwell, I.H. Studies of the binding of tetracycline to ribosomes *in vitro*, *Molecular Pharmacology*, *4*, 25–37 (1968).
110. Maxwell, R.E. and Nickel, V.S. The antibacterial activity of the isomers of chloramphenicol, *Antibiotics and Chemotherapy*, *4*, 289–295 (1954).
111. McGowan, J.C. A substance causing abnormal development of fungal hyphae produced by *Penicillium janczewskii* ZAL. II. Preliminary notes on the chemical and physical properties of "curling factor", *The British Mycological Society Transactions*, *29*, 188 (1946).
112. Misato, T., Ishii, I., Asakawa, M., Okimoto, Y. and Fukunaga, K. Antibiotics as protectant fungicides against rice blast. II. The therapeutic action of blasticidin S, *Annals of the Phytopathological Society of Japan*, *24*, 302–306 (1959).
113. Misato, T., Ishii, I., Asakawa, M., Okimoto, Y. and Fukunaga, K.

Antibiotics as protectant fungicides against rice blast. III. Effect of blasticidin S on respiration of *Piricularia oryzae*, *Annals of the Phytopathological Society of Japan*, *26*, 19–24 (1961).
114. Misato, T., Okimoto, Y., Ishii, I., Asakawa, M. and Fukunaga, K. Antibiotics as protectant fungicides against rice blast. IV. Effect of blasticidin S on the metabolism of *Piricularia oryzae*, *Annals of the Phytopathological Society of Japan*, *26*, 25–30 (1961).
115. Mitani, M. and Inoue, Y. Antagonists of antifungal substance polyoxin, *Journal of Antibiotics*, *21*, 492–496 (1968).
116. Nathans, D. and Lipmann, F. Amino acid transfer from aminoacyl-ribonucleic acids to protein on ribosomes of *Escherichia coli*, *Proceedings of National Academy of Sciences U.S.A.*, *47*, 497–504 (1961).
117. Nioh, T. and Mizushima, S. Effect of validamycin on the general growth and morphology of *pellicularia sasakii*, *Journal of General and Applied Microbiology*, *20*, 373–383 (1974).
118. Nirenberg, M.W. and Matthaei, J.H. Dependence of cell-free protein synthesis in *E. coli* upon naturally occurring or synthetic polyribonucleotides, *Proceedings of National Academy of Sciences U.S.A.*, *47*, 1588–1602 (1961).
119. Ohmori, K. Studies on characters of *Piricularia oryzae* made resistant to kasugamycin, *Journal of Antibiotics*, Ser. *A*, 20, 109–114 (1967).
120. Okimoto, Y. and Misato, T. Antibiotics as protectant bactericide against bacterial leaf blight of rice plant. 2. Effect of cellocidin on growth, respiration, and glycolysis of *Xanthomonas oryzae*, *Annals of the Phytopathological Society of Japan*, *28*, 209–215 (1963a).
121. Okimoto, Y. and Misato, T. Antibiotics as protectant bactericide against bacterial leaf blight of rice plant. 3. Effect of cellocidin on TCA cycle, electron transport system, and metabolism of protein in *Xanthomonas oryzae*, *Annals of the Phytopathological Society of Japan*, *28*, 250–257 (1963b).
122. Okuda, T., Suzuki, M., Egawa, Y. and Ashino, K. Studies on cycloheximide and its new stereoisomeric antibiotics, *Chemical and Pharmaceutical Bulletin*, *6*, 328–330 (1958).
123. Okuda, T., Suzuki, M., Egawa, Y. and Ashino, K. Studies on streptomyces antibiotic, cycloheximide. II. Naramycin-B, an isomer of cycloheximide, *Chemical and Pharmaceutical Bulletin*, *7*, 27–30 (1959).
124. Onuma, S., Nawata, Y. and Saito, Y. An X-ray analysis of blasticidin S monohydrobromide, *Bulletin of Chemical Society of Japan*, *39*, 1091 (1966).
125. Ōtake, N., Takeuchi, S., Endo, T. and Yonehara, H. Chemical studies on blasticidin S. III. The structure of blasticidin S, *Agricultural and Biological Chemistry*, *30*, 132–141 (1966).
126. Oxford, A.E., Raistrick, H. and Simonart, P. Studies on the biochemi-

stry of microorganisms. 60. Griseofulvin, $C_{17}H_{17}O_6Cl$, a metabolic product of *penicillium griseofulvum* DIECK, *Biochemical Journal, 33*, 240–248 (1939).

127. Ozaki, M., Mizushima, S. and Nomura, M. Identification and functional characterization of the protein controlled by streptomycin-resistant locus in *E. coli, Nature, 222*, 333–339 (1969).
128. Paget, G.E. and Alcock, S.J. Griseofulvin and Colchicine. Lack of Carcinogenic action, *Nature, 188*, 867 (1960).
129. Paget, G.E. and Walpole A.L. Some cytological effects of griseofulvin, *Nature, 182*, 1320–1321 (1958).
130. Pestka, S. Inhibitors of ribosomal functions, *Annual Review of Microbiology, 25*, 487–562 (1971).
131. Pestka, S. Peptidyl-puromycin synthesis on polyribosomes from *Escherichia coli, Proceedings of National Academy of Sciences U.S.A., 69*, 624–628 (1972).
132. Polglase, W.J. Contribution of the cationic groups of dihydro-streptomycin to biological activity, *Nature, 206*, 298–299 (1965).
133. Rebstock, M.C., Crooks, H.M. Controulis, J. and Bartz, Q. Chloramphenicol (chloromycetin). IV. Chemical studies, *Journal of American Chemical Society, 71*, 2458–2462 (1949).
134. Rendi, R. The effect of chloramphenicol on the incorporation of labeled amino acids into proteins by isolated subcellular fractions from rat liver, *Experimental Cell Research, 18*, 187–190 (1959).
135. Sagawa, T., Hirano, S., Takahashi, H., Tanaka, N., Oishi, H. Ando, K. and Togashi, K. Tetranactin, a new miticidal antibiotic. III. Miticidal and other biological properties, *Journal of Economic Entomology, 65*, 372–375 (1972).
136. Sasaki, S., Ohta, N., Eguchi, J., Furukawa, Y., Akashiba, T., Tsuchiyama, T. and Suzuki, S. Studies on polyoxins, antifungal antibiotics. VIII. Mechanism of action on sheath blight of rice plant, *Annals of the Phytopathological Society of Japan, 34*, 272–279 (1968).
137. Sasaki, S., Ohta N., Yamaguchi, I., Kuroda, S. and Misato, T. Studies on polyoxin action. Part I. Effect on respiration and synthesis of protein, nucleic acids and cell-wall of fungi, *Nippon Nogei-Kagaku Kaishi, 42*, 633–638 (1968).
138. Sakata, K., Sakurai, A. and Tamura, S. Structures of exomycins A_1 and A_2, *Tetrahedron Letters*, 4327–4330 (1974).
139. Sakata, K., Sakurai, A. and Tamura, S., Degradative studies on ezomycins A_1 and A_2, *Agricultural and Biological Chemistry, 39*, 885–892 (1975).
140. Sakata, K., Sakurai, A. and Tamura, S. Isolation of novel antifungal antibiotics, ezomycins A_1, A_2, A_3, B_1 and B_2, *Agricultural and Biological Chemistry, 38*, 1883–1890 (1974).

141. Sakata, K., Sakurai, A. and Tamura, S. Structures of ezomycins B_1, B_2, C_1, C_2, D_1 and D_2, *Tetrahedron Letters*, 3191–3194 (1975).
142. Schatz, A., Bugie, E. and Waksman, S.A. Streptomycin, a substance exhibiting antibiotic activity against Gram-positive and Gram-negative bacteria, *Proceedings of the Society for Experimental Biology and Medicine*, *55*, 66–69 (1944).
143. Seino, A., Sugawara, H., Shirato, S. and Misato, T. Aabomycin A, a new antibiotic. III. Taxonomic studies on the aabomycin producing strain, *Streptomyces hygroscopicus* subsp. *aabomyceticus* SEINO subsp. Nov., *Journal of Antibiotics*, *23*, 204–209 (1970).
144. Seto, H., Ōtake, N. and Yonehara, H. Studies on the biosynthesis of blasticidin S. Part II. Leucylblasticidin S, a metabolic intermediate of blasticidin S biosynthesis, *Agricultural and Biological Chemistry*, *32*, 1299–1305 (1968).
145. Seto, H., Yamaguchi, I., Ōtake, N. and Yonehara, H. Biogenesis of blasticidin S, *Tetrahedron Letters*, 3793–3799 (1966).
146. Seto, H., Yamaguchi, I., Ōtake, N. and Yonehara, H. Studies on the biosynthesis of blasticidin S. Part I. Precursors of blasticidin S biosynthesis, *Agricultural and Biological Chemistry*, *32*, 1292–1298 (1968).
147. Shunk, C.H., Stammer, C.H., Kaczka, E.A., Walton E., Spencer, C.F., Wilson, A.N., Richter, J.W., Holly, F.W. and Folkers, K. Nobociocin. II. Structure of novobiocin, *Journal of American Chemical Society*, *78*, 1770–1771 (1956).
148. Sisler, H. D. and Siegel, M.R. Cycloheximide and other glutarimide antibiotics, in *Antibiotics I. Mechanism of Action* (Edited by D. Gottlieb and P.D. Shaw), Springer-Verlag, Berlin, Heidelberg, New York, (1967).
149. Sinha, R.C. and Peterson, E.A. Uptake and oxytetracycline in ester plants and vector leafhoppers in relation to inhibition of clover phyllody agent, *Phytopathology*, *62*, 377–383 (1972).
150. Smith, D.H. and Davis, B.D. Inhibition of nucleic acid synthesis by novobiocin, *Biochemical and Biophysical Research Communications*, *18*, 796–800 (1965).
151. Smith, D.H. and Davis, B.D. Mode of action of novobiocin in *Escherichia coli*, *Journal of Bacteriology*, *93*, 71–79 (1967).
152. Smith, C.G., Dietz, A., Sokolski, W.T. and Savage, G. Streptonivicin, a new antibiotic, *Antibiotics and Chemotherapy*, *6*, 135–142 (1956).
153. So, A.G. and Davie, E.W. The incorporation of amino acids into protein in a cell-free system from yeast, *Biochemistry*, *2*, 132–136 (1963).
154. Stern, J.L. Barner, H.D. and Cohen, S.S. The lethality of streptomycin and the stimulation of RNA synthesis in the absence of protein synthesis, *Journal of Molecular Biology*, *17*, 188–217 (1966).
155. Suarez, G. and Nathans, D. Inhibition of aminoacyl-sRNA binding to

ribosomes by tetracycline, *Biochemical and Biophysical Research Communications, 18*, 743–750 (1965).
156. Suhara, Y., Maeda, K. and Umezawa, H. Chemical studies on kasugamycin. V. The structure of kasugamycin, *Tetrahedron Letters, 12*, 1239–1244 (1966).
157. Suhara, Y., Sasaki, F., Maeda, K., Umezawa, H. and Ohno, M. The total synthesis of kasugamycin, *Journal of American Chemical Society, 90*, 6559–6560 (1968).
158. Suzuki, S., Isono, K., Nagatsu, J., Mizutani, T., Kawashima, Y. and Mizuno, T. A new antibiotic, polyoxin A, *Journal of Antibiotics*, Ser. *A, 18*, 131 (1965).
159. Suzuki, S., Isono, K., Nagatsu, J., Kawashima, Y., Yamagata, K., Sasaki, K. and Hashimoto, K. Studies on polyoxins, antifungal antibiotics. Part IV. Isolation of polyoxins C, D, E, F, and G, new components of polyoxin complex, *Agricultural and Biological Chemistry, 30*, 817–819 (1966).
160. Suzuki, S., Nakamura, G., Okuma, K. and Tomiyama, Y. Cellocidin, a new antibiotic, *Journal of Antibiotics*, Ser. *A, 11*, 81–83, (1958).
161. Suzuki, S. and Okuma, K. The structure of cellocidin, *Journal of Antibiotics*, Ser. *A, 11*, 84–86 (1958).
162. Tanaka, N., Matsukawa, H. and Umezawa, H. Structural basis of kanamycin for miscoding activity, *Biochemical and Biophysical Research Communications, 26*, 544–549 (1967).
163. Tanaka, N., Nishimura, T., Yamaguchi, H., Yamamoto, G., Yoshida, Y., Sashikata, K., and Umezawa, H. Mechanism of action of Kasugamycin, *Journal of Antibiotics*, Ser. *A, 18*, 139–144 (1965).
164. Tanaka, N., Sakagami, Y., Yamaki, H. and Umezawa, H. Activity of cytomycin and blasticidin S against transplantable animal tumors, *Journal of Antibiotics*, Ser. *A, 14*, 123–126 (1961).
165. Tanaka, N., Sakaguchi, K., Otake, N. and Yonehara, H. An improved screening method for inhibitors of nucleic acid synthesis, *Agricultural and Biological Chemistry, 32*, 100–103 (1963).
166. Tanaka, N., Yamaguchi, H. and Umezawa, H. Mechanism of kasugamycin action on polypeptide synthesis, *Journal of Biochemistry, 60*, 429–434 (1966).
167. Tanaka, N., Yoshida, Y., Y. Sashikata, K., Yamaguchi, H. and Umezawa, H. Inhibition of polypeptide synthesis by kasugamycin, an aminoglycosidic antibiotic, *Journal of Antibiotics*, Ser. *A, 19*, 65–68 (1966).
168. Takeuchi, Y. and Buerger, M.J. The chemical structure of terramycin hydrochloride, *Proceedings of National Academy of Sciences U.S.A., 46*, 1366–1370 (1960).
169. Takeuchi, S., Hirayama, K., Ueda, K., Sasaki, H. and Yonehara, H.

Blasticidin S, a new antibiotic, *Journal of Antibiotics*, Ser. *A*, *11*, 1–5 (1958).

170. Thyagarajan, T.R., Srivastava, O.P. and Vora, V.V. Some cytological observations on the effect of griseofulvin on dermatophytes, *Naturwissenschaften*, *50*, 524–525 (1963).
171. Umezawa, H., Okami, Y., Hashimoto, T., Suhara, Y., Hamada, M. and Takeuchi, T. A new antibiotic, Kasugamycin, *Journal of Antibiotics*, Ser. *A*, *18*, 101–103 (1965).
172. Vazquez, D. The binding of chloramphenicol by ribosomes from *Bacillus megaterium*, *Biochemical and Biophysical Research Communications* *15*, 464–468 (1964).
173. Vogel, Z., Vogel, T., Zamir, A. and Elson, D. Ribosome activation and binding of dihydrostreptomycin. Effect of polynucleotides and temperature on activation, *Journal of Molecular Biology*, *54*, 379–386 (1970).
174. Wakae, O. and Matsuura, K., Characteristics of validamycin as a fungicide for *Rhizoctonia* disease control, *Review of Plant Protection Research*, *24*, 107–110 (1971).
175. Wakae, O. and Matsuura, K. Validamycin. A new antibiotic for *Rhizoctonia* disease control, *Proceedings of the First International Congress of IAMS*, Vol. 3, Science Council of Japan (1975).
176. Wakimoto, S. and Mukoo, H. National occurrences of streptomycin resistant *Xanthomonas oryzae*, the causal bacteria of leaf blight disease of rice, *Annals of the Phytopathological Society of Japan*, *28*, 153–158 (1963).
177. Welch, J.J. Rodent control. A review of chemical repellents for rodents, *Journal of Agricultural and Food Chemistry*, *2*, 142–149 (1954).
178. Whiffen, A.J. The production, assay, and antibiotic activity of actidione, an antibiotic from *Streptomyces griseus*, *Journal of Bacteriology*, *56*, 283–291 (1948).
179. Whiffen, A.J. The activity *in vitro* of cycloheximide (actidione) against fungi pathogenic to plants, *Mycologia*, *42*, 253–258 (1950).
180. Wilkins, J.R., Lewis, C. and Barbiers, A.R. Streptonivicin, a new antibiotic. III. *In vitro* and *in vivo* evaluation, *Antibiotics and Chemotherapy*, *6*, 149–156 (1956).
181. Wintersberger, E. Protein-synthese in isolierten Hefemitochondrien, *Biochemische Zeitschrift*, *341*, 409–419 (1965).
182. Wisseman, C.L., Hahn, F.E. and Hopps, H.E. Chloramphenicol inhibition of protein synthesis, *Federation Proceedings*, *12*, 466 (1953).
183. Wolf, A.D. and Hahn, F.E. Mode of action of chloramphenicol. IX. Effects of chloramphenicol upon a ribosomal amino acid polymerization system and its binding to bacterial ribosome, *Biochimica et Biophysica Acta*, *95*, 146–155 (1965).

184. Yamaguchi, I., Taguchi, R., Huang, K.T. and Misato, T. Biological studies on aabomycin A, *Journal of Antibiotics*, *22*, 463–466 (1969).
185. Yamaguchi, I., Takagi, K. and Misato, T. The sites for degradation of blasticidin S, *Agricultural and Biological Chemistry*, *36*, 1719–1727 (1972).
186. Yamaguchi, H. and Tanaka, N. Inhibition of protein synthesis by blasticidin S. II. Studies on the site of action in *E. coli* polypeptide synthesizing systems, *Journal of Biochemistry*, *60*, 632–642 (1966).
187. Yonehara, H. and Ōtake, N. Absolute configuration of blasticidin S, *Tetrahedron Letters*, 3785–3791 (1966).
188. Yonehara, H., Seto, H., Aizawa, S., Hidaka, T., Shimazu, A. and Otake, N. The detoxin complex, selective antagonists of blasticidin S, *Journal of Antibiotics*, *21*, 369–370 (1968).
189. Yoneyama, K., Koike, M., Sekido, S., Ko, K. and Misato, T. Effect of myomycin on plant bacterial diseases, *Journal of Pesticide Society*, *31*, 359–364 (1978).
190. Yoneyama, K., Sekido, S. and Misato, T. Antagonistic mechanism of sulfhydryl compounds on cellocidin activity, *Journal of Antibiotics*, *31*, 1065–1066 (1978).
191. Yukioka, M., Hatayama, T. and Morisawa, S. Affinity labelling of the ribonucleic acid component adjacent to the peptidyl recognition center of peptidyl transferase in *Escherichia coli* ribosomes, *Biochimica et Biophysica Acta*, *390*, 192–208 (1975).

SECTION C

New Strategies in Bioconversion

N.S. SUBBA RAO

19. Utilization of Farm Wastes and Residues in Agriculture

Animal excreta and residues from crops after harvest can be recycled in agriculture not only to conserve energy but also to minimize pollution. Intensive agriculture in the last two decades has no doubt increased food production but the disposal of plant residues has posed fresh problems. In the United States of America agricultural and food wastes and manure account for $600^{10^{-6}}$ t/year. Similarly, farm animals (beef cattle, dairy cattle, swine, sheep broilers, laying hens) provide annually more than 1386 billion kg of excreta having N.P.K. content approximately greater than 9.9, 2.4 and 7.2 billions of kg., respectively (57).

The net productivity of plants by photosynthesis is about 155.2 billion tonnes of dry biomass per year. Barring one-third of this biomass coming from the oceans, the remainder comes from terrestrial ecosystems. About 15 per cent of the rest of the two-thirds biomass can be ascribed to grasslands and cultivated lands, of which approximately 24 billion tonnes of the biomass generated happens to be either wastes or residues (10). The types of plants which provide residues after harvest are straw from rye, wheat, barley, rice, oat and grass, stover from sorghum, soybean and corn, sugar cane bagasse, cotton gin trash, husks from grains and vegetable or fruit peelings.

While quantitative estimates of available wastes and residues from advanced countries are readily available, similar estimates for developing countries are not so easily available. Apart from the considerations of availability, the mode and economics of collection, processing and utilization of wastes are not generally taken into account when one thinks of harnessing wastes in agriculture in developing countries. Nevertheless, the estimates of the available and realizable plant nutrient potential from the residues of principal crops in a developing country such as India, have been made by Bhardwaj (11) which are reproduced in Table 1. Similarly, the sources, estimated yield and chemical composition of animal organic wastes in India have been recently provided by Ranjhan (52) (Tables 2 and 3).

Table 1. Estimates of the available and realizable plant nutrient potential from the residues of principal crops in India [Bhardwaj (11)]

Crop	Residue/economic yield ratio	Thousand tonnes		Nutrient content (%)			Thousand tonnes		
		yield of economic component (1978–79)	Residue yield	N	P	K	Total NPK potential	Utilizable NPK potential (one-third)	Fertilizer equivalent value (50%)
Rice	1.5	53829	80744	0.70	0.09	1.15	1566.4	522.1	261.0
Wheat	1.3	34982	44987	0.48	0.07	0.98	688.3	229.4	114.7
Sorghum	1.0	11563	11563	0.85	0.12	1.21	253.0	84.3	42.1
Maize	1.0	6219	6219	0.82	0.09	1.25	134.3	44.8	22.4
Pearl miilet	1.5	5515	8283	0.45	0.07	0.95	121.6	40.5	20.2
Barley	1.5	2120	3180	0.52	0.08	1.25	58.8	19.6	9.8
Sugar cane	0.1	156450	15645**	0.45	0.08	1.20	270.7	270.7*	135.3
Potato	0.5	10125	5062	0.52	0.09	0.85	73.9	73.9*	36.9
Groundnut	1.5	6387	9580	1.65	0.12	1.23	277.3	277.3*	138.6

*Entire nutrient potential in the residues of the three crops could be taken as utilizable.

**Only trash.

Table 2. Sources and estimated yield of various animal organic wastes [Ranjhan (52)]

Animal organic waste	Estimated yield/year on dry matter basis	
	million tonnes	percentage
Bovine dung	267.00	92.62
Sheep and goat faeces	15.53	5.387
Pig faeces	1.94	0.673
Poultry droppings	1.26	0.437
Unhatched eggs and dead embryos	0.0012	—
Feather meal	0.017	—
Egg shell	0.0006	—
Dead animals		
Meat meal	0.920	0.319
Bone meal	0.575	0.199
Animal fat	0.690	0.239
Total	288.28	

Table 3. Chemical composition of various animal organic wastes on dry matter basis (%) [Ranjhan (52)]

Wastes	Crude protein	Crude fibre	Organic matter	Ash	Ca	P
Bovine dung	12–15	30–50	80–85	15–20	2–3	2–2.5
Sheep and goat faeces	15–20	25–35	70–80	20–30	3–4	2–3
Swine faeces	15–25	10–20	80–85	15–20	3–4	2–3
Poultry droppings	25–32	10–15	70–80	20–30	5–7	4–5
Poultry litter	15–25	20–25	65–75	25–35	4–6	3–4
Digested slurry from gobar gas plant	13–16	20–25	75–80	20–25	1–1.5	0.75–1.0

Organic matter decomposition

An age-old agricultural practice has been that a basal dressing of organic manure by way of well-decomposed compost is often added to the soil by the cultivator to restore soil structure and ensure proper root growth. Even under natural conditions when plant and animal residues reach the soil, they are acted upon by microorganisms when the processes of decomposition begin. Such residues comprise carbohydrates, proteins, fats, oils, alcohols, aldehydes, ketones, organic acids, lignin, phenols, tannins, hydrocarbons, alkaloids, pyrimidines and purines, enzymes and pigments (8). The bulk of plant residue is made up of lignocelluloses. Lignin is not easily degradable whereas cellulose is attacked by a host of microorganisms (Table 4). The enzymes involved are collectively known as cellulases.

Table 4. Genera of microorganisms which exhibit cellulolytic activity (collected from several sources)

Fungi:	*Acremoniella, Acremonium, Allesciaeizia, Arthrobotris, Aspergillus, Bispora, Botryotrichum, Catenularia, Ceratocystis, Chaetomium, Chloridium, Chrysosporium, Coniothyrium, Cordana, Coryne, Dictyosporium, Doratomyces, Fusarium, Gliocladium, Gonatobotrys, Graphium, Humicola, Myrothecium, Oidiodendron, Penicillium, Petriellidium, Phialocephala, Phoma, Pseudorotium, Rhinocladiella, Scytalidium, Sporotrichum, Stachybotrys, Trichoderma, Wardomyces, Xylogone*
Bacteria and Actinomycetes:	*Actinomyces, Angiococcus, Bacillus, Cellulomonas, Clostridium, Polyangium, Sporangium, Streptomyces*

When fresh plant residues reach the soil, under suitable conditions, bacteria, fungi and actinomycetes multiply with the formation of carbon dioxide. The rate of liberation of this gas reflects the extent of decomposition. Several physical and chemical factors influence organic matter decomposition. The finer the particle size of residues the better is the extent of decomposition. Other factors such as the moisture level, pH, available nutrients, C/N status, qualitative nature of microflora, temperature, aeration and the presence or absence of inhibitory substances (tannins, excess salts, pesticides and antibiotics) also influence the rate of decomposition.

Lignin constitutes a major portion of plant tissue material, next only to cellulose and hemicelluloses. Very little is known about the microbiology of lignin decomposition. Many basidiomycetous fungi have the ability to degrade lignin and they belong to the genera *Agaricus, Armillaria, Clavaria, Clitocybe, Coprinus, Cortinellus, Ganoderma, Lenzites, Marasmius, Mycena, Panus, Pholiota, Polystictus, Schizophyllum, Stereum, Ustulina, Collybia, Fomes, Pleurotus, Polyporus, Poria* and *Trametes* (7). Some of these fungi which degrade lignin can also break down glucose. Lignin degradation is considered to be significant in humus formation.

Mineralization and immobilization processes

Organic materials reaching the soil contain nutrients in a bound form which have to be rendered into an inorganic form for easy assimilation by plants. The conversion of organic forms of carbon, nitrogen and phosphorus to the easily assimilable inorganic state is known as mineralization. Plants growing in natural ecosystems depend upon organic matter for their nutrient requirements and since microorganisms are largely involved in nutrient transformations in soil, due consideration has to be given to the mineralization and immobilization processes in the soil during organic matter decomposition.

Nitrogen mineralization involves ammonification which is a microbiological process by which ammonium is formed from organic compounds mediated by diverse genera of microorganisms. Some of the frequently investigated fungi and bacteria in this connection are *Alternaria*, *Aspergillus*, *Mucor*, *Penicillium*, *Rhizopus*, *Pseudomonas*, *Bacillus*, *Clostridium*, *Serratia* and *Micrococcus*. Nitrification, a process by which ammonium is converted to nitrate is a consequence of mineralization. The magnitude of mineralization of nitrogen in soils depends upon several environmental factors.

Nitrogen is needed for microbial growth in the same way as crops need it for their growth. Ammonium salts are the most readily available sources of nitrogen which are depleted in the process of organic matter decomposition to build up the microbial biomass in soil. Nitrogen immobilization is a process opposite in magnitude to that of mineralization and locks up inorganic nitrogen in microbial cells, a process which is rapid in such soils to which crop residues poor in nitrogen are added. Nitrogen immobilization followed by additions of nitrogen-poor crop residues to soil is undesirable for crop productivity because nitrogen becomes a limiting nutrient to crop growth owing to the competition of microorganisms with the higher plant for the same element. It is in this context that the addition of nitrogen-rich legume residues is advocated for immediate benefits to crop growth (9, 17, 30, 34).

Phosphorus mineralization and immobilization are also processes similar to those of nitrogen. Microorganisms convert the organic source of phosphorus into inorganic forms. The organic phosphorus is derived from plant tissues and microbial protoplasm. Environmental conditions influence phosphorus mineralization which is more rapid in uncultivated rather than cultivated soils. The enzymes which take part in the release of phosphorus from organic sources are collectively termed Phosphatases and species of *Aspergillus*, *Penicillium*, *Rhizopus*, *Cunninghamella*, *Arthrobacter*, *Streptomyces*, *Pseudomonas* and *Bacillus* are known to synthesize one of the enzymes, phytase (7). Assimilation of phosphorus into the cytosol of microorganisms takes place during organic matter decomposition and hence in phosphate deficient soils, the addition of fresh crop residues may depress crop yields due to the immobilization of this scarce element in plant cytosol. Therefore, the addition of crop residues which are rich in both nitrogen and phosphorus is desirable in impoverished soil (19, 36, 46, 56, 59).

Organic matter provides carbon for microorganisms. Assimilation of this element into the microbial biomass under aerobic conditions is accompanied by the release of carbon dioxide. The efficiency of carbon assimilation is reflected in the release of this gas—more carbon assimilation results in less carbon dioxide liberation and *vice versa*. The assimilation of carbon is also accompanied by the immobilization of other nutrients into the microbial biomass, and hence carbon assimilation is linked with the nitrogen, phosphorus, potassium and sulphur status of soils as well as the quality and quantity of added organic matter (14, 15, 19, 27, 35, 47, 48).

Humus

The organic fraction of the soil, often termed humus, is a product of the synthetic and decomposing activities of the microflora (7). It is necessary to distinguish between the degradation of freshly added plant residues and the decomposition of organic matter native to any given soil. The native soil organic matter is subject to continuous mineralization processes since it consists of essential nutrients in an organic form. Microbial decomposition of freshly added organic matter, after a series of degradations involved in decomposition results in brown or black organic complexes which remain in a dynamic state. Humus is sparingly soluble in water but is rendered into solution by alkali. It contains amino acids, purines, pyrimidines, aromatic molecules, uronic acids, amino sugars, pentose and hexose sugars, sugar alcohols, methyl sugars, aliphatic acids and probably other unknown substances. Humus can be fractionated into humins, humic acids, fulvic acids and hymetomelanic acids. Although resistant to microbial attack, many soil bacteria, fungi and actinomycetes have been shown to degrade humic acid in pure culture. The genera of microorganisms involved in humic acid degradation are *Bacillus*, *Pseudomonas*, *Streptomyces*, *Aspergillus* and *Penicillium*.

There is plenty of evidence to show that humus or its fractions improve plant growth and nutrient uptake (12, 13, 18, 20, 23, 24, 26, 38, 39, 42–44, 49, 50, 55, 61). These and other evidence point out to the multipronged benefits of humus on plants but the mode of action still remains a matter of conjecture.

Composting

Composting is an age-old process (40). It is a process essentially meant to utilize solid wastes of animal and plant origin in agriculture. The concept of farm composting has now recently been enlarged to recycle city wastes, sewage and night soil of human origin. Much of the early work on composting was related to evolving measures to achieve quick composting by the addition of chemicals and determining the factors involved in optimum composting. Artificial farmyard manure fortified with chemicals was known in the early part of this century (54). Similarly the role of nitrogenous starters (0.7 to 0.8 g mineral nitrogen per 100 g composting materials) and the importance of C/N ratio of composting material as factors in decomposition were also recognized early (32). At the same time, it was soon realized that fungi were involved in collulose breakdown (53, 60, 63).

Some of the important points which are relevant in scientific composting are:

(1) Shredding of materials to smaller pieces.

(2) The initial C/N ratio of the material because a ratio of 30 to 35 is ideal for composting. A low C/N ratio encourages the ammonia to escape while a

Plate 1. Composting on the farm: (A) Digging the pit; (B) Layering of crop wastes and bovine excreta (urine and dung); (C) Filling with soil; and (D) Sealing with mud-plaster (courtesy: Farm Information Directorate, Government of India).

Plate 2. Biogas plants on the farm: (A) Closed or below ground level type of biogas plant in China; (B) A family size Chinese plant which has been exposed in the process of preparation for below ground level operation (photos, courtesy: P.R. Hesse, UNDP/FAO); (C and D) Above ground level type of village biogas plants which are operated in India (photos, courtesy: Division of Agricultural Extension, I.A.R.I., New Delhi).

high C/N ratio is not congenial for microbial equilibrium.

(3) Frequent turning of compost material for proper decomposition and controlling fly breeding and lessening noxious odours.

(4) Maintenance of optimum temperature at 50–60°C which helps in destroying harmful pathogens and *Ascaris* eggs (1–4, 16, 21, 22, 28, 29, 41, 45, 51, 58, 62, 64–66).

Although composting has been practised even from Biblical times, scientific procedures have been envolved only during this century (1, 5, 25, 31). In the aerobic method, a pit (approximately 3 feet deep and 6–8 feet wide) is dug near the cattle house on a site free from waterlogging. Plant residues are shredded and piled up in the pit with excreta of cattle in layers. The layering of plant residues and cattle excreta is done alternately so as to fill the pit completely. The residues are turned once a fortnight and a good quality compost is made (Plate 1) in about three months. In the anaerobic method, the layers in the pit are sealed with plaster of mud. Due to the anaerobic conditions, the temperature inside the pit rises. Variations of this process involve the opening of the mud plaster periodically, followed by the turning over of the contents and the re-sealing of the pit by mud plaster. This leads to an alteration of the aerobic and anaerobic conditions. Individual preferences, practices and needs dictate the adoption of any of these methods. In China (FAO report) various devices are adopted to combine the anaerobic and aerobic methods in circular or rectangular pits filled with aquatic weeds, rice straw, animal dung, silt, shredded crop residues, night soil, urine etc. The turning over of the residues at intervals and the resealing of the mud plaster of the pits are essential ingredients of the Chinese methods. Hollow bamboo poles are driven into the sealed mud plaster to provide aeration which can easily be removed and the holes plastered again with mud to provide anaerobic conditions.

Acceleration of composting

Essentially, composting is a microbiological process. The addition of nitrogen-rich materials such as dried blood, cowdung, oil-cakes, ammonium sulphate, sodium nitrate, urea or ammonia can activate this process by increasing microbial activity. It is possible to accelerate the composting processes by adding selected cellulolytic microorganisms into the compost pit. In a study (67) undertaken to find out the role of added microorganisms such as *Trichoderma viride*, *Chaetomium abuanse*, *Myrothecium roridum*, *Aspergillus niger*, *A. terreus*, *Cellulomonas* sp. and *Cytophaga* sp., it was observed that after microbial treatment, the C/N ratio and the weight of the biomass were not only drastically lowered depending on the nature of the microorganism but the treatment also brought about an increase in the humic acid content of the composted material (Table 5). Similar results have also been obtained at different centres of the All India Coordinated Project of the Indian Council of Agricultural Research on Organic Matter Decomposition.

Table 5. Analysis of compost at the end of 12 weeks (average of three replications) [Yadav and Subba Rao (67)]

Name of the isolate	% organic carbon	% total nitrogen	C : N ratio	% loss in weight	% humic acid
Control	49.11	1.05	46.8	35.7	4.0
A. niger GF5	48.33	1.26	38.5	45.7	7.3
A. terreus BE2	48.04	1.26	45.5	42.9	6.4
Chaetomium abuanse COF7	47.62	1.25	38.1	47.7	7.5
Trichoderma viride BF8	38.21	2.22	17.7	62.9	8.3
Cellulomonas CAB3	37.82	2.40	15.8	63.6	9.5
Cytophaga BHB2	48.25	1.49	32.5	50.0	6.6
Mixture of above fungi	39.52	1.74	22.6	56.3	7.6
Mixture of above bacteria	40.31	1.93	20.9	57.7	7.3
Mixture of above fungi and bacteria	41.78	1.64	25.1	53.4	6.9

Initial: Organic carbon 53.08%; Total nitrogen 0.7%; C : N ratio 75.83.

Biogas production

Under anaerobic conditions, organic matter is degraded by primary colonizers into simple sugars. The genus *Clostridium* is very active in converting cellulose into organic acids such as acetic, formic, lactic, succinic and butyric acids and alcohols. Methane bacteria (*Methanococcus*, *Methanosarcina*, *Methanobacillus* and *Methanobacterium*) which are secondary colonizers break down organic acids into CH_4 and CO_2 (see also the next chapter by Hobson in this volume). The principle of anaerobic digestion of cowdung with agricultural wastes and residues is utilized at the village level to generate methane gas for domestic use in what are known as 'biogas plants' in India, China, Taiwan, Korea, Uganda and Bangladesh (Plate 2). The residue or slurry left behind after gas production is rich in nutrients (Table 6) and is useful as an organic fertilizer (6, 33, 37, 45).

Table 6. Typical chemical composition of biogas slurry in comparison with other manurial sources [Khandelwal (37)]

	Cattle dung	Digested slurry	Farmyard manure
Total solids (%)	20.0	8.2	30.8
C%	47.8	41.6	38.4
N%	1.55	2.24	1.99
P%	0.76	1.20	1.02
C : N	30.8	18.6	19.3

Future prospects

There is an urgent need for the developing countries to accurately assess the availability of farm wastes and the economics of recycling of such wastes both on the small as well as big farms so as to obtain a realistic picture of the type of wastes which remain abundant on the farm for inexpensive recycling operations. The two well-recognized methods of utilizing farm wastes are composting and biogas production. Realistically speaking, organized scientific composting is yet to be taken up on a large scale in developing countries whereas the biogas technology is catching up increasingly due to the massive efforts to popularize biogas plants. Obviously, more extension work is needed to inculcate the concept of total recycling of wastes on the farm, which is indeed the need of the hour.

References

1. Acharya, C.N. Studies on the hot fermentation processes for the composting of town refuse and other waste material, *Indian Journal of Agricultural Science*, *9* (6), 817–833 (1939).
2. Acharya, C.N. Composts and soil fertility, *Indian Farming*, *1* (2), 66–68 (1940a).
3. Acharya, C.N. Composts and soil fertility, *Indian Farming*, *1* (3), 121–125 (1940b).
4. Acharya, C.N. Relation between nitrogen conservation and quantity of humus obtained in manure preparation, *Indian Farming*, *7* (2), 66–67 (1946).
5. Acharya, C.N. Preparation of compost manure from town wastes, *Miscellaneous Bulletin, Imperial Council of Agricultural Research*, *60*, 3rd edition (1949).
6. Acharya, C.N. *Preparation of Fuel Gas and Manure by Fermentation of Organic Material, Indian Council of Agricultural Research Technical Bulletin Series No. 15*, Krishi Bhavan, New Delhi (1961).
7. Alexander, M. *An Introduction to Soil Microbiology*, 2nd edition, John Wiley and Sons Inc., New York (1977).
8. Allison, F.E. *Soil Organic Matter and its Role in Crop Production*, Elsevier Scientific Publishing Company, Amsterdam (1973).
9. Allison, F.E. and Klein, C.J. Rates of immobilization and release of nitrogen following additions of carbonaceous materials and nitrogen to soil, *Soil Science*, *93*, 383–386 (1962).
10. Bassham, J.A. General considerations, p. 9, Cellulose as a chemical and energy resource, in *Proceedings, Biotechnology and Bioengineering symposium No. 5* (Edited by C.R. Wilke), Inter Science, New York (1975).
11. Bhardwaj, K.K.R. Potentials and problems in the recycling of farm and city wastes on the land, pp. 57–75 in *Recycling Residues of Agriculture*

and Industry, Proceedings of a Symposium at the Punjab Agricultural University, Ludhiana, India (Edited by M.S. Kalra), (1980).
12. Bhardwaj, K.K.R. and Gaur, A.C. Growth promoting effect of humic substances on *Rhizobium trifolii*, *Indian Journal of Microbiology*, *12*, 19–21 (1972).
13. Burk, D., Lineweaver, H. and Harner, C.K. Iron in relation to stimulation of growth by humic acids, *Soil Science*, *33*, 413–451 (1932).
14. Chandra, P. and Bollen, W.B. Effect of nitrogen sources, wheat straw, saw dust on nitrogen transformations in sub-humid soils under green house conditions, *Journal of Indian Society of Soil Science*, *7*, 115–122 (1959).
15. Chandra, P. and Bollen, W.B. Effect of wheat straw, nitrogenous fertilizers and C/N ratio on organic decomposition in a sub-humid soil, *Journal of Agricultural Food Chemistry*, *8*, 19–24 (1960).
16. Chang, Y. and Hudson, H.J. The fungi of wheat straw compost. I. Ecological studies, *Transactions of the British Mycological Society*, *50* (4), 649–666 (1967).
17. Colon, J. and McCalla, T.M. The decomposition of partridge pee and its influence on nitrification, *Soil Science Society of America Proceedings*, *16*, 208 (1952).
18. De Kock, P.C. Influence of humic acid on plant growth, *Science*, *121*, 473–474 (1955).
19. Debnath, N.C. and Hazra, J.N. Transformation of organic matter in soil in relation to mineralization of carbon and nutrient availability, *Journal of Indian Society of Soil Science*, *20*, 95–102 (1972).
20. Dyakonova, K.V. Iron humus complexes and their role in plant nutrition, *Pochvovedenie*, *7*, 19–25 (1962).
21. Eastwood, D.J. The fungus flora of composts, *Transactions of the British Mycological Society*, *35* (3), 215–220 (1952).
22. Elliot, L.F. and Travis, T.A. Detection of carbonyl sulfide and other gases emanating from beef cattle manure, *Soil Science Society of America Proceedings*, *37* (5), 700–702 (1973).
23. Fernandez, V.H. The action of humic acid of different sources on the development of plants and their effect on increasing concentration of nutrient solution, pp. 805–856 in *Study Week on Organic Matter and Soil Fertility*, John Wiley and Sons Inc., New York (1968).
24. Flaig, W. Uptake of organic matter substances from soil organic matter by plant and their influence on metabolism, pp. 723–776 in *Study Week on Organic Matter and Soil Fertility*, John Wiley and Sons Inc., New York (1968).
25. Fowler, G.J. Recent experiments on the preparation of organic manure, A review, *Indian Agriculture and Live Stock*, *7*, 711–712 (1930).
26. Gaur, A.C. and Mathur, R.S. Stimulating influence of humic substances

on nitrogen fixation by *Azotobacter*, *Science and Culture*, *32*, 319 (1966).

27. Giddens, J. Rate of loss of carbon and Georgia soils, *Soil Science Society of America Proceedings*, *21*, 513–515 (1957).
28. Golucke, C.G. Composting: A review of rationale, principles and public health, *Compost Science*, *17* (3), 20–28 (1976).
29. Gray, K.R., Sherman, K. and Biddlestone, A.J. A review of composting—Part I, *Process Biochemistry*, 32–36 (1971).
30. Hiltbold, A.E., Bartholomew, W.V. and Werkman, C.H. The use of tracer technique in the simultaneous measurement of mineralization and immobilization of nitrogen in soil, *Soil Science Society of America Proceedings*, *15*, 166 (1951).
31. Howard, A. The waste products in agriculture, *Journal of the Royal Society of Arts*, 84–120 (1935).
32. Hutchinson, H.B. and Richards, E.H. Artificial farm yard manure, *Journal of the Ministry of Agriculture*, *28*, 398–411 (1921).
33. Idnani, M.A. and Vardarajan, S. Preparation of fuel gas and manure by anaerobic fermentation of organic material, *Indian Council of Agricultural Research Technical Bulletin in Agriculture No. 46*, Krishi Bhavan, New Delhi (1974).
34. Jansson, S.L. and Clark, F.E. Losses of nitrogen during decomposition of plant materials in the presence of inorganic nitrogen, *Soil Science Society of America Proceedings*, *16*, 330 (1952).
35. Jones, M.J. The maintenance of soil organic matter under continuous cultivation at Samaru, Nigeria, *Journal of Agricultural Science*, *77* (3), 473–482 (1971).
36. Kaila, A. Biological absorption of phosphorus, *Soil Science*, *68*, 279–289 (1949).
37. Khandelwal, K.C. Biogas systems and production of manure, pp. 183–190 in *Compost Technology Project Document No. 13 of Improving Soil Fertility through Organic Recycling, FAO/UNDP Regional Project RAS/75/004* (1980).
38. Khandelwal, K.C. and Gaur, A.C. Influence of humate on the growth of *moong* (*Phaseolus aureus*), *Science and Culture*, *36*, 110–111 (1970).
39. Khristeva, L.A. About the nature of physiologically active substances of soil humus and organic fertilizers and their agricultural importance, pp. 701–721 in *Study Week on Organic Matter and Soil Fertility*, John Wiley and Sons Inc., New York (1968).
40. King, F.H. *Farmers of Forty Centuries on Permanent Agriculture in China, Korea and Japan*, Jonathan Cape, London (1911).
41. Kochtitzky, O.W., Seaman, W.K. and Wiley, J.S. Municipal composting research at Johnson City, Tennessee, *Compost Science*, *9* (4), 5–16 (1969).
42. Kononova, M.M. *Soil Organic Matter, Its Nature, Its Properties, Its Role in Soil Formation and Soil Fertility*, Pergamon Press Ltd., Oxford (1966).

43. Kononova, M.M. and Alexandrova, I.V. The effect of humus substances on the utilization of mineral nitrogen by plants, p. 497 in *Transactions of International Symposium on Humus and Plant*, V, Prague (1971).
44. Kozar, D.G. and Savon, A.S. The effect of physiologically active substances and humus substances on the oil formation in sun flower seeds, p. 591 in *Transactions of International Symposium on Humus and Plant*, V, Prague (1971).
45. Krishna Murthy. *A Manual on Compost and Other Organic Manures*, Today and Tomorrow's Printers and Publishers (1978).
46. Lockett, J.L. Nitrogen and phosphorus changes in the decomposition of rye and clover at different stages of growth, *Soil Science*, *45*, 13–24 (1938).
47. Mariakulandi, A. and Thyagarajan, S.R. Long-term manurial experiment at Coimbatore, *Journal of the Indian Society of Soil Science*, *7*, 263–272 (1959).
48. McGill, W.B., Shields, J.A. and Paul, E.A. Relation between C and N turnover in soil organic fraction of microbial origin, *Soil Biology and Biochemistry*, *7*, 57–63 (1975).
49. Mishustin, E.N. and Sinha, M.K. Humic substances as a factor mobilizing inorganic phosphorus compounds in the soil, p. 491 in *Transactions of International Symposium on Humus and Plant*, V, Prague (1971).
50. Mockeridge, F. The occurrence and nature of plant growth promoting substances in organic manurial composts, *Biochemical Journal*, *18*, 550–554 (1924).
51. Poincelot, R.P. A scientific examination of the principles and practices of composting, *Compost Science*, *15*, 24–31 (1975).
52. Ranjhan, S.K. Nutritional potential of animal organic wastes for live stock feeding, pp. 6–17 in *Recycling Residues of Agriculture and Industry, Proceedings of a Symposium at the Punjab Agricultural University, Ludhiana, India* (Edited by M.S. Kalra) (1980).
53. Rege, R.D. Bio-chemical decomposition of cellulosic material with special reference to the action of fungi, *Annals of Applied Biology*, *14*, 1–44 (1927).
54. Russel, E.J. and Richards, E.H. The changes taking place during the storage of farmyard manure, *Journal of Agricultural Science*, *8*, 495–563 (1917).
55. Rypacek, V. Biological activity of isolated humic acids in relation to the time and method of storing, p. 288 in *Transactions of International Symposium on Humus and Plant*, IV, Prague (1967).
56. Schwartz, S.M. and Martin, W.P. Influence of soil organic acids on soluble phosphorus in Miami and Wooster silt loam soils, *Soil Science Society of America Proceedings*, *19*, 185 (1955).
57. Shuler, M.L. An introduction to the role of agricultural wastes and resi-

dues as resources, pp. 1–17 in *Utilization and Recycle of Agricultural Wastes and Residues* (Edited by M.L. Shuler), CRC Press Inc. (1980).

58. Stutzenberger, F.J., Kaufman, A.J. and Lossin, R.D. Cellulolytic activity in municipal solid waste composting, *Canadian Journal of Microbiology, 16* (7), 553–560 (1970).
59. Swenson, R.M., Cole, C.V. and Sieling, D.H. Fixation of phosphate by iron and aluminum and replacement by organic and inorganic ions, *Soil Science, 67*, 3–22 (1949).
60. Tenney, F.G. and Waksman, S.A. Composition of natural organic material and their decomposition in the soil, IV. The nature and rapidity of decomposition of the various organic complexes in different plant materials under aerobic condition, *Soil Science, 28*, 55–84 (1929).
61. Titova, N.A. Iron humus complexes in some soils, *Pochvovedenie, 12*, 38–43 (1962).
62. Waksman, S.A., Cordon, T. and Hulpoi, H. Influence of temperature upon the microbiological population and decomposition process in composts of stable manures, *Soil Science, 47*, 83–114 (1939).
63. Waksman, S.A. and Stevens, R.K. Contribution to the chemical composition of peat, 1. Chemical nature of organic complexes in peat and method of analysis, *Soil Science, 26*, 113–137 (1928).
64. Webley, D.M. The microbiology of composting, I. The behaviour of the aerobic mesophilic bacterial flora of composts and its relation to other changes taking place during composting, *Proceedings of the Society of Applied Bacteriology, 2*, 83–89 (1947).
65. Wiley, J.S. Progress report on high rate composting studies, p. 17 in *Proceedings of the 12th Industrial Wastes Conference, Purdue University Engineering Bulletin*, 596–603 (1957).
66. Wiley, J.S. and Spillance, J.T. Refuse sludge composting in windows and bins, *Compost Science, 2* (4), 18–25 (1962).
67. Yadav and Subba Rao, N.S. Use of cellulolytic microorganisms in composting, pp. 267–273 in *Recycling Residues of Agriculture and Industry, Proceedings of a Symposium at the Punjab Agricultural University, Ludhiana, India* (Edited by M.S. Kalra) (1980).

P.N. HOBSON

20. Production of Biogas from Agricultural Wastes

Introduction

Anaerobic digestion of agricultural wastes has come to the fore in the last few years because of the realisation that costs of conventional fuels such as oil, coal and natural gas will continue to rise. The supplies of these fuels can be influenced for political reasons at present, but in the future such fuels, particularly oil and gas, will become less available as resources get exhausted. Previous experiments with automated digesters using agricultural wastes, during war time for instance in Germany, were abandoned when the need for biogas fuel as a substitute for scarce petrol declined with the onset of peace, reopening of trade and the return of cheap oil. These experiments did not result in an efficient digester and it soon became apparent that although sewage works continued to use digesters as part of the sewage treatment, with gas for fuel for the works produced as a byproduct, building a digester purely as a source of fuel was not economic in countries where electricity and other fuels were not only cheap but universally distributed. In various places where conventional fuels, even wood, were scarce some digesters were built, but the running of these involved much hand labour and they were generally built in warm climates where digester heating was scarcely required. These digesters were generally a 'one-off' job built by some enthusiast, the exception probably being the Indian 'Gobar Gas' digester which originated in the 1950s.

However, before the 'fuel crisis' beginning about 1973, another factor arose which led to experiments in the West on anaerobic digestion. This was the change from the former 'extensive' systems of farming to the 'intensive' units where hundreds or thousands of animals and millions of birds are housed all the year round in one group of buildings or small areas of 'feed-lots'. Pollution caused by these animals became, and still is, a problem and hence the experiments on anaerobic digestion were designed to control the pollution. The biogas was possibly a useful byproduct of these experiments. Pollution control is still a major consideration in the use of digesters, but the increasing fuel prices make the production of biogas and the use of this to run engines and boilers economically viable.

While the microbiological aspects of digestion in the gobar digester and the heated, automated, intensive-farm digester are the same, the engineering problems however, are quite different, and because of the need for heat and automatic working in cold climate, the large digester has to be more efficient than the small one in converting feedstock to gas. Large digesters, of some hundreds of cubic metres capacity are now being built and run in Britain and other countries using animal excreta from intensive farms or wastes from agricultural industries as feedstocks. These are developmental plants and mechanical difficulties are not only being found and overcome but also experience on the running of digesters and ancillary equipment is being gained all the time. The engineering design and construction aspects of the digesters are outside the scope of this chapter. However, they will be considered only insofar as microbiological implications are concerned. Similarly, mathematical modelling of the digesters based on biochemistry and microbiology cannot also be considered here. For more detailed information on these aspects including feedstocks and their use, the reader may consult original papers, recent reviews and books on the subject (6, 9, 20, 21, 48).

Digester feedstocks

As digestion is a process carried out by bacteria the 'medium' in which the bacteria grow must contain an energy source, or energy sources, and sources of carbon and nitrogen for cell synthesis, as well as the trace elements and other ions needed for bacterial metabolism. These are usually provided in the animal excreta, vegetable matter and factory wastes used as feedstocks. Although most wastes do contain all the nutrients required by the bacteria, generally the main constituents, the energy and nitrogen sources, will not be in the correct ratio for optimum utilization of each. Faecal wastes, for instance, are generally too high in nitrogen content and some factory wastes (e.g., starch or sugar solutions) may have too little nitrogen. The C : N ratio of the former might be adjusted by the addition of a carbon source (e.g., potatoes or straw where these are seasonally available). Experiments on optimizing C : N ratios in feedstocks based on animal excreta have been made (15). Pure analysis, particularly if it is for materials such as cellulase, will not enable true C : N ratios to be calculated. What is required is a knowledge of the C and N available to the bacteria, and this may be only a fraction of the amount of substance determined by analysis. For many vegetable and vegetable byproduct feedstocks an approximate idea of their potential for biogas production can be obtained from the tables of digestibilities of the materials as ruminant feeds. But this does not say what detention time is needed to obtain breakdown. For instance, barley straw was little degraded with pig waste in a digester running at a 10-day detention time. When the detention time was increased to 20 days, approximately 35 per cent of the straw was digested to gas (16). Barley straw is said to contain about 76 per cent 'carbohydrate' and 'fibre' (58), so only some 50 per cent of the potential substrate was gassified (Tables 1 and 2). The digestible nutrients for ruminants are given in the same tables as about 41 per cent. On the other hand rotten potatoes were easily gassified in a pig waste digester running at 10-day detention time (49) and approximate calculations showed that about 82 per cent of the potato starch was gassified, so some 64 per cent of the dry matter would be degraded (using the analysis for potato sludge in the tables quoted). The 'carbohydrate' and fibre available to animals are given in the tables as about 62 per cent of the dry matter.

These results illustrate the point about available C and N. The starch of potato is easily attacked by amylolytic bacteria and enzymes, the cellulose and hemicellulose of straw is not. Pig excreta has a high ammonia content

Table 1. Gas production from animal excreta

Piggery waste. Slurry from fattening pigs on dry barley feed.

Detention time 10–15 days.	*Gas* 0.300 $m^3\ kg^{-1}$ TS*	
7 days	0.284	,,
5 days	0.240	,,
3 days	0.170†	,,

Temp. above, 35°C. At 30°C gas 0.300 m^3; 40°C (10 day detention) gas 0.360 m^3; 44°C gas 0.420 m^3. Below 25°C and above 45°C gas production falls off rapidly. Total Solids in slurry 2–6.5 per cent.

Poultry waste. From caged layers, no litter, slurried with water to required TS.

Detention time 20 days.	*Gas* 0.380 $m^3\ kg^{-1}$ TS*	
15 ,,	0.362	,,

Temp. above, 35°C. Total Solids in slurry *ca.* 6 per cent.
At 4 per cent TS slurry, gas 0.480 $m^3\ kg^{-1}$. TS: 6 per cent, gas 0.379 m^3; 12 per cent, gas 0.291 m^3. Probably due to NH_3 inhibition at higher TS.

Fattening cattle waste. Slurry from cattle on variety of mixed feeds.

Detention time 20 days.	*Gas* 0.215 $m^3\ kg^{-1}$ TS*	
10 ,,	0.195	,,

Temp. above, 35°C. Total Solids in slurry *ca.* 6 per cent.
At 5 per cent TS slurry, gas 0.189 $m^3\ kg^{-1}$. TS: 8 per cent, gas 0.258 m^3; 10 per cent, gas 0.264 m^3. Possibly due to low NH_3 in slurry of low TS.

Dairy cattle waste. Slurry from cows on silage-concentrate feed.

Detention time 21 days.	*Gas* 0.206 $m^3\ kg^{-1}$ TS* at 35°C.
20 ,,	0.172 $m^3\ kg^{-1}$ TS at 25°C

For all these wastes Volatile Solids was about 70–72 per cent of Total Solids. Gas composition 65–70 per cent CH_4, usually 70 per cent, 30 per cent CO_2 traces of H_2S and other gas, for pig and poultry wastes. For cattle wastes gas 55–60 per cent CH_4. Except for † where methane content and gas production were falling off rapidly.

Table 2. Biogas production from some vegetable wastes (Batch cultures)

Substrate	Gas ($m^3\ kg^{-1}$ TS)	Methane (%)
Grass hay	0.462	54
Kale	0.440	60
Sugar beet leaves	0.380	66
Maize	0.500	65
Oats	0.470	54
Wheat straw	0.412	58
Lake weed	0.380	56

Slurry 5 per cent TS, temp. 37° C, total reaction time 17–36 days. Data from Badger *et al.* (1).

which does not change during digestion (4, 50) so nitrogen is in excess in these feedstocks. The addition of straw or potatoes might be expected to give a more favourable C : N ratio.

However, it should be remembered that in some factory wastes a nutrient other than C or N, perhaps phosphate, may be limiting bacterial growth.

To get a proper estimate of the potential of some material as a digester feedstock experimentation is needed. Full experiments with a continuous-flow digester must be done at some time, but these take a long time and a quicker method is to carry out small-scale batch tests. These are usually on 1–2 litre scale and the apparatus consists of a flask, with gas outlet and some form of intermittent shaking or stirring, and a water bath or other means of heating the flask to the desired temperature. An inoculum of digester bacteria is usually supplied by adding a small amount (say 10 per cent) of digesting sludge from a working digester to a solution or, usually a slurry, of the proposed feedstock of perhaps 5 per cent Total Solids. Gas from the digestion is measured for volume and should be analyzed for methane and carbon dioxide. Digestion is allowed to proceed until gas production ceases, maybe in 30 or 40 days. However, the procedure may be more complicated than this and pH adjustment initially or at intervals may, perhaps be needed.

The gas production per unit weight of feedstock can then be calculated and the average gas composition found, and an approximate idea of the rate of digestion and so the detention time in a continuous-flow digester can be obtained. Badger and colleagues (1), for example, used this method to determine digestibilities of various vegetable matters.

The main materials used as digester feedstocks are wastes containing human and animal excreta. The principal use of digestion has been in domestic sewage treatment. Domestic sewage consists of a suspension and solution of faeces and urine in water. In industrial cities, and small towns and villages where mains water is available, this suspension and solution is very dilute, as the water-closet uses some 10 litres or more of water to flush away the excreta of one person, and wash basins, baths, street drains, and in some cases industrial factories, add more water. On entering the sewage work, grit and large debris are removed from the water and then the coarser particles of faeces, toilet papers, food scraps, etc., are allowed to sediment out to form a sludge which is removed. The relatively clean water then passes on for aerobic treatment which generates a sludge of coagulated debris and aerobic microorganisms which, again, has to be removed before the purified water is allowed to flow to a river or lake. Some of the aerobic sludge may be returned to the plant, depending on the system used, but there is always excess. This excess and the first sludge removed form a semi-liquid mass which is self-polluting and rapidly putrefies further if left untreated. The most common method of treating these combined sludges is anaerobic digestion. The digestion removes the degradable organic material which causes pollution and converts it to gas which can be used to fuel boilers and engines to provide power for the aerobic system and for pumping sewage and so on. It is also used to heat the digesters. The digested sludge is stable and virtually odourless, and much less polluting than the original sludges and can be used as liquid fertilizer, or dewatered to give a dry fertilizer, or dumped at sea or on land without giving offence.

The sludges used as domestic digester feedstocks are usually about 4 per cent 'Total Solids' (TS), although they may be 'thickened' to 6 or 8 per cent TS, and the solids are some 40 per cent degradable, but this varies with the sludge characteristics.

These domestic sewage sludges are more akin to the excreta from animals, particularly the slurries from intensive farms, than the sewage water of low BOD treated in the aerobic plant and so it seemed that digestion might be a better system for treating farm animal excreta than an aerobic system. Digestion cannot of itself produce a water low enough in pollutants to be put into a river, but it can reduce the smell of the pollutants sufficiently, for the excreta to be used as a liquid fertilizer without causing offence, while still retaining the NPK fertilizer value of the original material. The digestion process can also reduce pathogenic microorganisms and weed seeds in the original excreta. Smell, particularly from the slurries of excreta produced in vast amounts from large, intensive farm units, is a principal source of public complaint in Britain and other countries, and reducing the smell of excreta can go a long way towards obviating public complaints about farming operations.

There have been many estimates of the amounts of farm animal excreta produced in various countries and the gas production from these. In most cases these are only of academic interest as they ignore the availability of the wastes and the size of the units. The latter consideration is most important in the care of digesters in cold climates where automation is needed because labour is scarce and also costly. Although a digester can be built in any size, the cost of a small automated digester is proportionately higher than that of a large one, and the gas energy produced may not be much higher than that needed to run the digester. If pollution control is a major consideration then this does not matter so much, as pollution treatment will be costly whatever method is used. If energy production is the consideration then size of plant and capital costs must be taken into consideration. For reasonale energy production in Britain or a country similar to it, farm digester should be designed to treat excreta from 1500–2000 fattening pigs or more and at least some 75–100 cattle.

Some figures for gas production from farm animal excreta are given in Table 1. These are from the author's laboratory (4, 18, 22, 50, unpublished) but others have found similar results. The decrease in gas production with increasing percentage of TS in the feed with poultry slurry is most likely due to ammonia inhibition.

Although these size considerations apply in building some digesters, they do not apply to all. For example, in a warm climate and where energy requirements of the farm are small (simple cooking and lighting in the farm house) digesters of only 2 or 3 m^3 for four or five cattle or pigs may be quite economical. Even here, economy in building, and better control over running, may be obtained by building one larger, communal, digester for a number of farms rather than one small digester for each farm.

In a country such as Britain, with high energy requirements, a large population and a comparatively small area of farming land, a cold climate and so relatively low crop yields, the 'energy farm' concept may not be viable as a worthwhile contribution towards much of the country's energy needs. But in a country with a small population compared to its area of farmable land, and a warm climate and high crop yields, the digestion of vegetation might contribute considerably towards the country's energy. In the former class of countries, of course, the digestion of waste vegetable matter may help considerably towards the energy supply of a particular farm in winter by boosting the gas production from a digester already running on animal excreta. This was the subject of the previously-mentioned, and unpublished, experiments in adding rotten and waste potatoes and silage liquid to a piggery-waste digester. Half a per cent (TS) of potato pulp added to a 3.3 per cent TS pig slurry increased gas production by 27 per cent, while 10 per cent (v/v) silage liquid added to the same pig slurry increased gas by 16 per cent and also contributed to the purification of this highly-polluting liquid (3, 50).

In New Zealand experiments have been done on laboratory and large pilot-plant scale (for later expansion to a full system) where part of the arable land of farms would be used as 'energy farm'. Laboratory-scale batch digestions gave the figures shown in Table 2 for gas production from some vegetation (1). The gas production from straw seems to be much better than that obtained by the author and colleagues and some other workers.

Various factory wastes from food processing, alcohol production and other processes have been tested on laboratory or larger scale. References to these will be found in the books and reviews previously cited, but one waste that is being used on a large scale is that from palm-oil factories, which is a highly-polluting liquid. Gas production is substantial and reduction in pollution is considerable (38).

This has been only a short review of the kinds of feedstooks now being used or considered for anaerobic digestion. In most cases pollution control is a major factor, along with, or even to the exclusion of, gas energy production. As an illustration the reductions in pollution of pig waste by digestion are on and an average (10-day detention): in VFA 92.6 per cent, in BOD 82.5 per cent, in COD 53.2 per cent, in ammonia 15.6 per cent and odour very much reduced (45). These figures are for the whole digested sludge. If the sludge is allowed to settle, only 4 per cent of the original BOD is in the supernatant liquid, while flocculation of the sludge with a polyelectrolyte leaves only 2 per cent of the original BOD in a water with only a faint brownish tinge. Because of the very high initial BOD, however, the residual BOD is still far above the British standards for the addition of waste waters to a river. But for fertilizer use, this reduction in smell and pollutants is valuable when farming is carried out near villages or towns, and the digested sludge is not objectionable when stored: in fact the very slow digestion continuing in stored sludge tends to further decrease pollutants (22). Detailed results of different secondary treatments of digested pig waste are given by Summers and Bousfield (50).

The microbiology of anaerobic digestion

General considerations and building up the flora

Anaerobic digestion is a microbiological process that has been adapted to the agricultural industry, but unlike other industrial microbiological processes it does not use pure culture or sterile conditions. Obviously pure culture manufacturing processes can be carried out only where there is skilled staff and a high degree of automation coupled with culture vessels designed for sterile working and a system for sterilizing feedstocks. There must also be a backup service to maintain cultures and monitor the plant for sterility. Such a process is expensive and can be used only when the value of the product is high. Presently, the values of the products of anaerobic digestion, gas and fertilizer, are low. The value of pollution control is in many cases unquantifiable in monetary terms to the farmer, unless he is faced with legal action over pollution. But, even here, the farmer controlling pollution in some way cannot pass on the costs in higher prices for his products than the farmer who does not have pollution-control problems. So the plant must be relatively inexpensive in capital cost and maintenance. It has, for the intensive farm unit of the Western countries, to be automated and require only a short daily checking by a farm worker unskilled in microbiology but able to look after tractors and similar pieces of machinery. With the gobar and similar digesters the capital costs must be commensurate with a peasant-farm income and automation must be nil.

The use of excreta as feedstock means that unlike most industrial feedstocks the input to a digester contains vast numbers of bacteria and the cost of sterilizing such slurries would be prohibitive. However, digestion is a process carried out by a consortium of bacteria and these bacteria are contained in the feedstock, although not in the correct numbers. So the feedstock itself acts as an initial inoculum for the process and also as a continued source of bacteria. This latter could help to stabilize the digester flora, because if strain degeneration occurs new strains from the feed can replace those in the digester. The bacterial flora of a digester is extremely mixed, in fact we do not know the number of different species and strains in the mixture. Some of the bacteria may not be necessary for the digestion, but many, although not concerned in the biochemical pathways leading to methane, may provide growth factors for the biochemically-active bacteria or scavenge toxic byproducts or have some other necessary action. Although there must be variations with time in a digester flora, overall the flora is very stable and self-balancing, being little affected by variations in feed or many other factors.

Anaerobic digestion is, as the name implies, a process carried out by anaerobic bacteria growing under anaerobic conditions. Amongst the principal bacteria, and particularly the methanogens, are the most exacting anaerobes known, needing special culture techniques for growth in pure culture in the laboratory. These techniques [described for instance by Hungate (29) and Wolfe (57)] are essentially those used for the rumen anaerobes and involve total exclusion of air from media during preparation and from cultures during growth and manipulations of the bacteria. But in the natural systems where they are found (decaying vegetation under water, muds, wet soils, the rumen etc.) the bacteria of the anaerobic digestion process do not grow in pure culture. Indeed, even on the basis of nutrients some of the bacteria could not grow in natural pure culture as their substrates are found only as metabolic products of other bacteria. The mixed culture provides not only bacterial substrates but also provides anaerobiosis, as part of the flora consists of aerobes or facultative anaerobes which, under conditions where ingress of air is limited, can reduce any oxygen present and lower the Eh of the habitat to some –300 mV, at which Eh the methanogens and other anaerobes can grow and metabolize.

About half of the bacteria in digesters treating animal excreta are facultative anaerobes but the proportion is less in sewage digesters (25, 27, 33, 51). These bacteria do not have hydrolytic or other properties which can make them noteworthy in the digestion process, but could metabolise sugars formed by the hydrolytic bacteria, or other substrates, coincidentally using up any oxygen leaking into the digester. Examples of the ability of pig-waste digester contents to withstand and take up oxygen are given by Hobson, Bousfield and Summers (20). Many of these bacteria may, however, be non-metabolizing or drying off in the digester, keep up their numbers only by continued inoculation from the feed. For instance Hobson and Shaw (25) found 10^8 facultative and aerobic bacteria per ml of the feed slurry added to a continuous-flow pig-waste digester, but only 10^6 of these bacteria in the digester population. But that the bacteria are influenced by the feed was shown by the fact that the facultative population of a sewage digester was composed principally of *Escherichia coli*, a predominant bacterium of the human gut, but when a digesting sewage sludge was adapted to pig-waste digestion, the predominant facultative bacteria became the streptococci of the pig gut (25).

These results came from investigations of mesophilic digesters although the bacterial types may differ in thermophilic digesters, but presumably similar reactions occur. In general, there is a lack of information about digester microbiology and microbial ecology. The reactions of digestion are known and pure or mixed cultures of bacteria carrying out many of these reactions have been obtained. While these are 'representative' bacteria, it is not known how many species of bacteria are involved in each reaction, whether species are common to all types of digesters and feedstocks, in what numbers many of these species occur, and their exact interrelationships. There is so much to learn about the bacteria involved in digester reactions although some work has been done, mainly on mesophilic digestions.

'Mesophilic' and 'thermophilic' digestions were mentioned earlier. Like any other microbial reaction the rate of digestion is influenced by temperature. The rate drops off rapidly below about 25°C, so that unheated digestions in a cool climate like Britain are extremely slow, with a detention time of three months or more for sewage sludge stabilization and an uneconomic rate of gas production. Between about 25° and 45°C the rate of digestion increases slowly and above 45°C the rate rapidly falls off as the limit of growth of mesophilic bacteria is attained (50, 55). Under correct conditions, thermophilic bacterial populations can be developed which are active from about 50° to 65°C although the limiting and optimum temperatures are not well defined. Although thermophilic populations have been developed from mesophilic digesters or rumen contents, it is not clear how the changeover occurred—whether by adaptation or by selection mechanisms. Varel *et al.* (56) have described small-scale experiments on thermophilic digestion of cattle wastes.

Most commercial digesters run at mesophilic temperatures, usually between 30° and 40°C. There could be advantages in the rate of reaction in running thermophilic digestions, although some lack of stability has been reported in thermophilic sewage digestions, but a big drawback to the thermophilic digester is the energy required in a cold climate to keep the digester at a temperature some 30° higher than a mesophilic digester, unless the feedstock is a very hot factory waste.

As mentioned before, the digester bacteria are present in faecal matter, soil, water and air. If faecal or other wastes are left undisturbed in tanks or lagoons they will rapidly go anaerobic a few centimetres below the liquid surface and a population of bacteria and biochemical reactions similar to those of the anaerobic digester will develop. These reactions can be a source of trouble; for instance, cases have been known of explosions, or poisoning of animals or men, by the gases given off from stored animal excreta, especially when this is suddenly disturbed and the gas in the liquid is rapidly given off. The reactions and the bacteria involved may not be in the balance necessary for successful digestion, but storing the waste will have encouraged growth of many of the bacteria necessary for digestion and so such stored wastes can be used as an initial inoculum (10 or 20 per cent (v/v) perhaps) for a digester for similar wastes, or even as a complete first filling of the digester.

The first reaction in the chain which leads to biogas is fermentation of carbohydrates. The products of these fermentations are then converted to methane. Hydrolysis and fermentation of carbohydrates gives volatile fatty and other acids, hydrogen and carbon dioxide. The fermentative bacteria develop first in the waste and unless methanogenesis from the acids can take place the acids will accumulate, the pH will drop (in spite of the buffering activity of ammonium and other ions) and the microbial activity will, more or less, cease entirely. The methanogenic bacteria have a short pH range for growth near pH 7 and the fibre-degrading bacteria (on which much of digestion of animal excreta and vegetable wastes depends) also have little activity below a pH of about 6.5. In addition, propionic acid, which is liable to accumulate under these conditions, is itself toxic to hydrogen-using methanogenic bacteria and so to the digestion in general (26, 35).

When starting up a continuous-flow digester the inoculum should be of dilute sludge (about 2 per cent TS) and if a stored waste (as previously mentioned) is used it should, if possible, be ascertained that the 'Volatile Fatty Acid' (VFA) content is not too high, or will not be too high when the inoculum is diluted with fresh waste (*ca.* 1000 mg l^{-1}, as acetic). The fresh waste used as feedstock initially should itself be only about 2 per cent TS. The digester and contents are heated to working temperature, and if possible mixed. The feed input can then be set at a low daily volume (equivalent to perhaps 40–60 day detention time) with the same low-solids waste. If analytical facilities are available analyses for VFA concentrations can be carried out every few days. But gas production is often the only parameter that can be measured. Gas production will be low for some days and then should begin to slowly increase. As gas production rises, then perhaps at two weeks intervals the feed rate can be slightly increased. All being well the digester will be running at its working detention time after perhaps six or eight weeks, and the concentration of the feed can then be increased over another week or two to the working range (*ca.* 6 per cent TS for pig slurry, or 8–10 per cent TS for cattle slurry, for example). These figures are only approximate and taken from the starting of animal-waste or sewage sludge digesters, but for all digesters the principle holds good that whether the digestion is being started from water, dilute feedstock, with or without an inoculum, initial loading should be low. Daily gas production will vary slightly during start-up. If it falls for some days or ceases then it is most likely that acid concentrations have become too high. Often, if loading is stopped the acids will gradually be metabolized and gas production will start again after some days. If the digester has become very acid ('sour') then neutralization with lime may restart digestion, otherwise the digester will have to be partially emptied and refilled with dilute slurry or water to restore pH and acid concentration for a fresh start.

If an inoculum (say 10 per cent or more by volume) of sludge from a digester working on the same feedstock can be obtained, a digester may be brought to stability rather quicker, but the same principles of slow build-up to working loadings still apply.

During this slow start-up of digestion, the various groups of bacteria develop to their final populations. In a pig waste digester starting from water, Hobson and Shaw (25) showed that amylolytic bacteria, present in large numbers in the pig waste, were detectable from one week after loading started in numbers $>4 \times 10^3$ ml^{-1} and by three weeks had attained a stable population greater than 4×10^4 ml^{-1}. Cellulolytic bacteria were not detectable in the pig waste and did not attain a population of 4×10^3 ml^{-1} until the fifth week after starting. Formate- or hydrogen-using methanogens were not detectable in the pig waste and a population of 10^4 ml^{-1} did not develop until the fourth week. The development of these methanogens was paralleled by that of butyrate-using methanogens. Acetate- and propionate-utilizing bacteria were not detected but this may have been due to limitations in the time of incubation used for the cultures (20). One week after the digester had started a general count of 2×10^6 anaerobic bacteria and 2.2×10^7 facultative and aerobic bacteria was obtained. By about four to six weeks these counts had stabilized at 10^6–10^7 ml^{-1} for each group, the aerobic count being somewhat lower than the anaerobic count.

In a laboratory-scale digester initially filled with digesting sewage sludge, cellulolytic bacteria were present only at 4×10^3 ml^{-1}. As the digester was slowly changed to a pig-waste feed a cellulolytic population of 4×10^5 ml^{-1} developed in the course of time. The cellulose content of sewage sludge is low compared with that of pig waste, and the increase in numbers of cellulolytic bacteria reflected the change in substrate available to the bacteria (24, 25).

Fermentations, nitrogen metabolism and some other reactions

Except for some factory wastes, digester feedstocks contain little, if any, simple sugars. Carbohydrate is in the form of polymers, usually the cellulose, hemicellulose and other vegetable polysaccharides not digested in the gut of farm animals. These fibrous polysaccharides are usually heavily lignified and not easily hydrolyzed by the digester bacteria, and hydrolysis of fibres can be a rate-limiting step in digestion. The increase in detention time needed to digest lignified straw fibres compared with the barley residues in pig-excreta has already been mentioned, as has the fact that only part of the cellulose can be degraded. The microbiology of fibre breakdown in digesters has not yet been worked out; investigations however show that 'cellulolytic' activity, as in the rumen by bacteria able to degrade prepared celluloses in culture media, is not the only requisite and a consortium of bacteria is probably involved. Hobson and Shaw (25) found 11 types of cellulolytic bacteria in a pig-waste digester all but one Gram-negative. Maki (34) isolated 10 types from a sewage digester, and Hungate (28) isolated and classified three types. These were from mesophilic digesters. So far only one thermophilic cellulolytic bacterium [*Clostridium thermocellum* (40)] has been isolated. Besides the 'cellulolytic' bacteria (i.e., those which degrade powdered filter paper or similar material in a culture medium), bacteria degrading hemicelluloses must also be important. Hobson and Shaw (25) cultured two types of hemicellulose-degrading bacteria, the principal type being identified with the rumen *Bacteroides ruminicola*, the other a Gram-negative rod. Other bacteria of the rumen have been found in digesters, *Methanobacterium ruminantium* is another example, but there are many digester bacteria which do not form any major part of the rumen population.

Much more work remains to be done on fibre degradation in digesters, and particularly the actions of mixed cultures. Maki (34) found that culturing a cellulolytic bacterium from a digester together with a non-cellulolytic *Clostridium* from the same digester more than doubled the rate of cellulose degradation shown by the former bacterium in pure culture. The cause of this effect was not determined, but there must be many similar interactions in the digester population.

The carbohydrate fermentations provide energy for the growth of the bacteria and the feedstock and bacterial reactions provide nitrogen. Sewage sludge and animal-excreta feedstocks contain nitrogen in the form of the constituents of intestinal bacteria, shed intestinal epithelial cells and secretions, food residues and urea and other non-protein-nitrogen compounds of urine. There is also ammonia, and the breakdown of urea and other compounds in animal excreta standing in collecting channels can give high concentrations. Poultry excreta is particularly high in ammonia, values of 4900 mg l^{-1} being found in a 12 per cent TS slurry (4).

In digesters for animal excreta the breakdown of proteins and deamination of amino acids seems to about balance out the utilization of ammonia for bacterial synthesis, as ammonia concentrations change little, if at all, during digestion (see, e.g. 4, 50). Proteolytic activity can be detected in digesters and the most numerous proteolytic bacteria isolated from sewage-sludge and piggery-waste digesters have been clostridia (25, 45). Deaminative activity has been ascribed to the previously-mentioned *B. ruminicola* (25), but other bacteria are probably involved.

Ammonia is probably a major source of nitrogen for digester bacteria. Hobson and Shaw (25) found the fermentative bacteria could, overall, use amino acids or ammonia, while the methanogens use only ammonia (6).

The proteolytic bacteria probably use mainly sugars formed from polysaccharide degradation by the hydrolytic bacteria as energy sources, and there is a heterogeneous population of bacteria in digesters which must derive energy in the same way. But there are also bacteria which can ferment lactate if this is formed as a product of sugar fermentation (20). Lactate or succinate cannot be detected in digester contents (24), either because they are immediately fermented to acetic and propionic acids, or because they are not produced (see later). Lactate-fermenting bacteria were in numbers of 3×10^7 ml^{-1} in a piggery-waste digester.

One of the lactate fermenters was a *Desulphovibrio*, and sulphate-reducing bacteria can be generally found in digesters. Torien, Thiel and Hattingh (53) counted 3 to 5×10^4 sulphate-reducing bacteria per ml of digesting sewage sludge. Thermophilic sulphate-reducing bacteria have also been isolated (61). These bacteria form sulphides, but in addition sulphides can be formed from sulphur amino acids by some of the fermentative bacteria.

Some sulphide is found in all digesters, but high concentrations are undesirable. The sulphide gets into the biogas as hydrogen sulphide where it can be nauseating or poisonous to anyone smelling the gas and can cause problems with gas burners or engines. It is also toxic to the digester bacteria and can inhibit digestion (26, 35). The sulphide is removed from contact with the bacteria by precipitation with metals in the digester feedstock and this is the reason why heavy metals themselves may not be toxic in digesters and perhaps why iron additions sometimes increase digester activity. No problems have been found, for instance, with copper in pig excreta. The copper content of waste from pigs being fed 200 ppm copper in a barley feed was 850 ppm of the dry solids, and most of this was insoluble (24). The sulphide is also removed from the digester liquid as hydrogen sulphide, and vigorous gassing in a well-functioning digester can remove much of the free sulphide in this way.

Lignin was previously mentioned as a barrier to fibre breakdown in digesters and polymerized lignin is not metabolized by anaerobic bacteria. On the other hand aromatic monomers of the lignin molecules can be metabolized to methane by mixed digester bacteria (12).

Methanogenesis

The fermentative bacteria produce in pure cultures a mixture of volatile fatty acids, lactic and succinic acids, ethanol, hydrogen and carbon dioxide. Hydrogen, with carbon dioxide, is a general substrate for methanogenesis and the majority of methanogenic bacteria so far characterized use only hydrogen and carbon dioxide, or the equivalent formic acid, as a substrate. Counts of 10^4–10^6 ml^{-1} and 10^5–10^8 ml^{-1} of hydrogen-utilizing bacteria were determined for pig-waste and sewage-sludge digesters respectively (25, 27, 39, 46). The utilization of hydrogen in a mixed-culture fermentation such as a digester 'pulls' the fermentations from production of reduced acids and ethanol towards production of acetic acid and hydrogen, as has been demonstrated with a number of rumen bacteria [see (6), (19), (20) for a full discussion and references].

Acetate thus forms the main, or only, residual acid in properly functioning digesters (24, 50), and its concentration depends on the detention time of the digester, as does the concentration of any limiting substrate in a continuous culture (23). The detection of propionate or butyrate as a high proportion of the digester acids means that digestion is failing, and propionate, in particular, is toxic to hydrogen-utilizing digester bacteria (26), so that the failure tends to be autocatalytic (20).

Early experiments with labelled acetate showed that acetate was the main ultimate precursor of methane in digesters (47). Acetate is the main fermentation product in the digester, but it is formed from lipids and volatile fatty acids as well. In sewage digesters fats can form a large part of the digestible feedstock (perhaps 30 per cent of the TS); in animal wastes the proportion is lower (5–15 per cent of the TS). Much of this fat disappears during digestion, and experiments with digesting sewage sludges (9, 14) showed that acetate was produced. This is an energetically unfavourable reaction and recent experiments have shown that two bacteria are required. Although a bacterium degrading long-chain fatty acids (e.g., stearic) has not yet been found, there has been demonstrated a bacterium degrading C_4 to C_8 acids to either acetate and hydrogen or acetate, propionate and hydrogen (depending on whether the starting acid is odd- or even-numbered) and a bacterium degrading propionic acid to acetate, hydrogen and carbon dioxide, both in co-culture with a methanogenic or a sulphate-reducing bacterium which removes the hydrogen. Only by removal of the hydrogen can the degradation of the fatty acid take place (2, 36).

Butyric and propionic acids formed by carbohydrate fermentation can thus be converted to methane. These reactions were, of course, shown previously by dilution counts without isolation of the bacterial consortia. For instance, 2×10^4 and 6×10^2 to 25×10^6 butyrate-utilizing bacteria per ml of pig- and sewage-digester fluids were demonstrated (13, 27). The figures for the pig-waste digester were lower by a factor of 10 or 10^2 than those for hydrogen-utilizing methanogens.

The long-chain fatty acids are produced by hydrolysis of glycerol esters of the feedstock lipids. Hobson, Bousfield and Summers (20) counted 10^4 to 10^5 long-chain-glyceride-hydrolysing bacteria per ml of pig-digester fluid, and Torien (51) also isolated lipolytic bacteria from a sewage digester by enrichment techniques. Unsaturated long-chain fatty acids are also hydrogenated in digesters (14).

Although a number of mesophilic hydrogen-utilizing methanogens are known, only one thermophile has so far been obtained, *Methanobacterium thermoautotrophicum* (60). Rose and Pirt (42) have isolated a hydrogen-utilizing mycoplasma from a consortium of two mycoplasmas, one of which ferments sugars with production of hydrogen. This consortium was obtained by enrichment from a mesophilic sewage digester. On the other hand only one mesophilic bacterium producing methane from acetate has so far been obtained in unquestionable pure culture and that is *Methanosarcina barkerii* (44), and strains have been isolated by a number of workers. However, cultural and microscopical evidence suggests that acetate-utilizing rods may be more numerous constituents of the digester flora (8, 20, 54). A thermophilic acetate-utilizing sarcina has been isolated (62).

The acetate-utilizing methanogens are slow-growing, and conversion of acetate to methane is a rate-limiting reaction in digester design and running. Hobson and McDonald (23) showed that acetate-utilizers in pig-waste digestions had a maximum growth rate of about 0.4 d^{-1} at 35°C, and at a three-day detention time a pig-waste digestion was producing little gas and on the point of complete breakdown (50). A washout dilution rate of 0.49 d^{-1} was calculated by Ghosh and Klass (10) for the acetate-utilizing stage of a two-phase digester and Cohen *et al.* (8) used a detention time of 100 h for fatty acid utilization in the second stage of their laboratory two-stage digestion.

The digester reactions can then be summarized as a first step of hydrolysis of polysaccharides to mono- or disaccharides, and hydrolysis of lipids to glycerol (and presumably galactose from galactosyl glycerides) and long-chain fatty acids. The sugars, including glycerol are fermented with production mainly of acetic acid, hydrogen and carbon dioxide, and long-chain acids and volatile fatty acids higher than acetic are also degraded to the acetic acid and gases. These latter are then converted to methane and carbon dioxide or water. The fermentations and methane production give rise to ATP used for growth of the bacteria, the main nitrogen source being ammonia, either from the feedstock or from hydrolysis of proteins and deamination of amino acids, or hydrolysis of urea and other non-protein nitrogen compounds to ammonia. Sulphur, trace elements and other ions for cell synthesis are contained in the feedstock, and cell carbon must come from carbohydrate, carbon dioxide, volatile fatty acids (including acetic for methanogens) (6).

Calculations show that complete degradation of carbohydrate gives a 50 : 50 mixture of methane and carbon dioxide. Degradation of lipid gives more methane. These are principal sources of gas, the contribution of protein is unknown but probably small with most feedstocks.

The composition of the biogas given off from the digester varies with feedstock; cattle wastes give 55–60 per cent methane, pig poultry and human excreta near 70 per cent and meat wastes about 80 per cent. The percentage of methane is influenced by the composition of the feedstock, but is always higher than calculated, because while methane is virtually insoluble an appreciable amount of carbon dioxide dissolves in the digesting sludge.

Growth of the bacteria concerned in the different steps of the complex reactions leading to methane influences the rates of digestion and the design of digesters for different substrates, as discussed in the following section. In the final analysis, much more needs to be done to characterize digester bacteria and their interactions, but actual isolation of bacteria is not needed to understand the reactions occurring.

Design of digesters for different feedstocks

The batch and single-stage digester

A digester is a microbial culture and, so far as is known, a bacterial culture: the flagellates and other organisms occasionally noted in digesters for animal and other excreta are probably only 'passenger' organisms brought in with the feed. As a bacterial culture it can be run, as any other culture, on a batch or continuous basis. The time course of bacterial activity and gas production follows that of bacterial growth in a batch culture, with a lag phase, a phase of maximum production and a declining and then a death phase. Starting of batch digestions relies on either development of a methanogenic flora from animal excreta used as substrate or development from an inoculum left by only partially emptying a working digester. The lag phase of a batch digester can be long and the development of the digester flora uncertain, and the gas may vary in composition. In practice to obtain a steady gas production for use as a fuel, and to deal with a continuous flow of feedstock, a number of digesters running out of phase are needed so that one is at optimum production when others are starting up or slowing down in production. To run commercial digesters in this way with large volumes of feed to load and unload requires quite extensive mechanization in cranes, etc. Batch digesters have been tested, in France, for instance, but these were more for the production of a solid fertilizer from a mixture of animal excreta and straw than for gas production, and detention times were long.

Jewel (30) suggested the use of very large batch digesters with a solid substrate of animal excreta and vegetable matter with a running time of up to a year. He calculated that such a digester could be selfheating but the previously-mentioned problems would occur.

For most purposes some form of continuous-culture digester is the best type of plant and nearly all research and development work is on this type of digester.

Two possible rate-limiting reactions have previously been pointed out and for the digester running on animal wastes, vegetation, and most sewage sludges, the breakdown of cellulose and fibrous materials is the overall limitation. These feedstocks are also sludges or slurries with a high content of suspended particles. The single-stage, mixed, continuous-flow digester, with a long detention time is the best system for these feeds. For details of construction, the reader should consult general books on digestion (21, 48) or specialist papers and books (5, 7, 43) and as some general descriptions can be given here.

As previously mentioned, temperature plays an important role in the rate of digestion, and the overall running temperature of the digester should be fixed early in the design stage. The general range of temperature can be mesophilic or thermophilic, and most digesters run in the mesophilic range, at about 30–40°C; as previously explained, below about 25°C digestion is very slow. In any temperature range what is most important is to ensure that temperature is kept uniform, in actual value and over the digester contents. Sudden changes in digester temperature (that is changes over a few hours) of even 5°C can cause drastic oscillations in bacterial metabolism and gas production, and the bacteria can take days to recover (20). There are three reasons for mixing digester contents. One is to ensure that incoming feed is distributed evenly and quickly into the mass of active bacteria; a second is to ensure that all the suspended material and bacteria remain homogeneous. This latter ensures that the substrates are properly degraded and that there are no masses of drying, floating debris which are microbially inactive and can cause blockage problems in digester and overflow, and also that there are no masses of grit and heavy debris at the bottom of the digester which can also be inactive and which by accumulation can seriously alter the effective digester volume. A third reason is to ensure uniformity of temperature over the digester contents. If temperatures are not uniform then, obviously, bacterial activities will vary, and if flow is not maintained over internal heat-exchangers then not only is there danger of killing the bacteria by local overheating (even to 50°C), but the heat-exchanger surfaces may become encrusted with drying sludge and so become ineffective. Poor performance of a number of sewage digesters in Britain could be ascribed to poor mixing with the consequent results (5), and poor performance of climatically-heated small digesters might often be due to temperature fluctuations between day and night or between days. A slow change in temperature, perhaps 1°C per day or less, will allow the bacteria to adapt to some new temperature. So a digester can be allowed to cool naturally if being shut down, say, in a cold climate and after slow heating later to a working temperature, will attain its normal production. A climatically-heated digester will adapt to winter temperatures as the climate changes and although gas production will be less at the lower temperature the bacteria will not be harmed.

The heating of a 'mechanically'-heated digester should be such as to keep the temperature within about 1°C of the working temperature and such that no change in temperature of more than a degree or so takes place as digester contents pass over or through heat exchangers. Although the metabolic heat of bacteria in a big batch digester could theoretically maintain a mesophilic temperature, metabolic heat from digestion of the usual feedstock slurries is not sufficient to compensate for heat losses in a cold climate and also heat up the input slurry to digester temperature. In a cold climate bacterial metabolism may be noted as an indication of decrease in rate of cooling of an unheated digester over the rate of cooling with non-metabolizing slurry. Although, for financial and other reasons the very small, peasant-farm, type of digester used in India, Taiwan, China and other countries has to be only climatically heated, very large, commercial digesters in hot countries which must treat a waste uniformly and have uniform gas production from week to week may need to be fitted with heating equipment to ensure uniformity of temperature, even if this is little used, and they must certainly be fitted with adequate mixing systems to ensure uniformity of temperature within the digester contents. Even in Britain, summer heating from bright sunshine on the one side of a digester can cause variations in skin temperature and internal temperature if mixing is not adequate.

Digester heating can be achieved by an external, water heat-exchanger through which the digester contents are pumped as required by a signal from a digester thermostat. This system is used on one of the Aberdeen 13 m^3 experimental agricultural-waste digesters (22), and on some sewage digesters. However, it requires a large pump to circulate the sludge and the commercial digesters now being built for farm use have some form of internal water-circulation heat exchangers which require only a small pump to circulate hot water. The heat exchangers can be flat 'radiator' types, or coiled tubes. These need adequate gas mixers underneath to keep the sludge circulating and to prevent fouling. Or, the heating and mixers can be combined in double-skin draught tubes with the hot water circulating through the double skin. Digester mixing in the large farm digesters is by gas recirculation from the head space of the digester to outlets under draught tubes or giving free-rising gas bubbles. Operation of mixers, like heaters, is intermittent, but the mixer programme needs to be decided by experience with the particular digester and sludge, although some general guides to volumes of gas per volume of digester are available (37).

With the simple, small digester, mixing is carried out by hand turning of a paddle attached to the gas dome, or to a rope, or is non-existent and only natural mixing by gas evolution is relied on.

The single-stage digester, then, consists of a tank made from steel or concrete or brick, and commercial glass-enamelled slurry-storage tanks have formed the basis of many farm-waste digesters in Europe and America. Steel tanks, particularly, are usually insulated with some form of plastic. The size of the tank is calculated from the daily volume of waste to be treated and the detention time required. This latter varies from about 10 to 25 days with a heated digester working on agricultural wastes at 35° to 40°C, but may be 50 or 60 days for climatically-heated digesters on the same feeds. Size varies from about 2 m^3 to 1 to 3000 m^3 and between about 300 and 600 m^3 is the size used for the intensive pig or cattle units of the British farming systems (i.e., from about 4000 to 10000 pigs or 100 to 300 cattle, or mixtures of these animals, sometimes with poultry). But whatever the size, the tanks are usually cylindrical with either a fixed roof and a separate gas holder or a floating roof which also acts as a gas holder. The gas is stored and used at only a few inches water-gauge pressure. One exception is the Chinese digester which is a below-ground tank with an upper dome of fixed volume where gas collects under pressure, and this system unlike the floating roof or water-sealed separate gas holder, delivers gas at a varying pressure. This is of no consequence for simple gas usage, but an engine or sophisticated gas boiler needs gas at a constant pressure. In the few cases where a butyl rubber digester top, which acts as a 'balloon' gas holder, has been used on large digesters, a gas-pressure regulator is needed.

The theoretical continuous-culture requires a continuous feed of fresh medium and a continuous outflow of spent medium and bacteria. This is almost impossible to obtain with a single-stage digester with a slurry feedstock and using a pumped inlet system. Pumps which can handle slurries are usually of too great a pumping volume to continuously feed the relatively small hourly volumes of feedstock demanded by long detention times. Input is usually intermittent, by timed pump, but the shorter the intervals between pump-on periods, the more efficient is the digestion. An hour or so between inputs is of the order of timings in many cases.

It may be possible to produce a continuous feed input by gravity flow, if the digester is below ground for instance. This is possible even with a big digester. The smallest digesters have either this type of input or an intermittent manual input.

Outflow in small digesters is usually by gravity flow over some form of 'weir' which keeps a constant head of liquid to balance internal gas pressure, and most large digesters have this kind of overflow, although pumped output is sometimes used.

In a cold climate, energy can be conserved by transferring heat from the output to the input. Such a heat-exchange is technically very difficult to construct and run. One form of such heat-exchanger on a 13 m^3 experimental plant in Aberdeen was described by Mills (37), but whether this could be used on a large scale remains to be seen. Some other heat-exchangers tested have not been successful.

The digester output usually goes to storage tanks for later distribution as fertilizer. More recent experiments and developmental work on separation of farm slurries before digestion (to get over problems of fibres clogging digesters and pipes) and on use of digested sludges as animal feedstuffs, and so on, cannot be considered here but are discussed elsewhere (6, 9, 20, 21, 48).

The two-stage and two-phase digester

The single-stage digester behaves as a continuous culture and residual limiting substrate levels for the various bacterial reactions depend basically on the K_M for the reaction and the detention time of the system, i.e., the growth rate of bacteria. Concentrations of substrates in excess, e.g. ammonia in animal-waste digesters, will depend on input levels. The term 'substrates' was used here because the statements, of course, apply to bacterial substrates in solution, not potential substrates in suspended form and unavailable to the bacteria, for instance lignified cellulose fibres. The K_M and hence residual substrates for sugar fermentations are low as single-stage digesters are never run at detention times short enough to be near maximum growth rates of fermentative bacteria, but residual acids can be high at low detention times (23) and these form the main dissolved pollutants (other than ammonia) in digested sludges.

Theoretical considerations show that a second stage to a continuous culture, carrying out the same reactions as the first stage but running at a different detention time, can reduce residual substrates from the first stage. Again, in theory, for a simple reaction, say a sugar fermentation, a two-stage reactor can have a lower overall detention time for better substrate utilization than a single stage, as the second stage can be run at a rate greater than the maximum growth rate of the bacteria without washout. However, theory becomes complicated for a reaction of many steps and with slowly-degraded solid substrates such as the fibres in digesters. If the first stage has to be of long detention time to obtain degradation of solids then residual acids will be low in any case. A second stage then might be possible as a 'polishing' stage (together with solids separation) as an alternative to aerobic or other treatment if a final water of low acid content (and hence low BOD) were required, but for most digesters a second stage would seem hardly worth the extra monetary expense and energy cost. A second stage could be a stirred tank, as suggested, or a 'plug-flow' digester or in some cases of low solids feedstocks, an anaerobic filter, but all require extra tanks, pumps, heating and in some cases stirring.

Not to be confused with the two-stage digester, is the so-called 'two-phase' digester which is a mechanically similar system of two stirred-tank digesters. This is based on the suggestion that the natural equilibrium conditions of Eh and pH (about 7.2) obtaining in sewage and agricultural waste digesters are not optimum for both fermentative and methanogenic bacteria. If the two 'phases', fermentation and methanogenesis could be separated, both could be run under optimum conditions. As digestion is a non-sterile system with, in most cases, a continued inoculation of methanogenic and other bacteria from the feedstock, the only way of running a two-phase system is to make conditions so unsuitable in the first digester tank that methanogens can grow only in the second tank. However, the conditions of pH and Eh for growth of hydrolytic and fermentative bacteria are so close that the only possible practical way of separating the two phases is by having different detention times. To prevent growth and ensure washout of methanogens in the first tank the detention time must be a matter of hours. This is possible if the first phase is fermentation of a simple sugar, as used by Cohen *et al.* (8) in a laboratory two-phase digester with glucose as the primary substrate. The fermentor tank had a detention time of 10 hours, while the methanogenic tank had a detention time of 100 hours. But if the first step is hydrolysis of polymers before fermentation of sugars as with agricultural feedstocks then the hydrolysis requires a detention time of days and methanogens must then grow in the first-phase tank. So a two-phase digester is not possible. A small point about two-phase digesters is that unless the hydrogen produced in the fermentative phase can be collected and transferred to the methanogenic phase a loss of potential methane occurs. Cohen *et al.* (8) calculated they lost 12 per cent of the methane as hydrogen.

A kind of two-phase system being tested in Ireland uses an anaerobic filter as a second phase or stage (Dunican, private communication). Piggery wastes stored in outside tanks or in the channels under the piggery undergo slow decomposition with the production of carbon dioxide, hydrogen and some methane, but the acid fermentation products tend to accumulate to higher levels than in a properly-controlled digestion. The Irish system allows the pig waste to stand in such a way and the solids are allowed to settle in the tanks and the supernatant, acid-containing, liquid is then run off through an anaerobic filter where methanogenesis takes place. In short, a detention time, longer than the time needed in a normal digester is required and so the holding tank volumes need to be large. If the tanks are in the open and unheated then variations in reaction rate will occur with change of season. The anaerobic filter needs to be heated if its reaction rate is to be rapid and uniform in the cold climate. The system does not require the mixers of the stirred-tank digester, but does need pumps. It seems that this system could be a substitute for the stirred-tank digester on a small farm, but it requires more testing and development and more experiments to assess performance under different conditions. Its use with cattle slurries might be impossible as such slurries do not separate, even after digestion, as pig slurries do.

The tubular or plug-flow digester

A continuous-flow system which is theoretically a kind of batch-digester is the tubular, plug-flow, digester. In a plug-flow fermentor a medium enters at one end of a tube and flows (filling the tube) along it and out at the other end. If an inoculum is added to a volume of the medium entering the tube, then as this volume flows along the tube the bacteria grow as in a batch culture and if the rate of passage of the medium is correct, by the time this volume reaches the end of the tube the bacteria will have used up the substrate and ceased to grow. In theory there is no mixing in the tube and each volume of medium passes as a discreet mass. It can thus be seen that no inoculation from the bacteria growing to the medium coming in behind can occur. For a continuous operation the medium entering the tube must be continuously inoculated, and this can be done by separating some of the bacteria flowing from the end of the tube and returning them to the medium at the beginning.

If animal excreta are the feedstock, then these contain the necessary inoculum bacteria and, provided the time of passage of feed through the tube is sufficient, each 'plug' of feed entering the fermentor should develop a 'digester' flora, and this flora should digest the feed substrates as they pass along the tube.

In actual fact a plug-flow fermentor cannot be made. Wall friction, heating of the fermentor, gassing in the case of a digestion, all tend to mix the liquid flowing along the tube. With a feedstock such as pig waste, particulate matter will settle out in the tube. Cattle waste does not settle so easily and a laboratory plug-flow digester has been tested and this was followed by a 36 m^3 volume digester using cattle waste (11). The digester is a concrete trough with a butyl-rubber cover which also acts as a gas collector, as the liquid level is only to the top of the semi-circular-section trough, and is heated by water pipes. It seems a possible system for a small farm, but in Britain, at least, rubber bags or digester tops would be expensive. A large tubular digester (some hundreds of cubic metres) would seem impossible to build and operate and overall the pumps and equipment needed are similar to that of a stirred-tank system.

The detention time of a tubular digester with a farm-waste feedstock must be similar to that of a stirred-tank digester to allow fibre degradation, and must be even longer if time for the development of a digester flora in the incoming feed has to be allowed.

The anaerobic filter, feedback and sludge-blanket digester

In the digesters for farm wastes just described the size of the digester is governed by the necessity of a long detention time for the degradation of solids. The same considerations would apply to digesters for spent mash and other wastes of high solids content from distilleries. But in some cases, such as those of factories processing agricultural products such as fruits and vegetables or meat, the waste-waters are of low suspended-solids content but contain dissolved sugars and other substances. The rate of production of farm-animal excreta is commensurate with operating a digester, or digesters, of long detention time without them being too large. However, factory waste-waters are often produced at hundreds of gallons a minute. In calculating detention time the limiting factor in the case of the easily-fermentable factory wastes is not the degradation of fibres, but the growth rate of the methanogens, previously seen to need a doubling time of some four to six days. With the flow rates of wastes just mentioned even a six-day detention would mean an impossibly large digester. Therefore, some method of retaining the methanogens in the digester while allowing a liquid detention time of only a few hours, is required.

The simplest way of doing this is to allow the bacteria to attach to a solid support around which the waste-water can flow. This is the principle of the anaerobic filter previously mentioned and first tested on a laboratory scale by Young and McCarty (59). The water flows upward through a tank packed with stones or other support matrix and gas is collected from the top. The system can be at ambient temperature or heated, or hot waste-water may be used. The mass of bacteria on the matrix rapidly degrades the substrates in the water and so the void volume need be equivalent to a liquid detention time measured in hours rather than days. The bacterial mass tends to break off at intervals, but its overall detention time in the digester may be perhaps 20 days which certainly provides enough time for growth of the methanogens. Once a suitable flora has developed to a sufficient degree, it always remains in the digester to ensure continuous growth.

Another form of digester in which the bacteria are retained is the sludge-blanket digester. In this the bacteria grow naturally as a flocculent mass in an upward-flowing waste-water. The bacterial 'blanket' is retained by its own mass and by baffles in the top of the tank while gas and liquid can escape from the top. The bacterial mass does break up to some extent and so bacteria are lost in the outflow, but as in the filter the mean detention time of the bacteria is long enough to permit a dense mass of methanogenic bacteria to grow while liquid detention times are, again, only a few hours. This type of digester is being developed on a commercial scale in Holland (32).

Another variant on this theme is the fluidized-bed digester where the bacteria are attached not to a fixed support as in the filter, but to small glass spheres or similar particles which are freely suspended in the up-flowing liquid like the bacterial flocs of the sludge blanket. Hayes *et al.* (11) described some experiments with this type of digester.

Commercial farm digesters and gas use

In the previous sections a brief description of digester microbiology and types of digesters has been given. Many laboratory and pilot-plant experiments have been now done and the optimum conditions for, and results obtained from, digestion of many possible substrates (mainly wastes) determined. With no problems of aeration to consider, results of small-scale digestions can be duplicated on any scale provided digester conditions are duplicated (22). The biochemical and microbiological data for designing digesters and calculating pollutant degradation and gas available for use are available. What is now being done in Britain and other countries is to translate these data into large, working digesters. What has not been available is large-scale experience with collecting and pumping slurries, mixing and heating digesters, and other engineering matters, as little of the technology of the large-scale sewage digester can, for various reasons, be used in the farm digester. This experience is now being obtained in the commercial farm digesters now built, or being built, in Britain and elsewhere. Microbiological problems have been few, if any; problems that have occurred and are being solved are caused by the necessity to automate a system using animal excreta containing fibres, feathers, stones and other material and even without these of a very intractable nature. Theory can do little in the design of equipment for these slurries, only practical experience is of use. This is now being obtained and problems are being overcome.

On the smaller 'Gobar' and similar scales, although automation is non-existent, digesters still need some attention to work properly and the development of simple digesters and the training of farmers are continuing.

The small farm digester uses the gas for lighting and cooking with simple gas stoves. The large, intensive-farm requires much energy in the form of heat for animals, electricity for ventilating fans, feed mixers and conveyers, muck scrapers, milking machines, and so on. The idea of the intensive-farm digester is to provide this energy, so the digester gas is fed to automatic boilers which supply hot water to heat the digester as well as the farm, and it is also fed to engines which drive electricity generators. These latter are usually converted spark-ignition, car-type engines fitted with heat exchangers for cylinders and exhaust which supply hot water to heat the digester and farm buildings. In this way 80–90 per cent of the gas energy can be recovered for use. Experience in long-term running of these engine-generators, of the 20 to 100 kVA output range, is now being obtained.

The next few years should see the development of these, and factory digesters providing power for the factory in similar ways and to be generally used in systems for pollution control and 'alternative' energy.

References

1. Badger, D.M., Bogue, M.J. and Stewart, D.J. Biogas production from crops and organic wastes, 1. Results of batch digestions, *New Zealand Journal of Science*, *22*, 11–20 (1979).
2. Boone, D.R. and Bryant, M.P. Propionate-degrading bacterium, *Syntrophobacter wolinii* sp. nov. gen. nov., from methanogenic ecosystems, *Applied and Environmental Microbiology*, *40*, 626–632 (1980).
3. Bousfield, S., Hobson, P.N. and Summers, R. Pilot-plant high-rate digestion of piggery and silage wastes, *Journal of Applied Bacteriology*, *37*, xi (1974).
4. Bousfield, S., Hobson, P.N. and Summers, R. A note on anaerobic digestion of cattle and poultry wastes, *Agricultural Wastes*, *1*, 161–163 (1979).
5. Brade, C.E. and Noone, G.P. Anaerobic sludge digestion—need it be expensive? I. Making more of existing resources, Institute of Water Pollution Control, University of Aston (1979).
6. Bryant, M.P. Microbial methane production—theoretical aspects, *Journal of Animal Science*, *48*, 193–201 (1979).
7. *Chinese Biogas Manual*, Translated by M. Crook, Intermediate Technology Publications Ltd., King Street, London (1979).
8. Cohen, A., Zoetmeyer, R.J., van Deursen, A. and van Andel, J. G. Anaerobic digestion of glucose with separated acid production and methane formation, *Water Research*, *13*, 571–580 (1979).
9. Chynoweth, D.P. and Mah, R.A. Volatile acid formation in sludge digestion, Advances in Chemistry Series, American Chemical Society, No. 105.
10. Ghosh, S. and Klass, D. Two-phase anaerobic digestion, *Process Biochemistry*, *13* (4), 15–24 (1978).
11. Hayes, T.D., Jewell, W.J., Dell'Orto, S., Fanfori, K.J., Leuschner, A.P. and Sherman, D.F., Anaerobic digestion of cattle manure, in *Anaerobic Digestion* (see ref. 48) (1981).
12. Healy, J.B. and Young, L.Y. Anaerobic biodegradation of eleven aromatic compounds to methane, *Applied and Environmental Microbiology*, *38*, 84–89 (1979).
13. Heukelekian, H. and Heinemann, B. Studies on the methane-producing bacteria, 1. Development of a method for enumeration, *Sewage Works Journal*, *11*, 426–428 (1939).
14. Heukelekian, H. and Mueller, P. Transformation of some lipids in anaerobic sludge digestion, *Sewage and Industrial Wastes*, *30*, 1108–1110 (1958).
15. Hills, D.J. Effect of carbon : nitrogen ratio on anaerobic digestion of dairy manure, *Agricultural Wastes*, *1*, 267–278 (1979).
16. Hobson, P.N. Straw as feedstock for anaerobic digesters, pp. 217–224 in

Straw Decay and its Effect on Disposal and Utilisation (Edited by E. Grossbard), John Wiley, London (1979).

17. Hobson, P.N. Production of methane from wastes and crops, pp. 1–8 in *Processes for Chemicals from Some Renewable Raw Materials*, The Institution of Chemical Engineers, London (1979).
18. Hobson, P.N. Biogas production—agricultural wastes, pp. 37–44 in *Energy from the Biomass*, 5, The Watt Committee on Energy, London (1979).
19. Hobson, P.N. Microbial pathways and interactions in the anaerobic treatment process, in *Mixed Culture Fermentation*, Society for General Microbiology (in press) (1981).
20. Hobson, P.N., Bousfield, S. and Summers, R. Anaerobic digestion of organic matter, *Critical Reviews in Environmental Control*, *4*, 131–191 (1974).
21. Hobson, P.N., Bousfield, S. and Summers, R. *Methane Production from Agricultural and Domestic Wastes*, Applied Science Publishers, Barking, England, pp. 295 (1980).
22. Hobson, P.N., Bousfield, S., Summers, R. and Mills, P.J. Anaerobic digestion of piggery and poultry wastes, in *Anaerobic Digestion* (see ref. 48) (1981).
23. Hobson, P.N. and McDonald, I. Methane production from acids in piggery-waste digesters, *Journal of Chemical Technology and Biotechnology*, *30*, 405–408 (1980).
24. Hobson, P.N. and Shaw, B.G. The anaerobic digestion of waste from an intensive pig unit, *Water Research*, *7*, 437–449 (1973).
25. Hobson, P.N. and Shaw, B.G. The bacterial population of piggery waste anaerobic digesters, *Water Research*, *8*, 507–516 (1974).
26. Hobson, P.N. and Shaw, B.G. Inhibition of methane production by *Methanobacterium formicicum*, *Water Research*, *10*, 849–852 (1976).
27. Hobson, P.N. and Shaw, B.G. The role of strict anaerobes in the digestion of organic material, pp. 103–121 in *Microbial Aspects of Pollution* (Edited by G. Sykes and F.A. Skinner), Academic Press, London (1971).
28. Hungate, R.E. The anaerobic mesophilic cellulolytic bacteria, *Bacteriological Reviews*, *14*, 1–49 (1950).
29. Hungate, R.E. A roll-tube method for cultivation of strict anaerobes, pp. 117–132 in *Methods in Microbiology*, Vol. 3B, Academic Press, London (1969).
30. Jewell, W.J. Future trends in digester design, in *Anaerobic Digestion* (see ref. 48) (1981).
31. Kirsch, E.J. and Sykes, R.M. Anaerobic digestion in biological waste treatment, *Progress in Industrial Microbiology*, *9*, 155–237 (1971).
32. Lettinga, G., van Velsen, A.F.M., Hobma, S.W., de Zeeuw, W. and Klapwijk, A. Use of the upflow sludge blanket (USB) reactor concept for biological wastewater treatment, especially for anaerobic treatment,

Biotechnology and Bioengineering, XXII, 699–734 (1980).

33. Mah, R.A. and Susman, C. Microbiology of anaerobic sludge fermentation, 1. Enumeration of the non-methanogenic bacteria, *Applied Microbiology, 16*, 358–361 (1968).
34. Maki, L.R. Experiments on the microbiology of cellulose decomposition in a municipal sewage plant, *Antonie van Leeuwenhoek, 20*, 185–200 (1954).
35. McCarty, P.L. Anaerobic waste treatment fundamentals, part three, toxic materials and their control, *Public Works*, U.K., November, 91–94 (1964).
36. McInerney, M.J., Bryant, M.P. and Pfenning, N. Anaerobic bacterium that degrades fatty acids in symtrophic association with methanogens, *Archives of Microbiology, 122*, 129–135 (1979).
37. Mills, P.J. Minimisation of energy-input requirements of an anaerobic digester, *Agricultural Wastes, 1*, 57–66 (1979).
38. Morris, J.E. The digestion of crop residues—an example from the Far East, in *Anaerobic Digestion* (see ref. 48) (1981).
39. Mylroie, R.L. and Hungate, R.E. Experiments on the methane bacteria in sludge, *Canadian Journal of Microbiology, 1*, 55–59 (1954).
40. Ng, T.K., Weimer, P.J. and Zeikus, J.G. Cellulolytic and physiological properties of *Clostridium thermocellum*, *Archives of Microbiology, 114*, 1–7 (1977).
41. Robertson, A.M., Burnett, G., Bousfield, S., Hobson, P.N. and Summers, R. Bioengineering aspects of anaerobic digestion of piggery waste, pp. 544–548 in *Managing Livestock Wastes, Proceedings of the Third International Symposium on Livestock Wastes* ASAE (1975).
42. Rose, C.S. and Pirt, S.J. The roles of two mycoplasmal agents in the conversion of glucose to fatty acids and methane, *Society for General Microbiology Quarterly, 8* (1), 40 (1980).
43. Sathianathan, M.A. *Bio-Gas, Achievements and Challenges*, AVARD, New Delhi, pp. 192 (1975).
44. Schnellen, C.G.T.P. Onderzoekingen over de methaangisting, Dissertation, Technische Hoogeschool, Delft, Holland (1947).
45. Siebert, M.L. and Torien, D.F. The proteolytic bacteria present in the anaerobic digestion of raw sewage sludge, *Water Research, 3*, 241–250 (1969).
46. Smith, P.H. The microbial ecology of sludge methanogenesis, *Developments in Industrial Microbiology, 7*, 156–159 (1966).
47. Smith, P.H. and Mah, R.A. Kinetics of acetate metabolism during sludge digestion, *Applied Microbiology, 14*, 368–371 (1966).
48. Stafford, D.A., Wheatley, B.I. and Hughes, D.E. (Editors). *Anaerobic Digestion* (*Proceedings of the First International Symposium*), Applied Science Publishers, Barking, England (1980).

49. Summers, R. and Bousfield, S. Practical aspects of anaerobic digestion, *Process Biochemistry*, *11* (5), 3–6 (1976).
50. Summers, R. and Bousfield, S. A detailed study of piggery-waste anaerobic digestion, *Agricultural Wastes*, *2*, 61–78 (1980).
51. Torien, D.F. Enrichment culture studies on aerobic and facultative anaerobic bacteria found in anaerobic digesters, *Water Research*, *1*, 147–155 (1967).
52. Torien, D.F., Siebert, M.L. and Hattingh, W.H.J. The bacterial nature of the acid-forming phase of anaerobic digestion, *Water Research*, *1*, 497–507 (1967).
53. Torien, D.F., Thiel, P.G. and Hattingh, W.H.J. Enumeration, isolation and identification of sulphate-reducing bacteria of anaerobic digestion, *Water Research*, *2*, 505–509 (1968).
54. Van den Berg, L., Patel, G.B., Clark, D.S. and Lentz, C.P. Factors affecting rate of methane formation from acetic acid by enriched methanogenic cultures, *Canadian Journal of Microbiology*, *22*, 1312–1319 (1976).
55. Van Velsen, A.F.M., Lettinga, G. and Ottelander, D. den. Anaerobic digestion of piggery waste, 3. Influence of temperature, *Netherlands Journal of Agricultural Science*, 255–267 (1979).
56. Varel, V.H., Isaacson, H.R. and Bryant, M.P. Thermophilic methane production from cattle waste, *Applied and Environmental Microbiology*, *33*, 298–307 (1977).
57. Wolfe, R.E. Microbial formation of methane, *Advances in Microbial Physiology*, *6*, 107–146 (1971).
58. Woodman, H.E. *Rations for Livestock*, Bulletin 48, Her Majesty's Stationery Office, London (1954).
59. Young, J.C. and McCarty, P.L. The anaerobic filter for waste treatment, *Journal of the Water Pollution Control Federation*, *41*, R160 (1969).
60. Zeikus, J.G. and Wolfe, R.S. *Methanobacterium thermoautotrophicus*, an anaerobic, autotrophic, extreme thermophile, *Journal of Bacteriology*, *109*, 707–713 (1972).
61. Zeikus, J.G. Microbial populations in anaerobic digesters, in *Anaerobic Digestion* (see ref. 48) (1981).
62. Zinder, S.H. and Mah, R.A. Isolation and characterization of a thermophilic strain of *Methanosarcina* unable to use H_2-CO_2 for methanogenesis, *Applied and Environmental Microbiology*, *38*, 996–1008 (1979).

D.S. CHAHAL

21. Bioconversion of Lignocelluloses into Food and Feed Rich in Protein

Introduction

Malthus was not in error when he predicted that the human population would ultimately increase at a more rapid rate than its capacity to increase the rate of food production. The world population increased from 3.67 billion in 1970 to 4.03 billion in 1975 and is expected to reach 5.85 billion in the year 2000, almost doubling after 30 years. Cereal production increased from 1.245 billion metric tonnes in 1970 to 1.553 billion metric tonnes in 1979, an increase of only 24.7 per cent in nine years, in spite of the green revolution (FAO Yearbook, 1979).

Keeping in view the ever-increasing human population and the lack of a corresponding increase in food production, the importance of microbial sources of food/feed rich in protein has been raised a number of times since World War II. The success of the green revolution during the last two decades diminished the interest in the development of projects for microbial food/feed. Now, it seems that the situation has changed entirely because of the shortage of fossil fuels in the world and the diversion of food/feed grains for alcohol production.

The use of grains for fuel ethanol production would continue until the gap between food supply and demand is widened to such an extent that human starvation becomes a worldwide menace. Diversion of grains for fuel ethanol production will also continue until a technology is developed to convert crop and forest residues into fuel ethanol. Recently, the study entitled "Power Alcohol: Impact on Agribusiness Sectors, 1980–1990" indicated that use of corn in ethanol production may lead to widespread feed shortages by 1990, if not checked by the U.S. government (Biofuels Report, January 5, 1981). However, the technology for the bioconversion of crop and forest residues into microbial food/feed rich in protein is available now and it may solve the food/feed problem and also spare grains for the production of fuel ethanol.

Why microorganisms as a source of food and feed?

The use of microorganisms for the conversion of lignocellulose into food and feed rich in protein seems to be the wise choice in developing uncoventional source of protein as food/feed because of the following characteristics of microorganisms:

1) Microorganisms have a very fast growth rate (0.05 to 1.0h^{-1}). A protein doubling time as short as two hours has been recorded for *Neurospora crassa* (Zalokar, 1959), and the same rate of protein production has been maintained by the growth of mycelium in a 3,000-litre fermentor (Worgan, 1973).

2) They can be easily modified genetically for growth on a particular substrate and under particular cultural conditions.

3) Their protein content is quite high, varying from 35 to 60 per cent.

4) They can be grown in slurry or solid-state fermentation.

5) Their nutritional values are as good as those of other conventional foods rich in protein (56, 108).

During the first International Conference on Microbial Protein, convened in 1967 at the Massachusetts Institute of Technology (MIT), Cambridge, Massachusetts, U.S.A., a new generic term "Single-Cell Protein" (SCP) was coined to replace the supposedly less aesthetic term "Microbial Protein or Petroprotein" (91). This decision was most appropriate at that time, when the majority of the microorganisms used in protein production were single-celled (yeasts, bacteria and algae). By the time the Second International Conference was convened in 1973, again at MIT, some filamentous fungi and actinomycetes were reported to produce protein from various substrates (99). Since then, more and more reports on the use of filamentous fungi for protein production from starchy and lignocellulosic materials are appearing. Thus the use of the term single-cell protein is not a logical one when an organism is filamentous.

Filamentous fungi have been used for protein production since the 1920's (101) and more recently during the 1960's and 1970's (15, 39, 40, 74, 82, 108). The term "Fungal Protein" has been extensively used by many workers in the past, and now a new term "Mycoprotein" has been introduced by Ranks Hovis McDougall (RHM) in the United Kingdom, for protein produced on glucose or starchy substrates. The U.K. Ministry of Agriculture, Fisheries and Food has allowed the use of mycoprotein produced by the filamentous fungus, *Fusarium graminearum* A35, for human consumption (77).

The use of filamentous fungi for protein production from various substrates is becoming popular because of the following reasons:

1) Some of the filamentous fungi grow as fast as most single-celled organisms (1).

2) The finished product of the filamentous fungi is fibrous in nature and

can be easily converted into various-textured foods. In comparison, protein is extracted from single celled organisms and spun into fibrous form.

3) Filamentous fungi have a greater retention time in the digestive system than single-celled organisms.

4) Protein content can be as high as 35–50 per cent with comparatively less nucleic acid than single-celled organisms.

5) Digestibility and net protein utilization (NPU) without any pretreatment is higher than that of single-celled organisms.

6) The overall cost of protein production from filamentous fungi is more economical when compared to that of the single-celled organisms.

7) Filamentous fungi have greater penetrating power into insoluble substrates and are therefore more suitable for solid-state fermentation of lignocellulosic materials (Chahal, Moo-Young and Vlach, unpublished).

8) Most of the filamentous fungi have a faint mushroom-like odour and taste which may be more readily acceptable as a new source of food than the yeasty odour and green colour associated with yeasts and algae respectively.

9) The biomass produced by filamentous fungi can be used as such without any further processing because it provides carbohydrates, lipids, minerals and vitamins as well as protein. In addition, the nucleic acid content of fungal protein is lower than that of yeasts and bacteria.

Calories and protein supply

The present food supply in terms of calories and protein, per capita, per day (Table 1), indicates that the world is short of not only good quality animal protein but also in calorie (energy) requirement. Only developed countries have sufficient supplies of protein of animal origin; while in most developing countries, especially India, Bangladesh and Indonesia, many people suffer from protein-calorie malnutrition. Some other developing countries like Brazil, Mexico, Pakistan, South Africa and Cuba are on a marginal scale for energy requirements, but are short of animal protein. Fortunately, countries like India, Bangladesh, Indonesia, Pakistan and Brazil are quite rich in crop residues (straw). Moreover, India, Brazil and Cuba also produce large quantities of sugar cane bagasse that can be used for the production of mycoprotein.

Crop residues are composed of 30–45 per cent cellulose, 16–27 per cent hemicellulose and 3–13 per cent lignin (94). Wood harvesting and wood-processing residues are composed of 45–56 per cent cellulose, 10–25 per cent hemicelluloses and 18–30 per cent lignin (106). The average yield of crop residues (cereals) is about 2,041 kg/hectare and bagasse yields are 14,000 kg/hectare. If on an average, 70 per cent of the carbohydrates (cellulose and hemicelluloses) available from crop and forest residues are taken into consideration, then 1,428 kg of carbohydrates from crop residues and 9,000 kg of carbohydrates from bagasse per hectare would be available for bioconversion into microbial biomass rich in protein. Thus, additional food/feed in the form of microbial protein of about 357 kg—a factor of 0.25 of the total carbohydrate portion from crop residues and 2,450 kg from bagasse per hectare can be obtained to alleviate food and feed shortages in countries rich in such crop residues. The yield of microbial protein from crop residues is much higher (357 kg) than that obtained from cereal grains per hectare (244 kg based on approximately 12 per cent protein in cereal grains). The amount of protein produced from bagasse will be quite significant, i.e., 2,450 kg hectare.

Table 1. World food supply: Calories per capita and protein per capita[1]

Country	Calories[2] available per capita per day	Protein[3] available per capita per day	Protein (animal) available per capita per day	Protein (vegetable) available per capita per day
World	2,590	69.3	24.4	44.8
USA	3,537	106.2	72.1	33.5
Canada	3,346	101.3	65.6	35.6
Bangladesh	1,945	42.4	5.7	36.7
Brazil	2,522	61.2	23.2	38.0
Cuba	2,636	68.8	33.6	35.1
India	1,949	48.4	5.2	43.2
Indonesia	2,115	43.7	5.1	38.6
Mexico	2,668	66.1	20.3	45.8
Pakistan	2,255	62.0	15.6	46.3

[1]Food and Agriculture Organization. *Production Yearbook*, Rome, Italy, Volume 33 (1979).

[2]Estimated average calories required per capita=app. 3,000.

[3]Estimated average protein required per capita=app. 64 g.

Pretreatment of substrates

Unfortunately, lignin in the plant cell wall not only encrusts the cellulose microfibrils in a sheath-like manner, but is bonded physically and chemically to the plant polysaccharides (51). Lignin-carbohydrate bonds form metabolic blocks that greatly limit the action of microbial hemicellulases and cellulases. Physically, lignin forms a barrier, suppressing the penetration by polysaccharide-digesting enzymes (57). Unless the lignin is depolymerized, solubilized, or removed, the cellulose and the hemicelluloses cannot be utilized for conversion into microbial biomass by most microorganisms lacking lignin-degrading enzymes. White-rot fungi (Basidiomycetes) are known to utilize lignin as well as cellulose (21), but they are slow-growing organisms and so far none have shown any promise for microbial protein production (9, 24, 59). *Phanerochaete chrysosporium* Bursdal, a basidiomycete, had been shown to possess lignin-degrading enzymes (29); however, Eriksson and Larsson (30) reported only 13.8 per cent dry weight (DW) crude protein in the final product from alkali-pretreated tree bark (washed), without any degradation of lignin by *P. chrysosporium.* Recently, Reid (86) reported about 50 per cent removal of lignin from water-extracted aspen wood by growing *P. chrysosporium* for a long time (30 days). Thus, the pretreatment of lignocellulosic materials will be necessary for the more efficient bioconversion of carbohydrates into microbial food and feed.

The effect of physical and chemical pretreatments to increase the availability of carbohydrates from lignocellulosic materials as feedstock for animals, saccharification and microbial protein production have been discussed in detail by Tarkow and Feist (100) and Millet *et al.* (70). The most common pretreatments to break the lignocellulosic bonds are discussed as follows:

1) *Alkali*

Dilute alkali or liquid ammonia increases the fibre saturation point and the swelling capacity of lignocellulosics. The increase in swelling capacity results from the saponification or ammonolysis of esters of 4-0-methylglucuronic acid attached to xylan chains. In the natural state, the esters act as cross-links, limiting the swelling or dispersion of polymer segments in water. Such treatments increase the accessibility of plant polysaccharides to cellulolytic microorganisms (100).

2) *Steam*

Steam treatment under high pressure (4, 48, 79) makes the lignocellulosic material easily accessible to hydrolytic enzymes.

The basic principle of the Iotech process is to pressurize the substrate (wood chips, straw, etc.) with steam in a pressure-tight reactor, then suddenly release the pressure, extruding the cooked substrate through a nozzle. The effect of steam hydrolysis on various fractions of lignocellulosic materials can be represented by the following first order reactions:

a) Cellulose $\xrightarrow{k_1}$ Enzyme-accessible low DP cellulose $\xrightarrow{k_2}$ Degradation products (hydroxymethyl furfural)

b) Hemicelluloses $\xrightarrow{k_1'}$ Water soluble monomers and oligomers (mainly xylose) $\xrightarrow{k_2'}$ Degradation products (furfural)

c) Lignin $\xrightarrow{k_1''}$ Low molecular weight reactive lignin $\xrightarrow{k_2''}$ High molecular weight condensed lignin

The chemical changes in steam-treated wood depend on the temperature, pressure, and time of exposure to steam. The hemicelluloses are hydrolyzed to soluble sugars by organic acids, mainly acetic acid derived from acetylated polysaccharides present in wood (8). Under more drastic conditions, secondary reactions occur which result in the formation of furfural, hydroxymethyl furfural, and their precursors by dehydration of pentoses and hexoses (52). Campbell *et al.* (7), reported that phenolic-like compounds increased from 0.43 to 5.3 per cent in steam-pretreated bagasse at 27.58 bar (ca. 400 psig) for 45 minutes. Since phenolic-like compounds and furfurals are usually toxic to most microorganisms (47, 61), such pretreated lignocellulosics may not be good substrates for SCP production by most cellulolytic microorganisms. This necessitated determining the optimum temperature, pressure, and time combination for steam explosion of wood.

Iotech has developed a process where optimum conditions for steam-explosion pretreatment of wood and crop residues have been worked out to have large k_1/k_2 ratios to minimize the degradation of final products. This can be achieved by either an increase in the reaction temperature or an increase in the acidity of the substrate. However, too high a temperature may make it difficult to obtain uniform heat distribution within wood chips and the carbohydrate may be degraded. Adding a small amount of strong acid (e.g., HCl, H_2SO_4, SO_2) to wood chips will substantially increase the steam hydrolysis rate and the k_1/k_2 ratio. The application of mineral acid catalysts is particularly useful for certain wood species with a high lignin content and/or a low native-acid content (acetic, uronic acids) as they are less responsive to steam pretreatment alone (78, 79). High yields of glucose and xylose are obtained as well as a high-quality lignin. The Iotech lignin residue possesses all of the qualities necessary for synthesis of lignin formaldehyde resin and many other chemicals. The glucose is utilized for the production of fuel ethanol and xylose is used for the production of fungal protein, xylitol, furfurals, and other chemicals. A detailed engineering design of the Iotech process has an estimated operating cost of $7.50 per ton of biomass. At this low cost, the Iotech process achieves many important pretreatment goals. The substrate is sterilized, and pulverized into a fibrous or powdered form; the cellulose and hemicelluloses are made easily accessible to the polysaccharide-digesting enzymes for the production of simple sugars and SCP; and the process provides excellent yields of high-quality lignin. Plate 1 (A, B, C) presents a comparison of structural features of aspen wood processed into chips, ground with a Wiley mill, and exploded with steam.

3) *Sodium chlorite* ($NaClO_2$)

Sodium chlorite, a strong oxidizing agent, has long been used for removing lignin during the preparation of "holocellulose", the total carbohydrate portion of lignocellulose (41). Goering *et al.* (36) demonstrated that *in vitro* digestibility of straws is increased with $NaClO_2$ treatment. Chahal *et al.* (13) reported that protein productivity increased considerably on delignified wheat straw when fermented with *Cochliobolus specifer* Nelson. Delignification of lignocelluloses by $NaClO_2$ seems to be useful for increasing the production of protein per unit weight of original substrate because hemicelluloses, a potential carbon source, remain intact with cellulose, and both the substrates (hemicelluloses and cellulose) are utilized concurrently. However, this process of delignification is not economically attractive because of the high cost of chemicals involved (35).

Choice of substrate

One of the earliest processes for SCP production from lignocelluloses was the production of yeast from wood hydrolysate. During World Wars I and II, Germany developed a process to produce food yeast from wood sugars obtained by acid hydrolysis (64). Later, commercial plants were set up at various places in the United States of America and other parts of the world (34, 46, 103). SCP production from wood hydrolysate could not survive because of the high costs of production. Hydrocarbons then became attractive substrates and were investigated extensively for SCP production by a number of oil companies (31, 38, 42, 65, 95, 96, 109). The dream of abundant and low-cost hydrocarbons as substrates for SCP has been shattered recently because of the world energy crisis.

Keeping in view that lignocelluloses are constantly being replenished through photosynthesis, the attention of researchers all over the world has now been diverted to the direct utilization of this carbon source for SCP production. From time to time, various cellulolytic organisms (actinomycetes, bacteria, and fungi) have been recommended for SCP production from lignocelluloses, in slurry and solid-state fermentation.

Thus, the choice of substrate for microbial protein production will depend on the availability of suitable substrate in a particular country. Crop residues (straw) may be the best choice for India, Bangladesh, Mexico, Pakistan, and Indonesia; and sugar cane bagasse for Brazil, Cuba and India (Table 2). On the other hand, forest residues may be the best choice for some of the African countries. However, hydrocarbons may still be the main and cheapest carbon source for SCP production in the Middle East where other carbon sources (non-hydrocarbons) are in limited supply (42).

Various processes for food/feed production

A number of processes are now available for the bioconversion of agricultural, forestry and animal wastes into food/feed rich in protein:

Table 2. Crop residues in the world[1]

Country	Cereal grains 10^{-3} MT	Cereal[2] straw 10^{-3} MT	Sugar cane 10^{-3} MT	Bagasse[3] 10^{-3} MT
World	1,553,076	1,553,076	754,130	188,532
USA	299,257	299,257	25,112	6,278
Canada	36,381	36,381	—	—
Bangladesh	19,902	19,902	—	—
Brazil	27,134	27,134	138,325	34,581
Cuba	596	596	70,000	17,500
India	129,603	129,603	156,450	39,112
Indonesia	29,550	29,550	—	—
Mexico	16,501	16,501	35,415	8,853
Pakistan	16,462	16,462	—	—

[1]Food and Agriculture Organization. *Production Yearbook*, Rome, Italy, Volume 33 (1979).

[2]Calculated on grain : straw ratio of 1 : 1.

[3]Calculated on sugar cane : bagasse ratio of 1 : 0.25.

1) *Yeast protein from wood sugars*

In this process wood is hydrolyzed with acid. The sugars produced in the initial stage are decomposed in the following stage when the most resistant cellulose is being hydrolyzed. Careful control of the hydrolysis conditions is therefore essential to avoid considerable destruction of sugars. In practice, two types of processes have been developed to minimize these losses: the Bergius process where concentrated acid is used at normal temperature; and the Scholler process where dilute acid is used at temperatures of 160–180°C.

The Scholler process was operated on a large scale in Germany during World War II. About 6.8 kg of yeast protein was produced from 100 kg (dry weight) wood (64). Microbial protein production from wood hydrolysate could not survive because of the high cost of production. However, in Russia a process for the production of *Candida utilis* on wood hydrolysates is still considered to be efficient enough to justify production on a large scale, with a predicted output of 900×10^6 kg yield (6).

2) *Fungal protein from carob sugars*

The carob (*Ceratonia siligna* L.) is a member of Leguminosae and is grown throughout the Mediterranean region, Rhodesia, parts of the U.S.A., and South America (53). The shelled pod has almost no commercial value but contains about 55 per cent sugars (60 per cent sucrose and 40 per cent reducing sugars) on a dry-weight basis (20).

A simple process has been devised by Imrie and Vlitos (53) for the extraction of the sugars from the shelled pods and the cultivation of *Aspergillus niger* Ml for fungal protein production. Carob extract is a complex medium, capable of providing all the necessary nutrients for the growth of microorganisms. To optimize the yield of mycelium and protein content in the final product, supplementation with a small amount of nitrogen and phosphorous is necessary. The fermentation is carried on at 30–36°C. Imrie and Vlitos (53) have reported a yield of 45 per cent mycelial biomass on the initial sugars and a cell density of 31.5 g/l in a 3,000 litre fermentor. The biomass contained 35 per cent crude protein. The amino acid composition of fungal protein meets the FAO reference standards except for a deficiency in sulphur-containing amino acids, cystine and methionine. Preliminary feeding trials with chicken showed no pathological signs.

3) *Yeast protein from starch—The Symba process*

The Symba process was developed by Tveit, Jarl, Skogman and Bergkvist at the Swedish Sugar Corporation (107). In this process, starch is hydrolyzed into maltose and glucose with *Endomycopsis fibuligera*, and then *Candida utilis* is grown on the sugars thus produced. The concurrent enzymatic starch hydrolysis and yeast cell multiplication at a pH of about 5, is very favourable from the point of view of process technology. Enzymatic hydrolysis of starch is preferred to mineral acid hydrolysis because the latter requires highly corrosion-resistant equipment. The process also results in the formation of compounds which can only be partially assimilated by the cells of *C. utilis*. Thus, acid hydrolysis produces low yields of cell biomass when calculated on the basis of total starch supplied in the fermentation medium.

In this symbiotic culture system, the growth of *E. fibuligera* is kept at low levels so that the final product will contain more cells of *C. utilis* which has greater nutritional value. If the cells of *E. fibuligera* are increased, the total protein in the final product will decrease because *E. fibuligera* cells are low in protein. However, the great disadvantage of keeping the growth of *E. fibuligera* restricted is that it would limit the potential for cyclic-batch or continuous fermentation. In every subsequent cyclic-batch, or, after each residence time in continuous fermentation, the cell population of *E. fibuligera* would keep on declining affecting the final yield and ultimately ending the process. Under these conditions, the inoculum of *E. fibuligera* should be strengthened at every cyclic-batch or after every residence time in continuous fermentation.

According to Wiken (107), the Symba yeast (*C. utilis*) is slightly straw-coloured and has a pleasant taste. The protein digestibility is 80–90 per cent as determined by the Pepsin-hydrochloric acid method. On supplementation with 0.1 per cent methionine, its protein value will become equivalent to that of milk protein. So far, the product has been used mainly as feed for chicken and pigs, and to a lesser extent for human consumption.

4) *Fungal protein from starch*

Cassava (*Manihot esculenta* Crantz) is a starch-producing root crop cultivated extensively in tropical regions in Africa, Asia and South America as a staple food. Its yield, in terms of calories per acre, has been reported to be among the highest of any cultivated plant (67). However, its protein content is very low, and cases of malnutrition have been reported in places where cassava is the staple constituent of the diet (2). Thus, a process for upgrading the protein values of cassava was developed by the University of Guelph, Guelph, Ontario, Canada in collaboration with the Centro Internacional de Agricultura Tropical (CIAT), Cali, Colombia (40).

In this system an amylolytic fungus, *Aspergillus fumigatus* I-21 (85) is used. This organism has the ability to produce sufficient amylases to hydrolyse starch into glucose and ultimately glucose is used by the same organism to synthesize fungal biomass rich in protein. This process has an advantage over the Symba process where an additional organism (*E. fibuligera*) is required for the pre-hydrolysis of starch for the growth of *C. utilis*.

The cassava roots are washed to remove dirt and sand. The rasped and grated cassava roots are dumped into a fermentor containing water at 70°C. The temperature is maintained at 70°C for 10 minutes to gelatinize the starch and also to prevent the development of a fungistatic activity in the mash. The fungistatic activity is believed to result from the release of HCN from the glucoside linamarin due to the action of the enzyme linamarase and both the glucoside and the enzyme are present in cassava roots (85). More water is added to make a final carbohydrate concentration of 4 per cent. The pH is adjusted to 3.5 with H_2SO_4, which also provides a supply of sulphur. Urea (3.5 g/l) is added as a nitrogen source and KH_2PO_4 (0.5 g/l) as a source of potassium and phosphorus. The temperature of the fermentation medium is maintained at 45–47°C. The medium is inoculated with 7 per cent vol/vol mycelial broth of *Aspergillus fumigatus* I-21 produced on the same medium.

The final product contains 37 to 44 per cent crude protein. The nutritional value of *A. fumigatus* proved to be inferior to the reference diet containing casein (55). When the fungal diets were supplemented with methionine and all the ratios contained 10 per cent 'true' protein, the net protein ratio (NPR) values were only slightly inferior to the values obtained with casein.

Aspergillus fumigatus is associated with aspergillosis, a lung infection caused by the inhalation of its spores. For this reason, a stable (non-revertible) asporogenous mutant designated as *A. fumigatus* I-21A has been obtained by gamma radiation. Data presented on this organism by Khor *et al.* (55) based on feeding trials with rats, indicated that weight gain, feed efficiency, protein efficiency ratio and the net protein ratio was never equal to that of the control diet of casein.

5) *Microbial protein from ryegrass sugars*

Han *et al.* (45) again revived interest in microbial protein production on sugars obtained by acid hydrolysis of crop residues. Such a process could not survive about 30 years ago because of the high cost of production. They grew *Aureobasidium (pullularia) pullulans*, a yeast-like fungus, on the acid hydrolysate of rye-grass straw (treated with 3 per cent wt/wt H_2SO_4 at 121°C for 30–40 minutes) containing 6 g sugars/l supplemented with 1.25 g of yeast extract. A yield of only 1.5 g cells/l was obtained. The final product contained 42.6 per cent protein. Therefore, only 0.64 g protein/l was obtained, even after supplying 1.25 g of yeast extract/l, a very expensive constituent in the medium. Similarly, they have reported 1.35 g cells/l by growing *C. utilis* under similar conditions. The results indicated that they obtained 0.64 g protein by providing 0.62 g protein/l in the form of yeast extract. There would be a net economic loss if SCP was produced by this system. Moreover, *A. pullulans* is pathogenic to plants and occasionally has been reported to be a human pathogen (22).

6) *Fungal protein from hemicellulose sugars*

Hemicelluloses constitute 20–25 per cent of lignocellulosic materials. The hemicelluloses are easily hydrolysed to their monomeric sugars (xylose, mannose, arabinose, galactose, glucose and also uronic acids) by dilute acid hydrolysis at moderate temperature and atmospheric pressure (60). Iotech Corporation has developed a pretreatment process for lignocellulosic materials whereby yields of almost 89 per cent of hemicellulose sugars are attained by high steam pressure treatment (79). The cellulose obtained from these treatments is used for fuel ethanol production. At present these sugars (mainly xylose) have little value because they are not fermented into ethanol by common yeasts (32). Chahal, McGuire, Pikor, and Nole (unpublished) tried to utilize these sugars for fungal protein production. They reported 8.7 g fungal biomass/l by growing a fungus *Chaetomium cellulolyticum* on 15.9 g hemicellulose sugars/l obtained from aspen (*Populus tremuloides* Michx.) wood pretreated by the Iotech process.

The fungal biomass contained 44–46 per cent crude protein. They reported synthesis of 4.4 g crude potein/l within 22 hours of fermentation with a protein productivity of 200 mg/l.h. This is the highest protein productivity ever recorded in such a system. About 100 kg of fungal food/feed rich in protein (approximately 45 per cent) could be produced for every 1,000 kg of wood. The residual cellulose (400–500 kg) is used for ethanol production and the residual lignin (220 kg) could be used for making adhesives and other compounds (Noble, G., personal communication). When compared to the yield of yeast on wood sugars (64) and the yield of *A. pullulans* on ryegrass straw hydrolysate (45), production of fungal protein with *C. cellulolyticum* on hemicellulose sugars seems to be the most suitable and economical system at present (37, 90).

7) *Bacterial protein from cellulose*

A process was developed by the State University of Louisiana for SCP production from sugar cane bagasse using a new bacterial species, *Cellulomonas* sp. ATCC No. 21399 (27). This new species was isolated, defined and its preliminary growth studies were carried out by Srinivasan and Han (97). Dunlap (26) has reported that this organism is C_x (endo β-glucosidase) positive but lacks C_1 (exo β-glucocellobiose-hydrolase). Moreover, this organism has low β-glucosidase activity. Thus, this system needs another organism, *Alcaligenes faecalis* ATCC No. 21400, to remove the accumulated cellobiose which inhibits cellulase activity (66). Since this organism lacks C_1 activity, a drastic pretreatment of the substrate with alkali is necessary to remove lignin and to swell the cellulose. Dunlap (26) has recommended a pretreatment of sugar cane bagasse with 30 g NaOH/100 g substrate with a system pressure of 30 psig for four hours. It has been estimated from the data presented by Dunlap (26), that there is a loss of 32 g in the form of solubilized hemicelluloses and lignin, from 100 g of sugar cane bagasse during such pretreatment. When the 68 g of washed residual cellulose was fermented with a mixture of *Cellulomonas* sp. and *A. faecalis* at 34°C and pH of 6.6–6.8, only 40 g of cellulose was utilized and a bacterial cell mass of 18–20 g was obtained. Thus, a yield of only 10 g of protein was obtained from 100 g of original sugar cane bagasse. Theoretically, 100 g of sugar cane bagasse containing about 70 g of fermentable sugars will yield 35 g of bacterial cell mass which represents about 18 g of protein. If the protein yield of 10 g obtained in this process is compared to that of the theoretical yield of 18 g, this process does not seem to be an economically attractive alternative. Moreover, in this process hemicellulose sugars are not utilized which would create a disposal problem because of their very high BOD.

In another report, Han and Callihan (44) reported the production of 1.76 g protein/1 by growing a mixed culture of *Cellulomonas* sp. and *A. faecalis* for five days on a 1–5 per cent slurry of alkali-treated rice straw. They also reported that alkali treatment (3 per cent NaOH) of sugar cane bagasse under very high pressure (260 psig, 5 minutes) gave 7.2 g protein/1. The treated substrate was washed with water to remove solubilized lignin and hemicelluloses. The process seems to be attractive but it was difficult to estimate its efficiency since the concentration of the substrate used in the medium was not given. The disposal problems of solubilized hemicelluloses would still exist.

8) *Fungal protein from lignocelluloses*

The earliest work on fungal protein production was done by Pringsheim and Lichtenstein in 1920 when they fermented straw with *Aspergillus fumigatus* for animal feed (62). It is only during the last decade that a large number of papers began to appear on fungal protein production from cellulosic materials. Rogers *et al.* (87), reported 13.3 per cent DW crude protein after four days of growth of *Aspergillus fumigatus* on alkali-treated cellulose. Peitersen (82) obtained 21–26 per cent DW crude protein by growing *Trichoderma viride* on alkali-treated barley straw for two to four days. Romanelli *et al.* (88), reported 60 per cent utilization of fine Solka Floc powder (B.W. 200) in three days by *Sporotrichum thermophile*. Eriksson and Larsson (30) obtained a product with 6 per cent DW crude protein from powdered cellulose, 13.8 per cent from waste fibres, and 32 per cent from highly amorphous cellulose by growing the lignocellulolytic organism, *Sporotrichum pulverulentum* for six days. Recently, Miller and Srinivasan (69) switched over from the bacterium *Cellulomonas* sp. to the fungus *Aspergillus terreus* for fungal protein production from cellulose. They reported 32.9 per cent DW crude protein on pure cellulose (Solka Floc) treated with 1N NaOH solution (80 g NaOH/100 g cellulose) at 121°C for 15 minutes. The quantity of NaOH used in the pretreatment of cellulose is too high to be considered economically feasible.

A new cellulolytic fungus

During the time when other microorganisms were being tested for SCP production, a new cellulolytic fungus, *Chaetomium cellulolyticum* Chahal and Hawksw. (ATCC 32319), entered into the field of fungal protein production from lignocelluloses. Its taxonomic and morphological characteristics along with the preliminary physiological studies were reported by Chahal and Hawksworth (10) Chahal and Wang (19) later reported growth behaviour and production of protein on pure cellulose. They reported that long fibres of cellulose are broken into shorter ones during the early phases of fermentation, followed by longitudinal splitting of fibres into fibrils and ultimately, almost complete utilization of the substrate (Plate 1, D, E, F). This organism does not form any pellets during fermentation; mycelium is always found dispersed uniformly with cellulose in the fermentation medium. This characteristic makes it more desirable for growth in submerged culture because oxygen often becomes limited in a central biomass of pellets (83). Moreover, the pelleted mycelium may have less cellulase, as shown by Kobayashi and Suzuki (58) for α-galactosidase activity in large pellets of *Mortierella vinacea* Dixon-Stewart. Pelleting can also create problems in large-scale production of biomass, because of poor mass transfer. Another favourable characteristic of *C. cellulolyticum*, reported by Moo-Young *et al.* (75), is that it can utilize all the major hemicellulose sugars (xylose and mannose) as well as cellobiose. Cellobiose inhibits cellulase activity when accumulated in large quantity during cellulose degradation (66). Chahal, Vlach and Moo-Young (unpublished) also found that hemicelluloses and cellulose were utilized concurrently during SCP production from alkali-pretreated corn stover slurry containing solubilized hemicelluloses and lignin. The presence of solubilized lignin was found not to interfere with the normal metabolism of the organism.

Recently, Chahal and Moo-Young and Vlach (unpublished) have shown that the hyphae of *C. Cellulolyticum* enter into the cell lumen through natural openings, mechanical breaks, or spaces in the cell wall of plant materials created by the solubilization of hemicelluloses and lignin during alkali treatment. Once inside the cell lumen, the hyphae start digesting the cell wall from the inside towards the outside ultimateiy consuming the whole cell wall. They have emphasized that because of the good intrusion power of the hyphae, *C. cellulolyticum* can penetrate deep into the substrate for maximum conversion into fungal biomass. Deep penetration of the hyphae into the substrate makes *C. cellulolyticum* the most suitable organism for solid-state fermentation when compared to single-celled organisms (yeasts, bacteria) which lack such power of penetration. All of these characteristics indicate that *C. cellulolyticum* is a potentially strong candidate for fungal food/feed production from lignocelluloses.

Fungal protein production with *Chaetomium cellulolyticum*

1) *Crop residues*

First reports (19, 74) have shown that *C. cellulolyticum* was better for SCP production on pure cellulose than the other organisms previously reported. A final product with up to 40–45 per cent DW crude protein was recorded (Table 3). Chahal and Wang (19) indicated that *C. cellulolyticum* seems to be the most suitable organism for cyclic-batch and continuous fermentations. Chahal *et al.* (15) reported high protein productivity (69.7 mg/1/h) on alkali-treated wheat straw with a final product containing 40 per cent crude protein. This was the highest protein productivity (with a high percentage of protein) ever achieved when compared to other cellulolytic microorganisms used to ferment alkali-treated lignocellulosics (Table 3).

In most of the studies compared in Table 3, the lignocellulosic substrates were treated with alkali and washed to remove lignin. Such treatments not only remove lignin but also hemicelluloses, a potential carbon source for SCP production. The lignin and hemicelluloses thus removed present waste disposal problems. Moo-Young *et al.* (75) proved that *C. cellulolyticum* can utilize xylose and other hemicellulose sugars. Chahal *et al.* (16) retained the pretreatment liquor in the fermentation medium from alkali-treated corn stover. The authors obtained a final product with 35.7 per cent DW crude protein and a protein productivity of 146.0 mg/1/h in batch fermentation and approximately 400 mg/1/h in continuous fermentation. This is the highest productivity of protein ever reported from fermentation of a 1 per cent slurry of lignocelluloses.

Moo-Young and Chahal (73) reported that *C. cellulolyticum* can convert cattle manure (containing pretreatment liquor) into animal feed rich in protein (37 per cent). They have also shown that anaerobic liquor, obtained from methane-generating anaerobic fermentation of cattle manure was a good nutrient source (N, P, and other chemicals) for the bioconversion of corn

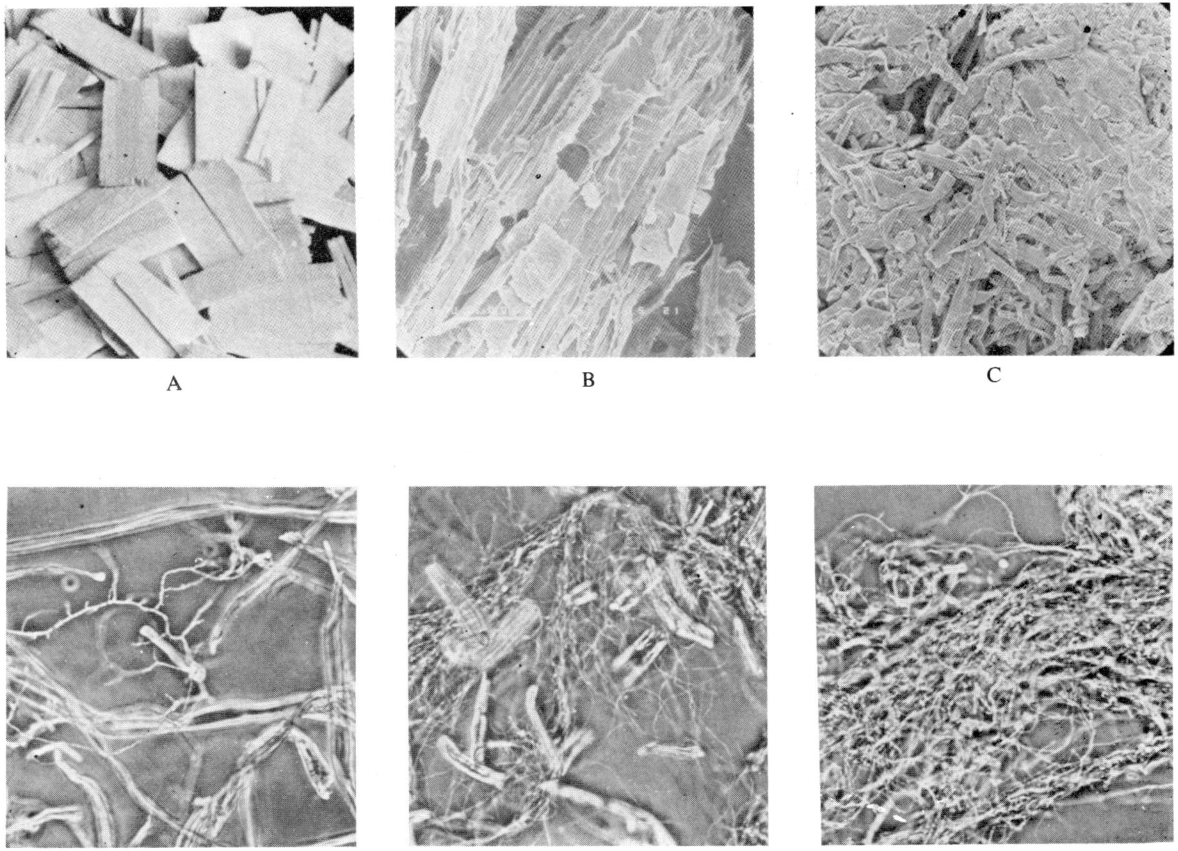

A B C D E F

Plate 1.

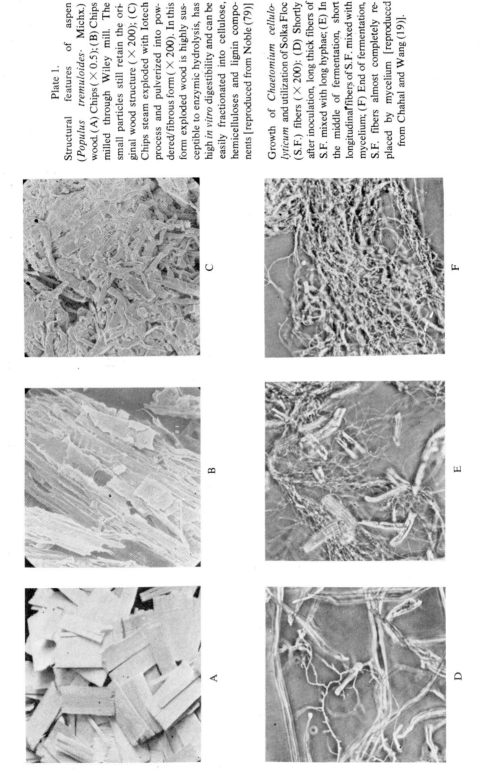

Structural features of aspen (*Populus tremuloides* Michx.) wood. (A) Chips (× 0.5); (B) Chips milled through Wiley mill. The small particles still retain the original wood structure (× 200); (C) Chips steam exploded with Iotech process and pulverized into powdered/fibrous form (× 200). In this form exploded wood is highly susceptible to enzymic hydrolysis, has high *in vitro* digestibility and can be easily fractionated into cellulose, hemicelluloses and lignin components [reproduced from Noble (79)]

Growth of *Chaetomium cellulolyticum* and utilization of Solka Floc (S.F.) fibers (× 200): (D) Shortly after inoculation, long thick fibers of S.F. mixed with long hyphae; (E) In the middle of fermentation, short longitudinal fibers of S.F. mixed with mycelium; (F) End of fermentation, S.F. fibers almost completely replaced by mycelium [reproduced from Chahal and Wang (19)].

Table 3. Comparison of SCP production on various lignocellulosics with various cellulolytic microorganisms (crop residues)

Microorganism[a]	Substrate type	Substrate concn. (w/v%)	Max. crude protein (% D.W.)	Max. crude protein (g/l)	Culture time (h)	Crude protein productivity (mg/l h)	Reference
1	2	3	4	5	6	7	8
Sporotrichum pulverulentum	Pure powdered cellulose	1	4.3	0.34	168	2.0[b]	Eriksson & Larsson (30)
Tricohoderma viride (QM 9123)*	NaOH treated barley straw	1	23.8	1.84	48	38.3[b]	Peitersen (82)
Cellulomonas sp. + *Alcaligenes faecalis*	4% NaOH, 100°C, 15 min. rice straw	1–5	—	1.76	120	14.3[b]	Han and Callihan (44)
Cellulomonas sp. + *Alcaligenes faecalis*	3% NaOH, (260 psig, 5 min) sugar cane bagasse	—	—	7.2	—	—	Han and Callihan (44)
*Aspergillus terreus** (ATCC 20514)	4% NaOH, 121°C for 15 minutes Solka-Floc	1	32.9	1.73	30	57.6[b]	Miller and Srinivasan (69)
C. cellulolyticum	Solka-Floc	1	40.3	2.7	36	75.0	Moo-Young *et al.* (74)
*C. cellulolyticum**	Solka-Floc	1	45	1.8	24	80.0[b]	Chahal and Wang (19)
C. cellulolyticum	untreated wheat straw	1	4.6	0.43	48	8.9	Chahal *et al.* (15)
C. cellulolyticum	1% NaOH treated wheat straw	1	40.6	2.51	36	69.7	Chahal *et al.* (15)
*C. cellulolyticum**[c]	1% NaOH, 100°C, 1 h, corn stover	1	35.7	1.8	12	146.0	Chahal, Moo-Young and Vlach (unpubl.)

(Contd.)

1	2	3	4	5	6	7	8
*C. cellulolyticum**[c]	1% NaOH, 100°C, cattle manure	1	37.0	1.52	11	108.2	Moo-Young and Chahal (73)
*C. cellulolyticum**[c]	Anaerobic liquor + corn stover treated with 1% NaOH, 100°C, 1 h	1.5	22.0	3.62	14	178.0	Moo-Young and Chahal (73)
*C. cellulolyticum**[c]	1% NaOH, 100°C, 1 h, wheat straw + swine manure	1	27.8	1.52	24	63.3	Chahal *et al.* (18)
*C. cellulolyticum**[c]	Anaerobic liquor + wheat straw treated with 1% NaOH, 100°C, 1 h	1.5	25.0	4.21	21	200.4	Chahal *et al.* (18)

[a]Data marked with asterisk from stirred tank fermentor; all other data on shake-flasks.
[b]Calculated from the given data.
[c]Pretreatment liquor retained in the fermentation medium.

Table 4. Comparison of SCP production on various lignocellulosics with various cellulolytic microorganisms (wood and wood residues)

Microorganism[a]	Substrate type	Substrate concn. (w/v%)	Max. crude protein (%D.W.)	Max. crude protein (g/l)	Culture time (h)	Crude protein productivity (mg/l h)	Reference
Rhizoctonia solani	Wood pulp and glucose	1.5 0.1	23.2	2.3	96	23.5[b]	Chahal *et al.* (14)
Myrothecium verrucaria*	Ball-milled newspaper	4	9.8	3.3	144	22.9[b]	Updegraff (104)
*Thermomonospora fusca**	Cellulose fibre	0.5	30.0	0.40	96	4.1[b]	Crawford *et al.* (23)
Brevibacterium sp.*	Mesquite wood	1	—	0.53	72	7.3[b]	Fu and Thayer (33)

Phanerochaete chrysosporium	1% NaOH and dilute HCl, maple bark	1	12–16	1.43	—	46.1	Daugulis and Bone (24)
*C. cellulolyticum**	1% NaOH, 100°C, 1 h, saw dust	1	40	1.8	36	50	Moo-Young *et al.* (74)
*C. cellulolyticum**[c]	Steamed at 100°C, 1 h, aspen	1	3.6	0.32	48	6.6	Chahal *et al.* (17)
*C. cellulolyticum**[c]	1% NaOH, 121°C, 30 minutes, aspen	1	16.2	1.1	17	64	Chahal *et al.* (17)
*C. cellulolyticum**[c]	Strong acid hydrolysis, high pressure, oak [Jelks (54)]	1	21	1.4	32	43	Moo-Young *et al.* (75)
*C. cellulolyticum**[c]	Steamed at 280 psig 4 minutes, aspen (Stake)[1]	1	21.4	1.5	20	75	Chahal *et al.* (17)
C. cellulolyticum	Iotech processed aspen	2	23.5–27.7	3.37	12	177.4	Chahal, McGuire, Pikor and Noble (unpublished)
*C. cellulolyticum**	$NaClO_2$, aspen	1	37.9	2.1	17	126	Chahal *et al.* (17)

[a]Data marked with asterisk from stirred tank fermentor; all other data on shake-flasks.
[b]Calculated from the given data.
[c]Pretreatment liquor retained in the fermentation medium.
[1]Stake Technology Ltd., Ottawa, Canada. Material was in the form of thick fibres. It was dried and ground to 1 mm size before preparation of medium [Bender (4)].

stover into protein-rich animal feed. Similarly, Chahal *et al.* (18) have found successful bioconversion of mixtures of swine manure and wheat straw and anaerobic liquor of swine manure and wheat straw. The final product contained 25–28 per cent protein (Table 3) without any obnoxious odour of swine manure.

2) *Wood and wood residues*

One of the early reports concerning utilization of wood by various fungi was by Chahal and Gray (9). They reported that only 12 out of 44 fungi tried could be grown on wood pulp. *Trichoderma* sp. and *Rhizoctonia* sp. were the best. Chahal *et al.* (14) later reported achieving approximately 23 per cent crude protein in the final product by growing a new isolate of *Rhizoctonia solani*, a soil fungus, on wood pulp. Updegraff (104) obtained approximately 10 per cent crude protein by growing *Myrothecium verrucaria* on ball-milled newspaper for six days. Crawford *et al.* (23) reported a 30 per cent crude protein product from a 0.5 per cent cellulose fibre slurry by growing *Thermomonospora fusca*, an actinomycete, in four days. Fu and Thayer (33) fermented mesquite wood with *Brevibacterium* sp. with limited success (Table 4).

There are a few reports about the degradation of tree bark (59, 80). Updegraff and Grant (105) tried to grow pure cultures of 200 fungi, bacteria, and yeasts on bark. Only six fungi were able to grow and these produced too low a protein product to achieve an economical process. Recently, Daugulis and Bone (24) used *Phanerochaete chrysosporium*, a white-rot fungus, for fermentation of alkali-treated (washed) maple bark and reported 12–16 per cent crude protein in the final product.

Considering its relatively good performance in fungal protein production from crop residues, *C. cellulolyticum* was tried on wood and wood residues. The first attempt with alkali-treated sawdust (washed) (74) produced a final product containing 40 per cent DW crude protein and *C. cellulolyticum* gave a higher protein productivity (50 mg/l/h) than other cellulolytic organisms (Table 4).

As *C. cellulolyticum* can simultaneously utilize hemicelluloses and cellulose, it was grown on wood containing pretreatment liquor in the medium to maximize the SCP production per unit of substrate weight. The data presented by Chahal *et al.* (17) (Table 4) on fermentation of aspen wood with *C. cellulolyticum*, indicates that very little protein was produced when wood was steamed at 100°C for one hour before fermentation. Due to its high lignin content and high degree of crystallinity, wood requires strong pretreatments so that its polysaccharides are easily accessible to hydrolytic enzymes. Alkali-pretreatment at 121°C for 30 minutes improved the performance of the organism and a final product containing 16 per cent protein was obtained. The protein content increased to approximately 21 per cent when wood was pretreated with steam (4) or the strong acid process (54) under high-pressure steam.

The preliminary study of fungal protein production using Iotech-processed wood indicated that *C. cellulolyticum* produced the same protein content (21 per cent) in just 15 hours (16). The growth almost stopped after 14 hours because the pH of the medium dropped as low as 3.9. When Iotech-processed wood was fermented under controlled pH conditions, a product of high protein content (23.5–27.7 per cent) with a protein productivity of 177.4 mg/1/h was obtained (16).

Food, feed and fuel production from steam-exploded lignocelluloses

An integrated plan for production of food, feed, and fuel from lignocelluloses is shown in Fig. 1. In this plan the lignocelluloses (wood and crop residues) are steam exploded under high pressure.

The steam-exploded lignocelluloses have a very high rumen digestibility (82 per cent) and can be fed as such to animals as an energy source (79). In cases where a complete animal feed rich in protein is required, the exploded wood can be fermented by *C. cellulolyticum.* The final product contains 20–27 per cent protein (16) which will supply the energy and protein requirements of the animal. An animal feed with various protein contents (10–15 per cent) can be prepared by mixing products in such ratios that a final mixture with a desired quantity of protein is obtained.

The steam-exploded lignocelluloses are easily fractionated into water-soluble hemicelluloses and a water-insoluble mixture of cellulose and lignin. Water-soluble hemicelluloses are fermented with *C. cellulolyticum* into high-quality fungal food containing 45 per cent protein (16) or they can be converted into fuel ethanol through pentose fermentation with a special yeast, *Pachysolen tannophilus* (90).

The lignin can be used to synthesize adhesives and other chemicals (79). The residual cellulose is hydrolyzed into glucose with cellulases produced by *Trichoderma reesei* on lignocelluloses (17). The glucose thus produced is fermented to fuel ethanol or used to synthesize various pharmaceuticals and other chemicals. During the enzyme production the residual fungal biomass is also used as animal feed.

In another phase of this whole process, the steam-exploded lignocelluloses are hydrolyzed with cellulases which yield a mixture of hexoses, pentoses and lignin. Lignin is separated by precipitation and filtration. The fractionated lignin is used to make adhesives and other chemicals. The mixture of soluble hexoses and pentoses is fermented with a common yeast (*Saccharomyces cerevisiae*) into ethanol. The unfermented pentoses are converted by *C. cellulolyticum* into a high-quality fungal food containing 45 per cent protein (16), or into fuel ethanol with *P. tannophilus*, or, evaporated to molasses to supplement the animal feed.

In this integrated plan, every fraction of the lignocelluloses is utilized and

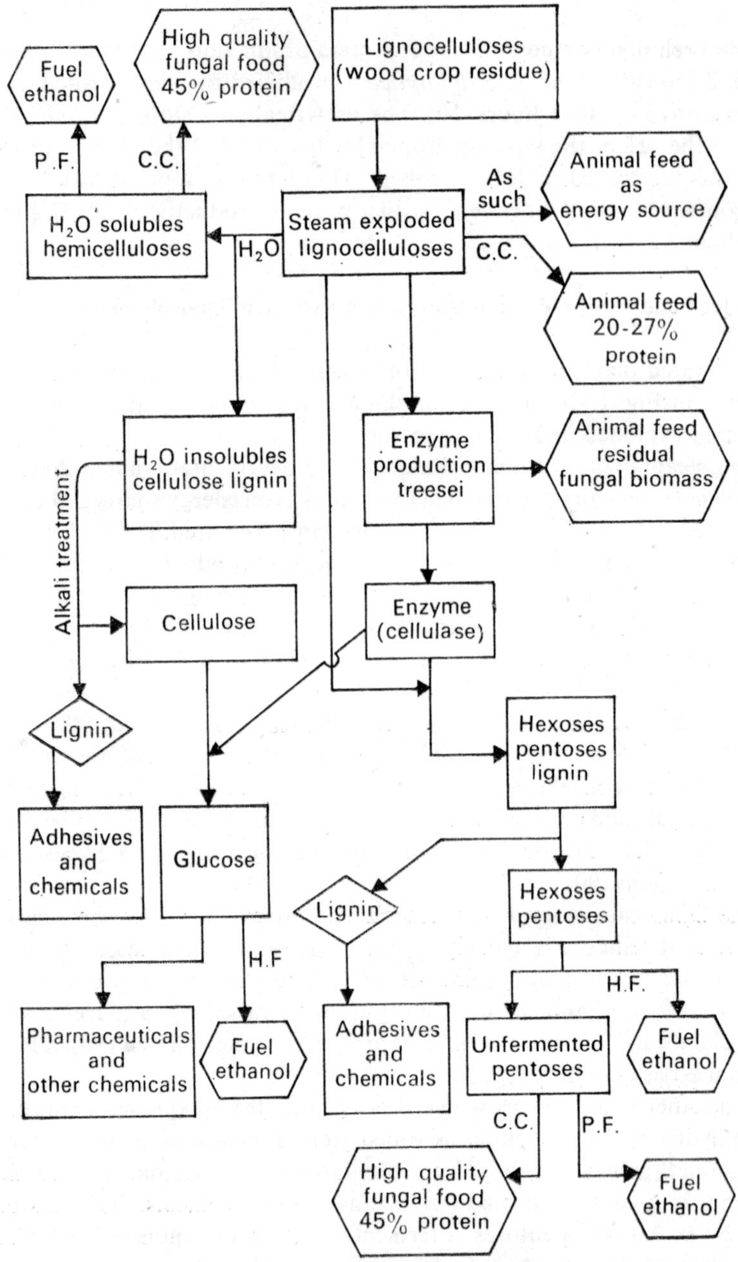

Fig. 1. An integrated plan for production of food, feed, and fuel from steam-exploded lignocelluloses.
C.C.—*C. cellulolyticum*; P.F.—Pentose fermentation;
H.F.—Hexose fermentation; H_2O—Hydrolysis.

no harmful chemicals or residues with high BOD are generated. Thus, no pollution problems of any kind are present in this process.

Solid-state fermentation of lignocelluloses

Solid-state fermentation, unlike slurry-state, requires no complex fermentation controls (50). This method is now being used for aflatoxin production (49, 92); for enzyme production (93, 102) and for upgrading the values of existing foods—especially oriental foods (49). More recently, researchers have turned their attentions to the bioconversion of lignocelluloses into protein-rich animal feed, using solid-state fermentation, because of its various advantages (low technology, low cost of dewatering of the final product, etc.) over slurry-state fermentation.

Detroy *et al.* (25) reported that fermentation of wheat straw with *Pleurotus ostreatus* in solid-state for 50 days, modified the substrate for enzymatic hydrolysis, which resulted in 72 per cent conversion of the residual cellulose component into glucose. They have attributed this to the utilization of 32 per cent lignin and modification of residual lignin by the organism. During the fermentation, however, there was almost a total loss of hemicelluloses and a 40 per cent loss of the cellulose from the original straw fermented. Zadrazil (111) reported that during solid-state fermentation of wheat straw for 120 days with *Stropharia rugosannulata* or *Pleurotus cornucopiae*, *in vitro* digestibility increased up to 60–70 per cent.

In both these processes the increase in digestibility—or increase in susceptibility to hydrolytic enzymes—is attained after a long time (40–120 days) of fermentation and loss of a considerable portion of the carbohydrates. Large-scale solid-state fermentation by these processes to improve the feed values of straw may not be economical because of these two factors.

Barnes *et al.* (3) obtained 6.5 per cent DW crude protein in six days by growing *Sporotrichum thermophile* on ground newsprint in solid-state fermentation. Pamment *et al.* (81) obtained only 8 per cent DW crude protein in the final product when alkali pretreated maple (*Acer saccharum*) sawdust was fermented in solid-state fermentation with *C. cellulolyticum* for nine days. The crude-protein content rose to 11 per cent DW when fermentation was continued up to 20 days.

Han *et al.* (45), based on their previous findings (43, 110), developed the following two processes for large-scale production of animal feed through solid-state fermentation of straw:

1) *Process 1*

Figure 2 shows a schematic diagram of the first process. The scheme can be applied as a continuous or batch process. The straw is first chopped to 1/4 inch to 1 inch lengths using a hammer-mill, knife grinder or attrition mill and then conveyed to a pressure cooker. During transportation, three parts of 0.5N H_2SO_4 solution are sprayed on one part of the straw. The addition of a minimal amount of liquid is essential to this process for reducing the cost of

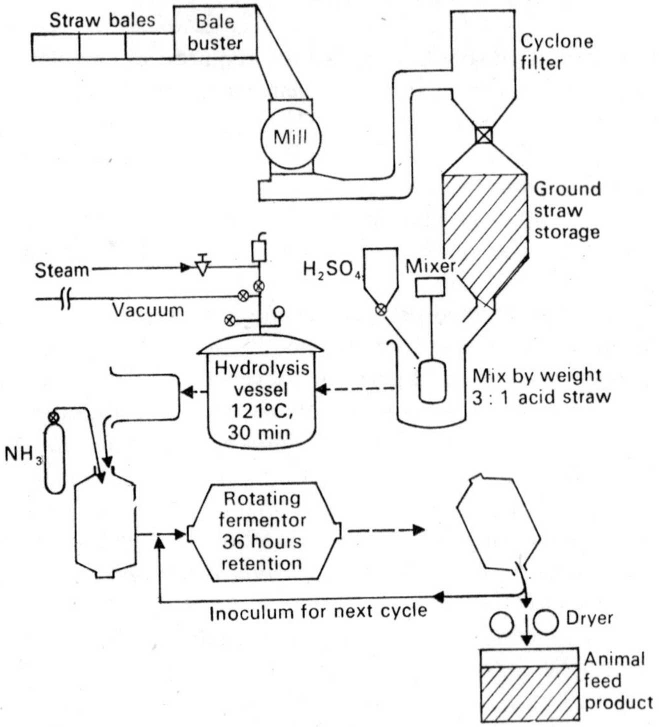

Fig. 2. Schematic diagram of solid-state fermentation [reproduced from Han *et al.* (45)].

operation and maintaining the product yield. The straw is hydrolyzed in the pressure cooker under 15 lb steam for 30 minutes. The hydrolyzed straw is treated with ammonia or ammonium hydroxide to raise the pH to 4.5, and is then conveyed to a fermentation chamber. Acid treatment and subsequent neutralization with ammonia produces a straw containing 20 per cent fermentable sugars and 2.3 per cent nitrogen. This gives optimal conditions for yeast fermentation. The pretreated straw is then inoculated with a suitable organism, such as *Aureobasidium pullulans*, *Candida utilis*, or *Trichoderma viride*. A liquid culture or a small portion of recycled fermented straw can be used as an inoculum.

The fermentation chamber should be capable of maintaining constant temperature and humidity and should be sterilizable with steam or gases, although sterilization should not be necessary in most cases. The fermentor must provide a constant tumbling motion to the straw or permit the free exchange of air. For continuous fermentation, the chamber should be large enough to provide up to 36 hours of residence time for the substrate straw. For a batch process, the chamber size is not critical. At the end of fermentation, the straw can be dried with hot air. Owing to the low pH, the limited level of nutrients and the large inoculum size, this process can be operated without strict adherence to aseptic conditions.

2) *Process 2*

The second process involves treatment of the straw with sodium hydroxide (4 per cent dry weight) and growth of cellulolytic organisms on a semi-solid substrate. The alkali-treated straw is neutralized and a nitrogen source such as ammonium sulphate is added. The cellulolytic organisms (mixed culture of *Cellulomonas* sp. and *Alcaligenes faecalis*) are areobic, so the fermentation requirements are similar to those for yeast cultivation on acid-hydrolyzed straw. The aseptic technique is unnecessary because only few organisms can grow on cellulose.

Crude protein contents of 14, 12.4, and 10.9 per cent DW were reported when pretreated (NH_3-neutralized) ryegrass straw was fermented for two to three days with *A. pullulans*, *C. utilis* and *T. viride* respectively. In the other process crude protein content up to 6.8 per cent was noticed when alkali-pretreated straw was fermented with a mixed culture of *Cellulomonas* sp. and *A. faecalis* for two to three days. The acid-hydrolyzed, yeast-fermented (*A. pullulans* or *C. utilis*) straw had *in vitro* rumen digestibility of 46.7 per cent whereas it was 55.9 per cent for alkali-treated, bacteria-fermented (*Cellulomonas* sp. and *A. faecalis*) straw.

Solid-state fungal protein production with *Chaetomium cellulolyticum*

The literature surveyed indicates that only limited success in microbial protein production has been achieved with solid-state systems. Thus, *C. cellulolyticum* was tried on wheat straw and corn stover in solid-state fermentation by Chahal *et al.* (16), to upgrade their protein values. *Chaetomium cellulolyticum* gave a solid product containing 9.8, 10.7, and 19.7 per cent DW crude protein on untreated, cold-NH_3-pretreated, and washed-alkali-pretreated chopped wheat straw respectively within 10 to 12 days culture time. In another experiment, about 19.2 per cent DW crude protein was obtained from alkali-pretreated straw (unwashed) containing all the solubilized hemicelluloses and lignin with five days of fermentation when mineral salts of Czapek medium (19) were replaced with those of Chahal and Gray (9) medium.

Corn stover proved to be better than wheat straw as a substrate for fungal protein production in solid-state fermentation. About 17.5 per cent DW of crude protein was found in the final product within three to four days culture time from cold- or hot-NH_3-treated stover without additional nitrogen or other mineral salts. The highest percentage of protein (20–24 per cent of DW) was found when alkali-pretreated stover containing solubilized hemicelluloses and lignin was fermented with 0.118 g $(NH_4)_2SO_4$/g substrate for five days. In this case, the success in solid-state fermentation may be attributed to the fact that *C. cellulolyticum* has better penetrating power into the substrate as compared to other organisms, especially single-celled organisms as explained earlier. As well, *C. cellulolyticum* has a greater ability to utilize hemicellulose and cellulose concurrently (16). *Chaetomium cellulolyticum* again seems to be better than all the organisms for solid-state fermentation (Table 5).

Table 5. Comparison of SCP production on various lignocellulosics with various cellulolytic microorganisms in solid-state fermentation

Reference/Organism	Substrate (treatment)	Crude protein in product (% DW)	Incubation time (day)
1. Barnes *et al.* (3) *Sporotrichum thermophile*	Newsprint (untreated)	6.5	6
2. Han and Anderson (43) (a) *Aurobasidium pullulans*	Ryegrass (0.5 N H_2SO_4, 121°C, 30 min.)	14.0	2–3
(b) *Candida utilis*	,,	12.4	2–3
(c) *Trichoderma viride*	,,	10.9	2–3
3. Yu *et al.* (110) (a) *Cellulomonas* sp. + *Alacaligenes faecalis*	Ryegrass (4% NaOH, app. 25°C)	6.8	2–3
(b) —do—	Ryegrass (NH_3)	9.5	2–3
4. Pamment *et al.* (81) *Chaetomium cellulolyticum*	Saw-dust (20% NaOH washed)	7.7 11.2	9 20
5. Chahal *et al.* (16) (a) *Chaetomium cellulolyticum*	Wheat straw (4% NaOH)	19.0	5
(b) *Chaetomium cellulolyticum*	Corn stover (4% NaOH)	20–24	4–5
(c) *Chaetomium cellulolyticum*	Corn stover (3% NH_3)	17	~4

Nutritive values of microbial food and feed

The nutritive value of microbial food/feed of a given species of microorganism varies with the substrate used and the environment in which it is grown. It has been shown that the growth rate decreases with the increase in glucose units of the substrate (1), i.e., from 0.28 h^{-1} to 0.22 h^{-1} on maltose (two glucose units) and further reduced to 0.18 h^{-1} on maltotriose (three glucose units).

The microbial food produced on pure solubilized carbohydrates would contain more protein than that produced on agricultural waste materials. The low protein content in biomass produced from agricultural wastes is mainly due to the presence of unutilized cellulose and lignin in the final product. The microbial biomass produced on such substrates would be good for animal feed.

The protein and nucleic acid content varies with the species of microorganism as well as with the growth rate. In general, bacteria and yeasts have more protein (55–65 per cent) and also more nucleic acids (10–20 per cent). On the other hand, some of the filamentous fungi have high protein contents (65 per cent in *Neurospora sitophila*; 50–55 per cent in *Fusarium graminaerum*) and low nucleic acids (1). The most extensively used microbial protein for food and feed purposes is from *C. utilis*. Results of various feeding trials have shown that *C. utilis* cells are not nutritionally equivalent to animal protein. Net protein utilization (NPU) of 38 per cent has been reported by Miller (68) and only 23.6 per cent by Mitsuda *et al.* (72). The low NPU of *C. utilis* is due to its hard cell wall which reduces the digestibility. Mitsuda *et al.* (72) reported that the digestibility could be increased to 95 per cent and the NPU to 57.7 per cent, by isolating the protein from the cells and supplementing it with methionine. On the other hand, the NPU of some filamentous fungi is as high as 64.9 per cent for *Boletus edulis* (84); and 60 per cent for *Fusarium semitectum* (71) without any pretreatment of the final product. The NPU can be increased to 76 per cent by supplementing with methionine (108).

Table 6. Essential amino acid composition[a] of protein in *C. cellulolyticum*, other cellulolytic organisms of SCP interest, alfalfa, soya, and FAO reference protein

Amino acid	*C. cellulolyticum*[1]	*Fusarium graminaerum*[2]	*T. viride*[3]	*Cellulomonas*[4]	Alfalfa[5]	Soybean[6]	FAO reference
Threonine	6.14	5.1	4.9	4.7	5.12	4.0	2.8
Valine	5.76	7.2	4.4	6.79	6.70	5.0	4.2
Cystine	0.31	0.87	1.45(?)	0.41	1.40	1.4	2.0
Methionine	2.33	2.17	1.35	1.69	1.96	1.4	2.2
Isoleucine	4.70	4.3	3.5	4.12	5.54	5.4	4.2
Leucine	7.54	6.2	5.8	8.66	8.43	7.7	4.8
Tyrosine	3.26	4.1	3.3	2.41	3.72	2.7	2.8
Phenylalanine	3.77	4.4	3.7	3.69	5.75	5.1	2.8
Lysine	6.80	7.5	4.4	8.0	6.70	6.5	4.2
Tryptophan	NA	NA	NA	NA	NA	1.5	1.4

[a]In per cent total true protein, NA means no answer.
[1]Moo-Young *et al.* (74).
[2]Anderson *et al.* (1).
[3]Peitersen (82).
[4]Han and Anderson (43); Han and Callihan (44).
[5]Livingston *et al.* (63).
[6]Shacklady (89).
[7]Duthie *et al.* (28).

The *in vitro* N-digestibility of the pure fungal biomass of *C. cellulolyticum* was 73 per cent (98).

The amino acids composition of various microorganisms used to produce food/feed from agricultural wastes and other carbohydrates is compared with that of alfalfa—a common animal fodder, soybean—a common protein source, and the FAO reference and shown in Table 6. The comparison indicates that the amino acid composition of all these organisms meets the FAO reference requirement except for sulphur-containing amino acids. Anderson *et al.* (1) have also reported similar findings by comparing the amino acid composition of various filamentous fungi grown on soluble sugars or starch. The deficiency of sulphur amino acids can be easily met by supplementing the microbial protein with methionine.

Feeding trials with various cellulolytic organisms (including *C. cellulolyticum*) used to produce microbial food/feed are still in their infancy. However, most of the feeding trials have indicated that up to 20–40 per cent of total protein requirement can be replaced with the protein from these microorganisms without any pathological problems (43, 82, 98). Rank, Hovis, McDougall Research Ltd. (RHM) of the United Kingdom has gone ahead and developed a process to produce fungal food for human consumption following 17 years of research. In this process, the fungus *Fusarium graminaerum* A3/5 is grown on starch or glucose. Extensive feeding trials have indicated that this fungal food is quite suitable for human consumption (28). The composition of essential amino acids is quite favourable except for the deficiency of sulphur amino acids which can be corrected by supplementing with methionine. Recently, the British Government has approved this fungal food for human consumption (*Globe* and *Mail*, Toronto, June 3, 1981). *Chaetomium cellulolyticum* is very similar to *F. graminaerum* in respect to amino-acid composition of protein. Unlike *F. graminaerum*, *C. cellulolyticum* can be grown successfully on cheap carbohydrates obtained from lignocellulosic materials and also from manures (15, 16–18, 73). Although preliminary feeding trials on rats have not shown any toxicity or pathological symptoms, extensive feeding trials are to be arranged before it can be recommended for human or animal consumption.

Acknowledgements

The author wishes to thank Geoffrey Noble for his sincere encouragement, Heather Pikor for her critical reading of the manuscript and Dianne Steele for typing.

References

1. Anderson, C., Longton, J., Maddix, C., Scammell, G.W., Solomons, G.W. and Solomons, G.L. The growth of microfungi on carbohydrates, pp. 314–329 in *Single-cell Protein II* (Edited by S.R. Tannenbaum and D.I.C. Wang), MIT Press, Cambridge, Mass. (1975).
2. Bailey, K.V. Rural nutrition studies in Indonesia III, Epidemiology of hunger edema in the cassava area, *Tropical Geographical Medicine*, *13*, 289 (1961).
3. Barnes, T.G., Eggins, H.O.W. and Smith, E.L. Preliminary stages in the development of a process for the microbial upgrading of waste paper, *International Biodeterioration Bulletin*, *8* (3): 112–116, (1972).
4. Bender, F. Method of treating lignocellulose materials to produce ruminant feed, U.S. patent No. 4,136,207 (1979).
5. Bender, F., Heaney, D.P. and Bowden, A. Potential of steamed wood as a feed for ruminants, *Forest Product Journal*, *20* (4): 36–41 (1970).
6. Bunker, H.J. Sources of single-cell protein: perspective and prospect, p. 64 in *Single-cell Protein* (Edited by R.I. Mateles and S.R. Tannenbaum), MIT Press, Cambridge, Mass. (1968).
7. Campbell, C.M., Wayman, O., Stanley, R.W., Kamstra, L.D., Olbrich, S.E., Hoa, E.B., Nakayama, T., Kohler, G.O., Walker, H.G. and Graham, R. Effects of pressure treatment of sugar cane bagasse upon nutrient utilization, *Proceedings Western Section American Society of Animal Science*, *24*, 173–184 (1973).
8. Casebier, R.L. Hamilton, J.K. and Hergert, H.L. Chemistry and mechanism of water prehydrolysis on southern pine wood, *Tappi*, *52* (12), 2369–2377 (1969).
9. Chahal, D.S. and Gray, W.D. Growth of selected cellulolytic fungi on wood pulp, pp. 584–593 in *Biodeterioration of Materials—Microbiological and Allied Aspects* (Edited by A.H. Walter and J.S. Elphic), Elsevier Publishing Co., Barking, Essex, England (1968).
10. Chahal, D.S. and Hawksworth, D.L. *Chaetomium cellulolyticum*, a new thermotolerant and cellulolytic *Chaetomium*, *Mycologia*, *68*, 600–610 (1976).
11. Chahal, D.S., McGuire, S., Pikor, H. and Noble, G. Bioconversion of cellulose and hemicelluloses of wood into fungal food and feed with *Chaetomium cellulolyticum*, *Biomass* (1981a—in press).
12. Chahal, D.S., McGuire, S., Pikor, H. and Noble, G. Production of cellulase-complex by *Trichoderma reesei* Rut-C30 on lignocellulose and its hydrolytic potential, *Biomass* (1981b—in press).
13. Chahal, D.S., Moo-Young, M. and Dhillon, G.S. Bioconversion of wheat straw and wheat straw components into single-cell protein, *Canadian Journal of Microbiology*, *25*, 793–797 (1979).

14. Chahal, D.S., Munshi, G.D. and Cheema, S.P.S. Fungal protein production by *Rhizoctonia solani* when grown on cellulose as the sole source of organic carbon, *Proceedings of the National Academy of Sciences, India, Section B, 39*, 287–291 (1969).
15. Chahal, D.S., Swan, J.E. and Moo-Young, M. Protein and cellulose production by *Chaetomium cellulolyticum* grown on wheat straw, *Developments in Industrial Microbiology, 18*, 433–442 (1977).
16. Chahal, D.S., Vlach, D., and Moo-Young, M. Upgrading the protein feed value of lignocellulosic materials using *Chaetomium cellulolyticum* in solid-state fermentation. Presented at the 30th Annual Meeting of the Canadian Society of Microbiologists, Guelph, Ontario, June 15–19 (1980a).
17. Chahal, D.S., Vlach, D. and Moo-Young, M. Effect of physical and physiochemical pretreatments of wood for SCP production with *Chaetomium cellulolyticum*. Presented at VI International Fermentation Symposium, London, Ontario, Canada, July 20–25 (1980b).
18. Chahal, D.S., Vlach, D., Stickeny, B. and Moo-Young, M. Pollution control of swine manure and straw by conversion to *Chaetomium* SCP stuff. Presented at VI International Fermentation Symposium, London, Ontario, Canada, July 20–25 (1980c).
19. Chahal, D.S. and Wang, D.I.C. *Chaetomium cellulolyticum*, growth behavior on cellulose and protein production, *Mycologia, 70*, 160–170 (1978).
20. Charalambous, J. and Papaconstantinou, J. *The Composition and Uses of Carob Bean* (Edited by J. Charalambous), Cyprus Agricultural Research Institute, Nicosia, Cyprus (1966).
21. Cochrane, V.W. *Physiology of Fungi*, John Wiley & Sons, Inc., New York (1958).
22. Cooke, W.B. An ecological life history of *Aureobasidium pullulans* (de Bary) Arnaud, *Mycopathologia et Mycologia Applicata, 12*, 1–45 (1959).
23. Crawford, D.L., McCoy, E., Harkin, J.M. and Jones, P. Production of microbial protein from waste cellulose by *Thermomonospora fusca*, a thermophilic actinomycete, *Biotechnology & Bioengineering, 15*, 833–843 (1973).
24. Daugulis, A.J. and Bone, D.H. Production of microbial protein from tree bark by *Phanerochaete chrysosporium*, *Biotechnology & Bioengineering, 20*, 1639–1649 (1978).
25. Detroy, R.W., Lindenfelser, L.A., St. Julian Jr., G. and Orton, W.L. Saccharification of wheat-straw cellulose by enzymatic hydrolysis following fermentative and chemical pretreatment, *Biotechnology & Bioengineering Symposium, 10*, 135–148 (1980).
26. Dunlap, C.E. Production of single-cell protein from insoluble agricultural wastes by mesophiles, pp. 244–262 in *Single-cell Protein II* (Edi-

ted by S.R. Tannenbaum and D.I.C. Wang), MIT Press, Cambridge, Mass. (1975).
27. Dunlap, C.E., Callihan, C.D. Single-cell protein from waste cellulose, Final report on grant EP00328–4 to the Federal Solid Waste Management Program, U.S. Environmental Protection Agency (1973).
28. Duthie, I.F. Animal feeding trials with a microfungal protein, pp. 505–554 in *Single-cell Protein II* (Edited by S.R. Tannenbaum and D.I.C. Wang), MIT Press, Cambridge, Mass. (1975).
29. Eriksson, K.E. Enzyme mechanisms involved in fungal degradation of wood components, pp. 195–201 in *Proceedings of Bioconversion Symposium*, Indian Institute of Technology, New Delhi (1977).
30. Eriksson, K.E. and Larsson, K. Fermentation of waste mechanical fibers from a newsprint mill by the rot fungus *Sporotrichum pulverulentum*, *Biotechnology & Bioengineering*, *17*, 327–348 (1975).
31. Evans, G.H. Industrial production of single-cell protein from hydrocarbons, pp. 243–254 in *Single-cell Protein* (Edited by R.I. Mateles and S.R. Tannenbaum), MIT Press, Cambridge, Mass. (1968).
32. Flickinger, M.C. Current biological research in conversion of cellulosic carbohydrates into liquid fuel: how far have we come? *Biotechnology & Bioengineering*, *22*, Supplement 1, 27–48 (1980).
33. Fu, T.T. and Thayer, D.W. Comparison of batch and semi-continuous cultures for production of protein from mesquite wood by *Brevibacterium* sp. J.M. 98A, *Biotechnology & Bioengineering*, *17*, 1749–1760 (1975).
34. Gilbert, N., Hobbs, I.A. and Levine, J.D. Hydrolysis of wood, *Industrial Engineering Chemistry*, *44*, 1712–1720 (1952).
35. Goering, H.K. and van Soet, P.J. *In vitro* digestibility of lignified materials ensiled with sodium chlorite, *Journal of Dairy Science*, *51*, 974 (1968).
36. Goering, H.K., Smith, L.W., van Soet, P.J. and Gordon, C.H. Digestibility of roughage materials ensiled with sodium chlorite, *Journal of Dairy Science*, *56*, 233 (1973).
37. Gong, C.S., Chen, I.F., Flickinger, M.C., Chang, L.C. and Tsao, G.T. Production of ethanol from D-xylose by using D-xylose isomerase and yeasts, *Applied Environmental Microbiology*, *41*, 430–436 (1981).
38. Gradova, N.B., Kruchkova, A.P., Redionova, G.S., Mikkhaylova, V.V. and Dikanskaya, E.M. Microbiological and physiological principles in the biosynthesis of protein and fatty substances from hydrocarbons, pp. 99–104 in *Proceedings of 2nd International Congress on Global Impacts of Microbiology* (Edited by E.L. Gaden Jr.), Addis Ababa, Interscience Publ., New York (1969).
39. Gray, W.D. Microbial protein for the space age, *Developments in Industrial Microbiology*, *3*, 63–71 (1962).

40. Gregory, K.F., Reade, A.E., Khor, G.L., Alexander, J.C., Lumsden, J.H. and Losos, G. Conversion of carbohydrates to protein by high temperature fungi, *Food Technology*, 30–35, March (1976).
41. Green, J.W. Wood cellulose, *Methods Carbohydrate Chemistry*, *3*, 9–20 (1963).
42. Hamer, G. Technical aspects of single-cell protein production from natural gas (methane), *Microbial Bioconversion Systems for Food and Fodder Production and Waste Management* (Edited by F.G. Overmire), KISR (1977).
43. Han, Y.W. and Anderson, A.W. Semi-solid fermentation of rye grass straw, *Applied Microbiology*, *30*, 930–934 (1975).
44. Han, Y.W., and Callihan, C.D. Cellulose fermentation: Effect of substrate treatment on microbial growth, *Applied Microbiology*, *27*, 159–165 (1975).
45. Han, Y.W., Cheeke, P.R., Anderson, A.W. and Lekprayoon, C. Growth of *Aurobasidium pullulans* on straw hydrolysate, *Applied Environmental Microbiology*, *32*, 799–802 (1976).
46. Harris, E.E. and Belinger, E. Madison wood sugar process, *Industrial Engineering Chemistry*, *38*, 890–895 (1946).
47. Harris, E.E., Hajny, G.J., Hannan, M., Rogers, S.C. Fermentation of douglas fir hydrolysate by *S. cerevisiae*, *Industrial Engineering Chemistry* *38*, 896–904 (1946).
48. Heaney, D.P. and Bender, F. The feeding value of steamed aspen for sheep, *Forest Product Journal*, *20* (9), 98–102 (1970).
49. Hesseltine, C.W. A millennium of fungi, food, and fermentation, *Mycologia*, *57*, 149–197 (1965).
50. Hesseltine, C.W. Solid-state fermentations, *Biotechnology and Bioengineering*, *14*, 517–532 (1972).
51. Higuchi, T. Fermentation and biological degradation of lignin, *Advanced Enzymology*, *34*, 207–277 (1971).
52. Hosaka, H., Suzuki, H., Nunomura, A., Uesugi, T., Takahashi, H., Hongo, M. and Hasegawa, I. *Report of the Hokkaido Forest Products Research Institute, No. 15*, 32 (1959).
53. Imrie, F.K.E. and Vlitos, A.J. Production of fungal protein carob (*Ceratonia siliqua* L.), pp. 223–243 in *Single-cell Protein II* (Edited by S.R. Tannenbaum and D.I.C. Wang), MIT Press, Cambridge, Mass. (1975).
54. Jelks, J.W. Process for oxidizing and hydrolyzing plant organic matter particles to increase the digestibility thereof by ruminants, U.S. Patent No. 3,939,286 (1976).
55. Khor, G.L., Alexander, J.C., Santos-Nunez, J., Read, A.E. and Gregory, K.F. Nutritive value of thermotolerant fungi grown on cassava, *Journal of Institute of Canadian Science Technology*, *9* (3), 139–143 (1976).

56. Kilberg, R. The microbe as a source of foods, *Annual Review of Microbiology, 26*, 423–466 (1972).
57. Kirk, T. and Haskin, J.M. Lignin biodegradation and the bioconversion of wood, *American Society of Chemical Engineering Symposium Series, 69*, 124–126 (1973).
58. Kobayashi, H. and Suzuki, H. Studies on decomposition of raffinose by α-galactosidase of mold II, Fermentation of mold pellet and its enzyme activity, *Journal of Fermentation Technology, 50*, 625–632 (1972).
59. Kuhlman, E.G. Decomposition of loblolly pine bark by soil- and root-inhabiting fungi, *Canadian Journal of Botany, 48*, 1787–1793 (1970).
60. Lee, Y.Y., Lin, C.M., Johnson, T. and Chambers, R.P. Selective hydrolysis of hardwood hemicellulose by acids, *Biotechnology & Bioengineering Symposium, No. 8*, 75–88 (1978).
61. Leonard, R.H. and Hajny, G.J. Fermentation of wood sugars to ethyl alcohol, *Industrial Engineering and Chemistry, 37*, 390–395 (1945).
62. Litchfield, J. The production of fungi, pp. 304–329 in *Single-cell Protein* (Edited by R.I. Mateles and S.R. Tannenbaum), MIT Press, Cambridge, Mass. (1968).
63. Livingston, A.L., Allis, M.E. and Kohler, G.O. Amino acid stability during alfalfa dehydration, *Journal of Agriculture and Food Chemistry, 19*, 947–953 (1971).
64. Looke, E.G., Saeman, J.F. and Dickerman, G.K. *U.S. Dept. Comm. Off. Public Board Report*, 7736 (1945).
65. Malek, I. Production of SCP from hydrocarbons: Czechoslovakia, pp. 268–270 in *Single-cell Protein* (Edited by R.I. Mateles and S.R. Tannenbaum), MIT Press, Cambridge, Mass. (1968).
66. Mandels, M. and Reese, E.T. Induction of cellulase in *Trichoderma viride* as influenced by carbon sources and metals, *Journal of Bacteriology, 73*, 269–278 (1957).
67. Martin, F.W. Cassava in the world of tomorrow, pp. 51–58 in *Proceedings of Second International Symposium. Tropical Root and Tuber Crops* (Edited by D.L. Plucknell), University of Hawaii, Honolulu (1970).
68. Miller, D.S. Some nutritional problems in the utilization of nonconventional protein for human feeding, in *Recent Advances in Food Science* (Edited by J.M. Leitch and D.N. Rhodes), Volume 3, Butterworths, London (1963).
69. Miller, T.F. and Srinivasan, V.R. Studies on continuous cultivation of a cellulolytic strain of *Aspergillus terreus*, Presented at ACS Division Microbiology and Biochemical Technology Meeting, Washington, D.C., September 9-14 (1979).
70. Millet, M.A., Baker, A.J. and Sattar, L.D. Physical and chemical pretreatments for enhancing cellulose saccharification, *Biotechnology & Bioengineering Symposium, No. 6*, 125–153 (1976).

71. Mitrakos, K., Sekari, K., Dtouliscos, N. and Georgi, M. Microbial protein from carob, Paper read at 3rd International Congress of Food Science and Technology, Washington, D.C., August (1970).
72. Mitsuda, H., Yasumoto, K. and Nakamura, H. A new method for obtaining protein isolates from *Chlorella* alga, *Torula* yeasts and other microbial cells, *Chemical Engineering Progress Symposium Series, 65*, 93–103 (1969).
73. Moo-Young, M. and Chahal, D.S. Utilization of cattle manure for single-cell protein production with *Chaetomium cellulolyticum*, *Animal Feed Science Technology, 4*, 199–208 (1979).
74. Moo-Young, M., Chahal, D.S., Swan, J.E. and Robinson, C.W. SCP production by *Chaetomium cellulolyticum*, a new thermotolerant fungus, *Biotechnology & Bioengineering, 19*, 527–538 (1977).
75. Moo-Young, M., Chahal, D.S. and Vlach, D. Single-cell protein from various chemically pretreated wood substrates using *Chaetomium cellulolyticum*, *Biotechnology & Bioengineering, 20*, 107–118 (1978).
76. Myhre, D.V. and Smith, F.J. Constituents of the hemicellulose of alfalfa (*Medicaga sativa*), Hydrolysis of hemicellulose and identification of neutral and acidic compounds, *Journal of Agriculture and Food Chemistry, 8*, 359–364 (1960).
77. Newmark, P. Fungal food, *Nature*, September (1980).
78. Nguyen, Q. and Noble, G. Preparation of powdered feedstock from biomass with steam, Paper presented at Specialists' Workshop on Fast Pyrolysis of Biomass, Sponsored by the Solar Energy Research Institute, Copper Mountain, Colorado, U.S.A., October 19–22 (1980).
79. Noble, G. Optimization of steam explosion pretreatment, Final report, Submitted to U.S. Dept. of Energy, Fuels from Biomass Program, Iotech Corporation Ltd., Ottawa, Ontario, Canada (1980).
80. Nordstrom, U.M. Bark degradation by *Aspergillus fumigatus*, Growth Studies, *Canadian Journal of Microbiology, 20*, 283–298 (1974).
81. Pamment, N., Robinson, C.W., Hilton, J. and Moo-Young, M. Solid-state cultivation of *Chaetomium cellulolyticum* on alkali-pretreated sawdust, *Biotechnology & Bioengineering, 20*, 1735–1744 (1978).
82. Peitersen, N. Production of cellulase and protein from barley straw by *Trichoderma viride*, *Biotechnology & Bioengineering, 17*, 361–374 (1975).
83. Phillips, D.H. Oxygen transfer into mycelial pellets, *Biotechnology & Bioengineering, 8*, 456–460 (1966).
84. Rafalski, J., Slam, J., Kluszcynska, Z. and Switonick, T. *Proceedings of Second International Congress of Food Science and Technology*, Warsaw (1966).
85. Reade, A.E. and Gregory, K.F. High temperature protein-enriched feed from cassava by fungi, *Applied Microbiology, 30*, 897–907 (1975).
86. Reid, I.D. The influence of nutrient balance on lignin degradation by

the white-rot fungus *Phanerochaete chrysosporium, Canadian Journal of Botany, 57*, 2050–2058 (1979).
87. Rogers, D.J., Coleman, E., Spino, D.F. and Purcell, T.C. Production of fungal protein from cellulose and waste cellulosics, *Environmental Science Technology, 6*, 715–719 (1972).
88. Romanelli, R.A., Houston, C.W. and Barnett, R.M. Studies on thermophilic cellulolytic fungi, *Applied Microbiology, 30*, 276–281 (1975).
89. Shacklady, C.A. Value of SCP for animals, pp. 489–502 in *Single-cell Protein II* (Edited by S.R. Tannenbaum and D.I.C. Wang), MIT Press, Cambridge, Mass. (1975).
90. Schneider, H., Wang, P.Y., Chan, Y.K. and Maleszka, R. Conversion of D-xylose in ethanol by the yeast *Pachysolen tannophiles, Biotechnology Letters*, February (1981).
91. Scrimshaw, N.S. Introduction, pp. 3–7 in *Single-cell Protein* (Edited R.I. Mateles and S.R. Tannenbaum), MIT Press, Cambridge, Mass. (1968).
92. Shotwell, D.L., Hesseltine, C.W., Stubblefield, R.D. and Sorenson, W.G. Production of aflatoxin on rice, *Applied Microbiology, 14*, 425–428 (1966).
93. Silman, R.W. Enzyme formation during solid-substrate fermentation in rotating vessels, *Biotechnology & Bioengineering, 22*, 410–420 (1980).
94. Sloneker, J.H. Agricultural residues, including feedlot wastes, *Biotechnology & Bioengineering Symposium, 6*, 235–250 (1976).
95. Snez, J.C. A technological and economical survey of industrial single-cell protein, *Proceedings of Symposium of Bioconversion in Food Technology*, Feb. 8–10, Helsinki (1978).
96. Snyder, H.E. Microbial sources of protein, *Advances in Food Research, 18*, 85–140 (1970).
97. Srinivasan, V.R. and Han, Y.W. Utilization of bagasse, *Advances in Chemistry Series, 95*, 447–460 (1969).
98. Srivastava, V.K., Mowat, D.N., Moo-Young, M., Daugulis, A.J. and Chahal, D.S. Preliminary evaluation of the feed quality of product from the Waterloo SCP process, Paper presented at VIth International Fermentation Symposium, London, Ontario, July 20–25 (1980).
99. Tannenbaum, S.R. and Wang, D.I.C. (Editors). *Single-cell Protein*, MIT Press, Cambridge, Mass. (1975).
100. Tarkow, H. and Feist, W.D. A mechanism for improving the digestibility of lignocellulosic materials with dilute alkali and liquid ammonia, *Advances in Chemistry Series, 95*, 197–218 (1969).
101. Thatcher, F.S. Food and feeds from fungi, *Annual Review of Microbiology, 3*, 449–472 (1954).
102. Toyama, N. Feasibility of sugar production from agricultural and urban cellulosic wastes with *Trichoderma viride cellulase, Biotechnology &*

Bioengineering Symposium, *6*, 207–219 (1976).

103. Underkofler, L.A. and Hickey, R.J. *Industrial Fermentation*, Vol. 1, Chemical Publishing Co., New York (1954).
104. Updegraff, D.M. Utilization of cellulose from waste paper by *Myrothecium verucaria*, *Biotechnology & Bioengineering*, *13*, 77–97 (1971).
105. Updegraff, D.M. and Grant, W.D. Microbial utilization of *Pinus radiata* bark, *Applied Microbiology*, *30*, 722–726 (1975).
106. Wenzel, H.F.J. *The Chemical Technology of Wood*, Academic Press, New York (1970).
107. Wiken, T.O. Utilization of agricultural and industrial wastes by cultivation of yeasts, pp. 569–576 in *Proceedings of IV International Fermentation Symposium*, *Fermentation Technology Today* (Edited by G. Terui), (1972).
108. Worgan, J.T. Protein production by microorganisms from carbohydrate substrates, pp. 339–361 in *The Biological Efficiency of Protein Production* (Edited by J.G.W. Jones), Cambridge University Press, U.K. (1973).
109. Yamada, K., Takahashi, J., Kawabata, Y., Okada, T. and Onihara, T. SCP from yeast and bacteria grown on hydrocarbons, pp. 193–207 in *Single-cell Protein* (Edited by R.I. Mateles and S.R. Tannenbaum), MIT Press, Cambridge, Mass. (1968).
110. Yu, P.L., Han, Y.W. and Anderson, A.W. Semi-solid fermentation of alkali-treated straw, *Proceedings Western Section American Society of Animal Science*, *27*, 189–191 (1976).
111. Zadrazil, F. The conversion of straw into feed by Basidiomycetes, *European Journal of Applied Microbiology*, *4*, 273–281 (1977).
112. Zalokar, M. Enzyme activity and cell differentiation in *Neurospora*, *American Journal of Botany*, *46*, 555–559 (1959).

D.S. CHAHAL and R.P. OVEREND

22. Ethanol Fuel from Biomass

Introduction

In a world of dwindling fossil fuel reserves, production of liquid fuel from biomass, a renewable resource, is an attractive proposition. The term biomass refers to all matters of plant and animal origin excluding fossil fuels. The term biomass used here will refer to agricultural crops and residues, wood, and forest industry residues. The plant biomass may also be termed as 'phytomass'. Of the two liquid fuels which can be produced from biomass—ethanol and methanol—ethanol (C_2H_5OH, ethyl alcohol) is an obvious choice when considering agricultural crops because the conversion of sugars and starch to ethanol is simple, well established, has known production costs and is already an established industry in Brazil. Other countries with surplus cane sugar production (for example, the Philippines) are contemplating similar programmes. The burgeoning U.S. gasohol programme based on corn starch is another evidence to show that, whatever the future may hold, the biomass-derived liquid fuel of today is ethanol (121).

Methanol (CH_3OH, methyl alcohol) known as wood alcohol can be synthesized from a variety of sources—biomass, natural gas and coal. In case of biomass and coal, the substrate must be gasified before synthesis. In the production of methanol from wood biomass, three basic steps are involved: gasification of wood, cleanup and modification of the gas produced, and liquefaction of the gas. As the methanol production is a non-microbiological process, it will not be discussed here.

Methane (CH_4), also known as biogas, is produced by anaerobic digestion of biomass. In this process various types of bacteria degrade organic material in the absence of air to produce a gaseous mixture composed predominantly of methane and carbon dioxide in varying proportions. The organisms which cause the breakdown may already be present in the feedstock or may be added by way of an inoculum (see also Chapter 21).

Ethanol

Ethanol (ethyl alcohol) production from biomass was known since the earliest days of recorded history. Processes to brew beer are depicted by the Mesopotamians and the Egyptians as early as 2500 B.C. Pictures of grape presses in Egyptian towns exist since 2000 B.C. The art of ethanol production from biomass (grains) was also known to the Chinese since 3000 B.C. Alcohol (wine) production was also known to Aryans about 2000 B.C. This is evident from references to 'Soma' (wine) in the ancient Vedas written in the Indian subcontinent during 1000 to 500 B.C. The medicinal values of ethanol were known to the ancient Aryans of the Indian subcontinent. As a matter of fact, ethanol was the first candidate as commercial fuel when automobiles were introduced at the end of the last century (86). However, with the large discoveries of petroleum sources and the steady decline in the delivered cost of petroleum products until the beginning of the last decade, alcohol lost markets to petroleum-based products such as gasoline, diesel, naphtha, fuel oil and ethylene. This situation has currently changed because a tenfold increase in petroleum prices occurred during the last decade. Moreover, the increasing fears about the adequacy of future petroleum supplies have resulted in renewed interest in alcohol production from biomass sources which is also based on economic and strategic reasons (5).

Ethanol can be manufactured by the two major processes: (a) Direct hydration of ethylene gas (synthetic alcohol), and (b) Fermentation of carbohydrates and distillation of the beer (fermented alcohol). There is basically no chemical difference between ethanol produced from the two processes; however, the alcohol manufactured for human consumption is required by law to be made by the second method. Synthetic alcohol (from ethylene) is mostly used for the preparation of various chemicals.

Synthetic alcohol

Synthetic ethanol is produced by the reaction of water with ethylene. Ethylene is derived from both natural and coke oven gases, and the waste gases released in refining petroleum to produce gasoline. The pyrolysis of ethane or propane provides a major source of ethylene in Canada (109). Scrubbing of such gases (of suitable and relatively high ethylene content and semipurified by removal of other interfering hydrocarbons) with concentrated sulphuric acid under pressure removes the ethylene as ethyl hydrogen sulphate, which in turn is easily hydrolyzed by water to ethyl alcohol. Yields of alcohol obtained are about 90 per cent of theoretical value (7). Some ether (5 per cent of total yield) is formed as a byproduct. The sulphuric acid may be regenerated re-use or recovered as ammonium sulphate. Catalyzed hydration of ethylene for by water vapour, without the use of acid is also employed in commercial production.

Synthetic alcohol production in the U.S.A. reached 74 million gallons in 1949 almost equalling the production rate of alcohol from molasses and grains. In 1976, the production increased to 202.9 million gallons compared to only 50 million gallons of ethanol from fermented sugars and 30 million gallons for beverages from grains (34, 54).

Fermented alcohol

All the land area of the planet earth (150×106 km^2) is not capable of supporting biomass because a significant fraction of land lies permanently under snow and ice. Moreover, the length of crop growing seasons and the availability of water impose additional constraints on raising a good biomass. The data of Rodin *et al.* (101) in Table 1 give biomass productivity and annual production for five generalized terrestrial zones and the oceans. The oceans represent 70 per cent of the earth's surface but fix only about 25 per cent of the annual carbon budget. Thus most of the carbon is fixed in the biomass produced on the terrestrial zones.

Table 1. Earth—area and phytomass [from Rodin *et al.* (101)]

Thermal zones	Area/10^6 km^2	%	Standing/Pg	%	Production/Pg	%
Polar	8.05	1.6	13.8	0.6	1.3	0.6
Boreal	23.20	4.5	439.1	18.3	15.2	6.5
Sub-boreal	22.53	4.5	278.7	11.5	17.0	7.7
Sub-tropical	24.26	4.8	323.9	13.5	34.6	14.8
Tropical	55.85	10.8	1347.1	56.1	102.5	44.2
Terrestrial total	133.4	26.2	2402.5	100	171.54	73.8
Ice covered	13.9	2.7	0	0	0	0
Lakes and rivers	2.0	0.4	0.04	0.01	1.0	0.4
Oceans	361.0	70.7	0.17	0.01	60.0	25.8
Earth	510.3	100	2402.7	100	232.54	100

The process responsible for the production of biomass is photosynthesis which was also originally responsible for the ultimate deposition of fossil fuel, and would still constitute the only process by which massive amounts of solar energy are being currently captured and stored. The net reaction of photosynthesis or carbon fixation is as follows:

$$CO_2 + H_2O \xrightarrow{\text{sunlight}} (CH_2O)_n + O_2$$

The amount of carbon fixed per year on a world basis is in the order of 30 times greater than the total world energy production of 15.2×10^{10} KJ (14.4×10^{10} BTU) or 7.0×10^9 metric tons of coal equivalent in 1970. Thus biomass seems to be the best choice for production of liquid fuel energy (10, 61).

Fermented alcohol can be obtained from three main types of raw materials: Sugar-bearing materials (such as sugar cane, molasses, sweet sorghum, etc.) which contain carbohydrates in sugar form; Starchy materials (such as cassava, corn, potato, etc.) which contain carbohydrates in starch form; and Lignocelluloses (such as wood, agricultural crop residues, etc.) which contain more complex forms of carbohydrates i.e. cellulose and hemicelluloses.

Ethanol from sugar-bearing materials

Sugar cane is the most attractive biomass since it involves the simplest conversion process of sugar into ethanol and it also provides more than adequate energy for generating the steam and power needed for crushing, fermentation and the distillation processes. A ton of sugar cane, with an average sugar content of 12.5 per cent, gives about 70 litres of ethanol through direct fermentation of the juice. Sugar cane gives one of the highest ethanol yields per hectare of crop land as compared to other biomass candidates such as cassava, sweet potatoes, babassu and corn (Table 2). The next best substrate for high ethanol yields per hectare is wood and possibly cassava depending on the improved production technology. On the other hand, the ethanol yield per ton of biomass is highest for corn and molasses and possibly for cassava.

Table 2. Ethanol yields of main biomass raw materials [from Anonymous (6)]

Raw material	Ethanol yield (litres/ton)	Raw material yield[a] (ton/h)	Ethanol yield (litres/h/yr)
Sugar cane	70	50.0	3,500
Molasses	280	N.A.	N.A.
Cassava	180	12.0	2,160
		(20.0)[b]	(3,600)[b]
Sweet sorghum	86	35.0[c]	3,010[c]
Sweet potatoes	125	15.0	1,875
Babassu	80	2.5	200
Corn	370	6.0	2,220
Wood	160	20.0	3,200

N.A.—Not applicable.

[a]Based on current average yields in Brazil, except for corn which is based on the average in the U.S.A.

[b]Potential with improved production technology.

[c]Tons of stalks/h/crop. Two crops per year may be possible in some locations.

Sugar cane cultivation is currently devoted largely to sugar production and any future large-scale production of alcohol from this source would depend on: (a) a choice between export of sugar and conversion of sugar to ethanol as a substitute for imported petroleum, and (b) competition of good land between food crops and sugar cane in third world countries.

Cane molasses (blackstrap molasses) is a byproduct of sugar production from sugar cane. Every ton of sugar produced gives approximately 190 litres of molasses. It contains between 50 and 55 per cent fermentable sugar (mainly sucrose, glucose and fructose) and about 280 litres of ethanol can be produced per ton of molasses. Molasses is also used as animal feed and for human consumption. It is estimated that in 1978–79 the total world production of (cane and beet) molasses was 33.5 million tons, of which about 22 million tons were consumed in the countries of origin and another 6.6 million tons entered world trade mainly for use as animal feed. The remaining 5 million tons (15 per cent of total) was most probably disposed of as valueless waste (5). If this waste molasses (5 million tons) is converted, it would yield only 1.35 billion litres of ethanol. And, if the total molasses of 33.5 million tons were diverted to ethanol production it would be equivalent to 9 billion litres or 1 per cent of the world gasoline demand in 1978, and 9 per cent of the gasoline demand from the developing countries. However, the major use of molasses in the developing countries is mostly for fermentation of beverage alcohol.

Fermentation technology for molasses and sugars (or sugars obtained from starch or wood) is already well established (135), though there is much scope for improvement in strains of yeasts used and in the distillation technology. However, ethanol production from sugar cane for fuel purposes in Brazil has made a tremendous impact on the world. Ethanol production in Brazil reached 1.5 billion litres in 1977. By the end of 1978 over 200 projects were approved by the National Alcohol Committee of Brazil to increase capacity by 3.9 billion litres per year. By 1985 fermented ethanol may supply about 2 per cent of Brazil's energy requirement (65, 145).

Sweet sorghum, which contains a mixture of sucrose and glucose, is increasingly considered as another attractive biomass material for ethanol production. Two crops can be raised in certain areas of the world. This crop can also be grown on marginal lands.

Ethanol from starch-bearing materials

The main starchy materials of interest for ethanol production are cassava and corn. Other starchy materials such as wheat, barley, grain sorghum, oats, rice and potatoes can be used depending on their availability in preference to being items of food. Cassava is a root crop grown extensively as a subsistence crop in a large number of developing countries. It is one of the most efficient converters of solar energy into biomass, can be grown on marginal land, can withstand adverse weather conditions, and the annual yield could be expected up to 20 tons/ha. However, ethanol production from cassava will have to face a tough competition from an equally good demand of the crop as a source of calories for the rural poor people who have habitually consumed it for generations as a staple diet. Corn is already being used for ethanol production especially for whisky in the U.S.A. since a long time but during the last few years, there has been an increasing interest in the U.S.A. to use corn for ethanol production to blend with gasoline.

The process for producing ethanol from corn yields a number of valuable byproducts which contribute to the economic feasibility of this process. The process can augment rather than deplete animal feed supplies because one bushel of corn (56 pounds) yields not only 2.5 U.S. gallons of liquid fuel, but also produces 17 pounds of animal feed with the same nutrient content as that contained in the original 56 pounds of corn. In the corn wet milling process, employed by Archer Daniels Midland, Decatur, Illinois, U.S.A. 56 pounds of corn contains 47.3 pounds of dry solids that yield:

—9.2 pounds of corn gluten feed, containing 21 per cent protein,
—2.7 pounds of gluten meal, containing 60 per cent protein,
—3.5 pounds of corn germ, and
—31.5 pounds of starch which is converted to 2.5 gallons (9.46 litres) of anhydrous ethanol.

If 31.5 pounds (11.74 kg) of corn starch is converted into ethanol according to methods employed in Table 3, the yield should be 7 litres instead of 9.46 litres. Accordingly one ton of corn will yield 387 litres of ethanol. However, for short term, corn may be a good feedstock for ethanol production without any visible effect on food/feed shortage in the world but long term use will offset the food/feed balance as there is also a need of starch (calories) for feeding the starving millions of people in the world. The only alternative will be the use of agricultural and forest residues (lignocelluloses) for ethanol production.

Ethanol from lignocelluloses from wood

The yields of ethanol from wood (lignocelluloses) given in Table 2 are low. The low yields are because of the old technology used to hydrolyze wood and also due to non-conversion of pentoses into ethanol. However, the expected ethanol yields from lignocelluloses given in Table 3 are quite high

Table 3. Expected ethanol[1] yields from lignocelluloses

Components (average)	Carbohydrates per ton lignocelluloses	Monomer[2] sugars	Ethanol[3] yields by weight	Ethanol[4] yields by volume
(i) Cellulose=50%	500 kg	495 kg	239.8 kg	303.5 litres
(ii) Hemicelluloses=25%	250 kg	247.5 kg	119.9 kg	151.7 litres
			83.3[5] kg	105.4[5] litres
(iii) Lignin=25%	—	—	—	—
			Total expected yields=455.2 litres	
			Total present yields=408.9 litres	

[1]Anhydrous ethanol.
[2]Conversion factor 1.1 and 90% efficiency.
[3]Conversion factor 0.51 and 95% efficiency.
[4]Conversion factor 0.79 (specific gravity).
[5]Present pentose conversion efficiency 66% [based on Slininger *et al.* (116)].

(455.2 litres/ton) and are based on the present pretreatments available, high hydrolytic power of cellulase-complex, and the improved knowhow of pentose fermentation into ethanol. Even on the basis of technology available today about 409 litres of ethanol can be produced from one ton of lignocelluloses, which is equivalent to that of corn (397 litres). Moreover, the production cost of lignocelluloses is much less than that of corn and the net energy gain will also be higher than that of corn.

Structural features and chemical composition of lignocelluloses in relation to cellulase activity

The lignocelluloses (biomass) are composed of cellulose, hemicelluloses, lignin, protein and various extraneous materials. The composition of various lignocellulosic materials in comparison with that of cotton is presented in Table 4. The lignocelluloses are the major constituents of the plant cell wall.

Structure of plant cell wall

The cotton fibre is unicellular and is an outgrowth on the seed of the cotton plant and it is the purest form of cellulose in nature. But the cellulose of lignocellulosic materials (wood and crop residues) is mixed with lignin and hemicelluloses and is a part of the cell wall of plant tissues. Chemically there are no differences between the cellulose of cotton and of the other lignocelluloses.

Table 4. Chemical composition of various lignocellulosic materials and cotton (per cent dry weight)

Constituents	Cotton	Birch[1] angiosperm	Spruce[1] gymnosperm	Crop[2] residues	Wood[3] residues
Cellulose	89.0	44.9	46.1	30–45	45–56
Hemicelluloses	5.0	32.7	24.6	16–27	10–25
Lignin	0.0	19.3	26.3	3–13	18–30
Protein (1N × 6.25)	1.3	0.5	0.2	3.6–7.2	N.A.
Extractable extraneous materials	2.5	2.3	2.5	N.A.	N.A.
Ash	1.2	0.3	0.3	N.A.	N.A.

N.A.=Not available.
[1]Cowling (28).
[2]Sloneker (117).
[3]Wenzel (139).

Both the cotton and plant cells have a thin primary wall that surrounds the relatively thick secondary wall (Fig. 1). The primary wall, which is only 0.1 to 0.2 μm in thickness contains randomly and loosely organized network

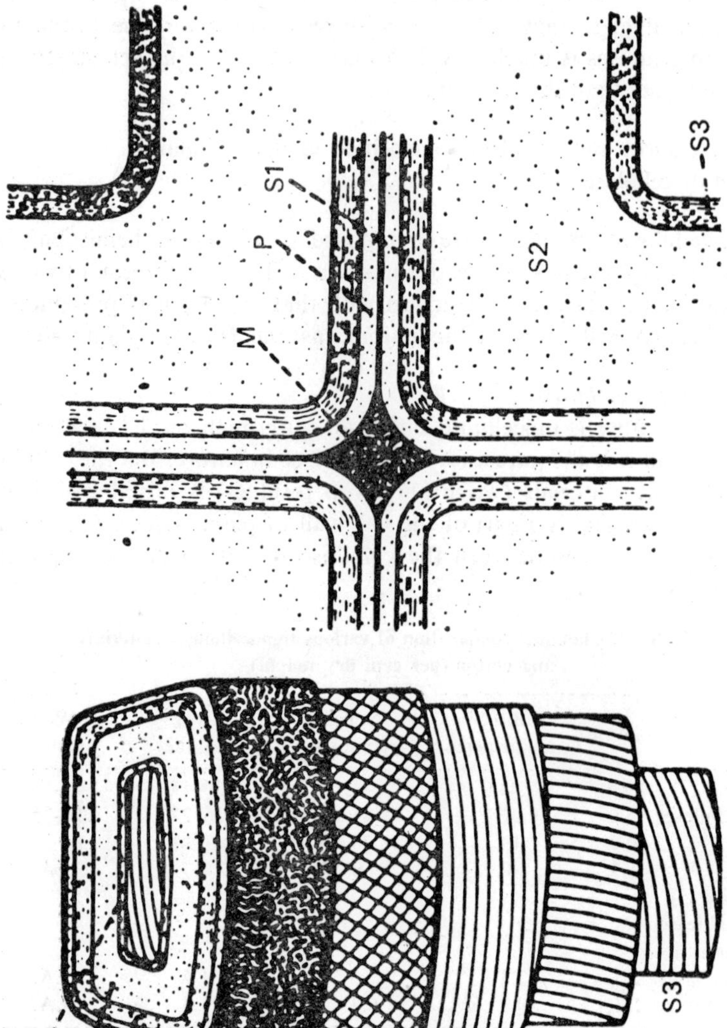

Fig. 1. Diagrammatic representation of various layers of cell wall. The intercellular materials—middle lamella (M) and adjacent primary walls (P) of contiguous cells comprise the compound middle lamella. The secondary walls are composed of outer (S_1), middle (S_2), and inner (S_3) layers. Each layer has different orientation of microfibrils.

of cellulose microfibrils. The outer layer of the secondary wall, S_1, has a crossed fibrillar structure. In the S_2-layer, the main portion of the secondary wall (1 to 5 μm thick), the microfibrils are oriented almost parallel to the lumen axis. In the thin S_3-layer (0.1 μm) the microfibrils form a flat helix. The innermost portion of the cell wall consists of the so-called warty layer, probably formed from protoplasmic debris. The central empty portion, formed after the disintegration of protoplasm at the time of ageing, is called the lumen. The cell lumen contains some nutrients of disintegrated protoplasm. These nutrients are good for the growth of various wood rotting fungi (Basidiomycetes). The primary wall is mostly composed of cellulose and pectic compounds. The primary walls of two adjacent cells are cemented together with pectic compounds and lignin. The space between two adjacent cells is called the middle lamella.

Cellulose

The cellulose in each layer of cell wall occurs as long slender bundles composed of long chains of β-D-glucopyranose residues linked by (1→4)-glycosidic bond (cellulose molecules) called elementary fibrils with diameter of 35 Å. The length of cellulose molecules in elementary fibril varies from less than 15 β-D-glucopyranose residues in gamma-cellulose to as many as 10,000 to 14,000 residues per molecule in alpha-cellulose. The length of the cellulose molecule is measured as degree of polymerization (DP) i.e. number of β-D-glucopyranose residues. Within each elementary fibril the cellulose molecules are bound laterally with adjacent molecules running in opposite direction, by hydrogen bonds. They are associated in various degrees of parallelism—regions that contain highly oriented molecules are called crystalline whereas those of lesser order are called paracrystalline or amorphous regions (Fig. 2). A number of elementary fibrils form microfibrils of 100–200 Å wide.

There are four recent concepts about the structure of microfibril (Fig. 2). According to one concept (92) the microfibril is 50 × 100 Å in cross-section and consists of a "crystalline core" surrounded by "amorphous sheath" which in cotton contains mainly cellulose molecules but in case of other lignocelluloses it also contains hemicelluloses and lignin. But according to another concept (52) the microfibril is composed of 15–40 cellulose molecules. At certain places the microfibrils are less ordered (amorphous or paracrystalline region) along the length of the microfibrils. Still in another concept (77) the cellulose molecules exist in a folded chain lattice formed as ribbon which in turn is wound in tight helix. The most recent speculation (102) on structure is that the microfibril at certain length has strain-distorted tilt and twist regions which are easily accessible to the enzymes. Whatever may be the different opinions about the structure of the microfibril, it is certain that there is some crystalline and some amorphous regions. Several microfibrils when joined laterally form a macrofibril. The microfibrils or macrofibrils are differently

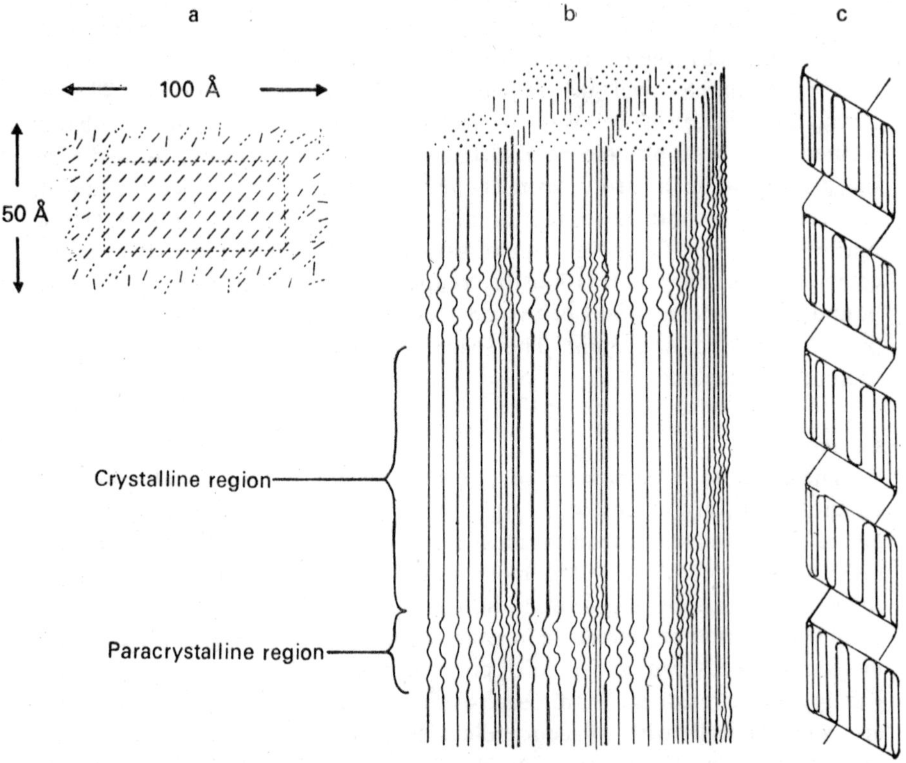

Fig. 2. Diagrammatic representation of different models of cellulose fibrillar ultrastructure.

a) According to Preston and Cronshaw (92) the microfibril strand is about 100 Å wide and about 50 Å thick. The solid strokes represent the planes of the glucose residues in the cellulose chain molecules. The broken strokes indicate the position of other sugars or sugar derivatives in noncellulosic chains. The central crystalline core is surrounded by a paracrystalline sheath.

b) According to Hess *et al.* (52), the microfibril contains a number of elementary fibrils which are segmented into crystalline and paracrystalline (amorphous) regions.

c) According to Manley's (77) proposal of the elementary fibril, the cellulose molecule is first folded into a ribbon, which is subsequently wound as a tight helix.

oriented in each layer of secondary wall to give a structural strength to the cell wall. To hold the microfibrils together in the cell wall, they are almost covered with lignin and the interspaces are impregnated with hemicelluloses and other extraneous materials. Thus in nature the cellulose in the lignocelluloses is very well protected from the hydrolytic enzymes.

Hemicelluloses

Hemicelluloses are low-molecular weight polysaccharides, most of which are water-insoluble but they are easily solubilized by alkali. They are more readily acid hydrolyzed than cellulose. Although the hemicelluloses are usually considered to be structural polysaccharides, it is convenient to include among them a few other plant polymers, such as the arabinogalactans, which obviously have other functions. Hemicelluloses are built up from relatively few sugar residues, the most common of which are D-xylose, D-mannose, D-galactose, D-glucose, L-arabinose, 4-0-methyl-D-glucuronic acid, D-galacturonic acid, D-glucuronic acid. Among the more rare constituents are L-rhamnose, L-fucose, and various methylated neutral sugars.

Xylan and mannan are the two major polysaccharides. The complete formula of hardwood xylan (0-acetyl-4-0-methylglucurono-xylan) is given in Fig. 3 (131). The polysaccharide framework consists of approximately 200 β-D-xylopyranose residues, linked together (1→4)-glucosidic bonds. Some of the xylose units carry a single, terminal side chain consisting of a 4-0-methyl-α-D-glucuronic acid residue, attached directly to the 2-position of the xylose. Seven out of ten xylose residues contain an 0-acetyl group at C-2 or more frequently at C-3. The basic framework of softwood xylan (arabino-4-0-methyl-glucurono-xylan) is same as that of the hardwood. However, the softwood xylan also contains α-L-arabinofuranose residues directly linked to C-3 of the xylose.

Mannan (galacto-glucomannan) is predominant in softwood. The framework consists of (1→4)-linked β-D-glucopyranose and β-D-mannopyranose residues distributed at random (Fig. 4). Some of the hexose units carry a terminal residue of α-D-galactopyranose attached to C-6. It is probably that all the galacto-glucomannans are acetylated in their native state. The acetyl groups are attached to the mannose residues.

Lignin

The biological role of lignin in living plant is to form, together with the cellulose and other carbohydrates of the cell wall, a tissue of excellent strength and durability. Lignin is concentrated mainly in the spaces between the cells (middle lamella) and in S_2-layer of cell wall where it is deposited during the lignification of the plant tissue although some lignin is also in other layers. The completion of the lignification process usually coincides with stoppage of the living function of the cells. Lignin is a complex three-dimensional polymer formed from coniferyl alcohol units in case of gymnosperms and both coniferyl and syringyl units in the case of angiosperms. The latest structure of lignin was proposed by Freudenberg (38).

Extraneous materials in cell wall

The small quantity of extraneous materials deposited in fine capillaries of the cell wall may affect degradation and hydrolysis in five different ways (29): 1) growth promoting substances such as vitamins (thiamin) and certain

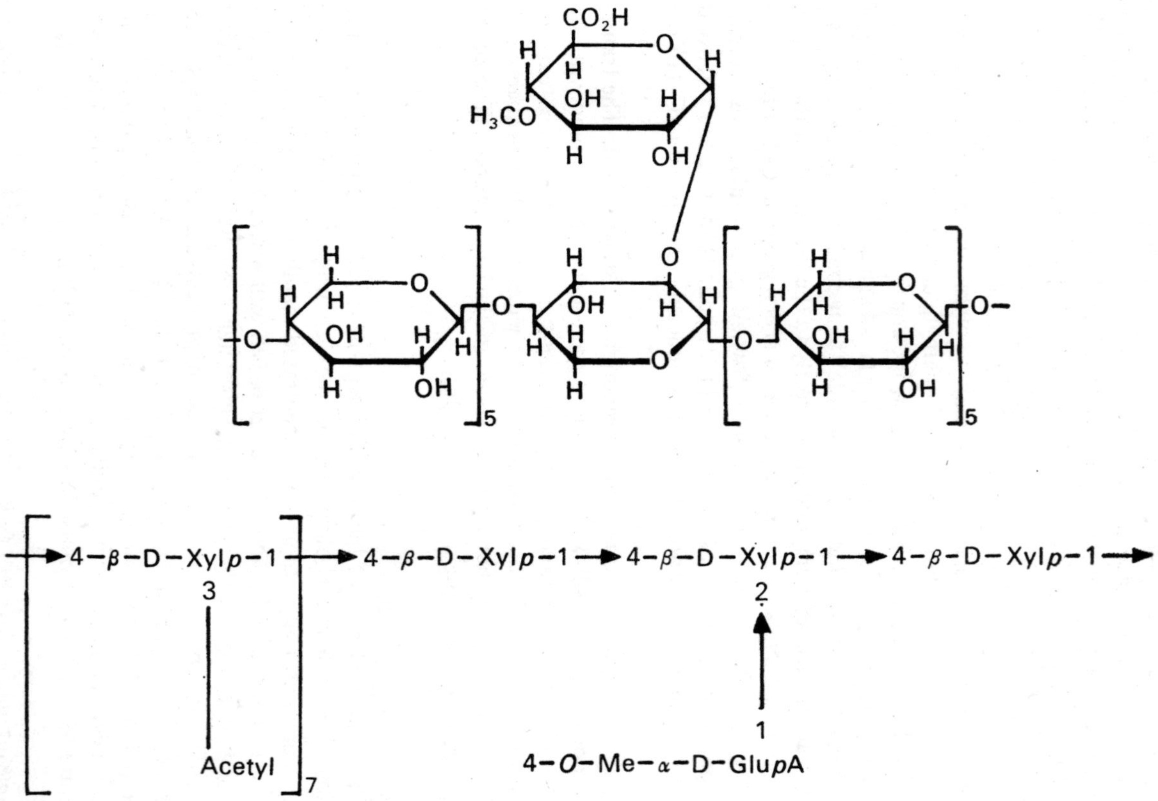

Fig. 3. Top: Partial structure of a 4-0-methylglucurono-xylan.
Bottom: Structural formula of 0-acetyl-4-0-methylglucurono-xylan [reproduced from Timell (131)].

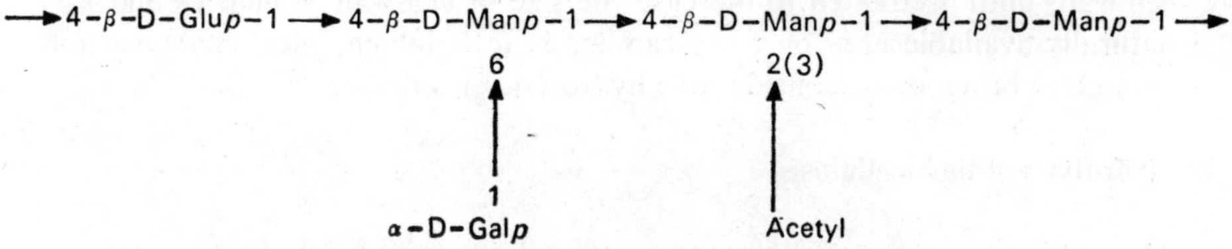

Fig. 4. Structural form of 0-acetyl-galacto-glucomannans [reproduced from Timell (131)].

soluble carbohydrates provide substrates for rapid growth and development of cellulolytic micro-organisms within cellulose fibres; 2) poisonous materials, particularly toxic phenolic materials (106) inhibit the normal growth and development of cellulolytic organisms and activity of cellulolytic enzymes; 3) various substances are deposited within the fine capillary structure of the cell wall and thus reduce the accessibility of the cellulose to extracellular enzymes; 4) certain specific enzyme inhibitors act directly to reduce the rate or extent of enzymatic hydrolysis of the cellulose.

Relationship between structural features of cell wall and cellulolytic enzymes

The most important structural feature in a cell wall is the capillary area, whose dimensions and shape control the enzymatic hydrolysis of lignocelluloses. The cell lumen itself is the biggest capillary in a cell followed by pit apertures and the pores in pit-membrane ranging from 200 Å to 10 or more microns in diameter. The cell-wall capillaries are the space between microfibrilis and cellulose molecules in the amorphous regions. Most of the area of such capillaries is filled with hemicelluloses and lignin, thus making most of the cellulose inaccessible to enzymes although some capillaries remain unfilled with such materials. Most of the cell-wall capillaries are closed when dry but open when moisture is absorbed. When the cell wall is fully saturated with water, the cell-wall capillaries attain their maximum dimensions. Any pretreatment of lignocelluloses which depolymerizes, solubilizes or removes lignin and hemicelluloses from the cell wall, will considerably increase the accessibility of cellulose to enzymatic hydrolysis. A detailed account of pretreatment processes (1, 16) has been given in the previous chapter of this book (Chapter 21).

The cellulolytic enzyme molecules of various microorganisms are water-soluble proteins of high molecular weight. It has been estimated (137, 140) that if they are spherical they range from 24 to 77 Å in diameter with an average of 59 Å. If they are ellipsoids with an axil ratio of about 6, the sizes range from 13 × 79 Å to 42 × 252 Å in width and length, respectively. The comparison of the sizes of capillary and enzymes indicated that cellulase enzyme molecules are quite big and will not be able to diffuse in cotton and wood cell walls until pretreated to increase the size of cell-wall capillaries and the naturally available area of gross capillaries (cell lumen, pits, etc.) are not enough to bring about considerable hydrolysis of cellulose.

Hydrolysis of lignocelluloses

The utilization of lignocelluloses for ethanol production involves two essential steps: 1. the hydrolysis of lignocelluloses into its monomer sugars; and 2. the fermentation of all these sugars into ethanol. There are two major methods for hydrolyzing the lignocelluloses—acids and enzymes.

Acid hydrolysis

Acid hydrolysis of lignocelluloses was discovered about 150 years ago. Since then there have been persistent efforts to improve upon the yields by acid hydrolysis. Two processes were developed in Germany in which high yields of ethanol were claimed (105). In one, the Scholler process (108) a dilute sulphuric acid of 0.2 to 1 per cent concentration is circulated through layers of sawdust or wood chips, under pressure at a temperature of 121–180°C. The sugar produced in the hydrolysis is easily destroyed by the acid, hence it must be quickly and continuously removed. A solution containing only 3 per cent sugar is obtained. The acid is neutralized with lime before the solution is fermented. Yields of about 50 gallons of alcohol are obtained per ton of dry wood.

In another method, known as the Bergius process, concentrated hydrochloric acid is used. This process requires special acid-resistant vessels and also necessitates the recovery of acid. For these reasons the Bergius process was not as attractive as the Scholler process.

The Scholler process was modified at the Forest Products Laboratory, Madison, Wisconsin, and was known as the Madison Wood-Sugar Process. By this process, in 3 to 5 hours, a bark-free wood can be converted into solution containing 5 to 6 per cent reducing sugars which on fermentation gave 50 to 65 gallons of ethanol per ton wood. A full-scale plant based on this process was built at Springfield, Oregon, U.S.A. in 1945, but it did not reach full production before it was shut down because of several operating problems. These problems were later resolved (43) and a satisfactory method was developed which produced a 6 per cent reducing sugar solution with a yield of 75 per cent of theoretical. But by that time ethanol was being produced chemically from ethylene much cheaper than that produced by this process of hydrolysis and fermentation. Hence, this process could not stand competition with other processes.

A great disadvantage of acid hydrolysis is that the hemicelluloses and amorphous cellulose are hydrolyzed faster than the crystalline cellulose, the soluble sugars thus produced are degraded during the subsequent exposure to acid and temperature when reaction is continued to hydrolyze the crystalline cellulose. To avoid such degradation of sugars, a 2-stage process has been used, the first-stage being relatively mild to hydrolyze the hemicelluloses (which is then removed from the reaction vessel) and the second being more severe to hydrolyze cellulose (26, 47, 62). The extraction of hemicellulose sugars during the first stage of acid hydrolysis make the remaining cellulose more susceptible to enzymatic hydrolysis (62). The hemicellulose sugars obtained could be fermented into ethanol separately by suitable organisms because of domination of xylose—a pentose, in this solution. Optimum yields of xylose and glucose from hemicelluloses of hardwood are obtained with 0.1 to 0.2 per cent H_2SO_4 at 170°C for one to two hours with solid-to-liquid ratio of 1:10. Under these conditions xylose yield of 24 g/100 g dry wood was obtained. Taking the hemicellulose content of southern red oak as 22.74 per cent, the yield of xylose is calculated to be 94 per cent of the theoretical maximum (accounting for water addition in hydrolysis). When the spent wood was hydrolyzed with enzymes the reaction rate was close to that of filter paper.

A continuous plug flow reactor was built for acid hydrolysis of lignocelluloses (26). A single pass hydrolysis at 240°C with 1 per cent acid gives 50 to 57 per cent glucose yield for a slurry concentration of 5 to 13.5 per cent solids. For high xylose recovery, a two-stage hydrolysis is needed, the first stage at low temperature (180–200°C) and the second stage at high temperature (230–240°C).

Partial acid hydrolysis was studied as a pretreatment to enhance enzymatic hydrolysis (59). Such a treatment was carried out in a continuous flow reactor on oak, corn stover, newsprint, and Solka Floc at temperatures ranging from 160 to 220°C, acid concentrations ranging from 0–1.2 per cent and a fixed treatment time of 0.22 minutes. The resulting slurries and solids were then hydrolyzed with *T. reesei* QM9414 cellulase at 50°C for 48 hours. In the case of newsprint and corn stover, 100 per cent of the potential glucose content of the substrates was converted to glucose after 24 hours of enzymatic hydrolysis whereas 90 and 81 per cent conversion was obtained in case of oak and Solka Floc respectively even after 48 hours.

The mild acid pretreatment followed by enzymatic hydrolysis seems to hold good promise for further developments in ethanol production from lignocelluloses. The mild acid treatment seems to be very similar to that of steam explosion of wood by Iotech and Stake process as described under pretreatments of lignocelluloses in the previous chapter.

Enzymatic hydrolysis

SOURCES OF CELLULASE. The cellulase system is present and used for different purposes (140) almost throughout animal and plant kingdoms, for example: in guts of snails (*Helix pomatia*); termites (*Termopsis angusticollis*,

Reticulitermes flavipes); in protozoa (*Endoplodinum neglectum, Trichomonas termopsidus*); in germinating rye and barley grains (malt); and many fungi, actinomycetes and bacteria (40).

HISTORICAL DEVELOPMENT. In the U.S.A., several organizations within the Armed Services set up laboratories including the Quartermaster Corps (now U.S. Army Natick Development Center) to investigate the nature of rotting, the causal organisms, their mechanisms of action, and the development of methods of control not requiring the use of fungicides (94). This resulted in the Quartermaster (QM) collection of over 14,000 fungi active in degradation of materials such as wool, leather, cellulose, and other polysaccharides. This collection is now housed at the University of Massachusetts in Amherst, Massachusetts. Recently Dr. Toyama's cultures of *Trichoderma* strains BIA (QM9973) and LE (QM9974) have been added to this collection (75).

In the late 1940's, Dr. Elwyn Reese noted that although many fungi degrade cellulose in nature, very few produce culture filtrate active against insoluble cellulose which led him to the multiple enzyme (C_1 and C_x) concept in 1950 (99) preceding by many years the isolation of the endo- and exo-glucanases as recognized today. In 1964 Mandels and Reese (74) were able to separate C_1 and C_x components of the cellulase complex. In 1968 Katz and Reese (56) reported the production of 30 per cent glucose syrup by using concentrated cellulase-complex on ball milled cellulose pulp.

SELECTION OF CELLULASE-PRODUCING ORGANISMS. The cellulolytic fungi and actinomycetes can easily be grown on simple media and hence are the best choice for cellulase production. Most of the cellulolytic fungi and actinomycetes produce cellulase extracellularly when grown on cellulose as a source of carbon, and it is easy to separate the enzyme from the microbial biomass.

The ability of an organism to degrade cellulosic substrates and to convert the substrate into microbial biomass does not necessarily mean that its cell-free cellulase system will be good to hydrolyze other cellulosic materials. The analysis of data given by Mandels and Weber (76) indicate that *Trichoderma viride* produced high C_1 and C_x units and this cellulose system was able to give highest hydrolysis of cotton (Table 5). On the other hand *Chrysosporium pruinosum* and *Penicillium pusillum* produced less C_1 units but high C_x units and their cellulase system was poor in hydrolysis of cotton because their cellulase systems were unstable as compared to that of *T. viride*. All the other organisms produced quite high C_x activity but low C_1 activity and thus were very poor in hydrolysis of cotton. There was one exception, *Fusarium moniliforme* which produced very low C_x activity (C_1 activity not tested) but still this enzyme-system was better than that of *C. pruinosum* and *P. pusillum*. It is concluded from their data that C_1 plays a more important role than C_x in the hydrolysis of cellulosic materials. Thus high activity of C_x is no direct indication of the hydrolytic potential of a cellulase-system.

It has been observed (21, 22) that the cellulase activity of *T. viride* broth

Table 5. Production of cellulase by various fungi [reproduced from Mandels and Weber (76)]

QM No.	Preparation	C_1 units	C_x units	Hydrolysis of cotton, %	
				Sugar	Wt. loss
	Buffer (0.05M acetate)	0	0	0	1
6a	*Trichoderma viride*	50.0	50.0	58	53
826	*Chrysosporium pruinosum*	30.0	70.0	11	19
137g	*Penicillium pusillum*	27.0	110.0	22	23
1224	*Fusarium moniliforme*	N.T.	3.5	39	39
72f	*Aspergillus terreus*	5.0	36.0	28	24
806	*Basidiomycete*	5.0	75.0	15	23
94d	*Stachybotrys atra*	1.0	8.0	5	6
B814	*Streptomyces* sp.	0.7	40.0	9	12
38g	*Fusarium roseum*	0.7	10.0	9	10
381	*Pestalogiopsis westerdijkii*	0.7	60.0	4	8
460	*Myrothecium verrucaria*	0.4	28.0	2	2
459	*Chaetomium globosum*	0.2	0.5	N.T.	N.T.

Cultures grown on Solka Floc, except *Penicillium pusillum* 137g, grown on cotton duck.

C_1 units—action on cotton sliver for 24 hours at 40°C.

C_x units—action on carboxymethyl cellulose.

N.T.—Not tested.

Hydrolysis of 1% cotton sliver 35 days at 29°C (4 changes of enzyme).

was higher than that of *Chaetomium cellulolytium,* but on the other hand *C. cellulolytium* was able to produce more protein and mycelial biomass than *T. viride* when grown on sigmacell-20 (cellulose). This showed that high levels of cellulase activity in the broth need not necessarily mean high conversion of cellulose into microbial biomass. *Chaetomium cellulolytium* was low in cellulase production but the enzyme system was perfect in releasing the monomer sugars from cellulose substrate and simultaneously incorporating them into the mycelial protein. On the other hand the cellulase system of *T. viride* was also perfect in converting the cellulose substrate into its monomer sugars and their simultaneous conversion to enzyme protein rather than to the mycelial protein. Thus an organism which will be able to convert most of the cellulose substrate into enzymic protein with high hydrolytic potential will be the most suitable organism for cellulase production.

Unfortunately very few fungi are able to produce an enzyme system which can degrade native cellulose. Such fungi are: *Trichoderma viride, T. lignorum* and *T. koningii* (13, 143, 144), *Sporotrichum pulverulentum* (33, 35, 122), *Penicillium funiculosum* (111), *P. iriensis* (15), *Polyporus adustus* (33), *Myrothecium verrucaria* (136), *Fusarium solani* (143), and *Chaetomium thermophile* var. *dis-*

situm (44). There are many more fungi and bacteria which can produce cellulases but their enzyme system can hydrolyze only pretreated cellulose or carboxymethyl cellulose (CMC), but not crystalline cellulose (32, 127). Some gliding bacteria, (Gram-negative and Gram-positive ones) and some actinomycetes are also able to produce cellulase (44, 115). Cellulase activity is also found in facultative anaerobes (*Bacillus*, *Cellulomonas*) as well as in anaerobes (*Clostridium*).

Recently thermophilic organisms have been studied to obtain a cellulase-system which could be more stable at higher temperature (50–50°C), rendering them most suitable for hydrolysis. *Chaetomium thermophile* var. *dissitum* is a typical thermophilic fungus which is able to produce a cellulase-system and hydrolyze native cellulose (44). *Chaetomium thermophile*, *Sporotrichum thermophilium* and *Thermoasus aurantiacus* grow on and decompose cellulose very rapidly but their cellulase activity is very low (67). Similarly *Thermomonospora curvata* produce both endo- and exo-glucanases but their enzymes hydrolyze cotton only up to 1 per cent (124). However, it has been observed that cellulases from thermophilic organisms need not necessarily be more heat-stable than those from mesophiles (67).

From the hydrolysis point of view, *Trichoderma viride* Persoon and Fries is the most suitable organism to produce a stable and potential cellulase-system (48, 57, 74, 97, 113, 132, 142). Now, this organism is recognized as a new species, *Trichoderma reesei* Simmon (114). For the sake of uniformity this latest name has been used throughout the text even where *T. viride* was used in the old research work referred to in this chapter.

HEALTH HAZARD. *Thermonospora curvata*, a thermophilic actinomycete is used for degradation of cellulose during high temperature composting of municipal wastes (123, 126). It has been proved (125) that activation of alternate pathway complement (APC) by extracellular products of *T. curvata* is akin to other organisms known to cause hypersensitivity pneumonitis (HP) a "flue-like" respiratory disease. Repeated attack causes narrowing of the terminal airways and ultimately leads to the development of obliterative bronchiolitis and interstitial fibrosis (6, 31, 79, 100). Therefore, in our anxiety to improve the cellulolytic activity of fungi by genetic engineering, care should be taken to avoid strains of microorganisms which may prove to be health hazards.

Production of cellulases

Medium

The basic medium for cellulase production was developed by Mandels and Weber at the U.S. Army Natick Development Center, Natick, Massachusetts, U.S.A. (76). The composition of the medium is given in Table 6. There have been various changes in the constituents especially on protein source and surfactants like Tween 80 (Polyoxyethylene sorbitan mono-oleate) to reduce the cost of the medium as well as to increase the yields of cellulases.

Table 6. ***Trichoderma reesei*** **medium for cellulase production [from Mandels and Weber (76)]**

	g/1		mg/1
$(NH_4)_2SO_4$	1.4	$FeSO_4 \cdot 7H_2O$	5.0
KH_2PO_4	2.0	$MnSO_4 \cdot H_2O$	1.6
Urea	0.3	$ZnSO_4 \cdot 7H_2O$	1.4
$CaCl_2$	0.3	$CoCl_2$	2.0
$MgSO_4 \cdot 7H_2O$	0.3		

Cellulose 0.75–1.0%, Proteose peptone 0.075–0.1%, Tween 80 0.1–0.2%, Initial pH 5.0–6.0.

Inducers

These inducers are glucans of mixed linkage including the B (1→4) and a few oligosaccharides (71–73). The true inducers of cellulases for a fungus growing on cellulose are the soluble hydrolysis products of the cellulose, especially cellobiose (72). The role of cellobiose is complex. Cellobiose is an inducer at low concentrations but is an inhibitor at high concentrations (95). Concentrations of 0.5–1.0 per cent of cellobiose or other rapidly metabolizable carbon sources such as glucose or glycerol strongly repress cellulase formation (72, 73) until these carbon sources are consumed by the organism. If these sugars are added at these concentrations (0.5 to 1.0 per cent) to a culture already producing cellulase, they inactivate the enzyme already formed (72). If the culture is still in its logarithmic stage, the enzyme activity will reappear when these sugars are utilized and the fungal metabolism is shifted from soluble sugars to insoluble substrate, the cellulose.

Mandels and Weber (76) reported that if the rapid metabolism of the sugar (cellobiose) is slowed down by suboptimum temperature and aeration, deficiencies of mineral nutrients such as calcium, magnesium, or trace metals, and excess of a mineral such as cobalt (72), *T. reesei* will produce as much cellulase on sugar as on cellulose. But it has been noticed that a new hypercellulase producing strain Rut-C30 produces about one-third cellulase when grown on glucose as compared to that on cellulose under similar cultural conditions (19).

Addition of small amounts (0.1 per cent) of readily available substrates such as glucose, glycerol or peptone along with cellulose reduces the lag phase and increases the cellulase production (72). Similarly, it was noticed that addition of small amounts (400 mg C/1) of any sugar in β-form (β-methyl-D-glucoside), β-1, 4 linkage (cellobiose) or even simple hexose (glucose) supported the early phase of growth and ultimately increased total protein synthesis in *Myrothecium verrucaria*, *Chaetomium globosum*, *Rhizoctonia solani* and *Tri-*

choderma sp. Although carboxymethyl cellulose is a soluble cellulose derivative it did not support the initial growth in these fungi except in *M. verrucaria* which could grow even without any additional sugar (17). Thus it is clear that addition of small amounts of easily metabolizable sugar in cellulase fermentation helps the organism to pick up growth quickly and get acclimatized during this period to switch over the metabolism to insoluble substrate—the cellulose.

These findings were exploited to increase the cellulase productivity (39). A new strain of *T. reesei* MCG77 was first grown on glucose for mycelial growth. A 50 per cent of this mycelial broth was used to inoculate 4 per cent roll-milled and ball-milled cellulose. A productivity of 72 filter paper units/litre/hour on the 84-hour batch cycle was obtained. A very high level of cotton activity (17 mg/ml at 84 hour) was also achieved. Steam exploded wood (81, 129) contains about 15–20 per cent soluble sugars mainly xylose, glucose, galactose, etc. These soluble sugars when retained in the fermentation media proved to increase the cellulase productivity (18). It was noticed that when these sugars were retained in a medium of 1.4–1.9 per cent cellulose from steam exploded wood (H60) a cellulase productivity of 37–45 filter paper units/litre/hour was obtained in 52–60 hours of cyclic-batch fermentation with a new strain of *T. reesei*, Rut-C30. In this medium no proteose peptone was used. These results were much better than those obtained by another improved strain of *T. reesei*, MCG77 when the quantity of cellulose substrate used was the criterion for comparison. Thus adding soluble sugars to the steam exploded wood medium greatly reduces the cost of cellulase fermentation when compared to the use of pure cellulose, proteose peptone and heavy inoculum grown on glucose. Sophorose, a B (1→2) glucoside, was a very powerful inducer of cellulase for *T. reesei* (76). Lactose, a B (1→4) galactoside, is another cellulase inducer. These two sugars are the only known cellulase inducers that do not have a B (1→4) glucosidic linkage.

Replacement of proteose peptone

Proteose peptone in small concentration is necessary to obtain high yields of cellulases. But proteose peptone is more costly than other ingredients in the cellulase fermentation medium. Optimum peptone concentration is 0.1 to 0.2 per cent depending on the cellulose concentration and concentrations higher than 0.5 per cent peptone strongly inhibit the cellulase production whatever may be the cellulose concentration (76). Peptone can be replaced with proflo (cottonseed flour), phyton, casein hydrolysate and yeast extract (76) and corn steep liquor (78) with slight decrease in cellulase production. The spent mycelium (0.25–0.3 per cent) from a cellulase fermentation was also an excellent replacement for proteose peptone (2). Hot water, acid or alkali extracts of the mycelium or fresh undried or frozen mycelium were equal to the freezedried mycelium. It was also found (18) that retention of soluble sugars available in steam exploded wood medium did not need proteose peptone in cyclic-batch cellulase fermentation.

Effect of surfactants

Surfactants are known to be useful in the fermentation industry for increasing growth rates and metabolite production. Addition of Tween 80 and Tween 40 doubled the yield of cellulase in *T. reesei* (96). The mechanism of enhancement in cellulase yield is not well understood, but may be related to an increase in the permeability of the cell membrane of the fungus, allowing more rapid secretion of the enzymes.

pH effects

The initial pH of the medium at the time of inoculation is usually about 5.6. The pH starts falling with increase in growth and utilization of NH_4^+ ions from $(NH_4)_2SO_4$ by the organism and formation of H_2SO_4 (22). The growth of *T. reesei* was rapid when the medium was controlled at pH 5, and decreased with a fall in pH to 3.5 although the production of cellulase was just in the reverse order. In media controlled at pH 3.5 from the time of inoculation, the growth of the fungus and enzyme production were markedly retarded; the media controlled at pH 4.5 or 4 were similar in cellulase production to those where the pH was allowed to fall and then maintained at pH 4.5 or 4.

Age and volume of inoculum

Spore inoculum (10^4 spores/ml in the fermentation medium) gives long lag phase and produces about half the amount of enzyme as compared to that with the use of a three day-old mycelial inoculum. The level of inoculum, 1 or 5 per cent v/v, had little effect on the yield of the cellulase (2). But most of the studies have shown that inoculum level higher than 5 per cent v/v was most suitable for cellulase production (18, 39, 84, 103). But the use of inoculum by volume may not be a good criterion as the amount of mycelial biomass produced varies with the concentration of carbon substrate in the inoculum medium. However the use of a 10 per cent v/v inoculum of mycelial broth grown on the same type of substrate and at same concentration is suggested in cellulase production. For example, if the cellulase production medium contains 2 per cent cellulose (Solka Floc), the inoculum should also be grown in the same medium with 2 per cent cellulose (Solka Floc) for reducing the lag phase and for enhancing the cellulase yield.

Temperature

Normally the fermentation is carried out at 29°C (optimum for enzyme production) while the growth rate continues to increase up to temperatures as high as 35°C. A considerable increase in cellulase production was recorded when the temperature of fermentation was maintained at 33–34°C in the early phase of the growth (about 24 hours) followed by reduction to 29°C throughout the rest of the fermentation period (85).

Substrate

The data on cellulase production (70) given in Table 7 indicate that the cellulosic substrates differ slightly when their filter paper activity was considered but the difference was high when C_1 activity was compared. In general highest C_1 activity was noticed in the case of crude cellulosic substrate-newspaper, but crystalline cellulosic substrates, Avicel pH 105 and absorbent cotton, produced lowest C_1 activity although Avicel pH 105 produced highest filter paper activity. Pure cellulosic substrates (SW 40, SW 200, Sweco 270, Avicel pH 105, and absorbent cotton) produced highest C_x activity as compared to the crude cellulosic substrates. Unfortunately the hydrolytic activity of these cellulase-systems has not been evaluated. Had this been done the real role of C_1 activity in hydrolysis could have been determined. There are some indications (20) that the C_1 activity of *T. reesei* increases with an increase

Table 7. Effect of cellulose growth substrate on enzyme production by *Trichoderma reesei* QM 9414 [from Mandels *et al.* (70)]

Substrate growth (1%)	Soluble protein (mg/ml)	C_x CMC (μ/ml)	Cellulase FP (μ/ml)	C_1 Cotton (mg glucose/ml)
SW 40	1.84	152	1.48	7.6
BW 200	1.40	89	1.11	2.4
Sweco 270	1.56	102	1.30	3.5
Avicel pH 105	1.68	144	2.04	3.5
Absorbent cotton	1.84	85	1.30	4.6
Jay bee newspaper	1.28	24	0.93	7.6
NEP 40	1.44	48	1.48	7.7
Sweco newspaper	1.44	56	1.11	8.6
Milled computer paper	0.46	22	0.74	5.9

Grown for 13 days on *T. reesei* medium with 1% cellulose, 0.1% proteose peptone, 0.2% Tween 80. Soluble protein and enzyme activities were determined on the culture filtrates.

in the crystallinity of the cellulosic substrate. It appears that the native crystallinity found in wheat straw cellulose was more favourable for production of C_1 activity than in "Sigmacell", a regenerated crystalline cellulose, or "Solka Floc", a purified form of cellulose with lots of amorphous portions. On the other hand, the crystallized substrates (wheat straw and sigmacell) gave almost the same filter paper activity but Solka Floc produced more filter paper activity. The cellulase-system produced on wheat straw which had high C_1 activity was found to be more active in hydrolyzing all the three substrates (wheat straw, sigmacell, and Solka Floc). The ratio of C_1 and filter paper activity in this enzyme system was about 3 : 1. The cellulase-system produced on Solka Floc with C_1 and filter paper activity of about 1 : 1 was poor in hydrolysis of all the three substrates (Tables 8 and 9). It is inferred from the data available that the potential of cellulase-system to hydrolyze various cellulosic substrates depends on the nature of the substrate used to produce the enzyme system. Recently it has been reported that the enzyme produced by the Rut-C 30 mutant of *T. viride* on steam exploded wood was faster and more effective in hydrolysis of wood as compared to the commercial enzyme—Novo enzyme (18). This might be due to the fact explained earlier that the enzyme was produced on wood with native crystallinity as observed in case of wheat straw.

Concentration of substrate

In most of the early studies the concentration of cellulose used for cellulase production was 0.5 to 1 per cent (21, 76, 119). A cellulose concentration

Table 8. Cellulase activities of broths [from Chahal *et al.* (20)]

Growth substrate	CMCase (μ/ml)	Filter paper activity (μ/ml)	Cotton activity (μ/ml)
T. reesei			
Wheat straw	2.54	1.94	7.17
Sigmacell	2.34	1.42	3.31
Solka Floc	2.85	2.84	2.87

Table 9. Hydrolysis of different cellulosic materials by cellulases produced from different cellulosic substrates [from Chahal *et al.* (20)]

Substrate for enzyme production	Substrate for enzyme reaction	mg. RS/g substrate/ml broth enzyme Reaction time		
		1 hour	6 hours	24 hours
T. reesei				
Wheat straw	Wheat straw	20.68	36.56	50.04
Sigmacell	Wheat straw	17.20	34.72	36.76
Solka Floc	Wheat straw	13.56	29.08	36.80
Wheat straw	Sigmacell	7.28	9.36	21.16
Sigmacell	Sigmacell	0.24	8.28	17.28
Solka Floc	Sigmacell	2.72	12.48	20.24
Wheat straw	Solka Floc	3.04	12.96	16.96
Sigmacell	Solka Floc	2.96	11.60	12.44
Solka Floc	Solka Floc	1.88	8.16 (?)	5.52 (?)

of 0.75 per cent was reported to be the optimum concentration for cellulase production by *T. reesei* and higher cellulose concentrations resulted in marked decrease in final enzyme level (76). The decrease in enzyme production with increase in cellulose concentration was because there was no corresponding increase in the nutrient salts. This fact becomes very evident when the data

of Sternberg (119) are examined in the light of the above statement. He obtained 0.6 filter paper units with 0.75 per cent cellulose and filter paper units increased correspondingly to 1.6 units when the cellulose concentration was increased to 2 per cent along with an increase in nutrient salts. The low enzyme yields obtained at 2 per cent cellulose by Mandels and Weber (76) has been attributed by Sternberg (119) to a fall in pH to 2.4 where the organism cannot grow properly and cellulase becomes inactivated. It was also

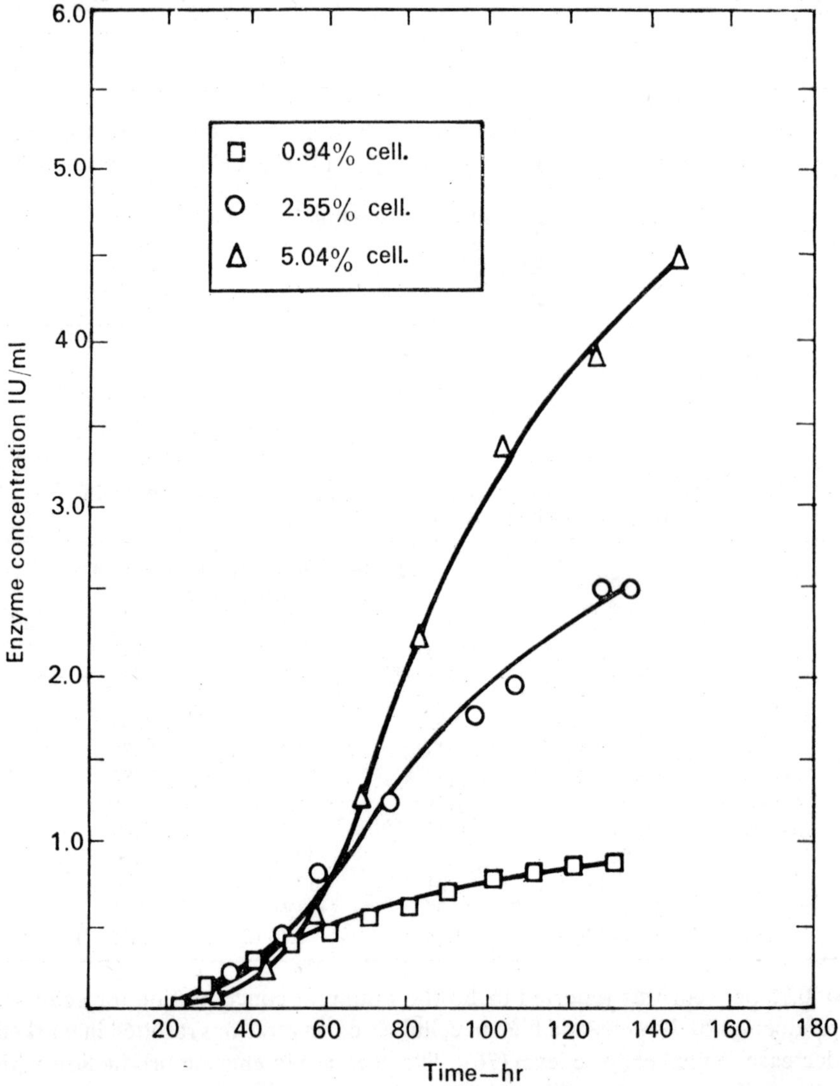

Fig. 5. Effect of concentration of cellulose on cellulase production [from Nystrom and Dilca (85)].

claimed by him that increase in enzyme yield on 2 per cent cellulose was mainly due to pH control at 3 but failed to note that such an increase may also be due to increase in nutrient salts. This fact has been confirmed by growing *T. reesei* in various cellulose concentrations (0.94, 2.55 and 5.04 per cent) with corresponding increase in nutrient salts (85). There was corresponding increase in cellulase yields with increase in cellulose concentration when the media were also enriched with salts accordingly (Fig. 5). It has also been pointed out that optimum C : N ratio should be close to 8 to get optimum cellulase yields and the supply of N seems to be most critical.

Kinetics of cellulase production

Cellulase synthesis is believed to be related to the growth of the organism (120) although contrary reports (12, 41, 42) indicate that the enzyme is synthesized near or at the end of the growth phase and suggest a negative correlation between specific enzyme synthesis and the growth rate of the organism.

The kinetics of growth and cellulase production of *T. reesei* postulated here, is based on the data obtained from Ryu and Mandels (104), Andreotti *et al.* (2), Gallo *et al.* (39), Chahal *et al.* (18–20), Peitersen (87) and the data and speculations of Chahal (unpublished). The postulated kinetics of cellulase production given in Fig. 6, are explained as follows:

i) The initial growth of the organism starts immediately on the utilization of soluble sugars or proteins (proteose peptone, etc.) present in the growth medium. This phase of growth is faster than the subsequent stage.

ii) The soluble sugars or proteins are consumed by the organism within a short time (12–24 hours) depending on their concentrations. During this time almost no cellulose is utilized.

iii) As soon as about 80–90 per cent of sugars and proteins are consumed the organism shifts its metabolism to utilization of cellulose. There may be some signs of diauxic type of growth which may be very clear or may not be visible at all depending on the crystallinity of the cellulose and particle size of cellulose. Thereafter, there will be continuous consumption of cellulose till it reaches a plateau. There may be another shift in cellulose utilization if the substrate has some amorphous portion as found in Solka Floc SW40. The cellulose utilization curve shows that some cellulose is always left unutilized at the end of the fermentation which is attributable to the cellulose positive pure mycelium of *T. reesei* (19).

iv) Mycelium synthesized is very difficult to estimate especially when it is mixed with cellulose. This is the main reason that most studies do not show growth parameters in cellulase fermentation thus making it more difficult to interpret the cellulase kinetics. Wherever such growth parameters were given they were based on indirect measurements, usually the protein content of the total biomass (mycelium + unutilized cellulose). The growth curve of the organism given here is based on such estimations of biomass protein at

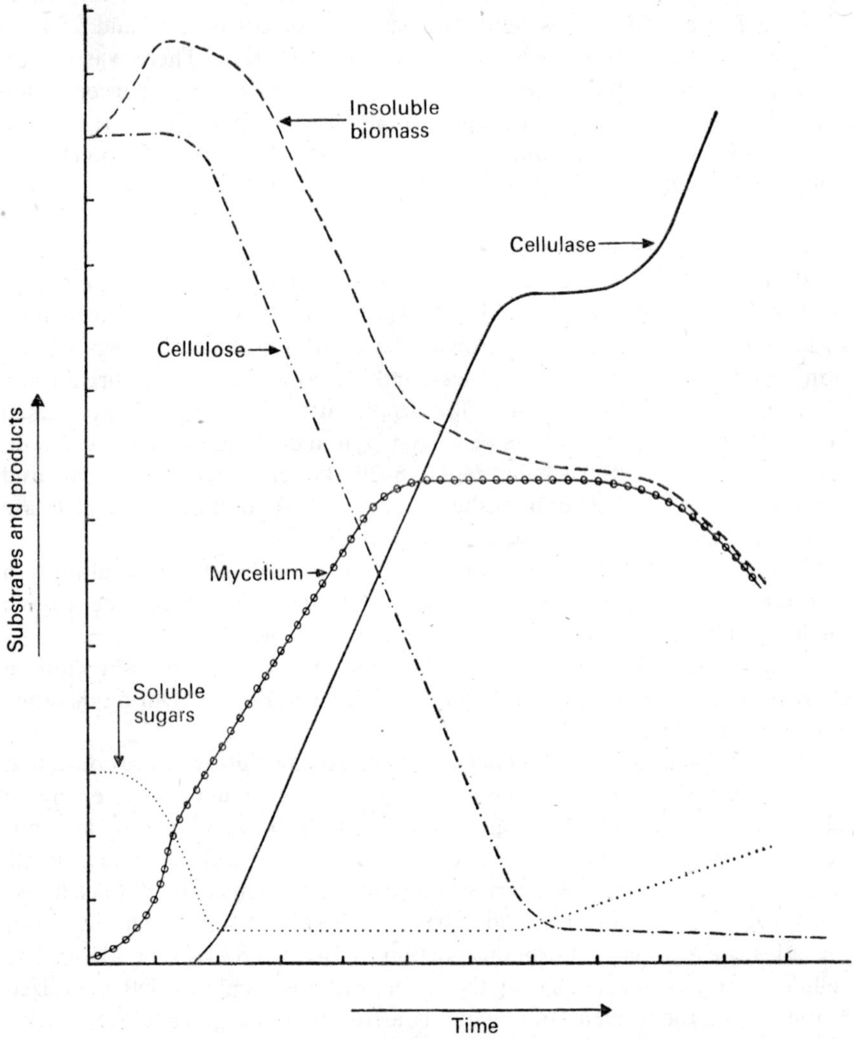

Fig. 6. Kinetics of growth and cellulase production of *Trichoderma reesei*.
— · — Cellulose.
— — — Insoluble biomass (mixture of mycelium and unutilized cellulose).
. . . . . . Soluble sugars added or already present in the substrate (lignocelluloses) produced during pretreatment with acid at high pressure or with steam at high pressure.
○-○-○ Mycelium. Calculated on the basis of insoluble protein synthesized by fungal biomass (mixed with unutilized cellulose).
——— Cellulase as measured in the cell (biomass) free of fermentation broth.

certain times of growth. There is a slight lag phase followed by fast exponential growth on soluble sugars and proteins and followed by another slow exponential growth on insoluble substrate, cellulose. Thereafter, there is a stationary phase. During this stage the cellulose is still being consumed which indicates that some mycelium is being synthesized while the older mycelium may be autolyzing thereby providing for growth in the stationary phase. The stationary phase is followed by the declining phase because there is no more cellulose available for further growth and due to autolysis of the older mycelium. Autolysis of mycelium is a common phenomenon in fungi (25).

v) The weight of insoluble biomass (mixture of mycelium and unutilized cellulose) increases a little during the first few hours of growth of an organism on soluble sugars and proteins. This increase in weight is due to the newly synthesized mycelium and the presence of insoluble cellulose which is not utilized during this stage. Thereafter continuous decline in weight of insoluble biomass is seen because for every unit of cellulose utilized about half of it is converted into biomass and the other half is consumed as a source of energy for various metabolic processes of the organism and synthesis of cellulase. Then there is the levelling of the curve just parallel to the stationary phase of the growth. Thereafter the decline in weight of insoluble biomass starts again which is due to the autolysis of the mycelium of the organism.

vi) Cellulase activity is not noticed until almost all the soluble sugars and proteins are consumed by the organism and the growth is started on cellulose. There might be some cellulase activity during the initial growth of the organism on soluble sugars or proteins and some cellulose, but it is not detected because most of the synthesized cellulase is adsorbed on the substrate-cellulose. Thereafter, the quantity of cellulase synthesized is continuously increased and runs almost parallel to the exponential growth phase. Cellulase production continues even during the stationary phase when some cellulose is also being consumed. The cellulase production levels off when the substrate-cellulose is completely consumed. Thereafter, in most of the fermentations more cellulase appears in the broth even when the growth of the organism ceases and cellulose is depleted in the fermentation medium. This increase in cellulase at the end of the fermentation is mainly due to the release of the enzyme from the mycelial cell walls due to autolysis but not due to the *de novo* synthesis as postulated by Ghose (41). The autolysis of the mycelium is indicated by the decline in the weight of the insoluble substrate biomass, the mycelial biomass and the increase output in soluble sugars released by the autolysis of fungal cells. Some protein is also released from the disintegration of this protoplasm during autolysis. The specific enzyme activity will be low at this stage because some of the protein released from the autolyzing cells is of non-enzymic nature. The specific activity of the enzyme at this stage will have insignificant value in determining its hydrolytic potential.

Strain improvements

There are three primary enzymes elaborated by *T. reesei* which act synergetically to hydrolyze cellulose: 1,4-β-D-glucan 4-glucanohydrolase (EC3.2.1.4) (endoglucanase), 1,4-β-D-glucan cellobiohydralase (EC3.2.1.91) (cellobiohydrolase), and β-glucosidase (EC3.2.1.21) (cellobiase). Strain improvement in *T. reesei* was done keeping the following mechanisms controlling the synthesis and activity of the cellulase complex of enzymes in view (78):

Biochemical mechanisms controlling cellulase synthesis and activity in *T. reesei* [from Montenecourt and Eveleigh (78)]

Enzyme	End-product inhibitor	Catabolite repressor	Inducer
cellobiohydrolase	cellobiose	glucose or	cellulose
endoglucanase	cellobiose	metabolites of	sophorose
cellobiase	glucose	glucose	cellobiose

Montenecourt and Eveleigh (78) devised selective screening techniques which allow for the isolation of mutant strains that can specifically overcome those catabolite repression and end-product inhibition regulatory mechanisms and at the same time hyperproduce one or all of the individual enzymes in the cellulase complex. Utilizing these screening techniques roughly 800,000 colonies were screened and approximately 100 mutants were isolated. Of these only Rut-NG14 and Rut-C30 have been fully investigated to date.

Yields of cellulase of 15 units/ml under controlled fermentor conditions have been achieved with both Rut-NG14 and Rut-C30. Quantitative reaction of Rut-NG14 enzyme preparation with purified antibodies to cellobiohydrolase shows that in this mutant, the cellobiohydrolase is specifically hyperproduced relative to the rest of the enzymes in the cellulase complex. Rut-C30, which was derived from Rut-NG14, shows resistance to catabolite repression for filter paper hydrolyzing enzymes, endoglucanase, and cellobiase and is capable of producing these enzymes during growth on cellulose. The future goal is to isolate constitutive mutants and end-product resistant strains so as to increase overall cellulase yields.

At the U.S. Army Natick Research and Development Command, Natick, Massachusetts (39), Strain MCG77 of *T. reesei* was selected based on its ability to clear cellulose on an agar plate containing 8 per cent glycerol and its near freedom from glycerol repression in submerged culture (Fig. 7). The mutants QM9414 and NG14 have high cellulase productivities but are subject to catabolite repression. Strain MCG77 gives yields similar to the other enhanced mutant strains QM9414 and NG14 in regular cellulase fermentations. However, strain MCG77 grows rapidly on soluble substrates without subsequent restrictions in cellulase synthesis and has a faster rate of metabolism and enzyme production. It seems that MCG77 of Natick is similar to that of C30 of Rutgers as both are resistant to catabolite repression. The comparison of parent strain (QM6a) with the other improved strains is given in Table 10.

Table 10. Cellulase production by mutant strains of *Trichoderma reesei* [from Ryu and Mandels (104)]

Strain	CMC (units/ml)	Filter paper (units/ml)	β-glucosidase (units/ml)	Productivity (FPU/1/h)	Soluble protein (mg/ml)
QM6a (parent)	88	5	0.3	15	7
QM9414 (Natick)	109	10	0.6	30	14
MCG77 (Natick)	104	11	0.9	33	16
C30 (Rutgers)	150	14	0.3	42	19
NG14 (Rutgers)	133	15	0.6	45	21

Cultures grown 14 days in 10–1 fermenters on 6% 2 roll-milled cotton with pH control greater than 3.0 using 2 N NH_4OH.

Enzyme units = μmol glucose produced per min in standard assay.

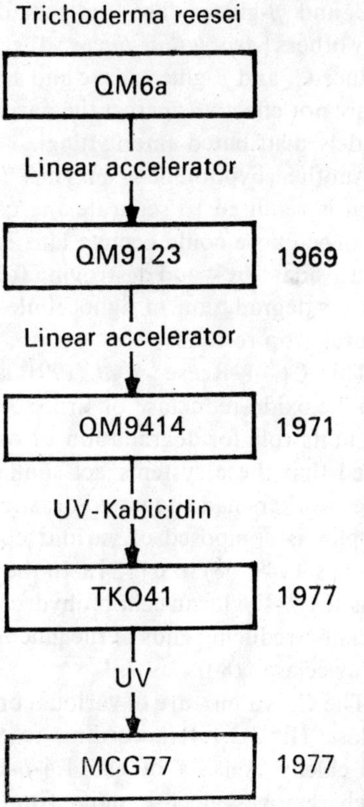

Fig. 7. Genealogy of high-yielding cellulase mutants developed at the United States Army Natick Research and Development Command, Natick, Massachusetts. The years when the mutants were isolated are indicated and the types of mutagen used are also given. Kabicidin is a fungicide [adapted from Gallo *et al.* (39)].

This comparison indicates that Rut-C30 and Rut-NG14 are good for filter paper activity but poor for β-glucosidase while Natick QN9414 and Natick MCG77 are slightly poor for filter paper activity and slightly better for β-glucosidase. But the real evaluation lies in their ability to hydrolyze the cellulosic substrates. Unfortunately this comparison is not available at present.

Nature of the cellulase-complex system and mechanism of lignocellulose breakdown

About thirty years ago the first concept of degradation of cellulose was postulated by Reese *et al.* (99), when it was proposed that C_1 factor acts on hydrogen bonds and/or on van der Waals forces to release long chains of glucose polymers from crystalline cellulose for further degradation with C_x enzymes into a mixture of glucose and cellobiose. The cellobiose is hydrolyzed into glucose units with β-glucosidase (Fig. 8). They also suggested that C_1 is essential to work with C_x for complete hydrolysis of crystalline cellulose. There are a few fungi [*Myrothecium verrucaria, Aspergillus terreus, A. fumigatus, A. niger* QM877, and *Trichoderma reesei* (added later)] which produce C_1, C_x and β-glucosidase and can degrade native cellulose while there are many others [*Aspergillus flavus, A. sydowi, A. tamari, A. niger* (QM455)] which produce C_x and β-glucosidase and are able to degrade only modified cellulose but are not effective against the native cellulose. The activity of β-glucosidase is widely distributed among fungi.

Another hypothetical enzyme "X" has been added to this scheme (27) which is required to separate the cellulose from lignin before any degradation of cellulose could occur. This enzyme is produced along with C_1, C_x and β-glucosidase by wood destroying fungi (*Poria monticola, Polyporus versicolor*, etc.) for degradation of lignocellulosic materials as found in wood and agricultural crop residues.

This C_1 of Reese *et al.* (99) is very much comparable to cellobiose: quinone oxido-reductase of Eriksson (34) and to H_2O_2 : Fe system of Koenig (60) in its role for degradation of native crystalline cellulose. Both had suggested that these systems act similar to C_1 on hydrogen bond in cellulose. Other workers had suggested drastic modifications of this concept. This C_1-C_x complex is composed of various components varying from 3 (74, 112) to as many as 4 (49, 58) to 6 (134). In the recent concept (143) C_1 has been described as 1, 4-β-D-glucan cellobiohydrolase (EC3.2.1.91) which releases cellobiose from non-reducing ends of the glucose polymer chains. It has also been named as "avicelase" (80).

The C_x is a mixture of various components of 1,4-β-D-glucan 4-glucanohydrolase (EC3.2.1.4) according to various workers (53, 58, 89). The mixture chiefly consists of exo 1,4-β-D-glucan glucohydrolase (exo-glucanase) which removes glucose units from non-reducing end of glucose polymer

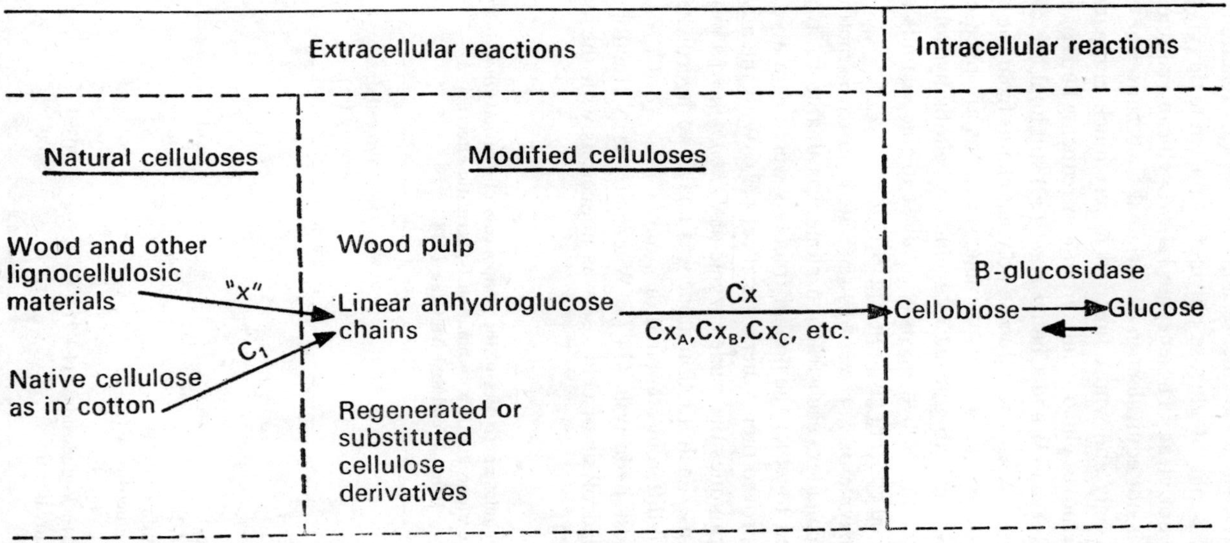

Fig. 8. Breakdown of cellulose in lignocellulosic materials, adapted from the concept postulated by Reese *et al.* (98). C_1 acts on the hydrogen bonds and/or van der Waals forces to release long chains of glucose polymers from crystalline portion of cellulose. Cx—a complex enzyme system acts on linear anhydrous glucose chains, modified cellulose (wood pulp), regenerated or substituted cellulose derivatives and releases mixture of cellobiose and glucose. The cellobiose is hydrolyzed by β-glucosidase into glucose. An unnamed ("X") enzyme has been added here in this system. According to Cowling (27) "X" is necessary to release cellulose from lignin. The "X", C_1 and Cx are extracelluelar enzymes which must be released by the cellulolytic organisms to act on lignocelluloses to release soluble dimers (cellobiose) and monomers (glucose) which can penetrate into the cells. β-glucosidase is an intracellular enzyme which hydrolyzes cellobiose into glucose.

chains and endo 1,4-β-D-glucan glucanohydrolase (endo-glucanase) which randomly attacks glucose polymer chains to break them into smaller chains (oligomers) and also release some glucose units during this reaction. However, the mode of first split of glucose polymer chain from the crystalline portion of cellulose is not clear. The enzyme 1,4-β-D-glucan cellobiohydrolase (the so-called C_1) is not equivalent to the original C_1 because the present C_1 releases cellobiose only and whose function is very much comparable to exo-glucanase which removes glucose units from non-reducing ends of glucose polymer chain (58). Reese (93) and Chahal *et al.* (20) still hold the old concept of C_1 that it helps to release the glucose polymer chains from the crystalline portion of cellulose and as soon as the glucose chains are released they are attacked by endo- and exo-glucanases including cellobiohydrolase and β-glucosidase to give complete hydrolysis of cellulose to glucose units according to the recent modified concept given in Fig. 9.

When the crude enzyme of *T. reesei* was separated into two fractions (C_1 and C_x) on DEAE sephadex, each fraction alone was unable to hydrolyze cotton but when mixed together the pooled fractions were again able to hydrolyze cotton which proved their synergistic effect (72). With the advent of more sophisticated techniques the crude enzyme of *T. koningii* was separated into 5 different fractions. Each fraction alone was unable to hydrolyze cellulose but when tested in different combinations, only $C_1+C_{x\,(1)}+C_{x\,(2)}$ combination gave the highest hydrolysis (24%). When all the 5 fractions were combined, as much hydrolysis of cellulose was noticed as with the original enzyme (Table 11) (144).

Table 11. Relative cellulase activities of the components of *Trichoderma koningii* cellulase alone and in combination (synergistic effect) [from Wood and McCrae (144)]

Enzyme	Relative cellulase activity (%)
C_1	1
$C_{x\,(1)}$	1
$C_{x\,(2)}$	1
β-glucosidase-1	nil
β-glucosidase-2	nil
$C_1+\beta$-glucosidase (1+2)	5
$C_x(1+2)+\beta$-glucosidase (1+2)	4
$C_1+C_{x\,(1)}+C_{x\,(2)}$	24
$C_1+C_x\,(1+2)+\beta$-glucosidase (1+2)	103
Original culture filtrate	100

Cellulase activity=Hydrolysis of cotton. Components added at levels equal to original filtrate.

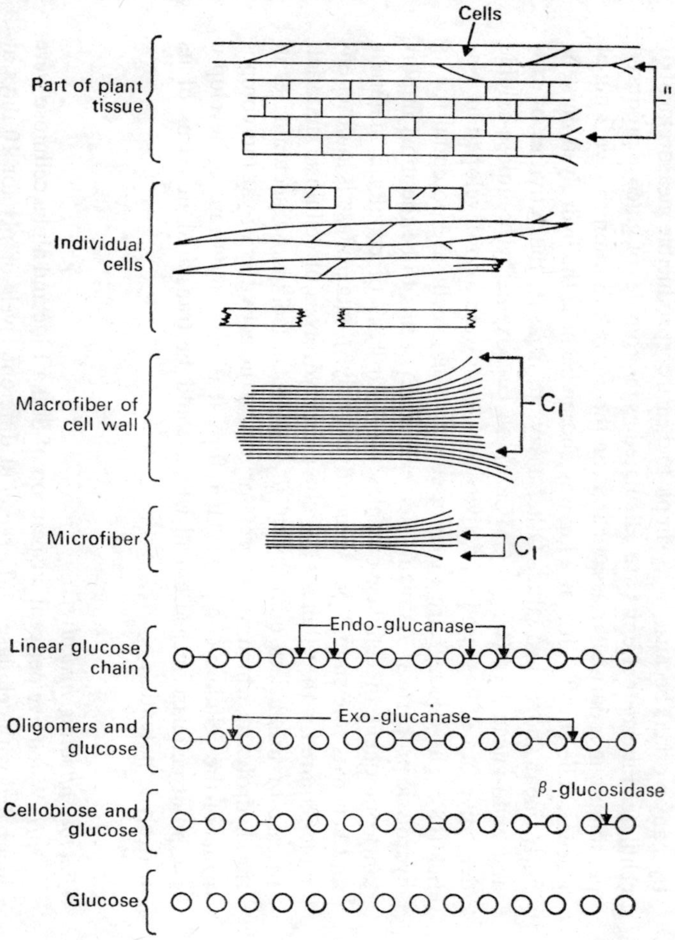

Fig. 9. Diagrammatic representation of breakdown of cellulose in lignocellulosic materilas (based on recent findings). Top to bottom:

—Part of plant tissue showing different types of cells and the action of an unnamed enzyme "X" on middle lamella.

—Individual cells released by the action of "X". Some of the cells show broken ends while others show some cracks in their cell walls.

—C_1 acts on cellulose of cell wall and releases macrofibres (macrofibrils) and it continues to work on them to release microfibres (microfibrils). During this process single linear anhydrous glucose chains are also released.

—C_1 releases single linear anhydrous glucose chains from microfibre (microfibril).

—Endo-glucanase acts on the linear anhydrous glucose chain at random to release oligomers. During this reaction some single glucose units may also be released.

—Exo-glucanase acts on non-reducing ends of long linear anhydrous glucose chains and also on oligomers to release cellobiose and glucose units. The exo-glucanase which releases cellobiose from non-reducing ends is called cellobiohydrolase.

—*β*-glucosidase acts on cellobiose to yield glucose.

—All the reactions given above occur simultaneously and aiso synergistically on cellulose to release glucose units.

A cellulase-system containing hemicellulases will also hydrolyze the hemicelluloses to release various sugars—xylose, mannose, galactose, arabinose, etc.

Hydrolysis

The broth from cellulase fermentation obtained after filtration or centrifugation can be used directly as a source of enzyme for hydrolysis of various cellulosic materials. The enzyme broth can be concentrated by the use of Amico or Abcor ultrafiltration membranes of 10,000–30,000 molecular weight cutoffs with little loss of activity. It can also be precipitated by 66 per cent acetone without any loss in its activity (75). The enzyme broth can be easily stored in a refrigerator (1–3°C) using toluene, thymol and azide at 0.01 per cent as preservatives (98). Merthiolate when used at 0.01 mg/ml as a preservative inactivates the cellulase.

Measurement of cellulases

In industry a well-defined unit of cellulase is needed to determine the operation cost of fermentation and also quantify yield of sugars from certain cellulosic materials. But this is not an easy task because a bewildering array of substrates, enzyme actions, units, and activities have been used to measure cellulase. Added to this, every worker develops his own assay method and consequently the results from one laboratory cannot be compared with those of others.

The filter paper assay (68) as developed at Natick has been used widely by many workers because of its simple and easily reproducible measurement of cellulase. Filter paper units are calculated as micromoles of glucose produced per minute in a particular assay based on the enzyme dilution to give 2 mg of glucose. The cutoff value of 2 mg was chosen because the hydrolysis curve is fairly linear above this value. Other assays to measure the activities of exo- and endo-glucanases (C_x or CMCase) on carboxymethyl cellulose-soluble derivative of cellulose and C_1 activity on cotton were also developed at Natick (68). These assays are rarely used to evaluate cellulase potential for its hydrolytic properties. As mentioned earlier, C_1 seems to be the most important component of cellulase-complex for hydrolysis of cellulosic materials and hence its role should be evaluated properly. β-glucosidase is another enzyme required with cellulases for complete hydrolysis of cellulosic material. In case of insufficient quantity of β-glucosidase, cellobiose accumulates in the hydrolyzate which will repress or inhibit the activities of various components of the cellulase-complex. Therefore, it is very important to develop a foolproof cellulase measurement which could be translated in terms of its hydrolytic potential.

Factors affecting hydrolysis

pH. When five per cent suspension of Solka Floc and alpha cellulose were incubated with cellulase of *T. reesei* at different levels of pH for 10 days at

50°C, maximum hydrolysis of both the substrates was recorded at pH of 4.8 (76).

TEMPERATURE. On the other hand at higher temperatures of 55 and 60°C the initial reaction was very high and after six hours of hydrolysis the enzyme became inactivated as there was almost no increase in sugar yields. The temperature of 50°C proved to be the most suitable for hydrolysis as most of the hydrolysis was complete within the first six hours and sugar yields continued to increase on further incubation (75) (Table 12).

Table 12. Effect of enzyme and substrate concentration on hydrolysis of hydropulped paper waste [from Mandels and Sternberg (75)]

Substrate conc. (wt %)	Enzyme conc. FP units/ml	Temperature (°C)	Reducing sugar at 24 hr (mg/ml)	Saccharification at 24 hr (mg/ml)
2.5	0.5	50	16	64
5.0	0.5	50	23	43
7.5	0.5	50	26	33
2.5	1.0	50	20	77
5.0	1.0	50	33	60
7.5	1.0	50	39	47
2.5	1.5	50	21	79
5.0	1.5	50	39	67
7.5	1.5	50	46	53
5.0	1.0	45	28	54
5.0	1.5	45	33	59
5.0	1.0	55	30	54
5.0	1.5	55	33	61

One litre stirred tank reactor at pH 4.8.

Enzyme *Trichoderma* QM9414 Culture Filtrate. Specific activity 0.62 FP units/mg protein.

Substrate hydropulped paper waste from the Pentagon.

$$\% \text{ Saccharification} = \frac{\text{glucose mg/ml} \times 0.9}{\text{substrate mg/ml}}.$$

CONCENTRATION OF SUBSTRATE AND ENZYME. The effect of different concentrations of substrate (hydropulped paper waste) and *T. reesei* enzyme was investigated extensively in a series of one litre stirred tank reactors (Table 13) (75). As the substrate concentration was increased the yield of sugar was increased, but the percentage of conversion was correspondingly decreased. Similarly when enzyme concentration was increased the yield of sugar increased, but the amount of sugar produced per unit of enzyme was correspondingly decreased. From these data, Mandels and Sternberg (75) concluded that the fastest rates are attained by using high substrate and enzyme concentrations for short hydrolysis times, but such conditions lead to only 20–30 per cent conversion, yield dilute sugar syrups and consume large quantities of enzyme. Longer hydrolysis time yields more concentrated syrups but requires long residence time which increases capital costs.

A critical analysis of these data shows that 60 units of enzyme/g of substrate were used for 2.5 per cent slurry of hydropulp paper whereas only half the amount (30 units/g) of enzyme was used to hydrolyze double the quantity of substrate (@ 1.5 units/ml for 2.5, 5.0 and 7.5 per cent substrate). Similarly in another study (5) 20 units/g substrate were used for 10 per cent slurry of milled newspaper NEP200 and the quantity was reduced to 13.3 units/g substrate when 15 per cent slurry was used. Had equal amounts of enzyme units per g of substrate been used at different concentrations of the substrate, there might have been an entirely a new picture on the rate of hydrolysis and the amount of substrate hydrolyzed. From these observations it is hypothesized that the use of high concentration of cellulose and high concentration of enzyme will yield a sugar syrup of sufficient concentration for economical fermentation of ethanol.

Effect of structural features of substrates and pretreatments

These constraints are due to insolubility, high crystallinity and coating of lignin over the cellulose microfibrils.

Crystallinity

The influence of the degree of crystallinity on the susceptibility of cellulose to enzymatic hydrolysis has been studied by many workers (83, 98). It was observed (83) that cellulases readily degrade the more readily accessible amorphous portions of regenerated cellulose but are unable to attack the less accessible crystalline portion. The residual crystalline portion will not be hydrolyzed until the cellulase-complex is rich in C_1 factor. It is the crystallinity which affects hydrolysis the most, the degree of polymerization of cellulose molecules is of limited significance in determining the susceptibility of cellulose to hydrolysis (29). It was found that the hydrolysis rate is mainly dependent upon the fine structural order of cellulose which can best be represented by the crystallinity rather than the simple surface area. The rate of hydrolysis of microcrystalline cellulose was much slower than that of Solka Floc because of its high crystallinity index (CrI). When the CrI of Solka Floc was reduced by ball-milling, the rate of hydrolysis increased considerably (37). Similarly, Mandels and Sternberg (75) had shown that ball-milling increased the hydrolysis of various pure and waste cellulosic materials to a great extent (Table 13).

Effect of lignin

The effect of delignification of bagasse and rice straw on hydrolysis becomes very clear by a comparison of the data in Table 13 with those in Table 14. Bagasse as such gave only 6 per cent hydrolysis which increased to 48 per cent with ball-milling (Table 13) and to 88 per cent when delignified with alkali and paracetic acid (Table 14). These results show that ball-milling is not as effective as delignification because in ball-milling the particle size and crystallinity are reduced but some of the cellulose still remains attached to lignin and hence is not easily accessible to the enzyme.

Table 13. Hydrolysis of cellulose by *Trichoderma reesei* cellulase [adatped from Mandels and Sternberg (75)]

Substrate	% Saccharification			
	1 hr	4 hr	24 hr	48 hr
(a) *Pure cellulose*				
Cotton-fibrous	1	2	6	10
Cotton-ball milled	14	26	49	55
Cellulose pulp-SW40	5	13	26	37
Cellulose pulp-ball milled	23	44	74	92
(b) *Waste cellulose*				
Bagasse	1	3	6	6
Bagasse-ball milled	14	29	42	48
Newspaper-shredded	10	24	31	70
Newspaper-ball milled	18	49	65	70

QM9414 cellulase 1.2 FP units per ml.
Saccharification at 50° pH 4.8.

Table 14. Saccharification of delignified bagasse and rice straw with cellulase Onozuka TvCL7 [from Toyama and Ogawa (133)]

Substrate	Substrate conc. (%)	After 24 h incubation		After 48 h incubation	
		sugar (%)	decom. (%)	sugar (%)	decom. (%)
Bagasse	10	9.69	87.2	9.79	88.1
	15	12.57	71.2	15.07	85.4
Rice straw	10	8.58	77.2	9.38	84.4
	15	11.15	63.2	13.59	77.0

Substrates were delignified by boiling with a 1 per cent NaOH solution for 3 h in a 100 ml Erlenmeyer flask; 10 g of delignified bagasse of rice straw and 90 ml of a 3 per cent cellulase Onozuka TvCL7 solution or 15 g of the same substrate and 85 ml of the same enzyme solution were incubated at pH 5.0, 45°C for 24 to 48 h.

Alkali treatment (treated in 2 per cent NaOH at 70°C for 90 minutes, washed and neutralized) of newspaper also proved to be a more effective means of increasing susceptibility provided the product was undried (70). Analysis showed little or no decrease in lignin content so the change must have been due to the swelling and hydration of the fibres. Alkali-treated newspaper was more susceptible than hammer-milled paper but poorer than the ball-milled sample. However, alkali-treated newspaper when dried lost its susceptibility but when dried and ball-milled proved to be the best in hydrolysis with about 60 per cent saccharification (70). When 80-mesh spruce wood treated with 2N NaOH was hydrolyzed with cellulase, 80 per cent total carbohydrates was converted to sugars (90). It is concluded that the substrate ground to about 80 mesh and then treated with alkali is most susceptible to enzymatic hydrolysis.

Steam explosion of lignocellulosic materials is another pretreatment which has proved to be the best for hydrolysis (19, 114). It has been reported that hydrolysis of steam-treated (190°C) poplar shavings increased four times the rate of hydrolysis in control samples (114). Similarly about 89 per cent theoretical conversion of glucose was obtained when aspen wood steam exploded with Iotech process was hydrolyzed with *T. reesei* Rut-C30 enzyme (19). Aspen wood treated with Stake process gave more than 90 per cent of theoretical saccharification with cellulase from *T. reesei* (129).

Role of β-glucosidase

The end-products depend on the nature of cellulose hydrolyzed and the nature of the enzyme. In the case of pure cellulose and incomplete enzyme system the main constituents of the hydrolysate are glucose and cellobiose (119). In crude cellulosic substrates in addition to these sugars other sugars like xylose (75) are also found in the hydrolysate. The cellobiose levels generally vary between 10 and 70 per cent of the total sugars depending on the nature of the enzyme and the concentration of the substrate. Cellulase is moderately inhibited by glucose and strongly inhibited by cellobiose (69). For both sugars the extent of inhibition increases with increasing resistance of the cellulose. β-glucosidase is inhibited by glucose (14, 46). As glucose accumulates in the hydrolysate the β-glucosidase is inhibited resulting in cellobiose accumulation ultimately affecting the hydrolysis by inhibiting the cellulases. *Trichoderma reesei* cellulase preparations usually have ~0.02–0.1 units of β-glucosidase for every filter paper unit which is not sufficient for complete hydrolysis of cellobiose produced especially at high concentrations of cellulose. There have been some efforts to increase the β-glucosidase activity in cellulase system of Rut-C30 strains by changing temperature and pH profiles during fermentation (128). But the potential of such enzyme-systems in hydrolysis of hammer milled and acid-treated corn stover seems to be not promising as only 60 per cent theoretical glucose was obtained at 5 per cent concentration of substrate and it decreased to 43 per cent when concentration was increased to 25 per cent (88). It has been reported that ~1–1.5 β-glucosidase units per filter paper unit (19, 114) is sufficient for complete hydrolysis. In the case of a cellulase-system deficient in β-glucosidase addition of this enzyme greatly increases the glucose, total sugars and percentages of hydrolysis. A good source of β-glucosidase is *Aspergillus phoenicis* (1).

Fermentation of hydrolysate into ethanol

Once glucose is obtained in solution, whether from sugar, starch or cellulosic materials many microorganisms especially brewer's yeast (*Saccharomyces cerevisiae*) can ferment it into ethanol anaerobically:

$$C_6H_{12}O_6 \rightarrow 2C_2H_5OH + 2CO_2.$$

Recently, work is being done on the utilization of a bacterium, *Zymomonas mobilis* (23) which may prove to be more productive and more tolerant to high glucose and ethanol concentrations.

The maximum theoretical conversion of glucose to ethanol is 51 per cent by weight, and this value can be attained very closely in practice through the use of immobilized yeast cells because in normal fermentation some sugar is utilized for the multiplication of yeast cells. The actual yield is therefore about 95 per cent, i.e. 48–49 per cent overall.

Batch fermentation has been the traditional method of producing ethanol for thousands of years. The simplicity of the batch method makes it most adaptable to small-scale operation. From an industrial point of view, there are two major disadvantages of this system. Firstly, there is a period of down time between batches when the fermentor is emptied, cleaned, sterilized, refilled with new substrate and inoculated with a starter culture for subsequent fermentation run. The second disadvantage is the lag phase of microbial growth and the occurrence of death phase, as the concentration of ethanol in the "beer" rises, resulting in further loss of time in alcohol production. On the other hand the continuous fermentation method, where beer is continuously drawn out as new substrate is added, offers the advantage of keeping the microbes in the active stage with rapid production of ethanol. The continuous operation is periodically shut down and restarted to ensure that no contamination has taken place, but the frequency of this is much less than in the batch method. The use of immobilized cells for fermentation is becoming the most important factor for reducing the cost of fermentation (24, 121).

Ethanol production is severely suppressed when ethanol concentration in the beer rises above 7 to 10 per cent. At the University of California an improved fermentation technology has been developed for achieving a twelvefold increase in ethanol production rate over conventional continuous fermentation (30). Such an increase in productivity would decrease the required fermentor volume over batch and conventional continuous fermentation to approximately 1/26 and 1/8, respectively. In this process (still at laboratory scale), fermentation proceeds under vacuum (6.7 KPa) so that ethanol is continuously boiled off at fermentation temperature of 35°C, thereby eliminating end-product inhibition. A portion of fermentation broth is continually bled off and the cells are recovered by centrifugation. These cells are then washed to get rid of toxic non-volatile compounds and recycled to maintain a very high cell density in the fermentor. Finally a small amount of pure oxygen is sparged into the fermentor to increase yeast viability. The ethanol condensate is delivered to the still at a concentration of 16 to 21 per cent, thus reducing distillation requirements. The compressor employed in the vacuum fermentation is driven by high pressure steam (4.135 MPa), the exhaust from which supplies 63 per cent of the distillation energy requirements resulting in considerable economy in the energy involved in the process.

Ethanol production from pentoses

Acid or enzymatic hydrolysis of lignocelluloses yields hexoses (glucose, mannose, galactose, etc.) and pentoses (xylose, arabinose). Hexoses are easily fermented into ethanol by common yeasts (*Saccharomyces* spp.) or by bacteria (*Zymomonas* spp.). But pentoses are not fermented by these organisms. The pentoses are about 25–30 per cent of total sugars released from the hydrolysis of carbohydrate from lignocellulosic materials. Thus from every 1000 kg of total sugar 250–300 kg of pentoses (mostly xylose) will be left unutilized, which would create a disposal problem.

The residual pentoses (xylose) can be concentrated into molasses for animal feed or can be fermented into animal feed with *Chaetomium cellulolyticum* (15). The final product contains about 45 per cent protein. But keeping in view the demand for ethanol fuel in the world various research laboratories are trying to develop a process for converting xylose—a major pentose sugar—into ethanol.

As early as 1940's *Fusarium* sp. was used for ethanol production from pentoses in sulphuric acid-derived wood mill waste liquors (64, 66), Douglas fir hydrolysate (82), corn stalk syrup (64), wheat mashes, and wheat hydrolysates (82). In this process yeast was used to convert glucose to ethanol and then inoculated with *Fusarium* to produce additional ethanol from pentoses (66, 110). Recently (11) *Fusarium oxysporum* has been reviewed as a potential system for ethanol production from xylose in combination with *S. cerevisiae*.

Within the last couple of years a few papers on xylose fermentation into ethanol have appeared. The published literature indicates that there are many yeasts which can ferment pentoses aerobically into yeast cell biomass but not into ethanol (9, 127). But Wang *et al.* (138) reported that most yeasts especially *Schizosaccharomyces pombe* can ferment xylose, a keto-form of xylose, into ethanol. They reported that fermentation of xylose by xylose-oxidizing yeast indicates that a control operates under conditions of low oxygen tension which prevents the catabolism of xylose to xylulose. Subsequently the same authors (138) reported that xylose is fermented into ethanol in the presence of glucose/xylose isomerase (xylose is converted to xylulose) by *S. pombe* but only ~10 per cent of theoretical ethanol yield was obtained. Meanwhile another paper (45) appeared which described a very similar process for ethanol production by using xylose isomerase to convert xylose to xylulose and then fermentation with a common yeast *Saccharomyces cerevisiae.* They got only 30 per cent conversion of total xylose supplied and only 60 per cent of conversion of utilized xylose into ethanol. This indicated that some of the xylose was utilized for synthesis of cells and some for the production of other metabolites. Xylitol is one of the major intermediate metabolites of xylose fermentation (9).

Recently Schneider *et al.* (107) reported a yeast, *Pachysolen tannophilus* which can convert 52 per cent xylose to ethanol under semi-anaerobic conditions without using isomerase. At the same time another group led by Bothast (116) also reported that *P. tannophilus* converted ~66 per cent of xylose into ethanol at 32°C and initial pH of 4.5. In contrast to Schneider's work the Bothast group reported that aerobic conditions were required for cell growth but not for ethanol production. Meanwhile yet another report appeared (55) claiming that *Candida tropicalis* can convert xylose into ethanol under aerobic conditions. As calculated from their data only 12 per cent theoretical ethanol yield could be obtained.

The overall conversion of xylose into ethanol is very low, therefore, the process is not economical at this stage. There is also a lot of controversy about the aerobic/anaerobic conditions for the fermentation of xylose into ethanol. Moreover, the work reported here was done on pure xylose and also in low concentrations (1–5 per cent) and there could be a number of unexpected problems when xylose, obtained from partial or complete hydrolysis of wood with acid (62) or from steam exploded wood (81, 129) containing toxic phenolic compounds (51, 63), is fermented at higher concentrations.

Biochemistry of fermentation of hexoses and pentoses

The carbohydrates (cellulose, hemicelluloses and some starch) of lignocelluloses yield the following sugars on hydrolysis with acids or enzymes:

(a) cellulose: (i) hexoses = glucose
(b) hemicelluloses: (i) hexoses = glucose, fructose, mannose, galactose
(ii) pentoses = xylose, arabinose
(iii) uronic acids
(c) starch: (i) hexoses = glucose.

Although the intermediary metabolism of carbohydrates is described in standard textbooks of biochemistry, the main steps are given here (9), particularly to make clear how the various sugars obtained from lignocellulosic or starchy materials are catabolized into ethanol. A yeast fails to catabolize a particular sugar because: (i) the sugar does not enter the cells, (ii) the yeast lacks one or more enzymes necessary to convert the sugar into an intermediary metabolite of a central pathway, or (iii) the appropriate central pathway is inoperative from lack of one or more enzymes that control its reaction. The interrelationships of the central pathways are shown in Fig. 10.

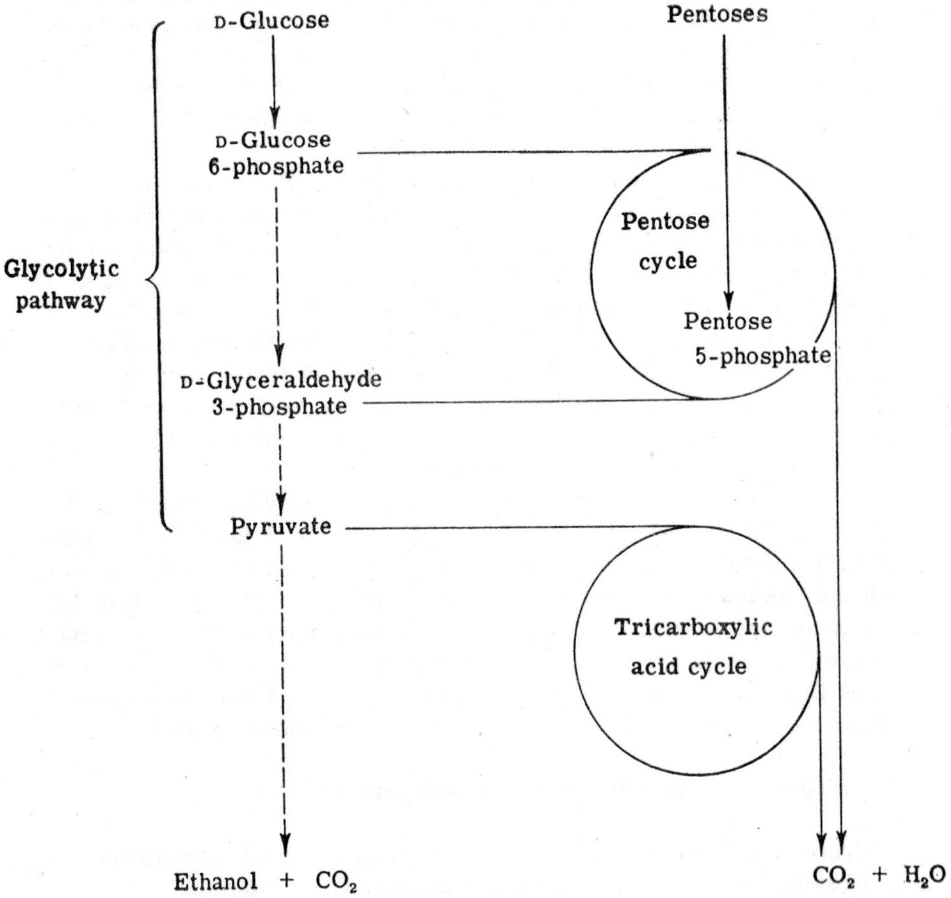

Fig. 10. Interrelationships of the central pathways of carbohydrate metabolism in yeasts [from Barnett (9)].

Catabolism of hexoses

The hexoses other than glucose must be converted into D-glucose or D-glucose 6-phosphate and pentoses into pentose 5-phosphate before they could enter into central pathways. D-fructose is phosphorylated with hexokinase to D-fructose 6-phosphate which is an intermediate of both glycolytic pathway and the pentose cycle. D-mannose is also phosphorylated with hexokinase to

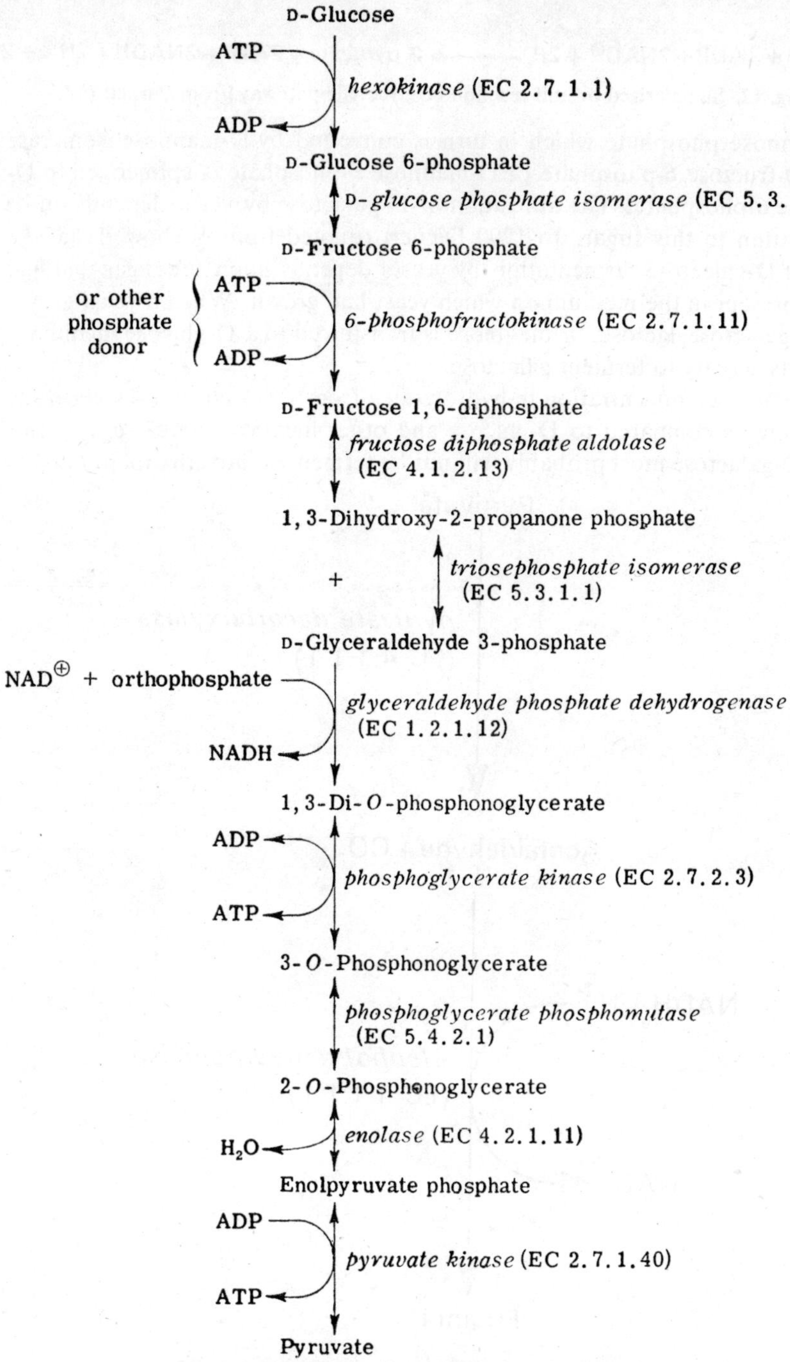

Fig. 11. The glycolytic pathway [from Barnett (9)].

$$\text{D-Glucose} + 2\,\text{ADP} + 2\,\text{NAD}^{\oplus} + 2\,\text{P}_i \longrightarrow 2\ \text{pyruvate} + 2\,\text{ATP} + 2\,\text{NADH} + 2\,\text{H}^{\oplus} + 2\,\text{H}_2\text{O}$$

Fig. 12. Summarized overall reaction of glycolytic pathway [from Barnett (9)].

D-mannose phosphate which in turn is converted by D-mannose isomerase into D-fructose 6-phosphate or D-mannose 6-phosphate is epimerized to D-glucose 6-phosphate. The utilization of D-galactose by yeast depends on its adaptation to this sugar. In 1900 Dienert (quoted from 9) showed that the rate of D-galactose fermentation by yeasts depends upon the sugar that had been present in the medium on which yeast had grown. When a yeast grown on D-galactose, lactose, or melibiose is transferred to a D-glucose medium it loses its activity to ferment galactose.

Galactose concentration in hydrolysates from lignocelluloses is generally very low as compared to D-glucose and other hexoses, therefore, it seems that D-galactose most probably will not be fermented into ethanol in most of

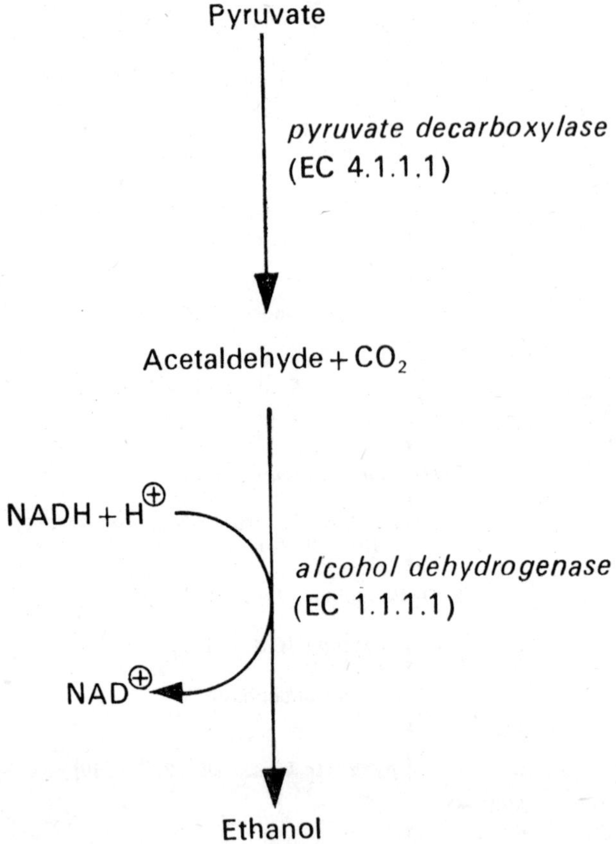

Fig. 13. Conversion of pyruvate into ethanol [from Barnett (9)].

the cases. However, a yeast grown on galactose will be able to ferment D-galactose. Under such conditions D-galactose is converted into D-glucose 6-phosphate after passing through a number of reactions, before it can enter into glycolytic pathways. The most commonly accepted glycolytic pathway and further conversion of pyruvate into ethanol are given in Figs. 11–13.

Catabolism of pentoses

The pentoses and alditols are not easily catabolized by most of the yeasts because (i) pentoses and pentitols may be reversibly interconverted by dehydrogenases; (ii) catabolic routes may be shared; and (iii) many of the alditol dehydrogenases have particularly wide substrate-specificity, so that a single enzyme may act on pentitols and pentoses, as well as on hexitols and hexoses (9).

In principle, the first step by which yeasts could convert an aldopentose into an intermediate of the pentose cycle might be: (a) an epimerization, (b) a conversion into the corresponding ketose by way of the enediol, (c) conversion by oxidation and reduction, or (d) phosphorylation. D-xylose is expected to be catabolized initially, either by isomerization to D-threo-pentulose or reduction to xylitol which would then be oxidized to D-threo-pentulose (D-xylulose). D-threo-pentulose is phosphorylated to give D-threo-pentulose 5-phosphate (D-xylulose 5-phosphate) which can enter into the pentose cycle. Possible routes of pentitol and pentose catabolism by yeasts have been postulated by Barnett (8, 9) in Fig. 14. *Pachysolen tannophilus* (107, 116) and

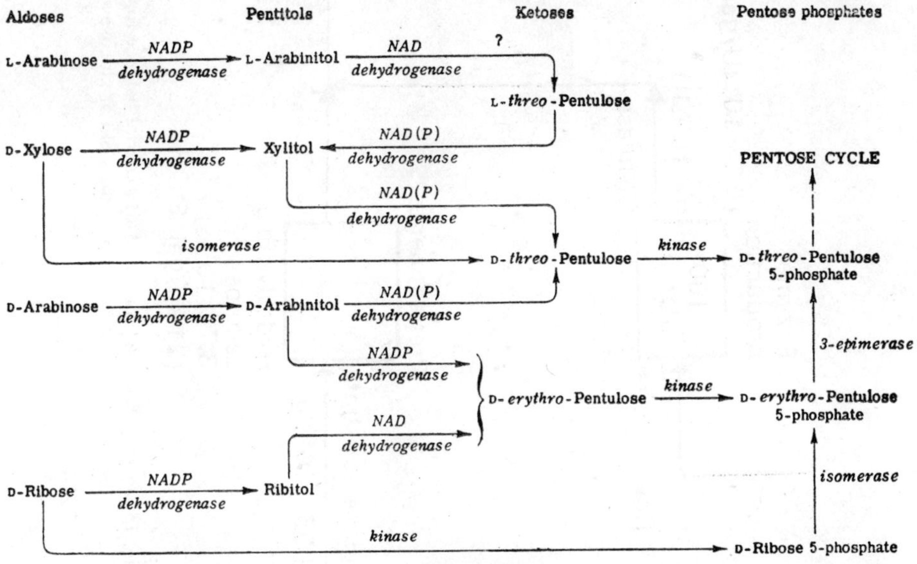

Fig. 14. Possible routes of pentitol and pentose catabolism by yeasts [from Barnett (8)].

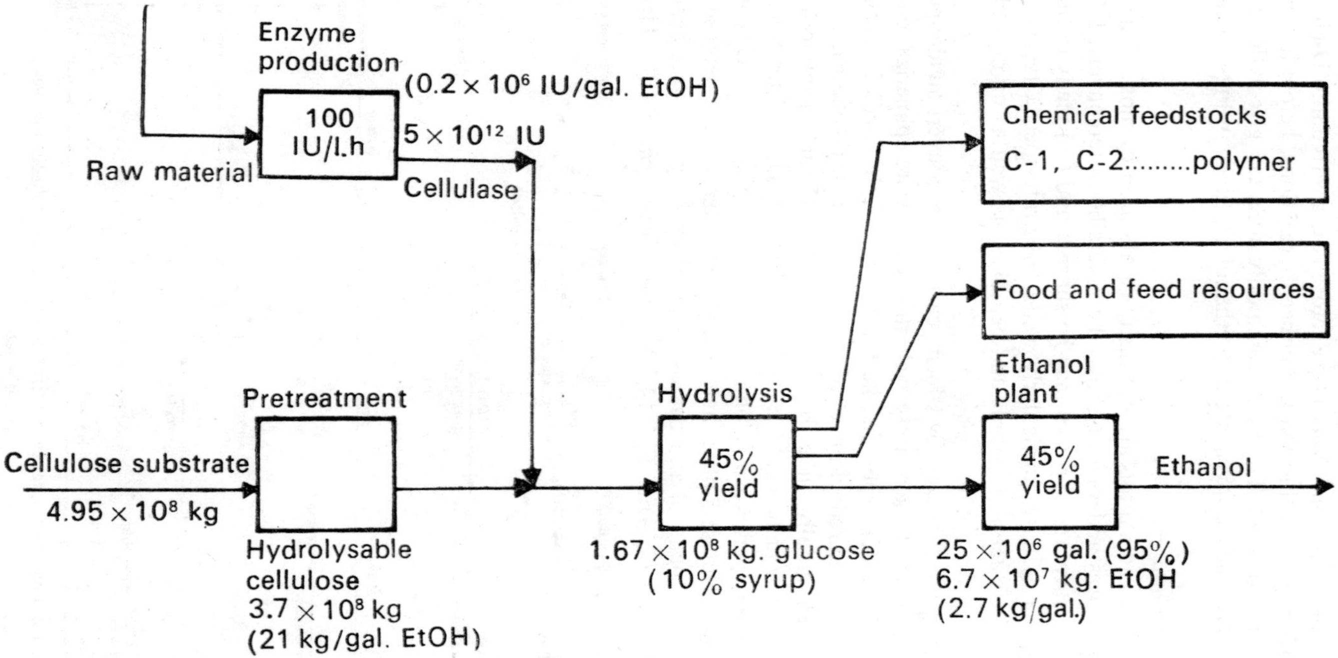

Fig. 15. Process flowsheet and overall material balance for production of ethanol (25×10^6 gallons per year) from cellulose [from Spano *et al.* (118)]. Compare the yield of glucose and ethanol obtained in this process with that given in Table 3.

Candida tropicalis (55) seem to catabolize xylose through this pathway. The yields are so low that a lot of work is to be done to optimize these conditions to obtain higher conversion of xylose and arabinose into ethanol.

Distillation

Following fermentation, the cells are separated from the liquid, leaving "beer" (ethanol, water and small amounts of higher alcohols and ether). Ethanol is then separated from the water by one of two basic methods of distillation: vaporization of the ethanol/water mixture without reflux (flash distillation) or continuous distillation with reflux (rectification).

Economics of ethanol production from cellulose

A number of cost estimates have been made by several groups. Some of the studies were to evaluate cellulase production (89), enzymatic hydrolysis (141, 142), and ethanol production (118, 130). Most of the studies showed that enzyme production is the most expensive item in the whole process of fermentation of cellulose to ethanol. It was estimated that the production cost of cellulase is ~\$3.00/kg cellulase protein having a specific activity of 0.6 IU/mg protein (114). This enzyme production cost varies with the carbon source used, and it could be lowered to ~\$1.50/kg. The production cost of glucose from cellulose by enzymatic hydrolysis is about \$0.15/kg crude glucose syrup and that of ethanol is about \$1.43/gallon 95 per cent ethanol based on the current cellulase process technology without taking any byproduct credit (Table 15 and Fig. 15). By using 5×10^{12} units of cellulase and 4.95×10^8 kg of cellulose, 25×10^6 gallons of alcohol (95 per cent) or 6.7×10^7 kg ethanol could be produced by this process (Fig. 15).

A very good comparison of cost estimates from various workers for ethanol production has been compiled by Stone and Marshall (121) which varies from \$0.27 to \$1.12/litre ethanol.

Table 15. Relative cost factor analysis for production of ethanol from cellulose [from Spano *et al.* (118)]

	Unit cost (cent/gal 95% ETOH)	Per cent cost
Enzyme production	57.33	43.4
Pretreatment	30.38	23.0
Hydrolysis	13.03	10.0
Ethanol production	31.07	23.6
Total	131.81	100.0

References

1. Allen, A. and Sternberg, D. β-glucosidase production by *Aspergillus phoenicis* in stirred tank fermentors, Presented at the second symposium on Biotechnology in Energy Production and Conservation, Gatlinburg, Tennessee, U.S.A. (1979).
2. Andreotti, R.E., Mandels, M. and Roche, C. Effect of some fermentation variables on growth and cellulase production by *Trichoderma* QM9414, pp. 249–267 in *Proceedings of Bioconversion Symposium*, Indian Institute of Technology, Delhi (1977).
3. Anonymous. U.S. Department of the Treasury, Internal Revenue Service, Annual Report of the Commissioner, for the fiscal year ending June 30 (1956).
4. Anonymous. U.S. Department of the Treasury, Bureau of Alcohol, Tobacco and Firearms, Summary Statistics, fiscal year ending June 30 (1976).
5. Anonymous. *Alcohol Production from Biomass in the Developing Countries*, World Bank, 1818 H Street, N.W. Washington, D.C. 20433, U.S.A. (1980).
6. Anonymous. *American Lung Association Bulletin*, *63* (3), 7 (1977).
7. Aries, R.S. Synthetic alcohol from petroleum, *Oil and Gas Journal*, *46* (46), 108–110 (1948).
8. Barnett, J.A. Biochemical differentiation of taxa with special reference to the yeasts, pp. 557–595 in *The Fungi—An Advanced Treatise* Vol. 3 (Edited by G.C. Ainsworth and A.S. Sussman), Academic Press, New York (1968).
9. Barnett, J.A. The utilization of sugars by yeasts, *Advances in Carbohydrate Biochemistry*, *32*, 125–234 (1976).
10. Bassham, J.A. Feed and Food from Desert Environments, in *The Biosaline Concept* (Edited by A. Hollaender), Plenum Press, New York (1979).
11. Batter, T.R. and Wilke, C.R. Report of Lawrence Berkeley Laboratory, University of California, LBL-6351 (1977).
12. Berg, B. and Pettersson, G. Location and fermentation of cellulases in *Trichoderma viride*, *Journal of Applied Bacteriology*, *42*, 65–75 (1977).
13. Berghem, L.E.R. and Pettersson, L.G. The mechanism of enzymatic cellulose degradation. Purification of cellulolytic enzyme from *Trichoderma viride* active on highly ordered cellulose, *European Journal of Biochemistry*, *37*, 21–32 (1973).
14. Bissett, F. and Sternberg, D. Immobilization of *Aspergillus* Beta-glucosidase on chitosan, *Applied Environmental Microbiology*, *35*, 750–755 (1978).
15. Boretti, G., Garafano, L., Montecucci, P. and Spalla, C. Cellulase pro-

duction with *penicillium iriensis* (n. sp.), *Archives Mikrobiologie, 92,* 189–200 (1973).

16. Chahal, D.S. Bioconversion of lignocelluloses into food and feed rich in protein, pp. 551–584 in *Advances in Agricultural Microbiology* (Edited by N.S. Subba Rao), Oxford and IBH Publishing Co., New Delhi (1982).
17. Chahal, D.S. and Gray, W.D. Growth of cellulolytic fungi on wood pulp. (ii) Effect of different sugars and different levels of nitrogen on production of fungal protein, *Indian Phytopathology, 23,* 74–79 (1970).
18. Chahal, D.S., McGuire, S., Pikor, H. and Noble, G. Production of cellulase-complex by *Trichoderma reesei* Rut-C30 on lignocellulose and its hydrolytic potential. Paper presented in a symposium: *Fuel and Chemicals from Biomass,* American Chemical Society Annual Meeting, held at New York, August (1981).
19. Chahal, D.S., McGuire, S., Pikor, H. and Noble, G. Production of cellulase-complex by *Trichoderma reesei* Rut-C30 on lignocellulose and its hydrolytic potential, *Biomass* (1982) (in press).
20. Chahal, D.S., Swan, J.E. and Moo-Young, M. Role of C_1 enzyme of *Trichoderma viride* and *Chaetomium cellulolyticum* in degradation of cellulose. Paper presented at the 69th Annual Meeting of American Institute of Chemical Engineering, Chicago, Illinois, U.S.A., November 14–18 (1976).
21. Chahal, D.S., Swan, J.E. and Moo-Young, M. Protein and cellulase production by *Chaetomium cellulolyticum* grown on wheat straw, *Developments in Industrial Microbiology, 18,* 433–442 (1977).
22. Chahal, D.S. and Wang, D.I.C. *Chaetomium cellulolyticum,* growth behavior on cellulose and protein production, *Mycologia, 70,* 160–170 (1978).
23. Chase, J. Jr., Eveleigh, D.E., Pincus, R., and Montenecourt, B.S. Ethanol production, presented at the VI International Fermentation Symposium, July 20–25 (1980).
24. Chibata, I. and Tosa, T. *Immobilized Microbial Cells and their Applications,* TIBS, Elsevier/North-Holland Biomedical Press (1980).
25. Cochrane, V.W. *Physiology of Fungi,* John Wiley and Sons, Inc., New York, p. 524 (1958).
26. Converse, A.O. and Grethlein, H.E. Acid hydrolysis of cellulosic biomass, pp. 91–95 in *Proceedings of Third Annual Biomass Energy Systems Conference (SERI, DOE),* Golden, Co., U.S.A. (1979).
27. Cowling, E.B. A review of literature on the enzymatic degradation of cellulose and wood, *Forest Products Laboratory Report No. 2116,* Madison, Wisconsin, U.S.A., pp. 26 (1958).
28. Cowling, E.B. Physical and chemical constraints in hydrolysis of cellulose and lignocellulosic materials, *Biotechnology & Bioengineering Symposium, 5,* 163–181 (1975).

29. Cowling, E.B. and Kirk, T.K. Properties of cellulose and lignocellulosic materials as substrate for enzymatic conversion process, *Biotechnology & Bioengineering Symposium, 6*, 95–123 (1976).
30. Cysewski, G.R. and Wilke, C.R. Rapid ethanol fermentation using vacuum and cell recycle, *Biotechnology & Bioengineering, 19*, 1125–1143 (1977).
31. Ellis, E.F. *Journal of Respiratory Diseases, 1* (2), 3 (1980).
32. Enari, T.-M., Markkanen, P. and Korhonen, E. Cellulase production by *Aspergillus awamori*, pp. 171–180 in *Symposium on Enzymatic Hydrolysis of Cellulose*, SITRA, Helsinki, Finland (1975).
33. Eriksson, K.-E. Enzymic mechanisms involved in the degradation of wood components, pp. 263–280 in *Proceedings of Symposium on Enzymatic Hydrolysis of Cellulose* (Edited by M. Bailey, T.M. Enari, M. Linko), Aulanko, Finland, SITRA, Helsinki (1975).
34. Eriksson, K.-E. Enzyme mechanisms involved in fungal degradation of wood components, *Proceedings of Bioconversion Symposium*, Indian Institute of Technology, New Delhi (1977).
35. Eriksson, K.-E. and Pettersson, B. Extracellular enzyme system utilized by the fungus *Sporotrichum pulverulentum* (*Chrysosporium lignorum*) for the breakdown of cellulose, *European Journal of Biochemistry, 51*, 193–206 (1975).
36. Eriksson, K.-E., Pettersson, B. and Westermak, U. Oxidation: An important enzyme reaction in fungal degradation of cellulose, *PEBS Letters, 49* (2), 282–285 (1974).
37. Fan, L.T., Lee, Y.-H. and Beardmore, D.H. Mechanism of the enzymatic hydrolysis of cellulose: Effects of major structural features of cellulose on enzymatic hydrolysis, *Biotechnology & Bioengineering, 22*, 177–199 (1980).
38. Freudenberg, K. Lignin: Its constitution and formation from P-hydroxycinnamyl alcohols, *Science, 148*, 595–600 (1965).
39. Gallo, B.J., Andreotti, R., Roche, C., Ryu, D. and Mandels, M. Cellulase production by a new mutant strain of *Trichoderma reesei* MCG77, *Biotechnology & Bioengineering Symposium, 8*, 89–101 (1978).
40. Gascoigne, J.A. and Gascoigne, M.M. *Biological Degradation of Cellulose*, Butterworth, London, pp. 264 (1960).
41. Ghose, T.K. Cellulase biosynthesis and hydrolysis of cellulosic substances, *Advances in Biochemical Bioengineering, 6*, 39–76 (1977).
42. Ghose, T.K., Pathak, A.N. and Bisaria, V.S. Kinetic and dynamic studies of *Trichoderma viride* cellulase production, pp. 111–136 in *Proceedings of Symposium on Enzymatic Hydrolysis of Cellulose*, Aulanko, Finland (1975).
43. Gilbert, N., Hobbs, I.A. and Levine, J.D. Hydrolysis of wood, *Industrial and Engineering Chemistry, 44*, 1712 (1952).

44. Goksoyr, J., Edisa, G., Eriksen, J. and Osmundsvag, K. A comparison of cellulases from different microorganisms, pp. 217–230 in *Proceedings of Symposium on Enzymatic Hydrolysis of Cellulose*, Aulanko, Finland, SITRA, Helsinki (1975).
45. Gong, D.S., Chen, L.-F., Flickinger, M.C., Chiang, L.-C. and Tsao, G.T. Production of ethanol from D-xylose by using D-xylose isomerase and yeasts, *Applied and Environmental Microbiology, 41*, 430–436 (1981).
46. Gong, D.S., Ladisch, M.R. and Tsao, G.T. Cellobiase from *Trichoderma viride:* Purification, properties, kinetics and mechanism, *Biotechnology and Bioengineering, 19*, 959–981 (1977).
47. Grethlein, H.E., Knappert, D.R. and Converse, A.O. Partial acid hydrolysis of cellulosic materials to increase the rate of enzymatic hydrolysis, *Final Report of National Science Foundation Grant ENG 75–17969*, Thayer School of Engineering, May (1979).
48. Halliwell, G. Solubilization of native and derived forms of cellulose by cell-free microbial enzymes, *Biochemical Journal, 100*, 315–320 (1966).
49. Halliwell, G. Mode of action of components of the cellulase complex in relation to cellulolysis, pp. 319–336 in *Proceedings of Symposium on Enzymatic Hydrolysis of Cellulose*, Aulanko, Finland, SITRA, Helsinki (1975).
50. Harris, E.E. and Belinger, E. Madison wood sugar process, *Industrial and Engineering Chemistry, 38*, 890–895 (1946).
51. Harris, E.E., Hajny, G.J., Hannan, M. and Rogers, S.C. Fermentation of douglas fir hydrolysate by *S. cerevisiae, Industrial and Engineering Chemistry, 38*, 896–904 (1946).
52. Hess, K., Mahl, H. and Gutter, E. Elektronenmikroskopische Darstellung grosser Langsperioden in Zellulosefasern und ihr Vergleich mit den Perioden anderer Faserarten, *Kolloid-Zeitschrift, 155*, 1–18 (1954).
53. Hofsten, B.Y. Topological effects in enzymatic and microbial degradation of highly ordered polysaccharides, pp. 281–295 in *Proceedings of Symposium on Enzymatic Hydrolysis of Cellulose*, Aulanko, Finland, SITRA, Helsinki (1975).
54. Jacob, P.B. *Industrial Alcohol*, United States Department of Agriculture, Miscellaneous Publication No. 695, Washington, D.C., U.S.A. (1950).
55. Jeffries, T.W. Conversion of xylose to ethanol under aerobic conditions by *Candida tropicalis, Biotechnology Letters* (1981) (in press).
56. Katz, M. and Reese, E.T. Production of glucose by enzymatic hydrolysis of cellulose, *Applied Microbiology, 16*, 419–420 (1968).
57. King, K.W. Enzymatic attack on highly crystalline hydrocellulose, *Journal of Fermentation Technology* (Japan), *43*, 79–94 (1965).
58. King, K.W. and Vessal, M.I. Enzymes of the cellulase complex, *Advances in Chemistry Series, No. 95*, 7–25 (1969).
59. Knappert, D. Grethlein, H. and Converse, A. Partial acid hydrolysis

of cellulosic materials as a pretreatment for enzymatic hydrolysis, *Biotechnology & Bioengineering*, *22*, 1449–1463 (1980).

60. Koenigs, J.W. Hydrogen peroxide and iron: A microbial cellulolytic system, *Biotechnology & Bioengineering Symposium*, *5*, 151–159 (1975).
61. Kosaric, N., Ng, D.C.M., Russell, I., and Stewart, G.S. Ethanol production by fermentation: An alternative liquid fuel, *Advances in Applied Microbiology*, *26*, 147–227 (1980).
62. Lee, Y.Y., Liu, C.M., Johnson, T. and Chambers, A.P. Selective hydrolysis of hardwood hemicellulose by acids, *Biotechnology & Bioengineering Symposium*, *8*, 75–88 (1978).
63. Leonard, R.H. and Hajny, G.J. Fermentation of wood sugars to ethyl alcohol, *Industrial and Engineering Chemistry*, *37*, 390–395 (1945).
64. Letcher, H. and Willigan, J.J. Biochemistry of plant diseases, VIII Alcoholic fermentation of *Fusarium lini*, *Phytopathology*, *16*, 941–949 (1926).
65. Lindeman, L.R. and Rocchiccioli, C. Ethanol in Brazil: Brief summary of the state of the industry in 1977, *Biotechnology & Bioengineering*, *21*, 1107–1119 (1979).
66. Loughran, G.A., Soodak, M. and Nord, F.F. Fermentation of wood hydrolysate by yeast and Fusaria, *Archives of Biochemistry*, *6*, 163–164 (1945).
67. Mandels, M. Microbial sources of cellulase, *Biotechnology & Bioengineering Symposium*, *5*, 81–106 (1975).
68. Mandels, M., Andreotti, R. and Roche, C. Measurement of saccharifying cellulase, *Biotechnology & Bioengineering Symposium*, *6*, 21–33 (1976).
69. Mandels, M., Dorval, S. and Madeiros, J. Cellulases, pp. 627 in *Proceedings of 2nd Annual Symposium on Fuels from Biomass* (Edited by W.W. Shuster), John Wiley and Sons, New York (1978).
70. Mandels, M., Hontz, L. and Nystrom, J. Enzymatic hydrolysis of waste cellulose, *Biotechnology & Bioengineering*, *16*, 1671–1693 (1974).
71. Mandels, M. and Reese, E.T. Induction of cellulase in *Trichoderma viride* as influenced by carbon sources and metals, *Journal of Bacteriology*, *73*, 269–278 (1957).
72. Mandels, M. and Reese, E.T. Induction of cellulase in fungi by cellobiose, *Journal of Bacteriology*, *79*, 816–826 (1960).
73. Mandels, M. and Reese, E.T. Sophorose as an inducer of cellulase in *Trichoderma viride*, *Journal of Bacteriology*, *83*, 400–408 (1962).
74. Mandels, M. and Reese, E.T. Fungal cellulases and the microbial decomposition of cellulosic fabric, *Developments in Industrial Microbiology*, *5*, 5–20 (1964).
75. Mandels, M. and Sternberg, D. Recent advances in cellulase technology, *Journal of Fermentation Technology*, *54* (4), 267–286 (1976).
76. Mandels, M. and Weber, J. The production of cellulases, pp. 391–414

in *Advances in Chemistry Series, 95* (1969).
77. Manley, R. St. J. Fine structure of native cellulase microfibrils, *Nature, 204*, 1155–1157 (1964).
78. Montenecourt, B.S. and Eveleigh, D.E. Selective screening methods for the isolation of high yielding cellulase mutants of *Trichoderma reesei, Advances in Chemistry Series, No. 181*, 289–301 (1979).
79. Mue, S., Ise, T., Ono, Y. and Akaska, K. A study of western red cedar sensitivity: Workers allergy reactions and symptoms, *Annals of Allergy, 35*, 148–152 (1975).
80. Nisizawa, K., Tomita, Y. and Kanda, T. Substrate specificity of C_1 and C_x cellulase components from fungi, pp. 719–725 in *Proceedings of IV International Fermentation Symposium, Fermentation Technology Today*, (1972).
81. Noble, G. Optimizations of steam explosion pretreatment, *Final Report Submitted to U.S. Department of Energy, Fuels from Biomass Program*, Iotech Corporation Limited, Ottawa, Ontario, Canada (1980).
82. Nord, F.F. and Mull, R.P. Recent progress in the biochemistry of Fusaria, *Advances in Enzymology, 5*, 165–205 (1945).
83. Norkrans, B. Influence of cellulolytic enzymes from hymenomycetes on cellulose preparations of different crystallinity, *Physiologia Plantarum, 3*, 75–78 (1950).
84. Nystrom, J.M. and Allen, A.L. Pilot scale investigations and economics of cellulase production, *Biotechnology & Bioengineering Symposium, 6*, 55–74 (1976).
85. Nystrom, J.M. and Dilca, P.H. Enhanced production of *Trichoderma* cellulase on high levels of cellulose in submerged culture, pp. 293–304 in *Proceedings of Bioconversion Symposium*, Indian Institute of Technology, Delhi (1977).
86. Paul, J.K. *Ethyl Alcohol Production and Use as a Motor Fuel*, Noyes Data Corporation, Park Ridge, New York (1979).
87. Peitersen, N. Production of cellulase and protein from barley straw by *Trichoderma viride, Biotechnology & Bioengineering, 17*, 361–374 (1975).
88. Perez, J., Wilke, C.R. and Blanch, H.W. Economics of sugar production with *Trichoderma reesei* Rutgers C-30. Presented at the Second Congress of the North American Continent, Los Vegas, Nevada, U.S.A. (1980).
89. Pettersson, L.G. The mechanism of enzymatic hydrolysis of cellulose by *Trichoderma viride*, pp. 255–261 in *Proceedings of Symposium on Enzymatic Hydrolysis of Cellulose*, Aulanko, Finland, SITRA, Helsinki, March 12–14 (1975).
90. Pew, J.C. and Weyna, P. Fine grinding, enzyme digestion, and lignin-cellulose bond in wood, *Tappi, 45* (3), 247–256 (1962).

91. Prescot, S.C. and Dunn, C.G. *Industrial Microbiology*, McGraw-Hill Book Company, Incorporation, New York, pp. 945 (1959).
92. Preston, R.D. and Cronshaw, J. Constitution of the fibrillar and non-fibrillar components of the walls of *Valonia ventricosa*, *Nature*, *181*, 248–250 (1958).
93. Reese, E.T. Polysaccharases and hydrolysis of insoluble substrates, pp. 165–181 in *Biological Transformation of Wood by Microorganisms* (Edited by W. Liese), Springer-Verlag, Heidelberg, Germany (1975).
94. Reese, E.T. History of the cellulase program at the U.S. Army Natick Development Center, *Biotechnology & Bioengineering Symposium*, *6*, 9–20 (1976).
95. Reese, E.T., Gilligan, W. and Norkrans, B. Effect of cellobiose on the enzymatic hydrolysis of cellulose and its derivatives, *Physiologia Plantarum*, *5*, 379–390 (1952).
96. Reese, E.T. and Maguire, A. Increase in cellulase yields by addition of surfactants to cellobiose cultures of *Trichoderma viride*, *Developments in Industrial Microbiology*, *12*, 212–217 (1971).
97. Reese, E.T. and Mandels, M. Stability of the cellulase of *Trichoderma reesei* under use conditions, *Biotechnology & Bioengineering*, *22*, 323–335 (1980).
98. Reese, E.T., Segal, L. and Tripp, V.W. The effect of cellulase on the degree of polymerization of cellulose and hydrocellulose, *Textile Research Journal*, *27*, 626–632 (1957).
99. Reese, E.T., Siu, S.G.H. and Levinson, H.S. The biological degradation of soluble cellulose derivatives and its relationship to the mechanism of cellulose hydrolysis, *Journal of Bacteriology*, *59*, 485–489 (1950).
100. Roberts, R.C. and Moore, V.L. Immunopathogenesis of hypersensitivity pneumonitis, *American Review in Respiratory Diseases*, *116*, 1075–1090 (1977).
101. Rodin, L.E., Basilevich, N.I., Rozov, N.N. Productivity of world ecosystem, in *Proceedings of a Symposium on Productivity of World Ecosystems*, National Academy of Sciences, Washington, D.C. (1976).
102. Rowland, S.P. and Roberts, E.J. The nature of accessible surfaces in the microstructure of cotton cellulose, *Journal of Polymer Science, Part A-1*, *10*, 2447 (1972).
103. Ryu, D., Andreotti, R., Mandels, M., Gallo, B. and Reese, E.T. Studies on quantitative physiology of *Trichoderma reesei* with two-stage continuous culture for cellulase production, *Biotechnology & Bioengineering*, *21*, 1887–1903 (1979).
104. Ryu, D.D.Y. and Mandels, M. Cellulases: Biosynthesis and applications, *Enzyme Microbial Technology*, *2*, 91–102 (1980).
105. Saeman, J.F., Locke, E.G. and Dickerson, G.K. The production of wood sugar in Germany and its conversion to yeast and alcohol, *U.S.*

Joint Intelligence Objectives Agency, FIAT Final Report No. 499, Washington, D.C. (1945).

106. Scheffer, T.C. and Cowling, E.B. Natural resistance of wood to microbial deterioration, *Annual Review of Phytopathology, 4*, 147–170 (1966).
107. Schneider, H., Wang, P.Y., Chan, Y.K. and Maleszka, R. Conversion of xylose into ethanol by the yeast *Pachysolen tannophilus, Biotechnology Letters, 3* (2), 89–92 (1981).
108. Scholler, H. and Walter, K. Process of converting cellulose and the like into sugar with dilute acids under pressure, U.S. Patent No. 1,990,097.
109. Schutt, H.C. Production of ethylene from ethane-propane, *Chemical Engineering Progress Transactions Section, 43* (3), 103–116 (1947).
110. Sciarini, L.J. and Wirth, J.C. Ethanol production from wheat hydrolysates and stillage by yeast and Fusaria, *Cereal Chemistry, 22*, 11–21 (1945).
111. Selby, K. Mechanism of biodegradation of cellulose, pp. 62–78 in *Biodeterioration of Materials*, Vol. 1 (Edited by A.H. Walter and J.J Elphick), Applied Science Publication Limited, London (1968).
112. Selby, K. The purification and properties of the C_1 component of cellulase complex, *Advances in Chemistry Series, No. 95*, 34–52 (1969).
113. Selby, K. and Maitland, C.C. The cellulase of *Trichoderma viride, Biochemical Journal, 104*, 716–724 (1971).
114. Simmon, E.G. Abstracts of Second International Mycological Congress, Tampa, Florida, p. 618 (1977).
115. Siu, R.G.H. *Microbial Decomposition of Cellulose*, Reinhold Publishing Corporation, New York (1951).
116. Slininger, P.J., Bothast, R.J., Van Cauwenberge, J.E. and Kurtzman, C.P. Conversion of D-xylose to ethanol by the yeast *Pachysolen tannophilus*, Personal communication (1981).
117. Sloneker, J.H. Agricultural residues including feedlot wastes, *Biotechnology & Bioengineering Symposium, 6*, 235–250 (1976).
118. Spano, L., Allen, A., Tassinari, T., Mandels, M. and Ryu, D. *Proceedings of 2nd Annual Symposium on Fuels from Biomass* (Edited by W.W. Shuster), John Wiley and Sons, New York (1978).
119. Strenberg, D. Production of cellulase by *Trichoderma, Biotechnology & Bioengineering Symposium, 6*, 35–53 (1976).
120. Sternberg, D. and Mandels, G. Induction of cellulolytic enzymes in *Trichoderma reesei* by sophorose, *Journal of Bacteriology, 139*, 761–769 (1979).
121. Stone, J.E. and Marshall, H.B. *Analysis of Ethanol Production Potential from Cellulosic Feedstocks*, John Stone and Associates Limited, 19 Saguenay Drive, Aylmer, Quebec, Canada (1980).
122. Streamer, M., Eriksson, K.E. and Pettersson, B. Extracellular enzyme

system utilized by the fungus *Sporotrichum pulverulentum* (*Chrysosporium lignorum*) for the breakdown of cellulose, *European Journal of Biochemistry*, *59*, 607–613 (1975).

123. Stutzenberger, F.J. Cellulase production by *Thermomonospora curvata* isolated from municipal solid waste compost, *Applied Microbiology*, *22*, 147–152 (1971).
124. Stutzenberger, F.J. Cellulolytic activity of *Thermomonospora curvata*: Nutritional requirements for cellulase production, *Applied Microbiology*, *24*, 77–82 (1972).
125. Stutzenberger, F.J. and Bowden, M.W. Potential health hazard in cellulose bioconversions by thermophilic actinomycetes, *Biotechnology & Bioengineering*, *22*, 2443–2447 (1980).
126. Stutzenberger, F.J., Kaufman, H.J. and Lossin, R.D. Cellulolytic activity in municipal solid waste composting, *Canadian Journal of Microbiology*, *16*, 553–560 (1970).
127. Suomalainen, H. and Oura, E. Yeast nutrition and solute uptake, in *The Yeasts* (Edited by A.H. Rose and J.S. Harrison), Academic Press, New York (1971).
128. Tangnu, K.S., Blanch, H.W. and Wilke, C.R. Enhanced production of cellulase, hemicellulase and β-glucosidase by *Trichoderma reesei* (Rut-C30), Lawrence Berkeley Laboratory, University of California, Preprint 11074 for U.S. Department of Energy, Contract W-7405-ENG-48 (1980).
129. Taylor, J.D. Continuous high pressure steaming as an efficient pretreatment in the hydrolysis of biomass, Presented at Organization for Economic Co-operation and Development Workshop held at Amersfoort, The Netherlands, October 8–10 (1980).
130. *The Report on the Alcohol Fuel Policy Review*, U.S. Department of Energy (1979).
131. Timell, T.E. Recent progress in the chemistry of wood hemicelluloses, *Wood Science and Technology*, *1*, 45–70 (1967).
132. Toyama, N. Degradation of foodstuffs by cellulase and related enzymes, pp. 235–253 in *Advances in Enzymatic Hydrolysis of Cellulose and Related Material* (Edited by E.T. Reese), Pergamon Press, London (1963).
133. Toyama, N. and Ogawa, K. Sugar production from agricultural woody wastes by saccharification with *Trichoderma viride* cellulase, *Biotechnology & Bioengineering Symposium*, *5*, 225–244 (1975).
134. Toyama, N. and Ogawa, K. Utilization of cellulosic wastes by *Trichoderma viride*, pp. 743–757 in *Proceedings of Fourth International Fermentation Symposium, Fermentation Technology Today* (Edited by G. Terui), Society of Fermentation Technology, Osaka (1972).
135. Underkofler, L.A. and Hickey, R.J. *Industrial Fermentations*, Vol. I, Chemical Publishing Company, Incorporation, New York (1954).

136. Updegraff, D.M. Utilization of cellulose from waste paper by *Myrothecium verrucaria*, *Biotechnology & Bioengineering*, *13*, 77–79 (1971).
137. Walseth, C.S. Influence of fine structure of cellulose on the action of cellulases, *Tappi*, *35*, 233–238 (1952).
138. Wang, P.V., Shopsis, C. and Schneider, H. Fermentation of a pentose by yeasts, *Biochemical and Biophysical Research Communications*, *94* (1), 248–254 (1980).
139. Wenzel, H.F.J. *The Chemical Technology of Wood*, Academic Press, New York (1970).
140. Whitaker, D.R. Cellulase, pp. 273–290 in *The Enzymes*, 5, Academic Press, Inc., New York (1971).
141. Wilke, C.R. and Mitra, G. Process development studies on the enzymatic hydrolysis of cellulose, *Biotechnology & Bioengineering Symposium*, *5*, 253–274 (1975).
142. Wilke, C.R., Yang, R.D. and Van Stockar, U. Preliminary cost analyses for enzymatic hydrolysis of newsprint, *Biotechnology & Bioengineering Symposium*, *6*, 155–175 (1976).
143. Wood, T.M. The C_1 component of the cellulase complex, pp. 711–718 in *Proceedings of Fourth International Fermentation Symposium, Fermentation Technology Today* (Edited by G. Terui), Society of Fermentation Technology, Osaka (1972).
144. Wood, T.M. and McCrae, S.I. The purification and properties of the C_1 components of *Trichoderma koningii* cellulose, *Biochemical Journal*, *128*, 1183–1192 (1972).
145. Yand, V. and Trindade, S. Brazil's gasohol program, *Chemical Engineering Progress*, *75*, 11–19 (1979).

M. KOBAYASHI

23. The Role of Phototrophic Bacteria in Nature and Their Utilization

Phototrophic bacteria are widely distributed in nature (10, 12) and play an important role in carbon dioxide assimilation and nitrogen fixation by utilizing solar energy. Moreover, phototrophic bacteria have been found to contribute greatly to the purification of the environment (6, 11). In large amounts, the bacteria have been used as feed for small animals, fish and shell-fish. The bacteria are also beneficial to rice cultivation (7).

Classification

Van Niel (17) classified phototrophic bacteria as photosynthetic bacteria. However, with the accumulation of evidence on their physiological activity, the name "phototrophic bacteria" was considered more appropriate, and N. Pfennig and H.G. Trüper have classified this group of bacteria under phototrophic bacteria since 1974 (16). The classification is shown in Table 1.

According to Van Niel's classification Rhodospirillales are non-sulphur purple bacteria and belong to the Athiorhodaceae; Chromatiaceae are sulphur purple bacteria and belong to the Thiorhodaceae; and Chlorobiaceae are sulphur green bacteria and belong to the Chlorobacteriaceae.

Chromatiaceae and Chlorobiaceae favour anaerobic conditions, and grow by utilizing solar energy with sulphide, hydrogen or organic materials such as lower fatty acids as a hydrogen donor.

The general equation of this reaction is as follows:

$$CO_2 + 2H_2S \xrightarrow{\text{light}} (CH_2O) + H_2O + 2S \qquad 1$$

$$S + CO_2 + 3H_2O \xrightarrow{\text{light}} (CH_2O) + H_2SO_4 + H_2 \qquad 2$$

$$2CO_2 + Na_2S_2O_3 + 3H_2O \xrightarrow{\text{light}} 2(CH_2O) + Na_2SO_4 + H_2SO_4 \qquad 3$$

It is said that Chromatiaceae contain sulphur particles in cells but Chlorobiaceae do not accumulate such sulphur particles inside cells.

Table 1. Classification of phototrophic bacteria

Order Rhodospirillales [Pfennig and Trüper (16)]
- Family 1. Rhodospirillaceae
  - Genus 1. *Rhodospirillum*
    - Species: *rubrum, tenue, fulvum, molischianum, photometricum*
  - Genus 2. *Rhodopseudomonas*
    - Species: *palustris, viridis, acidophila, gelatinosa, capsulata, sphaeroides*
  - Genus 3. *Rhodomicrobium*
    - Species: *vannielii*
- Family 2. Chromatiaceae
  - Genus 1. *Chromatium*
    - Species: *okenii, weissei, warmingii, buderi, minus, violascens, vinosum, gracile, minutissimum*
  - Genus 2. *Thiocystis*
    - Species: *violacea, gelatinosa*
  - Genus 3. *Thiosarcina*
    - Species: *rosea*
  - Genus 4. *Thiospirillum*
    - Species: *sanguineum, jenense, rosenbergii*
  - Genus 5. *Thiocapsa*
    - Species: *roseopersicina, pfennigii*
  - Genus 6. *Lamprocystis*
    - Species: *roseopersicina*
  - Genus 7. *Thiodictyon*
    - Species: *elegans, bacillosum*
  - Genus 8. *Thiopedia*
    - Species: *rosea*
  - Genus 9. *Amoebobacter*
    - Species: *roseus, pendens*
  - Genus 10. *Ectothiorhodospira*
    - Species: *mobilis, shaposhnikovii, halophila*
- Family 3. Chlorobiaceae
  - Genus 1. *Chlorobium*
    - Species: *limicola, vibrioforme, phaeobacteroides, phaeovibrioides*
  - Genus 2. *Prosthecochloris*
    - Species: *aestuarii*
  - Genus 3. *Chloropseudomonas*
    - Species: *ethylica*
  - Genus 4. *Pelodictyon*
    - Species: *clathratiforme, luteolum*
  - Genus 5. *Clathrochloris*
    - Species: *sulphurica*

Rhodospirillales utilize mainly organic materials such as lower fatty acids, favourably. According to Van Niel, the general equation for this reaction is:

$$CO_2 + 2\ H_2.\ \text{Acceptors} \xrightarrow{\text{light}} (CH_2O) + H_2O + 2\ \text{Acceptors},$$

assuming that the fatty acids alone are used as hydrogen donors. However, besides the utilization of lower fatty acids as hydrogen donors, these bacteria utilize butyric acid as a carbon source as seen in the following equation:

$$C_4H_7O_2-Na+2\,H_2O+2\,CO_2 \xrightarrow{\text{light}} 5(CH_2O)+NaHCO_3.$$

These bacteria grow in both anaerobic and aerobic conditions as heterotrophic bacteria if organic materials such as yeast extract and peptone are added to the media in the absence of lower fatty acids and CO_2. Furthermore, these bacteria can grow in dark conditions, i.e., without utilizing solar energy (1). Such characteristics of phototrophic bacteria are summarized in Table 2.

Table 2. Characteristics of phototrophic bacteria

Phototrophic bacteria	Colour	Hydrogen donors effective for growth: Inorganic[a] materials	Hydrogen donors effective for growth: H_2	Hydrogen donors effective for growth: Organic[b] materials	Requirement of growth factors	Growing ability under dark condition	Ability of nitrogen fixation
Chromatiaceae	purple reddish	+ (well available)	+	+	—	+ Growth under dark*	+
Chlorobiaceae	green	+ (well available)	+	— (hardly available)	—	—	+
Rhodospirillaceae	purple reddish brownish	+ (weakly available)	+	+ (well available)	+	+	+

[a] H_2S, $H_2S_2O_3$, etc., [b] Organic acids, alcohol, etc.
*The observation of the present author.

Rhodospirillales can use both inorganic and organic materials as hydrogen donors and are capable of growing even in dark conditions, but require one or more of the following growth factors; biotin, *p*-amino-benzoic acid, thiamine and nicotinic acid. The kind and number of the factors required vary depending on the strain of bacteria. On the other hand, Chromatiaceae and Chlorobiaceae do not require a growth factor, but are said not to grow in the absence of solar energy. However, in our studies on the ecological distribution of Chromatiaceae, growth in complete darkness and under anaerobic conditions was observed (see Table 2). Higher plants and algae evolve large amounts of oxygen during photosynthesis, but a characteristic feature of phototrophic bacteria is that oxygen is not evolved. For example, if lower fatty acids are utilized, the equation for hydrogen evolution is as follows:

$$C_4H_6O_5+H_2O \xrightarrow{\text{light}} (CH_2O)_2+2\,CO_2+2\,H_2\uparrow.$$

Another characteristic of these bacteria is that they fix atmospheric nitrogen, and play an important role in the nitrogen and carbon cycles of nature.

Distribution

Phototrophic bacteria are widely distributed, and are found in most submerged areas such as paddy fields, ditches, soils in river beds and seashores, and sewage disposal plants, etc. A few extreme examples are that they have been found in the soil from the seashores of the Antarctic continent under ice where the water temperature was 0°C and in hot springs where the temperature is about 90°C. They have also been found in sea water at a depth of 2000 m. They are fairly tolerant to salt, and can grow well in saline conditions of 10 per cent and survive in salt lake with a salt content of 30 per cent. Phototrophic bacteria have also been confirmed to be present in the water of red tide sea (5).

Chromatiaceae and Chlorobiaceae of the phototrophic bacteria produce sulphur. Production of sulphur using the activity of these bacteria has been industrialized in many lakes around the world (production is said to be 100 tons per 20 km^2 per year). Phototrophic bacteria are abundant in the soil of various curative thermal waters.

Though in nature, green algae generally live only in areas where light penetrates, red coloured phototrophic bacteria (Rhodospirillales, Chromatiaceae) are abundant even in complete darkness, 8 cm deep from the surface soil layer and have been assumed to play some role. Even though green-coloured Chlorobiaceae are also phototrophic bacteria, these bacteria can live only in light conditions on the surface of the soil.

Table 3. Cell counts of phototrophic bacteria (number per g) in various samples

Ditch (B.O.D. 250 ppm)	10^6–10^7
Lake (B.O.D. 10 ppm)	10^2–10^3
River (B.O.D. 1.0 ppm)	+ –10
Aeration tank (Activated sludge method) (B.O.D. 150 ppm)	10^6–10^7
Soil of paddy field	10^5–10^6
Soil of seaside	10^3–10^4

Table 3 shows that phototrophic bacteria are abundant in dirty areas with high BOD values such as submerged ditches and sewage treatment plants.

The microorganisms which occur in a paddy field in Southeast Asia are shown in Table 4. Since no chemical nitrogen fertilizer was applied to the paddy field, the nitrogen supplied was mainly by the nitrogen fixers such as nitrogen-fixing algae and phototrophic bacteria. The soils of Tg. Kalang and Province Wellesley in Malaya were more fertile than the other samples.

Phototrophic bacteria grow well in the rhizosphere of rice plants from the

Table 4. Cell counts of nitrogen-fixing and other microorganisms in paddy soil of Southeast Asia

Soils	Nitrogen-fixing algae	Phototrophic bacteria: Chromatiaceae	Phototrophic bacteria: Rhodospirillaceae	Nitrogen-fixing heterotrophs	Nitrifiers	Protein decomposers
Thailand						
Bangken	10^4–10^5	10 –10^2	10^4–10^5	10^5–10^6	10^6–10^7	10^6–10^7
Rangsit	10^4–10^5	10 –10^2	10^3–10^4	10^6–10^7	10^5–10^6	10^6–10^7
Khon Kaen	10^4–10^5	10 –10^2	10^3–10^4	10^5–10^6	10^5–10^6	10^5–10^6
Malaya						
Tg. Kalang	10^6–10^7	10 –10^2	10^5–10^6	10^7–10^8	10^4–10^5	10^6–10^7
Province Wellesley	10^6–10^7	+ –10	10^5–10^6	10^4–10^5	10^2–10^3	10^7–10^8
Philippines						
Maahas clay	10^3–10^4	+ –10	10^4–10^5	10^5–10^6	10^3–10^4	10^5–10^6
Luisiana soil	10^4–15^5	10^4–10^5	10^6–10^7	10^5–10^6	10^5–10^6	10^6–10^7
Taiwan						
Taipei	10^5–10^6	+ –10	10^4–10^5	10^4–10^5	10^3–10^4	10^6–10^7

seedling stage to the reproductive stage, but their growth gradually becomes poor at more advanced stages. The growth of phototrophic bacteria in the rhizosphere of rice plants (Kyoto-Asahi species) starts to increase rapidly at the reproductive stage (early August in Japan), and reaches maximum at the ear budding stage and gradually declines with the draining of the paddy fields as shown in Fig. 1. This suggests that phototrophic bacteria may influence the yield of rice (2).

Nitrogen fixation

Nitrogen fixation of isolated phototrophic bacteria (*Rhodopseudomonas capsulata*) was examined in the following conditions: anaerobic-light, aerobic-light, aerobic-dark, and anaerobic-dark. The results of this study (15) have been shown in Table 5. A characteristic of phototrophic bacteria was that their nitrogen-fixing activity reached maximum under light and anaerobic condition. The question is whether such an anaerobic-light condition exists under the sun. Ordinarily it is assumed that no such condition would exist. However, higher nitrogen fixation was observed when some heterotrophs co-existed with other bacteria than when phototrophic bacteria in pure condition were subjected to anaerobic-light condition (13, 14).

R. capsulata can neither fix nitrogen nor maintain growth when cultured in a medium with glycerol as the sole substrate. However, even in such an extreme condition, nitrogen fixation takes place if the nitrogen-fixer, *Azoto-*

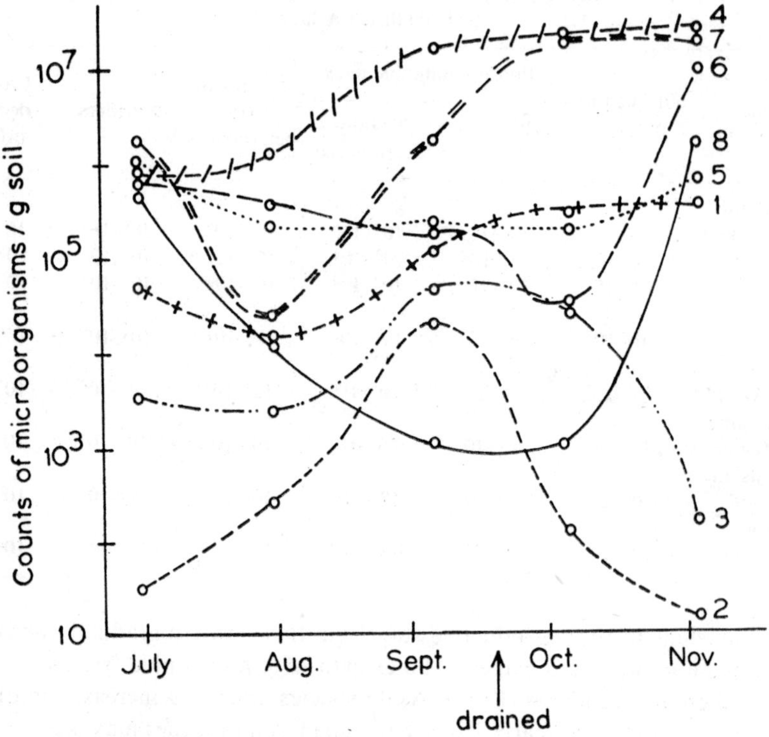

Fig. 1. Seasonal changes of microbial counts in paddy fields (Japan).

1. *Chlorella*, etc. (non-nitrogen fixing photoautotrophs)
2. Chromatiaceae } Phototrophic bacteria
3. Rhodospirillaceae } Phototrophic bacteria
4. Nitrogen fixers (Heterotrophs)
5. Non-uitrogen fixers (Heterotrophs)
6. Protein decomposers, etc.
7. Starch decomposers, etc.
8. Nitrifiers.

bacter, and non-nitrogen fixers *Bacillus megaterium* or *B. subtilis*, co-exist (7) with the *R. capsulata* (Table 6). Because phototrophic bacteria show nitrogen-fixing activity even in aerobic conditions when heterotrophs co-exist in a sort of symbiotic relation, the nitrogen fixation of phototrophic bacteria in nature can amount to a considerable extent.

Ecological variation of phototrophic bacteria in organic sewage (6, 11)

When rice straw and a small amount of rice field soil are placed in a glass bottle under waterlogged condition for a month at 25–30°C, air-bubbles are

Table 5 (a). Comparison of nitrogen fixation of phototrophic bacteria by using butyrate and pyruvate as substrate under various conditions

Conditions		A Butyrate*		B Pyruvate**
Aerobic	Light	16.8%	(1.01)	67.1%
	Dark	1.5%	(0.09)	40.2%
Anaerobic	Light	100.0%	(6.00)	100.0%
	Dark	0.7%	(0.04)	21.9%

*Incubated for 3 weeks.
**Incubated for 1 week.
In parenthesis: fixed N mg/100 ml.

Table 5 (b). Comparison of nitrogen-fixing activity under aerobic light and dark conditions by phototrophic bacteria and mixing cells of phototrophic bacteria with *Bacillus megaterium*

Organisms		PTB*	PTB-Bm**	PTB (^{15}N)	PTB-Bm Mix (^{15}N)
Substrate		Butyrate	Glycerol	Pyruvate	Pyruvate
Condition	Light	100	100	100	100
	Dark	8.9	22.2	60.0	67.8

*PTB: Phototrophic bacteria, *Rhodopseudomonas capsulata.*
**Bm: *Bacillus megaterium.*

Table 6. Nitrogen fixation in pure and mixed culture of *R. capsulata* with *Azotobacter agilis*, *Bacillus subtilis* or *Bacillus megaterium* (in shaking culture under aerobic condition with illumination)

	Fixed nitrogen: N mg/100 ml*
Rhodopseudomonas capsulata	0.00
B. megaterium	0.00
Azotobacter	2.69
B. subtilis	0.00
R. cap.—B. mega.—Mix	4.41
R. cap.—Azotobacter—Mix	4.95
R. cap.—B. sub.—Mix	0.54

*0.3 % glycerol medium—Nitrogen free.

B. megaterium, *Azotobacter* and *B. subtilis* can use glycerol as substrate but *R. capsulata* cannot.

B. megaterium and *B. subtilis* are non-nitrogen fixers; *R. capsulata* and *Azotobacter* are nitrogen fixers.

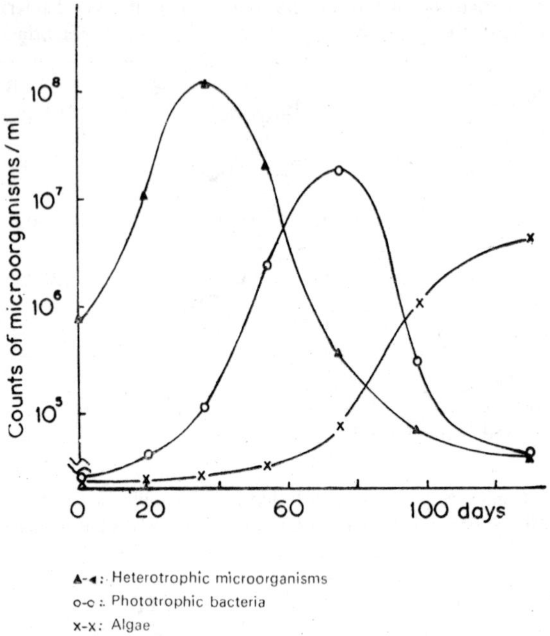

Fig. 2. Microbial changes in the decomposing process of straw powder of rice plant under submerged, natural condition.

produced due to the abundant growth of heterotrophs accompanied by decomposition and foul odour. With the decrease of the substrate and accumulation of the decomposed products the growth of the heterotrophs becomes poorer, and in such a condition the growth of phototrophic bacteria becomes rapid as shown in Fig. 2. In two or three weeks, phototrophic bacteria stop growing gradually and green algae appear. Examination of the change in the biochemical oxygen demand (BOD) during this process showed that the degree of putrefaction was high (several ten thousand ppm) when the heterotrophs were growing but the value decreased to several hundred ppm when phototrophic bacteria were growing. This indicates that purification of the medium progressed, and with the appearance of the green algae the purification could be accelerated even further (see Table 7).

Table 7. Purification of polluted water, including organic materials (animal excrement), under natural biological process

	Ammonia ppm	B.O.D. ppm
Heterotrophs growing stage	over 5,000	over 5,000
After growth of phototrophic bacteria	200–500	200–600
Photoautotrophs growing stage	10–50	10–60

Such succession of microorganisms was found to purify other organic sewage materials such as human excreta, livestock waste, sewage waters, effluents from food processing factories (confectionery, canned food, starch, beer, alcohol, antibiotics, amino acids) and other industrial effluents from oil and chemical plants.

Treatment of sewage using photosynthetic microorganisms

It was found that organic sewage is purified by the natural ecological change of microorganisms as mentioned above. Therefore, a method of purifying organic sewage in an artificial plant which simulates natural microbial processes of phototrophic bacteria and algae have been studied for more than 10 years, and it is now in practical use (3).

Figure 3 shows a flow diagram of the operative elements of a purifying plant. Solid matter is first filtered or separated by a vibrating screen. The concentration of the inflow waste solution should be adjusted so as to exceed 1,000 ppm BOD as the waste cannot be satisfactorily treated by the activated sludge method. The waste flowing into the first tank is aerated. This process facilitates decomposition by decarboxylation of the high molecular substances by the aerobes. This decomposition step prepares the substrate for further action by phototrophic bacteria and aerobes acting symbiotically. When BOD levels exceed 10,000 ppm, certain types of organic matter produce a foam. If this occurs, a defoaming propeller should be applied at the surface.

After 24 hours, aeration of the first tank is discontinued, and the contents are allowed to precipitate for 30 minutes to 1 hour. Then 80 per cent of the supernatant is transferred to a second tank containing phototrophic bacteria. A precipitating tank may be used for continuous treatment. The first tank is again filled and the aeration process repeated. In the second tank, the phototrophic bacteria are induced to multiply rapidly in symbiosis with other bacterial flora using controlled aeration. As these bacterial cells multiply, aeration is gradually increased for the purpose of increasing the dissolved oxygen level to 3 to 5 ppm. This process greatly enhances purifying efficiency. During this process the pigment of the bacteria changes to pink, and the concentration of phototrophic bacteria decreases.

When phototrophic bacteria grow well, a deep red colour develops, when dissolved oxygen levels are adjusted to 1 to 2 ppm using controlled aeration and illumination. Solar light is used during the daytime and artificial illumination at night. The amount of bacteria produced as a byproduct increases proportionally with the removal of nitrogen and phosphorus.

The "staying time" of phototrophic bacteria in the second tank is about three to five days. The capacity of the tank is determined by the BOD value of the original waste solution. Either a single tank or three to five tanks placed in series, may be utilized for the phototrophic bacterial process.

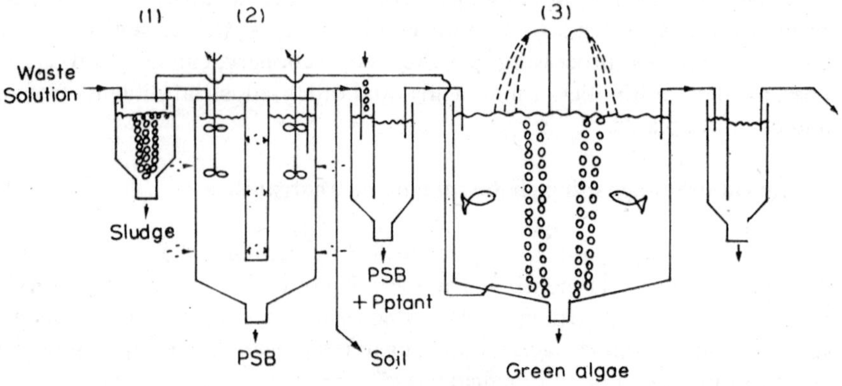

Fig. 3. A flow diagram of the operative elements in a purifying plant.

Tank 1. Receiving tank. The liquid waste is received in this tank and the suspension aerated for 20 h (even during the inflow of waste solution, e.g. sewage). The pH will vary according to aeration, especially for liquid waste containing protein. Higher aeration increases pH and vice versa. For liquid waste from pharmaceutic industry (penicillin and erythromycin production) aeration is not required. In this case Tank No. 1 is used as a storage tank for short periods. If storage is to be prolonged it is essential that aeration be carried out to prevent deterioration of the waste product by anaerobic fermentation. After initial aeration, the aerobic environment is eliminated by the stoppage of air supply. The large particles settle on the bottom of the tank and the supernatant is transferred to Tank No. 2. This conversion of the conditions from aerobic to anaerobic is essential in order that the accumulation of the sludge be avoided. New liquid waste is added and the aeration resumed. This tank is a closed system type and the air having the unpleasant odour is piped into Tank No. 3, or to an earth column to be purified.

Tank 2. Phototrophic bacteria (**PTB**) tank. This tank is illuminated continuously by combined day light and artificial light. The suspension is agitated but not aerated by mechanical propellers. In most cases the pH will remain at a suitable level; however, for waste water from penicillin production, it is advisable to adjust the pH (pH=7.0–7.5) with an alkaline solution. After 24 h 20–30 per cent of the liquid is transferred to a by-pass settling tank. The **PTB** cells are collected by simple precipitations with hydrolyzed chitin solution (0.01–0.001 per cent for liquid waste) and the supernatant is transferred to Tank No. 3. (If the separation of **PTB** cells is not required, the suspension can be used directly as fertilizer.) The 70–80 per cent remaining liquid is used as starter for the next purification.

Tank 3. Algal culture tank. Tank No. 3 is aerated by the gas produced from Tank No. 1, and by the air. If the liquid from Tank No. 2 is deficient in mineral elements (e.g. from erythromycin production), small quantities of N, P, K, Mg should be added to promote the growth of algae. The liquid is maintained for totally 5 days. The aeration is stopped every day and the algal cells are collected. The algal cells settle down to the bottom of the tank and can be readily harvested. (At this point in the process 20–30 per cent of the purified water can be released.) However, this harvesting operation can be obviated if nylon netting is set on the surface of the algal tank. By overflooding the culture, algal cells will be collected by the nylon net. The algal cells can then be used as animal and fish feed.

The effluent of the second tank is then transferred to a precipitation tank. Bacteria may be removed by dropwise addition of the precipitant (ex. chitosan). The precipitated phototrophic bacteria can ultimately be used as an animal feed or fertilizer. After precipitation is completed, the effluent is transferred to a tank containing algae, where it remains for a longer period of time (over two times bigger than phototrophic bacterial tank). The results obtained in purifying waste produced from livestock by this method is shown in Table 8.

Table 8. Example of purification of waste solution from swinery

	Original	Supernatant in precipitation tank after PTB* treatment	Discharged water
BOD	(ppm) 6600	380	15
COD	(ppm) 3364	354	64
SS	(ppm) 6540	450	17
Kjeldahl nitrogen (as N)	(ppm) 915	32.8	7.8
pH	6.8	7.3	7.1

*PTB: Phototrophic bacteria.

In many parts of Japan temperatures are often so low in winter that the use of an algae tank is impractical. Thus, as a substitute, a contact oxidation tower could be used (see Fig. 4). Solid materials are removed and the effluent is aerated in the first tank. After the waste is solubilized, the supernatants are transferred to a tank containing phototrophic bacteria. The pH is adjusted to 7 if necessary. Again, precipitated bacteria are utilized as animal feed supplements. The addition of chitinous substances such as crab shells would make it possible to utilize the precipitated bacterial cells as livestock feed. However, care should be taken in the use of such feeds, since some chemosynthetic, high

Table 9. Example of purification of waste solution from bean cake factory

	Original	Supernatant in precipitation tank after PTB* treatment	Discharged water
BOD	(ppm) 11300	340	15
COD	(ppm) 9800	270	17
SS	(ppm) 3930	23	5
Kjeldahl nitrogen (as N)	(ppm) 3850	280	11
pH	6.4	7.8	7.2

*PTB: Phototrophic bacteria.

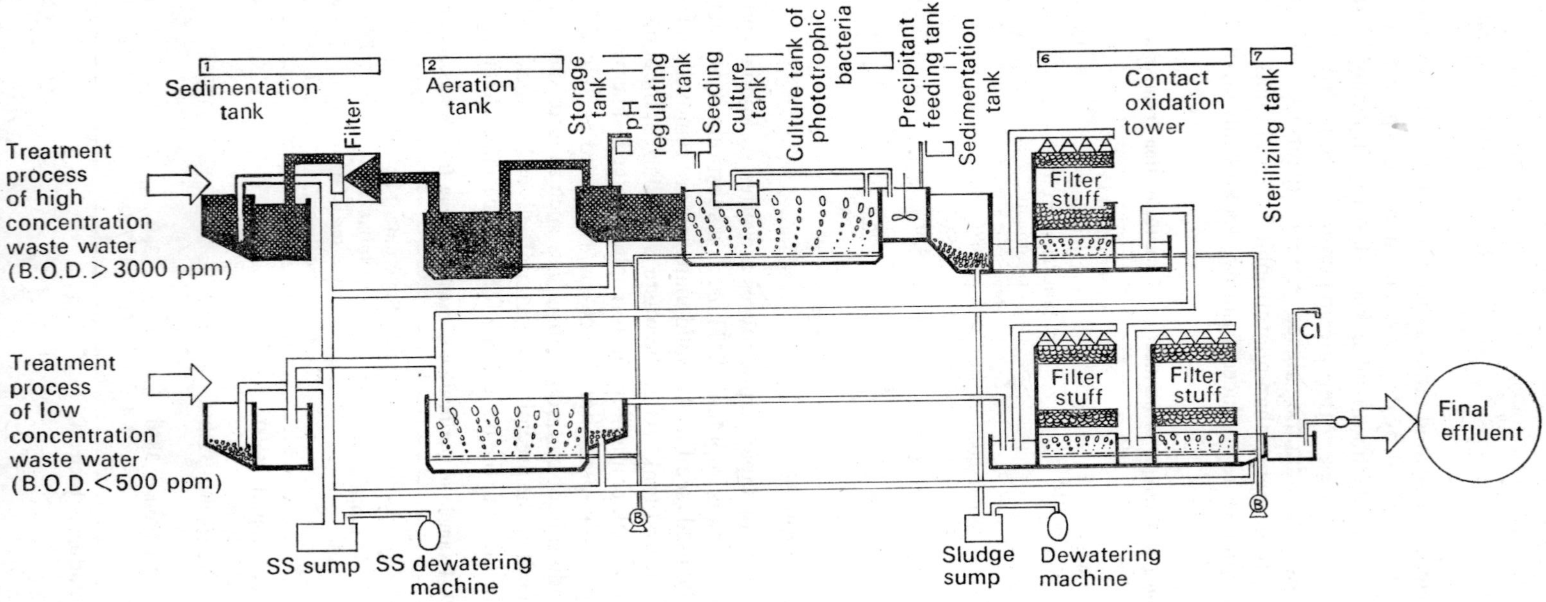

Fig. 4. Flow sheet of purification of organic waste water by using phototrophic bacteria.

molecular precipitants may be toxic. The BOD values of the supernatant may be reduced to very low ppm levels by passing the effluent through an exudation tower, as illustrated in Fig. 4. The data obtained in the purification of bean cake are shown in Table 9. This process of purification by using phototrophic bacteria may be effective for a variety of waste products (Table 10).

Table 10. Source of waste materials purified by phototrophic bacteria

Various kinds of microbial industry (beer, antibiotics, amino acids, nucleic acids, etc. fermentations)
Various kinds of chemical industry (synthetic fibres, synthetic resins, chemical fertilizers, chemicals, etc.)
Various kinds of food industry (canned food, bottled food, cakes, *miso*, *tofu:* bean cake, etc.)
Petroleum industry
Starch and wool industry
Others: activated sludge, excrement, other organic materials

The advantages of this system over the activated sludge process are: (a) dilution is not required and hence the system can be used in areas where water is a scarce resource; (b) unlike the activated sludge method which creates a secondary sludge disposal problem, this system produces phototrophic bacteria and green algae which are useful byproducts; and (c) because very little dilution is required and much less space is required for this process in comparison with the activated sludge method.

The use of byproducts

The use of byproducts (microbial cells) in pisciculture and poultry industry is of significance as outlined in Tables 11 and 12. The survival rate of carp fry is significantly higher in the group fed with phototrophic bacteria, but the size of individual fish was lowered (Table 11).

Table 11. Effect of phototrophic bacteria on survival of young fry of crucian carps*

	Survival numbers after 1 month	Survival ratio (%)
Control	2772	69.3
Treatment (with 0.1% PTB cells)**	3860	96.5

*The experiment is carried out in a tank of 2 tons capacity with an initial number of 4000 fry.

**The PTB cells were obtained from waste treatment plant of fish meat industry. The bacterial powder contains only 50 per cent of PTB cells, the other half is made up of heterotrophic bacteria (contaminants).

It is possible that the higher survival rate reduced the living space in the tank, which may be responsible for the lower weight of individual fish. The food value of phototrophic bacterial cells has also been demonstrated in the culture of brine shrimp (*Artemia salina*) and in plankton production.

The effect of phototrophic bacteria on egg laying by hens is commercially important. In six months, the treated group of birds produced 3708 more eggs (or 243 kg) than the control group (8). Such an increase would have substantial benefit to the poultry industry (Table 12).

Table 12. Effect of phototrophic bacteria on egg laying by hen

	Total number of eggs/6 months	Total weight of eggs/6 months (kg)	Average number of eggs laid by 200 birds	Average weight of egg (g)
Control	24,408	1486.5	136±15	60.9±1.5
Treatment (+0.01% PTB*)	28,116	1729.1	156±7	61.5±0.3

*See text.

The value of phototrophic bacterial cells as organic fertilizer (4, 9)

a) *Experiment with persimmon*

As shown in Table 13, those plants to which phototrophic bacterial cells were applied, showed a remarkable effect. The quantity of output and their sugar content increased and there was a substantial improvement in fruit colour. Subsequently, we fractionated and studied the carotenoid pigments of the fruit skin by means of column chromatography. It was confirmed that

Table 13. Weight and chemical components of persimmon fruit

	Total No. of fruit	Total Wt. of fruit kg.	Average Wt./fruit g	Fruit* colour H.C.C.	Water contents %	Degree of sugar in Brix	Acid contents %	Reducing sugar contents %	Non-reducing sugar contents %	Total sugar contents %
Control (Inorganic fertilizer)	32	7.1	222	12	84.7	14.7	0	10.82	2.34	13.16
Treatment (Organic fertilizer)**	43	8.2	191	13	83.3	16.4	0	12.56	2.57	15.13

*Fruit colour: 12 Orange, 13 Saturn red; **Phototrophic bacterial cells.

the fruits from plants receiving an application of phototrophic bacteria contained a larger quantity of carotene than the fruits from control plants to which inorganic fertilizer liquid was applied (Table 14). It is especially interesting that the fruits from treated plants had a large quantity of lycopene pigment, which is present in large quantities in phototrophic bacterial cells. We,

Table 14. Contents of carotenoid pigment in persimmon peel (mg/100 g fresh weight)

	β-carotene	Lycopene	Cryptoxanthin	Zeanxathin	Total contents of carotenoid pigment
Control					
(Inorganic fertilizer)	3.108	2.773	13.018	7.682	26.581
Treatment					
(Organic fertilizer*)	2.929	4.237	15.667	8.970	31.803

*Phototrophic bacterial cells.

therefore, studied the movement of the pigment by labelling the cells utilizing ^{14}C. The results are shown in Table 15 which confirmed that the compound was taken not only into the entire fruit but also into the pigment itself. It is possible that the phenomenon was caused by the $^{14}CO_2$ generated by the decomposition of the phototrophic bacterial cells by other soil microorganisms, which was then assimilated by photosynthesis into the plant. However, if it is

Table 15. Incorporation of ^{14}C in persimmon fruit from ^{14}C-labelled phototrophic bacterial cells

	^{14}C-activity incorporated (cpm)
Fruit	3.0×10^4 (100)
Pigments	3.4×10^3 (11.3)

Remark: Contents of carotenoid pigments in persimmon fruit are 0.026–0.032% (see Table 14).

assumed that this was the result of the assimilation of carbon dioxide, then it should follow that the pigment would contain the activity of 0.024 per cent of the ^{14}C-activity (3×10^4 cpm) taken into the fruit, inasmuch as the fruit's pigment content ratio is 0.024 per cent, while there was, in reality, a very high activity of 11.3 per cent. This proves that apart from the assimilation of carbon dioxide, there were certain materials which it is assumed to be the precursor of pigment, absorbed directly from the soil, which were involved in the synthesis of the pigments.

b) *Experiment with tomato plants*

As shown in Table 16, the vitamin B and vitamin C content was larger in plants treated with phototrophic bacterial cell suspension than control plants which did not receive the bacteria.

Table 16. Contents of vitamins B_1 and C of tomato fruit in culture supplied with inorganic and organic fertilizers

	Vitamin (mg %)	
	B_1	C
Sand culture		
Control		
(Inorganic fertilizer: I.F.)	0.09 (100)	25.6 (100)
Treatment		
(I.F.+organic materials*)	0.12 (133)	28.6 (111)
Soil culture		
Control (I.F.)	0.16 (100)	30.2 (100)
Treatment		
(I.F.+organic materials*)	0.18 (112)	32.6 (108)

*Organic materials: Phototrophic bacteria.

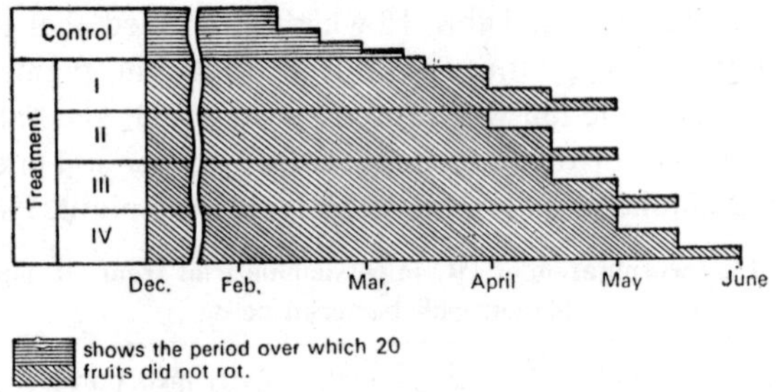

Fig. 5. Effect of organic fertilizer on the storage period of mandarin oranges. *Control:* supplied inorganic fertilizer only. *Treatment:* supplied with organic fertilizer (PTB cells I: in July; II: in Aug.; III: in Sept.; IV: in July, Aug. and Sept.)

c) *Experiments with mandarin oranges*

The mandarin fruits were harvested and stored at room temperature and the length of time to the onset of their decay was measured. As shown in Fig. 5, the control, which had been grown with inorganic fertilizer only, had almost completely rotted within a few months after harvesting. The treatments in which the organic fertilizer was applied, however, showed a remarkable resistance to such decay. Treatment IV in particular (application of the organic fertilizer in July, August and September) indicated practically no decay at all after five months following the harvest.

Use of the bacterial mass obtained from treatment of waste solution

The sewage purification using phototrophic bacteria is useful for treating waste water from palm oil and molasses. When phototrophic bacterial cells were collected as the secondary product along with other co-existing microbial cells and utilized as an organic fertilizer, the growth of plants was promoted as shown in Table 17. The analysis of microflora in soil indicated that beneficial actinomycetes multiply better in the presence of organic fertilizer when compared to the application of chemical fertilizer as shown in Table 18. It was observed that when phototrophic bacteria are added to the medium, *Streptomyces fradiae*, antagonistic to *Fursarium oxysporum* multiplies and causes lysis of *F. oxysporum*, a pathogenic organism.

Phototrophic bacteria inoculated on activated sludge and digested sludge

Table 17. Tomato plant growth and the fruit yield by supply of phototrophic and symbiotic heterotrophic microbial cells

	Leaves (dry weight, g)	Roots (dry weight, g)	Total fruits number	Total fruits (fresh weight, g)
Sand culture/pot				
Control (I.F.)	57.7	15.0	14	729
Treatment (I.F.+PTB, etc.*)	51.3	20.5	16	805
Soil culture/pot				
Control (I.F.)	58.2	22.1	16	1049
Treatment (I.F.+PTB, etc.*)	55.1	30.3	22	1408

***PTB, etc.**: Phototrophic and symbiotic heterotrophic microbial cells collected from purification of waste solution. I.F.—Inorganic fertilizer

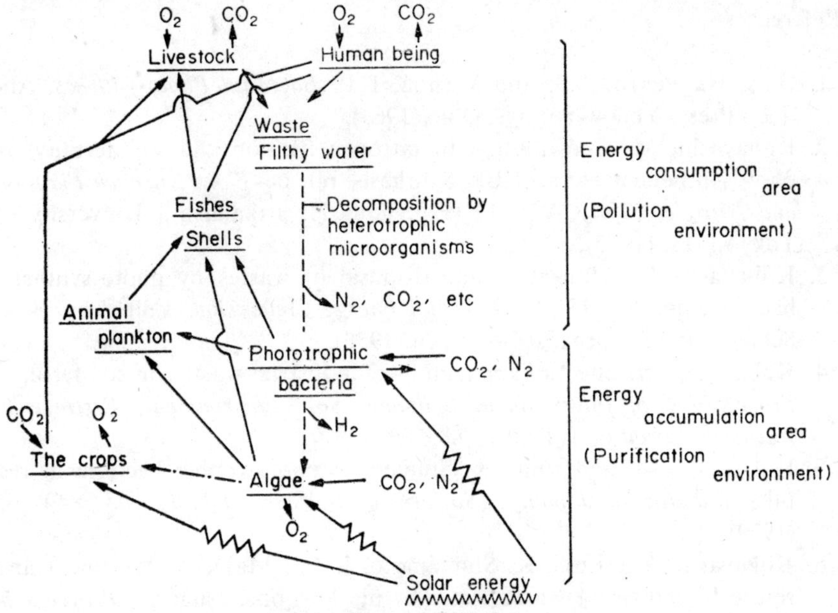

Fig. 6. Diagram showing role of phototrophic bacteria in natural environment.

Table 18. Comparison of microbial counts in tomato plant culture supplied with inorganic and organic (PTB, etc. cells) fertilizers

(after 3 months)

	Bacteria (counts/g)	Actinomycetes (counts/g)	Fungi (counts/g)	Act./Fungi ratio
Sand culture				
Control (I.F.)	3.8×10^5	1.0×10^5	18.0×10^4	0.56
Treatment (I.F. + PTB, etc.*)	9.0×10^5	1.2×10^5	4.0×10^4	3.0
Soil culture				
Control (I.F.)	8.5×10^6	5.2×10^5	5.0×10^5	1.0
Treatment (I.F. + PTB, etc.*)	16.3×10^6	23.0×10^5	15.0×10^5	1.5

*PTB, etc.: Phototrophic and symbiotic heterotrophic microbial cells. I.F.—Inorganic fertilizer.

grow well when agitated under illumination. One has to only remove hydrogen sulphide, amines and other smelling substances (ex. mercaptan, etc.) present in these sludges, to change them into the useful organic fertilizer. Thus phototrophic bacteria are excellent organisms for recycling organic wastes (3). A diagrammatic representation of the role of phototrophic bacteria in natural environment is shown in Fig. 6.

References

1. Gest, H., Pietro, A.S. and Vernon, L.P. *Bacterial Photosynthesis*, Antioch Press, Yellow Springs, Ohio (1963).
2. Kobayashi, M. Contribution to nitrogen fixation and soil fertility by photosynthetic bacteria, JIBP, Synthesis, pp. 66–77 in *Nitrogen Fixation and Nitrogen Cycle*, Vol. 12 (Edited by H. Takahashi), University of Tokyo Press (1975).
3. Kobayashi, M. Utilization and disposal of wastes by photo-synthetic bacteria, pp. 443–453 in *Microbial Energy Conversion* (Edited by H.G. Schlegel and J. Barnea), Gottingen (1976).
4. Kobayashi, M. Effect of application of industrial wastes on soil fertility, *Proceedings of International Seminar, Soil Environment, Fertility in Intensive Agriculture* (Tokyo), 688–697 (1977).
5. Kobayashi, M. and Fujii, K. Studies on phototrophic bacteria in red tide, *Bulletin of Japanese Society of Science Fisheries*, *45*, 849–855 (1979).
6. Kobayashi, M., Fujii, K. Shimamoto, I. and Maki, T. Treatment and re-use of industrial waste water by phototrophic bacteria, *Progress in Water Technology* (England), *11*, 279–284 (1978).

7. Kobayashi, M. and Haque, M.Z. Contribution to nitrogen fixation and soil fertility by photosynthetic bacteria, *Plant and Soil* (Netherlands), Special Volume, 443–456 (1971).
8. Kobayashi, M. and Kurata, S. The mass culture and cell utilization of photosynthetic bacteria, *Process Biochemistry* (England), *13*, 27–30 (1978).
9. Kobayashi, M. and Maki, T. Production of useful materials by photosynthetic bacteria, Abstracts of XII International Congress of Microbiology (Munchen), Bacteriology Section, p. 24 (1978).
10. Kobayashi, M., Takahashi, E. and Kawaguchi, K. Distribution of nitrogen-fixing microorganisms in paddy soils of Southeast Asia, *Soil Science* (U.S.A.), *104*, 113–118 (1967).
11. Kobayashi, M. and Tchan, Y.T. Treatment of industrial waste solution and production of useful by-products using a photosynthetic bacterial method, *Water Research* (England), *7*, 1219–1224 (1973).
12. Kondrateva, E.N. Photosynthetic bacteria, Akademiya Nauk, SSSR, *Institut Mikrobiologii*, Moskva (1963).
13. Okuda, A. and Kobayashi, M. Production of slime substance in mixed cultures of *R. capsulata* and *Azotobacter vinelandii*, *Nature* (England), *192*, 1207–1208 (1961).
14. Okuda, A. and Kobayashi, M. Symbiotic relation between *R. capsulata* and *Azotobacter vinelandii*, *Mikrobiologiya* (USSR), *32*, 936–945 (1963).
15. Okuda, A., Yamaguchi, M. and Kobayashi, M. Nitrogen fixation by photosynthetic bacteria under various conditions, *Soil, Plant Food*, *5*, 73–76 (1959).
16. Pfenning, N. and Truper, H.G. The Phototrophic bacteria, pp. 24–75 in *Bergey's Manual of Determinative Bacteriology*, 8th edition (1974).
17. Van Niel, C.B. *Bergey's Manual of Determinative Bacteriology*, 7th edition, 35–67, Baltimore, William and Wilkins Co. (U.S.A.) (1957).

F.X. ROTH

24. Microorganisms as a Source of Protein for Animal Nutrition*

Introduction

In view of population trends and the current shortage of protein, it is estimated that protein production for human nutrition must be doubled in the next twenty years (14). One of the ways to augment protein production is not only to increase protein production through conventional sources but also to produce proteins by microbial fermentation. Basically, microbial fermentations result in quick growth of selected microorganisms which are rich in protein and the dried biomass of the end product is referred to, somewhat incorrectly, as single cell protein (SCP).

Production of microorganisms

There are a number of advantages in microbial production of protein when compared with conventional methods of producing protein containing food and feed. These advantages relate to:

1) Rapid succession of generations (algae, 2–6 hr; yeasts, 1–3 hr; bacteria, 0.5–2 hr);

2) Easily modifiable genetically (e.g. for composition of amino acids);

3) High protein content of 45 to 85 per cent in the dry mass;

4) Broad spectrum of the original raw material used for the production which also includes waste products;

5) Production in continuous cultures, consistent quality not dependent on climate, determinable amount, low land requirements, ecologically beneficial.

A number of microorganisms can be used for the production of microbial protein. The most interesting are the green algae *Chlorella* and *Scenedesmus*,

*Reproduced with permission from *Animal Research and Development*, *12*, 7–19 (1980), published by the Institute for Scientific Cooperation, Federal Republic of Germany.

and the blue-green alga *Spirulina*. The use of yeasts, particularly *Rhodotorula* and *Saccharomyces*, has long been established; recently *Candida* has also been used. The use of bacteria such as *Hydrogenomonas*, *Methanomonas* and *Methylomonas* was recently attempted for the first time with remarkable success in some cases. Various fungi have also been used for the development of appropriate procedures in SCP production.

The availability of necessary substrates is of considerable biological and economic importance for the production of SCP. While algae which contain chlorophyll do not require organic substrate because they utilize energy from sunlight and CO_2 from the air, yeasts and bacteria require a source of energy since they do not possess chlorophyll. In addition to the source of energy, the nutrient substrate must be enriched with appropriate minerals and organic nitrogen from an adequate source. Several substances and compounds can be used as sources of energy. The most important according to Schulz (14) are presented in Table 1. Of the two groups of substances listed in the table, those in group A have long been recognized as the most suitable

Table 1. Nutrient substrates and/or C sources for the production of yeasts and bacteria

A. Long-known nutrient substrates such as
   1. Starch-containing raw materials (e.g. grain, potatoes, tapioca and their by-products)
   2. Sugar-containing raw materials (sugar beet, sugar cane and their processed products)
   3. Hydrolysates of wood and annual plants
   4. Other raw materials such as whey and refuse from processed foods

B. Unconventional nutrient substrates
   1. Alkanes of various long chain hydrocarbons (particularly C_{10}-C_{23} chains)
   2. Alcohols such as methanol, ethanol, isopropanol
   3. Aldehydes
   4. Organic acids
   5. Processed refuse from animal production
   6. Municipal waste water
   7. Water + carbon dioxide

nutrient substrates for yeast production and they are still being used today. The compounds in group B are only interesting at present and they have only been used in intensive research during the last twenty years. The most important processes for the production are those which utilize waste products. At present, there are technologies which are ready for production in addition to those using the conventional sources, particularly for yeasts using hydrocarbons (alkane) as a substrate and bacteria using methanol as a base material. Large-scale production has been envisaged in England and Rumania with annual production of 60,000 tonnes of bacterial mass in England.

A procedure for bacterial biomass production is shown in Fig. 1, which consists of the following steps:

—Supply of a nutrient substrate

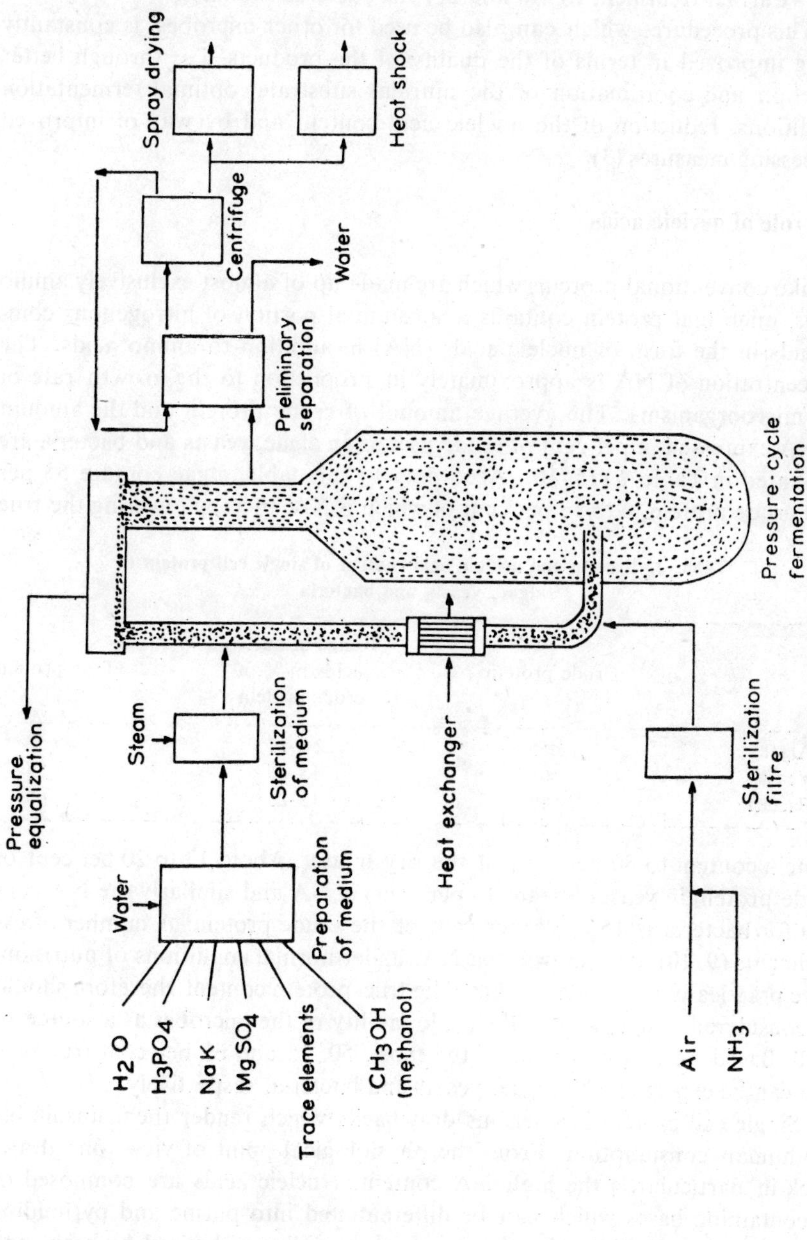

Fig. 1. Bacterial biomass production (ICI procedure)

—Formulation of a suitable medium
—Multiplication of microorganisms through fermentation
—Separation of cellular substance from the left-over medium
—Further treatment to kill and dry the bacterial biomass.

This procedure, which can also be used for other microbes, is constantly being improved in terms of the quality of the products, i.e. through better selection and coordination of the nutrient substrate, optimal fermentation conditions, reduction of the nucleic acid content and by way of improved processing measures (3).

The role of nucleic acids

Unlike conventional proteins which are made up of almost exclusively amino acids, microbial protein contains a substantial portion of nitrogenous compounds in the form of nucleic acids (NA) in addition to amino acids. The concentration of NA is approximately in proportion to the growth rate of the microorganisms. The average amount of crude protein and the amount of NA expressed in per cent of crude protein in algae, yeasts and bacteria are presented in Table 2 (15, 10). According to this table, algae contain 55 per cent crude protein, 6 to 13 per cent of which is NA, thereby lowering the true

Table 2. Protein and nucleic acid content of single cell protein of algae, yeasts and bacteria

	Crude protein	Range of nucleic acids in % of crude protein	True protein
Algae	55	6–13	50
Yeasts	65	13–20	55
Bacteria	80	15–25	65

protein content to 50 per cent of the dry matter. About 13 to 20 per cent of crude protein in yeasts (mean, 15 per cent) is NA and similarly the NA content for bacteria is 15 to 25 per cent of the crude protein. A number of experiments (9, 10) have shown that NA, under normal conditions of nutrition, have practically no protein value. The true protein content therefore should be considered as an index of the exploitability of the microbes as a source of SCP. Based on this criterion, at the most, 50, 55 and 65 per cent true protein can be expected with algae, yeasts and bacteria, respectively.

Single cell protein has serious drawbacks which render them unsuitable for human consumption. From the physiological point of view, one drawback in particular is the high NA content. Nucleic acids are composed of N-containing bases which can be differentiated into purine and pyrimidine bases. While pyrimidine bases are relatively well metabolized by both man and animal, purines present a health hazard for man. The metabolic breakdown of purines is shown in Fig 2. The purines in food are broken down into uric acid through several intermediary stages. In man, uric acid is the end-product of purine metabolism and the acid is excreted in urine. Under physiological pH values, uric acid is only sparingly soluble; increased uric acid

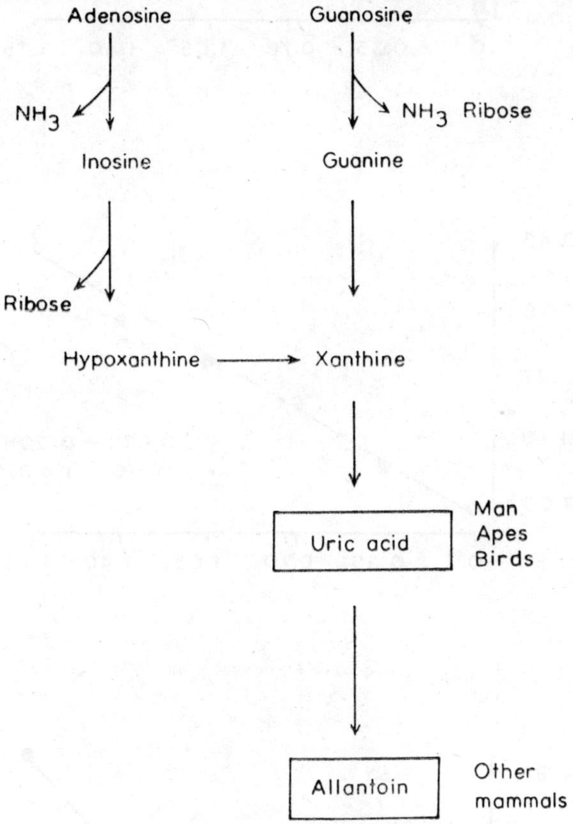

Fig. 2. Breakdown of purines in metabolism.

values in the blood result in the deposition of uric acid crystals in tissue and joints, resulting in painful attacks of inflammation (gout). Increased excretion of uric acid through the kidney can be responsible for kidney and bladder stones. Therefore, the maximum daily NA intake for man should not exceed 2 g.

On the contrary, mammals other than man are able to break down purine bases one stage further into freely soluble allantoin and excrete it. This was tested in many experiments with pigs, calves and sheep; allantoin was analyzed in each case as the most important end-product of the purine catabolism

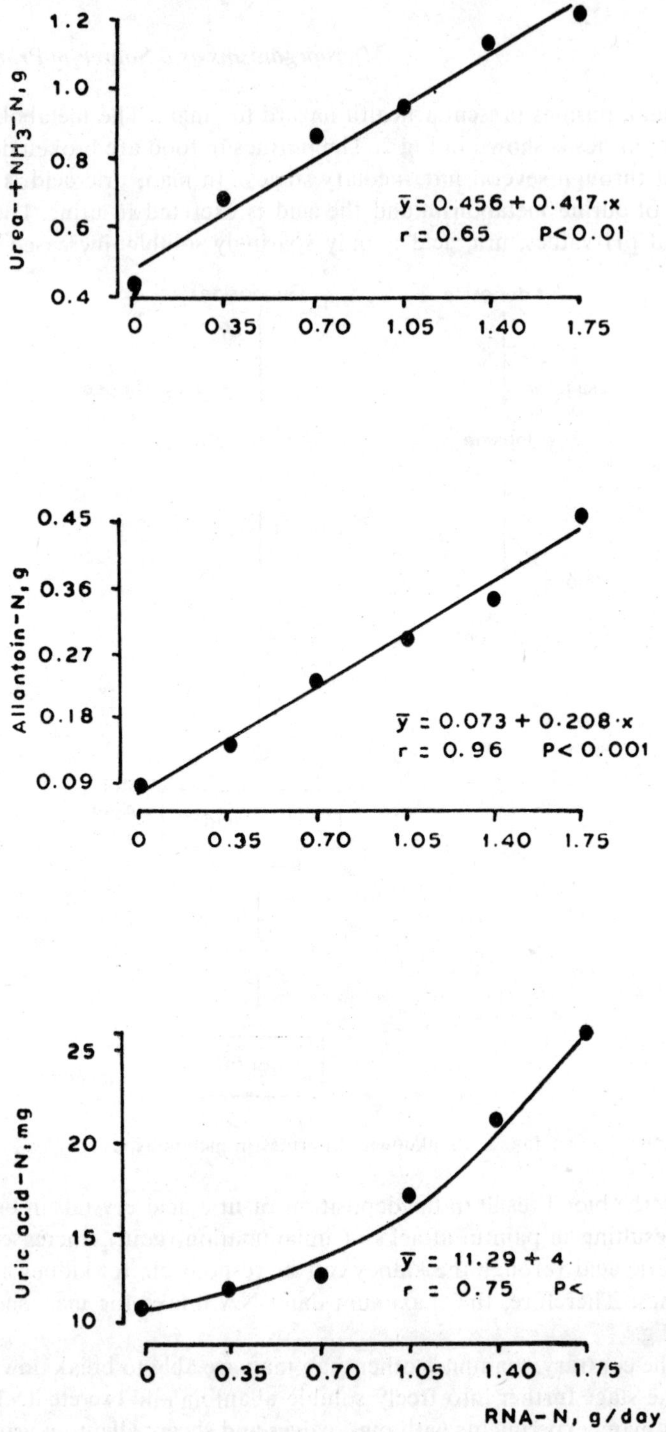

Fig. 3. Relationship between the intake of nucleic acid-N and daily excretion of N-fractions in the urine of piglets (●—mean of 4 animals)—for details see text.

(8, 9, 10, 11). The results presented in Fig. 3 show that increased NA intake in piglets results in increased excretion of allantoin. The elevated uric acid concentration is not quantitatively important since it is also apparent that a considerable portion of the NA-N is not utilized and is lost as urea + NH_3.

In view of the metabolic implications cited above, SCP is preferred as a feed for animals, including poultry. Poultry animals also form uric acid, but they possess an extremely effective excretory mechanism. If SCP is used as animal feed, conventional sources of protein could then be used for human consumption.

Crude nutrient content

If new feed materials are to be used for animal nutrition, the composition of the nutrients in the feeds is of primary importance. The nutrient content of microorganisms is dependent on a number of facts such as type of organism, the substrate on which they grow and the schedule of the fermentation process. The procedures involved in the preparation of the biomass after the fermentation process can also influence the chemical composition. From an extensive survey of the literature Schulz (14) lists the ranges of the nutrient content in algae, yeasts and bacteria (Table 3) which shows extreme variations (40–80 per cent) in the crude protein content of the microbial biomass

Table 3. Composition of microorganisms, per cent in dry matter

	Orgaic substrate (range)	Crude protein (range)	Crude fat (range)	Crude fibre (range)	NFE (range)
Algae	86–94	45–71	1.0–14	3.21	9–24
Yeasts in conventional substrate	90–93	45–60	0.2–8	0–3	28–46
Unconventional substrate	90–94	40–68	1.6–24	0–11	12–45
Bacteria (*Methylomonas*)	90–94	80–83	6.6–8	—	3–3.4

depending on the type of microorganisms. Ignoring the few extreme values, the mean crude protein in the dry matter for algae and yeasts on conventional substrate lies between 50 and 60 per cent and the same for alkane yeasts lies between 55 and 65 per cent and for bacteria at 80 per cent. A high content of NA-free protein is naturally extremely important for the economic efficiency of the procedure in SCP production.

Considerable differences in the crude fat contents microorganisms can also be seen from data in Table 3. In addition to the actual triglycerides and fatty acids, it may be pointed out that an extremely high portion of ether extract can contain unsaponifiable compounds. This unsaponifiable portion can be problematic, particularly with yeasts along with components of the nutrient medium if the microorganisms are grown on various hydrocarbons. The literature is extensive on this subject which should be consulted for further information not only on this problem but also on questions about the toxicity (6, 1). In addition, large amounts of unusable poly-β-hydroxybutyric acid can also accumulate in the medium with certain types of bacteria (2). Therefore, in SCP production an attempt is made to hold the upper limits of the ether extract at 8 to 10 per cent and to limit the unsaponifiable portion to half of the total fat content (14) particularly type C-14 to C-18 of the fatty acids. In yeasts, it has been observed that unsaturated fatty acids which are also partially polyunsaturated ones predominate.

Because of the high protein and fat content, the contribution of the carbohydrates to the nutritional value of SCP is not of prime importance. The amount of paraplastic substances from the cell walls in algae is considerable and therefore, it interferes with the digestibility of the cell contents especially if algae are dried very gently when the cell walls do not disintegrate.

The crude ash content is determined in particular by the nutrient salts of the fermentation medium. On the basis of the studies available, it can be concluded that while microorganisms, with the exception of algae, are poor sources of (1 to 2 g/kg dry matter), they are excellent sources of phosphorus because of the high NA content (20 to 30 g/kg dry matter) and phosphorus can be well utilized due to the nature of its chemical bonds. The high iron and zinc content in microorganisms should be emphasized particularly in considering the usefulness of SCP. Apparently these are easily stored during the development processes of the microorganisms.

Mention should be made of the high vitamin B contents of microorganisms. As has been known for some time, yeasts are the most important source of vitamin B for animal and human nutrition. In addition, the B-carotene content in algae is also very high (approx. 500 mg/kg dry matter).

Amino acids

Besides determining the crude nutrient content, protein feed should also be evaluated particularly for protein quality. Based on data provided by Schiller and co-workers (12), Beck and Gropp (1) and Schulz (14) quantitative figures for certain amino acids in SCP which are limiting factors in human diet are shown in Table 4 along with the amino acid content of fish meal and soybean meal as examples of the most important conventional protein components for comparative evaluation. The comparative figures show that yeasts are excellent source of lysine and quantitatively correspond to that of fish meal. Favourable lysine values however can also be seen in algae and bacteria. On the other hand, methionine and cystine content in algae and yeasts was unfavourable although somewhat more favourable in bacteria.

Table 4. Content of individual essential amino acids, g per 100 g crude protein

	Algae (*Scenedesmus*)	Yeasts (*Candida*)	Bacteria (*Methylomonas*)	Fish meal	Extracted soybean meal
Lysine	5.7	7.4	6.2	6.4	6.4
Methionine	1.6	1.5	2.4	2.7	1.5
Cystine	0.7	1.0	1.6	1.1	1.5
Threonine	5.2	4.9	4.6	4.5	4.0
Tryptophan	1.3	1.4	1.0	1.2	1.3

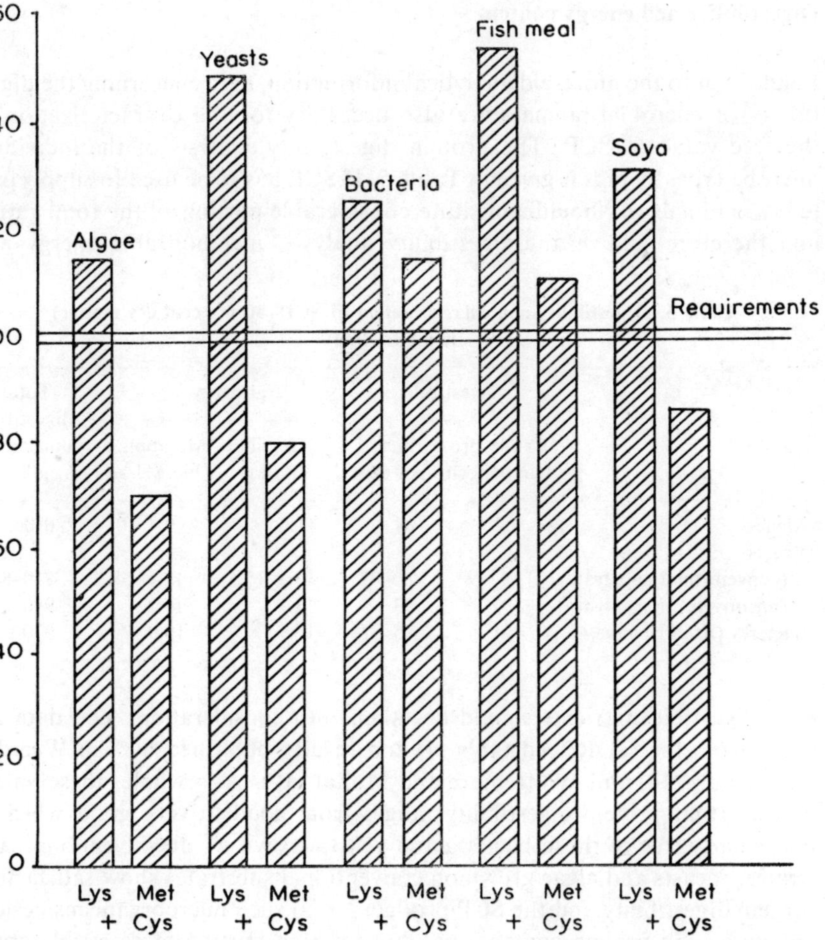

Fig. 4. Lysine, methionine and cystine contents of a few feed proteins compared with the requirements of pigs in the initial fattening phase.

Threonine and tryptophan contents vary little among the individual types of microorganisms and correspond to that of fish and soybean meal. If the lysine, methionine and cystine contents of these products are compared with the nutritional requirements of pigs in the initial fattening phase, it can be seen that algae and yeasts do not meet the requirements for methionine and cystine. As is well known, these amino acids limit the quality of protein. On the other hand, it is clear that like fish meal, yeasts can serve as a good complementary source of lysine. The supply of lysine, methionine and cystine is more balanced with bacterial protein although the complementary effect of lysine is lower.

Digestibility and energy content

In addition to the aforesaid analytical information, data concerning the digestibility of microbial biomass are also necessary for the characterization of the feed value of SCP. The protein digestibility analysis of the individual microbe types by pigs is given in Table 5. If SCP has to be used to supply protein to animals, it should constitute considerable portion of the total ration and therefore data on the digestibility analysis, metabolizable energy and

Table 5. Digestibility and energy content of SCP (90 per cent dry matter) in pig feeding

	Digestibility		Energy		Total digestible nutrients g/kg
	Crude protein per cent	Energy per cent	Digestible MJ/kg	Metabolizable MJ/kg	
Algae	82	65	11.2	9.4	610
Yeasts					
(conventional substrate)	82–90	80–97	13.7–15.5	12.6–14.2	740–820
(unconventional substrate)	92	93	18.1	15.2	980
Bacteria (*Methylomonas*)	91	85	16.9	15.9	920

total digestible nutrients should be taken into consideration. The data are taken from the GDR feed table (4) and results published by van Weerden (16), Schulz (14) and Whittemore and Moffat (19). As a whole, these values indicate that protein digestibility ranges from good to very good which is particularly true of those bacteria and yeasts grown on unconventional substrates. Yeasts and algae grown on conventional substrates show satisfactory protein digestibility and the SCP products from such microorganisms besides having a high protein content, also have a high portion of digestible crude protein. As far as yeasts and bacteria are concerned, favourable to very favourable values can be seen with regard to energy digestibility whereas values for algae decrease sharply because of the high content of cell wall components.

Use of SCP as feed for agricultural animals

Primarily monogastric animals and veal calves (considered because of their digestive function) come into consideration in the use of microorganisms as feed. The extensive tests described here were on poultry and pigs, to determine the extent to which the conventional protein sources, soybean and fish meal or both can be substitued by SCP without loss in performance. In calves, the substitution of skimmed milk by SCP powder was always considered to be of primary interest. Table 6 presents a statement summarizing the various test results for individual types of animals. Algae can only replace

Table 6. Replacement of total protein of the ration by means of SCP (per cent)

Broiler	10 through algae
	30 through yeasts and bacteria
Laying hens	30 through bacteria and 50 through yeasts
Piglets	30 through algae and yeasts
Fattened pigs	40 to 50 through algae, yeasts and bacteria
Calves	15 to 30 through yeasts
	20 to 40 through bacteria

10 per cent of the protein in the ration for broilers, because larger amounts showed a tendency to result in slower growth and a marked yellow staining of the slaughtered carcass (carotinoid) (18). Yeasts were well utilized and assimilated by broilers to the extent of 30 per cent of the crude protein. Compared to fish meal rations, large amounts of proteins (40 to 50 per cent of the protein) are required for adjustment in methionine and arginine levels of the feed. Large amounts of protein in the ration lead to a decrease in performance of the animals which cannot be corrected by adjusting the amino acids which are limiting factors for growth. If bacteria are used to replace this amount in the ration protein, the ration has to be given in the form of a pellet (1, 5, 17). Yeasts are the only source of protein feed for laying hens which do not influence the egg yield. The protein in the ration has to be 30 per cent of the crude protein when bacteria are used.

A number of studies have shown that 30 per cent protein replacement for piglets with algae and yeasts represents the upper limits for raising animals. The use of bacterial biomass with piglets should not exceed an intake of 100 g per animal per day (13). A quality of fattening and slaughter production equal to that obtainable by conventional sources can be obtained in the initial and final fattening phases for pigs by using good yeasts as the only source of protein feed. On the whole, bacteria (*Methylomonas*) is the only protein feed component for pigs if feeding is done on a lysine-equivalent basis and not according to the crude protein content (7). Veal calves do not fully utilize microbial proteins, particularly in the first fattening phase up to 100 kg live weight; the intake of yeast and bacteria by the animals therefore must be limited to 15 and 20 per cent equivalents of the lactoprotein, respectively. Similarly the appropriate intake values for the second fattening phase are 30 per cent for yeasts and 40 per cent for bacteria.

Conclusions

In light of the predicted protein shortage, microorganisms offer many possibilities for protein production. The biomass of microorganisms can be freed of unwanted cell contents and used as single cell protein (SCP) for feeding animals. Because of the high nucleic acid (NA) content, microorganisms are not directly suitable for human nutrition, but, as studies have shown they can be utilized as feed in animal nutrition.

To avoid production losses, basically a nucleic acid-free protein should be considered as the source of an animal feed. In spite of the partly very favourable amino acid composition, high vitamin-B and digestible nutrient contents, microorganisms are no "wonder drugs". They can, however, be used to replace, totally or partially, certain valuable amounts of conventional vegetable and animal protein feed. The technological procedures which utilize waste products should be of primary importance in the production of SCP. In addition to those procedures involving the use of conventional substrates, particularly hydrocarbons and methanol, other technologies are ready for production of SCP and large-scale production of SCP is envisaged in England and Rumania.

The extent to which microbial proteins can be competitive with the conventional protein sources cannot be determined at this point, since data on market prices are not available. The prices for yeasts and bacteria however would probably fall in the same range as fish meal. Certainly, the price would be cheaper than that of skim milk powder. Large-scale production and utilization of microorganisms as SCP in future will depend not only on price relationship but also on agricultural, developmental and commercial trends.

References

1. Beck, H. and Gropp, J. Alkanhefen in der Geflügelernährung, *Z. Tierphysiol., Tierernährg. u. Futtermittelkde.*, *33*, 158–176 and 305–323 (1974).
2. Brune, H. and Niemann, E. Über den Einsatz und die Verträglichkeit von Bakterieneiweiss (*Hydrogenomonas*) mit unterschiedlichem Gehalt an

Poly-β-hydroxibuttersäure in der Tierernährung, *Z. Tierphysiol., Tierernährg. u. Futtermittelkde.*, *38*, 13–22 and 81–93 (1977).
3. Davis, P. *Single Cell Protein*, Academic Press, London (1974).
4. DDR (GDR)-*Futtermitteltabellenwerk*, VEB Deutscher Landwirtschaftsverlag, Berlin (1970).
5. Gropp, J., Beck, H. and Erbersdobler, H. Alkanhefen in der Geflügelernährung, *Z. Tierphysiol., Tierernährg. u. Futtermittelkde.*, *34*, 141–163 and 181–199 (1975).
6. Kaemmerer, K. Bedeutung und Verträglichkeit von Alkanhefen als Nahrungseiweiss, *Züchtungskunde*, *46*, 56–72 and 139–156 (1974).
7. Peterson, U. and Oslage, H.J. Einsatz von biotechnischem Protein bei Schweinen, *Sonderh. Ber. Ldw.*, *192*, 622–643 (1975).
8. Roth, F.X. and Kirchgessner, M. Einfluss steigender Mengen alimentär zugeführter Ribonucleinsäure auf den N-Stoffwechsel beim Ferkel, *Z. Tierphysiol., Tierernährg. u. Futtermittelkde.*, *38*, 214–225 (1977).
9. Roth, F.X. and Kirchgessner, M. N-Verwertung alimentärer Ribonucleinsäure beim Ferkel, *Z. Tierphysiol., Tierernährg. u. Futtermittelkde.*, *40*, 315–325 (1978).
10. Roth, F.X. and Kirchgessner, M. Stoffwechselergebnisse von Kälbern bei parteillem Austausch von Milchprotein durch Bakterienmasse, *Z. Tierphysiol., Tierernährg. u. Futtermittelkde.*, *41*, 29–39 (1978).
11. Roth, F.X. and Kirchgessner, M. Verwertung alimentärer Ribonucleinsäure im N-Stoffwechsel des Kalbes, *Arch. Tierernährung*, *29*, 275–283 (1979).
12. Schiller, K., Simeček, K. and Oslage, H.J. Mikrobiell produzierte Eiweissfuttermittel in der Tierernährung, 1. Mitt. Untersuchungen zur ernährungsphysiologischen Bewertung des Proteins von Hefen, gewachsen auf Rohöl-, n-Paraffinen und anderen Substraten, *Z. Tierphysiol., Tierernährg. u. Futtermittelkde.*, *30*, 246–259 (1972).
13. Schneider, D. and Bronsch, K. Methanolbakterien (*Methylomonas methylotropha*) als Proteinquelle für die Ernährung von Ferkeln, *Z. Tierphysiol., Tierernährg. u. Futtermittelkde.*, *39*, 313–325 (1977).
14. Schulz, E. Mikroorganismen als Eiweissfuttermittel, *Übers. Tierernährung*, *3*, 177–206 (1975).
15. Schulz, E. and Oslage, H.J. Analytische und tierexperimentelle Untersuchungen zur ernährungsphysiologischen Qualität von biotechnischem Protein sowie dessen Ergänzungsmöglichkeiten, *Sonderh. Ber. Ldw.*, *192*, 607–621 (1975).
16. van Weerden, E.J. Digestibility and metabolisable energy value for pigs of BP Lavera and Grangemouth yeast, *ILOB-Report*, *233*, Wageningen, Holland (1970).
17. Waldroup, P.W. and Payne, J.R. Feeding value of methanol-derived single cell protein for broiler chicks, *Poultry Science*, *53*, 1039–1042 (1973).

18. Walz, O.P., Koch, F. and Brune, H. Untersuchungen zu einigen Qualitätsmerkmalen der Grünalge *Scenedesmus acutus* an Schweinen und Küken, *Z. Tierphysiol., Tierernährg. u. Futtermittelkde.*, *35*, 55–75 (1975).
19. Wittemore, C.T. and Moffat, I.W. The digestibility of dried microbial cells grown on methanol in diets for growing pigs, *Journal of Agricultural Science*, Cambridge, *86*, 407–410 (1976).

S.T. CHANG and S.F. LI

25. Mushroom Culture

Introduction

Misconceptions about mushroom culture are exceedingly common, especially in the developing countries. Although it is thought to be very simple, mushroom cultivation is in fact a complicated business. It involves a number of different operations including preparations of pure culture, spawn and compost as well as crop management and marketing. While it can be treated as a very primitive type of farming as is the cultivation of the straw mushroom, *Volvariella volvacea*, in the Southeast Asian countries (6, 7, 30, 32), it can also be a highly industrialized agricultural enterprise with a considerable capital outlay as is the cultivation of the *Agaricus* mushroom (32). In any case, it is the aim of mushroom growers and researchers to try to increase the yield from a given surface area, to shorten the length of time between spawning and fructification but to lengthen the cropping period, and to have many more flushes with a high yield each time. In this connection, a thorough understanding of the starting materials for growing and the proper preparation of the substrate, the selection of suitable media for spawn making, the breeding of high quality strains, as well as the improvement of bed caring and the prevention of the development of pests and mushroom diseases have all to be called to attention.

There has been extensive concern in recent years as regards the production of food protein from domestic, agricultural and industrial wastes (1). On the one hand, high and sophisticated technology in the yeast and algal cultures for single cell proteins demands complicated input requirements and large capital. The immediate products may even need to be processed before being accepted as human food. Nevertheless, such development is expected to have good prospects in the advanced and rich countries. On the other hand, the great value in promoting the cultivation of mushrooms lies in their ability to grow on cheap carbohydrate materials and to transform various waste materials which are inedible by man into a highly valued food protein for direct human consumption (4, 5, 11, 25). This is extremely important in the rural areas where there is an enormous quantity of wastes that have been found to be ideal as growing substrates for tropical mushrooms. Furthermore, the spent compost, which is the substrate left over after mushroom harvesting, can be converted into stockfeed (34) and plant fertilizer as a soil conditioner (9). It is obvious that mushroom cultivation opens the deadlock in the biological degradation of natural resources. Thus, its immediate potential contributions should be properly recognized.

At present there are about ten edible mushrooms being cultivated in industrialized or semi-industrialized ways throughout the world (8). The most established one is the white mushroom *Agaricus bisporus*, followed by the Shiitake mushroom *Lentinus edodes*. The third- and fourth-most important cultivated mushrooms are *Volvariella volvacea* and *Pleurotus* spp. respectively (12), but their cultivations are not very developed and can be regarded only as semi-industrialized. According to the recent data (Table 1), however, it is

Table 1. World production of some cultivated edible mushrooms in 1979

Species	Tonnage	Percentage
Agaricus bisporus/bitorquis	870	71.9
Lentinus edodes	170	14.1
Flammulina velutipes	60	5.0
Volvariella volvacea	49	4.1
Pleurotus spp.	32	2.7
Pholiota nameko	17	1.4
Tremella spp. *Auricularia* spp.	10	0.8
Other species	2	0.2

Source: Delcaire (13).

Table 2. Temperature requirement for the cultivation of some edible mushrooms

Species	Temperature (°C)	
	Spawn-run	Fruiting
Agaricus bisporus	20–27	10–20
Agaricus bitorquis	25–30	20–25
Lentinus edodes	20–30	12–20
Flammulina velutipes	18–25	3–8
Volvariella volvacea	35–40	30–35
Pleurotus ostreatus		
("low temp." strains)	20–27	10–15
("temp. tolerant" strains)	20–35	10–30
Pleurotus sajor-caju	25–35	20–30
Pholiota nameko	24–26	5–15
Tremella spp.	20–25	20–27
Auricularia spp.	20–35	20–30

Source: Kurtzman (26), Li (29).

found that *Flammulina velutipes* (the winter mushroom) has become the third most important variety and its world production is just above that of the straw mushroom. Cultivation of *F. velutipes* is, however, restricted mainly to some countries of cold climate or to those which can afford high production cost. This mushroom requires a very low temperature for fructification (Table 2) and an intensive cooling system in mushroom houses, if it is to be grown in the tropics/subtropics. Therefore, *V. volvacea* and *Pleurotus* are still the two most suitable candidates for mushroom cultures in areas of warm climate. Since the first two mushrooms have been treated very intensively in literatures (14, 15, 18, 19, 20, 21, 22, 24, 32), they are not discussed here. In this chapter, the general aspects of mushroom cultures are considered and *V. volvacea* and *Pleurotus sajor-caju* are used as working examples.

Biological nature

Good crops are impossible to get without good spawn. To obtain and maintain good strains for spawns, an understanding of the biological nature of the mushroom is very important. It is clear that mushrooms are fungi which are characterized by their heterotrophic mode of nutrition. Digestive enzymes are liberated from the fungal cells into the immediate environment where organic matter is simplified and the food molecules in turn pass into the fungal cells. There has been much dispute on which fungi should be taken as mushrooms. The term mushroom is most frequently applied to the gill-bearing fungi. According to Chang and Hayes (8), edible mushrooms refer to both epigeous and hypogeous fruiting bodies of macroscopic fungi that are already commercially cultivated or grown in half-culture processes or potentially implanted under controlled conditions (8). Most of them are Basidiomycetes while there are a few Ascomycetes, too.

Both *Pleurotus sajor-caju* and *Volvariella volvacea* are Basidiomycetes. The Basidiomycetes are united as a group by their formation of a basidium. The nuclei in the basidium fuse to form the only diploid cell in the life cycle. Meiosis follows and the resulting four nuclei then pass into exogenous spore initials and form four basidiospores. Whether one basidiospore is self-fertile (homothallism) or self-sterile but cross-fertile (heterothallism) is dictated by the different patterns of sexuality (31). *Pleurotus* is tetrapolar heterothallic (16), which means the compatibility among the basidiospores is governed by two genetic factors A and B, and there are four possible recombinations of them. Upon germination each basidiospore with a haploid nucleus initiates a haploid monokaryotic mycelium. The monokaryon is at first non-septate but later becomes divided into a number of uninucleate cells. Plasmogamy occurs by the fusion of haploid hyphae. When the two mates are with different A and B incompatibility factors, a dikaryon is formed. The vegetative mycelium and the tissues in the basidiocarp are dikaryotic (each cell is binucleate).

This binucleate condition is maintained by the formation of clamp connections. Under suitable environmental conditions, the mycelium may enter the reproductive phase by first forming a hyphal aggregate (primordia). Further differentiation and morphogenesis occur and finally a macroscopic fruit body with basidiospores is formed.

In contrast to the well-clarified life cycle of *Pleurotus*, much of that of *V. volvacea* remains unknown. Each basidiospore is also uninucleate when first formed but it gives rise to a multinucleate mycelium. There is no dikaryotic phase and no clamp connections have been observed. The next event which is known is that karyogamy and meiosis take place in the basidium and four basidiospores are produced. According to the phenomenon that some single spores alone may give rise to fruit bodies (i.e. self-fertile), it has been taken to be homothallic (2). Yet the great variation of many characteristics found among the single spores of one progeny (which is unexpected) cannot be properly explained in this way (28). So far there is no final conclusion on its life cycle and research work is still underway. It is hoped that much will be clarified soon since this is a great handicap on the breeding work and strain improvement of the straw mushroom.

Cultivation

General

Figure 1 is an overall summary of the major steps in mushroom culture. Details of each process are described as follows:

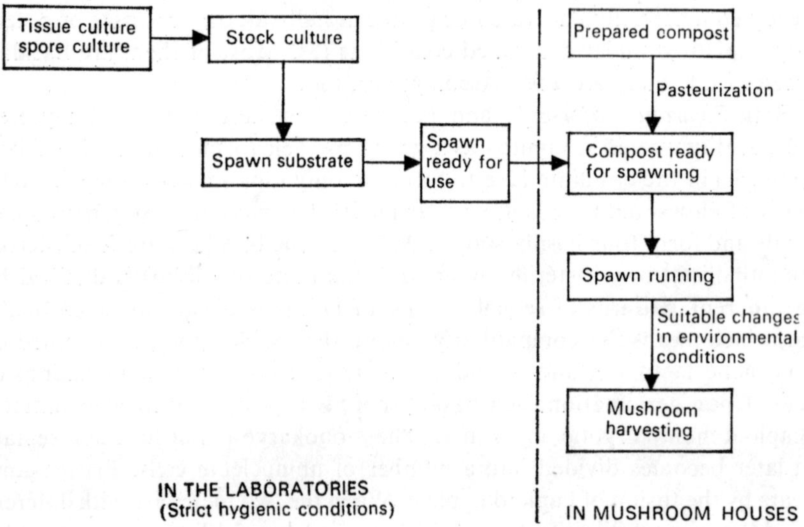

Fig. 1. Major steps in mushroom culture.

Plate 1. Fruit bodies of the straw mushroom, *Volvariella volvacea*. (A) "Button" and "egg" stages, good for harvesting; (B) "Mature" stage.

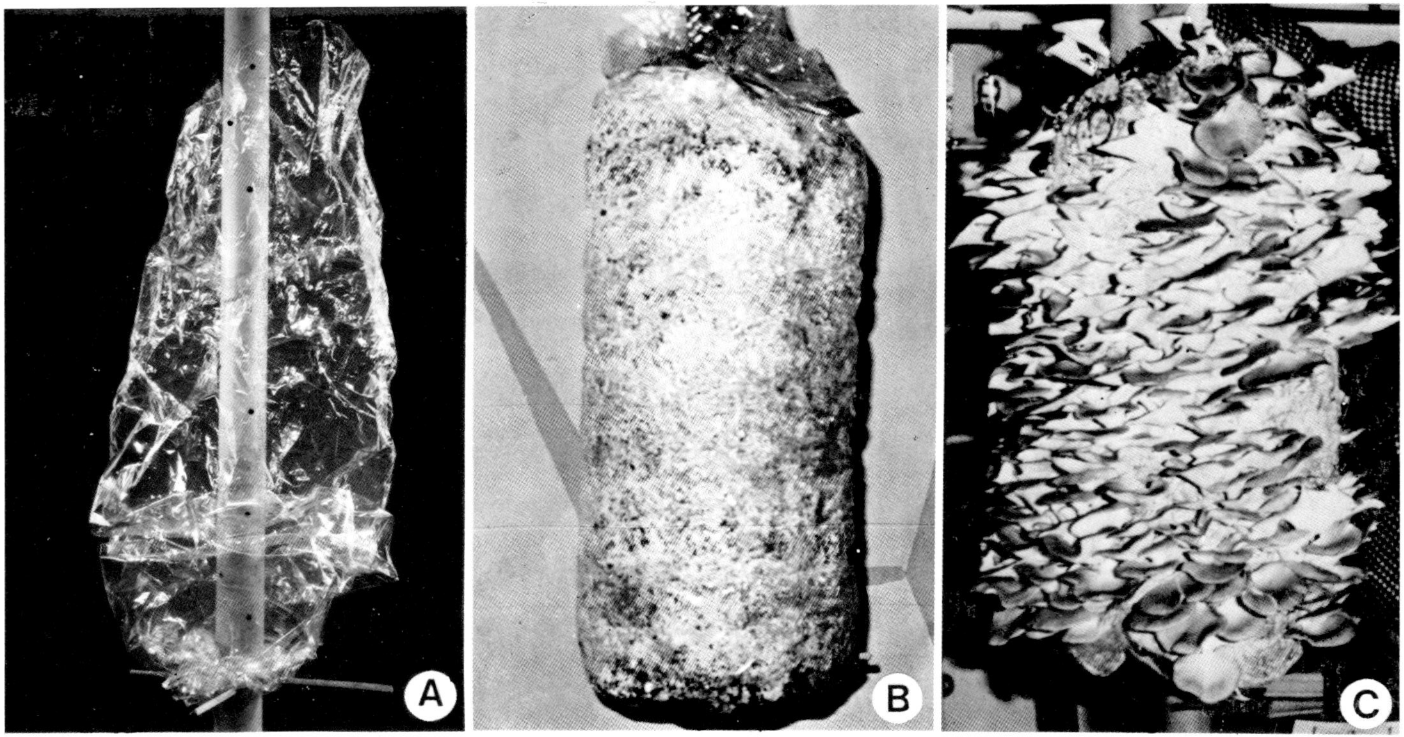

Plate 2. Cultivation of *Pleurotus sajor-caju* on cotton waste column in PP plastic sack. (A) Before filling: A PVC tubing (100 cm long, 4 cm in diameter) with holes with a cylindrical PP plastic sack which is tied near the bottom and supported by means of a pair of wooden sticks crossed to each other. (B) After spawn-running, before removal of the plastic wrapping: A column of cotton waste substrate thoroughly ramified by *Pleurotus* mycelium. (C) After removal of the plastic wrapping, pinheads formed and full growth of *Pleurotus* mushroom ready for harvesting.

1) PURE CULTURE. To be an honest grower pure cultures of known commercial strains which are patented should be purchased. However, pure cultures of mushrooms collected in nature can be established either by the tissue culture method, multispore culture method or single spore culture method. Aseptic techniques and considerable skill are required to accomplish this step without introducing contaminants. If this is not done properly, contaminating moulds and bacteria will overgrow the culture. It is preferable to have the manipulation being carried out in a sterile chamber.

a) *Culture media.* Generally, potato dextrose agar (PDA) medium or malt extract agar medium can be used for the growth and maintenance of mushroom pure cultures.

i) PDA medium: For 1 litre medium, ingredients required are 200 g potatoes, 20 g dextrose (glucose) and 15 g agar. Peel, wash and dice the potatoes. Cook them in approximately 1 litre distilled water until they are soft enough to be eaten (do not overcook!). Strain the decoction through a piece of cheese cloth, collect the liquid and add distilled water to have a final volume of 1 litre again. Add 20 g dextrose and 15 g agar, and heat to boil in a water bath. Stir occasionally until the agar is completely dissolved. Transfer 5 ml medium into 10 ml test tubes and plug with cotton wool (other glass containers like small bottles, flasks, etc. can also be used). Then sterilize them at 121°C for 15 minutes (e.g., by means of an autoclave or a pressure cooker). Slant the test tubes when hot but take care not to let the medium touch the cotton plug. To prepare agar plates, sterilize the medium in large Erlenmeyers flasks (e.g., 500 ml medium in 1 litre flask) and pour medium into the plates while still hot.

ii) Malt extract agar medium: Use 2 per cent malt extract and 1.5 per cent agar in distilled water. Sterilize and prepare in the same way as mentioned above.

b) *Tissue culture method.* Select a fresh mature mushroom (picked straight from the bed is the best). Wash it gently under running water to remove surface dirt. Blot dry. Wipe the surface gently with 70 per cent alcohol. Use a sterile scalpel (knife), cut a small slit at the bottom of the mushroom. Tear it into two halves and avoid touching the inner surface. Transfer a few pieces of tissue (pseudoparenchyma) from the centre of the mushroom to agar plates, one piece per plate. Incubate these plates for a few days. Then there should be mycelium growing out from the tissue onto the agar medium. If there is contamination, transfer some tiny bits of uncontaminated mycelium to another agar plate and incubate again.

c) *Spore culture method.* The mushroom spores serve for reproduction in the same manner as the seeds of higher plants. There are a number of satisfactory ways to collect spores in a sterile fashion. The surface of a mature (open) mushroom should be disinfected by wiping gently with 70 per cent alcohol. Remove the stipe and place the cap with gills downwards on a piece of sterile white paper. Preserve the whole arrangement in a sterile container. In a few minutes to one hour the spores drop from the gills onto the paper and make a pattern corresponding to the size and shape of the gills. Such a pattern is called a spore print. One cannot see a single spore with the naked eyes. The print, however, is made up of millions of spores and has a very striking appearance. Discard the gills and put the print into a clean petri dish. Spores remain viable for a long time when stored in a cool dry place.

i) Multispore culture: Cut a tiny piece of spore print and soak it in a few ml sterile tap water. Dip a sterile inoculating loop into this spore suspension and streak gently across the surface of an agar plate. Incubate the plate for a few days and a white mycelial mass resulting from the growth of germinated spores becomes evident.

ii) Single spore culture: Prepare a spore suspension as above. Make further dilutions with sterile tap water to get a very dilute one. Transfer one to two drops of it (contains 10–20 spores approximately) to an agar plate and spread evenly with a bent glass rod. After a few days' incubation, some distinct mycelial colonies appear. Transfer each colony to another agar plate and each plate is taken as a single spore culture. However, further checking has to be done before it is confirmed.

2) PREPARATION OF SPAWNS. Mushroom spawn can be simply defined as a medium impregnated with mushroom mycelium and serves as the 'seed' or 'inoculum' for mushroom cultivation. Pure culture spawn means the strain used is of known origin and free of contaminating organisms. The entire process of preparation should be aseptic from the beginning to the end.

The chief problem of the spawn manufacturers is the isolation of stable strains which will fulfil the expectation of the growers. A poor strain will be ultimately unsatisfactory no matter how ideal are the conditions in the spawn plant or in the growing house itself. There are certain principles on how to develop new strains and make appropriate selections (16, 31), but they will not be discussed here (see Fig. 2, later). Many different types of materials, which may or may not be the same substrate used in cultivation, alone or in various combinations, can be used as substrates for spawn making. To mention a few examples, rice straw cuttings, cotton waste, cotton seed hulls, rice hulls, sorghum grains, rye grains, etc. There has been a tendency to overemphasize the type of substrate the spawn is growing rather than the particular strain itself. Actually the substrate is merely the carrier or vehicle of the strain in a convenient medium which can be used to inoculate the beds. Nevertheless, the spawn substrate does influence the growth habit of the mycelium to some extent. Some spawns may grow (run) out more quickly and the beds fill out more rapidly than with other spawns. To choose a suitable spawn substrate the cost and availability of the raw material as well as the growth of mycelium on it should be considered. The details on the preparation of grain spawn and straw spawn are described here.

a) *Grain spawn* (*e.g. rye/sorghum/wheat*). Cook the grains in water until they swell but do not burst. Drain off excess water. Mix in 2 per cent (w/w) lime (calcium carbonate), fill loosely into glass bottles or PP plastic bags (three-quarters full only) and plug with cotton wool. Sterilize them in an autoclave for 30 minutes at 121°C and allow to cool. Inoculate the bottles with the prepared pure culture and incubate at the proper temperature. When the mycelium has run over the whole surface as well as permeated through the substrate, the spawn is ready for use.

b) *Straw spawn* (*e.g. paddy rice straw/wheat straw*). Cut the straw into approximately 5 cm long pieces. Soak in water for 5–10 minutes. Mix in 2 per cent (w/w) lime and proceed further as mentioned above.

Like handling the pure cultures, the inoculation and incubation of spawn should be done under strict hygienic conditions. The normal atmosphere is loaded with fungus spores and bacteria which float around freely in the air currents. If even one such spore happens to get into a bottle of fresh medium it will grow vigorously and render the spawn unusable. To avoid this, strict precautions are necessary.

3) PREPARATION OF SUBSTRATE. "Composting" is a process actually derived from the *Agaricus* mushroom growing industry which renders horse manure specific for the growth of mushrooms. It is difficult to say if the treatment of substrates for growing other mushrooms can be regarded as "composting" because, although the aim of both processes is the same, the starting materials and the accompanying changes in various lengths of time are incomparable. Therefore, only the purposes of composting and the possible general changes are discussed here. Any substrate material which is rich in readily available nutrients is not a satisfactory medium for growing mushrooms. This is not because the mycelium cannot grow in such material but will grow on them if they are first sterilized, which means that the mycelium cannot compete with other bacteria and moulds. Under natural circumstances when spawn is inoculated into raw substrate, the competing organisms quickly gain dominance and do not permit the mushroom mycelium to develop. The purpose of composting, then, is to prepare a medium of such characteristics that the growth of mushroom mycelium is promoted to the practical exclusion of other organisms. During composting certain chemical properties and physical qualities have to be developed. All of these are equally important and none are independent of the others.

Broadly speaking, the proper chemical state is one in which the best food materials able to serve the nutritional needs of the mushrooms are accumulated. These foodstuffs must be in a form available to the mushrooms. For instance, it would be unsatisfactory if all the nitrogen is changed to nitrate instead of protein since the mushrooms cannot use nitrates. Moreover, no toxic substances which inhibit the growth of spawn must be produced. The substrate must also have certain physical qualities. It must admit air freely, hold water without becoming waterlogged, have a proper pH and proper drainage. Biologically, the substrate must have a suitable population of microorganisms. The substrate is never sterile. It is literally teeming with millions of bacteria and fungi!

Composting is generally understood as the piling up of substrates for a certain period during which various changes occur so that the composted substrate is much different from the starting material. In the *Agaricus* industry, horse manure is the main starting material. Accordingly, any substrate which is composed of agricultural and chemical materials from any source except horse manure is called a "synthetic compost". Numerous formulations employing every manner of agricultural waste product and residue for the preparation of synthetic compost have been devised. Synthetic compost in this sense is in fact the general substrate used for growing mushrooms, including *Agaricus*.

To consider the straw or other plant wastes, their main composition is cellulose, hemicellulose and lignin. The hemicelluloses are generally found in rather close association with cellulose in plant fibre. Both cellulose and hemicellulose are carbohydrates which yield sugars after appropriate treatment. They are readily attacked by bacteria and are easily decomposed under suitable circumstances. Lignin, on the other hand, is a resistant substance which more or less impregnates plant fibre and is not readily attacked by bacteria. After composting, the easily decomposed carbohydrates, which serve as an excellent source of food for moulds and bacteria, tend to diminish and the substrate is no more favourable for these potential competitors. Proteins increase as a result of the activity of the microorganisms which convert simple nitrogenous materials such as ammonia and nitrates to complex proteins. The pH falls because of the growth of microorganisms. Although the principles of this complicated process are known and guidelines are available, in practice modifications are necessary which have to meet the various situations, e.g., the availability of raw materials, the facilities in the growing area and above all, the species of mushroom which is to be cultivated. There is no hope that the technology of the *Agaricus* cultivation industry can be uniformly followed for the growing of the straw mushroom *V. volvacea* in every situation.

When the substrates are ready, they are filled into trays, boxes or shelves to make the so-called "beds" and transferred into the room for pasteurization. This is a partial sterilization process operated at lower temperatures which justifies extreme care and effort in managing it properly. Carelessness at this stage may lead to a crop failure. The critical point is to achieve and maintain a satisfactory temperature range in all parts of the house. The objectives of pasteurization are twofold:

(a) To eradicate insects and pests carried in with the substrate and the destruction of spores of contaminating microorganisms; and

(b) To bring the substrate to a uniform temperature of around 50–55°C which promotes decomposition of the substrates by thermophilic micro-organisms. Through this final adjustment a more selective medium favouring the growth of the mushroom is accomplished.

Live steam generated from a water boiler is generally employed and it gives satisfactory results. It must be noted that it is the bed temperature which is critical and the object is to manage the air temperature in any way necessary to keep the bed temperature at the desired level. The entire process must be adapted to the specific circumstances existing in different mushroom houses. Generally speaking, it takes approximately two hours after the introduction of live steam to have an air temperature of 60–62°C. Maintain this temperature for two hours, then introduce a gentle stream of fresh air (by opening doors/windows or ventilators) to lower the temperature to 52°C. Maintain this temperature for the next six to eight hours. Then shut off the steam supply and let cool gradually to the desired temperature for spawning. How long it takes for this last step depends on the outdoor temperature.

4) SPAWNING, SPAWN-RUNNING AND CROPPING. Spawning is the process of planting spawn onto the bed materials. After pasteurization the bed temperature has to fall to a certain degree before spawning is begun because high temperature damages spawns. There is no strict rule on how much spawn is needed for a certain surface area. Larger amounts result in a more rapid filling out of the bed with mycelium, but the production cost will be higher.

Remove the spawn from the container. Break it into small pieces by crushing and crumbling with the fingers. Spawn pieces may be broadcast over the bed surface and be pressed down against the substrate to assure good contact, or they should be inserted 2–2.5 cm deep into the substrate.

During the spawn-running period, proper temperature and humidity should be maintained. Never let the bed surfaces dry out. As the spawn grows, it produces heat which contributes to the water loss. Water lightly with a fine water sprinklet. Improper environmental control is the common cause of a poor spawn run.

After the growing period and under suitable environmental conditions there is primordia formation and is followed by fruit bodies. Mushrooms appear in rhythmic cycles which are called "flushes" or "breaks". Different mushrooms are to be picked at different maturation stages. It depends on the consumers' preference and the market value. Care is required to pick mushrooms, twist them lightly and avoid disturbing the neighbouring smaller ones. During the cropping period suitable temperature, humidity and ventilation should still be maintained. This is important because it accounts for the total yield and the number of flushes which can be obtained.

Examples

(1) *Volvariella volvacea* (2, 3, 6) (Plate 1). The optimal range of temperatures for mycelial growth is 30–35°C, and 32°C is the best. Therefore, for pre-

paring pure cultures, spawns, as well as spore germination, the incubation temperature should be around 32°C. Mycelium of *V. volvacea* cannot tolerate low temperatures and becomes unviable at below 10°C. Therefore it is important to note that stock cultures should be kept at 20–25°C. A working example on growing *V. volvacea* with cotton waste as bedding material is given below.

Tear large pieces of cotton waste into smaller parts. Put the substrate inside a wooden frame (92 cm × 92 cm) and heap on concrete floors. Add sufficient water to get a moisture content of about 70 per cent, avoid overwetting and mix with 20 per cent (w/w) lime. Stamp on the mixture to enhance moistening. Prepare a pile of approximately 70–90 cm high. Cover it with a plastic sheet and let stand outdoors. After two days turn the pile thoroughly by hands or by a mechanical mixer. Add water when needed. Pile up the substrate and cover it with plastic sheets again. Let stand for another two days. When the compost is ready, transfer it into mushroom house to make up the beds. It does not matter whether shelf system or tray system is used; one layer of compost should be at least 10 cm thick. After the beds are done, pasteurization begins. Introduce live steam into the room until the temperature is 60–62°C and maintain this temperature for two hours. Turn on the ventilators (or open the windows) to lower the temperature to 50–52°C. Maintain this temperature for 8–10 hours. Then let it gradually cool down with controlled ventilation. When it is around 36–38°C spawning can begin. A spawning rate of 2 per cent (spawn to substrate, w/w) is sufficient. After spawning, cover the beds with plastic sheets to help keep the bed surface from drying out. Maintain the temperature at 32–34°C during the spawn running period. No light and watering are needed except a little ventilation. If the compost is prepared properly and the conditions are suitable, full growth may be achieved in three to four days. Then, remove the plastic sheets, turn on the light (white fluorescent lamps) and sprinkle some water onto the beds by means of a fine spray. Maintain the temperature at 28–30°C. On the fifth day after spawning, primordia of fruit bodies usually appear as white pinheads on the bed surface. After another three to four days, the first flush of mushrooms is ready for harvesting. For straw mushrooms, they are usually picked at the button stage, i.e., before the universal veil is ruptured.

(2) *Pleurotus sajor-caju. Pleurotus sajor-caju* is comparable to the high temperature strains of *Pleurotus ostreatus* in the temperature requirement for fructification (25, 28, 33). Their fruit body, morphology, colour, texture, etc., do not differ much either (28). It is however taken as another species because it is incompatible with the known *P. ostreatus* strains (16, 28, Li, unpublished data). Its origin is from India (22). This mushroom has a promising prospect in the tropical/subtropical areas because its cultivation is easy and demands no complicated requirement (22, 35). The cultivation method which has been tested to be successful is as follows: Cotton waste is used as the substrate. Tear large pieces into smaller parts. Add 2 per cent (w/w) lime and

mix with sufficient water to get a moisture content of about 70 per cent. Pile it up, cover with plastic sheets and let stand overnight. Load the substrate into small baskets or on shelves for pasteurization which is done in the same way as mentioned above. After cooling to approximately 25°C, mix around 2 per cent (w/w) spawn thoroughly with the substrate and pack into columns of 60 cm long which have hard plastic (PVC) tubings of 100 cm (4 cm in diameter) as central support (Plate 2), and with plastic sheets as outside wrapping.

Incubate these columns at around 24–28°C, preferably in the dark. When the mycelium of *Pleurotus* has ramified the entire column of substrate after three to four weeks, remove the plastic wrapping and switch on white light. Water occasionally to avoid the surface from drying. In around three days white primordia start to appear over the whole surface. After another two days, the *Pleurotus* mushrooms are ready for harvesting. During the cropping period watering is very important if many flushes are required.

Discussion and future prospects

The techniques used for the cultivation of edible mushrooms are derived from the disciplines of microbiology and agricultural engineering. When the basic principles are understood properly, its promotion should not be a difficult task. There are several essential aspects in the cultivation process which demand good attention and improvement (Fig. 2).

In addition to the significant role which it plays in the further utilization of organic waste materials, mushroom cultivation is in fact a way to help raising standard of living of the people in developing countries. It does not need to be a main business in farms. Farmers of horticultural crop plantation can also run mushroom cultivation as a sideline business. In this way, they can get from their farm wastes high quality food for themselves, or they might sell them in local markets and get some money for other foods. The main handicap for the farmers should be the hygienic conditions required for spawn production. It is suggested here that the government or scientific institutions of developing countries should give strong support to the mushroom business in rural areas by providing good spawns. The authorities should not overemphasize its large-scale development for export over domestic consumptions, but need to recognize the significance and the overall benefits of this small-scale business to the local people. As far as the nutritive values of these mushrooms produced from waste material is concerned, it is known that on a fresh weight basis the protein content of mushrooms is on an average twice as much as that of vegetables and four to twelve times higher than that of fruits (4). The quality of mushroom protein is valued better than that of cereal grains and legumes because all nine essential amino acids are present. In addition, mushrooms are also a source of some nutrients

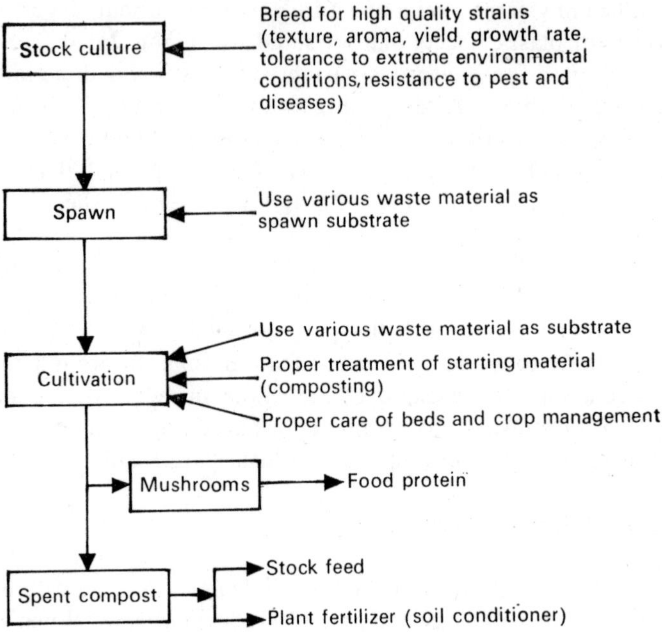

Fig. 2. Some alternatives suggested for the development of mushroom cultivation.

such as phosphorus, iron, thiamine (vitamin B1), riboflavin (vitamin B2) and niacin (11). It is true that proteins from animal sources are biologically still the best, but the people of developing countries are not able to afford food of animal origin which is not only expensive but also limited in supply. Although the mushroom protein is incomparable to the animal protein, its great value lies in the ability to grow on wastes and requires low production cost, and yet can contribute much to the human diets.

Because of the fact that the people of the developing countries eat too little protein (the average daily protein intake, 53.6 g, of each person in

Table 3. Comparison of the biological efficiency (yield of fresh mushrooms of one flush/dry weight of compost at spawning) of *Volvariella volvacea* and *Pleurotus sajor-caju*

Strain	Substrate	Biological efficiency
V. volvacea	cotton waste	35.90%
P. sajor-caju	cotton waste	69.41%
	cotton waste (after growing *V. volvacea*)	63.07%
	paddy straw	27.39%

Source: Chang and Yau (9).

Southeast Asia is only about half of that of people in the U.S.A. which is 104.3 g) (4), the significant role of mushroom cultivation in these areas becomes very obvious. Finally, it should be pointed out that the spent compost after harvesting the straw mushrooms has been proved to be a good substrate for growing the *Pleurotus* mushrooms (10, Table 3).

References

1. Birch, G.G., Paker, K.J. and Worgan, J.T. (Editors). *Food from Wastes*, Applied Science Publishers, London, p. 301 (1976).
2. Chang, S.T. *Volvariella volvacea*, pp. 573–603 in *The Biology and Cultivation of Edible Mushrooms* (Edited by S.T. Chang and W.A. Hayes), Academic Press, New York and London (1978).
3. Chang, S.T. Cultivation of *Volvariella volvacea* from cotten waste composts, *Mushroom Science, 10 (Part II)*, 609–618 (1979).
4. Chang, S.T. Mushrooms from waste, *Food Policy, 5*, 64–65 (1980).
5. Chang, S.T. Mushroom as human food, *Bio Science, 30* (6), 399–401 (1980).
6. Chang, S.T. Cultivation of *Volvariella* mushrooms in Southeast Asia, *Mushroom Newsletter for the Tropics, 1* (1), 5–10 (1980).
7. Chang, S.T. Mushroom production in Southeast Asia, *Mushroom Newsletter for the Tropics, 1* (2), 18–22 (1980).
8. Chang, S.T. and Hayes, W.A. (Editors). *The Biology and Cultivation of Edible Mushrooms*, 819 pp., Academic Press, New York and London (1978).
9. Chang, S.T. and Yau, C.K. Production of mushroom food and crop fertilizer from organic wastes, *The Proceedings of the Sixth International Conference of Global Impacts of Applied Microbiology (GIAM VI)*, Lagos, Nigeria (in press).
10. Chang, S.T., Lau, O.W. and Cho, K.Y. The cultivation and nutritional values of *Pleurotus sajor-caju, European Journal of Applied Microbiology and Biotechnology* (in press).
11. Crisan, E.V. and Sands, A. Nutritional value, pp. 137–168 in *The Biology and Cultivation of Edible Mushrooms* (Edited by S.T. Chang and W.A. Hayes), Academic Press, New York and London (1978).
12. Delcaire, J.R. Economics of edible mushrooms, pp. 727–793 in *The Biology and Cultivation of Edible Mushrooms* (Edited by S.T. Chang and W.A. Hayes), Academic Press, New York and London (1978).
13. Delcaire, J.R. Personal communication (1980).
14. Delmas, J. Cultivation in western countries: growing in caves, pp. 251–298 in *The Biology and Cultivation of Edible Mushrooms* (Edited by S.T.

Chang and W.A. Hayes), Academic Press, New York and London (1978).
15. Edwards, R.L. Cultivation in western countries: growing in houses, pp. 300–336 in *The Biology and Cultivation of Edible Mushrooms* (Edited by S.T. Chang and W.A. Hayes), Academic Press, New York and London (1978).
16. Eger, G. Biology and breeding of *Pleurotus*, pp. 497–519 in *The Biology and Cultivation of Edible Mushrooms* (Edited by S.T. Chang and W.A. Hayes), Academic Press, New York and London (1978).
17. Eger, G., Li, S.F. and Leal-Lara, H. Contribution to the discussion on the species concept in the *Pleurotus ostreatus* complex, *Mycologia, 71*, 577–588 (1979).
18. Hayes, W.A. Edible mushrooms, pp. 301–333 in *Food and Beverage Mycology* (Edited by L.R. Beuchat), AVI Publishing Co. Inc., Connecticut (1978).
19. Hayes, W.A. and Nair, N.G. The cultivation of *Agaricus bisporus* and other edible mushrooms, pp. 212–248 in *Filamentous Fungi, Vol. 1* (Edited by J.E. Smith and D.R. Berry), Edward Arnold, London (1975).
20. Hayes, W.A. and Wright, S.H. Edible mushrooms, pp. 141–176 in *Microbial Biomass* (Edited by A.H. Ross), Academic Press, New York (1979).
21. Ho, M.S. Cultivation in Asian countries: growing in subtropical areas, pp. 337–343 in *The Biology and Cultivation of Edible Mushrooms* (Edited by S.T. Chang and W.A. Hayes), Academic Press, New York and London (1978).
22. Ito, T. Cultivation of *Lentinus edodes*, pp. 461–473 in *The Biology and Cultivation of Edible Mushrooms* (Edited by S.T. Chang and W.A. Hayes), Academic Press, New York and London (1978).
23. Jandaik, C.L. Artificial cultivation of *Pleurotus sajor-caju* (Fr.) Sing., *Mushroom Journal, 22*, 405 (1974).
24. Kim, D.S. Cultivation in Asian countries: growing in temperate zones, pp. 345–363 in *The Biology and Cultivation Edible Mushrooms* (Edited by S.T. Chang and W.A. Hayes), Academic Press, New York and London (1978).
25. Kurtzman, R.H. Jr. Mushrooms convert wastes to food, *New Technology (Nov.-Dec.)*, 42, 45 (1976).
26. Kurtzman, R.H. Jr. Mushrooms: single cell protein from cellulose, pp. 305–309 in *Annual Report on Fermentation Processes, Vol. 3* (Edited by D. Perlman), Academic Press, New York (1979).
27. Li, S.F. and Eger, G. Characteristics of some *Pleurotus* strains from Florida, their practical and taxonomical importance, *Mushroom Science, 10 (Part I)*, 155–169 (1979).
28. Li, S.F. and Chang, S.T. Variation in the homothallic basidiomycete

Volvariella volvacea, *Mushroom Science*, *10* (*Part I*), 171–190 (1979).
29. Li, S.F. Studies on the tolerance to elevated temperatures in *Pleurotus ostreatus* (Jacq. ex Fr) Kummer—A contribution to taxonomy and the genetics of the fruiting process, *Bibliotheca Mycologica*, *76*, 1–86, J. Cramer, Vaduz (1980).
30. Lou, L.H. Cultivation of edible mushrooms in the tropical and subtropical regions of the Peoples' Republic of China, *Mushroom Newsletter for the Tropics*, *1* (2), 14–18 (1980).
31. Raper, C.A. Sexuality and breeding, pp. 83–117 in *The Biology and Cultivation of Edible Mushrooms* (Edited by S.T. Chang and W.A. Hayes), Academic Press, New York and London (1978).
32. Vedder, P.J.C. *Modern Mushroom Growing*, Educaboek-Culemborg, Netherlands (1978).
33. de Vries, C.A. Mushroom cultivation, *Tropical Abstracts*, *29*, 12 (1973).
34. Zadražil, F. The ecology and industrial production of *Pleurotus ostreatus*, *Pleurotus florida*, *Pleurotus cornucopiae* and *Pleurotus eryngii*, *Mushroom Science*, *9*, 621–652 (1974).
35. Zadražil, F. The conversion of straw into feed by basidiomycetes, *European Journal of Applied Microbiology*, *4*, 273–281 (1977).
36. Zadražil, F. Cultivation of *Pleurotus*, pp. 521–557 in *The Biology and Cultivation of Edible Mushrooms* (Edited by S.T. Chang and W.A. Hayes), Academic Press, New York and London (1978).

Index